EQUATIONS RELATING TO METHODS

Where numerical values are given, the following units are assumed: i (A), D_O(cm^2/sec), C_O^* (mol/cm^3), t (sec), A (cm^2), V (cm^3), v (V/sec), ΔE (V), ω (sec^{-1}). $T = 25$ °C

Controlled potential:

(Cottrell equation)

$$i(t) = \frac{nFAD_O^{1/2}C_O^*}{\pi^{1/2}t^{1/2}} \qquad (5.2.11)$$

(Ilkovic equation)

$$i_d = 708nD_O^{1/2}C_O^*m^{2/3}t^{1/6} \qquad (5.3.6)$$

(Integrated Cottrell equation)

$$Q_d = \frac{2nFAD_O^{1/2}C_O^*t^{1/2}}{\pi^{1/2}} \qquad (5.9.1)$$

(General voltammetric current-potential equation for a reversible wave)

$$E = E_{1/2} + \frac{RT}{nF}\ln\frac{i_d(t) - i(t)}{i(t)} \qquad (5.4.22)$$

[Linear potential sweep (peak current, reversible wave)]

$$i_p = (2.69 \times 10^5)n^{3/2}AD_O^{1/2}v^{1/2}C_O^* \qquad (6.2.19)$$

[Thin-layer equation (potential sweep, reversible reaction)]

$$i_p = (9.39 \times 10^5)n^2vVC_O^* \qquad (10.7.17)$$

$$i = i_p\frac{4\exp[nf(E - E^{0\prime})]}{\{1 + \exp[nf(E - E^{0\prime})]\}^2} \qquad (10.7.16)$$

Controlled current:

(Sand equation)

$$\frac{i\tau^{1/2}}{C_O^*} = \frac{nFAD_O^{1/2}\pi^{1/2}}{2} = 85500nAD_O^{1/2} \qquad (7.2.14)$$

Rotating disk electrode:

(Levich equation)

$$i_l = 0.620nFAD_O^{2/3}\omega^{1/2}\nu^{-1/6}C_O^* \qquad (8.3.22)$$

ac voltammetry:

(Peak current amplitude, reversible wave)

$$I_p = (9.39 \times 10^5)n^2A\omega^{1/2}D_O^{1/2}C_O^*\,\Delta E \qquad (9.4.12)$$

ELECTROCHEMICAL METHODS

Fundamentals and Applications

ALLEN J. BARD
Department of Chemistry
University of Texas

LARRY R. FAULKNER
Department of Chemistry
University of Illinois

John Wiley & Sons
New York • Chichester • Brisbane • Toronto • Singapore

Library of Congress Cataloging in Publication Data:

Bard, Allen J
Electrochemical methods.

Includes bibliographical references and index.
1. Electrochemistry. I. Faulkner, Larry R., 1944-
II. Title.
QD553.B37 541'.37 79-24712
ISBN 0-471-05542-5

Printed in the United States of America

10 9 8

Preface

Several, now classic, books appeared about 25 to 30 years ago, including Kolthoff and Lingane's *Polarography* (Interscience, 1952), Lingane's *Electroanalytical Chemistry* (Interscience, 1953), and Delahay's *New Instrumental Methods in Electrochemistry* (Interscience, 1954), which presented an analysis and discussion of many of the electrochemical methods of those times in terms of the underlying principles of thermodynamics, kinetics, mass transfer, and interfacial structure. These books also discussed the application of these methods to chemical problems, mainly in analytical chemistry, and presented a general strategy for dealing with electrochemical investigations. They were important stimuli for the dramatic development of electrochemical methodology from the mid-1950s into the early 1970s. Growth was abetted by the advent of sophisticated electronic instruments that greatly increased experimental flexibility, as well as by widespread use of digital computer techniques for data analysis and theoretical treatments. It is a tribute to these works that they could stimulate 20 years of rapid growth, reported in thousands of papers, and still remain among the better sources for learning the fundamentals.

Now electrochemists have turned most of their attention to chemical problems, rather than the methods themselves, and the methodology has become a mature body of knowledge. Electrochemical techniques are being recognized by nonelectrochemists as useful means for characterizing chemical systems. A greater diffusion of electrochemical practice into the larger scientific community seems inevitable. Thus there is a need for a textbook to teach the fundamentals and applications of electrochemical methods in an up-to-date, comprehensive, systematic fashion.

This book was written in pursuit of that goal. It is intended as a textbook and includes numerous problems and chemical examples. Illustrations have been employed to clarify presentations, and the style is pedagogical throughout. The book could be used in formal courses at the senior undergraduate and beginning graduate levels, but we have also tried to write in a way that would make self-study by interested individuals possible. A knowledge of basic physical chemistry is assumed, but the discussions generally begin at an elementary level and develop upward. We have sought to make the volume self-contained by developing almost all ideas of any importance to our subject from very basic principles of chemistry and physics. Much of the discussion depends on mathematics, but the specialized mathematical background is covered in Appendix A. The problems following each chapter have been devised as teaching tools. They often extend concepts introduced in the text or show how experimental data are reduced to fundamental results.

Our approach is to first give an overview of electrode processes (Chapter 1) showing the way in which the fundamental components of the subject come together in an electrochemical experiment. Then there are individual discussions of thermodynamics and potential, charge transfer kinetics, and mass transfer (Chapters 2 to 4). Concepts from these basic areas are integrated together in treatments of the various methods

(Chapters 5 to 10). The effects of homogeneous kinetics are treated separately (Chapter 11) in a way that provides a comparative view of the responses of different methods. A discussion of interfacial structure and the effects of adsorption follows (Chapter 12); then there is a taste of electrochemical instrumentation (Chapter 13) and a discussion of experiments involving spectrometric and photochemical aspects (Chapter 14). Appendix A teaches the mathematical background, and Appendix B provides an introduction to digital simulation.

References are selective, frequently to review series and papers. We make no attempt to document the history of the subject, and we have found it impractical to attempt a thorough listing of even the important papers in all of the areas covered here. Our hope is that the 800 citations that do exist will offer adequate entry to the specialized literature for an interested reader. To facilitate his or her effort, we have also included in Section 1.6 a survey of some important elements of the electrochemical literature.

The mathematical notation is uniform throughout the book and there is minimal duplication of symbols. The list of major symbols (pp. vii-xvi) and the list abbreviations (pp. xvii-xviii) offer definitions, dimensions, and section references. Usually we have adhered to the recommendations of the IUPAC Commision on Electrochemistry [R. Parsons et al., *Pure Appl. Chem.*, **37**, 503 (1974)]. Exceptions have been made where customary usage or clarity of notation seemed compelling.

Of neccessity, compromises have been made between depth, breadth of coverage, and reasonable size. "Classical" topics in electrochemistry, including many aspects of thermodynamics of cells, conductance, and potentiometry are not covered here. Usually they are treated in more elementary courses in analytical chemistry, although they may unfortunately be fast disappearing from the curriculum entirely. Similarly, we have not been able to accommodate discussions of many techniques that are useful but are not widely practiced. These include many different types of hydrodynamic electrochemical systems and some ac methods. The details of laboratory procedures, such as the design of cells, the construction of electrodes, and the purification of materials, are beyond our scope.

There is a relative neglect of dc polarography in this book by comparison to others of its type. We are mindful of the great contributions of Heyrovsky and his successors, and we appreciate that the impact of their work on the development of electrochemical methods cannot be overestimated. However, difficulties in dc polarography based on drawbacks of the dropping mercury electrode, such as the expanding-sphere mass transfer problem, lack of adsorption equilibrium, limited time domain, and the appearance of maxima, have led to a decrease in the general usefulness of the method with respect to other techniques. We have tried to treat it with that perspective.

We have been aided by many people in preparing this work. Colleagues including F. Anson, G. Christian, D. Evans, N. Good, K. Kadish, T. Kuwana, M. Morris, R. Murray, T. Nieman, R. Osteryoung, J. Saveant, R. Sioda, D. Smith, R. Van Duyne, and members of our research groups have read parts of the manuscript and offered many helpful critical comments. We are grateful for their care. Carolyn White and Glenna Wilsky are due special appreciation for the painstaking attention that they gave to the preparation of the final manuscript and the details of publication. Finally, we offer simple thanks to our families for affording us time for this endeavor.

Allen J. Bard
Larry R. Faulkner

Table of Contents

Major Symbols

The symbols listed below are used in several chapters or in large portions of a chapter. Symbols similar to some of these may have different meanings in a local context. In most cases, the usage follows the recent recommendations of the IUPAC Commission on Electrochemistry [R. Parsons et al., *Pure Appl. Chem.*, **37**, 503 (1974).]; however there are exceptions.

STANDARD SUBSCRIPTS

a	anodic
c	(a) cathodic
	(b) charging
D	disk
d	diffusion
dl	double layer
eq	equilibrium
f	(a) forward
	(b) faradaic
l	limiting
O	pertaining to species O in $O + ne \rightleftharpoons R$
p	peak
R	pertaining to species R in $O + ne \rightleftharpoons R$
R	ring
r	reverse

ROMAN SYMBOLS

Symbol	Meaning	Usual Dimensions	Section References
A	(a) area	cm^2	1.3.2
	(b) cross-sectional area of a porous electrode	cm^2	10.6.2
	(c) frequency factor in a rate expression	depends on order	3.1.2
	(d) open-loop gain of an amplifier	none	13.1.1
$\mathscr{A}$	absorbance	none	14.1.1

Symbol	Meaning	Usual Dimensions	Section References
a	internal area of a porous electrode	cm^2	10.6.2
a_j^α	activity of substance j in a phase α	usually, M	2.1.5
b	$\alpha n_a Fv/RT$	sec^{-1}	6.3.1
b_j	$\beta_j \Gamma_{j,s}$	mol/cm^2	12.4.3
C	capacitance	F, μF	9.1.2
C_d	differential capacitance of the double layer	μF, $\mu F/cm^2$	1.2.2, 12.2.2
C_i	integral capacitance of the double layer	μF, $\mu F/cm^2$	12.2.2
C_j	concentration of species j	M, mol/cm^3, mM	
C_j^*	bulk concentration of species j	M, mol/cm^3, mM	1.4.2, 4.3.3
$C_j(x)$	concentration of species j at distance x	M, mol/cm^3	
$C_j(x = 0)$	concentration of species j at the electrode surface	M, mol/cm^3	1.4.2
$C_j(x, t)$	concentration of species j at distance x at time t	M, mol/cm^3	4.3
$C_j(0, t)$	concentration of species j at the electrode surface	M, mol/cm^3	4.3.3
$C_j(y)$	concentration of species j at distance y below an RDE	M, mol/cm^3	8.3.3
$C_j(y = 0)$	surface concentration of species j at a rotating electrode	M, mol/cm^3	8.3.4
C_{SC}	space charge capacitance	F/cm^2, $\mu F/cm^2$	14.5.2
C_s	pseudocapacity	F, μF	9.1.3
c	speed of light in vacuo	cm/sec	14.1.2
D_j	diffusion coefficient of species j	cm^2/sec	1.4.1, 4.3
$\mathbf{D}_M$	model diffusion coefficient in simulation	none	B.1.3, B.1.8
d_j	density of phase j	g/cm^3	
E	(a) Potential of an electrode versus a reference	V	1.1, 2.1
	(b) emf of a reaction	V	2.1
$\dot{E}$	voltage or potential phasor	V	9.1.2
ΔE	(a) pulse height in differential pulse voltammetry	mV	5.8.3
	(b) amplitude (1/2 $p - p$) of ac excitation in ac voltammetry	mV	9.4.1
$\mathbf{E}$	electron energy	eV	14.5.2
E^0	(a) standard potential of an electrode	V	2.1.4
	(b) standard emf of a half-reaction	V	2.1.4
ΔE^0	difference in standard potentials for two couples	V, mV	
$E^{0'}$	formal potential of an electrode	V	2.1.6
E_A	activation energy of a reaction	kJ/mol	3.1.2
E_{ac}	ac component of potential	V, mV	9.1.1
$\mathbf{E}_b$	binding energy of an electron	eV	14.2.1
E_{eq}	equilibrium potential of an electrode	V	1.3.2, 3.5.1

Symbol	Meaning	Usual Dimensions	Section References
E_{dc}	dc component of potential	V, mV	9.1.1
$\mathbf{E}_F$	Fermi level	eV	14.5.2
E_{fb}	flat-band potential	V	14.5.2
$\mathbf{E}_g$	bandgap of a semiconductor	eV	14.5.2
E_i	initial potential	V	6.2.1
E_j	junction potential	mV	2.3.3, 2.3.4
E_m	membrane potential	mV	2.4
$E_{p/2}$	potential where $i = i_p/2$ in LSV	V	6.2.2
E_p	peak potential	V	6.2.2
ΔE_p	$\lvert E_{pa} - E_{pc} \rvert$ in CV	V, mV	6.5
E_{pa}	anodic peak potential	V	6.5
E_{pc}	cathodic peak potential	V	6.5
E_z	potential of zero charge	V	12.2.2
E_λ	switching potential for cyclic voltammetry	V	6.5
$E_{1/2}$	(a) measured half-wave potential in polarography or voltammetry	V	1.4.2, 5.4, 8.3.4
	(b) in derivations, the *reversible* half-wave potential, $E^{0\prime} + (RT/nF)\ \ln(D_R/D_O)^{1/2}$	V	5.4
$E_{1/4}$	potential where $i/i_d = \frac{1}{4}$	V	5.4.2
$E_{3/4}$	potential where $i/i_d = \frac{3}{4}$	V	5.4.2
$E_{\tau/4}$	quarter-wave potential in chronopotentiometry	V	7.3.1
$\boldsymbol{\mathscr{E}}$	electric field vector	V/cm	2.2.1
$\mathscr{E}$	electric field strength	V/cm	2.2.1
e	(a) quantity of charge on the electron	C	
	(b) voltage in an electric circuit	V, mV	9.1.1, 13.1
e_i	input voltage	V, mV	13.1.1
e_s	voltage across the input terminals of an amplifier	μV	13.1.1
e_0	output voltage	V, mV	13.1.1
$\mathrm{erf}(x)$	error function of x	none	A.2
$\mathrm{erfc}(x)$	error function complement of x	none	A.2
F	the faraday; charge on one mole of electrons	C	
f	(a) F/RT	V^{-1}	
	(b) frequency of rotation	r/sec	8.3
	(c) frequency of a sinusoidal oscillation	sec^{-1}	9.1.2
$f_i(j, k)$	fractional concentration of species i in box j after iteration k in a simulation	none	B.1.3
g	(a) gravitational acceleration	cm/sec^2	
	(b) interaction parameter in adsorption isotherms	$J\text{-}cm^2/mol^2$	12.4.2
	(c) g factor in ESR	none	14.3.1
G	Gibbs free energy	kJ	2.2.4
ΔG	Gibbs free energy change in a chemical process	kJ	2.1.2, 2.1.3
ΔG^0	standard Gibbs free energy change in a chemical process	kJ	2.1.2, 2.1.3

Symbol	Meaning	Usual Dimensions	Section References
$\Delta G^{\ddagger}$	standard Gibbs free energy of activation	kJ/mol	3.1.2
$\bar{G}$	electrochemical free energy	kJ	2.2.4
H	(a) $k_f/D_O^{1/2} + k_b/D_R^{1/2}$	$sec^{-1/2}$	5.5.1
	(b) enthalpy	kJ	2.1.2
ΔH	enthalpy change in a chemical process	kJ	2.1.2
ΔH^0	standard enthalpy change in a chemical process	kJ	2.1.2
$\Delta H^{\ddagger}$	standard enthalpy of activation	kJ/mol	3.1.2
h	Planck's constant	J-sec	
h_{corr}	corrected mercury column height at a DME	cm	5.3.4
I	amplitude of an ac current	A, μA	9.1.2
$I(t)$	convolutive transform of current; semi-integral of current	$C/sec^{1/2}$	6.7.1
$\dot{I}$	current phasor	A, μA	9.1.2
$\bar{I}$	diffusion current constant for average current	$\mu A\text{-}sec^{1/2}/mg^{2/3}\text{-}mM$	5.3.3
$(I)_{max}$	diffusion current constant for maximum current	$\mu A\text{-}sec^{1/2}/mg^{2/3}\text{-}mM$	5.3.3
I_p	peak value of ac current amplitude	A, μA	9.4.1
i	current	A, μA	1.3.2
δi	differential current output in differential pulse voltammetry	A, μA	5.8.3
$i(0)$	initial current in bulk electrolysis	A, μA	10.3.1
i_a	anodic component current	A, μA	3.2
i_c	(a) charging current	A, μA	5.3.5
	(b) cathodic component current	A, μA	3.2
i_d	(a) current due to diffusive flux	A, μA	4.1
	(b) diffusion-limited current	A, μA	5.2.1
$\bar{i}_d$	average diffusion-limited current flow over a drop lifetime at a DME	A, μA	5.3.1
$(i_d)_{max}$	diffusion-limited current at t_{max} at a DME (maximum current)	A, μA	5.3.1
i_f	faradaic current	A, μA	
i_j	current flow due to species j	A, μA	
i_k	kinetically limited current	A, μA	8.3.4
i_l	limiting current	A, μA	1.4.2
$i_{l,a}$	limiting anodic current	A, μA	1.4.2
$i_{l.}$	limiting cathodic current	A, μA	1.4.2
i_m	migration current	A, μA	4.1
i_p	peak current	A, μA	6.2.2
i_{pa}	anodic peak current	A, μA	6.5.1
i_{pc}	cathodic peak current	A, μA	6.5.1
i_r	current during reversal step	A, μA	5.7.1
i_0	exchange current	A, μA	3.5.1
$i_{0,t}$	true exchange current	A, μA	12.7.1
$\text{Im}(\omega)$	imaginary part of complex function ω		A.5

Symbol	Meaning	Usual Dimensions	Section References
$J_j(x, t)$	flux of species j at location x at time t	mol cm^{-2} sec^{-1}	1.4.1, 4.1
j	(a) current density	A/cm^2, μA/cm^2	1.3.2
	(b) box index in a simulation	none	B.1.2
	(c) $\sqrt{-1}$	none	A.5
j_0	exchange current density	A/cm^2, μA/cm^2	3.5.1
K	equilibrium constant		
k	(a) rate constant for a homogeneous reaction	depends on order	
	(b) iteration number in a simulation	none	B.1
	(c) extinction coefficient	none	14.1.2
$\mathscr{k}$	Boltzmann constant	J/K	
k^0	standard (intrinsic) heterogeneous rate constant	cm/sec	3.2, 3.3
k_b	(a) heterogeneous rate constant for oxidation	cm/sec	3.2
	(b) homogeneous rate constant for "backward" reaction	depends on order	3.1
k_b^0	heterogeneous rate constant for oxidation at $E = 0$ on the potential scale in use	cm/sec	3.3
k_f	(a) heterogeneous rate constant for reduction	cm/sec	3.2
	(b) homogeneous rate constant for "forward" reaction	depends on order	3.1
k_f^0	heterogeneous rate constant for reduction at $E = 0$ on potential scale in use	cm/sec	3.3
$k_{i,j}^{\text{pot}}$	potentiometric selectivity coefficient of interferent j toward a measurement of species i	none	2.4.2
k_t^0	true standard heterogeneous rate constant	cm/sec	12.7.1
L	length of a porous electrode	cm	10.6.2
$L\{f(t)\} = \bar{f}(s)$	Laplace transform of $f(t)$		A.1
$L^{-1}\{\bar{f}(s)\}$	Inverse Laplace transform of $\bar{f}(s)$		A.1
l	thickness of solution in a thin-layer cell	cm	10.7.2
$\mathbf{l}$	number of iterations corresponding to t_k in a simulation	none	B.1.4
m	mercury flow rate at a DME	mg/sec	5.3
$m(t)$	convolutive transform of current; semi-integral of current	C/sec$^{1/2}$	6.7.1
m_j	mass transfer coefficient of species j	cm/sec	1.4.2
N	collection efficiency at an RRDE	none	8.4.2
N_A	acceptor density	cm^{-3}	14.5.2
N_D	donor density	cm^{-3}	14.5.2
N_j	total number of moles of species j in a system	moles	10.3.1
n	(a) electrons per molecule oxidized or reduced; faradays per mole of substance electrolyzed	none	1.3.2

Symbol	Meaning	Usual Dimensions	Section References
n	(b) electron density in a semiconductor	cm^{-3}	14.5.2
	(c) refractive index	none	14.1.2
$\hat{n}$	complex refractive index	none	14.1.2
n^0	number concentration of each ion in a $z:z$ electrolyte	cm^{-3}	12.3.2
n_a	number of electrons involved in the rate-determining step	none	3.6
n_i	electron density in an intrinsic semiconductor	cm^{-3}	14.5.2
n_j	(a) number of moles of species j in a phase	moles	2.2.4, 12.1.1
	(b) number concentration of ion j in an electrolyte	cm^{-3}	12.3.2
n_j^0	number concentration of ion j in the bulk electrolyte	cm^{-3}	12.3.2
O	oxidized form of the standard system O + $ne \rightleftharpoons$ R; often used as a subscript denoting quantities pertaining to species O		
P	pressure	Pa, atm, torr	
p	(a) hole density in a semiconductor	cm^{-3}	14.5.2
	(b) $m_O A/V$, $m_R A/V$	sec^{-1}	10.3.1
p_i	hole density in an intrinsic semiconductor	cm^{-3}	14.5.2
Q	charge passed in electrolysis	C, μC	1.3.2, 5.9.1, 10.3.1
Q^0	charge required for complete electrolysis of a component by Faraday's law	C, μC	10.3.4
Q_d	chronocoulometric charge from a diffusing component	C, μC	5.9.1
Q_{dl}	charge devoted to double-layer capacitance	C, μC	5.9, 12.5.6
Q_r	chronocoulometric charge removed in a reversal step	C, μC	5.9.2
q^j	excess charge on phase j	C, μC	1.2, 2.2
R	reduced form of the standard system, O + $ne \rightleftharpoons$ R; often used as a subscript denoting quantities pertaining to species R		
R	(a) gas constant	$J\ mol^{-1}\ K^{-1}$	
	(b) resistance	Ω	9.1.2
	(c) fraction of substance electrolyzed in a porous electrode	none	10.6.2
	(d) reflectance	none	14.1.2
$R_{ct,j}$	charge transfer resistance for species j	Ω	1.3.3, 3.5.6
R_f	feedback resistance	Ω	13.1
$R_{mt,j}$	mass transfer resistance for species j	Ω	1.3.3, 3.5.6
R_s	(a) solution resistance	Ω	1.3.4
	(b) series resistance in an equivalent circuit	Ω	9.1.3
R_u	uncompensated resistance	Ω	1.3.4, 13.6
R_Ω	ohmic solution resistance	Ω	9.1.3, 13.4.1

Symbol	Meaning	Usual Dimensions	Section References
r	(a) radial distance from the center of a spherical electrode	cm	5.2.2
	(b) radial distance from the axis of rotation of a rotating electrode	cm	8.3.1
r_c	radius of a capillary	cm	5.3.3
r_0	radius of a spherical electrode	cm	5.2.2
r_1	radius of disk electrode	cm	8.3.5
r_2	inner radius of ring electrode	cm	8.4.1
r_3	outer radius of ring electrode	cm	8.4.1
Re	Reynolds number	none	8.2.1
$\mathrm{Re}(\omega)$	real part of complex function ω		A.5
ΔS	entropy change in a chemical process	kJ/K	2.1.2
ΔS^0	standard entropy change in a chemical process	kJ/K	2.1.2
$\Delta S^{\ddagger}$	standard entropy of activation	kJ mol^{-1} K^{-1}	3.1.2
$S_\tau(t)$	unit step function rising at $t = \tau$	none	A.1.7
s	(a) Laplace plane variable, usually complementary to t		A.1
	(b) specific area of a porous electrode	cm^{-1}	10.6.2
T	absolute temperature	K	
t	time	sec	
t'	transit time at an RRDE	sec	8.5.2
t_j	transference number of species j	none	2.3.3, 4.1
t_k	known characteristic time in a simulation	sec	B.1.4
t_{max}	drop time at a DME	sec	5.3.1
u_j	mobility of ion j	$\mathrm{cm}^2\ \mathrm{V}^{-1}\ \mathrm{sec}^{-1}$	2.3.3, 4.1
V_j	volume of phase j	cm^3	
v	(a) linear potential scan rate	V/sec	6.1
	(b) homogeneous reaction rate	mol $\mathrm{cm}^{-3}\ \mathrm{sec}^{-1}$	1.3.2, 3.1
	(c) heterogeneous reaction rate	mol $\mathrm{cm}^{-2}\ \mathrm{sec}^{-1}$	1.3.2, 3.2
	(d) linear velocity of solution flow, usually a function of position	cm/sec	1.4.1, 8.2
v_b	(a) "backward" homogeneous reaction rate	mol $\mathrm{sec}^{-1}\ \mathrm{cm}^{-3}$	3.1
	(b) anodic heterogeneous reaction rate	mol $\mathrm{sec}^{-1}\ \mathrm{cm}^{-2}$	3.2
v_f	(a) "forward" homogeneous reaction rate	mol $\mathrm{sec}^{-1}\ \mathrm{cm}^{-3}$	3.1
	(b) cathodic heterogeneous reaction rate	mol $\mathrm{sec}^{-1}\ \mathrm{cm}^{-2}$	3.2
v_j	component of velocity in the j direction	cm/sec	8.2.1
v_{mt}	rate of mass transfer to a surface	mol $\mathrm{cm}^{-2}\ \mathrm{sec}^{-1}$	1.4.1
X_C	capacitive reactance	Ω	9.1.2
X_j	mole fraction of species j	none	12.1.2
x	distance, usually from a planar electrode	cm	
x_1	distance of the IHP from the electrode surface	cm	1.2.3, 12.3.3
x_2	distance of the OHP from the electrode surface	cm	1.2.3, 12.3.3
Y	admittance	Ω^{-1}	9.1.2
$\mathbf{Y}$	admittance vector	Ω^{-1}	9.1.2
y	distance below an RDE or RRDE	cm	8.3.1
Z	(a) impedance	Ω	9.1.2

Symbol	Meaning	Usual Dimensions	Section References
Z	(b) dimensionless current parameter in simulation	none	B.1
$\mathbf{Z}$	impedance vector	Ω	9.1.2
Z_f	faradaic impedance	Ω	9.1.3
Z_{Im}	imaginary part of impedance	Ω	9.5.3
Z_{Re}	real part of impedance	Ω	9.5.3
Z_w	Warburg impedance	Ω	9.1.3
z	(a) charge on an ion in signed units of electronic charge	none	4.1
	(b) charge magnitude of each ion in a z:z electrolyte	none	12.3.2
z_j	charge on species j in signed units of electronic charge	none	4.1

GREEK SYMBOLS

Symbol	Meaning	Usual Dimensions	Section References
α	(a) transfer coefficient	none	3.2, 3.3
	(b) absorption coefficient	cm^{-1}	14.1.2
β	(a) geometric parameter for an RRDE	none	8.4.1
	(b) $1 - \alpha$	none	9.4.1
β_j	(a) $\partial E/\partial C_j(0, t)$	V-cm^3/mol	9.2.2
	(b) energy parameter in an adsorption isotherm for species j	none	12.4.2
Γ_j	surface excess of species j at equilibrium	mol/cm^2	12.1.2
$\Gamma_{j(r)}$	relative surface excess of species j with respect to component r	mol/cm^2	12.1.2
$\Gamma_{j,s}$	surface excess of species j at saturation	mol/cm^2	12.4.2
γ	surface tension	dyne/cm	
γ_j	activity coefficient for species γ_j	none	2.1.6
Δ	ellipsometric parameter	none	14.1.2
δ_j	"diffusion" layer thickness at an electrode fed by convective transfer	cm	1.4.2, 8.3.2
ε	(a) dielectric constant	none	12.3.1
	(b) optical-frequency dielectric constant	none	14.1.2
	(c) porosity	none	10.6.2
$\hat{\varepsilon}$	complex optical-frequency dielectric constant	none	14.1.2
ε_j	molar absorptivity of species j	$M^{-1}\ \mathrm{cm}^{-1}$	14.1.1
ε_0	permittivity of free space	$\mathrm{C}^2\ \mathrm{N}^{-1}\ \mathrm{m}^{-2}$	12.3.1
η	overpotential, $E - E_{eq}$	V, mV	1.3.2, 3.5.2
η_{conc}	concentration overpotential	V, mV	1.4.2
η_{ct}	charge transfer overpotential	V, mV	1.3.3, 3.5.6

Symbol	Meaning	Usual Dimensions	Section References
η_j	viscosity of fluid j	g cm^{-1} sec^{-1} = poise	8.2.2
$\eta_{mt,j}$	mass transfer overpotential for species j	V, mV	1.3.3, 3.5.6
θ	$\exp[(nF/RT)(E - E^{0\prime})]$	none	5.4.1
θ_j	fractional coverage of an interface by species j	none	12.4.2
κ	(a) conductivity of a solution	S/cm = Ω^{-1} cm^{-1}	2.3.3, 4.2
	(b) double-layer thickness parameter	cm^{-1}	12.3.2
Λ_j	equivalent conductivity of a solution	cm^2 Ω^{-1} equiv^{-1}	2.3.3, 4.1
λ	(a) $k_f\tau^{1/2}/D_O^{1/2}$	none	5.5.4
	(b) dimensionless homogeneous kinetic parameter, specific to mechanism	none	11.3
	(c) switching time in CV	sec	6.5
	(d) wavelength of light in vacuo	nm, Å	14.1.2
λ_j	equivalent ionic conductivity for ion j	cm^2 Ω^{-1} equiv^{-1}	2.3.3, 4.1
$\lambda_{0,j}$	equivalent ionic conductivity of ion j extrapolated to infinite dilution	cm^2 Ω^{-1} equiv^{-1}	2.3.3, 4.1
μ	(a) reaction layer thickness	cm	1.5.2
	(b) magnetic permeability	none	14.1.2
μ_j^α	chemical potential of species j in phase α	kJ/mol	2.2.4
$\mu_j^{0\alpha}$	standard chemical potential of species j in phase α	kJ/mol	2.2.4
$\bar{\mu}_e^\alpha$	electrochemical potential of electrons in phase α	kJ/mol	2.2.4, 14.5.2
$\bar{\mu}_j^\alpha$	electrochemical potential of species j in phase α	kJ/mol	2.2.4
μ	(a) kinematic viscosity	cm^2/sec	8.2.2
	(b) frequency of light	sec^{-1}	
ξ	$(D_O/D_R)^{1/2}$	none	5.4.1
ρ	resistivity	Ω-cm	4.2
σ	(a) nFv/RT	sec^{-1}	6.2.1
	(b) $(1/nFA\sqrt{2}\ [\beta_O/D_O - \beta_R/D_R]$	Ω-sec$^{1/2}$	9.2.3
σ^j	excess charge density on phase j	C/cm^2, μC/cm^2	1.2.3, 2.2
σ_j	parameter describing potential dependence of adsorption energy	none	12.5.4
τ	(a) transition time in chronopotentiometry	sec	7.2.2
	(b) sampling time in sampled-current voltammetry	sec	5.1, 5.8
	(c) forward step duration in a double-step experiment	sec	5.7.1
	(d) generally, a characteristic time defined by the properties of an experiment	sec	
τ'	start of potential pulse in normal and differential pulse voltammetry	sec	5.8.2, 5.8.3
ϕ	(a) electrostatic potential	V	2.2.1

Symbol	Meaning	Usual Dimensions	Section References
ϕ	(b) phase angle between two sinusoidal signals	degrees, radians	9.1.2
	(c) phase angle between $\dot{I}_{ac}$ and $\dot{E}_{ac}$	degrees, radians	9.1.2, 9.3
	(d) angular coordinate in a cylindrical system	radians	8.3.1
$\Delta\phi$	potential drop in the space charge region of a semiconductor	V, mV	14.5.2
ϕ^j	absolute electrostatic potential of phase j	V	2.2.1
ϕ_0	total potential drop across the solution side of the double layer	mV	12.3.2, 12.3.3
ϕ_2	potential at the OHP with respect to bulk solution	mV	1.2.3, 12.3.3
χ	$(12/7)^{1/2}k_f\tau^{1/2}/D_O^{1/2}$	none	5.5.5
$\chi(j)$	dimensionless distance of box j in a simulation	none	B.1.5
$\chi(bt)$	normalized current for a totally irreversible system in LSV and CV	none	6.3.1
$\chi(\sigma t)$	normalized current for sweep experiments with a reversible system	none	6.2.1
ψ	(a) ellipsometric parameter	none	14.1.2
	(b) dimensionless rate parameter in CV	none	6.5.2
ω	(a) angular frequency of rotation; $2\pi \times$ rotation rate	sec^{-1}	8.3
	(b) angular frequency of a sinusoidal oscillation; $2\pi f$	sec^{-1}	9.1.2
$\Delta\omega$	amplitude of modulated angular rotation rate	sec^{-1}	8.6

Standard Abbreviations

Abbreviation	Meaning	Section Reference
ASV	anodic stripping voltammetry	10.8
CB	conduction band	14.5.2
CE	homogeneous chemical process preceding heterogeneous electron transfer†	11.1.1
CV	cyclic voltammetry	6.1, 6.5
DME	(a) dropping mercury electrode	5.3.1
	(b) 1,2-dimethoxyethane	
DMF	*N,N*-dimethylformamide	
DMSO	dimethylsulfoxide	
DPP	differential pulse polarography	5.8.3
EC	heterogeneous electron transfer followed by homogeneous chemical reaction†	11.1.1
ECE	heterogeneous electron transfer, homogeneous chemical reaction, and heterogeneous electron transfer, in sequence†	11.1.1
ECM	electrocapillary maximum	12.2.2
emf	electromotive force	2.1.3
FFT	fast Fourier transform	A.6
FT	Fourier transform	A.6
GCS	Gouy-Chapman-Stern	12.3.3
GDP	galvanostatic double pulse	7.3.5
HMDE	hanging mercury drop electrode	6.2.3
IDE	ideal depolarized electrode	1.3.2
IHP	inner Helmholtz plane	12.3.3
IPE	ideal polarized electrode	1.2.1
ISE	ion-selective electrode	2.4
LSV	linear sweep voltammetry	6.1

Abbreviation	Meaning	Section Reference
MFE	mercury film electrode	10.8
NHE	normal hydrogen electrode = SHE	1.1.1
NCE	normal calomel electrode, $Hg/Hg_2Cl_2/KCl$ (1.0 M)	
NPP	normal pulse polarography	5.8.2
OHP	outer Helmholtz plane	12.3.3
OTE	optically transparent electrode	14.1.1
OTTLE	optically transparent thin-layer electrode	14.1.1
PC	propylene carbonate	
PZC	potential of zero charge	12.2.2
QRE	quasi-reference electrode, usually a metal wire immersed in the test solution, unpoised in a traditional sense	
RDE	rotating disk electrode	8.3
RRDE	rotating ring-disk electrode	8.4.2
SCE	saturated calomel electrode	1.1.1
SHE	standard hydrogen electrode = NHE	1.1.1
SMDE	static mercury drop electrode	5.8.3
SSCE	sodium-saturated calomel electrode, $Hg/Hg_2Cl_2/NaCl$ (sat'd)	
$TBABF_4$	tetra-*n*-butylammonium fluoborate	
TBAI	tetra-*n*-butylammonium iodide	
TBAP	tetra-*n*-butylammonium perchlorate	
TEAP	tetraethylammonium perchlorate	
THF	tetrahydrofuran	
UPD	underpotential deposition	10.2.1
VB	valence band	14.5.2

†Letters may be subscripted *i*, *q*, or *r* to indicate irreversible, quasi-reversible, or reversible reactions.

chapter 1

Introduction and Overview of Electrode Processes

1.1 INTRODUCTION

Scientists make electrochemical measurements on chemical systems for a variety of reasons. They may be interested in obtaining thermodynamic data about a reaction. They may want to generate an unstable intermediate such as a radical ion and study its rate of decay or its spectroscopic properties. The goal might be the analysis of a solution for trace amounts of metal ions or organic species. In these examples electrochemical methods are employed as tools in the study of chemical systems, in just the way spectroscopic methods are frequently applied. There are also investigations in which the electrochemical properties of the systems themselves are of primary interest, for example, in the design of a new power source or for the electrosynthesis of some product. A number of electrochemical methods have been devised for these investigations. Their application requires an understanding of the fundamental principles of electrode reactions and the electrical properties of electrode-solution interfaces.

In this chapter the terms and concepts employed in describing electrode reactions are introduced. In addition, before embarking on a detailed consideration of methods of studying electrode processes and the rigorous solutions of the mathematical equations that govern them, we will consider approximate treatments of several different types of electrode reactions to illustrate their main features. The concepts and treatments described here will be considered in a more complete and rigorous way in later chapters.

1.1.1 Electrochemical Cells and Reactions

We are constantly concerned with the processes and factors affecting the transport of charge across interfaces between chemical phases. Almost always, one of the two phases contributing to an interface of interest to us will be an *electrolyte*, which is merely a phase through which charge is carried by the movement of ions. Electrolytes may be liquid solutions or fused salts, or they may be ionically conducting solids, such as sodium β-alumina, which has mobile sodium ions. The second phase at the boundary might be another electrolyte, or it might be an *electrode*, which is a phase through which charge is carried by electronic movement. Electrodes can be metals or semiconductors, and they can be solid or liquid.

It is natural to think about events at a single interface, but we will find that one cannot deal experimentally with such an isolated boundary. Instead, one must study the properties of collections of interfaces called *electrochemical cells*. These systems are defined most generally as two electrodes separated by at least one electrolyte phase.

In general, there is a measurable difference in potential between the two electrodes whether the cell is passing a current or not. This difference is really a manifestation of the collected differences in electric potential between all of the various phases in the current path, and we will find in Chapter 2 that the transition in electric potential in crossing from one conducting phase to another usually occurs almost entirely at the interface itself. The sharpness of the transition implies that a very high electric field exists at the interface, and one can expect it to exert great effects on the kinetic behavior of charge carriers (electrons or ions) in the interfacial region. Also, the magnitude of the potential difference at an interface affects the relative energies of the carriers in the two phases; hence it controls the direction of charge transfer. Thus the measurement and control of *cell potentials* (the difference in potential across the electrodes of a cell) is one of the most important aspects of experimental electrochemistry.

Before we consider how these operations are carried out, it is useful to set up a shorthand notation for expressing the structures of cells. For example, the cell pictured in Figure 1.1.1*a* is written compactly as

$$Zn/Zn^{+2}, Cl^-/AgCl/Ag \qquad (1.1.1)$$

In this notation, a slash represents a phase boundary and a comma separates two components in the same phase. A double slash, not yet used here, represents a phase boundary whose potential is regarded as a negligible component of the overall cell potential. When a gaseous phase is involved, it is written adjacent to its corresponding conducting element. For example, the cell in Figure 1.1.1*b* is written schematically as

$$Pt/H_2/H^+, Cl^-/AgCl/Ag \qquad (1.1.2)$$

The overall chemical reaction taking place in a cell is made up of two independent *half-reactions*, which describe the real chemical changes at the two electrodes. Each half reaction responds to the interfacial potential difference at the corresponding electrode. Most of the time one is interested in only one of these reactions, and the electrode at which it occurs is called the *working* (or *indicator*) electrode. To focus on it, one standardizes the other half of the cell by using an electrode made up of phases having constant composition.

The internationally accepted primary reference is the *standard hydrogen electrode* (SHE), or *normal hydrogen electrode* (NHE), which has all components at unit activity:

$$Pt/H_2(a = 1)/H^+(a = 1, \text{aqueous}) \qquad (1.1.3)$$

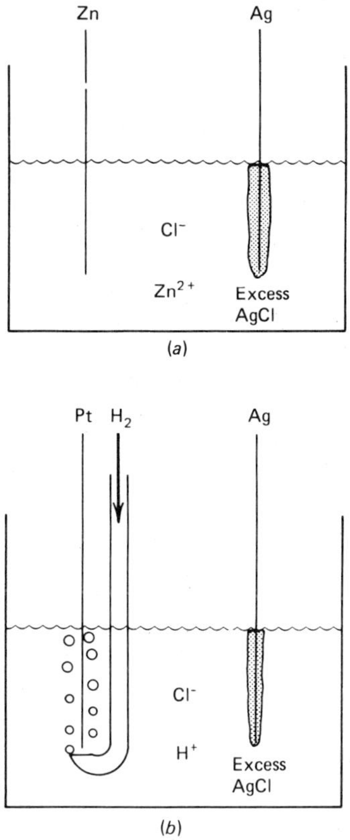

Figure 1.1.1
Typical electrochemical cells. (*a*) Zn metal and Ag wire covered with AgCl immersed in a $ZnCl_2$ solution. (*b*) Pt wire in a stream of H_2 and Ag wire covered with AgCl in HCl solution.

Potentials are often measured and quoted with respect to reference electrodes other than the NHE, which is not very convenient from an experimental standpoint. By far the most common reference is the *saturated calomel electrode* (SCE), which is

$$Hg/Hg_2Cl_2/KCl \text{ (sat'd in water)} \tag{1.1.4}$$

Its potential is 0.242 V *vs.* NHE.

Since the reference electrode has a constant makeup, its potential is fixed. Therefore, any changes in the cell are ascribable to the working electrode. We say that we observe or control the *potential* of the working electrode *with respect* to the reference, and that is equivalent to observing or controlling the energy of the electrons within the working electrode. By driving the electrode to more negative potentials the energy of the electrons is raised, and they will eventually reach a level high enough to occupy vacant states on species in the electrolyte. In this case, a flow of electrons from electrode to solution (a *reduction current*) occurs (Figure 1.1.2*a*). Similarly, the energy of the electrons can be lowered by imposing a more positive potential, and at some point electrons on solutes in the electrolyte will find a more favorable energy on the electrode and will transfer there. Their flow, from solution to electrode, is an *oxidation current* (Figure 1.1.2*b*). The critical potentials at which these processes occur are related to the *standard potentials*, E^0, for the specific chemical substances in the system.

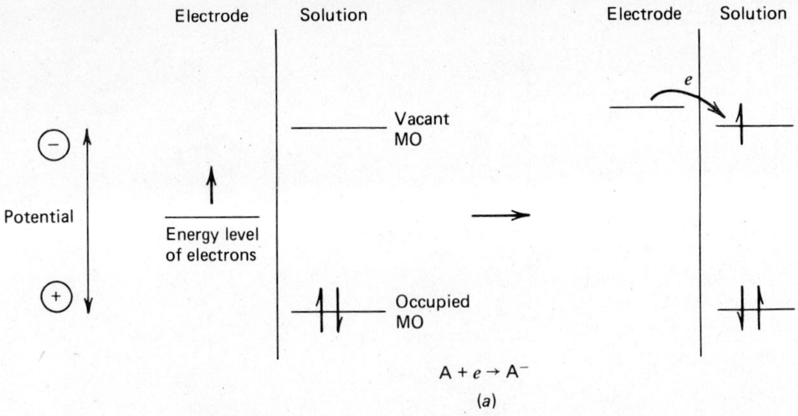

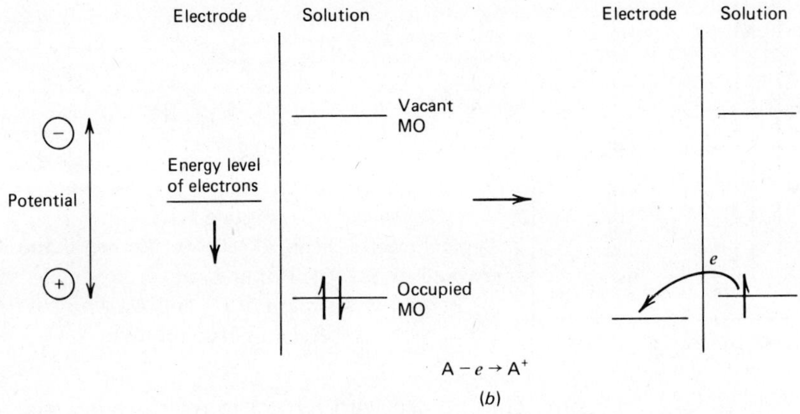

Figure 1.1.2
Representation of (*a*) reduction and (*b*) oxidation process of a species A in solution. The molecular orbitals (M.O.) of species A shown are the highest occupied M.O. and the lowest vacant M.O. As shown, these correspond in an approximate way to the E^0's of the A/A^- and A^+/A couples, respectively. The illustrated system could represent an aromatic hydrocarbon (e.g., 9,10-diphenylanthracene) in an aprotic solvent (e.g., acetonitrile) at a platinum electrode.

In general, when the potential of an electrode is moved from its equilibrium (or its zero-current) value toward more negative potentials, the substance that will be reduced first (assuming all possible electrode reactions are rapid) is the oxidant in the couple with the least negative (or most positive) E^0. For example, for a platinum electrode immersed in an aqueous solution containing 0.01 *M* each of Fe^{3+}, Sn^{4+}, and Ni^{2+}, in 1 *M* HCl, the first substance reduced will be Fe^{3+}, since the E^0 of this couple is most positive (i.e., Fe^{3+} is easiest to reduce) (Figure 1.1.3*a*). When the potential of the electrode is moved from its zero-current value toward more positive potentials, the substance that will be oxidized first is the reductant in the couple of least positive (or most negative) E^0. Thus for a gold electrode in an aqueous solution containing 0.01 *M* each of Sn^{2+} and Fe^{2+}, in 1 *M* HI, the Sn^{2+} will be first oxidized,

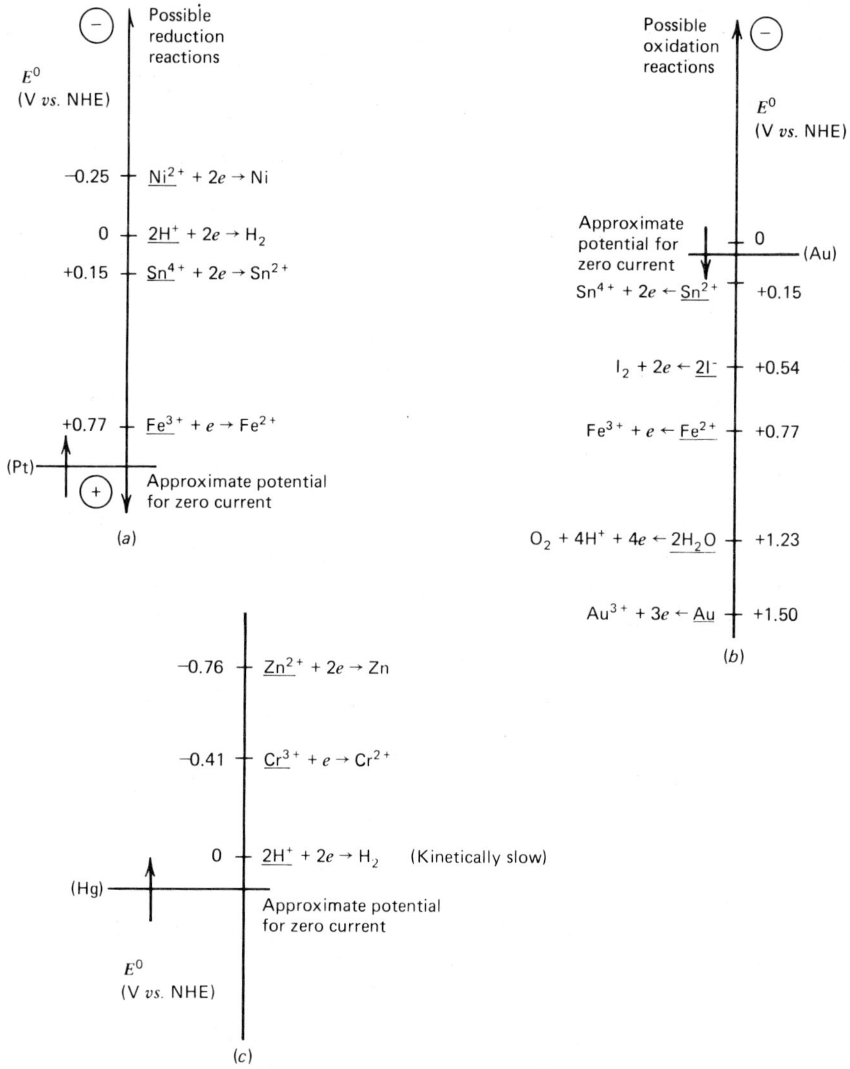

Figure 1.1.3

(*a*) Potentials for possible reductions at a platinum electrode, initially at ~1 V *vs.* NHE in a solution of 0.01 *M* each of Fe^{3+}, Sn^{4+}, and Ni^{2+} in 1 *M* HCl. (*b*) Potentials for possible oxidation reactions at a gold electrode, initially at ~ +0.1 V *vs.* NHE in a solution of 0.01 *M* each of Sn^{2+} and Fe^{2+} in 1 *M* HI. (*c*) Potentials for possible reductions at mercury electrode in 0.01 *M* Cr^{3+} and Zn^{2+} in 1 *M* HCl.

since the E^0 of this couple is least positive (i.e., Sn^{2+} is easiest to oxidize) (Figure 1.1.3*b*). Of course these predictions are based on thermodynamic considerations, and slow kinetics might prevent a reaction from occurring at a significant rate in a potential region where the E^0 would suggest the reaction was possible. Thus for a mercury electrode immersed in a solution of 0.01 *M* each of Cr^{3+} and Zn^{2+}, in 1 *M* HCl, the first reduction process predicted would be the reduction of H^+ (Figure 1.1.3*c*). This reaction, however, is very slow on mercury, and the first process actually observed is the reduction of Cr^{3+}.

1.1.2 Faradaic and Nonfaradaic Processes

Two types of processes occur at electrodes. One kind comprises those just discussed, in which charges (e.g., electrons) are transferred across the metal-solution interface. This electron transfer causes oxidation or reduction to occur. Since these reactions are governed by Faraday's law (i.e., the amount of chemical reaction caused by the flow of current is proportional to the amount of electricity passed), they are called *faradaic* processes. Electrodes at which faradaic processes occur are sometimes called *charge transfer* electrodes. Under some conditions a given electrode-solution interface will show a range of potentials where no charge transfer reactions occur because such reactions are thermodynamically or kinetically unfavorable. However, processes such as adsorption and desorption can occur, and the structure of the electrode-solution interface can change with changing potential or solution composition. These processes are called *nonfaradaic* processes. Although charge does not cross the interface under these conditions, external currents can flow (at least transiently) when the potential, electrode area, or solution composition changes. Both faradaic and nonfaradaic processes occur when electrode reactions take place. Although the faradaic processes are usually of primary interest in the investigation of an electrode reaction (except in studies of the nature of the electrode-solution interface itself), even here the effects of the nonfaradaic processes must be taken into account in using the electrochemical data to obtain information about the charge transfer and associated reactions. We discuss first, then, the simpler case of a system where only nonfaradaic processes occur.

1.2 NONFARADAIC PROCESSES AND THE NATURE OF THE ELECTRODE-SOLUTION INTERFACE

1.2.1 The Ideal Polarized Electrode

An electrode at which no charge transfer across the metal-solution interface can occur regardless of the potential imposed by an outside source of voltage is called an *ideal polarized electrode* (IPE). While no real electrode can behave as an IPE over the whole potential range available in a solution, some electrode-solution systems, over certain limited potential ranges, can approach ideal polarizability. For example, a mercury electrode in contact with a deaerated potassium chloride solution approaches the behavior of an IPE over a potential range greater than 1.5 V. At sufficiently positive potentials the mercury can oxidize in a charge transfer reaction:

$$\mathrm{Hg} + \mathrm{Cl^-} \rightarrow \tfrac{1}{2}\mathrm{Hg_2Cl_2} + e \ (\text{at} \sim +0.25\ \mathrm{V}\ \textit{vs.}\ \mathrm{NHE}) \tag{1.2.1}$$

and at very negative potentials K^+ can be reduced:

$$\mathrm{K^+} + \mathrm{Hg} + e \rightarrow \mathrm{K(Hg)} \ (\text{at} \sim -2.1\ \mathrm{V}\ \textit{vs.}\ \mathrm{NHE}) \tag{1.2.2}$$

However, in the potential range between these processes, no charge transfer reactions occur. The reduction of water:

$$H_2O + e \rightarrow \tfrac{1}{2}H_2 + OH^- \tag{1.2.3}$$

is thermodynamically possible in this region, but occurs at a very slow rate at a mercury surface unless quite negative potentials are reached. The only faradaic current that flows in this region is that due to charge transfer reactions of trace impurities (e.g., metal ions, oxygen, and organic species), and in clean systems this current is quite small.

1.2.2 Capacitance and Charge of an Electrode

Since charge cannot cross the IPE interface when the potential across it is changed, the behavior of the electrode-solution interface is analogous to that of a capacitor. A capacitor is an electrical circuit element, represented as in Figure 1.2.1*a*, composed of two metal sheets separated by a dielectric material whose behavior is governed by the equation

$$\frac{q}{E} = C \tag{1.2.4}$$

where q is the charge on the capacitor (in coulombs, C), E is the potential across the capacitor (in volts, V), and C is the capacitance (in farads, F). When a potential is applied across a capacitor, charge will accumulate on its metal plates until q satisfies equation 1.2.4. During this charging process, a current (called the *charging current*) will flow. The charge on the capacitor consists of an excess of electrons on one plate and a deficiency of electrons on the other (Figure 1.2.1*b*). For example, if a 2-V battery is placed across a 10 μF capacitor, current will flow [the magnitude of which depends on the resistance in the circuit (see Section 1.2.4)] until 20 μC have accumulated on the capacitor plates.

The electrode-solution interface has been shown experimentally to behave like a capacitor, and a model of the interfacial region somewhat resembling a capacitor can be given. At a given potential there will exist a charge on the metal electrode, q^M, and a charge in the solution, q^S (Figure 1.2.2). Whether the charge on the metal is negative or positive with respect to the solution depends on the potential across the interface

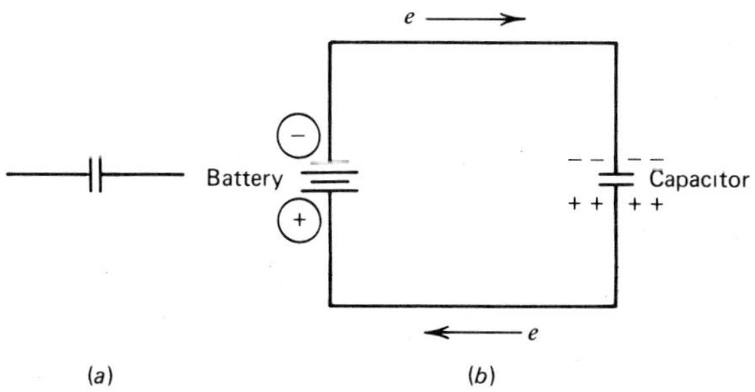

Figure 1.2.1
(*a*) A capacitor. (*b*) Charging a capacitor with a battery.

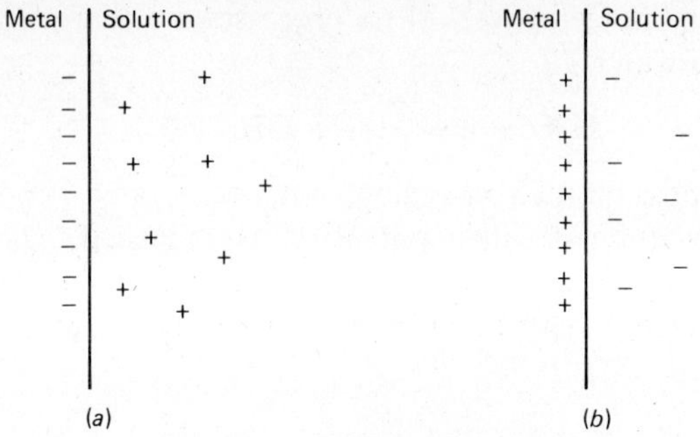

Figure 1.2.2
The metal-solution interface as a capacitor with a charge on the metal, q^M, (*a*) negative and (*b*) positive.

and the composition of the solution. At all times, however, $q^M = -q^S$. (Note that in an actual experimental arrangement, two metal electrodes, and thus two interfaces, would have to be considered; we concentrate our attention here on one of these and ignore what happens at the other.) The charge on the metal q^M represents an excess or deficiency of electrons and resides in a very thin layer (<0.1 Å) on the metal surface. The charge in solution q^S is made up of an excess of either cations or anions in the vicinity of the electrode surface. The charges q^M and q^S are often divided by the electrode area and expressed as charge densities, $\sigma^M = q^M/A$, usually given in μC/cm^2. The whole array of charged species and oriented dipoles existing at the metal-solution interface is called the *electrical double layer* (although its structure only very loosely resembles two charged layers, as we will see in Section 1.2.3). At a given potential the electrode-solution interface is characterized by a double-layer capacitance, C_d, typically in the range of 10 to 40 μF/cm^2; notice, however, that unlike real capacitors, whose capacitances are independent of the voltage across them, C_d is often a function of potential.†

1.2.3 Brief Description of the Electrical Double Layer

The solution side of the double layer is thought to be made up of several "layers." The layer closest to the electrode, the *inner layer*, contains solvent molecules and sometimes other species (ions or molecules) that are said to be *specifically adsorbed* (Figure 1.2.3). This inner layer is also called the *compact, Helmholtz*, or *Stern layer*. The locus of the electrical centers of the specifically adsorbed ions is called the *inner Helmholtz plane* (IHP) which is at a distance x_1. The total charge density (μC/cm^2) from specifically adsorbed ions in this inner layer is σ^i. Solvated ions can only approach the metal to a distance x_2; the locus of centers of these nearest solvated ions is called the *outer Helmholtz plane* (OHP). The interaction of the solvated ions with the

† In various equations in the literature and in this book, C_d may express the capacitance per unit area and be given in μF/cm^2, or it may express the capacitance of a whole interface and be given in μF. The usage for a given situation is always apparent from the context or from a dimensional analysis.

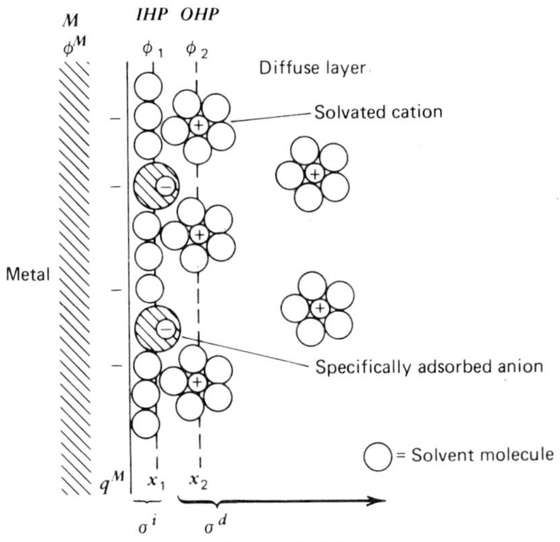

Figure 1.2.3
Proposed model of the electrode-solution, double-layer region.

charged metal involves only long-range electrostatic forces, so that their interaction is essentially independent of the chemical properties of the ions. These ions are said to be *nonspecifically adsorbed.* Because of thermal agitation in the solution, the nonspecifically adsorbed ions are distributed in a three-dimensional region, called the *diffuse layer*, which extends from the OHP into the bulk of the solution. The excess charge density in the diffuse layer is σ^d, so that the total excess charge density on the solution side of the double layer, σ^S, is given by

$$\boxed{\sigma^S = \sigma^i + \sigma^d = -\sigma^M} \tag{1.2.5}$$

The thickness of the diffuse layer depends on the total ionic concentration in the solution; for concentrations greater than 10^{-2} *M*, the thickness is less than 300 Å. The potential profile across the double-layer region is shown in Figure 1.2.4.

The structure of the double layer can affect the rates of electrode processes. Consider an electroactive species that is not specifically adsorbed. This species can only approach the electrode to the OHP, and the total potential it experiences is less than the potential between the electrode and the solution by an amount $\phi_2 - \phi^S$, which is the potential drop across the diffuse layer. For example, in 0.1 *M* NaF, $\phi_2 - \phi^S$ is -0.021 V at $E = -0.55$ V *vs.* SCE, but it has somewhat larger magnitudes at more negative and more positive potentials. Sometimes one can neglect double-layer effects in considering electrode reaction kinetics. At other times they must be taken into account. The importance of adsorption and double-layer structure is considered in greater detail in Chapter 12.

One usually cannot neglect the existence of the double layer-capacitance or the presence of a charging current in electrochemical experiments. Indeed, during electrode reactions involving very low concentrations of electroactive species, the charging current can be much larger than the faradaic current for the reduction or oxidation

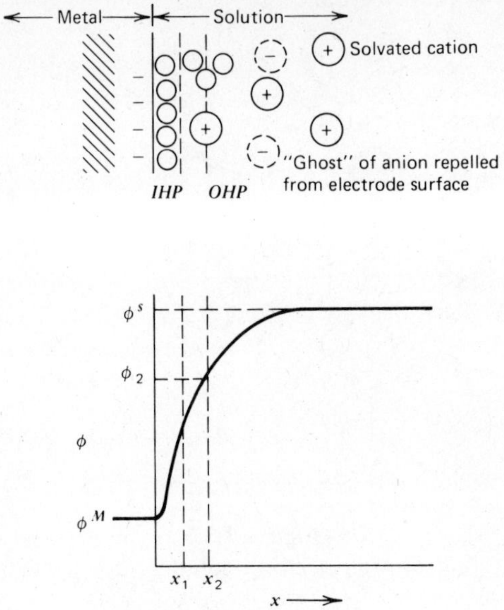

Figure 1.2.4
Potential (ϕ) profile across double-layer region. (A more quantitative representation is shown in Figure 12.3.6.)

reaction. For this reason we consider briefly the nature of the charging current at an IPE for several types of electrochemical experiments.

1.2.4 Double-Layer Capacitance and Charging Current in Electrochemical Measurements

A cell consisting of an IPE and an ideal reversible electrode, approximated by a mercury electrode in contact with a potassium chloride solution that also contacts a saturated calomel electrode (SCE) (represented by the cell formulation Hg/K^+, Cl^-/SCE), can be approximated by an electrical circuit of a resistor, R_s, representing the solution resistance, and a capacitor C_d representing the double layer at the Hg/K^+, Cl^- interface (Figure 1.2.5). (The capacitance of the SCE, C_{SCE}, should also be included but, since the series capacitance of C_d and C_{SCE}, $C_T = C_d C_{SCE}/[C_d + C_{SCE}]$, and usually $C_{SCE} \gg C_d$, then $C_T \simeq C_d$, and C_{SCE} can be neglected in the circuit.) However, C_d will generally be a function of potential so that the proposed model in terms of actual circuit elements is only useful for experiments where the overall cell potential does not change very much. Where it does, approximate results can be obtained using an "average" C_d over the potential range.

As we will see in later sections and chapters, information about an electrochemical system is gained by applying an electrical perturbation to the system and observing the resulting changes in the characteristics of the system. It is instructive to consider the behavior of the IPE system, represented by the circuit elements R_s and C_d in series, to several types of electrical perturbation.

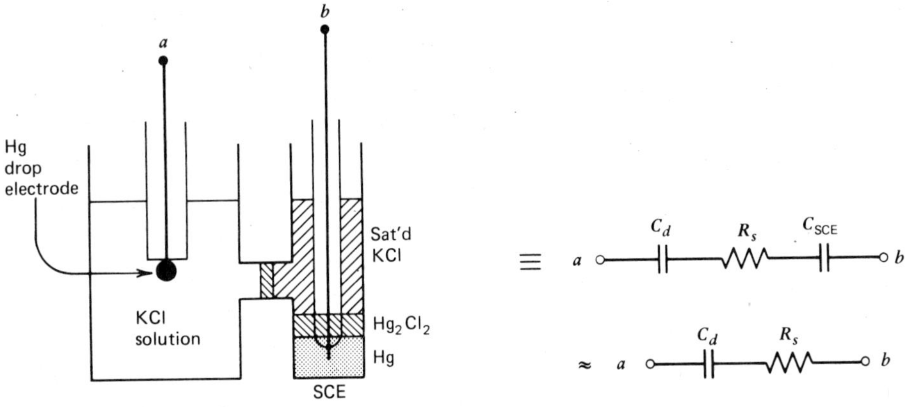

Figure 1.2.5
Two-electrode cell with an ideal polarized mercury drop electrode and an SCE, and the representation of this cell in terms of linear circuit elements.

(a) Voltage (or Potential) Step. The result of a potential step to the IPE is the familiar *RC* circuit problem (Figure 1.2.6). The behavior of the current, *i*, with time, *t*, when applying a potential step of magnitude *E*, is

$$\boxed{i = \frac{E}{R_s} e^{-t/R_sC_d}} \tag{1.2.6}$$

This equation is derived from the general equation for the charge *q*, on a capacitor as a function of the voltage across it, E_C:

$$q = C_d E_C \tag{1.2.7}$$

At any time the sum of the voltages across the resistor, E_R, and the capacitor must equal the applied voltage; hence

$$E = E_R + E_C = iR_s + \frac{q}{C_d} \tag{1.2.8}$$

Rearranging, and noting that $i = dq/dt$, yields

$$\frac{dq}{dt} = \frac{-q}{R_sC_d} + \frac{E}{R_s} \tag{1.2.9}$$

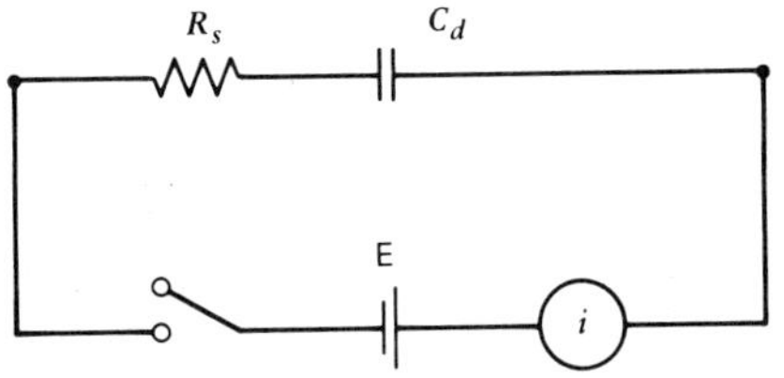

Figure 1.2.6
Potential step experiment for *RC* circuit.

If we assume that the capacitor is initially uncharged ($q = 0$ at $t = 0$), we obtain as a solution of (1.2.9)

$$q = EC_d\left[1 - \exp\left(\frac{-t}{R_sC_d}\right)\right] \tag{1.2.10}$$

By differentiating (1.2.10), equation 1.2.6 is obtained. Hence, for a potential step input, an exponentially decaying current is obtained, with a time constant, $\tau = R_sC_d$ (Figure 1.2.7). Note that for a potential step, the current for charging the double-layer capacitance drops to 37% of its initial value at $t = \tau$ and to 5% of its initial value at $t = 3\tau$. For example, if $R_s = 1\ \Omega$ and $C_d = 20\ \mu F$, $\tau = 20\ \mu$sec, or double-layer charging is 95% complete in 60 μsec.

(b) Current Step. When the R_sC_d circuit is charged by a constant current i (Figure 1.2.8), then again equation 1.2.8 applies. Since $q = \int i\,dt$, and i is a constant,

$$E = iR_s + \frac{i}{C_d}\int_0^t dt \tag{1.2.11}$$

or

$$\boxed{E = i(R_s + t/C_d)} \tag{1.2.12}$$

Hence, for a current step, assuming a constant C_d, the potential increases linearly with time (Figure 1.2.9).

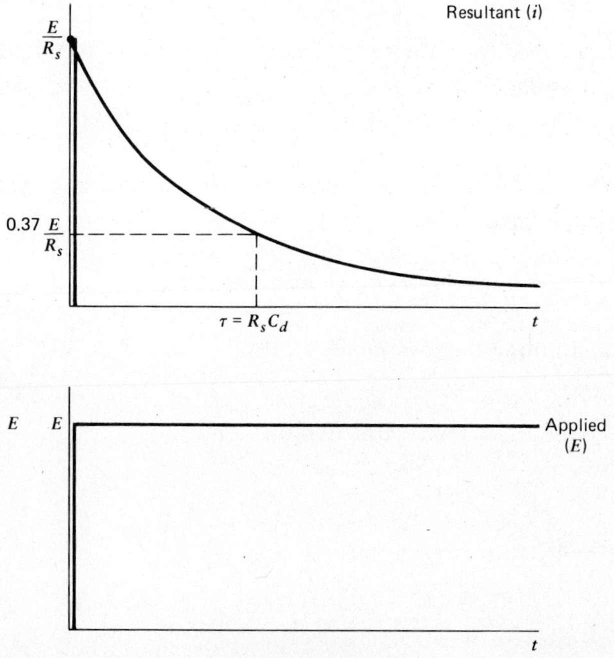

Figure 1.2.7
Current transient (i vs. t) resulting from potential step experiment.

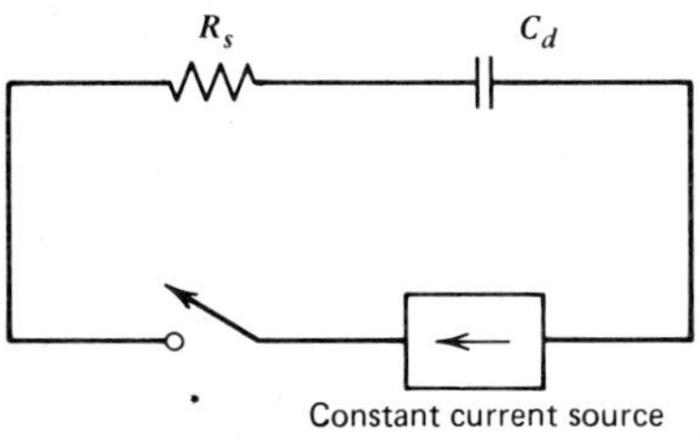

Figure 1.2.8
Current step experiment for *RC* circuit.

(c) Voltage Ramp (or Potential Sweep). A voltage ramp or linear potential sweep is a potential that increases linearly with time starting at some initial value, E_i, at a sweep rate, v (in volts per second) (see Figure 1.2.10*a*).

$$E = E_i + vt \tag{1.2.13}$$

If such a ramp is applied to the R_sC_d circuit, equation 1.2.8 still applies, or

$$E_i + vt = R_s \frac{dq}{dt} + \frac{q}{C_d} \tag{1.2.14}$$

Thus (for $q = 0$ at $t = 0$)

$$\boxed{i = vC_d + \left[\left(\frac{E_i}{R_s} - vC_d\right) \exp(-t/R_sC_d)\right]} \tag{1.2.15}$$

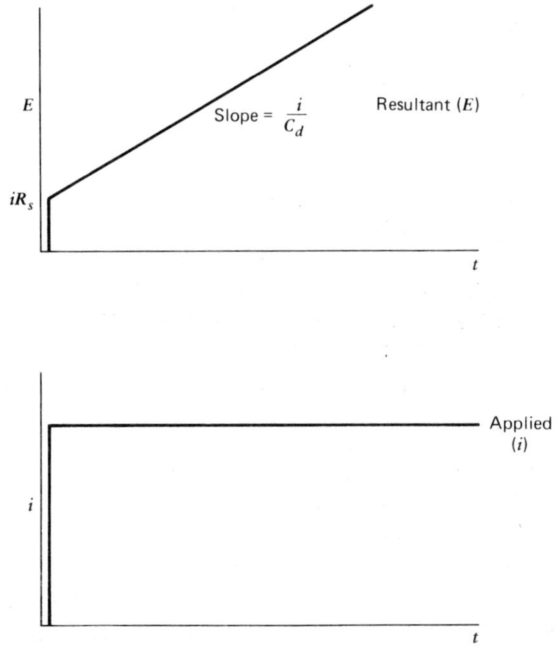

Figure 1.2.9
E-t behavior resulting from current step experiment.

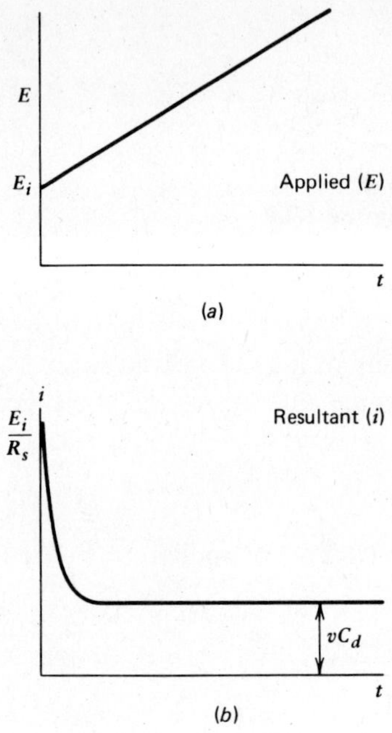

Figure 1.2.10
Current-time behavior resulting from a linear potential sweep applied to an *RC* circuit.

The current now contains a transient part (contained within the brackets), which dies away with a time constant, R_sC_d, and a steady-state current, vC_d (Figure 1.2.10*b*). If R_sC_d is small and v is constant, the instantaneous current can be used to measure C_d as a function of E.

If a triangular wave, that is, a ramp whose sweep rate switches from v to $-v$ at some potential, E_λ, is applied to the R_sC_d circuit, then the steady-state current changes from vC_d during the forward (increasing E) scan to $-vC_d$ during the reverse (decreasing E) scan. The result for a system with a very small time constant, and with C_d constant with E, is shown in Figure 1.2.11.

1.3 FARADAIC PROCESSES AND FACTORS AFFECTING RATES OF ELECTRODE REACTIONS

1.3.1 Electrochemical Cells—Types and Definitions

Electrochemical cells in which faradaic currents are flowing are classified as either *galvanic* or *electrolytic* cells. A *galvanic cell* is one in which reactions occur spontaneously at the electrodes when they are connected externally by a conductor (Figure 1.3.1*a*). These cells are often employed in converting chemical energy into electrical energy. Types of galvanic cells of commercial importance include primary (nonrechargeable) cells (e.g., the Leclanché Zn-MnO_2 cell), secondary (rechargeable) cells (e.g., a charged Pb-PbO_2 storage battery) and fuel cells (e.g., an H_2-O_2 cell). An *electrolytic cell* is one in which reactions are effected by the imposition of an external voltage greater than the reversible potential of the cell (Figure 1.3.1*b*). These cells are

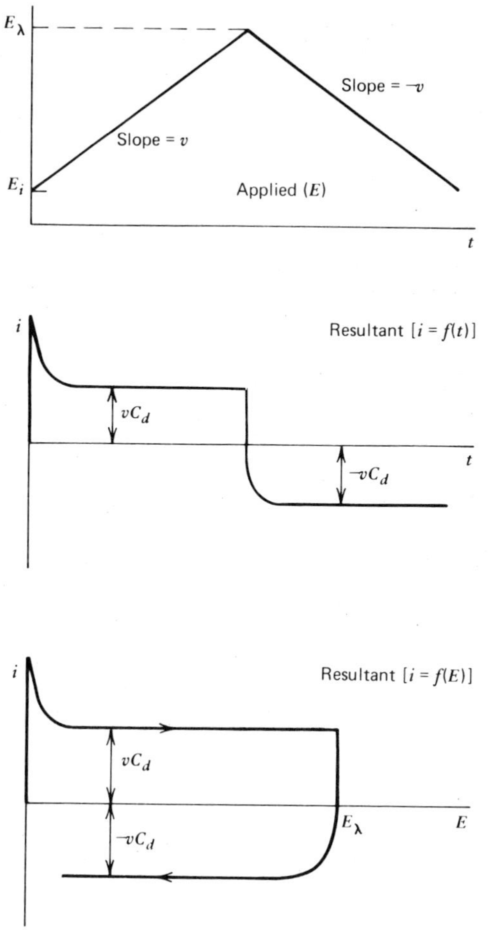

Figure 1.2.11
Current-time and current-potential plots from a cyclic linear potential sweep (or triangular wave) applied to an RC circuit.

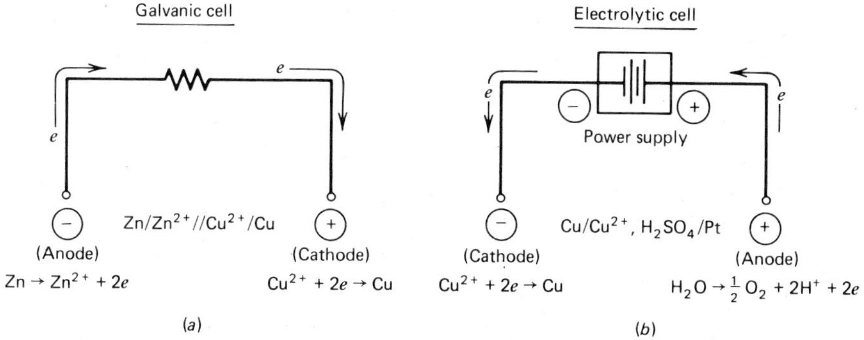

Figure 1.3.1
(a) Galvanic and (b) electrolytic cells.

frequently employed to carry out desired chemical reactions by expending electrical energy. Commercial processes involving electrolytic cells include electrolytic syntheses (e.g., the production of chlorine and aluminum), electrorefining (e.g., copper), and electroplating (e.g., silver and gold). The lead sulfate storage cell, when it is being "recharged," is an electrolytic cell.

Although it is often convenient to make a distinction between galvanic and electrolytic cells, we will most often be concerned with reactions occurring at only *one* of the electrodes. This simplifies the treatment by concentrating our attention on only one-half of the cell at a time. The behavior of a two-electrode cell can be ascertained later, if desired, by combining the individual half-cells. The fundamental nature of the reactions and behavior of a single electrode are independent of whether this electrode is part of a galvanic or electrolytic cell. For example, consider the cells in Figure 1.3.1. The nature of the reaction $Cu^{2+} + 2e \rightarrow Cu$ is the same in both cells. If one desires to plate copper, one could accomplish this in either a galvanic cell (using a counter half-cell with a more negative potential than that of Cu/Cu^{2+}), a so-called "internal electrolysis," or in an electrolytic cell (using any counter half-cell and supplying electrons to the copper electrode with an external power supply). Thus we generally define *electrolysis* as comprising the chemical changes accompanying faradaic reactions at electrodes in contact with electrolytes. In discussing cells, one calls the electrode at which reductions occur the *cathode* and the electrode at which oxidations occur the *anode*. Similarly a current in which electrons cross the interface from metal to a species in solution is termed a *cathodic current* while electron flow from a solution species into the metal is called an *anodic current*. In an electrolytic cell the cathode is negative with respect to the anode. In a galvanic cell the cathode is positive with respect to the anode.†

1.3.2 The Electrochemical Experiment and Variables in Electrochemical Cells

An investigation of the electrochemical behavior of a system consists of holding certain variables of the electrochemical cell constant and observing how other variables (usually current, potential, or concentration) vary with changes in the controlled variables. The parameters of importance in electrochemical cells are shown in Figure 1.3.2. For example, in *potentiometric* experiments, $i = 0$, and E is determined as a function of C. Since no current flows in this experiment, no net faradaic reaction occurs, and the potential is frequently (but not always) governed by the thermodynamic properties of the system. Many of the variables (electrode area, mass transfer, electrode geometry) do not affect the potential. Another way of visualizing an electrochemical experiment is in terms of the way in which the system under study responds to a perturbation. The electrochemical cell is considered as a "black box" to which a

† Because a cathodic current and a cathodic reaction can occur at an electrode which is either positive or negative with respect to another electrode (e.g., an auxiliary or reference electrode, see Section 1.3.4), it is poor usage to associate the term cathodic or anodic with potentials of a particular sign. For example, one should not say, "The potential shifted in a cathodic direction," when what is meant is, "The potential shifted in a *negative* direction." The terms anodic and cathodic refer to electron flow or current directions, not to potentials.

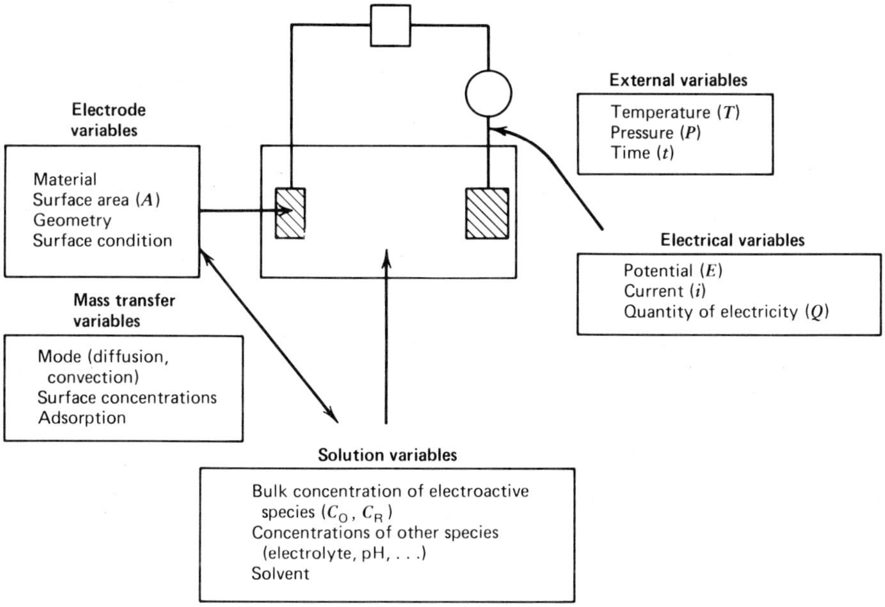

Figure 1.3.2
Variables affecting the rate of an electrode reaction.

certain excitation function (e.g., a potential step) is applied, and a certain response function (e.g., the resulting variation of current with time) is measured, with all other system variables held constant (Figure 1.3.3). The aim of the experiment is to obtain information (thermodynamic, kinetic, analytical, etc.) about the chemical system from observation of the excitation and response functions and a knowledge of appropriate models for the system. This same basic idea is used in many other types of experiments and tests, such as circuit testing or spectrophotometric analysis. In spectrophotometry, the excitation function is light of different wavelengths; the response function is the fraction of light transmitted by the system at these wavelengths; the system model is Beer's law or a molecular model; and the information content comprises the concentrations of absorbing species, their absorptivities, or their transition energies.

Before beginning an examination of simple models for electrochemical systems, let us consider more closely the nature of the current and potential of an electrochemical cell. Consider the cell composed of a cadmium electrode immersed in a 1 M $Cd(NO_3)_2$ solution and coupled through a salt bridge to an SCE (Figure 1.3.4). The reversible, open-circuit potential of the cell (neglecting activity coefficients and the liquid junction potential; see Chapter 2) is 0.64 V, with the copper wire attached to the cadmium electrode negative with respect to that attached to the mercury electrode. When the voltage applied by the external power supply, E_{appl}, is 0.64 V, $i = 0$. When E_{appl} is made larger (i.e., $E_{appl} > 0.64$ V, so that the cadmium electrode is made even more negative with respect to the SCE), the cell behaves as an electrolytic cell and a current flows. At the cadmium electrode, the reaction $Cd^{2+} + 2e \rightarrow Cd$ occurs;

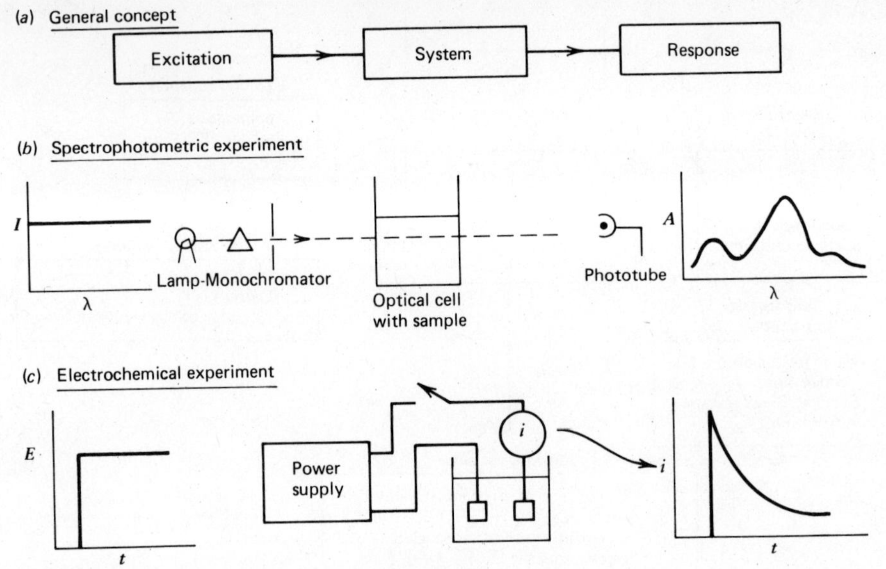

Figure I.3.3

(*a*) General principle of studying a system by application of an excitation (or perturbation) and observation of response. (*b*) In a spectrophotometric experiment the excitation is light of different wavelengths (λ) and the response is the absorbance (A) curve. (*c*) In an electrochemical (potential step) experiment, the excitation is the application of a potential step and the response is the observed *i-t* curve.

while at the SCE, mercury is oxidized to Hg_2Cl_2. A question of interest might be: "If $E_{appl} = 0.74$ V (i.e., if the potential of the cadmium electrode is made -0.74 V *vs.* the SCE), what current will flow?" Since i represents the number of electrons reacting with Cd^{2+} per second, or the number of coulombs of electric charge flowing per second, the question "What is i?" is essentially the same as "What is the rate (v) of

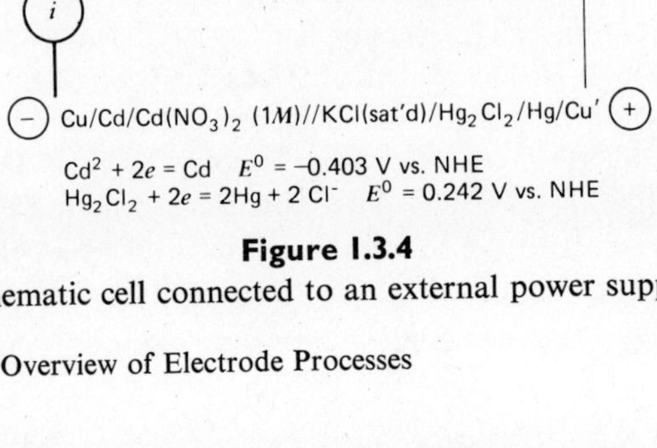

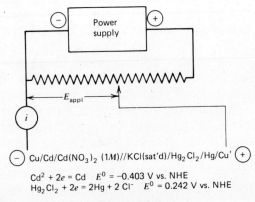

Figure I.3.4

Schematic cell connected to an external power supply.

the reaction $Cd^{2+} + 2e \rightarrow Cd$?" The following relations demonstrate the direct proportionality between faradaic current and electrolysis rate:

$$i \text{ (amperes)} = dQ/dt \text{ (coulombs/second)} \tag{1.3.1}$$

$$\frac{Q}{nF} \frac{\text{(coulombs)}}{\text{(coulombs/mole)}} = N \text{ (moles electrolyzed)} \tag{1.3.2}$$

$$\boxed{v\text{(moles/second)} = \frac{dN}{dt} = \frac{i}{nF}} \tag{1.3.3}$$

Interpreting the rate of an electrode reaction is often more complex than understanding a homogeneous reaction, because an electrode process is a heterogeneous reaction (occurring only at the electrode-electrolyte interface) with a rate that depends on mass transfer to the electrode and various surface effects, in addition to the usual kinetic variables. Since electrode reactions are heterogeneous, their reaction rates are usually described in units of mol/sec per unit area; that is,

$$\boxed{v(\text{mol sec}^{-1}\text{ cm}^{-2}) = \frac{i}{nFA} = \frac{j}{nF}} \tag{1.3.4}$$

where j is the current density (A/cm^2).

Information about an electrode reaction is often gained by determining current as a function of potential (*i-E* curves), and certain names are sometimes associated with features of the curves.† The departure of the electrode potential (or cell potential) from the reversible (i.e., nernstian or equilibrium) value upon passage of faradaic current is termed *polarization*. The larger this departure is, the larger the extent of polarization is said to be. We have seen that an *ideal polarized electrode* (Section 1.2.1) showed a very large change in potential upon the passage of an infinitesimal current; thus ideal polarizability is characterized by a horizontal region of an *i-E* curve (Figure 1.3.5*a*). A substance that tends to cause the potential of an electrode to be nearer its equilibrium value by virtue of its being oxidized or reduced is called a *depolarizer*.‡ An *ideal nonpolarizable electrode* (or *ideal depolarized electrode*) is thus an electrode whose potential does not change upon passage of current, that is, an electrode of fixed potential. Nonpolarizability is characterized by a vertical region on an *i-E* curve (Figure 1.3.5*b*). An SCE constructed with a large-area mercury pool would approach ideal nonpolarizability at small currents. The extent of polarization is measured by the *overpotential*, η, which is the deviation of the potential from the equilibrium value:

$$\boxed{\eta = E - E_{eq}} \tag{1.3.5}$$

† These names are often carry-overs from older electrochemical studies and models and, indeed, do not represent the best possible terminology. Their use is so ingrained in the electrochemical jargon that it seems wisest to keep them and attempt to define them as precisely as possible.

‡ The term depolarizer is also frequently used to denote a substance that is preferentially oxidized or reduced, in order to prevent an undesirable electrode reaction. Sometimes it is simply another name for an electroactive substance.

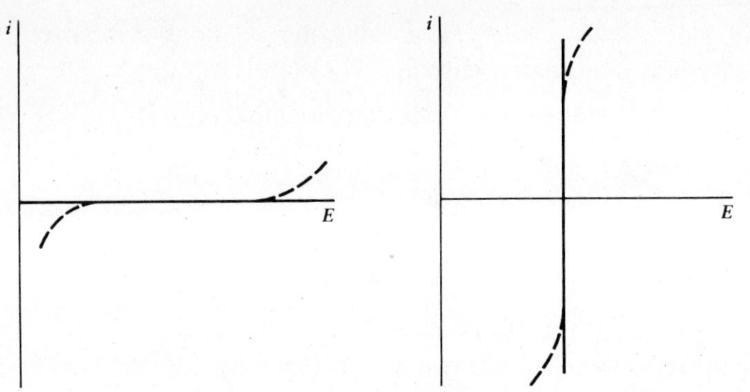

(a) Ideal polarizable electrode (b) Ideal nonpolarized electrode

Figure 1.3.5
Current potential curves for ideal (*a*) polarized and (*b*) nonpolarizable electrodes. Dashed lines show behavior of actual electrodes that approach the ideal behavior over limited current or potential ranges.

Current-potential curves, particularly those obtained under steady-state conditions, are sometimes referred to as *polarization curves*. When two or more faradaic reactions can occur simultaneously at an electrode, the fraction of the total current, (i_{total}) going to the rth process is called the instantaneous *current efficiency*:

$$\text{Instantaneous current efficiency for } r\text{th process} = \frac{i_r}{i_{\text{total}}} \tag{1.3.6}$$

A current efficiency of unity (or 100%) implies that only one process is occurring at an electrode. When one considers the result of an electrolysis over some period of time, the *overall current efficiency* represents the fraction of the total number of coulombs involved in the rth process:

$$\text{Overall current efficiency for } r\text{th process} = \frac{Q_{\text{r}}}{Q_{\text{total}}} \tag{1.3.7}$$

1.3.3 Factors Affecting Electrode Reaction Rate and Current

Consider an overall electrode reaction $\text{O} + ne \rightleftharpoons \text{R}$ composed of a series of steps that cause the conversion of the dissolved oxidized species, O, to a reduced form, R, also in solution (Figure 1.3.6). In general, the current (or electrode reaction rate) is governed by the rates of processes such as:

(1) Mass transfer (e.g., of O from the bulk solution to the electrode surface).

(2) Electron transfer at the electrode surface.

(3) Chemical reactions preceding or following the electron transfer. These might be homogeneous processes, such as protonations or dimerizations, or heterogeneous ones, such as catalytic decompositions on the electrode surface.

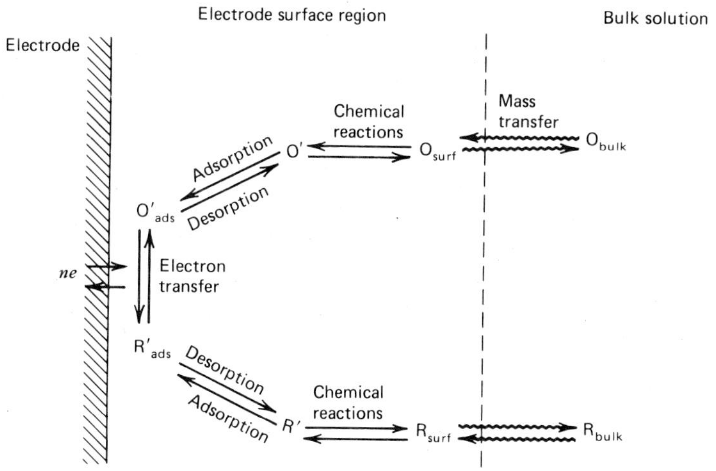

Figure 1.3.6
Pathway of a general electrode reaction.

(4) Other surface reactions, such as adsorption, desorption, or crystallization (electrodeposition).

The simplest reactions involve only mass transfer of a reactant to the electrode, heterogeneous electron transfer involving nonadsorbed species and mass transfer of the product to the bulk solution. A representative reaction of this sort is the reduction of an aromatic hydrocarbon (e.g., 9,10-diphenylanthracene, DPA) to the radical anion ($DPA^{\bar{\cdot}}$) in an aprotic solvent (e.g., *N,N*-dimethylformamide). More complex reaction sequences involving a series of electron transfers and protonations, branching mechanisms, parallel paths, or modifications of the electrode surface are quite common. When a steady-state current is obtained, the rates of all the reaction steps are the same. The magnitude of this current is often limited by the inherent sluggishness of one or more reactions called *rate-determining steps*. The more facile reactions are then held back from maximum rates by the slowness with which such a step disposes of their products or creates their participants.

At a given current density, j, there will be a certain overpotential η. This overpotential can be considered as a sum of different overpotential terms associated with the different reaction steps: η_{mt} (the mass transfer overpotential or concentration polarization), η_{ct} (the charge transfer overpotential or activation polarization), η_{rxn} (the overpotential associated with a preceding reaction or reaction polarization), etc. The electrode reaction can then be represented by a resistance, R, composed of a series of resistances (or more exactly, impedances) representing the various steps: R_{mt}, R_{ct}, etc. (Figure 1.3.7). A fast reaction step is characterized by a small resistance (or impedance) while a slow step is represented by a high resistance. However, except for very small current or potential perturbations, these impedances, unlike the analogous actual electrical elements they represent, are functions of E (or i). This picture of individual overpotentials is also useful in consideration of electrode reaction kinetics from the viewpoint of irreversible thermodynamics (1).

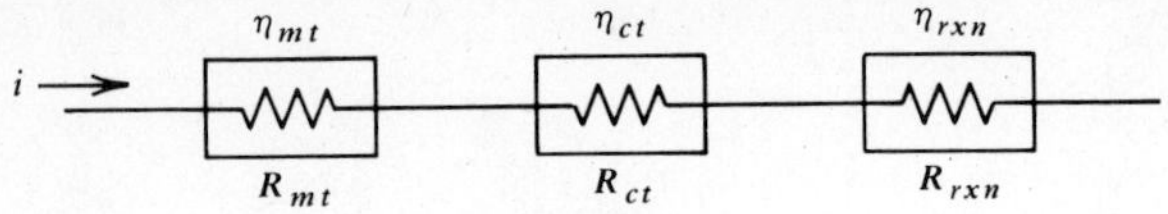

Figure 1.3.7
Processes in an electrode reaction represented as resistances.

1.3.4 Electrochemical Cells and Cell Resistance

When the potential of an electrode of interest is measured against a nonpolarizable reference electrode during the passage of current, a voltage drop equal to iR_s will always be included in the measured potential. Here R_s is the solution resistance between the electrodes which, unlike the pseudoimpedances describing the mass transfer and activation steps in the electrode reaction, behaves as an ideal resistance over a wide range of conditions. For example, consider the cell: ⊖Cd/Cd^{2+}// SCE⊕. At open circuit ($i = 0$) the potential of the cadmium electrode is the equilibrium value, E_{eq} (volts *vs.* SCE). By applying an external voltage of magnitude E_{appl} with a power supply, a current is forced through the cell and the potential of the cadmium electrode will shift to a new value E (volts *vs.* SCE). We assume that the SCE is essentially nonpolarizable at the extant current level and does not change its potential. Then:

$$\boxed{E_{appl} = E + iR_s = E_{eq} + \eta + iR_s} \tag{1.3.8}$$

The term iR_s is the ohmic potential drop in the solution. It is sometimes called the "ohmic polarization," but it should not be regarded as a form of overpotential, since it is characteristic of the bulk solution and not of the electrode reaction. Its contribution to the measured electrode potential can be minimized by proper cell design and instrumentation.

Consider a cell composed of two ideal nonpolarizable electrodes, for example, two SCEs immersed in a potassium chloride solution: SCE/KCl/SCE. The *i-E* characteristic of this cell would look like that of a pure resistance (Figure 1.3.8). These conditions

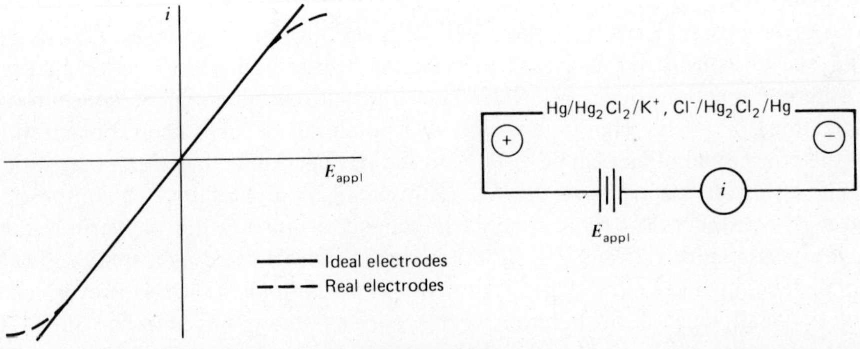

Figure 1.3.8
Current-potential curve for a cell composed of two electrodes approaching ideal non-polarizability.

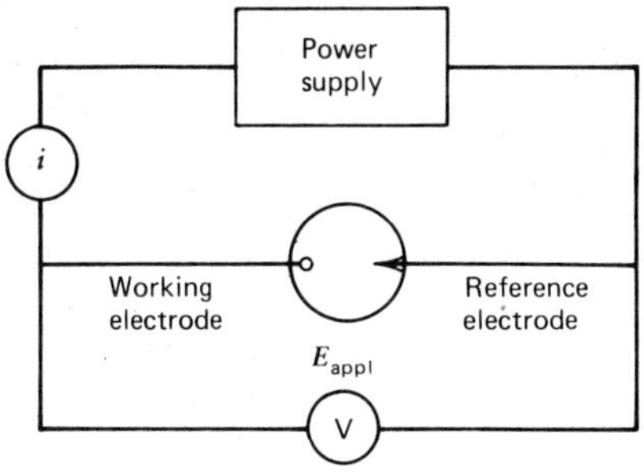

Figure 1.3.9
Two-electrode cell.

(i.e., nonpolarizable electrodes) are exactly those sought in measurements of solution conductivity. For any real electrodes (e.g., actual SCEs), at high enough current densities, mass transfer and charge transfer overpotentials would also become important.

Most of the time one is interested in reactions that occur at only one electrode, so that an experimental cell could be composed of the electrode system of interest, called the *working* (or *indicator*) electrode, coupled with an electrode that approaches an ideal nonpolarizable electrode of known potential (such as an SCE with a large-area mercury pool), called the *reference* electrode. If the passage of current does not affect the potential of the reference electrode, the E of the working electrode is given by equation 1.3.8. Under conditions when iR_s is small (say less than 1–2 mV), this *two-electrode cell* arrangement can be used to determine the i-E curve, with E either taken as equal to E_{appl} or corrected for the small iR_s drop (Figure 1.3.9). For example, in polarographic experiments in aqueous solutions, a two-electrode cell is often used. In this case $i < 10\ \mu A$, $R_s < 100\ \Omega$, so that $iR_s < (10^{-5}\ A)(100\ \Omega)$ or $iR_s < 1$ mV. On the other hand, when currents or solution resistances are high (e.g., in large-scale, electrolytic cells or galvanic cells or in experiments involving nonaqueous solutions with low conductivities), the iR_s term may be much larger.

In experiments where iR_s may be high, a *three-electrode cell* arrangement (Figure 1.3.10) is preferable. In this arrangement the current is passed between the working electrode and an *auxiliary* (or *counter*) *electrode*. The auxiliary electrode can be any

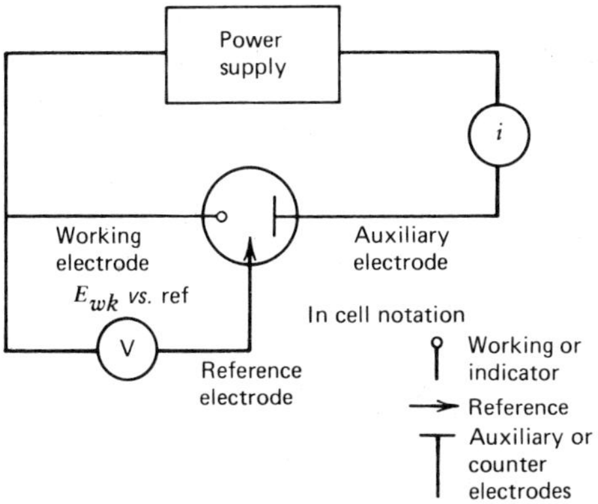

Figure 1.3.10
Three-electrode cell and notation for the different electrodes.

electrode desired, because its electrochemical properties do not affect the behavior of the electrode of interest; it is usually chosen to be an electrode that does not produce substances by electrolysis that will reach the working-electrode surface and cause interfering reactions there. It is frequently placed in a compartment separated from the working-electrode compartment by a sintered-glass disk or other separator. The potential of the working electrode is monitored relative to a separate reference electrode, positioned with its tip near the working electrode. The device used to measure or monitor the potential difference between the working electrode and the reference electrode has a high input impedance so that a negligible current is drawn through the reference electrode. Since essentially no current is passed through the reference electrode, its potential will remain constant and equal to its open-circuit value. Also, the iR_s contribution to the measured potential will be much smaller. This three-electrode cell arrangement is used in most electrochemical experiments; several practical cells are shown in Figure 1.3.11.

Note, however, that even in this arrangement not *all* of the iR_s term is removed from the reading of the potential-measuring device. Consider the potential profile in solution between the working and auxiliary electrodes, shown schematically in Figure 1.3.12. (The exact potential profile in an actual cell depends on the electrode shapes, geometry, solution conductance, etc.) The solution between the electrodes can be thought of as a potentiometer (not necessarily a linear one). For the reference electrode placed anywhere but exactly at the electrode surface, some fraction of iR_s (called iR_u, where R_u is the *uncompensated solution resistance*) will be included in the measured potential. Even when the tip of the reference electrode is designed for very close placement to the working electrode by use of a fine capillary tip called a *Luggin-Haber capillary*, some uncompensated resistance usually remains. This uncompensated potential drop can sometimes be taken into account, in steady-state measurements, for example, by measurement of R_u and point-by-point correction at any given i of the measured potentials. Modern electrochemical instrumentation frequently includes circuitry (such as positive-feedback circuits) for suppression of the iR_u term (see Chapter 13).

If a reference capillary with a tip diameter d is used, it can be placed as close as a distance $2d$ from the working electrode surface without causing appreciable shielding error. *Shielding* denotes a blockage of part of the solution current path at the working electrode surface, which causes nonuniform current densities to arise at the electrode surface (2,3).

For a plane electrode with uniform current density across its surface,

$$R_u = x/\kappa A \tag{1.3.9}$$

where x is the distance of the capillary tip from the electrode, A is the electrode area, and κ is the solution conductivity. The effect of iR_u can be particularly serious for spherical microelectrodes, such as the hanging mercury drop electrode or the dropping mercury electrode (DME). For a spherical electrode of radius r_0 (4),

$$R_u = \frac{1}{4\pi\kappa r_0}\left(\frac{x}{x + r_0}\right) \tag{1.3.10}$$

Note that in this case most of the resistive drop occurs close to the electrode; thus R_u

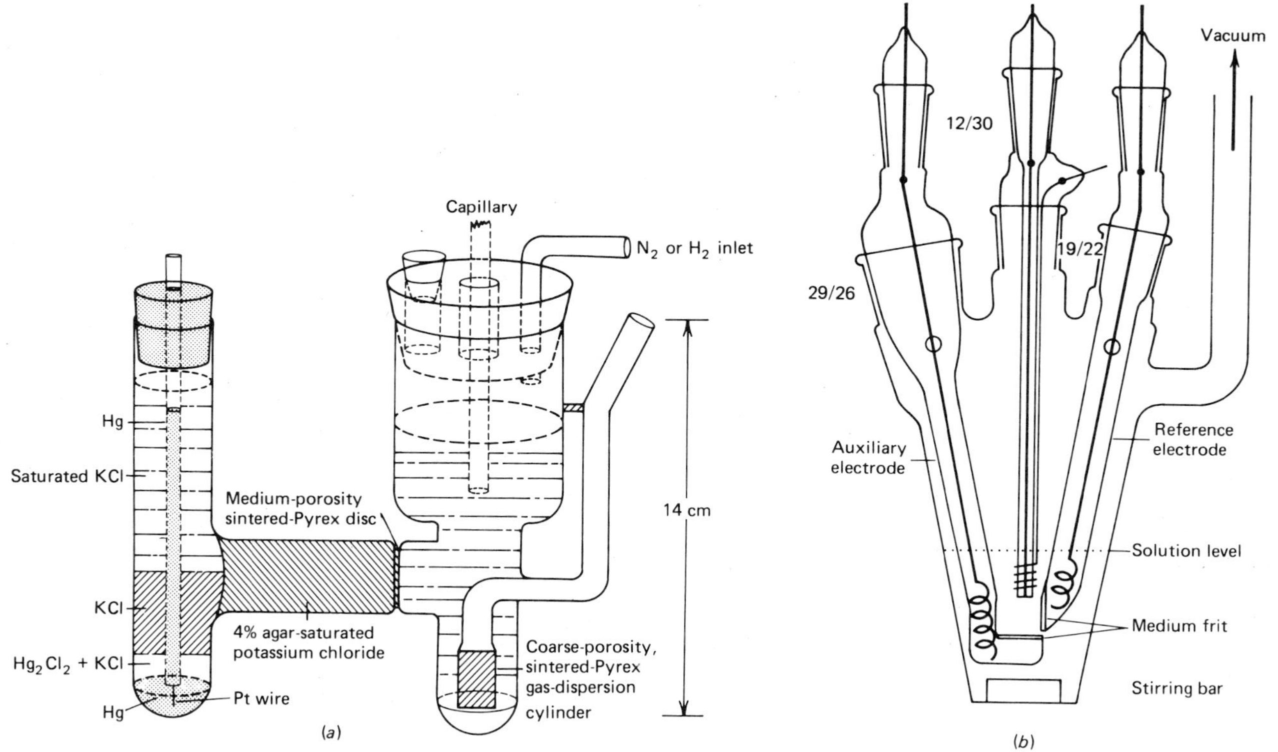

Figure 1.3.11

Typical two- and three-electrode cells used in electrochemical experiments. (*a*) Two-electrode cell for polarography. The working electrode is a dropping mercury electrode (capillary) and the N_2 inlet tube is for deaeration of the solution. (From L. Meites, *Polarographic Techniques*, 2nd ed., Wiley-Interscience, New York, 1965, with permission.) (*b*) Three-electrode cell designed for studies with nonaqueous solutions at a platinum-disk working electrode with provision for attachment to a vacuum line. (From A. Demortier and A. J. Bard, *J. Am. Chem. Soc.*, **95**, 3495 (1973), with permission.) Three-electrode cells for bulk electrolysis are shown in Figure 10.2.2.

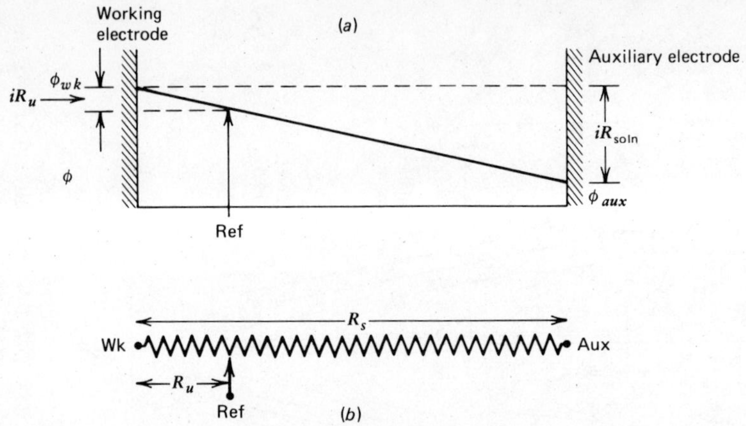

Figure 1.3.12
(*a*) Potential drop between working and auxiliary electrodes in solution and iR_u measured at reference electrode. (*b*) Representation of the cell as a potentiometer.

for the reference electrode tip placed one electrode radius away ($x = r_0$), is already one-half the value for the tip placed far away ($x \to \infty$). Any resistances in the working electrode itself (e.g., in thin capillaries of the DME, in semiconductor electrodes, or in resistive films on the electrode surface) will also appear in R_u.

1.4 INTRODUCTION TO MASS-TRANSFER-CONTROLLED REACTIONS

1.4.1 Modes of Mass Transfer

The simplest electrode reactions are those in which the kinetics of all electron transfer and associated chemical reactions are very rapid compared to those of the mass transfer processes. Under these conditions, the chemical reactions can usually be treated in a particularly simple way. If, for example, an electrode process involves only fast heterogeneous charge transfer kinetics and mobile, reversible homogeneous reactions, we will find below that (a) the homogeneous reactions can be regarded as being at equilibrium and (b) the *surface concentrations* of species involved in the faradaic process are related to the electrode potential by an equation of the Nernst form. The net rate of the electrode reaction, v, is then governed totally by the rate, v_{mt}, at which the electroactive species is brought to the surface by mass transfer. Hence, from equation 1.3.4,

$$v = v_{mt} = i/nFA \tag{1.4.1}$$

Those electrode reactions are often called *reversible* or *nernstian* because the chemical reactions obey thermodynamic relationships. Since mass transfer plays a big role in electrochemical dynamics, we consider here its three modes and begin a consideration of mathematical methods for treating them. Mass transfer, that is, the movement of material from one location in solution to another, arises either from differences in electrical or chemical potential at the two locations, or from movement of a volume element of solution. The modes of mass transfer are:

1. *Migration*. Movement of a charged body under the influence of an electric field (a gradient of electrical potential).
2. *Diffusion*. Movement of a species under the influence of a gradient of chemical potential (i.e., a concentration gradient).
3. *Convection*. Stirring or hydrodynamic transport. Generally fluid flow occurs because of natural convection (convection caused by density gradients) and forced convection, and may be characterized by stagnant regions, laminar flow, and turbulent flow.

Mass transfer to an electrode is governed by the *Nernst-Planck* equation, written for one-dimensional mass transfer along the x-axis as

$$\boxed{J_i(x) = -D_i \frac{\partial C_i(x)}{\partial x} - \frac{z_i F}{RT} D_i C_i \frac{\partial \phi(x)}{\partial x} + C_i v(x)} \tag{1.4.2}$$

where $J_i(x)$ is the flux of species i (mol sec^{-1} cm^{-2}) at distance x from the surface, D_i is the diffusion coefficient (cm^2/sec), $\partial C_i(x)/\partial x$ is the concentration gradient at distance x, $\partial \phi(x)/\partial x$ is the potential gradient, z_i and C_i are the charge and concentration of species i, respectively, and $v(x)$ is the velocity (cm/sec) with which a volume element in solution moves along the axis. This equation is derived and discussed in more detail in Chapter 4. The three terms on the right-hand side represent the contributions of diffusion, migration, and convection, respectively, to the flux.

While we will be concerned with particular solutions of this equation in later chapters, a rigorous solution is generally not very easy, and electrochemical systems are frequently designed so that one or more of the contributions to mass transfer are negligible. In this chapter we present an approximate treatment of steady-state mass transfer, which will provide a useful guide for these processes in later chapters and insight to electrochemical reactions unencumbered by the mathematical details.

1.4.2 Semiempirical Treatment of Steady-State Mass Transfer (5)

Consider the reduction of a species O at a cathode:

$$\mathrm{O} + ne \rightleftharpoons \mathrm{R} \tag{1.4.3}$$

In the absence of migration the rate of mass transfer is proportional to the concentration gradient at the electrode surface (where solution velocity is assumed to be zero):

$$v_{mt} \propto \left(\frac{\partial C_{\mathrm{O}}(x)}{\partial x} \right)_{x=0} \tag{1.4.4}$$

where x is distance from the electrode surface. This is approximated by the relation,

$$v_{mt} = m_{\mathrm{O}}[C_{\mathrm{O}}^* - C_{\mathrm{O}}(x = 0)] \tag{1.4.5}$$

where C_O^* is the concentration of O in the bulk solution (far from the electrode) and $C_O(x = 0)$ is the concentration at the electrode surface (Figure 1.4.1). The proportionality constant, m_O, is called the *mass transfer coefficient*; it has units of centimeters per second (which are those of a rate constant of a first-order heterogeneous reaction; see Chapter 3). These units follow from those of v and C_O, but can also be thought of as volume flow/sec per unit area ($cm^3\ sec^{-1}\ cm^{-2}$).† Thus from equations 1.4.1 and 1.4.5, taking a reduction current as positive,‡ [i.e., i is positive when $C_O^* > C_O(x = 0)$], we obtain

$$\frac{i}{nFA} = m_O[C_O^* - C_O(x = 0)] \tag{1.4.6}$$

Under the conditions of a net cathodic reaction, R is produced at the electrode surface, so that $C_R(x = 0) > C_R^*$ (where C_R^* is the bulk concentration of R). Therefore, with the convention of a cathodic current being positive,

$$\frac{i}{nFA} = m_R[C_R(x = 0) - C_R^*] \tag{1.4.7}$$

or for the particular case when $C_R^* = 0$ (no R in the bulk solution),

$$\frac{i}{nFA} = m_R C_R(x = 0) \tag{1.4.8}$$

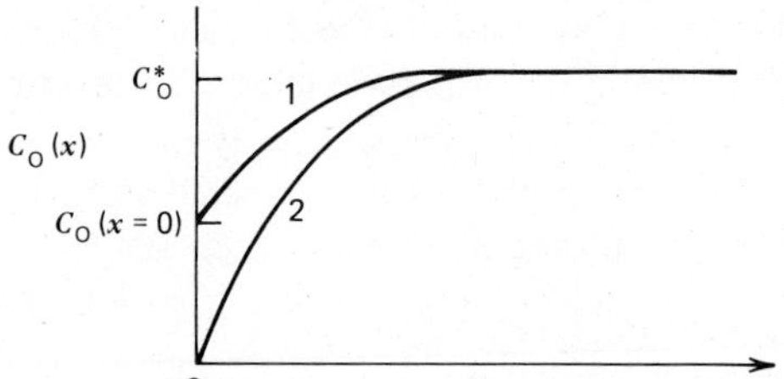

Figure 1.4.1
Concentration profiles at an electrode; $x = 0$ corresponds to the electrode surface. (1) At a potential where $C_O(x = 0)$ is about $C_O^*/2$; (2) At a potential where $C_O(x = 0) \approx 0$ and $i = i_l$.

† In approximate treatments of convective systems (see Section 1.4.3) m_O corresponds to D_O/δ_O where δ_O is the thickness of the hypothetical stagnant (diffusion) layer at the electrode surface. In more exact treatments the value of m_O can be specified without any arbitrary adjustable parameter like δ_O. For example, for the rotating disk electrode,

$$m_O = 0.617 D_O^{2/3} \omega^{1/2} \nu^{-1/6}$$

where ω is the angular velocity of the disk (i.e., $2\pi f$, with f as the frequency in revolutions per second) and ν is kinematic viscosity (i.e., viscosity/density, with units of cm^2/sec) (see Chapter 8).

‡ The convention of taking i positive for a cathodic current stems from the early polarographic studies, where reduction reactions were usually studied. This convention has continued among many analytical and electrochemists, even though oxidation reactions are now studied with equal frequency. Other electrochemists prefer to take an anodic current as positive. When looking over a derivation in the literature or examining a published *i-E* curve, it is important to decide first which convention is being used (i.e., "Which way is up?").

The values of $C_O(x = 0)$ and $C_R(x = 0)$ are functions of electrode potential, E. The largest rate of mass transfer of O occurs when $C_O(x = 0) = 0$ (or more precisely, when $C_O(x = 0) \ll C_O^*$, so that $C_O^* - C_O(x = 0) \simeq C_O^*$). The value of the current under these conditions is called the *limiting current*, i_l, where

$$\boxed{i_l = nFAm_OC_O^*} \qquad (1.4.9)$$

When the limiting current flows, the electrode process is occurring at the maximum rate possible for a given set of mass transfer conditions, because O is being reduced as fast as it can be brought to the electrode surface. Equations 1.4.6 and 1.4.9 can be used to obtain expressions for $C_O(x = 0)$:

$$\boxed{\frac{C_O(x = 0)}{C_O^*} = 1 - \frac{i}{i_l}} \qquad (1.4.10)$$

$$C_O(x = 0) = \frac{i_l - i}{nm_OFA} \qquad (1.4.11)$$

Thus the concentration of species O at the electrode surface is linearly related to the current and varies from C_O^*, when $i = 0$, to negligible values ($\simeq 0$), when $i = i_l$.

If the kinetics of electron transfer are rapid, the concentrations of O and R at the electrode surface can be assumed to be at their equilibrium values as governed by the Nernst equation for half-reaction (1.4.3):

$$E = E^{0'} + \frac{RT}{nF} \ln \frac{C_O(x = 0)}{C_R(x = 0)} \qquad (1.4.12)$$

assuming both O and R are soluble species. Such a process is called a *nernstian reaction*. We can derive the steady-state *i-E* curves for nernstian reactions under several different conditions.

(a) R Initially Absent. When $C_R^* = 0$, $C_R(x = 0)$ can be obtained from (1.4.8):

$$C_R(x = 0) = i/nFAm_R \qquad (1.4.13)$$

Then, combining equations 1.4.11 to 1.4.13, we obtain

$$E = E^{0'} - \frac{RT}{nF} \ln \frac{m_O}{m_R} + \frac{RT}{nF} \ln \left(\frac{i_l - i}{i}\right) \qquad (1.4.14)$$

An *i-E* plot is shown in Figure 1.4.2*a*. Note that when $i = i_l/2$,

$$\boxed{E = E_{1/2} = E^{0'} - \frac{RT}{nF} \ln \frac{m_O}{m_R}} \qquad (1.4.15)$$

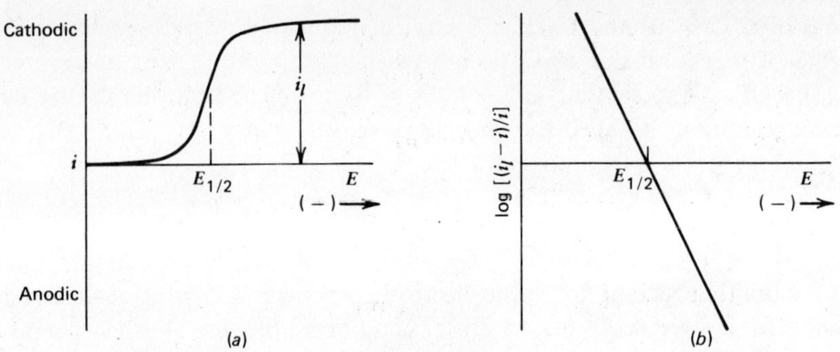

Figure 1.4.2
(*a*) Current-potential curve for a nernstian reaction involving two soluble species with only oxidant present initially. (*b*) $\log[(i_l - i)/i]$ *vs.* E for this system.

where $E_{1/2}$ is independent of the substrate concentration and is therefore characteristic of the O/R system. Thus

$$E = E_{1/2} + \frac{RT}{nF} \ln \left(\frac{i_l - i}{i}\right) \tag{1.4.16}$$

When a system conforms to this equation, a plot of E *vs.* $\log[(i_l - i)/i]$ is a straight line with a slope of $2.3\, RT/nF$ (or $59.1/n$ mV at 25°) (Figure 1.4.2*b*). When m_O and m_R have similar values, $E_{1/2} \simeq E^{0\prime}$.

(b) Both O and R Initially Present. In this case we must distinguish between a *cathodic* limiting current, $i_{l,c}$, when $C_O(x = 0) \simeq 0$, and an *anodic* limiting current, $i_{l,a}$, when $C_R(x = 0) \simeq 0$. We still have $C_O(x = 0)$ from (1.4.11), with i_l now specified as $i_{l,c}$. The limiting anodic current naturally reflects the maximum rate at which R can be brought to the electrode surface for conversion to O. It is obtained from (1.4.7):

$$i_{l,a} = -nFAm_RC_R^* \tag{1.4.17}$$

(The negative sign arises because of our convention that cathodic currents are taken as positive and anodic ones as negative.) Thus $C_R(x = 0)$ is given by

$$C_R(x = 0) = \frac{i - i_{l,a}}{nFAm_R} \tag{1.4.18}$$

$$\frac{C_R(x = 0)}{C_R^*} = 1 - \frac{i}{i_{l,a}} \tag{1.4.19}$$

The i-E curve is now given by

$$E = E^{0\prime} - \frac{RT}{nF} \ln \frac{m_O}{m_R} + \frac{RT}{nF} \ln \left(\frac{i_{l,c} - i}{i - i_{l,a}}\right) \tag{1.4.20}$$

A plot of this equation is shown in Figure 1.4.3. When $i = 0$, $E = E_{eq}$, the equilibrium potential of the system. Surface concentrations are then equal to bulk values. When current passes, the potential deviates from E_{eq} and the extent of this deviation is the concentration overpotential. (An equilibrium potential cannot be so defined when $C_R^* = 0$, of course.)

(c) R Insoluble. Species R is a metal, for example, and can be considered to be at essentially unit activity by having the electrode reaction take place on bulk R.† Thus, when $a_R = 1$, the Nernst equation is

$$E = E^{0'} + \frac{RT}{nF} \ln C_O(x = 0) \tag{1.4.21}$$

or, using the value of $C_O(x = 0)$ from equation 1.4.11,

$$\boxed{E = E^{0'} + \frac{RT}{nF} \ln C_O^* + \frac{RT}{nF} \ln \left(\frac{i_l - i}{i_l}\right)} \tag{1.4.22}$$

When $i = 0$, $E = E_{eq} = E^{0'} + (RT/nF) \ln C_O^*$ (Figure 1.4.4). If we define the concentration overpotential, η_{conc} (or mass transfer overpotential, η_{mt}), as

$$\eta_{\text{conc}} = E - E_{eq} \tag{1.4.23}$$

then

$$\eta_{\text{conc}} = \frac{RT}{nF} \ln \left(\frac{i_l - i}{i_l}\right) \tag{1.4.24}$$

Note that when $i = i_l$, $\eta_{\text{conc}} \rightarrow \infty$. Since η is a measure of polarization, this condition is sometimes called *complete concentration polarization*. Equation 1.4.24 can be written in exponential form:

$$1 - \frac{i}{i_l} = \exp\left(\frac{nF\eta_{\text{conc}}}{RT}\right) \tag{1.4.25}$$

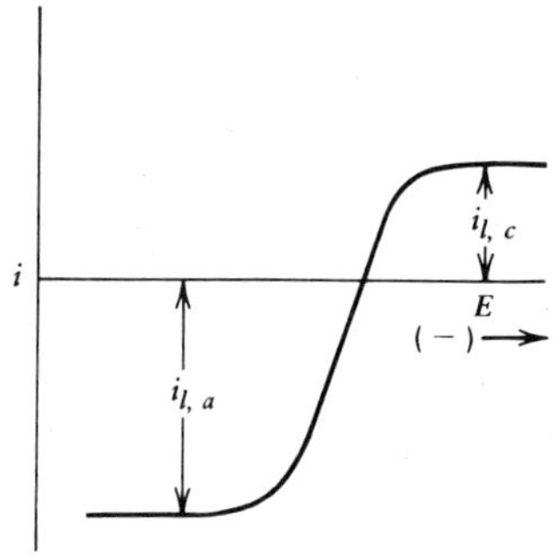

Figure 1.4.3
Current-potential curve for a nernstian system involving two soluble species with both forms initially present.

† This will not be the case for R plated onto an inert substrate (e.g., the substrate electrode being Pt and R being Cu) in amounts less than a monolayer. Under those conditions a_R may be considerably less than one (7).

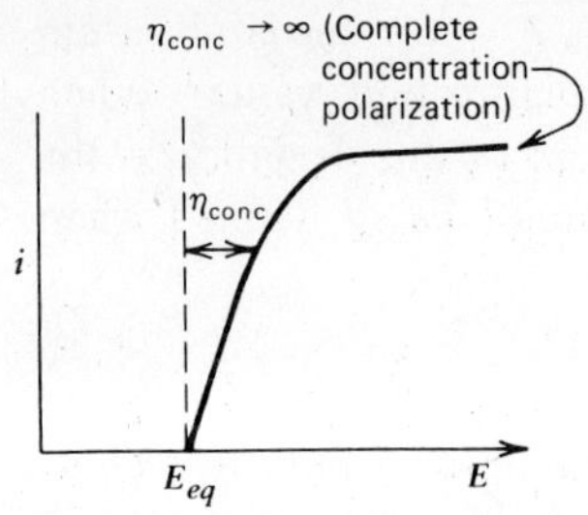

Figure 1.4.4
Current-potential curve for a nernstian system where the reduced form is insoluble.

The exponential can be expanded as a power series, and the higher-order terms can be dropped, if the argument is kept small; that is,

$$e^x = 1 + x + \frac{x^2}{2} + \cdots \simeq 1 + x \text{ (when } x \text{ is small)} \qquad (1.4.26)$$

Thus, under conditions of small deviations of potential from E_{eq}, the i-η_{conc} characteristic is linear:

$$\eta_{conc} = \frac{-RT}{nFi_l} i \qquad (1.4.27)$$

Since $-\eta/i$ has dimensions of resistance (i.e., ohms), we can define a "small signal" mass transfer resistance, R_{mt}, as

$$\boxed{R_{mt} = \frac{RT}{nF|i_l|}} \qquad (1.4.28)$$

Note that the mass-transfer-limited electrode reaction resembles an actual resistance element only at small overpotentials.

1.4.3 Semiempirical Treatment of Non-steady-State, Mass-Transfer-Controlled Reaction

The treatment in Section 1.4.2 involved a steady-state mass transfer condition in which the reaction rate was equal to the rate of mass transfer at the electrode surface, or the flux of species O to the electrode. In the *Nernst diffusion layer* treatment one assumes that a stagnant layer of thickness, δ_O, exists near the electrode surface (Figure 1.4.5). Outside this layer, convective transport maintains the concentration uniform

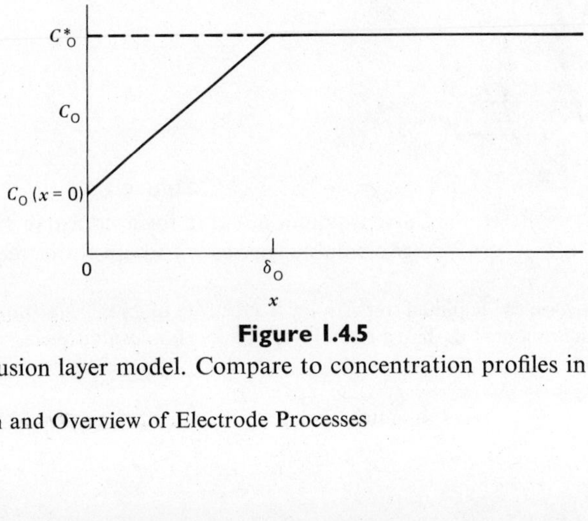

Figure 1.4.5
Nernst diffusion layer model. Compare to concentration profiles in Figure 1.4.1.

at the bulk concentration, C_O^*. Within the layer mass transfer occurs only by diffusion (i.e., $\partial\phi/\partial x$ and $v(x)$ are zero). Thus the flux of O at the electrode surface (assuming linear diffusion) from equation 1.4.2 is

$$-J_O(x = 0) = D_O\left(\frac{\partial C_O}{\partial x}\right)_{x=0} = \frac{i}{nFA} \tag{1.4.29}$$

The concentration gradient is thus

$$\left(\frac{\partial C_O}{\partial x}\right)_{0 \leq x \leq \delta} = \frac{C_O^* - C_O(x = 0)}{\delta_O} \tag{1.4.30}$$

$$\frac{i}{nFA} = \frac{D_O}{\delta_O}[C_O^* - C_O(x = 0)] \tag{1.4.31}$$

Comparison of equations 1.4.31 and 1.4.6 shows that they are equivalent for $m_O = D_O/\delta_O$.

This same concept can be used to give an approximate treatment of the time-dependent buildup of the diffusion layer either in a stirred solution or in an unstirred solution where the diffusion layer continues to grow with time. Consider what happens when a potential step of magnitude E is applied to an electrode immersed in a solution containing a species O. If the reaction is nernstian, the concentrations of O and R at $x = 0$ instantaneously adjust to the values governed by the Nernst equation (1.4.12). The thickness of the approximately linear diffusion layer, $\delta_O(t)$, grows with time (Figure 1.4.6), and an expression for $\delta_O(t)$ can be obtained by the following procedure: the volume of the diffusion layer is $A\delta_O(t)$, the current flow causes a depletion there of O, and the amount of O electrolyzed by this current is given by

$$\begin{matrix}\text{Moles of O electrolyzed} \\ \text{in diffusion layer}\end{matrix} \simeq [C_O^* - C_O(x = 0)]\frac{A\delta(t)}{2} = \int_0^t \frac{i\,dt}{nF} \tag{1.4.32}$$

or, by differentiation of (1.4.32) and use of (1.4.31),

$$\frac{[C_O^* - C_O(x = 0)]}{2}\frac{A\,d\delta(t)}{dt} = \frac{i}{nF} = \frac{D_O A}{\delta(t)}[C_O^* - C_O(x = 0)] \tag{1.4.33}$$

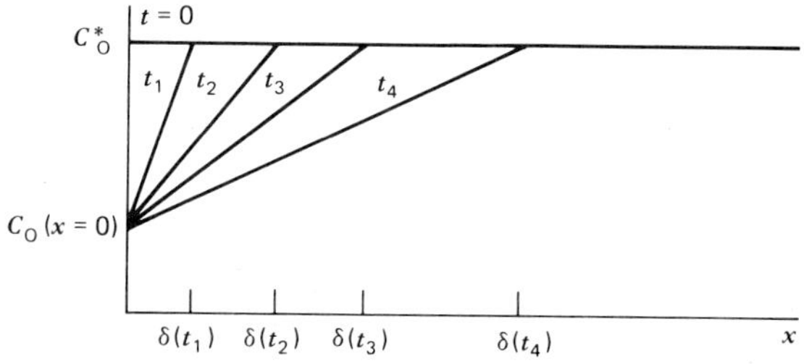

Figure 1.4.6
Growth of diffusion-layer thickness with time.

or

$$\frac{d\delta(t)}{dt} = \frac{2D_O}{\delta(t)} \tag{1.4.34}$$

Since $\delta(t) = 0$ at $t = 0$, the solution of (1.4.34) is

$$\delta(t) = 2\sqrt{D_O t} \tag{1.4.35}$$

and

$$\frac{i}{nFA} = \frac{D_O^{1/2}}{2t^{1/2}}[C_O^* - C_O(x = 0)] \tag{1.4.36}$$

Thus this approximate treatment predicts a diffusion layer that grows with $t^{1/2}$ and a current that decays with $t^{-1/2}$. In the absence of convection, the current continues to decay, but in a convective system it ultimately approaches the steady-state value characterized by $\delta(t) = \delta_O$ (see Figure 1.4.7). It is interesting that even this simplified approach approximates reality quite closely; equation 1.4.36 differs by only a factor of $2/\pi^{1/2}$ from the rigorous description of current arising for a nernstian system during a potential step (see Section 5.2).

1.5 SEMIEMPIRICAL TREATMENT OF NERNSTIAN REACTIONS WITH COUPLED CHEMICAL REACTIONS

1.5.1 Coupled Reversible Reactions

If a homogeneous process that is fast enough to be considered always in thermodynamic equilibrium (a reversible process) is coupled to a nernstian electron transfer reaction, a simple extension of the previously described treatment can be used to derive the *i-E* curve. Consider, for example, a species O involved in a reversible process that precedes the electron transfer reaction

$$A \rightleftharpoons O + qY \tag{1.5.1}$$

$$O + ne \rightleftharpoons R \tag{1.5.2}$$

For example, A could be a metal complex, MY_q^{+n}; O could be the free metal ion, M^{+n}; and Y could be the free ligand (see Section 5.4.5). The Nernst equation still applies for equation 1.5.2:

$$E = E^{0'} + \frac{RT}{nF}\ln\frac{C_O(x = 0)}{C_R(x = 0)} \tag{1.5.3}$$

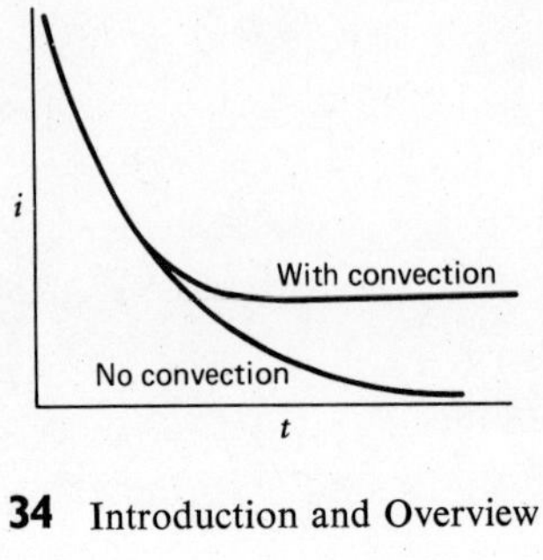

Figure 1.4.7
Current-time transient for a potential step to a stationary electrode (no convection) and to an electrode in stirred solution (with convection) where a steady-state current is attained.

and reaction (1.5.1) is assumed to be at equilibrium at all x:

$$\frac{C_O C_Y{}^q}{C_A} = K \qquad \text{(all } x\text{)} \tag{1.5.4}$$

Hence

$$E = E^{0'} + \frac{RT}{nF} \ln \left[\frac{K C_A(x = 0)}{C_Y{}^q(x = 0) C_R(x = 0)} \right] \tag{1.5.5}$$

Assuming (1) that at $t = 0$, $C_A = C_A^*$, $C_Y = C_Y^*$, and $C_R = 0$ (for all x); (2) that C_Y^* is so large compared to C_A^* that $C_Y(x = 0) = C_Y^*$ at all times; and (3) that $K \ll 1$; then at steady state

$$\frac{i}{nFA} = m_A[C_A^* - C_A(x = 0)] \tag{1.5.6}$$

$$\frac{i_l}{nFA} = m_A C_A^* \tag{1.5.7}$$

$$\frac{i}{nFA} = m_R C_R(x = 0) \tag{1.5.8}$$

Then, as previously,

$$C_A(x = 0) = \frac{(i_l - i)}{nFAm_A} \qquad C_R(x = 0) = \frac{i}{nFAm_R} \tag{1.5.9}$$

$$\boxed{E = E^{0'} + \frac{RT}{nF} \ln K + \frac{RT}{nF} \ln \frac{m_R}{m_A} - \frac{RT}{nF} q \ln C_Y^* + \frac{RT}{nF} \ln \left(\frac{i_l - i}{i} \right)} \tag{1.5.10}$$

$$E = E_{1/2} + (0.059/n) \log \frac{i_l - i}{i} \qquad (T = 25^\circ) \tag{1.5.11}$$

where

$$E_{1/2} = E^{0'} + \frac{0.059}{n} \log \frac{m_R}{m_A} + \frac{0.059}{n} \log K - \frac{0.059}{n} q \log C_Y^* \tag{1.5.12}$$

Hence the *i-E* curve (1.5.11) has the usual nernstian shape, but the $E_{1/2}$ is shifted in a negative direction (since $K \ll 1$) from that for (1.5.2) unperturbed. From the shift of $E_{1/2}$ with $\log C_Y$ both $q[= -(n/0.059)(dE_{1/2}/d \log C_Y^*)]$ and K can be determined. Although these thermodynamic quantities can be determined, no kinetic or mechanistic information can be obtained when both reactions are reversible.

1.5.2 Coupled Slow Chemical Reactions

When a slow chemical reaction is coupled to a nernstian electron transfer, a semiempirical treatment (based on the steady-state *reaction layer* concept) can be given. In this case the *i-E* curves can be used to provide kinetic and mechanistic information about the reaction. Consider a nernstian charge transfer reaction with an irreversible following reaction:

$$\text{O} + ne \rightleftharpoons \text{R} \tag{1.5.13}$$

$$\text{R} \xrightarrow{k} \text{T} \tag{1.5.14}$$

where k is the rate constant (in sec^{-1}) for the decomposition of R. (Note that k could be a pseudo-first-order constant, for example, in a buffered solution, where R reacts with protons and $k = k'C_{H^+}$.) As an example of this sequence, consider the oxidation of p-aminophenol in acid solutions.

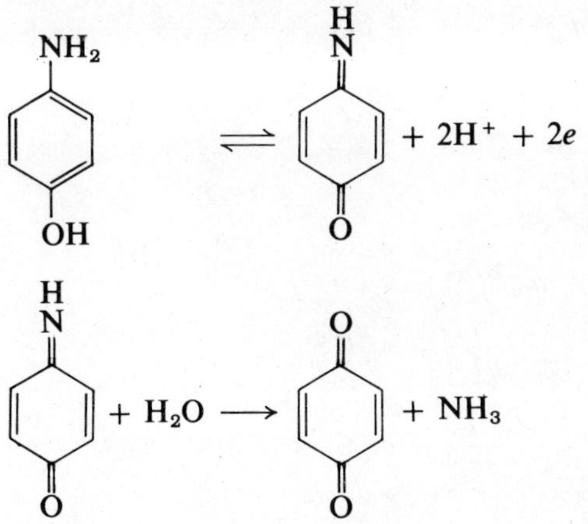

$$\text{HO-C}_6\text{H}_4\text{-NH}_2 \rightleftharpoons \text{O=C}_6\text{H}_4\text{=NH} + 2\text{H}^+ + 2e \quad (1.5.15)$$

$$\text{O=C}_6\text{H}_4\text{=NH} + \text{H}_2\text{O} \longrightarrow \text{O=C}_6\text{H}_4\text{=O} + \text{NH}_3 \quad (1.5.16)$$

The following reaction does not affect the reduction and mass transfer of O, so that (assuming $C_O = C_O^*$ and $C_R = 0$ at all x at $t = 0$) (1.4.6) and (1.4.9) still apply. At steady state, the rate of formation of R, given by (1.4.6), equals its rate of disappearance by diffusion [$(m_R C_R(x = 0)$; see (1.4.8)] and by reaction [$\mu k C_R(x = 0)$], so that

$$\frac{i}{nFA} = m_O[C_O^* - C_O(x = 0)] = m_R C_R(x = 0) + \mu k C_R(x = 0) \quad (1.5.17)$$

where μ is a so-called *reaction layer thickness* (in centimeters) analogous to δ considered above (8). One assumes that a reaction layer exists within which all molecules of R are candidates for reaction. Note that the product μk has dimensions of cm/sec as required; μ can be regarded as an adjustable parameter (like m_O). Thus, from (1.5.17)

$$C_O(x = 0) = \frac{i_l - i}{nFAm_O} \quad (1.5.18)$$

$$C_R(x = 0) = \frac{i}{nFA(m_R + \mu k)} \quad (1.5.19)$$

Substituting these values into the Nernst equation for (1.5.13) yields

$$\boxed{E = E^{0\prime} + \frac{RT}{nF} \ln \frac{m_R + \mu k}{m_O} + \frac{RT}{nF} \ln \left(\frac{i_l - i}{i}\right)} \quad (1.5.20)$$

or

$$E = E'_{1/2} + \frac{0.059}{n} \log \left(\frac{i_l - i}{i}\right) \quad \text{(at 25°)} \quad (1.5.21)$$

where

$$E'_{1/2} = E^{0'} + \frac{0.059}{n} \log\left(\frac{m_R + \mu k}{m_O}\right) \tag{1.5.22}$$

or

$$E'_{1/2} = E_{1/2} + \frac{0.059}{n} \log\left(1 + \frac{\mu k}{m_R}\right) \tag{1.5.23}$$

where $E_{1/2}$ is the half-wave potential for the kinetically unperturbed reaction. Two limiting cases can be examined: (a) when $\mu k/m_R \ll 1$, that is $\mu k \ll m_R$, the rate of the following reaction (1.5.14) is negligible compared to the rate of mass transfer of R away from the electrode surface and the unperturbed *i-E* curve results; and (b) when $\mu k/m_R \gg 1$, then

$$E'_{1/2} = E_{1/2} + \frac{0.059}{n} \log \frac{\mu k}{m_R} \tag{1.5.24}$$

In this case the effect of the following reaction is to shift the reduction wave in a *positive* direction without a change in shape. For the rotating disk electrode, where $m_R = 0.62\ D_R^{2/3}\omega^{1/2}\nu^{-1/6}$, (1.5.24) becomes [assuming $\mu \neq f(\omega)$]

$$E'_{1/2} = E_{1/2} + \frac{0.059}{n} \log \frac{\mu k}{0.62\ D_R^{2/3}\nu^{-1/6}} - \frac{0.059}{2n} \log \omega \tag{1.5.25}$$

so that an increase of rotation rate, ω, will cause the wave to shift in a *negative* direction (toward the unperturbed wave); a tenfold change in ω causes a $0.03/n$ V shift (Figure 1.5.1). A similar treatment can be given for other types of chemical reactions coupled to the charge transfer reaction (9). This approach is often useful in formulating a qualitative or semiquantitative interpretation of *i-E* curves. Notice, however, that unless explicit expressions for m_R and μ can be given in a particular case, the exact values of k cannot be determined. The rigorous treatment of electrode reactions with coupled homogeneous chemical reactions is discussed in Chapter 11.

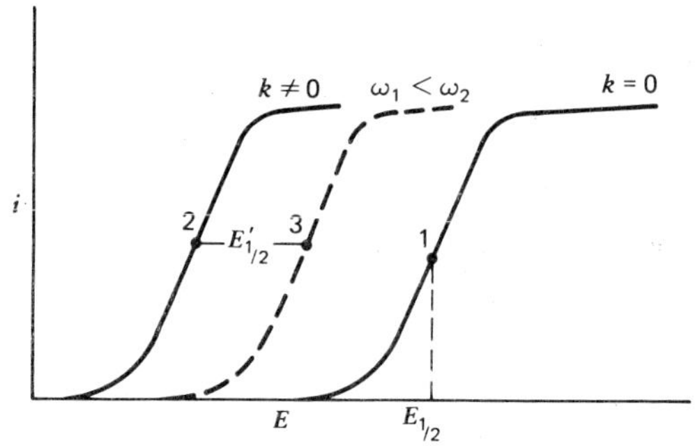

Figure 1.5.1

Effect of irreversible following homogeneous chemical reaction on nernstian *i-E* curves at rotating disk electrode. (1) Unperturbed curve. (2) and (3) Curves with following reaction at two rotation rates.

1.6 THE LITERATURE OF ELECTROCHEMISTRY

We now embark on more detailed and rigorous considerations of the fundamental principles of electrode reactions and the methods used to study them. At the outset we list the general monographs and review series in which many of these topics are treated in much greater depth. This listing is not at all comprehensive, but does represent the recent English-language sources on general electrochemical subjects. References to the older literature can be found in these. Monographs and reviews on particular subjects (e.g., polarography) are listed in the appropriate chapter. We also list the journals in which papers relating to electrochemical methods are published regularly.

1.6.1 Books and Monographs

(a) General Electrochemistry

Albery, W. J., "Electrode Kinetics," Clarendon Press, Oxford, 1975. (A short, very readable introduction to electrochemical systems; emphasizes kinetic aspects.)

Bockris, J. O'M., and A. K. N. Reddy, "Modern Electrochemistry," Plenum Press, New York, 1970 (2 vols.). (Wide-ranging description of fundamentals and applications of electrochemistry; principles rather than methodology.)

Conway, B. E., "Theory and Principles of Electrode Processes," Ronald Press, New York, 1965. (Kinetics, adsorption of reactants and intermediates.)

Eyring, H., D. Henderson, and W. Jost, Eds., "Physical Chemistry, An Advanced Treatise," Vols. IXA and IXB (*Electrochemistry*) Academic Press, New York, 1970. (Physical principles and aspects.)

Koryta, J., J. Dvořak, and V. Bohačková, "Electrochemistry," Methuen & Company, London, 1966. (Stress on electrolyte solution processes, but includes good section of heterogeneous electrochemical systems.)

MacInnes, D. A., "The Principles of Electrochemistry," Dover Publications, Inc., New York, 1961 (Corrected version of 1947 edition). (A very useful paperback on "classical" electrochemical topics, such as thermodynamics, activity, conductance, potentiometry, dielectric constants, and electrokinetic phenomena.)

Newman, J. S., "Electrochemical Systems," Prentice-Hall, Englewood Cliffs, N.J., 1972. (Emphasis on concepts of electrochemical engineering, especially those related to mass transport processes.)

Vetter, K. J., "Electrochemical Kinetics," Academic Press, New York, 1967. (Detailed compendium of theoretical and experimental aspects of the kinetics of electrode reactions, including discussion of many experimental systems, corrosion, and passivity.)

(b) Electrochemical Methodology

Adams, R. N., "Electrochemistry at Solid Electrodes," Marcel Dekker, New York, 1969. (A very readable discussion with many examples and experimental details.)

Charlot, G., J. Badoz-Lambling, and B. Tremillon, "Electrochemical Reactions," Elsevier, Amsterdam, 1962. (Electroanalytical chemistry with especially good discussions of the interpretation of current-potential curves.)

Damaskin, B. B., "The Principles of Current Methods for the Study of Electrochemical Reactions," McGraw-Hill, New York, 1967. (A brief paperback with clear, concise discussions of many modern electrochemical techniques.)

Delahay, P., "New Instrumental Methods in Electrochemistry," Interscience, New York, 1954. (Best review of voltammetry to 1954; this book was a major influence in the development of modern electroanalytical chemistry.)

Galus, Z., "Fundamentals of Electrochemical Analysis," Ellis Harwood, Ltd., Chichester, 1976. (An excellent treatment of voltammetric methods and coupled chemical reactions, with numerous references.)

Gileadi, E., E. Kirowa-Eisner, and J. Penciner, "Interfacial Electrochemistry—An Experimental Approach," Addison-Wesley, Reading, Mass., 1975. (A laboratory manual describing many useful experiments.)

Kolthoff, I. M., and P. J. Elving, Eds., "Treatise on Analytical Chemistry," Interscience, New York, 1963, Part I, Vol. 4. (Chapters by various authors on different methods; the ones by C. N. Reilley and R. W. Murray, also published separately as a paperback entitled "Electroanalytical Principles," make an especially good introduction to the field.)

Lingane, J. J., "Electroanalytical Chemistry," 2nd Ed., Interscience, New York, 1958. (Excellent introduction to the field, especially bulk electrolysis and coulometric methods; the instrumentation described is, of course, now out-of-date.)

Macdonald, D. D., "Transient Techniques in Electrochemistry," Plenum Press, New York, 1977. (Detailed discussion of the mathematics of electrochemical techniques with numerous references to the literature.)

Purdy, W. C., "Electroanalytical Methods in Biochemistry," McGraw-Hill, New York, 1965. (Simplified treatment with special emphasis on biochemistry.)

Sawyer, D. T., and J. L. Roberts, Jr., "Experimental Electrochemistry for Chemists," Wiley-Interscience, New York, 1974. (Extensive experimental details on cells, instrumentation, electrodes, and methods.)

(c) Descriptive Electrochemistry

Baizer, M. M., "Organic Electrochemistry," Marcel Dekker, New York, 1973. (Basic concepts and synthetic and mechanistic aspects; contains excellent bibliography to other literature in this field.)

Bard, A. J., and H. Lund, Eds., "Encyclopedia of the Electrochemistry of the Elements," Marcel Dekker, New York, 1973–80, Vols. 1–14. (Electrode potentials, rate constants, as well as descriptive and applied electrochemistry of inorganic and organic compounds reviewed in depth.)

Dryhurst, G., "Electrochemistry of Biological Molecules," Academic, New York, 1977. (A review of oxidation and reduction processes of purines, pyrimidines, flavins, pyrroles, porphyrins, and similar molecules of biochemical interest.)

Fry, A. J., "Synthetic Organic Electrochemistry," Harper & Row, New York, 1972. (Techniques and applications, classifications of synthetically useful electrode processes.)

Mann, C. K., and K. K. Barnes, "Electrochemical Reactions in Nonaqueous Systems," Marcel Dekker, New York, 1970. (Exhaustive review of reactions with numerous tables listing voltammetric potentials.)

Rifi, M. R., and F. H. Covitz, "Introduction to Organic Electrochemistry," Marcel Dekker, New York, 1974. (Basic principles and reactions; oriented to synthetic chemists interested in electrochemical methods.)

(d) Compilations of Electrochemical Data

Conway, B. E., "Electrochemical Data," Elsevier, Amsterdam, 1952.

Janz, G. J., and R. P. T. Tomkins, "Nonaqueous Electrolytes Handbook," Academic Press, New York, 1972 (2 volumes.)

Meites, L., and P. Zuman, "Electrochemical Data," Wiley, New York, 1974.

Parsons, R., "Handbook of Electrochemical Data," Butterworths, London, 1959.

1.6.2 Review Series

Five review series dealing exclusively with electrochemistry and related areas exist. Volumes are published every year or few years and contain chapters written by authorities in particular subject areas.†

Bard, A. J., Ed., "Electroanalytical Chemistry," Marcel Dekker, New York, from 1966. (11 volumes have appeared.)

Bockris, J. O'M., and B. E. Conway, Eds., "Modern Aspects of Electrochemistry," Plenum, New York, from 1954. (12 volumes have appeared.)

Delahay, P., and C. W. Tobias (from Vol. 10, H. Gerischer and C. W. Tobias), Eds., "Advances in Electrochemistry and Electrochemical Engineering," Wiley, New York, from 1961. (11 volumes have appeared.)

Specialist Periodical Reports, "Electrochemistry," G. J. Hills (Vols. 1–3) and H. R. Thirsk (from Vol. 4), Senior Reporters, The Chemical Society, London, from 1971. (6 volumes have appeared.)

Yeager, E., and A. J. Salkind, Eds., "Techniques of Electrochemistry," Wiley-Interscience, New York, from 1972. (3 volumes have appeared.)

Reviews on electrochemical topics also appear from time to time in the following:

Accounts of Chemical Research, The American Chemical Society, Washington.

† Articles in the first three series listed below are cited in this book, and often elsewhere in the literature, in journal reference format with the abbreviations *Electroanal Chem.*, *Mod. Asp. Electrochem.*, and *Adv. Electrochem. Electrochem. Eng.*, respectively. Note that the first should not be confused with *J. Electroanal. Chem.*

Analytical Chemistry (Annual Reviews), The American Chemical Society, Washington

Annual Reviews of Physical Chemistry, Annual Reviews, Inc., Palo Alto, Calif., from 1950.

Chemical Reviews, The Americal Chemical Society, Washington, D.C.

1.6.3 Journals

The following journals are primarily devoted to electrochemistry:

Electrochimica Acta

Electrokhimiya (Soviet Electrochemistry)

Journal of Applied Electrochemistry

Journal of Electroanalytical Chemistry

Journal of the Electrochemical Society

[Also *Journal of the Electrochemical Society of India*, *Journal of the Electrochemical Society of Japan*, *Journal of the Polarographic Society*, *Review of Polarography* (Polarographic Society of Japan)].

Papers of electrochemical interest also appear from time to time in many general journals including *Analyst*, *Anal. Chem.*, *Anal. Chim. Acta*, *Ber. Bunsenges. Phys. Chem.*, *Bioelectrochem. Bioenerg.*, *Bull. Chem. Soc. Japan*, *Chem. Letters*, *Coll. Czechoslov. Chem. Communs.*, *Inorg. Chem.*, *J. Am. Chem. Soc.*, *J. Chem. Soc.*, *J. Chem. Soc.* (*Faraday Trans.*), *J. Org. Chem.*, *J. Phys. Chem.*, *Nouv. J. Chim.*, *Talanta*, and *Z. Physik. Chem.*

1.7 REFERENCES

1. R. Guidelli, *Electroanal. Chem.*, **5**, 149 (1971).
2. R. Piontelli and G. Bianchi, *Proc. Intern. Comm. Electrochem. Thermodynam. and Kinetic* (CITCE), 2nd meeting, 1951, pp. 79–394; R. Aletti et al., CITCE, 3rd meeting, 1952, pp. 30–43; R. Piontelli, G. Bianchi, and R. Aletti, *Z. Elektrochem.*, **56**, 86 (1952).
3. S. Barnartt, *J. Electrochem. Soc.*, **108**, 102 (1961).
4. L. Nemec, *J. Electroanal. Chem.*, **8**, 166 (1964).
5. See, for example, J. Jordan, *Anal. Chem.*, **27**, 1708 (1955).
6. A. Demortier and A. J. Bard, *J. Am. Chem. Soc.*, **95**, 3495 (1973).
7. P. Delahay, "New Instrumental Methods in Electrochemistry," Wiley-Interscience, New York, 1954, pp. 278–281.
8. *Ibid.*, p. 92, *et seq.*
9. See, for example, G. J. Hoytink, J. VanSchooten, E. DeBoer, and W. Aalbersberg, *Rec. Trav. Chim.*, **73**, 355 (1954), for an application of this type of method to the study of reactions coupled to the reduction of aromatic hydrocarbons.

1.8 PROBLEMS

1.1 Consider the following electrode/solution interfaces and write the equation for the electrode reaction that occurs first when the potential is moved in (1) a negative direction and (2) a positive direction from the equilibrium potential. Next to each reaction write the approximate potential for the reaction in volts *vs.* SCE (assuming the reaction is reversible).
(a) Pt/Cu^{2+} (0.01 *M*), Cd^{2+} (0.01 *M*), H_2SO_4 (1 *M*)
(b) Pt/Sn^{2+} (0.01 *M*), Sn^{4+} (0.01 *M*), HCl (1 *M*)
(c) Hg/Cd^{2+} (0.01 *M*), Zn^{2+} (0.01 *M*), HCl (1 *M*)

1.2 For a rotating disk electrode, the treatment of steady-state, mass-transfer controlled electrode reactions applies, where the mass transfer coefficient m_O is given by

$$m_O = 0.62 D_O^{2/3} \omega^{1/2} \nu^{-1/6}$$

and D_O is the diffusion coefficient (cm^2/sec), ω is the angular velocity of the disk (sec^{-1}) ($\omega = 2\pi f$, where f is frequency of rotation in revolutions per second), and ν is the kinematic viscosity ($\nu = \eta/d$, η = viscosity, and d = density; for aqueous solutions $\nu \sim 0.010$ cm^2/sec). A rotating disk electrode of area 0.30 cm^2 is used for the reduction of 0.010 *M* Fe^{3+} to Fe^{2+} in a 1 *M* H_2SO_4. Given D_O for Fe^{3+} at 5.2×10^{-6} cm^2/sec, calculate the limiting current for the reduction for a disk rotation rate of 10 r/sec. (Include units on variables during calculation and give units of current in answer.)

1.3 A solution of volume 50 cm^3 contains 2.0×10^{-3} *M* Fe^{3+} and 1.0×10^{-3} *M* Sn^{4+} in 1 *M* HCl. This solution is examined by voltammetry at a rotating platinum disk electrode of area 0.30 cm^2. At the rotation rate employed, both Fe^{3+} and Sn^{4+} have mass transfer coefficients, m, of 10^{-2} cm/sec. (a) Calculate the limiting current for the reduction of Fe^{3+} under these conditions. (b) A current-potential scan is taken from +1.3 to −0.40 V *vs.* NHE under the above conditions. Make a labelled, quantitatively correct, sketch of the *i-E* curve that would be obtained. Assume that no changes in the bulk concentrations of Fe^{3+} and Sn^{4+} occur during this scan and that all electrode reactions are nernstian.

1.4 The conductivity of a 0.1 *M* KCl solution at 25 °C is 0.013 Ω^{-1} cm^{-1}. (a) Calculate the solution resistance between two parallel planar platinum electrodes of 0.1 cm^2 area placed 3 cm apart in this solution. (b) A reference electrode with a Luggin capillary is placed the following distances from a planar platinum working electrode (A = 0.1 cm^2) in 0.1 *M* KCl: 0.05, 0.1, 0.5, 1.0 cm. What is R_u in each case? (c) Repeat the calculations in part (b) for a spherical working electrode of the same area. [In parts (b) and (c) it is assumed that a very large counter electrode is employed.]

1.5 A 0.1 cm^2 electrode with C_d = 20 $\mu F/cm^2$ is subjected to a potential step under conditions where R_s is 1, 10, or 100 Ω. In each case what is the time constant, and what is the time required for the double-layer charging to be 95% complete?

1.6 For the electrode in Problem 1.5, what nonfaradaic current will flow (neglecting any transients) when the electrode is subjected to linear sweeps at 0.02, 1, 20 V/sec?

1.7 Consider the nernstian half-reaction:

$$A^{3+} + 2e \rightleftharpoons A^{+} \qquad E^{0\prime}_{A^{3+}/A^{+}} = -0.500 \text{ V } vs. \text{ NHE}$$

The *i-E* curve for a solution at 25 °C containing 2.00 mM A^{3+} and 1.00 mM A^{+}, in excess electrolyte shows $i_{l,c} = 4.00$ μA and $i_{l,a} = -2.40$ μA. (a) What is $E_{1/2}$ (volts *vs.* NHE)? (b) Sketch the expected *i-E* curve for this system. (c) Sketch the "log plot" (see Figure 1.4.2*b*) for the system.

1.8 Consider the system in Problem 1.7 under the conditions that a complexing agent, L^{-}, which reacts with A^{3+} according to the reaction

$$A^{3+} + 4L^{-} \rightleftharpoons AL_4^{-} \qquad K = 10^{16}\ M^{-4}$$

is added to the system. For a solution at 25 °C containing only 2.0 mM A^{3+} and 0.1 M L^{-} in excess inert electrolyte, answer parts (a), (b), and (c) in Problem 1.7. (Assume m_O is the same for A^{3+} and AL_4^{-}.)

1.9 Derive the current-potential relationship under the conditions of Section 1.4.2 for a system where only R is initially present at a concentration C_R^* (i.e., $C_O^* = 0$) with both O and R soluble. Sketch the expected *i-E* curve.

1.10 Suppose a mercury pool of 1 cm^2 area is immersed in a 0.1 M sodium perchlorate solution. How much charge (order of magnitude) would be required to change its potential by 1 mV? How would this be affected by a change in the electrolyte concentration to 10^{-5} M? Why?

chapter 2

Potentials and Thermodynamics of Cells

In Chapter 1, we sought to obtain a working feeling for potential as an electrochemical variable. Here we will explore the physical meaning of that variable in more detail. Our goal is to understand how potential differences are established and what kinds of chemical information can be obtained from them. At first, these questions will be approached through thermodynamics. That attack will show that potential differences are related to free energy changes involved in processes that may occur in an electrochemical system, and this discovery will open the way to the experimental determination of all sorts of chemical information through electrochemical measurements. Later in this chapter, we will explore the mechanisms by which potential differences are established. Those considerations will provide insights that will prove especially useful when we start to examine experiments involving the active control of potential in an electrochemical system.

2.1 BASIC ELECTROCHEMICAL THERMODYNAMICS

2.1.1 Reversibility

Since thermodynamics can strictly encompass only systems at equilibrium, the concept of *reversibility* is important in treating real processes thermodynamically. After all, the concept of equilibrium involves the idea that a process can move in either of two opposite directions from the equilibrium position. Thus the adjective *reversible* is an essential one. Unfortunately, it takes on several obliquely related meanings in the electrochemical literature, and we need to distinguish three of them now.

(a) Chemical Reversibility. Consider the electrochemical cell shown in Figure 1.1.1*b*:

$$Pt/H_2/H^+, Cl^-/AgCl/Ag \tag{2.1.1}$$

Experimentally one finds that the difference in potential between the silver wire and the platinum wire is 0.222 V when all substances are in their standard states. Furthermore, the platinum wire is the negative electrode and, when the two electrodes are shorted together, the following reaction takes place:

$$H_2 + 2AgCl \rightarrow 2Ag + 2H^+ + 2Cl^- \tag{2.1.2}$$

If one bucks out the cell voltage by opposing it with the output of a battery, the current flow through the cell will reverse, and the new cell reaction is

$$2Ag + 2H^+ + 2Cl^- \rightarrow H_2 + 2AgCl \tag{2.1.3}$$

Reversing the cell current merely reverses the cell reaction. No new reactions appear, and thus the cell is termed *chemically reversible.*

On the other hand, the system

$$Zn/H^+, SO_4^{2-}/Pt \tag{2.1.4}$$

is not chemically reversible. The zinc electrode is negative with respect to platinum, and discharging the cell causes the reaction

$$Zn \rightarrow Zn^{2+} + 2e \tag{2.1.5}$$

to occur there. At the platinum electrode, hydrogen evolves:

$$2H^+ + 2e \rightarrow H_2 \tag{2.1.6}$$

Thus the net cell reaction is†

$$Zn + 2H^+ \rightarrow H_2 + Zn^{2+} \tag{2.1.7}$$

By applying an opposing voltage larger than the cell voltage, the current flow reverses, but the reactions observed are

$$2H^+ + 2e \rightarrow H_2 \qquad \text{(Zn electrode)} \tag{2.1.8}$$

$$2H_2O \rightarrow O_2 + 4H^+ + 4e \qquad \text{(Pt electrode)} \tag{2.1.9}$$

$$2H_2O \rightarrow 2H_2 + O_2 \qquad \text{(Net)} \tag{2.1.10}$$

One has different electrode reactions as well as a different net process upon current reversal; hence this cell is said to be chemically irreversible.

(b) Thermodynamic Reversibility. A process is thermodynamically reversible when an infinitesimal reversal in a driving force causes the process to reverse its direction. Obviously this cannot happen unless the system feels only an infinitesimal driving force at any time; hence it must essentially be always at equilibrium. A reversible path between two states of the system is therefore one that connects a continuous series of equilibrium states. Traversing it would require an infinite length of time.

Note that a cell that is chemically irreversible cannot behave reversibly in a thermodynamic sense. A chemically reversible cell may or may not operate in a manner approaching thermodynamic reversibility.

† The net reaction will also occur without flow of electrons in the external circuit because H^+ in solution will attack the zinc. This "side reaction," which happens to be identical with the electrochemical process, is slow if dilute acid is involved.

(c) Practical Reversibility. Since all actual processes occur at finite rates, they cannot proceed with strict thermodynamic reversibility. However, they may in practice be carried out in such a manner that thermodynamic equations apply to any desired accuracy. Under these circumstances, one might term the processes reversible. *Practical reversibility* is not an absolute term; it includes certain attitudes and expectations an observer has toward the process.

A useful analogy involves the removal of a large weight from a spring balance. Carrying out this process strictly reversibly requires continuous equilibrium, and hence the "thermodynamic" equation that always applies is

$$kx = mg \tag{2.1.11}$$

where k is the force constant, x is the distance the spring is stretched when mass m is added, and g is the earth's gravitational acceleration. In the reversible process the spring never feels an urge to contract more than an infinitesimal distance.

Now if the same final state is reached by simply removing the weight all at once, equation 2.1.11 applies at no time during the process, which is characterized by severe disequilibrium and is grossly irreversible.

If one chose, one could remove the weight as pieces and, if there were enough pieces, the thermodynamic relation (2.1.11) would begin to apply a very large fraction of the time. In fact, one might not be able to distinguish the real (but slightly irreversible) process from the strictly reversible path. One could then legitimately label the real transformation as practically reversible.

In electrochemistry, one frequently relies on the Nernst equation:

$$E = E^{0'} + \frac{RT}{nF} \ln \frac{[\mathrm{O}]}{[\mathrm{R}]} \tag{2.1.12}$$

to provide a linkage between electrode potential E and the concentrations of participants in the electrode process:

$$\mathrm{O} + ne \rightleftharpoons \mathrm{R} \tag{2.1.13}$$

If an electrode system follows the Nernst equation or an equation derived from it, the electrode reaction is often said to be reversible (or nernstian).

Whether a process appears reversible or not depends on one's ability to detect the signs of disequilibrium. In turn, that ability depends on the time domain of the possible measurements, the rate of change of the force driving the observed process, and the speed with which the system can reestablish equilibrium. If the perturbation applied to the system is small enough, or the system can attain equilibrium rapidly enough compared to the measuring time, thermodynamic relations will apply. A given system may behave reversibly in one experiment and irreversibly in another, even of the same genre, if the experimental conditions have a wide latitude. This theme will be met again and again throughout this book.

2.1.2 Reversibility and Gibbs Free Energy

Consider three different methods (1) of carrying out the reaction $Zn + 2AgCl \rightarrow Zn^{2+} + 2Ag + 2Cl^-$:

(a) Suppose zinc and silver chloride are mixed directly in a calorimeter at constant, atmospheric pressure and at 25 °C. Assume also that the extent of reaction is so small that the activities of all species remain unchanged during the experiment. It is found that the amount of heat liberated, when all substances are in their standard states, is 233 kJ/mol of Zn reacted. Thus, $\Delta H^0 = -233$ kJ.†

(b) Suppose we now arrange the cell of Figure 1.1.1*a*, that is,

$$\text{Zn/Zn}^{2+}(a = 1), \text{Cl}^-(a = 1)/\text{AgCl/Ag} \tag{2.1.14}$$

and discharge it through a resistance R. Again assume that the extent of reaction is small enough to keep the activities essentially unchanged. During the discharge, heat will evolve from the resistor and from the cell, and we could measure the total heat change by placing the entire apparatus inside a calorimeter. We would find that the heat evolved is 233 kJ/mol of Zn, independent of R. That is, $\Delta H^0 = -233$ kJ, regardless of the rate of cell discharge.

(c) Let us now repeat the experiment with the cell and the resistor in separate calorimeters. Assume that the wires connecting them have no resistance and do not conduct any heat between the calorimeters. If we take Q_C as the heat absorbed by the cell and $-Q_R$ as that dissipated in the resistor, we find that $Q_C + Q_R = -233$ kJ/mol of Zn reacted, independent of R. However, the balance between these quantities does depend on the rate of discharge. As R increases, $|Q_C|$ decreases and $|Q_R|$ increases. In the limit of infinite R, Q_C approaches -43 kJ (per mole of zinc) and Q_R tends toward -190 kJ.

In this example, the energy $-Q_R$ was dissipated as heat, but it was obtained as electrical energy, and it might have been converted to light or mechanical work. In contrast, Q_C is an energy change that must appear as heat. Since discharge through $R \to \infty$ corresponds to a thermodynamically reversible process, the heat that must be absorbed in traversing a reversible path Q_{rev} is identified as $\lim_{R\to\infty} Q_C$. Now, since $\Delta S \equiv Q_{\text{rev}}/T$ (3), for our example, where all species are in their standard states,

$$T\Delta S^0 = \lim_{R\to\infty} Q_C = -43 \text{ kJ} \tag{2.1.15}$$

Since $\Delta G^0 = \Delta H^0 - T\Delta S^0$,

$$\Delta G^0 = -190 \text{ kJ} = \lim_{R\to\infty} Q_R \tag{2.1.16}$$

Note that we have now identified $-\Delta G$ with the maximum *net work* obtainable from the cell. Net work is defined as work other than PV work (2). For any finite R, $|Q_R|$ is *less* than the limiting value. Note also that the cell may absorb or evolve heat as it discharges. In the former case, $|\Delta G^0| > |\Delta H^0|$.

2.1.3 Free Energy and Cell emf

We found just above that if we discharged the electrochemical cell (2.1.14) through an infinite load resistance, the discharge is reversible. The potential difference is therefore always the equilibrium (open-circuit) value. Since the extent of reaction is supposed

† We adopt the thermodynamic convention in which absorbed quantities are positive.

to be small enough that all activities remain constant, the potential also remains constant. Then the energy dissipated in R is given by

$$|\Delta G| = \text{charge that passed} \times \text{reversible potential difference} \tag{2.1.17}$$

$$|\Delta G| = nF|E| \tag{2.1.18}$$

where n is the number of electrons passed per atom of zinc reacted, and F is the charge on a mole of electrons, which is about 96,500 C. Now, however, we realize that the free energy change has a *sign* associated with the *direction* of the net cell reaction. We can reverse the sign by reversing the direction. On the other hand, only an infinitesimal change in the overall cell potential is required to reverse the direction of the reaction, and hence E is essentially constant and independent of the direction of a (reversible) transformation. We have a quandary. We want to relate a direction-sensitive quantity (ΔG) to a direction-insensitive observable (E). This desire is the origin of almost all of the confusion that exists over sign conventions.

When we are interested in thermodynamic aspects of electrochemical systems, we have to rationalize this difficulty by inventing a thermodynamic construct called the *emf of the cell reaction*. Note that this quantity is assigned to the reaction (not to the physical cell); hence it has a directional aspect. To implement the plan, we associate a given chemical reaction with each cell schematic. For the one in (2.1.14), the reaction is

$$\text{Zn} + 2\text{AgCl} \rightarrow \text{Zn}^{2+} + 2\text{Ag} + 2\text{Cl}^- \tag{2.1.19}$$

The right electrode corresponds to reduction in the implied cell reaction, and the left electrode is identified with oxidation. Thus the reverse of (2.1.19) would be associated with the opposite schematic:

$$\text{Ag/AgCl/Cl}^-(a = 1), \text{Zn}^{2+}(a = 1)/\text{Zn} \tag{2.1.20}$$

The cell reaction emf, E_{rxn}, is then defined as the electrostatic potential of the electrode written on the right in the cell schematic with respect to that of the left.

For example, in the cell of (2.1.14), the measured potential difference is 0.985 V and the zinc electrode is negative; thus the emf of reaction (2.1.19) is +0.985 V. Likewise, the emf corresponding to (2.1.20) and the reverse of (2.1.19) is −0.985 V. By adopting this convention, we have managed to rationalize an (observable) *electrostatic* quantity (the cell potential difference), which is not sensitive to the direction of the cell's operation, with a (defined) *thermodynamic* quantity (the Gibbs free energy), which is sensitive to that direction. One can avoid completely the common confusion about sign conventions of cell potentials if one understands this formal relationship between electrostatic measurements and thermodynamic concepts (3, 4).

Since our convention implies a positive emf when a reaction is spontaneous,

$$\boxed{\Delta G = -nFE_{\text{rxn}}} \tag{2.1.21}$$

or as above, when all substances are at unit activity,

$$\boxed{\Delta G^0 = -nFE^0_{\text{rxn}}} \tag{2.1.22}$$

where E^0_{rxn} is called the *standard emf* of the cell reaction.

Other thermodynamic quantities can be derived from electrochemical measurements now that we have linked the potential difference across the cell to the free energy. For example, the entropy change in the cell reaction is given by the temperature dependence of ΔG:

$$\Delta S = -\left(\frac{\partial \Delta G}{\partial T}\right)_P \tag{2.1.23}$$

hence

$$\Delta S = nF\left(\frac{\partial E_{\text{rxn}}}{\partial T}\right)_P \tag{2.1.24}$$

and

$$\boxed{\Delta H = \Delta G + T\Delta S = nF\left[T\left(\frac{\partial E_{\text{rxn}}}{\partial T}\right)_P - E_{\text{rxn}}\right]} \tag{2.1.25}$$

The equilibrium constant of the reaction is given by

$$\boxed{RT \ln K_{\text{rxn}} = -\Delta G^0 = nFE^0_{\text{rxn}}} \tag{2.1.26}$$

Note that these relations are also useful for predicting electrochemical properties from thermochemical data. Several problems following this chapter illustrate the usefulness of that approach. Large tabulations of thermodynamic parameters exist in several sources (5–8).

2.1.4 Half-Reactions and Reduction Potentials

Just as the overall cell reaction comprises two independent half-reactions, one might think it reasonable that the cell potential could be broken into two individual electrode potentials. This view has experimental support—a self-consistent set of half-reaction emf's and half-cell potentials has been devised.

In order to establish the absolute potential of any conducting phase, according to definition, one must evaluate the work required to bring a unit positive charge, without associated matter, from the point at infinity to the interior of the phase. Although this quantity is not measurable by thermodynamically rigorous means, it can sometimes be estimated from a series of nonelectrochemical measurements and theoretical calculations if the demand for thermodynamic rigor is relaxed. Even if we could determine them, these absolute phase potentials would have limited utility because they would depend on magnitudes of the adventitious fields in which the phase is immersed (see Section 2.2). Much more meaningful is the difference in absolute phase potentials between an electrode and its electrolyte, for it is the chief factor determining the state of an electrochemical equilibrium. Unfortunately, we will find that it also is not rigorously measurable. Experimentally, we can find only the absolute potential difference between two electronic conductors. Even so, a useful scale results when one refers electrode potentials and half-reaction emf's to a standard reference electrode featuring a standard half-reaction.

The selected reference is the standard hydrogen electrode:

$$Pt/H_2(a = 1)/H^+(a = 1) \tag{2.1.27}$$

Its potential (the electrostatic standard) is taken as zero at all temperatures. Similarly, the standard emf's of the half-reactions:

$$2H^+ + 2e \rightleftharpoons H_2 \tag{2.1.28}$$

have also been assigned values of zero at all temperatures (the thermodynamic standard).

We can therefore record half-cell potentials by measuring them in whole cells against the NHE.† For example, in the system

$$Pt/H_2(a = 1)/H^+(a = 1)//Ag^+(a = 1)/Ag \tag{2.1.29}$$

the cell potential is 0.799 V and silver is positive. Thus the *standard potential of the* $Ag|Ag^+$ *couple* is +0.799 V *vs.* NHE. Moreover, the *standard emf of the* Ag^+ *reduction* is also +0.799 V *vs.* NHE, but that of the Ag oxidation is −0.799 V *vs.* NHE. One says that the *electrode potential* of Ag^+ is +0.799 V *vs.* NHE:

$$Ag^+ + e \rightleftharpoons Ag, \ E^0_{Ag^+/Ag} = +0.799 \text{ V } vs. \text{ NHE} \tag{2.1.30}$$

For the general system (2.1.13), the electrostatic potential of the R/O electrode (with respect to NHE) and the emf for the reduction of O always coincide. Therefore, one can condense the electrostatic and thermodynamic information into one list by tabulating electrode potentials and writing the half-reactions as reductions. Appendix C provides a listing of some frequently encountered potentials.

Tables of this sort are extremely useful because they feature much chemical and electrical information condensed into quite a small space. A few reduction potentials can characterize quite a number of cells and reactions. Since they really are indexes of free energies, they are also ready means for evaluating equilibrium constants, complexation constants, and solubility products. Also, they can often be taken in linear combinations to supply electrochemical information about additional half-reactions. One can tell from a glance at an ordered list of potentials whether or not a given redox process will proceed spontaneously.

2.1.5 emf and Concentration

Consider again a general cell in which the half-reaction at the right-hand electrode is

$$\nu_O O + ne \rightleftharpoons \nu_R R \tag{2.1.31}$$

The cell reaction is then

$$(n/2)H_2 + \nu_O O \rightarrow \nu_R R + nH^+ \tag{2.1.32}$$

† Note that an NHE is an ideal device and cannot be constructed. However, real hydrogen electrodes can approximate it, and its properties can be defined by extrapolation.

and its free energy is given from basic thermodynamics (2) by

$$\Delta G = \Delta G^0 + RT \ln \frac{(\mathrm{R})^{\nu_\mathrm{R}}(\mathrm{H}^+)^n}{(\mathrm{O})^{\nu_\mathrm{O}}(\mathrm{H}_2)^{n/2}} \tag{2.1.33}$$

Parenthesized quantities here represent activities.
Since $\Delta G = -nFE$ and $\Delta G^0 = -nFE^0$,

$$E = E^0 - \frac{RT}{nF} \ln \frac{(\mathrm{R})^{\nu_\mathrm{R}}(\mathrm{H}^+)^n}{(\mathrm{O})^{\nu_\mathrm{O}}(\mathrm{H}_2)^{n/2}} \tag{2.1.34}$$

but since $(\mathrm{H}^+) = (\mathrm{H}_2) = 1$,

$$\boxed{E = E^0 + \frac{RT}{nF} \ln \frac{(\mathrm{O})^{\nu_\mathrm{O}}}{(\mathrm{R})^{\nu_\mathrm{R}}}} \tag{2.1.35}$$

This relation, the Nernst equation, supplies the potential of the O/R electrode *vs.* NHE as a function of the activities of O and R. In addition, it defines the concentration dependence of the emf for the reaction (2.1.31).

It is now clear that the emf of any cell reaction, in terms of the reduction potentials of the two half-reactions, is

$$\boxed{E_{\mathrm{rxn}} = E_{\mathrm{right}} - E_{\mathrm{left}}} \tag{2.1.36}$$

where E_{right} and E_{left} refer to the cell schematic and are given by the appropriate Nernst equation. The cell potential is the magnitude of this value.

2.1.6 Formal Potentials

It is usually inconvenient to deal with activities in evaluations of half-cell potentials, because activity coefficients are almost always unknown. A device for avoiding them is the *formal potential*, $E^{0'}$. This quantity is the *measured* potential of the half-cell (*vs.* NHE) when (a) the species O and R are present at concentrations such that the ratio $[\mathrm{O}]^{\nu_\mathrm{O}}/[\mathrm{R}]^{\nu_\mathrm{R}}$ is unity and (b) other specified substances, for example, miscellaneous components of the medium, are present at designated concentrations. At the least, the formal potential incorporates the standard potential and some activity coefficients, γ_i. For example, consider

$$\mathrm{Fe}^{3+} + e \rightleftharpoons \mathrm{Fe}^{2+} \tag{2.1.37}$$

Its Nernst relation is simply

$$E = E^0 + \frac{RT}{nF} \ln \frac{(\mathrm{Fe}^{3+})}{(\mathrm{Fe}^{2+})} = E^0 + \frac{RT}{nF} \ln \frac{\gamma_{\mathrm{Fe}^{3+}}[\mathrm{Fe}^{3+}]}{\gamma_{\mathrm{Fe}^{2+}}[\mathrm{Fe}^{2+}]} \tag{2.1.38}$$

which is

$$E = E^{0'} + \frac{RT}{nF} \ln \frac{[\mathrm{Fe}^{3+}]}{[\mathrm{Fe}^{2+}]} \tag{2.1.39}$$

where

$$E^{0'} = E^0 + \frac{RT}{nF} \ln \frac{\gamma_{Fe^{3+}}}{\gamma_{Fe^{2+}}} \tag{2.1.40}$$

Since the ionic strength affects the activity coefficients, $E^{0'}$ will vary from medium to medium. One will find literature values listed for this couple in 1 M $HClO_4$, 0.1 M $HClO_4$, 1 M HNO_3, 1 M HCl, 1 M H_2SO_4, etc.

Actually $E^{0'}$ often contains factors related to complexation as well; as it does in fact for the Fe(III)/Fe(II) couple in HCl and H_2SO_4 solutions. Both iron species are complexed in these media, and hence (2.1.37) does not accurately describe the half-cell reaction. However, one can sidestep a full description of the complex competitive equilibria by using the empirical formal potentials. In such cases, $E^{0'}$ will also contain terms involving equilibrium constants and concentrations of some species involved in the equilibria.

2.1.7 Reference Electrodes

Many reference electrodes other than the NHE and the SCE have been devised for various studies of electrochemistry in aqueous and nonaqueous solvents. Several authors have provided extensive discussions on the subject (9–11).

Usually there are experimental reasons for the choice of a reference electrode. For example, the system

$$\text{Ag/AgCl/KCl (saturated, aqueous)} \tag{2.1.41}$$

has a smaller temperature coefficient of potential than an SCE and can be built more

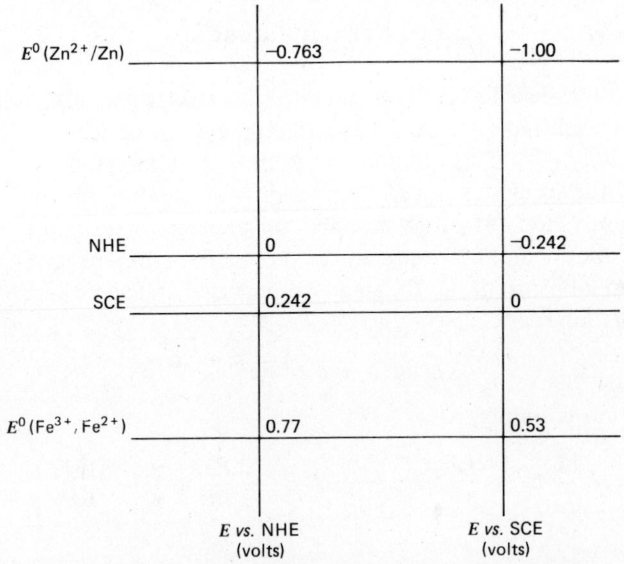

Figure 2.1.1

Relationship between potential scales based on the NHE and the SCE.

compactly. When chloride is not acceptable in the experimental system, the mercurous sulfate electrode may be used:

$$Hg/Hg_2SO_4/K_2SO_4 \text{ (saturated, aqueous)} \tag{2.1.42}$$

With a nonaqueous solvent, one may be concerned with the leakage of water from an aqueous reference electrode; hence a system like

$$Ag/Ag^+(0.01\ M \text{ in } CH_3CN) \tag{2.1.43}$$

might be preferred.

Since the potential of a reference electrode *vs.* NHE or SCE is nearly always specified in experimental papers, interconversion of scales can be accomplished easily. Figure 2.1.1 is a schematic representation of the relationship between the SCE and NHE scales. The inside back cover contains a tabulation of the potentials of the most common reference electrodes.

2.2 A MORE DETAILED VIEW OF INTERFACIAL POTENTIAL DIFFERENCES

2.2.1 The Physics of Phase Potentials

In considering the thermodynamic points of the previous section, we were not required to advance a mechanistic basis for the observable differences in potentials across certain phase boundaries. However, it is difficult to think chemically without a mechanistic model, and we may now find it helpful to consider the kinds of interactions between phases that could create these interfacial differences.

First, we must consider two prior questions: (1) Can we expect the potential within a phase to be uniform? (2) If so, what governs its value? One certainly can speak of the potential at any particular point within a phase. That quantity, $\phi(x, y, z)$, is defined as the work required to bring a unit positive charge, without material interactions, from an infinite distance to point (x, y, z). From electrostatics we have assurance that $\phi(x, y, z)$ is independent of the path of the test charge (12). The work is done against a coulombic field; hence we can express the potential generally as

$$\phi(x, y, z) = \int_{\infty}^{x,y,z} -\mathscr{E} \cdot d\mathbf{l} \tag{2.2.1}$$

where $\mathscr{E}$ is the electric field strength vector (force exerted on a unit charge at any point), and $d\mathbf{l}$ is an infinitesimal tangent to the path in the direction of movement. The integral is carried out over any path to (x, y, z). The difference in potential between points (x', y', z') and (x, y, z) is then

$$\phi(x', y', z') - \phi(x, y, z) = \int_{x,y,z}^{x',y',z'} -\mathscr{E} \cdot d\mathbf{l} \tag{2.2.2}$$

In general, the electric field is not zero everywhere between two points and the integral does not vanish over a path; hence some potential difference usually exists.

However, conducting phases have some special properties of great importance. Such a phase is one with mobile charge carriers, for example, a metal, a semiconductor, or an electrolyte solution. When no current passes through a conducting phase (i.e., there is no net movement of charge carriers), the electric field at all interior points must be zero. If it were not, the carriers would move in response to it in order to eliminate the field. From equation 2.2.2 one can see that the difference in potential between any two points in the interior of the phase must also be zero under these conditions, and thus the entire phase is an *equipotential volume.* We designate its potential as ϕ, which is known as the *inner potential* (or Galvani potential) of the phase.

How is the inner potential determined? A very important factor is the excess charge on the phase itself, since a test charge would have to work against its coulombic field. Other components of the potential would arise from the miscellaneous fields emanating from charged bodies outside the sample. As long as the charge distribution throughout the system is constant, the phase potential will remain constant, but changes in the phase potential will accompany alterations in charge distributions inside or outside the phase. Thus we have our first indication that differences in potential arising from chemical interactions between phases have some sort of charge *separation* as their basis.

An interesting question concerns the location of any excess charge on a conducting phase. The Gauss law from elementary electrostatics is extremely helpful here (13). It states that if we enclose a volume with an imaginary surface (a *Gaussian surface*), we will find that the net charge q inside the surface is given by an integral of the electric field over the surface:

$$q = \varepsilon_0 \oint \mathscr{E} \cdot d\mathbf{S} \tag{2.2.3}$$

where ε_0 is a proportionality constant† and $d\mathbf{S}$ is an infinitesimal vector normal outward from the surface. Now consider a Gaussian surface located within a conductor that is uniform in its interior (i.e., no voids or interior phases are present). If no current flows, $\mathscr{E}$ is zero at all points on the surface, hence the net charge within the boundary is zero. The situation is depicted in Figure 2.2.1. This conclusion applies to any Gaussian surface, even one situated just inside the phase boundary; thus we must conclude that the excess charge actually resides on the surface.‡

A view of the way in which phase potentials are established is now beginning to emerge:

1. Changes in the potential of a conducting phase can be effected by altering the charge distributions on or around the phase.

† The parameter ε_0 is called the permittivity of free space and has the value 8.85418×10^{-12} $C^2 N^{-1} m^{-2}$. See the footnote in Section 12.3.1 for a fuller explanation of electrostatic conventions followed in this book.

‡ There can be a finite thickness to this surface layer. The critical aspect is the size of the excess charge with respect to the bulk carrier concentration. If the charge must be established by drawing carriers from a significant volume, thermal processes will impede the compact accumulation of the excess strictly on the surface. Then the charged zone is called a *space charge region* because it has three-dimensional character. Its thickness can range from a few angstroms to several thousand angstroms in electrolytes and semiconductors. In metals it is negligibly thick. See Chapter 12 for more detailed discussion along this line.

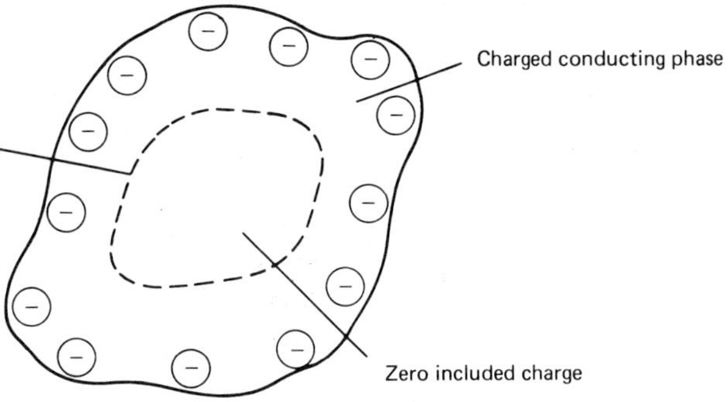

Figure 2.2.1
A three-dimensional conducting phase containing a Gaussian enclosure. Illustration that the excess charge resides on the surface.

2. If the phase undergoes a change in its excess charge, its charge carriers will move in a way such that the excess is wholly distributed over an entire boundary of the phase.
3. The surface distribution is such that the electric field strength within the phase is zero under null current conditions.
4. The entire phase features a constant potential ϕ.

The excess charge needed to change the potential of a conductor by electrochemically significant amounts is often not very large. Consider, for example, a spherical mercury drop of 0.5 mm radius. Changing its potential requires only about 5×10^{-14} C/V if it is suspended in air or in a vacuum (12).

2.2.2 Interactions Between Conducting Phases

When two conductors, for example a metal and an electrolyte, are placed in contact, the situation becomes complicated by the coulombic interaction between the phases. Charging one phase to change its potential tends to change the potential of the neighboring phase as well. This point is illustrated in the idealization of Figure 2.2.2, which portrays a charged metal sphere surrounded by an uncharged electrolyte. We know that the charge on the metal q^M resides on its surface. This unbalanced charge (negative in the diagram) will create an excess cation concentration near the electrode in the solution. What can we say about the magnitudes and distributions of the obvious charge imbalances in solution?

Consider the integral of equation 2.2.3 over the Gaussian surface shown in Figure 2.2.2. Since $\mathscr{E}$ at every point is zero, the net enclosed charge is also zero. We could place the Gaussian surface just outside the surface region bounding the metal and solution, and we would reach the same conclusion. Thus we know now that the excess positive charge in the solution at the interface, q^S, exactly compensates the excess metal charge. That is,

$$\boxed{q^S = -q^M} \tag{2.2.4}$$

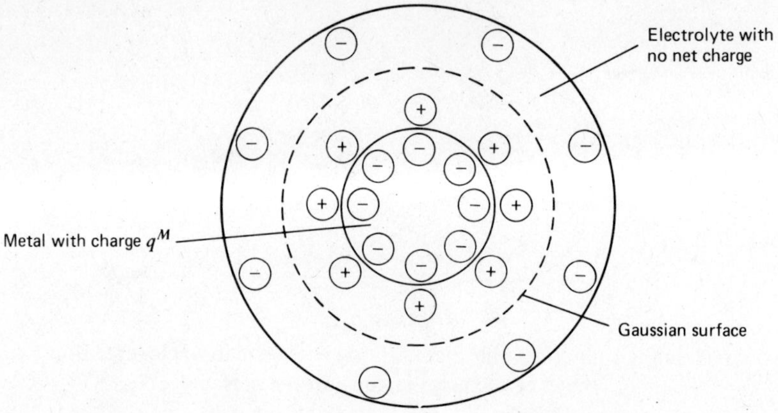

Figure 2.2.2
Interaction between a metal sphere and a surrounding electrolyte layer. The Gaussian enclosure is a sphere containing the metal phase and part of the electrolyte.

This fact is very useful in the treatment of these interfacial charge arrays, which we have already seen as *electrical double layers* (see Chapters 1 and 12).

Alternatively, we might move the Gaussian surface to a location just inside the outer boundary of the electrolyte. The enclosed charge must still be zero, yet we know that the net charge on the whole system is q^M. A negative charge equal to q^M must therefore reside on the outer surface of the electrolyte.

Figure 2.2.3 is a display of the potential versus distance from the center of this assembly. Increasing the negative charge on the metal would naturally lower ϕ^M, but

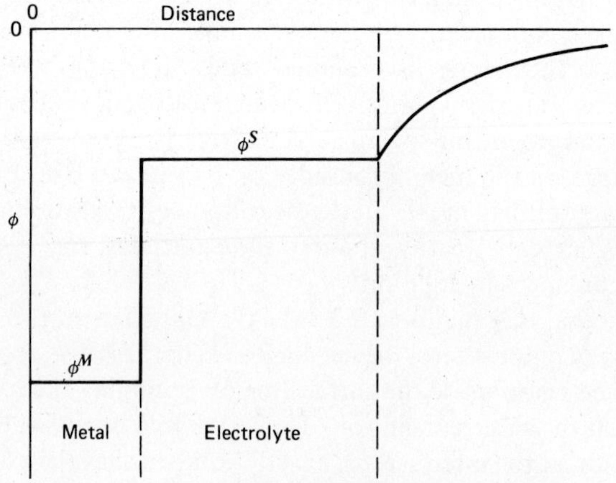

Figure 2.2.3
Potential profile through the system shown in Figure 2.2.2. Distance is measured radially from the center of the metallic sphere.

it would also lower ϕ^S because the excess negative charge on the outer boundary would increase. The difference $\phi^M - \phi^S$, called the interfacial potential difference, depends on the charge imbalance at the interface and the physical size of the interface. That is, it depends on the charge density (C/cm^2) at the interface. Making a 1-V change in this interfacial potential difference requires sizable alterations in charge density. For the spherical mercury drop considered above, now surrounded by 0.1 M strong electrolyte, one would need about 10^{-6} C for the change.

This is a much larger density than the insulated drop required. The difference appears because the coulombic field of any surface charge is counterbalanced to a very large degree by polarization in the adjacent electrolyte. In practical electrochemistry, metallic electrodes are partially exposed to an electrolyte and partially insulated. For example, one might use a 0.1 cm^2 platinum disk electrode attached to a platinum lead that is almost fully sealed in glass. It is interesting to consider the location of excess charge used in altering the potential of such a phase. Of course, the charge must be distributed over the entire surface, including both insulated and electrochemically active area. However, we have seen that the coulombic interaction with the electrolyte is so strong that essentially all the charge at any potential will lie adjacent to the solution, unless the percentage of the phase area in contact with electrolyte is really minuscule.

What real mechanisms are there for charging a phase at all? An important one is simply to pump electrons into or out of a metal or semiconductor with a power supply of some sort. In fact, we will find that this technique is the basis for our *control* over the kinetics of electrode processes.

In addition, there are chemical mechanisms. We know from experience, for example, that a platinum wire that is dipped into a solution containing ferricyanide and ferrocyanide will change its potential gradually toward a predictable equilibrium value given by the Nernst equation. This process occurs because the electron affinities of the two phases initially differ, hence there is a transfer of electrons from the metal to the solution or vice versa. Ferricyanide is reduced or ferrocyanide is oxidized. The transfer of charge continues until the resulting change in potential reaches the equilibrium point where the electron affinities of the solution and metal phases are equal. Only small amounts of charge need be transferred; hence the net chemical effects on the solution are usually not noticeable. The metal adapts to the solution and reflects its composition.

Electrochemistry is full of situations like this one, in which charged species, electrons, or ions cross interfacial boundaries. These processes generally create a net transfer of charge that sets up the equilibrium or steady-state potential differences that we observe. Considering them in more detail must follow the development of some additional concepts (see Section 2.3 and Chapter 3).

Actually, interfacial potential differences can develop without an excess charge on either phase. Consider an aqueous electrolyte in contact with an electrode. Since the electrolyte interacts with the metal surface (e.g., wetting it), the water dipoles in contact with the metal generally will have some preferential orientation. From a coulombic standpoint, this situation is equivalent to charge separation across the interface because the dipoles are not randomized with time. Since moving a test charge through the interface requires work, the interfacial potential difference is not zero (14).†

2.2.3 Measurement of Potential Differences

We have already noted that the difference in the inner potentials, $\Delta\phi$, of two phases in contact is a factor of primary importance to electrochemical processes occurring at the interface between them. Part of its influence comes from the high electric fields reflecting the large changes in potential in the boundary region. These fields can reach values as high as 10^7 V/cm. They are high enough to distort electroreactants so as to alter reactivity, and they can affect the kinetics of charge transport across the interface. Another aspect of $\Delta\phi$ is its direct influence over the relative energies of charged species on either side of the interface. In this way, $\Delta\phi$ controls the relative electron affinities of the two phases; hence it controls the direction of reaction.

Unfortunately, $\Delta\phi$ cannot be measured, because one cannot sample the electrical properties of the solution without introducing at least one more interface. It is characteristic of devices for measuring potential differences (e.g., potentiometers, voltmeters, or electrometers) that they can be calibrated only to register potential differences between two phases of the same composition, such as the two copper contacts available at most instruments. Now consider $\Delta\phi$ at the interface Zn/Zn^{2+}, Cl^-. The simplest approach one could make to $\Delta\phi$ using a potentiometric instrument with copper contacts is shown in Figure 2.2.4*a*. The measurable potential difference between the copper phases clearly includes interfacial potential differencces at the Zn/Cu interface and the Cu/electrolyte interface in addition to $\Delta\phi$. We might simplify matters by constructing a voltmeter wholly from zinc but, as shown in Figure 2.2.4*b*, the measurable voltage would still contain contributions from two separate interfacial potentials.

By now we realize that a measured cell potential is the sum of several interfacial differences, none of which we can evaluate independently. For example, we could represent the potential profile through the cell

$$Cu/Zn/Zn^{2+}, Cl^-/AgCl/Ag/Cu' \qquad (2.2.5)$$

according to Vetter's representation (15) as shown in Figure 2.2.5. (Although silver

Figure 2.2.4
Two devices for measuring the potential of a cell containing the Zn/Zn^{2+} interface.

† In some cases, it is useful to break the inner potential into two components called the *outer* (or *Volta*) potential and *surface* potential. The outer potential is represented by ψ and the surface potential is χ. Thus

$$\phi = \psi + \chi$$

There is a large, detailed literature on the establishment, the meaning, and the measurement of interfacial potential differences and their components. See, for example, References 14–17.

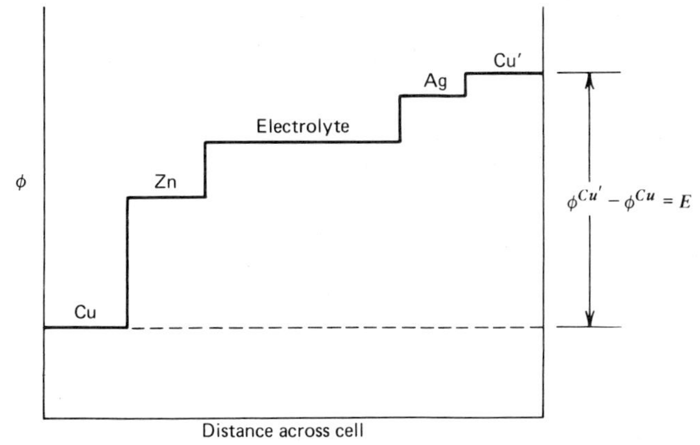

Figure 2.2.5
Potential profile across a whole cell at equilibrium.

chloride is a separate phase, it does not contribute to the cell potential E, because it does not physically separate silver from the electrolyte. In fact, it need not even be present; one merely requires a solution saturated in silver chloride to measure the same cell potential.)

Even through this complexity, it is still possible to focus on a single interfacial potential difference, such as that between zinc and the electrolyte in (2.2.5). If we can maintain constant interfacial potentials at all the other junctions in the cell, then any *change* in E must be wholly attributed to a *change* in $\Delta\phi$ at the zinc/electrolyte boundary. Keeping the other junctions at constant potential difference is not so difficult, for the metal-metal junctions always remain constant without attention, and the silver/electrolyte junction will be fixed if the activities of the participants in its half-reaction are fixed. When this idea is realized, the whole rationale behind half-cell potentials and the choice of reference electrodes becomes much clearer.

2.2.4 Electrochemical Potentials (14–17)

Consider again the interface Zn/Zn^{2+}, Cl^- (aqueous), and focus on zinc ions in metallic zinc and in solution. In the metal, Zn^{2+} is fixed in a lattice of positive zinc ions, with free electrons permeating the structure. In solution, zinc ion is hydrated and it may interact with Cl^-. The energy state of Zn^{2+} in any location clearly depends on the chemical environment, which manifests itself through short-range forces that are mostly electrical in nature. In addition, there is the energy required simply to bring the $+2$ charge, disregarding the chemical effects, to the location in question. This second energy is clearly proportional to the potential ϕ at the location; hence it depends on the electrical properties of an environment that is very much larger than the ion itself. Although one cannot experimentally separate these two components for a single species, the differences in the scales of the two environments responsible for them makes it possible to separate them mathematically. Butler (18) and Guggenheim

(19) developed the conceptual separation and introduced the *electrochemical potential*, $\bar{\mu}_i^\alpha$, for species i with charge z_i in phase α:

$$\boxed{\bar{\mu}_i^\alpha = \mu_i^\alpha + z_i F \phi^\alpha} \tag{2.2.6}$$

The term μ_i^α is merely the familiar chemical potential

$$\mu_i^\alpha = \left(\frac{\partial G}{\partial n_i}\right)_{T,P,n_{j \neq i}} \tag{2.2.7}$$

where n_i is the number of moles of i in phase α. Thus the electrochemical potential would be

$$\bar{\mu}_i^\alpha = \left(\frac{\partial \bar{G}}{\partial n_i}\right)_{T,P,n_{j \neq i}} \tag{2.2.8}$$

where the *electrochemical free* energy $\bar{G}$ differs from the *chemical* free energy G by inclusion of effects from the large-scale electrical environment.

(a) Properties of the Electrochemical Potential

1. For uncharged species: $\bar{\mu}_i^\alpha = \mu_i^\alpha$.
2. For any substance: $\mu_i^\alpha = \mu_i^{0\alpha} + RT \ln a_i^\alpha$ where $\mu_i^{0\alpha}$ is the standard chemical potential and a_i^α is the activity of species i in phase α.
3. For a pure phase at unit activity (e.g., solid Zn, AgCl, Ag, or H_2 at unit fugacity): $\bar{\mu}_i^\alpha = \mu_i^{0\alpha}$.
4. For electrons in a metal: $\bar{\mu}_e^\alpha = \mu_e^{0\alpha} - F\phi^\alpha$. Activity effects can be disregarded because the electron concentration never changes appreciably.
5. For equilibrium of species i between phases α and β: $\bar{\mu}_i^\alpha = \bar{\mu}_i^\beta$.

(b) Reactions in a Single Phase. Within a single conducting phase, ϕ is constant everywhere and exerts no effect on chemical equilibrium. The ϕ terms drop out of relations involving electrochemical potentials, and only chemical potentials will remain. Consider the acid-base equilibrium:

$$\text{HOAc} \rightleftharpoons \text{H}^+ + \text{OAc}^- \tag{2.2.9}$$

This requires that

$$\bar{\mu}_{\text{HOAc}} = \bar{\mu}_{\text{H}^+} + \bar{\mu}_{\text{OAc}^-} \tag{2.2.10}$$

$$\mu_{\text{HOAc}} = \mu_{\text{H}^+} + F\phi + \mu_{\text{OAc}^-} - F\phi \tag{2.2.11}$$

$$\mu_{\text{HOAc}} = \mu_{\text{H}^+} + \mu_{\text{OAc}^-} \tag{2.2.12}$$

(c) Reactions Involving Two Phases Without Charge Transfer. The solubility equilibrium:

$$\text{AgCl (crystal, } c) \rightleftharpoons \text{Ag}^+ + \text{Cl}^- \text{ (solution, } s) \tag{2.2.13}$$

can be treated in two ways. First, one can consider separate equilibria involving Ag^+ and Cl^- in solution and in the solid. Thus

$$\bar{\mu}_{Ag^+}^{AgCl} = \bar{\mu}_{Ag^+}^{s} \tag{2.2.14}$$

$$\bar{\mu}_{Cl^-}^{AgCl} = \bar{\mu}_{Cl^-}^{s} \tag{2.2.15}$$

Recognizing that

$$\bar{\mu}_{AgCl}^{AgCl} = \bar{\mu}_{Ag^+}^{AgCl} + \bar{\mu}_{Cl^-}^{AgCl} \tag{2.2.16}$$

one can obtain

$$\mu_{AgCl}^{0AgCl} = \bar{\mu}_{Ag^+}^{s} + \bar{\mu}_{Cl^-}^{s} \tag{2.2.17}$$

from the sum of (2.2.14) and (2.2.15). Expanding, we have

$$\mu_{AgCl}^{0AgCl} = \mu_{Ag^+}^{0s} + RT\ln a_{Ag^+}^{s} + F\phi^s + \mu_{Cl^-}^{0s} + RT\ln a_{Cl^-}^{s} - F\phi^s \tag{2.2.18}$$

and rearrangement gives

$$\mu_{AgCl}^{0AgCl} - \mu_{Ag^+}^{0s} - \mu_{Cl^-}^{0s} = RT\ln(a_{Ag^+}^{s}a_{Cl^-}^{s}) = RT\ln K_{sp} \tag{2.2.19}$$

where K_{sp} is the solubility product. A quicker route to the well-known result (2.2.19) is to write down (2.2.17) directly from the chemical equation (2.2.13).

Note that the ϕ^s terms cancelled in (2.2.18), and that an implicit cancellation of ϕ^{AgCl} terms occurred in (2.2.16). Since the final result depends only on chemical potentials, the equilibrium is unaffected by the potential difference across the interface. This is a general feature of interphase reactions without transfer of charge. When charge transfer does occur, the ϕ terms will not cancel and the interfacial potential difference strongly affects the chemical process. Thus we can use that potential either to probe or to alter the equilibrium position.

(d) Formulation of a Cell Potential. Consider now the cell (2.2.5) for which the cell reaction can be written:

$$Zn + 2AgCl + 2e(Cu') \rightleftharpoons Zn^{2+} + 2Ag + 2Cl^- + 2e(Cu) \tag{2.2.20}$$

At equilibrium,

$$\bar{\mu}_{Zn}^{Zn} + 2\bar{\mu}_{AgCl}^{AgCl} + 2\bar{\mu}_{e}^{Cu'} = \bar{\mu}_{Zn^{2+}}^{s} + 2\bar{\mu}_{Ag}^{Ag} + 2\bar{\mu}_{Cl^-}^{s} + 2\bar{\mu}_{e}^{Cu} \tag{2.2.21}$$

$$2(\bar{\mu}_{e}^{Cu'} - \bar{\mu}_{e}{}^{Cu}) = \bar{\mu}_{Zn^{2+}}^{s} + 2\bar{\mu}_{Ag}^{Ag} + 2\bar{\mu}_{Cl^-}^{s} - \bar{\mu}_{Zn}^{Zn} - 2\bar{\mu}_{AgCl}^{AgCl} \tag{2.2.22}$$

But,

$$2(\bar{\mu}_{e}^{Cu'} - \bar{\mu}_{e}^{Cu}) = -2F(\phi^{Cu'} - \phi^{Cu}) = -2FE \tag{2.2.23}$$

Expanding (2.2.22), we have

$$\begin{aligned} -2FE = {} & \mu_{Zn^{2+}}^{0s} + RT\ln a_{Zn^{2+}}^{s} + 2F\phi^s + 2\mu_{Ag}^{0Ag} + 2\mu_{Cl^-}^{0s} \\ & + 2RT\ln a_{Cl^-}^{s} - 2F\phi^s - \mu_{Zn}^{0Zn} - 2\mu_{AgCl}^{0AgCl} \end{aligned} \tag{2.2.24}$$

$$-2FE = \Delta G^0 + RT\ln a_{Zn^{2+}}^{s}(a_{Cl^-}^{s})^2 \tag{2.2.25}$$

where

$$\Delta G^0 = \mu_{Zn^{2+}}^{0s} + 2\mu_{Cl^-}^{0s} + 2\mu_{Ag}^{0Ag} - \mu_{Zn}^{0Zn} - 2\mu_{AgCl}^{0AgCl} = -2FE^0 \tag{2.2.26}$$

Thus we arrive at

$$E = E^0 - \frac{RT}{2F} \ln(a^s_{Zn^{2+}})(a^s_{Cl^-})^2 \tag{2.2.27}$$

which is the Nernst equation for the cell. This corroboration of an earlier result displays the general utility of electrochemical potentials for treating interfacial reactions with charge transfer. They are powerful tools. For example, they are easily used to consider whether the two cells:

$$Cu/Pt/Fe^{2+}, Fe^{3+}, Cl^-/AgCl/Ag/Cu' \tag{2.2.28}$$

$$Cu/Au/Fe^{2+}, Fe^{3+}, Cl^-/AgCl/Ag/Cu' \tag{2.2.29}$$

would have the same E values. Examination of this point is left to the reader as Problem 2.8.

2.3 LIQUID JUNCTION POTENTIALS

2.3.1 Potential Differences at an Electrolyte-Electrolyte Boundary

Up to this point, we have examined only systems at equilibrium, and we have learned that the potential differences in equilibrium electrochemical systems can be treated exactly by thermodynamics. However, many real cells are never at equilibrium because they feature different electrolytes around the two electrodes. There is somewhere an interface between the two solutions, and at that point mass transport processes attempt to mix the solutes until the two electrolytes are identical. Unless they are the same initially, the *liquid junction* will not be at equilibrium because net flows of mass occur continuously across it.

Such a cell is

$$\underset{\alpha}{Cu/Zn/Zn^{2+}}/\underset{\beta}{Cu^{2+}}/Cu' \tag{2.3.1}$$

for which we can depict the equilibrium processes as in Figure 2.3.1. The overall cell potential at null current is then

$$E = (\phi^{Cu'} - \phi^{\beta}) - (\phi^{Cu} - \phi^{\alpha}) + (\phi^{\beta} - \phi^{\alpha}) \tag{2.3.2}$$

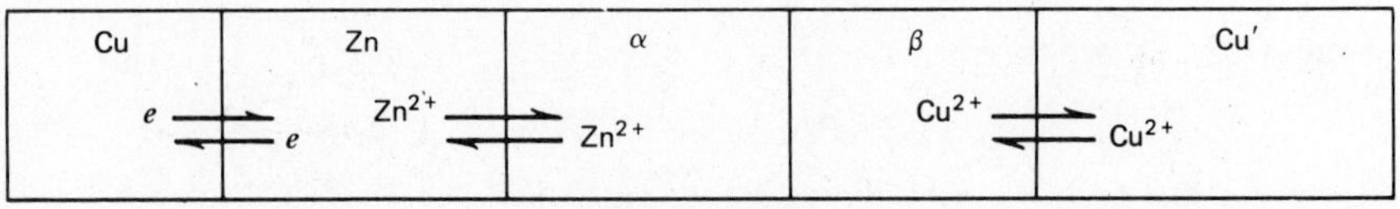

Figure 2.3.1

Schematic view of the phases in the cell (2.3.1). Equilibrium is established for certain charge carriers as shown, but at the liquid junction between the two electrolyte phases α and β, equilibrium is not reached.

Obviously the first two components of E are the expected interfacial potential differences at the copper and zinc electrodes. The third term shows that the measured cell potential depends also on the potential difference between the electrolytes, that is, on the *liquid junction potential.* This discovery is a real threat to our system of electrode potentials, because that system is based on the idea that all contributions to E can be assigned unambiguously to one electrode or to the other. How could the junction potential possibly be assigned properly? We must evaluate the importance of these phenomena.

2.3.2 Types of Liquid Junctions (3, 15, 20, 21)

The existence of junction potentials is easily realized by considering the boundary shown in Figure 2.3.2*a*. At the junction, there is a steep concentration gradient in H^+ and Cl^-; hence both ions tend to diffuse from right to left. Since the hydrogen ion has a much larger mobility than Cl^-, it penetrates the dilute phase initially at a higher rate. This process gives a positive charge to the dilute phase and a negative charge to the concentrated one, with the result that a boundary potential difference develops. The electric field then retards the movement of H^+ and speeds up the passage of Cl^-, until the two cross the boundary at equal rates. Thus, there is a detectable steady-state potential, which is not due to an equilibrium process. From its origin, this interfacial potential is sometimes called a *diffusion potential.*

Lingane (3) classifies liquid junctions into three types:

1. Two solutions of the same electrolyte at different concentrations; for example, Figure 2.3.2*a*, which is considered above.
2. Two solutions at the same concentrations with different electrolytes having an ion in common; for example, Figure 2.3.2*b*.
3. Two solutions not satisfying conditions 1 or 2; for example, Figure 2.3.2*c*.

We will find this classification useful in the treatments of junction potentials that follow.

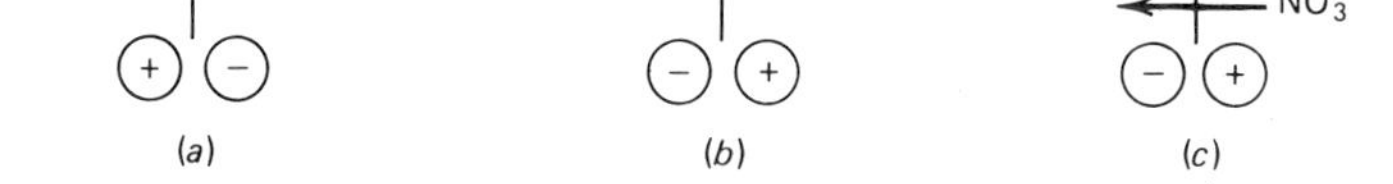

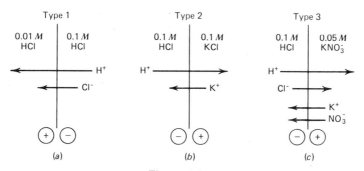

Figure 2.3.2
Types of liquid junction. Arrows show the direction of net transfer for each ion, and their lengths indicate relative mobilities. The polarity of the junction potential is indicated in each case by the circled signs. (Adapted from J. J. Lingane, "Electroanalytical Chemistry," 2nd ed., Wiley-Interscience, New York, 1958, p. 60, with permission.)

Even though the boundary region cannot be at equilibrium, it has a composition that is effectively constant over long time periods, and the reversible transfer of electricity through the region can be considered.

2.3.3 Conductance, Transference Numbers, and Mobility (21, 22)

When an electric current flows in an electrochemical cell, the current in solution is carried by the movement of ions. For example, take the cell:

$$\ominus \mathrm{Pt/H_2(1\ atm)}/\underset{(a_1)}{\overset{\alpha}{\mathrm{H^+, Cl^-}}}/\underset{(a_2)}{\overset{\beta}{\mathrm{H^+, Cl^-}}}/\mathrm{H_2(1\ atm)/Pt'}\oplus \tag{2.3.3}$$

where $a_2 > a_1$. When the cell operates galvanically, at the left electrode,

$$\mathrm{H_2} \rightarrow 2\mathrm{H^+}(\alpha) + 2e(\mathrm{Pt}) \tag{2.3.4}$$

and at the right electrode,

$$2\mathrm{H^+}(\beta) + 2e(\mathrm{Pt'}) \rightarrow \mathrm{H_2} \tag{2.3.5}$$

Therefore, there is a tendency to build up a positive charge in the α phase and a negative charge in β. This tendency is overcome by the movement of ions: H^+ to the right and Cl^- to the left. For one faraday of charge passed, 1 mol of H^+ is produced in α and 1 mol of H^+ is consumed in β. The total amount of H^+ and Cl^- migrating across the boundary between α and β must equal 1 mol.

The fractions of the current carried by H^+ and Cl^- are called their *transference numbers* (or *transport numbers*). If we let t_+ be the transference number for H^+ and t_- be that for Cl^-, then clearly,

$$t_+ + t_- = 1 \tag{2.3.6}$$

In general, for an electrolyte containing many ions, i,

$$\boxed{\sum_i t_i = 1} \tag{2.3.7}$$

Schematically, the process can be represented as shown in Figure 2.3.3. Initially the cell features a higher activity of hydrochloric acid (+ as H^+, − as Cl^-) on the right (Figure 2.3.3*a*), hence discharging it spontaneously produces H^+ on the left and consumes it on the right. Assume that five units of H^+ are reacted as shown in Figure 2.3.3*b*. For hydrochloric acid, since $t_+ \sim 0.8$ and $t_- \sim 0.2$, to maintain electroneutrality, 4 H^+ migrate to the right and 1 Cl^- to the left. This process is depicted in panel (*c*), and the final state of the solution is represented in Figure 2.3.3*d*.

A charge imbalance like that suggested in (*b*) could not actually occur because a very large electric field would be established. On a macroscopic scale, electroneutrality is always maintained throughout the solution. The migration represented in (*c*) occurs simultaneously with the electron transfer reactions.

Transference numbers are determined by the details of ionic conduction, which are understood mainly through measurements of the resistance to current flow in solution,

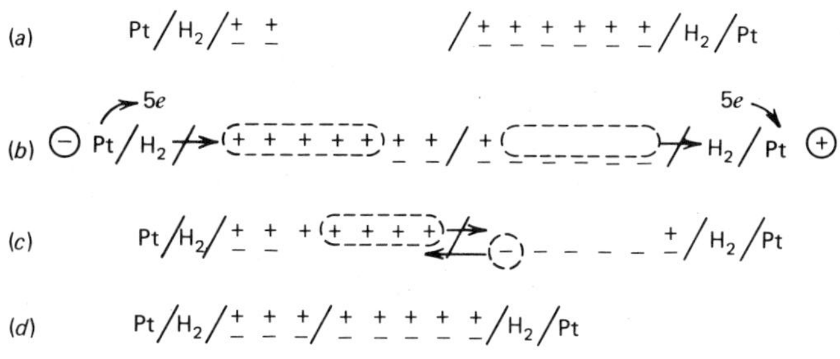

Figure 2.3.3

Schematic diagram showing the redistribution of charge during electrolysis of a system featuring a high concentration of hydrochloric acid on the right and a low concentration on the left.

or its reciprocal, the conductance, L. The value of L for a segment of solution immersed in an electric field is directly proportional to the cross-sectional area perpendicular to the field vector and it is inversely proportional to the length of the segment along the field. The proportionality constant is the *conductivity*, κ, which is an intrinsic property of the solution:

$$L = \kappa(A/l) \tag{2.3.8}$$

The conductance L is given in dimensions of siemens (S) (i.e., Ω^{-1}), and κ is expressed in siemens per centimeter (Ω^{-1} cm^{-1}).

Since the passage of current through the solution is accomplished by the independent movement of different species, κ is really the sum of contributions from all the ionic species i. It is intuitive that each component of κ is proportional to the concentration of the ion, the magnitude of its charge $|z_i|$, and some index of its intrinsic migration velocity.

That index is the *mobility*, u_i, which is the limiting velocity of the ion in an electric field of unit strength. Mobility usually carries dimensions of cm/sec per V/cm (or cm^2 V^{-1} sec^{-1}). When a field of strength $\mathscr{E}$ is applied to an ion, it will accelerate under the force imposed by the field until the frictional drag exactly counterbalances the electric force. Then the ion continues its motion at that terminal velocity. In Figure 2.3.4 this balance is represented. The force from the field is $|z_i|e\mathscr{E}$, where e is the electronic charge. The frictional drag can be approximated from the Stokes law as $6\pi\eta rv$, where η is the viscosity of the medium, r is the radius of the ion, and v

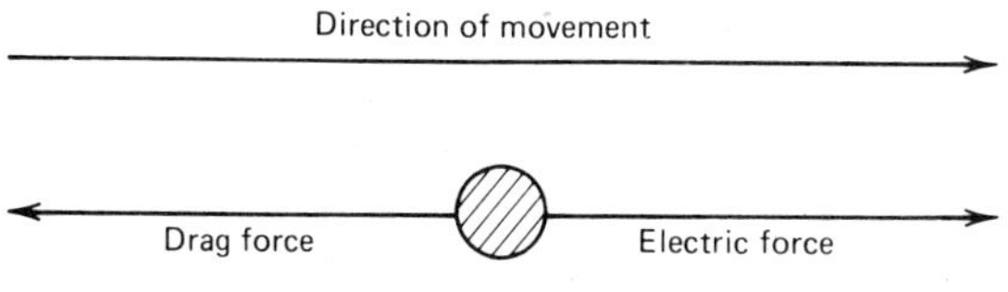

Figure 2.3.4

Forces on a charged particle moving in solution under the influence of an electric field. The forces balance at the terminal velocity.

is the velocity. When the terminal velocity is reached, we have by equation and rearrangement,

$$u_i = \frac{v}{\mathscr{E}} = \frac{|z_i|e}{6\pi\eta r} \tag{2.3.9}$$

The proportionality factor relating an individual ionic conductivity to charge, mobility, and concentration turns out to be the faraday; thus

$$\boxed{\kappa = F\sum_i |z_i|u_iC_i} \tag{2.3.10}$$

In general, then, the transference number for species i as merely the contribution to conductivity made by that species divided by the total conductivity:

$$\boxed{t_i = \frac{|z_i|u_iC_i}{\sum_j |z_j|u_jC_j}} \tag{2.3.11}$$

For solutions of simple, pure electrolytes (i.e., one positive and one negative ionic species), such as KCl, $CaCl_2$, and HNO_3, conductance is often quantified in terms of a parameter, Λ, the *equivalent conductivity*, which is defined by

$$\boxed{\Lambda = \frac{\kappa}{C_{eq}}} \tag{2.3.12}$$

where C_{eq} is the concentration of positive (or negative) charges. Thus Λ expresses the conductivity per unit concentration of charge. In these systems, since $C|z| = C_{eq}$ for either ionic species, one finds from (2.3.10) and (2.3.12) that

$$\Lambda = F(u_+ + u_-) \tag{2.3.13}$$

where u_+ refers to the cation and u_- to the anion. This relation suggests that Λ could be regarded as the sum of individual *equivalent ionic conductivities*:

$$\Lambda = \lambda_+ + \lambda_- \tag{2.3.14}$$

hence we find

$$\lambda_i = Fu_i \tag{2.3.15}$$

In these simple solutions, then, the transference number t_i is given by

$$t_i = \frac{\lambda_i}{\Lambda} \tag{2.3.16}$$

or, alternatively,

$$t_i = \frac{u_i}{u_+ + u_-} \tag{2.3.17}$$

Transference numbers can be measured by several approaches (21, 22), and numerous data for pure solutions appear in the literature. Table 2.3.1 displays a few values

Table 2.3.1

Cation Transference Numbers for Aqueous Solutions at 25 °C[a]

	Concentration, C_{eq}[b]			
Electrolyte	0.01	0.05	0.1	0.2
HCl	0.8251	0.8292	0.8314	0.8337
NaCl	0.3918	0.3876	0.3854	0.3821
KCl	0.4902	0.4899	0.4898	0.4894
NH_4Cl	0.4907	0.4905	0.4907	0.4911
KNO_3	0.5084	0.5093	0.5103	0.5120
Na_2SO_4	0.3848	0.3829	0.3828	0.3828
K_2SO_4	0.4829	0.4870	0.4890	0.4910

[a] From D. A. MacInnes, "The Principles of Electrochemistry," Dover, New York, 1961, p. 85 and references cited therein.
[b] Moles of positive (or negative) charge per liter.

for aqueous solutions at 25°. From results of this sort, one can evaluate the individual ionic conductivities, λ_i. Both λ_i and t_i depend on the concentration of the pure electrolyte, because interactions between ions tend to alter the mobilities as the concentration changes (21–23). Tabulations of λ values, such as that of Table 2.3.2, usually list λ_{0_i},

Table 2.3.2

Ionic Properties at Infinite Dilution in Aqueous Solutions at 25°

Ion	λ_0, cm^2 Ω^{-1} equiv^{-1}[a]	u, cm^2 sec^{-1} V^{-1}[b]
H^+	349.82	3.625×10^{-3}
K^+	73.52	7.619×10^{-4}
Na^+	50.11	5.193×10^{-4}
Li^+	38.69	4.010×10^{-4}
NH_4^+	73.4	7.61×10^{-4}
$\frac{1}{2}Ca^{2+}$	59.50	6.166×10^{-4}
OH^-	198	2.05×10^{-3}
Cl^-	76.34	7.912×10^{-4}
Br^-	78.4	8.13×10^{-4}
I^-	76.85	7.96×10^{-4}
NO_3^-	71.44	7.404×10^{-4}
OAc^-	40.9	4.24×10^{-4}
ClO_4^-	68.0	7.05×10^{-4}
$\frac{1}{2}SO_4^{2-}$	79.8	8.27×10^{-4}
HCO_3^-	44.48	4.610×10^{-4}
$\frac{1}{3}Fe(CN)_6^{3-}$	101.0	1.047×10^{-3}
$\frac{1}{4}Fe(CN)_6^{4-}$	110.5	1.145×10^{-3}

[a] From D. A. MacInnes, "The Principles of Electrochemistry," Dover, New York, 1961, p. 342.
[b] Calculated from λ_0.

which are data obtained by extrapolation to infinite dilution. In the absence of measured transference numbers, it is convenient to use these to estimate t_i for pure solutions by (2.3.16) or for mixed electrolytes by an equivalent to (2.3.11):

$$t_i = \frac{|z_i|C_i\lambda_i}{\sum_j |z_j|C_j\lambda_j} \tag{2.3.18}$$

2.3.4 Calculation of Liquid Junction Potentials

Imagine the concentration cell (2.3.3) connected to a power supply as shown in Figure 2.3.5. The voltage from the supply opposes that from the cell, and experimentally one finds that it is possible to oppose the cell voltage exactly so that no current flows through the galvanometer, G. If the magnitude of the opposing voltage is reduced very slightly, the cell operates spontaneously as described above, and current passes (as electrons) from Pt to Pt′ in the external circuit. The process occurring at the liquid junction is the passage of an equivalent negative charge from right to left. If the opposing voltage increases from the null point, the entire process reverses, including charge transfer through the interface between the electrolytes. The fact that an infinitesimal change in the driving force can reverse the direction of charge passage implies that the electrochemical free energy change for the whole process is zero.

These events can be divided into those involving the chemical transformations at the metal-solution interfaces:

$$\tfrac{1}{2}H_2 \rightleftharpoons H^+(\alpha) + e(\text{Pt}) \tag{2.3.19}$$

$$H^+(\beta) + e(\text{Pt}') \rightleftharpoons \tfrac{1}{2}H_2 \tag{2.3.20}$$

and that effecting charge transport at the liquid junction depicted in Figure 2.3.6:

$$t_+H^+(\alpha) + t_-Cl^-(\beta) \rightleftharpoons t_+H^+(\beta) + t_-Cl^-(\alpha) \tag{2.3.21}$$

Note also that (2.3.19) and (2.3.20) are at strict equilibrium under the null current condition; hence the free energy change for each of them individually is zero. Of course, this is also true for their sum:

$$H^+(\beta) + e(\text{Pt}') \rightleftharpoons H^+(\alpha) + e(\text{Pt}) \tag{2.3.22}$$

which describes the chemical change in the system. Likewise, the sum of this equation and the charge transport relation (2.3.21) describes the overall cell operation. However, since we have just learned that the free energy changes for both the overall

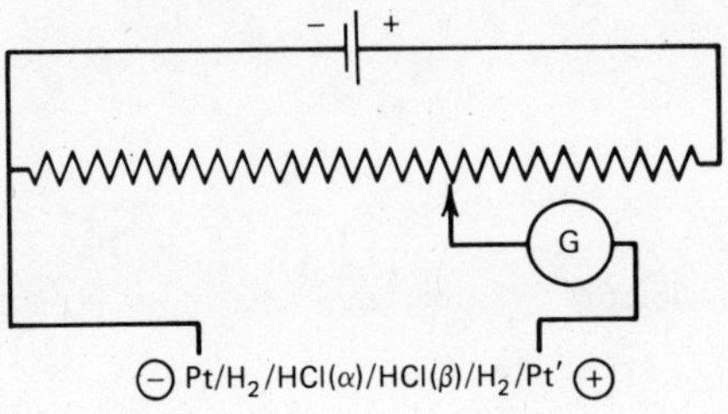

Figure 2.3.5
Experimental system for demonstrating reversible flow of charge through a cell with a liquid junction.

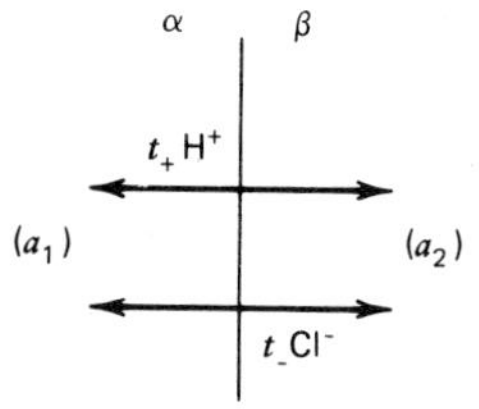

Figure 2.3.6
Diagram describing reversible charge transfer through the liquid junction in Figure 2.3.5.

process and (2.3.22) are zero, we must conclude that the free energy change for (2.3.21) is also zero. In other words, charge transport across the junction occurs in such a way that the electrochemical free energy change vanishes, even though it cannot be considered as a process at equilibrium. This important conclusion permits an approach to the calculation of junction potentials.

Focus first on the net chemical reaction (2.3.22). Since the electrochemical free energy change is zero,

$$\bar{\mu}_{\mathrm{H}^+}{}^{\beta} + \bar{\mu}_e^{\mathrm{Pt}'} = \bar{\mu}_{\mathrm{H}^+}{}^{\alpha} + \bar{\mu}_e^{\mathrm{Pt}} \tag{2.3.23}$$

$$FE = F(\phi^{\mathrm{Pt}'} - \phi^{\mathrm{Pt}}) = \bar{\mu}_{\mathrm{H}^+}{}^{\beta} - \bar{\mu}_{\mathrm{H}^+}{}^{\alpha} \tag{2.3.24}$$

$$E = \frac{RT}{F} \ln \frac{a_2}{a_1} + (\phi^\beta - \phi^\alpha) \tag{2.3.25}$$

The first component of E in (2.3.25) is merely the Nernst relation for the reversible chemical change, and $\phi^\beta - \phi^\alpha$ is the liquid junction potential. In fact, it is a general truth that for a chemically reversible system under null current conditions,

$$E_{\mathrm{cell}} = E_{\mathrm{Nernst}} + E_j \tag{2.3.26}$$

hence the junction potential is always an additive perturbation onto the nernstian response.

To evaluate E_j, we consider (2.3.21), for which

$$t_+\bar{\mu}_{\mathrm{H}^+}{}^{\alpha} + t_-\bar{\mu}^{\beta}_{\mathrm{Cl}^-} = t_+\bar{\mu}_{\mathrm{H}^+}{}^{\beta} + t_-\bar{\mu}^{\alpha}_{\mathrm{Cl}^-} \tag{2.3.27}$$

Thus

$$t_+(\bar{\mu}_{\mathrm{H}^+}{}^{\alpha} - \bar{\mu}_{\mathrm{H}^+}{}^{\beta}) + t_-(\bar{\mu}^{\beta}_{\mathrm{Cl}^-} - \bar{\mu}^{\alpha}_{\mathrm{Cl}^-}) = 0 \tag{2.3.28}$$

$$t_+\left[RT \ln \frac{a_{\mathrm{H}^+}{}^{\alpha}}{a_{\mathrm{H}^+}{}^{\beta}} + F(\phi^\alpha - \phi^\beta)\right] + t_-\left[RT \ln \frac{a^{\beta}_{\mathrm{Cl}^-}}{a^{\alpha}_{\mathrm{Cl}^-}} - F(\phi^\beta - \phi^\alpha)\right] = 0 \tag{2.3.29}$$

Activity coefficients for single ions cannot be measured with thermodynamic rigor (20, 24–26); hence they are usually equated to a measurable *mean ionic activity coefficient*. Under this procedure, $a_{\mathrm{H}^+}{}^{\alpha} = a^{\alpha}_{\mathrm{Cl}^-} = a_1$ and $a_{\mathrm{H}^+}{}^{\beta} = a^{\beta}_{\mathrm{Cl}^-} = a_2$. Since $t_+ + t_- = 1$, we have

$$\boxed{E_j = (\phi^\beta - \phi^\alpha) = (t_+ - t_-)\frac{RT}{F} \ln \frac{a_1}{a_2}} \tag{2.3.30}$$

for a type 1 junction involving 1:1 electrolytes.

Consider, for example, HCl solutions with $a_1 = 0.01$ and $a_2 = 0.1$. We can see from Table 2.3.1 that $t_+ = 0.83$ and $t_- = 0.17$; hence at 25°

$$E_j = (0.83 - 0.17)(59.1) \log\left(\frac{0.01}{0.1}\right) = -39.1 \text{ mV} \tag{2.3.31}$$

For the total cell,

$$E = 59.1 \log \frac{a_2}{a_1} + E_j = 59.1 - 39.1 = 20.0 \text{ mV} \tag{2.3.32}$$

thus the junction potential is a substantial component of the measured cell potential.

In the derivation above, we made the implicit assumption that the transport numbers were constant throughout the system. This is a very good approximation for junctions of type 1; hence (2.3.30) is not seriously compromised. For type 2 and type 3 systems, it clearly cannot be true. To consider these cases, one must imagine the junction region to be sectioned into an infinite number of volume elements having compositions that range smoothly from the pure α phase composition to that of pure β. Transporting charge across one of these elements involves every ion i in the element and, for each faraday passed, $t_i/|z_i|$ moles must move. Thus, the passage of positive charge from α toward β might be depicted as in Figure 2.3.7. One can see that the change in electrochemical free energy upon moving any species is $(t_i/z_i)\, d\bar{\mu}_i$ (recall that z_i is a signed quantity); therefore, the differential in free energy is

$$d\bar{G} = \sum_i \frac{t_i}{z_i}\, d\bar{\mu}_i \tag{2.3.33}$$

Integrating from the α phase to the β phase, we have

$$\int_\alpha^\beta d\bar{G} = 0 = \sum_i \int_\alpha^\beta \frac{t_i}{z_i}\, d\bar{\mu}_i \tag{2.3.34}$$

If μ_i^0 for the α phase is the same as that for the β zone,

$$\sum_i \int_\alpha^\beta \frac{t_i}{z_i} RT\, d \ln a_i + \left(\sum_i t_i\right) F \int_\alpha^\beta d\phi = 0 \tag{2.3.35}$$

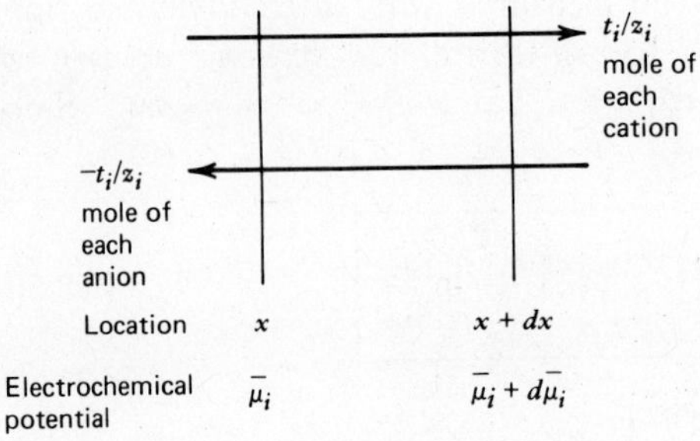

Figure 2.3.7
Transfer of net positive charge from left to right through an infinitesimal segment of a junction region. Each species must contribute t_i moles of charge per faraday transported; hence $t_i/|z_i|$ moles of that species must migrate.

Since $\sum t_i = 1$,

$$E_j = \phi^\beta - \phi^\alpha = \frac{-RT}{F} \sum_i \int_\alpha^\beta \frac{t_i}{z_i} \, d \ln a_i \tag{2.3.36}$$

which is the general expression for the junction potential.

It is easy to see now that (2.3.30) is a special case for type 1 junctions between 1:1 electrolytes having constant t_i. Note that E_j is a strong function of t_+ and t_-, and that it actually vanishes if $t_+ = t_-$. The value of E_j as a function of t_+ for a 1:1 electrolyte with $a_1/a_2 = 10$ is

$$E_j = 59.1\,(2t_+ - 1) \text{ mV} \tag{2.3.37}$$

at 25°. For example, the cell

$$\text{Ag/AgCl/KCl}(0.1\ M)\text{/KCl}(0.01\ M)\text{/AgCl/Ag} \tag{2.3.38}$$

has $t_+ = 0.49$; hence $E_j = -1.2$ mV.

While type 1 junctions can be treated with some rigor and are independent of the method of forming the junction or its geometry, type 2 and type 3 junctions have potentials that depend on the technique of junction formation (static or flowing) and can only be treated in an approximate manner. Apparently different approaches to junction formation lead to different profiles of t_i through the junction, which in turn lead to different integrals for (2.3.36).

Approximate values for E_j can be obtained by assuming that (a) concentrations of ions everywhere in the junction are equivalent to activities and (b) the concentration of each ion follows a linear transition between the two phases. Then (2.3.36) can be integrated to give the *Henderson equation* (15, 20):

$$E_j = \frac{\sum_i \frac{|z_i| u_i}{z_i} [C_i(\beta) - C_i(\alpha)]}{\sum_i |z_i| u_i [C_i(\beta) - C_i(\alpha)]} \frac{RT}{F} \ln \frac{\sum_i |z_i| u_i C_i(\alpha)}{\sum_i |z_i| u_i C_i(\beta)} \tag{2.3.39}$$

where u_i is the mobility and C_i is the molar concentration of species i. For type 2 junctions between 1:1 electrolytes, this equation collapses to the *Lewis-Sargent relation*:

$$E_j = \pm \frac{RT}{F} \ln \frac{\Lambda_\beta}{\Lambda_\alpha} \tag{2.3.40}$$

where the positive sign corresponds to a junction with a common cation in the two phases, and the negative sign applies to junctions with common anions. As an example, consider the cell

$$\text{Ag/AgCl/HCl}(0.1\ M)\text{/KCl}(0.1\ M)\text{/AgCl/Ag} \tag{2.3.41}$$

for which E_{cell} is essentially E_j. The measured value at 25° is 28 ± 1 mV, depending on the technique of junction formation (20), and the estimated value from (2.3.40) and the data of Table 2.3.2 is 26.8 mV.

2.3.5 Minimizing Liquid Junction Potentials

In most electrochemical experiments, the junction potential is an additional troublesome factor and attempts usually are made to minimize it, or one hopes that it is small or at least remains constant. A familiar method for minimizing E_j is to replace the junction, for example,

$$HCl(C_1)/NaCl(C_2) \tag{2.3.42}$$

with a system featuring a concentrated solution in an intermediate *salt bridge*, where the solution in the bridge has ions of nearly equal mobility. Such a system is

$$HCl(C_1)/KCl(C)/NaCl(C_2) \tag{2.3.43}$$

Table 2.3.3 lists some measured junction potentials for the cell:

$$Hg/Hg_2Cl_2/HCl(0.1\ M)/KCl(C)/KCl(0.1\ M)/Hg_2Cl_2/Hg \tag{2.3.44}$$

As C increases, E_j falls markedly because ionic transport at the two junctions is dominated more and more extensively by the massive amounts of KCl. The series junctions become more similar in magnitude and have opposite polarities; hence they tend to cancel. Solutions usually used in aqueous salt bridges contain KCl ($t_+ = 0.49$, $t_- = 0.51$) or, where Cl^- is deleterious, KNO_3 ($t_+ = 0.51$, $t_- = 0.49$). In many measurements, such as pH and other potentiometric determinations, it is sufficient if the junction potential remains constant between calibration (e.g., with a standard buffer or solution) and measurement. In general, however, variations in E_j of 1–2 mV can be expected, and should bc considered in any interpretations made from potential data.

Table 2.3.3

Effect of a Salt Bridge on Measured Junction Potentials[a]

Concentration of KCl, $C(M)$	E_j, mV
0.1	27
0.2	20
0.5	13
1.0	8.4
2.5	3.4
3.5	1.1
4.2 (sat'd)	<1

[a] See J. J. Lingane, "Electroanalytical Chemistry," Wiley-Interscience, New York, 1958, p. 65. Original data from H. A. Fales and W. C. Vosburgh, *J. Am. Chem. Soc.*, **40**, 1291 (1918); E. A. Guggenheim, *ibid.*, **52**, 1315 (1930); and A. L. Ferguson, K. Van Lente, and R. Hitchens, *ibid.*, **54**, 1285 (1932).

2.4 SELECTIVE ELECTRODES

2.4.1 Selective Interfaces

Suppose one could create an interface between two electrolyte phases across which only a single ion could penetrate. A selectively permeable membrane might be used as a separator to accomplish this end. Equation 2.3.34 would still apply; but it could be simplified by recognizing that the transference number for the permeating ion is unity and that for every other ion is zero. By integration, one obtains

$$\frac{RT}{z_i} \ln \frac{a_i^{\beta}}{a_i^{\alpha}} + F(\phi^{\beta} - \phi^{\alpha}) = 0 \tag{2.4.1}$$

where ion i is the permeating species. Rearrangement gives

$$E_m = -\frac{RT}{z_i F} \ln \frac{a_i^{\beta}}{a_i^{\alpha}} \tag{2.4.2}$$

Note that if the activity of species i is held constant in one phase, the potential difference between the two phases (i.e., the *membrane potential*, E_m) responds in a nernstian fashion to the ion's activity in the other phase.

This idea is the essence behind the operation of ion-selective electrodes. Measurements with these devices are essentially determinations of membrane potentials, which themselves comprise junction potentials between electrolyte phases. The performance of any single system is determined largely by the degree to which the species of interest can be made to dominate charge transport in part of the membrane. We will see below that real devices are fairly complicated, and that selectivity in charge transport throughout the membrane is rarely achieved and is actually unnecessary.

Many ion-selective interfaces have been studied, and several different types of electrode have been marketed commercially. We will examine the basic strategies for introducing selectivity by considering a few of them here. The glass membrane is our starting point because it offers a fairly complete view of the fundamentals, as well as the usual complications found in practical devices.

2.4.2 Glass Electrodes (15, 25, 27–34)

The ion-selective properties of glass/electrolyte interfaces were recognized early in this century, and glass electrodes have been used since then for measurements of pH and the activities of alkali ions. Figure 2.4.1 depicts the construction of a typical device. To make measurements, one immerses the electrode so that the thin membrane is fully in contact with the solution, and the potential of the electrode is registered with respect to a reference electrode such as an SCE. Thus the cell becomes,

$$\underbrace{\text{Hg/Hg}_2\text{Cl}_2\text{/KCl(sat'd)}}_{\text{SCE}} \Big/ \begin{matrix}\text{Test}\\ \text{solution}\end{matrix} \Big/ \underbrace{\begin{matrix}\text{Glass}\\ \text{Membrane}\end{matrix} \Big/ \underbrace{\text{HCl(0.1 }M\text{)/AgCl/Ag}}_{\text{Glass electrode's internal reference}}}_{\text{Glass electrode}} \tag{2.4.3}$$

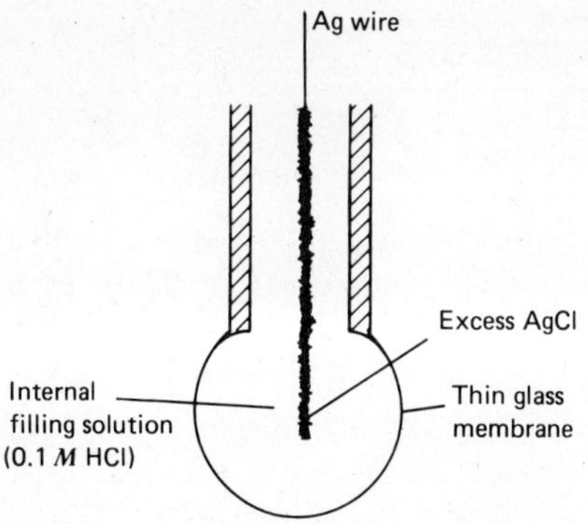

Figure 2.4.1
Schematic diagram of a typical glass electrode.

The properties of the test solution influence the overall potential difference of the cell at two points. One of them is the liquid junction between the SCE and the test solution. From the considerations of Section 2.3.5, we can hope that the potential difference there is small and constant. The remaining contribution from the test solution comes from its effect on the potential difference across the glass membrane. Since all of the other interfaces in the cell feature phases of constant composition, changes in the cell potential can be wholly ascribed to the junction between the glass membrane and the test solution. If that interface is selective toward a single species i, the cell potential is

$$E = \text{constant} + \frac{RT}{z_i F} \ln a_i^{\text{soln}} \tag{2.4.4}$$

where the constant term is the sum of potential differences at all of the other interfaces. The constant term is evaluated by "standardizing" the electrode, that is, by measuring E for a cell in which the test solution is replaced by a standard solution having a known activity for species i.†

Actually, the operation of the glass phase is rather complicated (15, 25, 27–31). The bulk of the membrane, which might be about 50 μm thick, is dry glass through which charge transport occurs exclusively by the mobile cations present in the glass. Usually these are alkali ions such as Na^+ or Li^+. Hydrogen ion from solution does not contribute to conduction in this region. The faces of the membrane in contact with solution differ from the bulk in that the silicate structure of the glass is hydrated. As shown in Figure 2.4.2, the hydrated layers are quite thin. Interactions between the glass and the adjacent solution, which occur exclusively in the hydrated zone between them, are facilitated kinetically by the swelling that accompanies the hydration.

The membrane potentials appear because the silicate network has an affinity for certain cations, which are adsorbed (probably at anionic fixed sites) within the structure. This action creates a charge separation that alters the interfacial potential difference. That potential difference, in turn, alters the rates of adsorption and desorption. Thus those rates are gradually brought into balance by a mechanism

† By the phrase "activity for species i," we mean the concentration of i multiplied by the mean ionic activity coefficient. See Section 2.3.4 for a commentary and references related to the concept of single-ion activities.

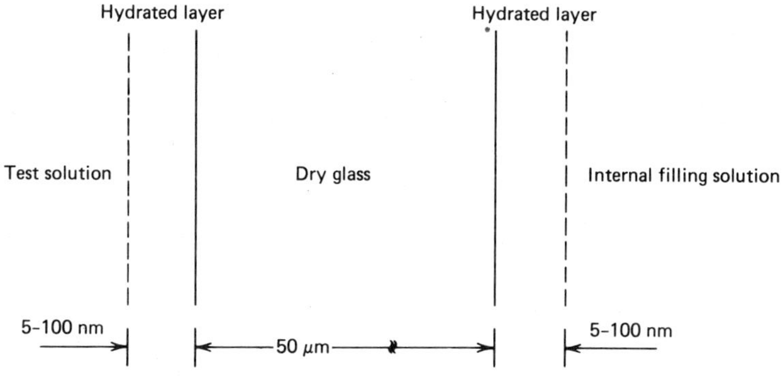

Figure 2.4.2
Schematic profile through a glass membrane.

resembling the one responsible for the establishment of junction potentials as discussed above.

It is clear that the glass membrane does not adhere to the simplified idea of a selectively permeable membrane. In fact, it may not be at all permeable to some of the ions of greatest interest, such as H^+. Thus, the transference number of such an ion cannot be unity throughout the membrane, and it may actually be zero in certain zones. Can we still understand the observed selective response to the ion of interest? The answer is yes, provided that the ion dominates charge transport in the interfacial regions of the membrane.

Let us consider a model for the glass membrane like that shown in Figure 2.4.3. The glass will be considered as comprising three regions. In the interfacial zones, m' and m'', there is rapid attainment of equilibrium with constituents in solution, so that each adsorbing cation has an activity reflecting its corresponding activity in the adjacent solution. The bulk of the glass is denoted by m, and we presume that conduction there takes place by a single species, which is taken as Na^+ for the sake of this argument. The whole system therefore comprises essentially five phases, and the overall difference in potential across the membrane is the sum of four contributions from the junctions between the various zones:

$$E_m = (\phi^\beta - \phi^{m''}) + (\phi^{m''} - \phi^m) + (\phi^m - \phi^{m'}) + (\phi^{m'} - \phi^\alpha) \qquad (2.4.5)$$

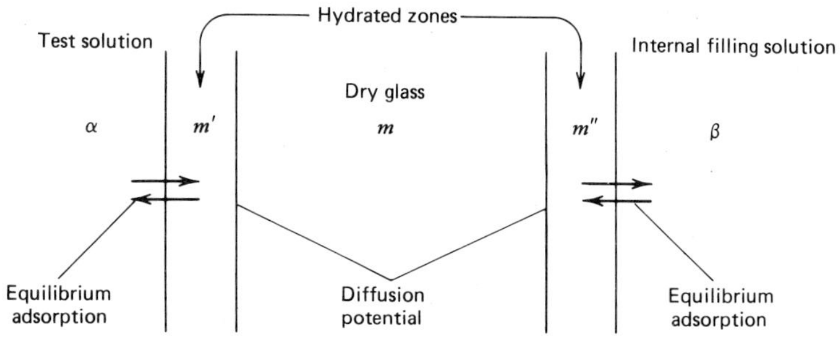

Figure 2.4.3
Model for treating the membrane potential across a glass barrier.

The first and last terms are interfacial potential differences arising from an *equilibrium balance* of selective charge exchange across an interface. This condition is known as *Donnan equilibrium* (15, 30). The magnitude of the resulting potential difference can be evaluated from electrochemical potentials. Suppose we have Na^+ and H^+ as interfacially active ions. Then,

$$\bar{\mu}_{H^+}{}^{\alpha} = \bar{\mu}_{H^+}{}^{m'} \quad (2.4.6)$$

$$\bar{\mu}_{Na^+}^{\alpha} = \bar{\mu}_{Na^+}^{m'} \quad (2.4.7)$$

Expanding (2.4.6), we have

$$\mu_{H^+}^{0\alpha} + RT \ln a_{H^+}{}^{\alpha} + F\phi^{\alpha} = \mu_{H^+}^{0m'} + RT \ln a_{H^+}^{m'} + F\phi^{m'} \quad (2.4.8)$$

and rearrangement gives

$$(\phi^{m'} - \phi^{\alpha}) = \frac{\mu_{H^+}^{0\alpha} - \mu_{H^+}^{0m'}}{F} + \frac{RT}{F} \ln \frac{a_{H^+}{}^{\alpha}}{a_{H^+}{}^{m'}} \quad (2.4.9)$$

An equivalent treatment of the interface between β and m'' gives

$$(\phi^{\beta} - \phi^{m''}) = \frac{\mu_{H^+}^{0m''} - \mu_{H^+}^{0\beta}}{F} + \frac{RT}{F} \ln \frac{a_{H^+}{}^{m''}}{a_{H^+}{}^{\beta}} \quad (2.4.10)$$

Note that $\mu_{H^+}^{0\alpha} = \mu_{H^+}^{0\beta}$ because both α and β are aqueous solutions. Similarly $\mu_{H^+}^{0m'} = \mu_{H^+}^{0m''}$. When we add (2.4.9) and (2.4.10) later in this development, these equivalences will cause the terms involving μ^0's to disappear.

The second and third components in (2.4.5) are junction potentials within the glass membrane. In the specialized literature they are called diffusion potentials because they arise from differential ionic diffusion in the manner discussed in Section 2.3.2. The chemical systems correspond to type 3 junctions, as defined there.

We can treat them through a variant of the Henderson equation, which was derived earlier in Section 2.3.4. The usual form (2.3.39) is derived from (2.3.36) by neglecting activity effects and assuming linear concentration profiles through the junction. Here we are interested only in univalent positive charge carriers, and hence we can specialize (2.3.39) for the interface between m and m' as

$$(\phi^{m} - \phi^{m'}) = \frac{RT}{F} \ln \frac{u_{H^+} a_{H^+}{}^{m'} + u_{Na^+} a_{Na^+}^{m'}}{u_{Na^+} a_{Na^+}^{m}} \quad (2.4.11)$$

where the concentrations have been replaced by activities. Also, for the interface between m and m'',

$$(\phi^{m''} - \phi^{m}) = \frac{RT}{F} \ln \frac{u_{Na^+} a_{Na^+}^{m}}{u_{H^+} a_{H^+}{}^{m''} + u_{Na^+} a_{Na^+}^{m''}} \quad (2.4.12)$$

Now let us add the component potential differences (2.4.9–2.4.12) as dictated by (2.4.5) to obtain the whole potential difference across the membrane:†

† Note that the diffusion term here is the same as that which would be predicted by the Henderson equation from the compositions of m' and m'' without considering m as a separate phase. Many treatments of this problem follow such an approach. We have added the phase m because the three-phase model for the membrane is more realistic with regard to the assumptions underlying the Henderson equation.

$$E_m = \frac{RT}{F} \ln \frac{a_{H^+}{}^{\alpha} a_{H^+}{}^{m''}}{a_{H^+}{}^{\beta} a_{H^+}{}^{m'}} \qquad \text{(Donnan term)}$$

$$+ \frac{RT}{F} \ln \frac{(u_{Na^+}/u_{H^+}) a_{Na^+}^{m'} + a_{H^+}{}^{m'}}{(u_{Na^+}/u_{H^+}) a_{Na^+}^{m''} + a_{H^+}{}^{m''}} \qquad \text{(Diffusion term)} \tag{2.4.13}$$

This is an unsatisfying result, and it does not provide confidence that we will find a direct, simple linkage between E_m and $a_{H^+}{}^{\alpha}$. However, some important simplifications can be made. First, we combine the two terms in (2.4.13) and rearrange the parameters to give

$$E_m = \frac{RT}{F} \ln \frac{(u_{Na^+}/u_{H^+})(a_{H^+}{}^{\alpha} a_{Na^+}^{m'}/a_{H^+}{}^{m'}) + a_{H^+}{}^{\alpha}}{(u_{Na^+}/u_{H^+})(a_{H^+}{}^{\beta} a_{Na^+}^{m''}/a_{H^+}{}^{m''}) + a_{H^+}{}^{\beta}} \tag{2.4.14}$$

Now consider (2.4.6) and (2.4.7) which apply simultaneously. Their sum must also be true:

$$\bar{\mu}_{Na^+}^{\alpha} + \bar{\mu}_{H^+}{}^{m'} = \bar{\mu}_{H^+}{}^{\alpha} + \bar{\mu}_{Na^+}^{m'} \tag{2.4.15}$$

This equation is a free energy balance for the ion-exchange reaction:

$$Na^+(\alpha) + H^+(m') \rightleftharpoons H^+(\alpha) + Na^+(m') \tag{2.4.16}$$

Since it does not involve net charge transfer, it is not sensitive to the interfacial potential difference [see Section 2.2.4(c)], and it has an equilibrium constant:

$$K_{H^+,Na^+} = \frac{a_{H^+}{}^{\alpha} a_{Na^+}^{m'}}{a_{H^+}{}^{m'} a_{Na^+}^{\alpha}} \tag{2.4.17}$$

An equivalent expression, involving the same numeric value of K_{H^+,Na^+}, would apply to the interface between phases β and m'. These relations can be substituted into (2.4.14) to give

$$E_m = \frac{RT}{F} \ln \frac{(u_{Na^+}/u_{H^+}) K_{H^+,Na^+} a_{Na^+}^{\alpha} + a_{H^+}{}^{\alpha}}{(u_{Na^+}/u_{H^+}) K_{H^+,Na^+} a_{Na^+}^{\beta} + a_{H^+}{}^{\beta}} \tag{2.4.18}$$

Since K_{H^+,Na^+} and u_{Na^+}/u_{H^+} are constants of the experiment, it is convenient to define their product as the *potentiometric selectivity coefficient*, $k_{H^+,Na^+}^{\text{pot}}$:

$$E_m = \frac{RT}{F} \ln \frac{a_{H^+}{}^{\alpha} + k_{H^+,Na^+}^{\text{pot}} a_{Na^+}^{\alpha}}{a_{H^+}{}^{\beta} + k_{H^+,Na^+}^{\text{pot}} a_{Na^+}^{\beta}} \tag{2.4.19}$$

If the β phase is the internal filling solution (of constant composition) and the α phase is the test solution, then the overall potential of the cell is

$$\boxed{E = \text{constant} + \frac{RT}{F} \ln (a_{H^+}{}^{\alpha} + k_{H^+,Na^+}^{\text{pot}} a_{Na^+}^{\alpha})} \tag{2.4.20}$$

This expression tells us that the cell potential is responsive to the activities of both Na^+ and H^+ in the test solution and that the degree of selectivity between these species is controlled by $k_{H^+,Na^+}^{\text{pot}}$. If it is small enough to make the product $k_{H^+,Na^+}^{\text{pot}} a_{Na^+}^{\alpha}$ much less than a_{H^+}, then the membrane responds essentially exclusively to H^+. When that condition applies, charge exchange between the phases α and m' is completely dominated by H^+.

We have formulated this problem in a manner that considers only Na^+ and H^+ as active species. Glass membranes also respond to other ions, such as Li^+, K^+, Ag^+, and NH_4^+, and the relative responses can be expressed through the appropriate potentiometric selectivity coefficients. They are controlled to a great extent by the composition of the glass, and different types of electrodes, based on different types of glass, are marketed. Rechnitz (32) has classified them broadly as (a) *pH electrodes* with a selectivity order $H^+ \ggg Na^+ > K^+, Rb^+, Cs^+ \gg Ca^{2+}$, (b) *sodium-sensitive* devices with the order $Ag^+ > H^+ > Na^+ \gg K^+, Li^+ \gg Ca^{2+}$, and (c) a more general *cation-sensitive* type with a narrower range of selectivities in the order $H^+ > K^+ > Na^+ > NH_4^+, Li^+ \gg Ca^{2+}$.

There is a large literature on the design, performance, and theory of glass electrodes. The interested reader is referred to it for more advanced discussions (25, 27–34).

2.4.3 Other Ion-Selective Membranes (29, 34)

The principles that we have reviewed above also apply to other types of commercial selective membranes. They fall generally into three categories.

(a) Solid-State Membranes. Like the glass membrane, which is a member of this group, the remaining common solid membranes are electrolytes having tendencies toward the preferential adsorption of certain ions on their surfaces.

Most prominent among them is the single-crystal LaF_3 membrane. It is doped with EuF_2 in order to create fluoride vacancies that allow ionic conduction by fluoride through the crystal. The surface selectively accommodates F^- ions to the virtual exclusion of every other species except OH^-.

Other devices are made from precipitates of insoluble salts, such as AgCl, AgBr, AgI, Ag_2S, CuS, CdS, and PbS. The precipitates are usually pressed into pellets or are suspended in polymer matrices. The silver salts conduct by mobile Ag^+ ions but, since the heavy metal sulfides are not very conductive, they are usually mixed with Ag_2S. The surfaces of these membranes are generally sensitive to the ions comprising the salts, plus other species that tend to form very insoluble precipitates with a constituent ion. For example, the Ag_2S membrane responds to Ag^+, S^{2-}, and Hg^{2+}. Likewise, the AgCl membrane is most sensitive to Ag^+, Cl^-, Br^-, I^-, CN^-, and OH^-.

(b) Liquid Ion Exchangers. Figure 2.4.4 is a diagram of the structure of a typical commercial device in which a hydrophobic liquid membrane is used as the sensing element. The liquid is stabilized physically between the aqueous internal filling solution and the aqueous test solution by allowing it to permeate a porous, lipophilic diaphragm. A reservoir contacting the outer edges of the diaphragm contains this liquid. Chelating agents with selectivity toward ions of interest are dissolved in it, and they provide the mechanism for selective charge transport across the boundaries of the membrane.

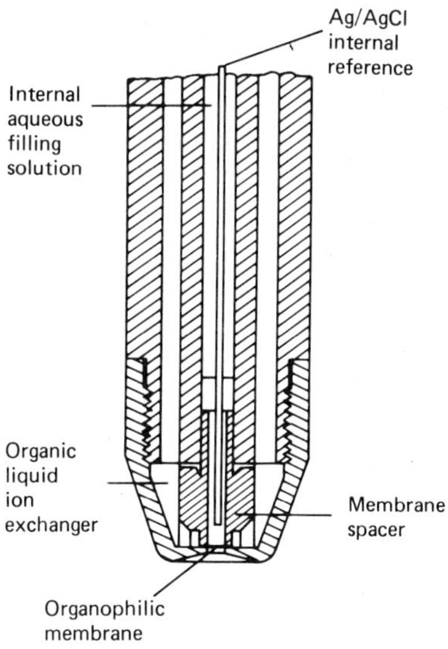

Figure 2.4.4
A typical electrode with a liquid membrane. (Courtesy of Orion Research, Inc.)

The most important commercial devices based on these principles are calcium-selective electrodes. The hydrophobic solvent may be one of several substances. A typical one is dioctylphenylphosphonate. The chelating agent is usually the sodium salt of an alkyl phosphate ester, $(RO)_2PO_2^-Na^+$, where R is an aliphatic chain having 8–18 carbons. The membrane is sensitive to Ca^{2+}, Zn^{2+}, Fe^{2+}, Pb^{2+}, Cu^{2+}, tetra-alkylammonium ions, and still other species to lesser degrees. "Water hardness" electrodes are based on the same agents, but are designed to show virtually equal responses to Ca^{2+} and Mg^{2+}.

Other commercial systems featuring liquid ion exchangers are available for anions such as NO_3^-, ClO_4^-, and Cl^-. Nitrate and perchlorate are sensed by membranes including alkylated 1,10-phenanthroline complexes of Ni^{2+} and Fe^{2+}, respectively. All three ions are active at other membranes based on quaternary ammonium salts.

Several of these electrodes based on liquid ion exchangers have very recently been offered in a form in which the chelating agent is immobilized in a hydrophobic polymer membrane. This design is more rugged and often seems to offer superior performance.

(c) Neutral Carriers. Liquid ion exchangers all feature charged chelating agents, and various ion-exchange equilibria play a role in their operation. Another type of device, also featuring a stabilized liquid membrane, involves uncharged chelating agents. These species enable the transport of charge from one side of the membrane to the other by selectively complexing certain ions; thus they are called *neutral carriers.*

The most important available devices are potassium-selective electrodes, which typically involve the natural macrocycle valinomycin as a neutral carrier in diphenyl ether. This membrane has a much higher sensitivity to K^+ than to Na^+, Li^+, Mg^+, Ca^+, or H^+; but Rb^+ and Cs^+ are sensed to much the same degree as K^+. The basis for selectivity seems to rest mostly on the fit between the size of the ion to be complexed and the volume of the complexing cavity in the macrocycle.

2.4.4 Gas-Sensing Electrodes (35)

Figure 2.4.5 is a schematic view of a typical potentiometric gas-sensing electrode. In general, such a device involves a glass pH electrode that is protected from the test solution by a polymeric diaphragm. Between the glass membrane and the diaphragm is a small volume of electrolyte. Small gaseous molecules, such as SO_2, NH_3, and CO_2, can penetrate the membrane and interact with the trapped electrolyte by reactions that produce changes in pH. The glass electrode responds to the alterations in acidity.

We should note here that the well-known Clark oxygen electrode differs fundamentally from these devices (11, 36). The Clark device is similar in construction to the apparatus of Figure 2.4.5, in that a polymer membrane traps an electrolyte against a sensing surface. However, the sensor is a platinum electrode, and the analytical signal is the steady-state *current* flow due to the faradaic reduction of molecular oxygen.

2.4.5 Enzyme-Coupled Devices (30, 32, 37, 38)

Recent years have seen rapidly growing interest in the prospects for exploiting the natural specificity of enzyme-catalyzed reactions as the basis for selective detection of analytes. One fruitful approach has featured potentiometric sensors with a structure similar to that of Figure 2.4.5, with the difference that the gap between the ion-selective electrode and the polymer diaphragm is filled with a matrix in which an enzyme is immobilized.

For example, urease might be held, together with a buffered electrolyte, in a cross-linked polyacrylamide gel. When the electrode is immersed in a test solution, there will be a selective response toward urea, which diffuses through the diaphragm into the gel. The response comes about because the urease catalyzes the process:

$$NH_2\text{—}\overset{\overset{\displaystyle O}{\|}}{C}\text{—}NH_2 + H^+ + 2H_2O \xrightarrow{\text{Urease}} 2NH_4^+ + HCO_3^- \qquad (2.4.21)$$

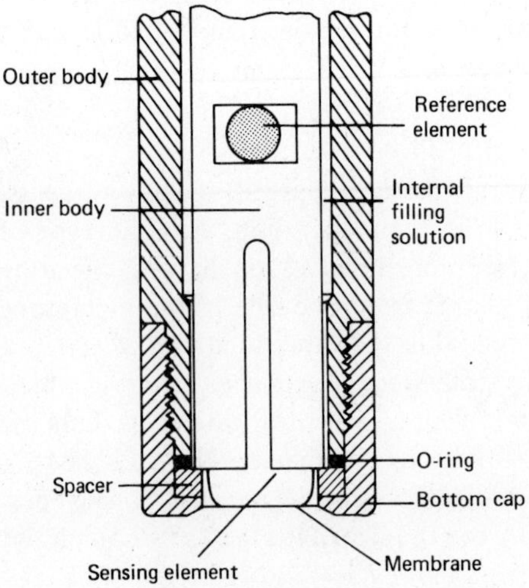

Figure 2.4.5
A gas-sensing electrode. (Courtesy of Orion Research, Inc.)

The resulting ammonium ions can be detected with a cation-sensitive glass membrane. Alternatively, one could use a gas-sensing electrode for ammonia in place of the glass electrode, so that interferences from H^+, Na^+, and K^+ are reduced.

The research literature features many examples of this basic strategy. Different enzymes allow selective determinations of single species, such as glucose (with glucose oxidase), or groups of substances such as the L-amino acids (L-amino acid oxidase). Recent reviews should be consulted for a more complete view of the field.

2.5 REFERENCES

1. The arguments presented here follow those given earlier by D. A. MacInnes ("The Principles of Electrochemistry," Dover, New York, 1961, pp. 110–113) and by J. J. Lingane ("Electroanalytical Chemistry," 2nd ed., Wiley-Interscience, New York, 1958, pp. 40–45). Experiments like those described here were actually carried out by H. Jahn (*Z. Physik. Chem.*, **18**, 399 (1895)).
2. I. M. Klotz, "Chemical Thermodynamics," Benjamin, New York, 1964.
3. J. J. Lingane, "Electroanalytical Chemistry," 2nd ed., Wiley-Interscience, New York, 1958, Chap. 3.
4. F. C. Anson, *J. Chem. Educ.*, **36**, 394 (1959).
5. W. M. Latimer, "Oxidation Potentials," 2nd ed. Prentice-Hall, Englewood Cliffs, N.J., 1952.
6. F. D. Rossini et al., "Selected Values of Chemical Thermodynamic Properties," *Nat. Bur. Stand. Circ.* 500, 1952, and succeeding publications from the Nat. Bur. Stand.
7. F. D. Rossini et al., "Selected Values of Physical and Thermodynamic Properties of Hydrocarbons and Related Compounds," American Petroleum Institute Project 44, Carnegie Press, Pittsburgh, 1953.
8. "JANAF Thermochemical Tables," Dow Chemical Co., Nat. Bur. Stand. Institute of Applied Technology, Washington, 1965–.
9. D. J. G. Ives and G. J. Janz, Eds., "Reference Electrodes," Academic Press, New York, 1961.
10. J. N. Butler, *Adv. Electrochem. Electrochem. Engr.*, **7**, 77 (1970).
11. D. T. Sawyer and J. L. Roberts, Jr., "Experimental Electrochemistry for Chemists," Wiley, New York, 1974.
12. D. Halliday and R. Resnick, "Physics," 3rd ed., Wiley, New York, 1978, Chap. 29.
13. *Ibid.*, Chap. 28.
14. J. O'M. Bockris and A. K. N. Reddy, "Modern Electrochemistry," Vol. 2, Plenum, New York, 1970, Chap. 7.
15. K. J. Vetter, "Electrochemical Kinetics," Academic Press, New York, 1967,
16. B. E. Conway, "Theory and Principles of Electrode Processes," Ronald, New York, 1965, Chap. 13.
17. R. Parsons, *Mod. Asp. Electrochem.*, **1**, 103 (1954).
18. J. A. V. Butler, *Proc. Roy. Soc., London*, **112A**, 129 (1926).
19. E. A. Guggenheim, *J. Phys. Chem.*, **33**, 842 (1929); **34**, 1540 (1930).

20. D. A. MacInnes, "The Principles of Electrochemistry," Dover, New York, 1961, Chap. 13.
21. J. O'M. Bockris and A. K. N. Reddy, *op. cit.*, Vol. 1, Chap. 4.
22. D. A. MacInnes, *op. cit.*, Chap. 4.
23. *Ibid.*, Chap. 18.
24. J. O'M. Bockris and A. K. N. Reddy, *op. cit.*, Chap. 3.
25. R. G. Bates, "Determination of pH," 2nd ed., Wiley-Interscience, New York, 1973.
26. R. M. Garrels in "Glass Electrodes for Hydrogen and Other Cations," G. Eisenman, Ed., Marcel Dekker, New York, 1967, Chap. 13.
27. M. Dole, "The Glass Electrode," Wiley, New York, 1941.
28. G. Eisenman, Ed., "Glass Electrodes for Hydrogen and Other Cations," Marcel Dekker, New York, 1967.
29. R. A. Durst, Ed., "Ion Selective Electrodes," Nat. Bur. Stand. Spec. Pub. 314, U.S. Government Printing Office, Washington, 1969.
30. J. Koryta, "Ion-Selective Electrodes," Cambridge University Press, Cambridge, 1975.
31. G. A. Rechnitz, *Chem. Engr. News*, **45** (25), 146 (1967).
32. R. P. Buck, *Anal. Chem.*, **50**, 17R (1978). This article is the latest in the series of biennial reviews on ion-selective electrodes published in *Analytical Chemistry*. Past articles in the series are also useful, and future contributions can be expected.
33. N. Lakshminarayanaiah in "Electrochemistry," (A Specialist Periodical Report), Vols. 2, 4, and 5; G. J. Hills (Vol. 2); and H. R. Thirsk (Vols. 4 and 5); Senior Reporters, The Chemical Society, London, 1972, 1974, and 1975.
34. H. Freiser, "Ion-Selective Electrodes in Analytical Chemistry," Plenum, New York, 1979.
35. J. W. Ross, J. H. Riseman, and J. A. Krueger, *Pure Appl. Chem.*, **36**, 473 (1973).
36. L. C. Clark, Jr., *Trans. Am. Soc. Artif. Intern. Organs*, **2**, 41 (1956).
37. G. G. Guilbault, *Pure Appl. Chem.*, **25**, 727 (1971).
38. G. A. Rechnitz, *Chem. Engr. News*, **53** (4), 29 (1975).

2.6 PROBLEMS

2.1 Devise electrochemical cells in which the following reactions could be made to occur. If liquid junctions are necessary, note them in the cell schematic appropriately, but neglect their effects.

(a) $H_2O \rightleftharpoons H^+ + OH^-$

(b) $2H_2 + O_2 \rightleftharpoons H_2O$

(c) $2PbSO_4 + 2H_2O \rightleftharpoons PbO_2 + Pb + 4H^+ + 2SO_4^{2-}$

(d) $An^{\overline{\cdot}} + TMPD^{\dot{+}} \rightleftharpoons An + TMPD$ (in acetonitrile, where An and $An^{\overline{\cdot}}$ are anthracene and its anion radical, and TMPD and $TMPD^{\dot{+}}$ are *N,N,N′,N′*-tetramethyl-*p*-phenylenediamine and its cation radical. Use anthracene potentials for DMF solutions given in Appendix C.2).

(e) $2Ce^{3+} + 2H^+ + BQ \rightleftharpoons 2Ce^{4+} + H_2Q$ (aqueous, where BQ is *p*-benzoquinone and H_2Q is *p*-hydroquinone.)
(f) $Ag^+ + I^- \rightleftharpoons AgI$ (aqueous)
(g) $3I_3^- + 2Fe \rightleftharpoons 2Fe^{3+} + 9I^-$ (aqueous)
(h) $Fe^{3+} + Fe(CN)_6^{4-} \rightleftharpoons Fe^{2+} + Fe(CN)_6^{3-}$ (aqueous)
(i) $Cu^{2+} + Pb \rightleftharpoons Pb^{2+} + Cu$ (aqueous)
(*j*) $An^{\overline{\cdot}} + BQ \rightleftharpoons BQ^{\overline{\cdot}} + An$ (in *N,N*,-dimethylformamide, where BQ, An, and $An^{\overline{\cdot}}$ are defined above and $BQ^{\overline{\cdot}}$ is the anion radical of *p*-benzoquinone. Use BQ potentials in acetonitrile given in Appendix C.2.)

What half-reactions take place at the electrodes in each cell? What is the standard cell potential in each case? Which electrode is negative? Would the cell operate electrolytically or galvanically in carrying out a net reaction from left to right? Be sure your decisions accord with chemical intuition in each case.

2.2 Several hydrocarbons and carbon monoxide have been studied as possible fuels for use in fuel cells. From thermodynamic data in references 6 to 9, derive E^0's for the following reactions at 25°:
(a) $CO(g) + H_2O(l) \rightarrow CO_2(g) + 2H^+ + 2e$
(b) $CH_4(g) + 2H_2O(l) \rightarrow CO_2(g) + 8H^+ + 8e$
(c) $C_2H_6(g) + 4H_2O(l) \rightarrow 2CO_2(g) + 14H^+ + 14e$
(d) $C_2H_2(g) + 4H_2O(l) \rightarrow 2CO_2(g) + 10H^+ + 10e$

Even though a reversible emf could not be established (Why not?), which half-cell would ideally yield the highest cell voltage when coupled with the standard oxygen half-cell (in acid solution)? Which of the fuels above could yield the highest net work per mole of fuel oxidized? Which would give the most net work per gram?

2.3 Devise a cell in which the following reaction is the overall cell process (T = 298K):

$$2Na^+ + 2Cl^- \rightarrow 2Na(Hg) + Cl_2 \text{ (aqueous)}$$

where Na(Hg) symbolizes the amalgam. Is the reaction spontaneous or not? What is the standard free energy change? Take the standard free energy of formation of Na(Hg) as -85 kJ/mol. From a thermodynamic standpoint another reaction should occur more readily at the cathode of your cell. What is it? It is observed that the reaction written above takes place with good current efficiency. Why? Could your cell have a commercial value?

2.4 What are the cell reactions and their emf's in the following systems? Are the reactions spontaneous? Assume that all systems are aqueous.
(a) $Ag/AgCl/K^+, Cl^-(1\ M)/Hg_2Cl_2/Hg$
(b) $Pt/Fe^{3+}(0.01\ M), Fe^{2+}(0.1\ M), HCl(1\ M)//Cu^{2+}(0.1\ M), HCl(1\ M)/Cu$
(c) $Pt/H_2(1\text{ atm})/H^+, Cl^-(0.1\ M)//H^+, Cl^-(0.1\ M)/O_2(0.2\text{ atm})/Pt$
(d) $Pt/H_2(1\text{ atm})/Na^+, OH^-(0.1\ M)//Na^+, OH^-(0.1\ M)/O_2(0.2\text{ atm})/Pt$
(e) $Ag/AgCl/K^+, Cl^-(1\ M)//K^+, Cl^-(0.1\ M)/AgCl/Ag$
(f) $Pt/Ce^{3+}(0.01\ M), Ce^{4+}(0.1\ M), H_2SO_4(1\ M)//Fe^{2+}(0.01\ M), Fe^{3+}(0.1\ M), HCl(1\ M)/Pt$

2.5 Consider the cell in part (f) of Problem 2.4. What would the composition of the system be at the end of a galvanic discharge to an equilibrium condition? What would the cell potential be? What would the potential of each electrode be *vs.* NHE? *Vs.* SCE? Take equal volumes on both sides.

2.6 Devise a cell for evaluating the solubility product of $PbSO_4$. Calculate the solubility product from the appropriate E^0 values (T = 298K).

2.7 Obtain the dissociation constant of water from the parameters of the cell constructed for reaction (a) in Problem 2.1. (T = 298K)

2.8 Consider the cell:

$$Cu/M/Fe^{2+}, Fe^{3+}, H^+//Cl^-/AgCl/Ag/Cu'$$

Would the cell potential be independent of the identity of M (e.g., graphite, gold, platinum) as long as M is chemically inert? Use electrochemical potentials to prove your point.

2.9 Given the half-cell of the standard hydrogen electrode,

$$Pt/H_2(a = 1)/H^+(a = 1) \text{ (soln)}$$
$$H_2 \rightleftharpoons 2H^+(\text{soln}) + 2e(\text{Pt})$$

Show that although the emf of the cell half-reaction is taken as zero, the potential difference between the platinum and the solution, that is, $\phi^{Pt} - \phi^{S}$, is not zero.

2.10 Devise a thermodynamically sound basis for obtaining the standard potentials for new half-reactions by taking linear combinations of other half-reactions (T = 298K). As two examples, calculate E^0 values for

(a) $CuI + e \rightleftharpoons Cu + I^-$

(b) $O_2 + 2H^+ + 2e \rightleftharpoons H_2O_2$

given	E^0, volts *vs. NHE*
$Cu^{2+} + 2e \rightleftharpoons Cu$	0.337
$Cu^{2+} + I^- + e \rightleftharpoons CuI$	0.86
$O_2 + 4H^+ + 4e \rightleftharpoons 2H_2O$	1.229
$H_2O_2 + 2H^+ + 2e \rightleftharpoons 2H_2O$	1.77

2.11 Calculate the individual junction potentials on either side of the salt bridge in equation 2.3.44 for the first two concentrations in Table 2.3.3. What is the sum of the two potentials in each case? How does it compare with the corresponding entry in the table? $T = 298K$

2.12 Estimate the junction potentials for the following situations ($T = 298K$):

(a) HCl(0.1 *M*)/NaCl(0.1 *M*)

(b) HCl(0.1 *M*)/NaCl(0.01 *M*)

(c) KNO_3(0.01 *M*)/NaOH(0.1 *M*)

(d) $NaNO_3$(0.1 *M*)/NaOH(0.1 *M*)

2.13 One often finds pH meters with direct readout to 0.001 pH unit. Comment on the accuracy of these readings in comparisons of pH from test solution to test solution. Comment on their meaning in measurements of small changes in pH in a single solution (e.g., during a titration).

2.14 The following values of $k^{\text{pot}}_{\text{Na}^+,i}$ are typical for interferents i at a sodium-selective glass electrode: K^+, 0.001; NH_4^+, 10^{-5}; Ag^+, 300; H^+, 100. Calculate the activities of each interferent that would cause a 10% error in the activity of Na^+ estimated to be 10^{-3} *M* from a potentiometric measurement.

2.15 Would Na_2H_2EDTA be a good ion exchanger for a liquid membrane electrode? How about Na_2H_2EDTA-R, where R designates a C_{20} alkyl substituent? Why or why not?

2.16 Comment on the feasibility of developing selective electrodes for the direct potentiometric determination of uncharged substances.

chapter 3

Kinetics of Electrode Reactions

In Chapter 1, we established a proportionality between the net rate of an electrode reaction v and the current. Specifically, $v = i/nFA$. We also know that for a given electrode process, current does not flow in some potential regions, yet it flows to a variable degree in others. The reaction rate is a strong function of potential, and thus we require potential-dependent rate constants for an accurate description of interfacial charge transfer dynamics.

In this chapter, our goal is to devise a theory that can quantitatively rationalize the observed behavior of electrode kinetics with respect to potential and concentration. Once constructed, the theory will serve us often as an aid for understanding kinetic effects in new situations. We begin with a brief review of certain aspects of homogeneous kinetics, because they provide both a familiar starting ground and a basis for the construction, through analogy, of the electrochemical kinetic theory.

3.1 REVIEW OF HOMOGENEOUS KINETICS

3.1.1 Dynamic Equilibrium

Consider two substances A and B, which are linked by simple unimolecular elementary reactions.†

$$A \underset{k_b}{\overset{k_f}{\rightleftharpoons}} B \tag{3.1.1}$$

Both elementary reactions are active at all times, and the rate of the forward process, v_f (M/sec), is

$$v_f = k_f C_A \tag{3.1.2}$$

† An *elementary reaction* describes an actual, discrete chemical event: Many chemical reactions, as written, are not elementary, because the transformation of products to reactants involves several distinct steps. These steps are the elementary reactions that comprise the *mechanism* for the overall process.

whereas the rate of the reverse reaction is

$$v_b = k_b C_B \tag{3.1.3}$$

The rate constants, k_f and k_b, have dimensions of $\sec^{-1}$, and one can easily show that they are the reciprocals of the mean lifetimes of A and B, respectively (Problem 3.6). The net conversion rate of A to B is

$$v_{\text{net}} = k_f C_A - k_b C_B \tag{3.1.4}$$

At equilibrium, the net conversion rate is zero; hence

$$\frac{k_f}{k_b} = K = \frac{C_B}{C_A} \tag{3.1.5}$$

The kinetic theory therefore predicts a constant concentration ratio at equilibrium, just as thermodynamics does.

Such agreement is required of *any* kinetic theory. In the limit of equilibrium the kinetic equations must collapse to relations of the thermodynamic form; otherwise the kinetic picture cannot be accurate. Kinetics describes the evolution of mass flow throughout the system, including both the *approach* to equilibrium and the *dynamic maintenance* of that state. Thermodynamics describes only equilibrium. Understanding of a system is not even at a crude level unless the kinetic view and the thermodynamic one agree on the properties of the equilibrium state.

On the other hand, thermodynamics has nothing to say about the mechanism required to maintain equilibrium, whereas kinetics describes the intricate balance quantitatively. In the example above, equilibrium features nonzero rates of conversion of A to B (and vice versa), but those rates are equal. Sometimes they are called the *exchange velocity* of the reaction, v_0:

$$v_0 = k_f (C_A)_{eq} = k_b (C_B)_{eq} \tag{3.1.6}$$

We will see below that the idea of exchange velocity plays an important role in treatments of electrode kinetics.

3.1.2 The Arrhenius Equation and Potential Energy Surfaces (1, 2)

It is an experimental fact that most rate constants of solution-phase reactions vary with temperature in a common fashion: Nearly always, $\ln k$ is linear with $1/T$. Arrhenius was first to recognize the generality of this behavior, and he proposed that rate constants be expressed in the form:

$$\boxed{k = Ae^{-E_A/RT}} \tag{3.1.7}$$

where E_A has dimensions of energy. Since the exponential factor is strongly reminiscent of the probability of using thermal energy to surmount an energy barrier of height E_A, that parameter has been known as the *activation energy*. The coefficient A is known generally as the *A factor* or the *frequency factor*. The latter name derives from the interpretation based on the energy barrier. If the exponential states the probability of surmounting the barrier, A must be related to the frequency of attempts on it. As usual, these ideas turn out to be oversimplifications, but they carry the essence of truth and are useful for casting a mental image of the ways in which reactions proceed.

The idea of activation energy has led to pictures of reaction paths in terms of potential energy along a *reaction coordinate*. An example is shown in Figure 3.1.1. In a simple unimolecular process, for example, the cis-trans isomerization of stilbene, the reaction coordinate might be an easily recognized molecular parameter, such as the twist angle about the central double bond in stilbene. In general, the reaction coordinate would express progress along a complex, favored path on the multidimensional surface describing potential energy as a function of all independent position coordinates in the system. One zone of this surface corresponds to the configuration we call "reactant," and another corresponds to the "product" structure. Both must occupy minima on the energy surface, because they are the only arrangements possessing a significant lifetime. Even though other configurations are possible, they must lie at higher energies and lack the energy minimum required for stability.

As the reaction takes place, the coordinates are changed from those of the reactant to those of the product. Since the path along the reaction coordinate connects two minima, it must rise, pass over a maximum, then fall into the product zone. Very often, the heights of the maximum above the two valleys are identified with the activation energies, $E_{A,f}$ and $E_{A,b}$, for the forward and backward reactions, respectively.

In another notation, we can express the idea that E_A is the change in internal energy in going from one of the minima to the maximum, which is called the *transition state* or *activated complex*. We might denote it as $\Delta E^\ddagger$. The enthalpy of activation, $\Delta H^\ddagger$, would then be $\Delta E^\ddagger + \Delta(PV)^\ddagger$, but $\Delta(PV)$ is usually negligible in a condensed phase reaction, so that $\Delta H^\ddagger \approx \Delta E^\ddagger$. Thus the Arrhenius equation could be recast as

$$k = A\,e^{-\Delta H^\ddagger/RT} \tag{3.1.8}$$

We are free also to factor the coefficient A into the product $A'e^{\Delta S^\ddagger/R}$, because the exponential, involving the standard activation entropy, $\Delta S^\ddagger$, is merely a dimensionless constant. Then

$$k = A'e^{-(\Delta H^\ddagger - T\Delta S^\ddagger)/RT} \tag{3.1.9}$$

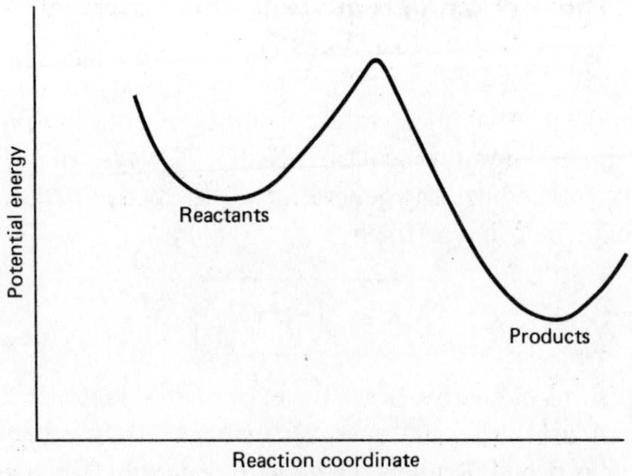

Figure 3.1.1
Simple representation of potential energy changes during a reaction.

or

$$\boxed{k = A' e^{-\Delta G^{\ddagger}/RT}} \tag{3.1.10}$$

where $\Delta G^{\ddagger}$ is the standard free energy of activation. This relation, like (3.1.8), is really an equivalent statement of the Arrhenius equation 3.1.7, which itself is an empirical generalization of reality. Equations 3.1.8 and 3.1.10 are derived from (3.1.7), but only by the *interpretation* we apply to the phenomenological constant E_A. Nothing we have written so far depends on a specific theory of kinetics.

3.1.3 Activated Complex Theory (1–3)

Many theories of kinetics have been constructed in order to permit greater illumination of the factors controlling reaction rates, and the prime goal of these theories is to predict the values of A and E_A for specific chemical systems in terms of quantitative molecular properties. An important general theory that often is adapted for electrode kinetics is the *activated complex theory*, which is also known as the *absolute rate theory* or the *transition state theory*.

Central to this approach is the idea that reactions proceed through a fairly well-defined activated complex as shown in Figure 3.1.2. The standard free energy change in going from the reactants to the complex is $\Delta G_f^{\ddagger}$, whereas the complex is elevated above the products by energy $\Delta G_b^{\ddagger}$.

Let us consider the system of (3.1.1), in which two substances A and B are linked by unimolecular reactions. First we focus on the special condition in which the entire system—A, B, and all other configurations—are at thermal equilibrium. For this

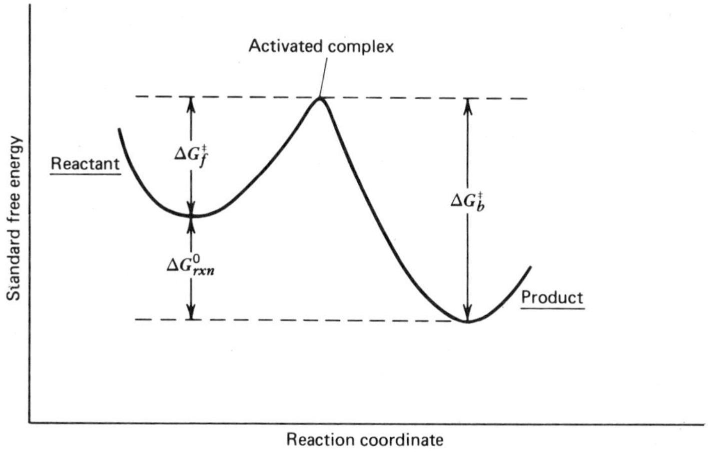

Figure 3.1.2
Free energy changes during a reaction. The activated complex is the configuration of maximum free energy.

situation, the concentration of complexes can be calculated from the free energies of activation in either of two ways:

$$\frac{[\text{Complex}]}{[\text{A}]} = e^{-\Delta G_f^{\ddagger}/RT} \tag{3.1.11}$$

$$\frac{[\text{Complex}]}{[\text{B}]} = e^{-\Delta G_b^{\ddagger}/RT} \tag{3.1.12}$$

These complexes decay into either A or B with a combined rate constant, k', and they can be divided into four fractions: (a) those created from A and reverting back to A, f_{AA}, (b) those arising from A and decaying to B, f_{AB}, (c) those created from B and decaying to A, f_{BA}, and (d) those arising from B and reverting back to B, f_{BB}. Thus the net rate of transforming A into B is

$$k_f[\text{A}] = f_{AB}k'[\text{Complex}] \tag{3.1.13}$$

and the net rate of transforming B into A is

$$k_b[\text{B}] = f_{BA}k'[\text{Complex}] \tag{3.1.14}$$

Since we require $k_f[\text{A}] = k_b[\text{B}]$ at equilibrium, f_{AB} and f_{BA} must be the same. In the simplest version of the theory, both are taken as $\frac{1}{2}$. This assumption implies that $f_{AA} = f_{BB} \simeq 0$; thus complexes are not considered as reverting to the source state. Instead, any system reaching the activated configuration is transmitted with unit efficiency into the product opposite the source. In a more flexible version, the fractions f_{AB} and f_{BA} are equated to $\kappa/2$, where κ, the *transmission coefficient*, can take a value from zero to unity.

Substitution for the concentration of the complex by (3.1.11) and (3.1.12) into (3.1.13) and (3.1.14), respectively, leads to the rate constants:

$$k_f = \frac{\kappa k'}{2} e^{-\Delta G_f^{\ddagger}/RT} \tag{3.1.15}$$

$$k_b = \frac{\kappa k'}{2} e^{-\Delta G_b^{\ddagger}/RT} \tag{3.1.16}$$

Statistical mechanics can be used to predict $\kappa k'/2$. In general, that quantity depends on the shape of the energy surface in the region of the complex, but for simple cases k' can be shown to be $2\,\mathscr{k}T/h$, where $\mathscr{k}$ and h are the Boltzmann and Planck constants. Thus the rate constants (equations 3.1.15 and 3.1.16) might both be expressed in the form:

$$\boxed{k = \kappa \frac{\mathscr{k}T}{h} e^{-\Delta G^{\ddagger}/RT}} \tag{3.1.17}$$

which is the equation most frequently seen for predicting rate constants by the activated complex theory.

To reach (3.1.17) we considered only a system at equilibrium. It is important to note now that the rate constant for an elementary process is fixed for a given temperature and does not depend on the reactant and product concentrations. Equation

3.1.17 is therefore a general expression. If it holds at equilibrium, it will hold away from equilibrium. The assumption of equilibrium, though useful in the derivation, does not constrain the equation's range of validity.

3.2 ESSENTIALS OF ELECTRODE REACTIONS (4–12)

We noted above that an accurate kinetic picture of any dynamic process must yield an equation of the thermodynamic form in the limit of equilibrium. For an electrode reaction, equilibrium is characterized by the Nernst equation, which links the electrode potential to the bulk concentrations of the participants. In the general case:

$$\mathrm{O} + ne \underset{k_b}{\overset{k_f}{\rightleftharpoons}} \mathrm{R} \tag{3.2.1}$$

this equation is

$$\boxed{E = E^{0'} + \frac{RT}{nF} \ln \frac{C_\mathrm{O}^*}{C_\mathrm{R}^*}} \tag{3.2.2}$$

where C_O^* and C_R^* are the bulk concentrations, and $E^{0'}$ is the formal potential. Any theory of electrode kinetics must predict this result for corresponding conditions.

We also require that the theory explain the observed dependence of current on potential under various circumstances. In Chapter 1, we saw that current is often limited wholly or partially by the rate at which the electroreactants are transported to the electrode surface. This kind of limitation does not concern a theory of interfacial kinetics. More to the point is the case of low currents and efficient stirring, in which mass transport is not a factor determining the current. Instead it is controlled strictly by the interfacial dynamics. Early studies of such systems showed that the current is often related exponentially to the overpotential η. That is,

$$i = a' \, e^{\eta/b'} \tag{3.2.3}$$

or, as given by Tafel in 1905,

$$\boxed{\eta = a + b \log i} \tag{3.2.4}$$

A successful model of electrode kinetics must explain the frequent validity of (3.2.4), which is known as the *Tafel equation.*

Let us begin by considering that reaction (3.2.1) has forward and backward paths as shown. The forward component proceeds at a rate v_f that must be proportional to the surface concentration of O. We express the concentration at distance x from the surface and at time t as $C_\mathrm{O}(x, t)$; hence the surface concentration is $C_\mathrm{O}(0, t)$. The constant of proportionality linking the forward reaction rate to $C_\mathrm{O}(0, t)$ is the rate constant k_f.

$$v_f = k_f C_\mathrm{O}(0, t) = \frac{i_c}{nFA} \tag{3.2.5}$$

Since the forward reaction is a reduction, there is a cathodic current i_c proportional to v_f. Likewise, we have for the backward reaction

$$v_b = k_b C_R(0, t) = \frac{i_a}{nFA} \tag{3.2.6}$$

where i_a is the anodic component to the total current. Thus the net reaction rate is

$$v_{net} = v_f - v_b = k_f C_O(0, t) - k_b C_R(0, t) = \frac{i}{nFA} \tag{3.2.7}$$

and we have

$$i = i_c - i_a = nFA[k_f C_O(0, t) - k_b C_R(0, t)] \tag{3.2.8}$$

Note that heterogeneous reactions are described differently than homogeneous ones. For example, reaction velocities in heterogeneous systems refer to unit interfacial area, and hence they have dimensions of mol sec^{-1} cm^{-2}. Thus heterogeneous rate constants must carry dimensions of cm/sec, if the concentrations on which they operate are expressed in mol/cm^3. Since the interface can respond only to its immediate surroundings, the concentrations entering rate expressions are always surface concentrations, which may differ from those of the bulk of the solution phase. It is clear also that any species participating in a heterogeneous redox reaction will have kinetic behavior that is strongly influenced by the interfacial potential difference.

For electrode reactions, that potential difference can be controlled, and we want to predict the precise way in which k_f and k_b depend on it. In the next two sections, two different approaches to this problem will be introduced. Both yield the same result for reaction (3.2.1); that is

$$k_f = k^0 \, e^{-\alpha n f(E - E^{0\prime})} \tag{3.2.9}$$

$$k_b = k^0 \, e^{(1-\alpha) n f(E - E^{0\prime})} \tag{3.2.10}$$

where $f = F/RT$, and k^0 and α are adjustable parameters called the *standard rate constant* and the *transfer coefficient*, respectively. These results and inferences derived from them are known generally as the *Butler-Volmer* formulation of electrode kinetics in honor of the pioneers in this area (13, 14).

3.3 MODEL BASED ON FREE ENERGY CURVES (7, 9, 15, 16)

We saw in Section 3.1 that reactions can be visualized in terms of progress along a reaction coordinate connecting a reactant configuration to a product configuration on an energy surface. This idea applies to electrode reactions too, but the shape of the surface turns out to be a function of electrode potential.

This point can be visualized by considering the reaction

$$Na^+ + e \underset{}{\overset{Hg}{\rightleftharpoons}} Na(Hg) \tag{3.3.1}$$

where Na^+ is dissolved in acetonitrile or dimethylformamide. We can take the reaction coordinate as the distance of the sodium nucleus from the interface. Then the free energy profile along the reaction coordinate would resemble Figure 3.3.1*a*. To the

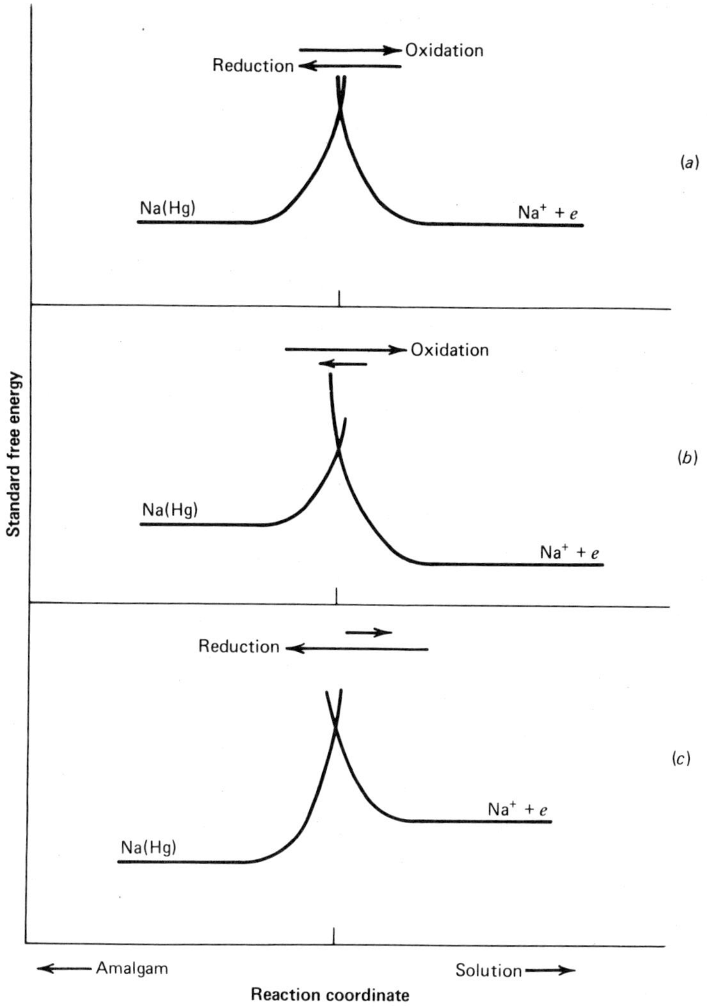

Figure 3.3.1

Simple representation of free energy changes during a faradic process. (a) At a potential corresponding to equilibrium. (b) At a more positive potential than the equilibrium value. (c) At a more negative potential than the equilibrium value.

right, we identify $Na^+ + e$. This configuration has an energy that depends little on the nuclear position in solution unless the electrode is approached so closely that the ion must be partially or wholly desolvated. To the left, the configuration corresponds to a sodium atom dissolved in mercury. Within the mercury phase, the energy depends only slightly on position, but if the atom leaves the interior, its energy rises as the favorable mercury-sodium interaction is lost. The curves corresponding to these reactant and product configurations intersect at the activated complex, and the heights of the barriers to oxidation and reduction determine their relative rates. When the rates are equal, as in Figure 3.3.1a, the system is at equilibrium, and the potential is E_{eq}.

Now suppose the potential is changed to a more positive value. The effect is to lower the relative energy of the electron, and hence the curve corresponding to $Na^+ + e$ drops with respect to that corresponding to Na(Hg), and the situation resembles that of Figure 3.3.1*b*. Since the barrier for reduction is raised and that for oxidation is lowered, the net transformation is conversion of Na(Hg) to $Na^+ + e$. Setting the potential to a value more negative than E_{eq} raises the energy of the electron and shifts the curve for $Na^+ + e$ to higher energies as shown in Figure 3.3.1*c*. Since the reduction barrier drops and the oxidation barrier rises, relative to the condition at E_{eq}, a net cathodic current flows. These arguments show qualitatively the way in which the potential affects the net rates and directions of electrode reactions. By considering the model a little more closely, we can establish a quantitative relationship.

Let us now deal with the general process

$$O + ne \underset{k_b}{\overset{k_f}{\rightleftharpoons}} R \tag{3.3.2}$$

and let us assume that the free energy profiles along the reaction coordinate have the general shapes shown in Figure 3.3.2. It is not important that we know the shapes of these profiles at distances far from the activated complex. Suppose further that the solid curves correspond to an electrode potential of 0 V on any convenient experimental scale. The cathodic and anodic activation energies are $\Delta G^{\ddagger}_{0c}$ and $\Delta G^{\ddagger}_{0a}$, respectively, at that potential.

A shift in potential to value E changes the relative energy of the electron resident on the electrode by $-nFE$; hence the O + ne curve moves up or down by that amount.

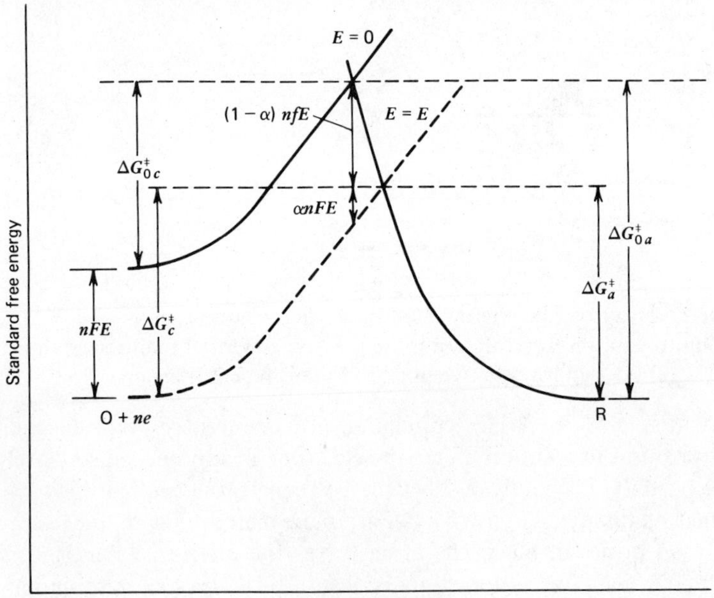

Figure 3.3.2
A detailed picture of the effects of a potential change on the free energies of activation for oxidation and reduction.

The dashed line in the figure shows the effect for a positive E. It is readily apparent that the barrier for oxidation, $\Delta G_a^\ddagger$, is now less than $\Delta G_{0a}^\ddagger$ by a fraction of the total energy change. Let us call that fraction $1 - \alpha$, where α can range from zero to unity, depending on the shape of the intersection region. Thus,

$$\Delta G_a^\ddagger = \Delta G_{0a}^\ddagger - (1 - \alpha)nFE \qquad (3.3.3)$$

A brief study of the figure will also reveal that the cathodic barrier $\Delta G_c^\ddagger$ at potential E is higher than $\Delta G_{0c}^\ddagger$ by αnFE; therefore,

$$\Delta G_c^\ddagger = \Delta G_{0c}^\ddagger + \alpha nFE \qquad (3.3.4)$$

Now let us assume that the rate constants k_f and k_b have an Arrhenius form that can be expressed:

$$k_f = A_f\, e^{-\Delta G_c^\ddagger/RT} \qquad (3.3.5)$$

$$k_b = A_b\, e^{-\Delta G_a^\ddagger/RT} \qquad (3.3.6)$$

Inserting the activation energies (3.3.3) and (3.3.4) gives

$$k_f = A_f\, e^{-\Delta G_{0c}^\ddagger/RT} e^{-\alpha nfE} \qquad (3.3.7)$$

$$k_b = A_b\, e^{-\Delta G_{0a}^\ddagger/RT} e^{(1-\alpha)nfE} \qquad (3.3.8)$$

where $f = F/RT$. The first two factors in each of these expressions form a product that is independent of potential and is equal to the rate constant at $E = 0$ on the scale in use. We designate it as k_f^0 or k_b^0; hence†

$$k_f = k_f^0\, e^{-\alpha nfE} \qquad (3.3.9)$$

$$k_b = k_b^0\, e^{(1-\alpha)nfE} \qquad (3.3.10)$$

Consider now the special case in which the interface is at equilibrium with a solution in which $C_O^* = C_R^*$. Thus $E = E^{0'}$ and $k_f C_O^* = k_b C_R^*$, which implies that $k_f = k_b$. That is,

$$k_f^0\, e^{-\alpha nfE^{0'}} = k_b^0\, e^{(1-\alpha)nfE^{0'}} = k^0 \qquad (3.3.11)$$

and we give the special name of *standard rate constant*‡ to k^0, which is the value of k_f and k_b at $E^{0'}$. Using equation 3.3.11 to substitute for k_f^0 and k_b^0 in (3.3.9) and (3.3.10) leads finally to the results

$$\boxed{k_f = k^0\, e^{-\alpha nf(E - E^{0'})}} \qquad (3.3.12)$$

$$\boxed{k_b = k^0\, e^{(1-\alpha)nf(E - E^{0'})}} \qquad (3.3.13)$$

which are identical to (3.2.9) and (3.2.10).

† In other electrochemical literature, k_f and k_b are designated as k_c and k_a or as k_{ox} and k_{red}. The corresponding symbols for rate constants at $E = 0$ on the scale in use are then k_c^0 and k_a^0 or k_{ox}^0 and k_{red}^0. Sometimes kinetic equations are written in terms of a complementary transfer coefficient, $\beta = 1 - \alpha$.

‡ The standard rate constant is also designated by $k_{s,h}$ or k_s in electrochemical literature. Sometimes it is also called the *intrinsic rate constant*.

Insertion of these relations into (3.2.8) yields the complete *current—potential characteristic:*

$$i = nFAk^0[C_O(0, t)\, e^{-\alpha nf(E - E^{0'})} - C_R(0, t)\, e^{(1-\alpha)nf(E - E^{0'})}] \tag{3.3.14}$$

This relation is very important. It, or a variation derived from it, is used in the treatment of every problem requiring an account of heterogeneous kinetics. Section 3.5 will consider some of its ramifications.

The physical interpretation of k^0 is straightforward. It simply is a measure of the kinetic facility of a redox couple. A system with a large k^0 will achieve equilibrium on a short time scale, but a system with small k^0 will be sluggish. The largest measured standard rate constants are in the range of 1 to 10 cm/sec and are associated with particularly simple electron transfer processes. For example, the reductions and oxidations of many aromatic hydrocarbons (such as substituted anthracenes, pyrene, and perylene) to the corresponding anion and cation radicals fall in this category (17–19). These processes involve only electron transfer and resolvation. There are no significant alterations in the molecular forms. Similarly, some electrode processes involving the formation of amalgams [e.g., the couples $Na^+/Na(Hg)$, $Cd^{2+}/Cd(Hg)$, and Hg_2^{2+}/Hg] are rather facile (20, 21). More complicated reactions involving significant molecular rearrangement upon electron transfer, such as the reduction of molecular oxygen to hydrogen peroxide or water, or the reduction of protons to molecular hydrogen, can be very sluggish (20–22). Many of these systems involve multistep mechanisms and are discussed more fully in Section 3.6. Values of k^0 significantly lower than 10^{-9} cm/sec have been reported (23–26); therefore, electrochemistry deals with a range of more than 10 orders of magnitude in kinetic facility.

Note that k_f and k_b can be made quite large, even if k^0 is small, by using a sufficiently extreme potential relative to $E^{0'}$. In effect, one drives the reaction by supplying the activation energy electrically. This idea is explored more fully in Section 3.5.

The transfer coefficient, α, is a measure of the symmetry of the energy barrier. This point can be amplified by considering α in terms of the geometry of the intersection region, as shown in Figure 3.3.3. If the curves are essentially linear, then the angles θ and ϕ are defined by

$$\tan\theta = \alpha nFE/x \tag{3.3.15}$$

$$\tan\phi = (1 - \alpha)nFE/x \tag{3.3.16}$$

hence

$$\alpha = \frac{\tan\theta}{\tan\phi + \tan\theta} \tag{3.3.17}$$

If the intersection is symmetrical, $\phi = \theta$, and $\alpha = \frac{1}{2}$. Otherwise $0 \le \alpha \le \frac{1}{2}$ or $\frac{1}{2} \le \alpha \le 1$, as shown in Figure 3.3.4. In most systems α turns out to lie between 0.3 and 0.7, and it can usually be approximated by 0.5 in the absence of actual measurements.

If the free energy curves are not linear over the domain corresponding to the potential range of interest, then α would be a potential-dependent factor, because the

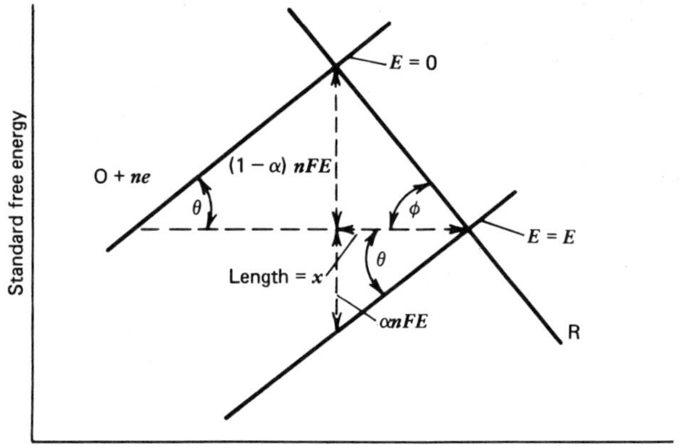

Figure 3.3.3
Relationship of the transfer coefficient to the angles of intersection of the free energy curves.

angles θ and ϕ would depend on the location of a particular intersection, which in turn is governed by the potential (see Section 6.7.3).

3.4 KINETIC MODEL BASED ON ELECTROCHEMICAL POTENTIALS (5, 8, 10, 27–29)

In this section, we pursue an alternate method for treating the kinetics of electrode processes. Even though the results of this approach are the same as those obtained in the previous section, the formulation can be more convenient for more complicated electrode processes, which may require the inclusion of double-layer effects or a sequence of interfacial reactions. Fuller discussions of the application of this method can be found in the cited literature.

We consider the simple process:

$$O^z + ne \underset{k_b}{\overset{k_f}{\rightleftharpoons}} R^{z'} \tag{3.4.1}$$

where z and z' are the charges on ions O and R, which are both dissolved in the electrolyte. Obviously, $z - n = z'$. As in Section 3.3, a postulate is made that the rate constants k_f and k_b follow the Arrhenius form and, in keeping with the literature, we add

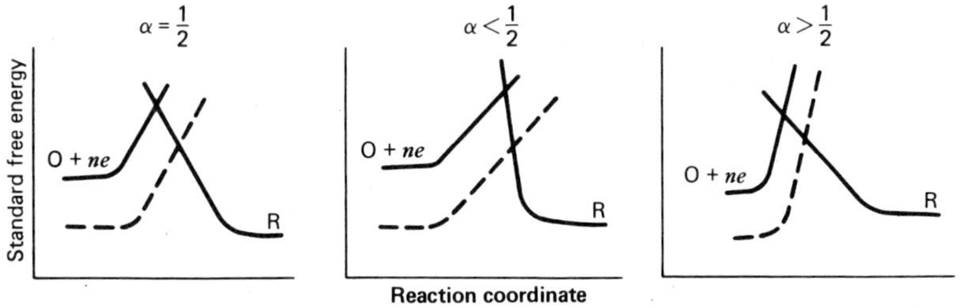

Figure 3.3.4
The transfer coefficient as an indicator of the symmetry of the barrier to reaction.

the further assumption that both rate constants adhere to the predictions of simplified activated complex theory (see Section 3.1). That is,

$$k_f = \frac{kT}{h} e^{-\Delta \bar{G}_f^{\ddagger}/RT} \tag{3.4.2}$$

$$k_b = \frac{kT}{h} e^{-\Delta \bar{G}_b^{\ddagger}/RT} \tag{3.4.3}$$

where the activation free energies $\Delta \bar{G}_f^{\ddagger}$ and $\Delta \bar{G}_b^{\ddagger}$ are written with an overbar in recognition of their dependence on the interfacial potential difference. In other words, they are *electrochemical* free energies of activation and they can be separated, as outlined in Section 2.2.4, into chemical and electrical components:

$$\Delta \bar{G}_f^{\ddagger} = \Delta G_f^{\ddagger} + (\Delta G_f^{\ddagger})_e \tag{3.4.4}$$

$$\Delta \bar{G}_b^{\ddagger} = \Delta G_b^{\ddagger} + (\Delta G_b^{\ddagger})_e \tag{3.4.5}$$

In fact, we can view the entire free energy surface connecting O + *ne* and R as being the sum of a chemical free energy surface and an electrical free energy surface, as illustrated in Figure 3.4.1.

We assume that the chemical free energies of activation $\Delta G_f^{\ddagger}$ and $\Delta G_b^{\ddagger}$ do not change with potential, so that the effects of potential are wholly manifested within the electrical components. To obtain an explicit relationship with potential, we assume further that the electrical components to the activation energies are fixed fractions of the overall electrical free energy change in going from O + *ne* to R. That is,

$$(\Delta G_f^{\ddagger})_e = \alpha(\Delta G^0)_e \tag{3.4.6}$$

$$(\Delta G_b^{\ddagger})_e = -(1 - \alpha)(\Delta G^0)_e \tag{3.4.7}$$

Our task now is to find $(\Delta G^0)_e$.

The standard electrochemical free energy (per mole) of the state corresponding to species O^z plus *n* electrons on metal M (state I) is

$$\bar{G}_I^0 = \bar{\mu}_O + n\bar{\mu}_e = \mu_O^{0S} + zF\phi^S + n\mu_e^{0M} - nF\phi^M \tag{3.4.8}$$

The equivalent free energy for species $R^{z'}$ (state II) is

$$\bar{G}_{II}^0 = \bar{\mu}_R = \mu_R^{0S} + z'F\phi^S \tag{3.4.9}$$

The desired energy component, $(\Delta G^0)_e$, is simply the difference between the electrical terms for states II and I; that is,

$$(\Delta G^0)_e = (z' - z)F\phi^S + nF\phi^M = nF(\phi^M - \phi^S) \tag{3.4.10}$$

Since the electrode potential E on any scale is related to the interfacial potential difference $\phi^M - \phi^S$ by a constant, K,

$$E = (\phi^M - \phi^S) + K \tag{3.4.11}$$

we have

$$(\Delta G_f^{\ddagger})_e = \alpha nF(E - K) \tag{3.4.12}$$

$$(\Delta G_b^{\ddagger})_e = -(1 - \alpha)nF(E - K) \tag{3.4.13}$$

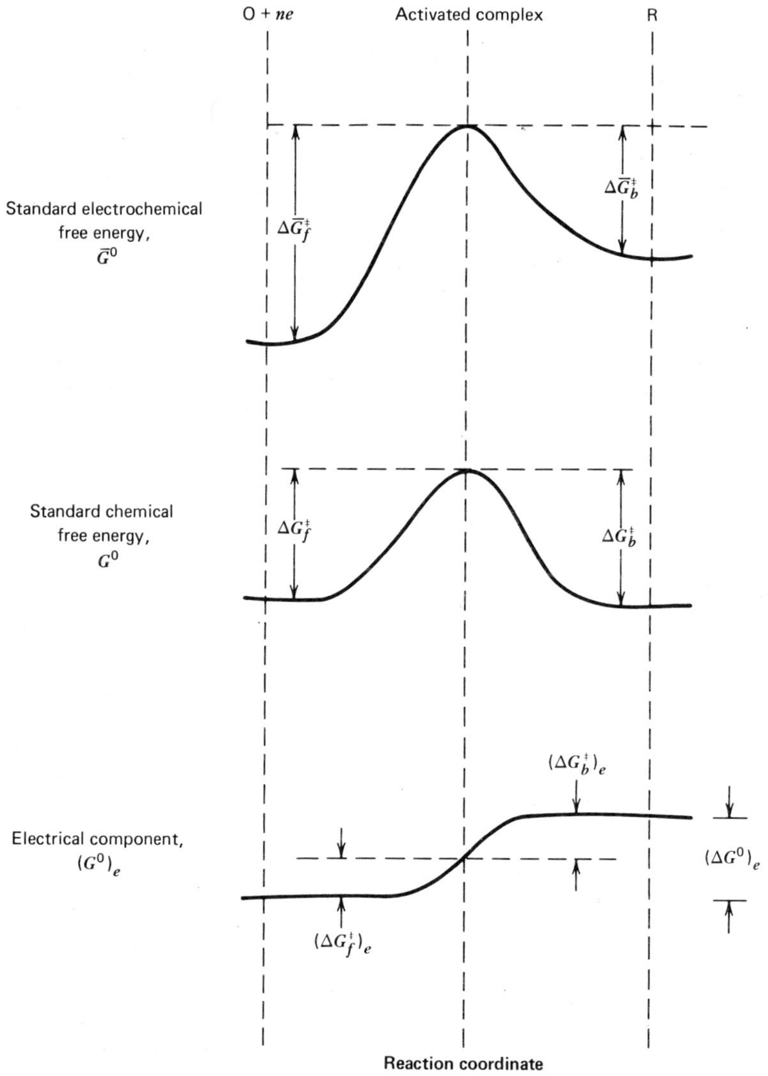

Figure 3.4.1
Separation of the electrochemical free energy (upper curve) into chemical and electrical components (lower two curves).

Substitution of (3.4.12) and (3.4.13) into (3.4.4) and (3.4.5) and then into (3.4.2) and (3.4.3) gives

$$k_f = \frac{kT}{h} e^{-\Delta G_f^\ddagger/RT} e^{\alpha nFK/RT} e^{-\alpha nFE/RT} \tag{3.4.14}$$

$$k_b = \frac{kT}{h} e^{-\Delta G_b^\ddagger/RT} e^{-(1-\alpha)nFK/RT} e^{(1-\alpha)nFE/RT} \tag{3.4.15}$$

All factors but the last exponential in each of these expressions are independent of potential, and they can be collected into the constants k_f^0 and k_b^0:

$$k_f = k_f^0 \, e^{-\alpha n f E} \tag{3.4.16}$$

$$k_b = k_b^0 \, e^{(1-\alpha) n f E} \tag{3.4.17}$$

where $f = F/RT$. These relations are exactly those derived in the last section, and the current-potential characteristic follows from them in just the manner shown above.

Even though we assumed the preexponentials in (3.4.2) and (3.4.3) to be $\mathscr{k}T/h$, the final result is not constrained by the appropriateness of that choice, because the factor $\mathscr{k}T/h$ has been gathered into k_f^0 and k_b^0 along with several unknowable factors. We could have used unspecified frequency factors A_f and A_b in (3.4.2) and (3.4.3), as we did in Section 3.3, and we would have arrived at the same final results.

The crucial assumptions of this treatment are (a) that $\Delta G_f^\ddagger$ and $\Delta G_b^\ddagger$ are independent of potential and (b) that $(\Delta G_f^\ddagger)_e$ and $(\Delta G_b^\ddagger)_e$ are constant fractions of $(\Delta G^0)_e$. The second assumption is a postulate that is integral to the treatment. The first is a convenience that is introduced in order to force all potential dependence into the electrical components of the activation energies. In general, it will hold provided that the *position* of the maximum on the electrochemical free energy surface (i.e., the location of the activated complex) does not move with potential. Such changes are inevitable in principle for a peak in standard chemical free energy superimposed on a potential-dependent sloping baseline of the electrical component; however, the shifts might not be large for small changes in potential. Such a change in position implies altered $\Delta G_f^\ddagger$ and $\Delta G_b^\ddagger$. The potential change does not alter the surface describing chemical free energy, but it may alter the contribution of that surface to the overall activation energies by changing the coordinates of the activated complex (30). Experimentally, this effect would probably appear as a potential dependence in α (see Section 6.7.3).

3.5 IMPLICATIONS OF THE CURRENT-POTENTIAL CHARACTERISTIC

3.5.1 Equilibrium Conditions. The Exchange Current (5–12)

At equilibrium, the net current is zero, and we require the electrode to adopt a potential based on the bulk concentrations of O and R as dictated by the Nernst equation. The current-potential characteristic derived above does indeed yield that thermodynamic relation as a special case. From equation 3.3.14 we have, at zero current,

$$nFAk^0 C_O(0, t) \, e^{-\alpha n f(E_{eq} - E^{0'})} = nFAk^0 C_R(0, t) \, e^{(1-\alpha) n f(E_{eq} - E^{0'})} \tag{3.5.1}$$

Since equilibrium applies, the bulk concentrations of O and R are found also at the surface; hence

$$e^{nf(E_{eq} - E^{0'})} = C_O^*/C_R^* \tag{3.5.2}$$

which is simply an exponential form of the Nernst relation:

$$E_{eq} = E^{0'} + \frac{RT}{nF} \ln \frac{C_O^*}{C_R^*} \tag{3.5.3}$$

Thus the theory has passed its first test of compatibility with reality.

Even though the net current is zero at equilibrium, we still envision balanced faradaic activity that can be expressed in terms of the *exchange current*, i_0, which is equal in magnitude to either component current i_c or i_a. That is,

$$i_0 = nFAk^0 C_O^* e^{-\alpha nf(E_{eq} - E^{0'})} \tag{3.5.4}$$

If both sides of (3.5.2) are raised to the $-\alpha$ power, we obtain

$$e^{-\alpha nf(E_{eq} - E^{0'})} = \left(\frac{C_O^*}{C_R^*}\right)^{-\alpha} \tag{3.5.5}$$

Substitution of (3.5.5) into (3.5.4) gives†

$$i_0 = nFAk^0 C_O^{*(1-\alpha)} C_R^{*\alpha} \tag{3.5.6}$$

The exchange current is therefore proportional to k^0, and it can often be substituted for k^0 in kinetic equations. For the particular case $C_O^* = C_R^* = C$,

$$i_0 = nFAk^0 C \tag{3.5.7}$$

Often the exchange current is normalized to unit area to provide the *exchange current density*, $j_0 = i_0/A$.

3.5.2 The Current-Overpotential Equation

An advantage of working with i_0 rather than k^0 is that the current can be described in terms of the deviation from the equilibrium potential, that is, the overpotential η, rather than the formal potential $E^{0'}$. Dividing (3.3.14) by (3.5.6), we obtain

$$\frac{i}{i_0} = \frac{C_O(0, t)\, e^{-\alpha nf(E - E^{0'})}}{C_O^{*(1-\alpha)} C_R^{*\alpha}} - \frac{C_R(0, t)\, e^{(1-\alpha)nf(E - E^{0'})}}{C_O^{*(1-\alpha)} C_R^{*\alpha}} \tag{3.5.8}$$

or

$$\frac{i}{i_0} = \frac{C_O(0, t)}{C_O^*} e^{-\alpha nf(E - E^{0'})} \left(\frac{C_O^*}{C_R^*}\right)^{\alpha} - \frac{C_R(0, t)}{C_R^*} e^{(1-\alpha)nf(E - E^{0'})} \left(\frac{C_O^*}{C_R^*}\right)^{-(1-\alpha)} \tag{3.5.9}$$

† The same equation for the exchange current can be derived from the anodic component current i_a at $E = E_{eq}$.

The ratios $(C_O^*/C_R^*)^{\alpha}$ and $(C_O^*/C_R^*)^{-(1-\alpha)}$ are easily evaluated from equations 3.5.2 and 3.5.5, and by substitution we obtain

$$\boxed{i = i_0\left\{\frac{C_O(0, t)}{C_O^*}\, e^{-\alpha n f \eta} - \frac{C_R(0, t)}{C_R^*}\, e^{(1-\alpha) n f \eta}\right\}} \tag{3.5.10}$$

where $\eta = E - E_{eq}$. This equation, known as the *current-overpotential equation*, will be used frequently in later discussions. Note that the first term describes the cathodic component current at any potential, and the second gives the anodic contribution.†

The behavior predicted by (3.5.10) is depicted in Figure 3.5.1. The solid curve shows the actual total current, which is the sum of the components i_c and i_a, shown as dashed traces. For large negative overpotentials, the anodic component is negligible; hence the total current curve merges with that for i_c. At large positive overpotentials, the cathodic component is negligible, and i_a is essentially the same as the total current. In going either direction from E_{eq}, the magnitude of the current rises rapidly because the exponential factors dominate behavior, but at extreme η, the current levels off. In these level regions, the current is limited by mass transfer rather than heterogeneous kinetics. The exponential factors in (3.5.10) are then kept under control by the factors $C_O(0, t)/C_O^*$ and $C_R(0, t)/C_R^*$, which manifest the reactant supply.

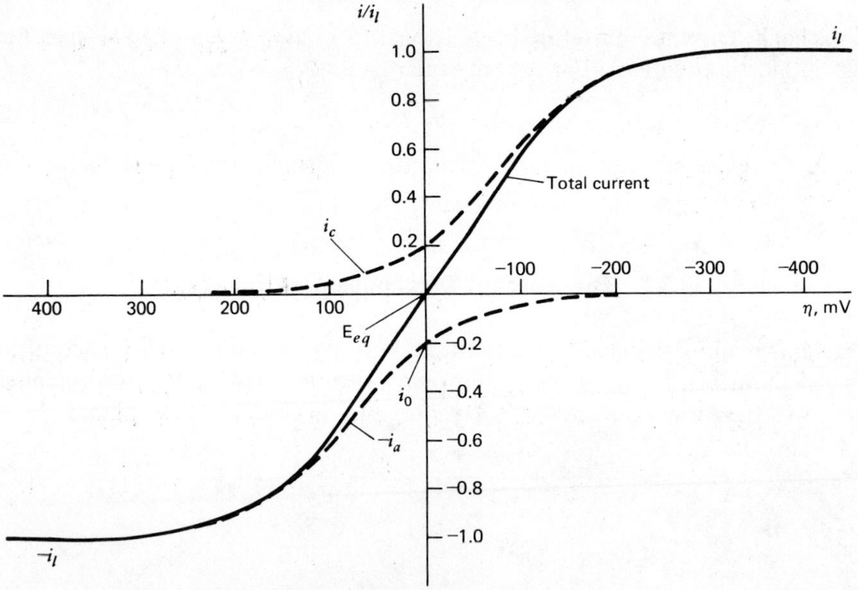

Figure 3.5.1

Current-overpotential curves for the system $O + ne \rightleftharpoons R$ with $\alpha = 0.5$, $n = 1$, $T = 298$ K, $i_{l,c} = -i_{l,a} = i_l$, and $i_0/i_l = 0.2$. The dotted lines show the component currents i_c and i_a.

† Since double-layer effects have not been included in this treatment, k^0 and i_0 are, in Delahay's nomenclature (5), *apparent* constants of the system. Both depend on double-layer structure to some extent and are functions of the potential at the outer Helmholtz plane relative to the solution bulk, ϕ_2. This point will be discussed in more detail in Chapter 12. See also References 5 to 12.

3.5.3 Approximate Forms of the *i*-η Equation

(a) No Mass Transfer Effects. If the solution is well stirred or currents are kept so low that the surface concentrations do not differ appreciably from the bulk values, then (3.5.10) becomes

$$\boxed{i = i_0[e^{-\alpha n f \eta} - e^{(1-\alpha)n f \eta}]} \tag{3.5.11}$$

which is known generally as the *Butler-Volmer equation.* This equation is a good approximation of (3.5.10) whenever i is less than about 10% of the smaller limiting current, $i_{l,c}$ or $i_{l,a}$. Equations 1.4.10 and 1.4.19 show that $C_O(0, t)/C_O^*$ and $C_R(0, t)/C_R^*$ will then be between 0.9 and 1.1.

The curves in Figure 3.5.2 show the behavior of (3.5.11) for different exchange current densities. In each case $n = 1$ and $\alpha = 0.5$. Figure 3.5.3 shows the effect of α in a similar manner. There $n = 1$ and the exchange current density is 10^{-6} A/cm^2. A notable feature of Figure 3.5.2 is the degree to which the inflection at E_{eq} depends on the exchange current.

Since mass transfer effects are not included here, the overpotential associated with any given current serves solely as an activation energy. It is required to drive the heterogeneous process at the rate reflected by the current. The lower the exchange current, the more sluggish are the kinetics; hence the larger is the *activation overpotential* for any particular net current. If the exchange current is very large, as for case (*a*) in Figure 3.5.2, then the system can supply large currents, perhaps even the mass transfer limited current, with insignificant activation overpotential. In that case, any observed overpotential is associated with changing surface concentrations of species O and R. It is called a *concentration overpotential* and can be viewed as an activation energy

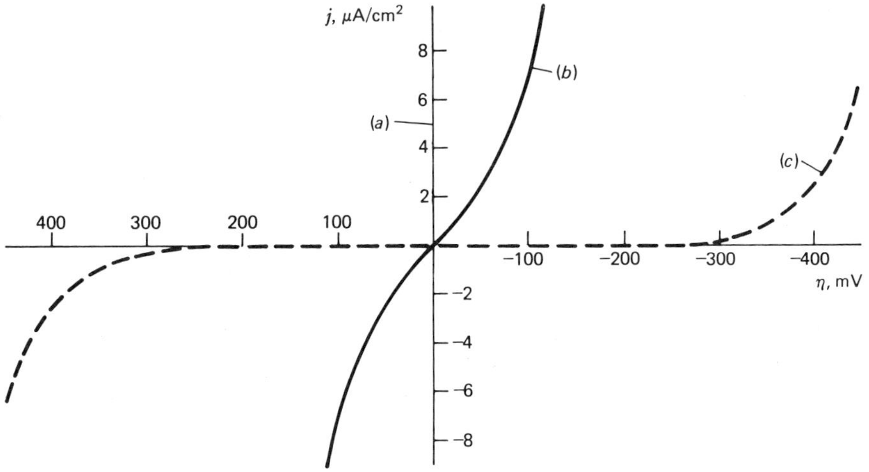

Figure 3.5.2

Effect of exchange current density on the activation overpotential required to deliver net current densities. (*a*) $j_0 = 10^{-3}$ A/cm^2, (*b*) $j_0 = 10^{-6}$ A/cm^2, (*c*) $j_0 = 10^{-9}$ A/cm^2. For all cases the reaction is $O + ne \rightleftharpoons R$ with $\alpha = 0.5$, $n = 1$, and $T = 298$ K.

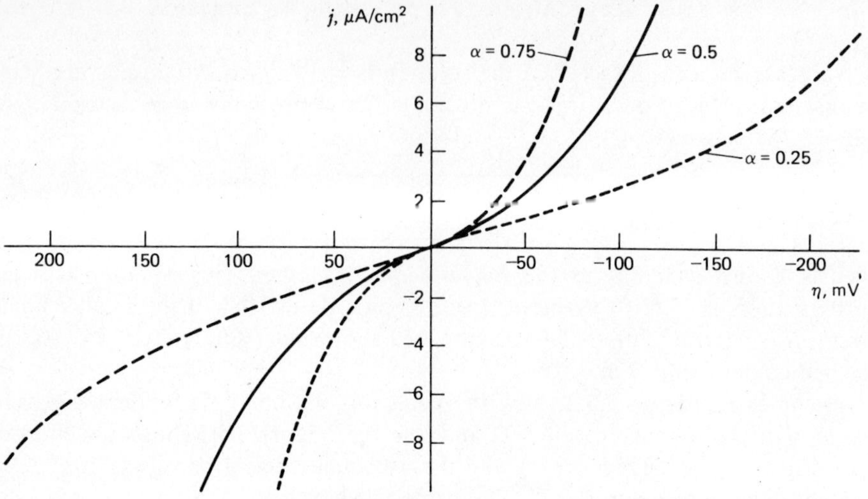

Figure 3.5.3
Effect of the transfer coefficient on the symmetry of the current-overpotential curves for $O + ne \rightleftharpoons R$ with $n = 1$, $T = 298$ K, and $j_0 = 10^{-6}$ A/cm².

required to drive mass transfer at the rate needed to support the current. If the concentrations of O and R are comparable, then E_{eq} will be near $E^{0'}$, and the limiting currents for both the anodic and cathodic segments will be reached within a few tens of millivolts of $E^{0'}$.

On the other hand, one might deal with a system with an exceedingly small exchange current because k^0 is very low, as for case (*c*) in Figure 3.5.2. In that circumstance, no significant current flows at all unless a very large activation overpotential is applied. At a sufficiently extreme potential, the heterogeneous process can be driven fast enough that mass transfer controls the current and a limiting plateau is reached. When mass transfer effects start to manifest themselves, then a concentration overpotential will also contribute, but the bulk of the overpotential is for activation of charge transfer. In this kind of system, the reduction wave occurs at much more negative potentials, and the oxidation wave lies at much more positive values than $E^{0'}$.

The exchange current can be viewed as a kind of "idle speed" for charge exchange across the interface. If we want to draw a net current that is only a small fraction of this bidirectional idle current, then a very small overpotential will be required to extract it. Even at equilibrium, the system is delivering charge across the interface at rates much greater than we require. The role of the slight overpotential is to unbalance the rates in the two directions to a small degree so that one of them predominates. On the other hand, if we ask for a net current that exceeds the exchange current, our job is much harder. We have to drive the system to deliver charge at the required rate, and we can only do that by applying a significant overpotential. From this perspective, we see that the exchange current is a measure of any system's ability to deliver a net current without a significant energy loss due to activation.

Exchange current densities in real systems reflect the wide range in k^0. They may exceed 10 A/cm² or they may be less than picoamperes/cm² (5–12, 23–26).

(b) Linear Characteristic at Small η. For small values of x, the exponential, e^x, can be approximated as $1 + x$; hence for sufficiently small η, equation 3.5.11 can be written:

$$\boxed{i = i_0(-nf\eta)} \tag{3.5.12}$$

which shows that the net current is linearly related to overpotential in a narrow potential range near E_{eq}. The ratio $-\eta/i$ has dimensions of resistance and is often called the *charge transfer resistance*, R_{ct}:

$$\boxed{R_{ct} = \frac{RT}{nFi_0}} \tag{3.5.13}$$

This parameter can be evaluated directly in some experiments, and it serves as a convenient index of kinetic facility. For very large k^0, it clearly approaches zero.

(c) Tafel Behavior at Large η. For large values of η (either negative or positive), one of the bracketed terms in (3.5.11) becomes negligible. For example, at large negative overpotentials, $\exp(-\alpha nf\eta) \gg \exp[(1 - \alpha)nf\eta]$ and (3.5.11) becomes

$$i = i_0\, e^{-\alpha nf\eta} \tag{3.5.14}$$

or

$$\boxed{\eta = \frac{RT}{\alpha nF}\ln i_0 - \frac{RT}{\alpha nF}\ln i} \tag{3.5.15}$$

Thus, we find that the kinetic treatment outlined above does yield a relation of the Tafel form, as required by observation, for the appropriate conditions. The empirical Tafel constants (see equation 3.2.4) can now be identified from theory as†

$$a = \frac{2.3RT}{\alpha nF}\log i_0 = \frac{0.0591}{\alpha n}\log i_0 \ (25\ ^\circ\mathrm{C}) \tag{3.5.16}$$

$$b = \frac{-2.3RT}{\alpha nF} = \frac{-0.0591}{\alpha n} \ (25\ ^\circ\mathrm{C}) \tag{3.5.17}$$

The Tafel form can be expected to hold whenever the back reaction (i.e., the anodic process when a net reduction is considered, and vice versa) contributes less than 1% of the current, or

$$\frac{e^{(1-\alpha)nf\eta}}{e^{-\alpha nf\eta}} = e^{nf\eta} \le 0.01 \tag{3.5.18}$$

which implies at 25 °C that

$$|\eta| > 0.118/n\ \mathrm{V} \tag{3.5.19}$$

† Note that for $\alpha = 0.5$ and $n = 1$, $b = 0.118$ V, a value that is sometimes quoted as a "typical" Tafel slope.

If the electrode kinetics are fairly facile, we will approach the mass transfer limited current by the time such an extreme overpotential is established. Tafel relationships cannot be observed for such systems, because they require the absence of mass transfer effects on the current. When electrode kinetics are sluggish and significant activation overpotentials are required, good Tafel relationships can be seen. This point underscores the fact that Tafel behavior is an indicator of *totally irreversible* kinetics. Systems in that category allow no significant current flow except at high overpotentials, where the faradaic process is effectively unidirectional and, therefore, chemically irreversible.

(d) Tafel Plots (5, 7–9, 31). A plot of log i *vs.* η, known as a *Tafel plot*, is a useful device for evaluating kinetic parameters. In general, there is an anodic branch with slope $(1 - \alpha)nF/2.3RT$ and a cathodic branch with slope $-\alpha nF/2.3RT$. As shown in Figure 3.5.4, both linear segments extrapolate to an intercept of log i_0. The actual current plots deviate sharply from linear behavior as η approaches zero, because the back reactions can no longer be regarded as negligible. The transfer coefficient α and the exchange current i_0 are obviously readily accessible from this kind of presentation, when it can be applied.

Some real Tafel plots are shown in Figure 3.5.5 for the Mn(IV/Mn(III) system in concentrated acid (32). The negative deviations from linearity at very large overpotentials come from limitations imposed by mass transfer. The region of very low overpotentials shows sharp falloffs for the reasons outlined just above.

An alternate method for plotting $i - \eta$ data, which allows the use of data obtained at low overpotentials, has been suggested by Allen and Hickling (33). Equation 3.5.11 can be rewritten as

$$i = i_0 e^{-\alpha n f \eta}(1 - e^{n f \eta}) \tag{3.5.20}$$

or

$$\log \frac{i}{1 - e^{n f \eta}} = \log i_0 - \frac{\alpha n F \eta}{2.3RT} \tag{3.5.21}$$

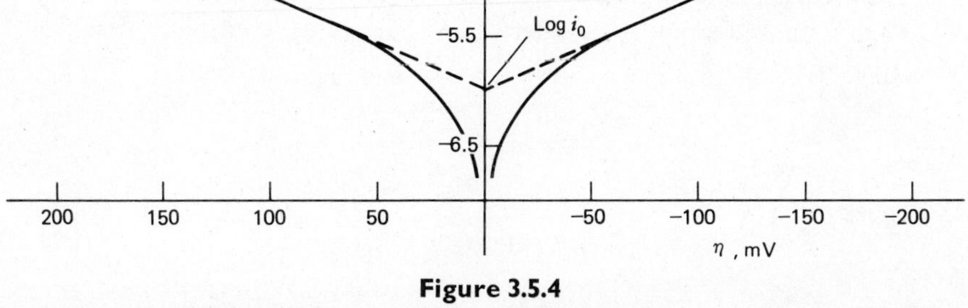

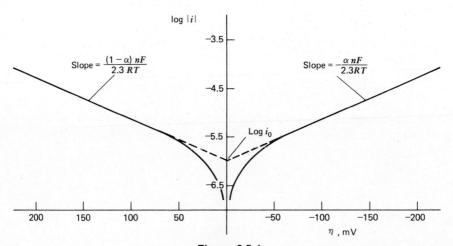

Figure 3.5.4
Tafel plots for anodic and cathodic branches of the current-overpotential curve for $O + ne \rightleftharpoons R$ with $n = 1$, $\alpha = 0.5$, $T = 298$ K, and $j_0 = 10^{-6}$ A/cm^2.

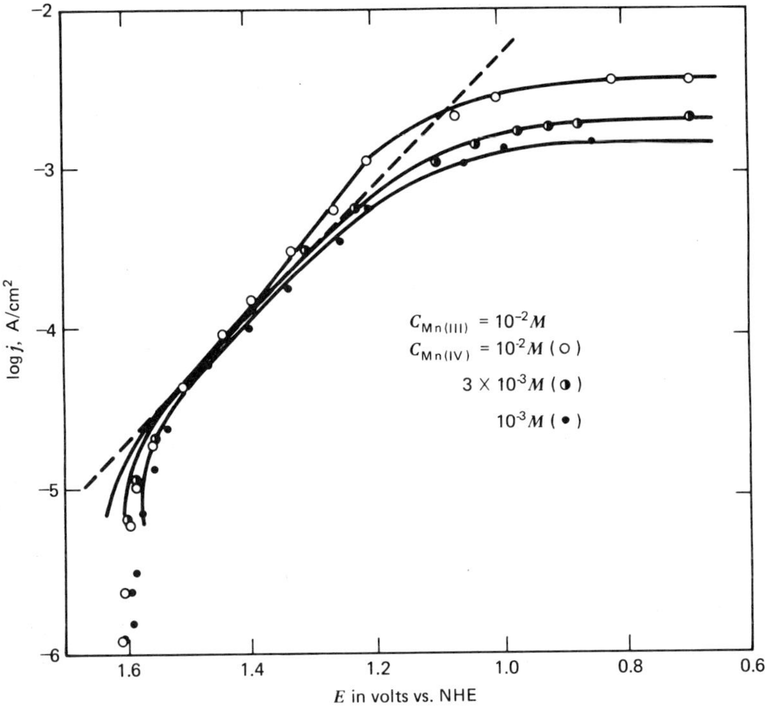

Figure 3.5.5

Tafel plots for the reduction of Mn(IV) to Mn(III) at Pt in 7.5 M H_2SO_4 at 298 K. The dashed line corresponds to $\alpha = 0.24$. [From K. J. Vetter and G. Manecke, *Z. Physik. Chem.* (Leipzig), **195**, 337 (1950), with permission.]

so that a plot of $\log[i/(1 - e^{nf\eta})]$ *vs.* η yields an intercept of $\log i_0$ and a slope of $-\alpha nF/2.3RT$. This approach has the advantage of being applicable to electrode reactions that are not totally irreversible, that is, those in which both anodic and cathodic processes contribute significantly to the currents measured in the overpotential range where mass transfer effects are not important. Such systems are often termed *quasi-reversible,* because the opposing charge transfer reactions must both be considered, yet a noticeable activation overpotential is required in order to drive a given net current through the interface.

3.5.4 Exchange Current Plots (5–12)

From equation 3.5.4 we recognize that the exchange current can be written as

$$i_0 = nFAk^0 C_O^* e^{-\alpha nf(E_{eq} - E^{0'})} \tag{3.5.22}$$

which can be restated:

$$\log i_0 = \log nFAk^0 + \log C_O^* + \frac{\alpha nF}{2.3RT} E^{0'} - \frac{\alpha nF}{2.3RT} E_{eq} \tag{3.5.23}$$

Therefore, a plot of $\log i_0$ *vs.* E_{eq} at constant C_O^* is linear with a slope of $-\alpha nF/2.3RT$. The equilibrium potential E_{eq} can be varied experimentally by changing the bulk concentration of species R, while that of species O is held constant. This kind of plot is useful for obtaining αn from experiments in which i_0 is measured essentially directly (see, for example, Chapter 9).

Careful inspection of relationships advanced earlier in this chapter for evaluating kinetic parameters will show that none of them permits a determination of α without knowledge of n. That is, one can only determine αn. In complex electrode processes involving more than one charge transfer step, the n value of the kinetically limiting step may not be readily apparent; hence it is useful to be able to obtain α separately. A means for doing so is suggested by rewriting (3.5.6) as

$$\log i_0 = \log nFAk^0 + (1 - \alpha) \log C_O^* + (\alpha) \log C_R^* \tag{3.5.24}$$

Thus we note that

$$\left(\frac{\partial \log i_0}{\partial \log C_O^*}\right)_{C_R^*} = 1 - \alpha \quad \text{and} \quad \left(\frac{\partial \log i_0}{\partial \log C_R^*}\right)_{C_O^*} = \alpha \tag{3.5.25}$$

An alternate equation, which does not require holding either C_O^* or C_R^* constant, is

$$\frac{d \log(i_0/C_O^*)}{d \log(C_R^*/C_O^*)} = \alpha \tag{3.5.26}$$

The last relation is easily derived from (3.5.6).

3.5.5 Very Facile Kinetics. Reversible Behavior

To this point, we have discussed in detail only those systems for which appreciable activation overpotential is observed. Another very important limit is the case in which the electrode kinetics require no driving force at all. As we noted above, that case corresponds to a very large exchange current, which in turn reflects a big intrinsic rate constant k^0. Let us rewrite the current-overpotential equation (3.5.10) as follows:

$$\frac{i}{i_0} = \frac{C_O(0, t)\, e^{-\alpha nf\eta}}{C_O^*} - \frac{C_R(0, t)\, e^{(1-\alpha)nf\eta}}{C_R^*} \tag{3.5.27}$$

and consider its behavior when i_0 becomes very large compared to any current of interest. The ratio i/i_0 then approaches zero, and we can rearrange the limiting form of equations 3.5.27 to

$$\frac{C_O(0, t)}{C_R(0, t)} = \frac{C_O^*\, e^{nf(E - E_{eq})}}{C_R^*} \tag{3.5.28}$$

and, by substitution from the Nernst equation in form (3.5.2), we obtain

$$\frac{C_O(0, t)}{C_R(0, t)} = e^{nf(E_{eq} - E^{0\prime})}\, e^{nf(E - E_{eq})} \tag{3.5.29}$$

or

$$\frac{C_O(0, t)}{C_R(0, t)} = e^{nf(E - E^{0\prime})} \tag{3.5.30}$$

This equation can be rearranged to the very important result:

$$E = E^{0'} + \frac{RT}{nF} \ln \frac{C_O(0, t)}{C_R(0, t)} \tag{3.5.31}$$

Thus we see that the electrode potential and the *surface* concentrations of O and R, regardless of the current flow, are linked by an equation of the Nernst form.

No kinetic parameters are present because the kinetics are so facile that no experimental manifestations can be seen. In effect, the potential and the surface concentrations are always kept in equilibrium with each other by the fast charge transfer processes, and the *thermodynamic* equation (3.5.31), characteristic of equilibrium, always holds. Net current flows because the surface concentrations are not at equilibrium with the bulk, and mass transfer continuously moves material to the surface where it must be reconciled to the potential by electrochemical change.

We have already seen that a system that is always at equilibrium is termed a *reversible* system; thus it is logical that an electrochemical system in which the charge transfer interface is always at equilibrium be called a *reversible* (or, sometimes, *nernstian*) *system*. These terms simply refer to cases in which the interfacial redox kinetics are so fast that activation effects cannot be seen. Many such systems exist in electrochemistry, and we will consider this case frequently under different sets of experimental circumstances. We will also see that any given system may appear reversible, quasi-reversible, or totally irreversible, depending on the time scale of the demands we make on the charge transfer kinetics.

3.5.6 Effects of Mass Transfer

A more complete $i - \eta$ relation can be obtained from (3.5.10), by substituting for $C_O(0, t)/C_O^*$ and $C_R(0, t)/C_R^*$ according to (1.4.10) and (1.4.19):

$$\frac{i}{i_0} = \left(1 - \frac{i}{i_{l,c}}\right) e^{-\alpha n f \eta} - \left(1 - \frac{i}{i_{l,a}}\right) e^{(1-\alpha) n f \eta} \tag{3.5.32}$$

This equation can be rearranged easily to give i as an explicit function of η over the whole range of η. In Figure 3.5.6, one can see $i - \eta$ curves for several ratios of i_0/i_l, where $i_l = i_{l,c} = -i_{l,a}$.

For small overpotentials, a linearized relation can be used. The complete Taylor expansion (Section A.2) of (3.5.27) gives, for $\alpha n f \eta \ll 1$,

$$\frac{i}{i_0} = \frac{C_O(0, t)}{C_O^*} - \frac{C_R(0, t)}{C_R^*} - \frac{nF\eta}{RT} \tag{3.5.33}$$

which can be substituted as above to give

$$\frac{i}{i_0} = -\frac{i}{i_{l,c}} + \frac{i}{i_{l,a}} - \frac{nF\eta}{RT} \tag{3.5.34}$$

or

$$\eta = -\frac{RT}{nF} i \left(\frac{1}{i_0} + \frac{1}{i_{l,c}} - \frac{1}{i_{l,a}}\right) \tag{3.5.35}$$

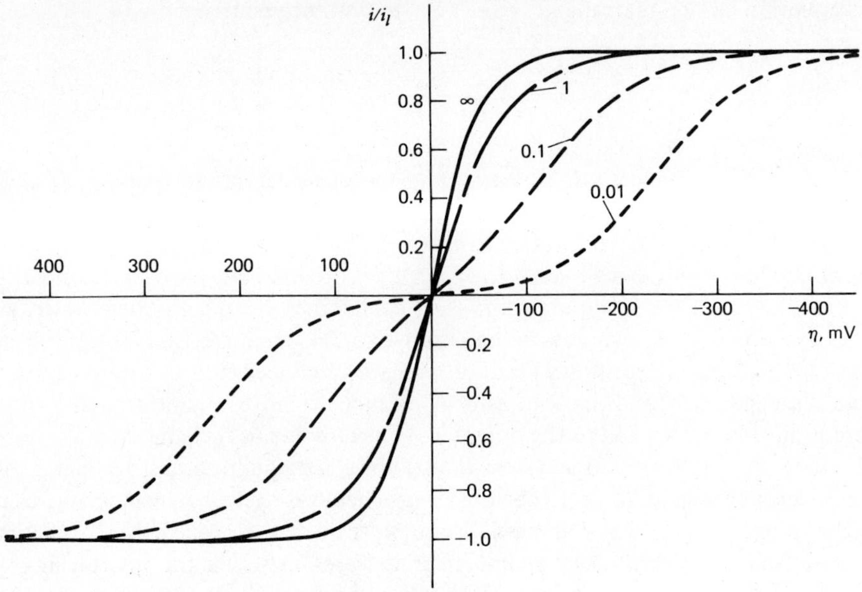

Figure 3.5.6
Relationship between the manifestation of an activation overpotential and net current demands relative to the exchange current. The reaction is $O + ne \rightleftharpoons R$ with $\alpha = 0.5$, $n = 1$, $T = 298$ K, and $i_{l,c} = -i_{l,a} = i_l$. Numbers by curves show i_0/i_l.

In terms of the polarization and mass transfer pseudoresistances defined in equations 1.4.28 and 3.5.13, this equation is

$$\boxed{\eta = -i(R_{ct} + R_{mt,c} + R_{mt,a})} \tag{3.5.36}$$

Here we see very clearly that when i_0 is much greater than the limiting currents, $R_{ct} \ll R_{mt,c} + R_{mt,a}$ and the overpotential, even near E_{eq}, is a concentration overpotential. On the other hand, if i_0 is much less than the limiting currents, then $R_{mt,c} + R_{mt,a} \ll R_{ct}$, and the overpotential near E_{eq} is due to activation of charge transfer. This argument is simply another way of looking at the points made above in Section 3.5.3(a).

In the Tafel regions, other useful forms of (3.5.32) can be obtained. For the cathodic branch at high η values, the anodic contribution is insignificant and (3.5.32) becomes

$$\frac{i}{i_0} = \left(1 - \frac{i}{i_{l,c}}\right) e^{-\alpha n f \eta} \tag{3.5.37}$$

or

$$\eta = \frac{RT}{\alpha nF} \ln \frac{i_0}{i_{l,c}} + \frac{RT}{\alpha nF} \ln \frac{(i_{l,c} - i)}{i} \tag{3.5.38}$$

This equation can be useful for obtaining kinetic parameters for systems in which the normal Tafel plots are complicated by mass transfer effects.

3.6 MULTISTEP MECHANISMS (9, 11, 12, 20, 21, 34)

The foregoing sections have concentrated on the potential dependences of the forward and reverse rate constants governing the simple one-step electrode reaction $O + ne \rightleftharpoons R$. By restricting our view in this way we have achieved a qualitative and quantitative understanding of the major features of electrode kinetics. Also, we have developed a set of relations that we can expect to fit a number of real chemical systems, for example,

$$Fe(CN)_6^{3-} + e \rightleftharpoons Fe(CN)_6^{4-} \tag{3.6.1}$$

$$Tl^+ + e \overset{Hg}{\rightleftharpoons} Tl(Hg) \tag{3.6.2}$$

$$\text{Anthracene} + e \rightleftharpoons \text{Anthracene}^{\overline{\cdot}} \tag{3.6.3}$$

On the other hand, we must now recognize that most electrode processes are mechanisms of several steps. For example, the important reaction

$$2H^+ + 2e \rightleftharpoons H_2 \tag{3.6.4}$$

even without consideration of experimental evidence, must clearly involve several elementary reactions. The hydrogen nuclei are separated in the oxidized form and combined by reduction. Somehow during reduction there must be sequential charge transfers and some chemical means for linking the two nuclei. Consider also the reduction

$$Sn^{4+} + 2e \rightleftharpoons Sn^{2+} \tag{3.6.5}$$

Is it realistic to regard two electrons as tunnelling simultaneously through the interface? Or must we regard the reduction and oxidation sequences as two one-electron processes proceeding through the ephemeral intermediate Sn^{3+}? Another case that looks simple at first glance is the deposition of silver:

$$Ag^+ + e \rightleftharpoons Ag \tag{3.6.6}$$

from a potassium nitrate electrolyte. However, there is evidence that this process involves at least a charge transfer step, which in reduction creates an adsorbed silver atom (adatom), and a crystallization step in which the adatom migrates across the surface until it finds a vacant lattice site. Electrode processes may also involve adsorption and desorption kinetics of primary reactants, intermediates, and products.

It is amply clear now that electrode reactions generally can be expected to show complex behavior, and for each mechanistic sequence one would obtain a different theoretical linkage between current and potential. That relation would have to take into account the potential dependences of all the steps and the surface concentrations of all the intermediates in addition to those of the primary reactants and products. A great deal of effort has been spent in developing such relations for important reaction sequences, and excellent expositions are available in the literature (5, 7–12, 20, 21, 34). We will not delve into specific cases here, except in Problems 3.8 and 3.9.

Nevertheless, we will note that complex mechanisms often show simple behavior and can be treated by the relations derived above. In one such case, we may find that every step in the mechanism is very facile with respect to the experimental time scale, so that no activation is needed. If so, the system is reversible, and (3.5.31) or an appropriate modification can be used. This situation applies, for example, to the polarographic reduction of the ethylenediamine (en) complex of cadmium (II):

$$Cd(en)_3^{2+} + 2e \overset{Hg}{\rightleftharpoons} Cd(Hg) + 3en \tag{3.6.7}$$

A second simplified situation is that in which a single step of the mechanism is much more sluggish than all the others, and the rate at which it operates controls the rate of the overall electrode reaction. It is therefore a *rate-determining step*. If it is a charge transfer reaction, we would expect the current and potential to be linked by relations like (3.3.14) and (3.5.10). However, they must be modified to allow a distinction between the overall n value of the electrode process, that is, the total number of electrons involved in converting one molecule of primary reactant into ultimate product, and the n value of the rate-determining step. The former quantity is usually designated as n, and the latter is n_a. Of course, $n_a \leq n$. For this kind of situation, the current-potential characteristic is

$$i = nFAk^0[C_O(0, t)\, e^{-\alpha n_a f(E - E^{0'})} - C_R(0, t)\, e^{(1-\alpha) n_a f(E - E^{0'})}] \tag{3.6.8}$$

The parameter n appears as a multiplier of the rate because the measured current is proportional to the total number of electrons per molecule of reactant converted, and n_a appears in the exponentials, because they express activation factors for the single rate-determining step. Many examples of this kind of behavior exist in the literature; one is the polarographic reduction of chromate in 0.1 M NaOH:

$$CrO_4^{2-} + 2H_2O + 3e \rightarrow CrO_2^- + 4OH^- \tag{3.6.9}$$

Despite the obvious mechanistic complexity of this system, it behaves as though it has a single, totally irreversible rate-determining reaction in which n_a is probably unity.

In closing this section, we note also that a complete electrode reaction may involve homogeneous chemistry in addition to the interfacial steps. Although the rate constants of these reactions are not dependent on potential, they affect the overall current-potential characteristic by their impact on the surface concentrations of species that are active at the interface. Some of the most interesting applications of electroanalytical techniques have been aimed at unraveling the homogeneous chemistry following the electrochemical production of reactive species, such as free radicals. Chapter 11 is devoted to the techniques for carrying out such studies.

3.7 MICROSCOPIC THEORIES OF CHARGE-TRANSFER KINETICS

What we have elaborated here is a generalized theory of electrode kinetics based on macroscopic concepts. We have described the basic features of behavior without attempting to make predictions about the impact of microscopic details such as the chemical structures of the electroreactants on the kinetic performance. Those specific aspects are lumped into the constants k^0 and α, which we treat as phenomenological parameters evaluated by experiment.

A great deal of work has gone into theoretical efforts to describe the effects of molecular structure and environment on k^0 and α. The goal, of course, is to make predictions that can be tested by experiments, so that we can understand the fundamental structural and environmental factors that cause reactions to be kinetically facile or sluggish. Then, we would have a firmer basis for designing superior new systems for many scientific and technological ends.

Major contributions in this area have been made by Marcus (35–38), Hush (39–41), Levich (42), Dogonadze (43), and others. Several comprehensive reviews are available (37, 42–44). The interested reader should refer to them for details. A full discussion of these theories is far outside the scope of this work, but we will consider two important aspects that can aid an intuitive grasp of the microscopic basis for charge transfer processes.

Electron transfer reactions, whether homogeneous or heterogeneous, are radiationless electronic rearrangements of a reacting complex. There are, accordingly, many common elements between theories of electron transfer and treatments of radiationless deactivation in excited molecules (37, 42–46). Since the transfer is radiationless, the electron must move from an initial state (on an electrode or a reductant) to a receiving state (on another solvated species or on an electrode) of the same energy. This demand for isoenergetic electron transfer is a fundamental aspect with important consequences throughout electrochemistry.

For example, an activation overpotential for reduction can sometimes be ascribed to a need to adjust the electrode potential so that the energies of the available electrons on the electrode match the energies where the density of possible receiving states (number of states per unit of energy) is high. Figure 3.7.1 is an illustration of such a case. Reduction would be allowed energetically at the potential corresponding to the lowest empty states, but it would be much faster at more negative potentials where many more receiving states are available. After the transfer, the electron could drop

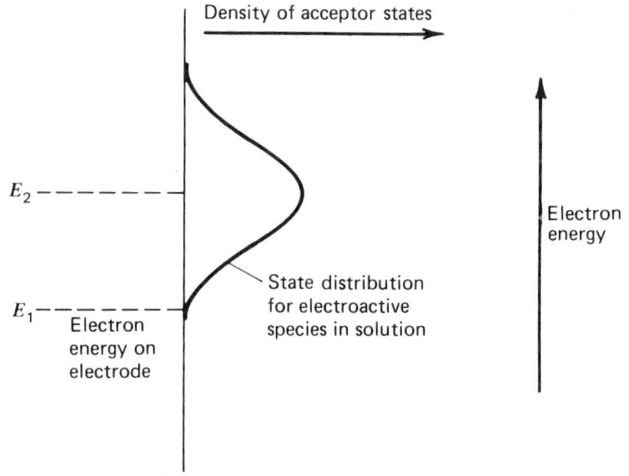

Figure 3.7.1

The effect of acceptor state density on the rate of electron transfer. The horizontal axis on the solution side shows the density of states. Reduction is possible at the potential E_1, but it may occur more rapidly at the more negative value E_2, where a higher density of acceptor states exists.

down into the lower states on the solute, so that the same final state is reached by either route. Ideas like these play a big role in the interpretation of electrochemistry at a semiconductor, which differs significantly from a metal in having a large range of energy (the *band gap*) over which the electrode has few, if any, states capable of mediating electron transfer (47–49).

A second important aspect of most microscopic theories of electron transfer is an assumption that the reactants and products do not change their configurations during the actual act of transfer. This idea is based essentially on the Franck-Condon principle (50), which says, in part, that nuclear momenta and positions do not change on the time scale of electronic transitions. If that tenet does apply here, then one must impose the condition that the reactants and products share a common nuclear configuration at the moment of transfer. Thus the rate constant for electron transfer has to include a factor accounting for the frequency with which a reactant molecule becomes distorted into a configuration that lies also on the energy surface describing the product. This configuration may be difficult to achieve if the reactant and product are structurally very different, and the rate constant for electron transfer may be correspondingly small. A complementary idea underlies the rationale for the very fast charge transfer rates observed for couples involving aromatic hydrocarbons and their radical ions (17–19). In those cases, very small structural changes arise upon oxidation and reduction. Accounting quantitatively for structural effects is complex, and differences in the approach to this aspect comprise some of the most important distinctions among the various microscopic theories.

3.8 REFERENCES

1. W. C. Gardiner, Jr., "Rates and Mechanisms of Chemical Reactions," Benjamin, New York, 1969.
2. H. S. Johnston, "Gas Phase Reaction Rate Theory," Ronald, New York, 1966.
3. S. Glasstone, K. J. Laidler, and H. Eyring, "Theory of Rate Processes," McGraw-Hill, New York, 1941.
4. J. Tafel, *Z. Physik. Chem.*, **50A**, 641 (1905).
5. P. Delahay, "Double Layer and Electrode Kinetics," Wiley-Interscience, New York, 1965, Chap. 7.
6. P. Delahay, "New Instrumental Methods in Electrochemistry," Wiley-Interscience, New York, 1954, Chap. 2.
7. B. E. Conway, "Theory and Principles of Electrode Processes," Ronald, New York, 1965, Chap. 6.
8. K. J. Vetter, "Electrochemical Kinetics," Academic Press, New York, 1967, Chap. 2.
9. J. O'M. Bockris and A. K. N. Reddy, "Modern Electrochemistry," Vol. 2, Plenum, New York, 1970, Chap. 8.
10. T. Erdey-Gruz, "Kinetics of Electrode Processes," Wiley-Interscience, New York, 1972, Chap. 1.

11. H. R. Thirsk, "A Guide to the Study of Electrode Kinetics," Academic, New York, 1972. Chap. 1.
12. W. J. Albery, "Electrode Kinetics," Clarendon, Oxford, 1975.
13. J. A. V. Butler, *Trans. Faraday Soc.*, **19**, 729, 734 (1924).
14. T. Erdey-Gruz and M. Volmer, *Z. Physik. Chem.*, **150A**, 203 (1930).
15. J. E. B. Randles, *Trans. Faraday Soc.*, **48**, 828 (1952).
16. C. N. Reilley, in "Treatise on Analytical Chemistry," Part I, Vol. 4, I. M. Kolthoff and P. J. Elving, Eds., Wiley-Interscience, 1963, Chap. 42.
17. M. E. Peover, *Electroanal. Chem.*, **2**, 1 (1967).
18. N. Koizumi and S. Aoyagui, *J. Electroanal. Chem.*, **55**, 452 (1974).
19. H. Kojima and A. J. Bard, *J. Am. Chem. Soc.*, **97**, 6317 (1975).
20. K. J. Vetter, *op. cit.*, Chap. 4.
21. T. Erdey-Gruz, *op. cit.*, Chap. 4.
22. P. Delahay, "Double Layer and Electrode Kinetics," *op. cit.*, Chap. 10.
23. N. Tanaka and R. Tamamushi, *Electrochim. Acta*, **9**, 963 (1964).
24. B. E. Conway, "Electrochemical Data," Elsevier, Amsterdam, 1952.
25. R. Parsons, "Handbook of Electrochemical Data," Butterworths, London, 1959.
26. A. J. Bard and H. Lund, "Encyclopedia of the Electrochemistry of the Elements," Marcel Dekker, New York, 1973–1980.
27. R. Parsons, *Trans. Faraday Soc.*, **47**, 1332 (1951).
28. J. O'M. Bockris, *Mod. Asp. Electrochem.*, **1**, 180 (1954).
29. D. M. Mohilner and P. Delahay, *J. Phys. Chem.*, **67**, 588 (1963).
30. P. Van Rysselbergh, *Electrochim. Acta*, **8**, 583, 709 (1963).
31. E. Gileadi, E. Kirowa-Eisner, and J. Penciner, "Interfacial Electrochemistry," Addison-Wesley, Reading, Mass., 1975, pp. 60–75.
32. K. J. Vetter and G. Manecke, *Z. Physik. Chem.* (Leipzig), **195**, 337 (1950).
33. P. A. Allen and A. Hickling, *Trans. Faraday Soc.*, **53**, 1626 (1957).
34. P. Delahay, "Double Layer and Electrode Kinetics," *op. cit.*, Chaps. 8–10.
35. R. A. Marcus, *J. Chem. Phys.*, **24**, 4966 (1956).
36. R. A. Marcus, *ibid.*, **43**, 679 (1965).
37. R. A. Marcus, *Ann. Rev. Phys. Chem.*, **15**, 155 (1964).
38. R. A. Marcus, *Electrochim. Acta*, **13**, 955 (1968).
39. N. S. Hush, *J. Chem. Phys.*, **28**, 962 (1958).
40. N. S. Hush, *Trans. Faraday Soc.*, **57**, 557 (1961).
41. N. S. Hush, *Electrochim. Acta*, **13**, 1005 (1968).
42. V. G. Levich, *Adv. Electrochem. Electrochem. Eng.*, **4**, 249 (1966) and references cited therein.
43. R. R. Dogonadze in "Reactions of Molecules at Electrodes," N. S. Hush, Ed., Wiley-Interscience, New York, 1971, Chapter 3 and references cited therein.
44. P. P. Schmidt in "Electrochemistry," A Specialist Periodical Report, Vols. 5 and 6, H. R. Thirsk, Senior Reporter, The Chemical Society, London, 1977 and 1978.
45. R. P. Van Duyne and S. F. Fischer, *Chem. Phys.*, **5**, 183 (1974).
46. S. F. Fischer and R. P. Van Duyne, *Chem. Phys.*, **26**, 9 (1977).

47. H. Gerischer, *Adv. Electrochem. Electrochem. Eng.*, **1**, 139 (1961).
48. H. Gerischer in "Physical Chemistry: An Advanced Treatise," Vol. 9A, H. Eyring, D. Henderson, and W. Jost, Eds., Academic Press, New York, 1970.
49. V. A. Myamlin and Yu V. Pleskov, "Electrochemistry of Semiconductors," Plenum Press, New York, 1967.
50. W. Kauzmann, "Quantum Chemistry," Academic Press, New York, 1957, pp. 664–665.

3.9 PROBLEMS

3.1 Consider the electrode reaction $O + ne \rightleftharpoons R$. Under the conditions that $C_R^* = C_O^* = 1\ mM$, $k^0 = 10^{-7}$ cm/sec, $\alpha = 0.3$, and $n = 1$ T = 298K:
(a) Calculate the exchange current density, $j_0 = i_0/A$, in $\mu A/cm^2$.
(b) Draw a current density-overpotential curve for this reaction for current densities up to 600 μA cm^2 anodic and cathodic. Neglect mass transfer effects.
(c) Draw $\log|j| - \eta$ curves (Tafel plots) for the current ranges in (b).

3.2 Consider one-electron electrode reactions for which $\alpha = 0.50$ and $\alpha = 0.10$. Calculate the relative error in current resulting from the use in each case of:
(a) The linear $i - \eta$ characteristic for overpotentials of 10, 20, and 50 mV.
(b) The Tafel (completely irreversible) relationship for overpotentials of 50, 100, and 200 mV.

3.3 According to G. Scherer and F. Willig [*J. Electroanal. Chem.*, **85**, 77 (1977)] the exchange current density j_0 for $Pt/Fe(CN)_6^{3-}$ (2.0 mM), $Fe(CN)_6^{4-}$ (2.0 mM), NaCl (1.0 M) at 25 °C is 2.0 mA/cm². The transfer coefficient α for this system is about 0.50. Calculate (a) the value of k^0; (b) j_0 for a solution 1 M each in the two complexes; (c) the charge transfer resistance of a 0.1 cm² electrode in a solution 10^{-4} M each in ferricyanide and ferrocyanide.

3.4 Berzins and Delahay [*J. Am. Chem. Soc.*, **77**, 6448 (1955)] studied the reaction

$$Cd^{2+} + 2e \overset{Hg}{\rightleftharpoons} Cd(Hg)$$

and obtained the following data with $C_{Cd(Hg)} = 0.40\ M$:

$C_{Cd^{2+}}$ (mM)	1.0	0.50	0.25	0.10
j_0 (mA/cm²)	30.0	17.3	10.1	4.94

(a) Calculate α.
(b) Calculate k^0.

3.5 The current-potential relationship near $E^{0'}$ for

$$Sn^{4+} + 2e \rightleftharpoons Sn^{2+}$$

looks like the curve in Figure 3.9.1 when both Sn^{2+} and Sn^{4+} are present at 1 mM. Explain its shape. Make an estimate of the exchange current relative to the limiting current. Sketch the curve expected for 0.5 mM Sn^{2+} and 2 mM Sn^{4+}

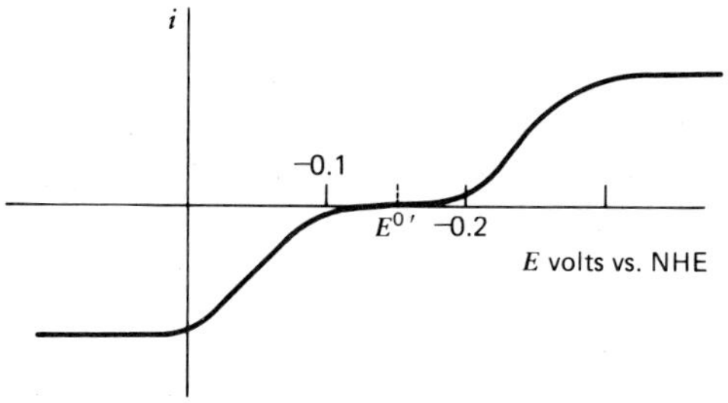

Figure 3.9.1

3.6 (a) Show that for a first-order homogeneous reaction:

$$\mathrm{A} \xrightarrow{kf} \mathrm{B}$$

the average lifetime of A is $1/k_f$.

(b) Derive an expression for the average lifetime of the species O when it undergoes the heterogeneous reaction:

$$\mathrm{O} + ne \xrightarrow{kf} \mathrm{R}$$

Note that only species within distance d of the surface can react. Consider a hypothetical system in which the solution phase extends only d (perhaps 10Å) from the surface.

(c) What value of k_f would be needed for a lifetime of 1 msec? Are lifetimes as short as 1 nsec possible?

3.7 Discuss the mechanism by which the potential of a platinum electrode becomes poised by immersion into a solution of Fe(II) and Fe(III) in 1 M HCl. Approximately how much charge is required to shift the electrode potential by 100 mV? Why does the potential become uncertain at low concentrations of Fe(II) and Fe(III), even if the ratio of their concentrations is held near unity? Does this experimental fact reflect thermodynamic considerations? How well do your answers to these issues apply to the establishment of potential at an ion-selective electrode?

3.8 A complex overall electrode process, such as the example in the next problem, may involve several steps, one of which is rate-determining. Consider the general overall reaction:

$$\nu_{\mathrm{O}}\mathrm{O} + \nu_{\mathrm{A}}\,\mathrm{A} + ne \rightleftharpoons (-\nu_{\mathrm{R}})\,\mathrm{R} + (-\nu_{\mathrm{Z}})\mathrm{Z}$$

where O and R are the oxidized and reduced species in the O/R couple, and A and Z are auxiliary reagents, such as acids, bases, or complexants. The ν's are signed *stoichiometric coefficients* (positive for oxidation products, negative for reduction products). A Nernst equation can be written for the overall process at equilibrium. Now consider the rate-determining steps of the redox reactions to be

$$z_{o,\mathrm{O}}\mathrm{O} + z_{o,\mathrm{A}}\mathrm{A} + z_{o,\mathrm{Z}}\mathrm{Z} + ne \rightleftharpoons z_{r,\mathrm{R}}\mathrm{R} + z_{r,\mathrm{A}}\mathrm{A} + z_{r,\mathrm{Z}}\mathrm{Z}$$

where the z's are *reaction orders* for oxidation and reduction. Here we consider a special case in which the n value for the rate-determining step is the same as that for the overall process. Also, we assume that the reaction above is rate-determining in both directions. However, the case is general in allowing A and Z to participate as reactants in both oxidation and reduction. Derive an expression equivalent to (3.3.14) for this situation; then derive two expressions for the exchange current by using the anodic and cathodic component currents. Show that

$$\frac{\partial \log i_0}{\partial \log C_A^*} = z_{o,A} - \alpha\nu_A$$

or

$$\frac{\partial \log i_0}{\partial \log C_A^*} = z_{r,A} + (1 - \alpha)\nu_A$$

Show also that these are equivalent expressions. Describe a set of experiments that could be used to obtain the reaction orders. Is a negative reaction order meaningful?

3.9 In ammoniacal solutions ($[NH_3] \sim 0.05\ M$), Zn(II) is primarily in the form of the complex ion $Zn(NH_3)_3(OH)^+$ [hereafter referred to as Zn(II)]. In studying the electroreduction of this compound to zinc amalgam at a mercury cathode, Gerisher [*Z. Physik. Chem.*, **202**, 302 (1953)] found that

$$\frac{\partial \log i_0}{\partial \log [\mathrm{Zn(II)}]} = 0.41 \pm 0.03 \qquad \frac{\partial \log i_0}{\partial \log [\mathrm{NH_3}]} = 0.65 \pm 0.03$$

$$\frac{\partial \log i_0}{\partial \log [\mathrm{OH^-}]} = -0.28 \pm 0.02 \qquad \frac{\partial \log i_0}{\partial \log [\mathrm{Zn}]} = 0.57 \pm 0.03$$

where [Zn] refers to a concentration in the amalgam. For this electrode reaction, $\alpha = 0.58$.

(a) Give the equation for the overall reaction, and define $\nu_{Zn(II)}$, ν_{NH_3}, ν_{OH^-}, and ν_{Zn}.

(b) Calculate the z_o and z_r values for all species.

(c) Identify the oxidized and reduced zinc species involved in the rate-determining step.

(d) Write chemical equations to give a mechanism consistent with the data.

3.10 The following data were obtained for the reduction of species R to R^- in a stirred solution at a 0.1 cm² electrode; the solution contained 0.01 M R and 0.01 M R^- T = 298K.

η (mV):	−100	−120	−150	−500	−600
i (μA):	45.9	62.6	100	965	965

Calculate: i_0, k^0, α, R_{ct}, i_l, m_O, R_{mt}

3.11 From results in Figure 3.5.5 for 10^{-2} M Mn(III) and 10^{-2} M Mn(IV), estimate j_0 and k^0. What is the predicted j_0 for a solution 1 M in both Mn(III) and Mn(IV)?

chapter 4

Mass Transfer by Migration and Diffusion

4.1 DERIVATION OF GENERAL MASS TRANSFER EQUATION

In this section we discuss the general partial differential equations that govern mass transfer; these will be used frequently in subsequent chapters for the derivation of equations appropriate to the different electrochemical techniques. As discussed in Section 1.4, mass transfer in solution occurs because of a gradient in electrochemical potential $\bar{\mu}$ (i.e., by diffusion and migration) and by convection. Consider a section of solution where, for a certain species j at two points in the solution, r and s (an infinitesimal distance from one another), $\bar{\mu}_j(r) \neq \bar{\mu}_j(s)$ (Figure 4.1.1). This difference of $\bar{\mu}_j$ over a distance (or gradient of electrochemical potential) can arise because of differences of concentration (or activity) of species j at r and s (a concentration gradient) or because of differences of ϕ at r and s (an electric field or potential gradient). In general, a flux of species j will occur to alleviate this difference of $\bar{\mu}_j$. The flux, $\mathbf{J}_j$ (mol sec^{-1} cm^{-2}), is proportional to the gradient, **grad** or ∇, of $\bar{\mu}_j$:

$$\mathbf{J}_j \propto \mathbf{grad}\,\bar{\mu}_j \qquad \text{or} \qquad \nabla\bar{\mu}_j \tag{4.1.1}$$

where **grad** or ∇ is a vector operator. For linear (one-dimensional) mass transfer $\nabla = \mathbf{i}(\partial/\partial x)$ (where $\mathbf{i}$ is the unit vector along the axis and x is distance), and for rectangular mass transfer

$$\nabla = \mathbf{i}\frac{\partial}{\partial x} + \mathbf{j}\frac{\partial}{\partial y} + \mathbf{k}\frac{\partial}{\partial z} \tag{4.1.2}$$

When the constant of proportionality is added to (4.1.1), we obtain

$$\mathbf{J}_j = -\left(\frac{C_j D_j}{RT}\right)\nabla\bar{\mu}_j \tag{4.1.3}$$

or for linear mass transfer,

$$J_j(x) = -\left(\frac{C_j D_j}{RT}\right)\frac{\partial\bar{\mu}_j}{\partial x} \tag{4.1.4}$$

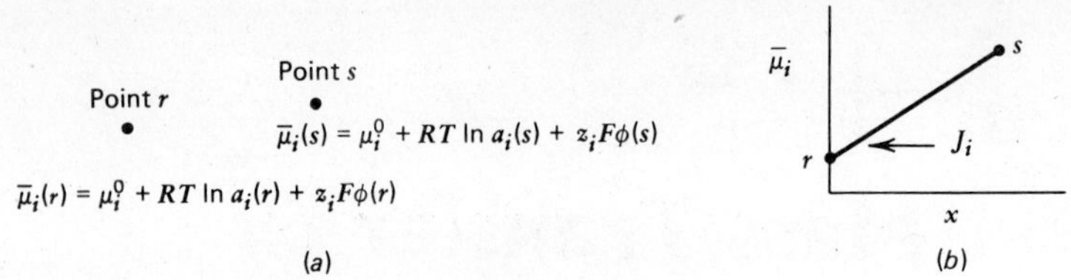

Figure 4.1.1
Electrochemical potential gradient.

The minus sign in (4.1.3) arises because the direction of the flux opposes the direction of increasing $\bar{\mu}_j$. If, in addition to this $\bar{\mu}$ gradient, the solution is moving so that an element of solution [with j at a concentration $C_j(s)$] moves from s with a velocity, $\mathbf{v}$, an additional term is added to the flux equation, yielding

$$\mathbf{J}_j = -\left(\frac{C_j D_j}{RT}\right) \nabla\bar{\mu}_j + C_j\mathbf{v} \tag{4.1.5}$$

or for linear mass transfer

$$J_j(x) = -\left(\frac{C_j D_j}{RT}\right)\left(\frac{\partial\bar{\mu}_j}{\partial x}\right) + C_j v(x) \tag{4.1.6}$$

For $a_j \simeq C_j$, we obtain the *Nernst-Planck* equations, which can be written as

$$J_j(x) = -\left(\frac{C_j D_j}{RT}\right)\left[\frac{\partial}{\partial x}(RT \ln C_j) + \frac{\partial}{\partial x}(z_j F\phi)\right] + C_j v(x) \tag{4.1.7}$$

$$\boxed{J_j(x) = -D_j \frac{\partial C_j(x)}{\partial x} - \frac{z_j F}{RT} D_j C_j \frac{\partial\phi(x)}{\partial x} + C_j v(x)} \tag{4.1.8}$$

or in general

$$\boxed{\mathbf{J}_j = -D_j \nabla C_j - \frac{z_j F}{RT} D_j C_j \nabla\phi + C_j\mathbf{v}} \tag{4.1.9}$$

In this chapter we are concerned with systems in which convection is absent. Under these conditions, that is, in an unstirred or stagnant solution with no density gradients, the solution velocity $\mathbf{v}$ is zero, and the general flux equation for species j, (4.1.9), becomes

$$\mathbf{J}_j = -D_j \nabla C_j - \frac{z_j F}{RT} D_j C_j \nabla\phi \tag{4.1.10}$$

(Solutions with convective mass transfer will be treated in Chapter 8.) For linear mass transfer (4.1.10) is given by

$$J_j(x) = -D_j\left(\frac{\partial C_j(x)}{\partial x}\right) - \frac{z_j F}{RT} D_j C_j \left(\frac{\partial\phi(x)}{\partial x}\right) \tag{4.1.11}$$

where the terms on the right-hand side of (4.1.11) represent the contributions of diffusion and migration, respectively, to the total mass transfer. If species j is charged, then the flux J_j is equivalent to a current density. Let us consider a linear mass-flow system with a cross-sectional area A normal to the axis of mass flow. Then J_j (mol sec^{-1} cm^{-2}) is equal to $-i_j/z_jFA$ (C/sec per (C mol^{-1} cm^2)), where i_j is the current component at any value of x arising from a flow of species j. Equation 4.1.11 can then be written:

$$-J_j = \frac{i_j}{z_jFA} = \frac{i_{d,j}}{z_jFA} + \frac{i_{m,j}}{z_jFA} \tag{4.1.12}$$

with

$$\frac{i_{d,j}}{z_jFA} = D_j \frac{\partial C_j}{\partial x} \tag{4.1.13}$$

$$\frac{i_{m,j}}{z_jFA} = \frac{z_jFD_j}{RT} C_j \frac{\partial \phi}{\partial x} \tag{4.1.14}$$

where $i_{d,j}$ and $i_{m,j}$ are *diffusion* and *migration currents* of species j, respectively. The factor $|z_j|FD_j/RT$ is the mobility u_j described in Section 2.3, so that

$$\frac{i_{m,j}}{|z_j|FA} = u_jC_j \frac{\partial \phi}{\partial x} \tag{4.1.15}$$

At any location in solution during electrolysis, the total current i is made up of contributions from all species; that is,

$$i = \sum_j i_j \tag{4.1.16}$$

and the current for each species is made up of a diffusional component arising from a concentration gradient and a migrational component arising from a potential gradient.

We now discuss migration and diffusion in electrochemical systems in a little more detail. Notice that the concepts and equations derived here date back to at least the work of Planck (1). Further details concerning mass transfer in electrochemical systems can be found in a number of reviews (2–6).

4.2 MIGRATION

The relative contributions of diffusion and migration to the flux of a species (and of the flux of that species to the total current) differ at a given time for different locations in solution. Near the electrode, the electroactive substance is, in general, transported by both processes. The flux of this substance at the electrode surface controls the rate of reaction and, therefore, the faradaic current flowing in the external circuit (see Section 1.3.2). That current can be separated into diffusion and migration currents reflecting the diffusive and migrational components to the flux of the electroactive species at the surface:

$$i = i_d + i_m \tag{4.2.1}$$

Note that i_m and i_d may be in the same or opposite directions, depending on the direction of the electric field and the charge on the electroactive species. Examples of three reductions: of a positively charged, a negatively charged, and an uncharged substance, are shown in Figure 4.2.1. The migrational component will always be in the same direction as i_d for cationic species reacting at cathodes and for anionic species reacting at anodes. It will oppose i_d when anions are reduced at cathodes and when cations are oxidized at anodes.

For many electrochemical studies of chemical systems, the mathematical treatments are simplified if the migrational component to the flux of the electroactive substance is made negligible. We discuss in this section the conditions under which that approximation holds. The topic is discussed in greater depth in references (7–10).

In the bulk solution (away from the electrode), concentration gradients are generally small, and the total current is carried mainly by migration. All charged species contribute. For species j in the bulk region of a linear mass transfer system having a cross-sectional area A, $i_j = i_{m,j}$, or,

$$i_j = |z_j| F A u_j C_j \frac{\partial \phi}{\partial x} \tag{4.2.2}$$

For a linear electric field,

$$\frac{\partial \phi}{\partial x} = \frac{\Delta E}{l} \tag{4.2.3}$$

where $\Delta E/l$ is the gradient (V/cm), arising from the change in potential ΔE over distance l. Thus,

$$i_j = \frac{|z_j| F A u_j C_j \, \Delta E}{l} \tag{4.2.4}$$

The total current in bulk solution, from (4.1.16), is given by

$$i = \sum_j i_j = \frac{FA\,\Delta E}{l} \sum_j |z_j| u_j C_j \tag{4.2.5}$$

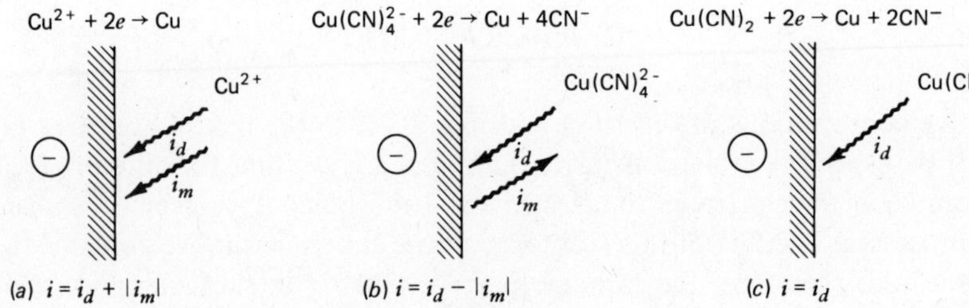

Figure 4.2.1
Examples of electrode reactions with different contributions of the migration current. All represent reductions at a negatively charged electrode. (*a*) Positively charged. (*b*) Negatively charged. (*c*) Uncharged reactants.

The conductance of the solution, $L(\Omega^{-1})$, which is the reciprocal of the resistance, $R(\Omega)$, is given by $i/\Delta E$ (Ohm's law); thus

$$L = \frac{1}{R} = \frac{i}{\Delta E} = \frac{FA}{l} \sum_j |z_j| u_j C_j = \frac{A}{l} \kappa \tag{4.2.6}$$

where κ, the *conductivity* (Ω^{-1} cm^{-1}), is (see also Section 2.3.3)

$$\kappa = F \sum_j |z_j| u_j C_j \tag{4.2.7}$$

Equally, one can write an equation for the solution resistance, in terms of ρ, the *resistivity* (Ω-cm), where $\rho = 1/\kappa$:

$$R = \frac{\rho l}{A} \tag{4.2.8}$$

Thus, in the bulk solution, current is carried by movement of the ions, each contributing to the total current i an amount i_j given by (4.2.4). The fraction of total current that a given ion j carries is t_j, the *transference number* of j, given by (see also equations 2.3.11 and 2.3.18)

$$\boxed{t_j = \frac{i_j}{i} = \frac{|z_j| u_j C_j}{\sum_k |z_k| u_k C_k} = \frac{|z_j| C_j \lambda_j}{\sum_k |z_k| C_k \lambda_k}} \tag{4.2.9}$$

4.2.1 Migration During Electrolysis

While migration carries the current in the bulk solution during electrolysis, in the vicinity of the electrodes concentration gradients of the electroactive species arise and diffusional transport also occurs. Indeed under some circumstances, as we will see, the total flux of electroactive species to the electrode is due to diffusion. To illustrate these effects let us consider the "balance sheet" approach (11) to transport in several examples.

EXAMPLE 4.1 Consider the electrolysis of a solution of hydrochloric acid at platinum electrodes (Figure 4.2.2*a*). Since the equivalent conductance of H^+, λ_+, and of Cl^-, λ_-, relate as $\lambda_+ \simeq 4\lambda_-$, then from (4.2.9), $t_+ = 0.8$ and $t_- = 0.2$. Assume that a total current equivalent to $10e$ per unit time is passed through the cell, producing 5 H_2 molecules at the cathode and 5 Cl_2 molecules at the anode. (Actually some O_2 could also be formed at the anode; for simplicity we neglect this reaction.) The total current is carried in the bulk solution by the movement of 8 H^+ toward the cathode and 2 Cl^- toward the anode (Figure 4.2.2*b*). To maintain a steady current, however, 10 H^+ must be supplied per unit time, so that an additional 2 H^+ diffuse to the electrode, bringing along 2 Cl^- to maintain electroneutrality. Similarly at the anode, to supply 10 Cl^- per unit time, 8 Cl^- must be supplied by diffusion, along with 8 H^+. Thus, the different currents (in arbitrary e-units) are: for H^+, $i_d = 2$, $i_m = 8$; for Cl^-, $i_d = 8$, $i_m = 2$. The total current i is 10. Equation 4.2.1 holds, with migration in this case being in the same direction as diffusion.

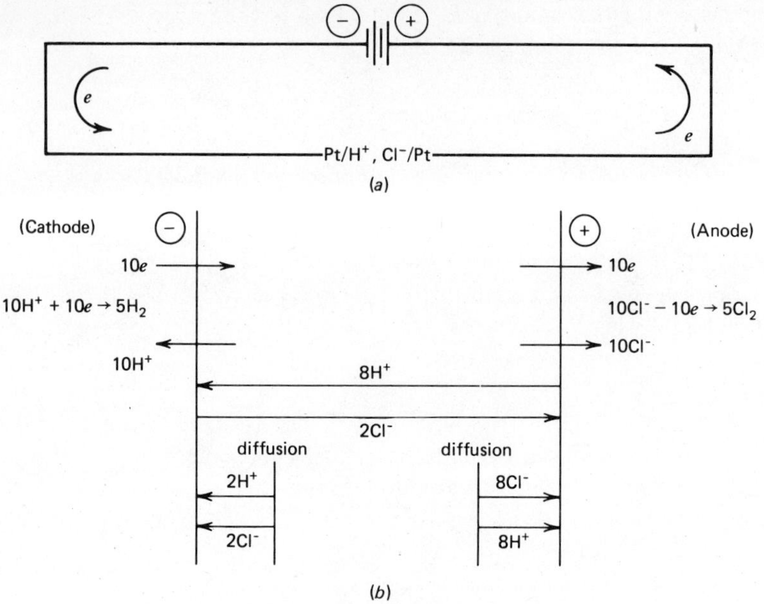

Figure 4.2.2

Balance sheet for electrolysis of hydrochloric acid solution. (*a*) Cell schematic. (*b*) Various contributions to the current assuming 10*e* passed in external circuit.

In general, for mixtures of charged species, the fraction of current carried by the jth species is t_j, so that of the total current i, the amount carried by the jth species is $t_j i$. The number of moles of jth species migrating per second is $t_j i/z_j F$. If the jth species is undergoing electrolysis, the moles electrolyzed per second are $|i|/nF$, while the moles arriving at the electrode per second by migration arc $\pm i_m/nF$, where the positive sign applies to reduction of j, and the negative sign pertains to oxidation. Thus

$$\pm \frac{i_m}{nF} = \frac{t_j i}{z_j F} \tag{4.2.10}$$

or

$$i_m = \pm \frac{n}{z_j} t_j i \tag{4.2.11}$$

Since, from equation 4.2.1,

$$i_d = i - i_m \tag{4.2.12}$$

$$i_d = i\left(1 \mp \frac{nt_j}{z_j}\right) \tag{4.2.13}$$

where the minus sign is used for cathodic currents and the positive sign is used for anodic currents. Note that both i and z_j are signed.

EXAMPLE 4.2 Consider the electrolysis of a solution of 10^{-3} M $Cu(NH_3)_4^{2+}$, 10^{-3} M $Cu(NH_3)_2^{+}$, and 3×10^{-3} M Cl^- in 0.1 M NH_3 at Hg electrodes (Figure

4.2.3*a*). Assuming the limiting equivalent conductances of all ions are equal, that is,

$$\lambda_{Cu(II)} = \lambda_{Cu(I)} = \lambda_{Cl^-} = \lambda \tag{4.2.14}$$

we obtain, using equation 4.2.9: $t_{Cu(II)} = \frac{1}{3}$, $t_{Cu(I)} = \frac{1}{6}$, and $t_{Cl^-} = \frac{1}{2}$. Thus, with an arbitrary current of 6*e* per unit time being passed, the bulk solution migration current is carried by movement of 1 Cu(II) and 1 Cu(I) toward the cathode, and 3 Cl^- toward the anode. The total balance sheet for this system is shown in Figure 4.2.3*b*. Thus, in this case, at the cathode, one-sixth of the current for the electrolysis of Cu(II) is provided by migration and five-sixths by diffusion. In calculating mass transfer of Cu(II), both modes must be considered. The NH_3, being uncharged, does not contribute to the carrying of the current but, in this case, only serves to stabilize the copper species in the +1 and +2 states. The cell resistance of this cell would also be quite large, since the total concentrations of ions in the solution is small.

4.2.2 Effect of Adding Excess Supporting Electrolyte

EXAMPLE 4.3 Let us consider the same cell as in Example 2, except now with the solution containing 0.10 *M* $NaClO_4$ as the supporting electrolyte (Figure 4.2.4*a*).

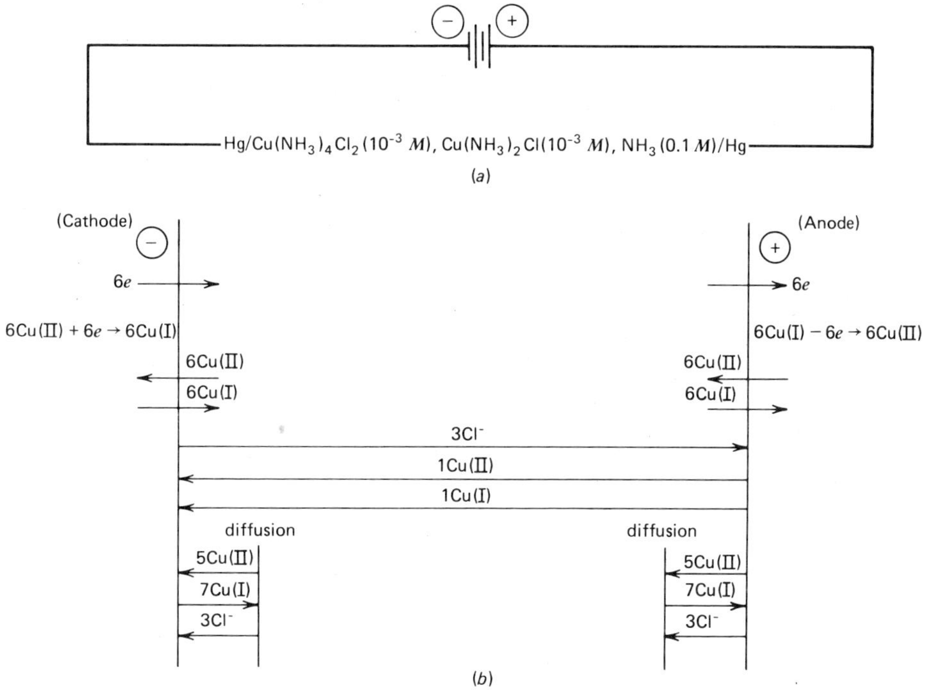

Figure 4.2.3

Balance sheet for electrolysis of Cu(II), Cu(I), NH_3 system. (*a*) Cell schematic. (*b*) Various contributions to the current assuming 6*e* passed in the external circuit; $i = 6$, $n = 1$. For Cu(II) at cathode, $|i_m| = (1/2)(1/3)(6) = 1$ (equation 4.2.11), $i_d = 6 - 1 = 5$ (equation 4.2.12). For Cu(I) at anode, $|i_m| = (1/1)(1/6)(6) = 1$, $i_d = 6 + 1 = 7$.

The Na^+ and ClO_4^- do not participate in the electron transfer reactions; but because their concentrations are high, they carry most of the current in the bulk solution. Assuming that $\lambda_{Na^+} = \lambda_{ClO_4^-} = \lambda$, and that the Cu(II), Cu(I), and Cl^- are at the same concentrations as in Example 4.2, then the following transference numbers are obtained: $t_{Na^+} = t_{ClO_4^-} = 0.485$, $t_{Cu(II)} = 0.0097$, $t_{Cu(I)} = 0.00485$, $t_{Cl^-} = 0.0146$. The balance sheet for this cell (Figure 4.2.4*b*) shows that most of the

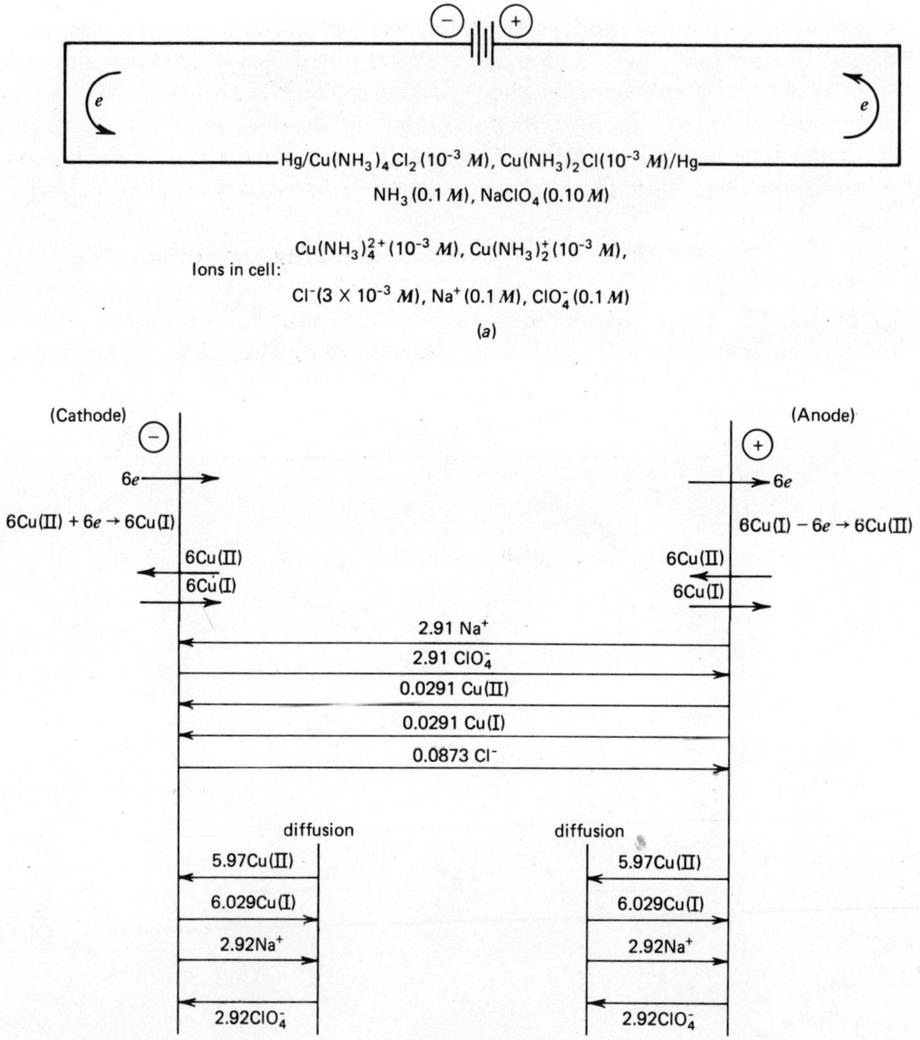

Figure 4.2.4

Balance sheet for system in Figure 4.2.3 containing excess $NaClO_4$ as supporting electrolyte. (*a*) Cell schematic. (*b*) Various contributions to the current assuming $6e$ passed in the external circuit, $i = 6$, $n = 1$. $t_{Cu(II)} = [(2 \times 10^{-3})\lambda/(2 \times 10^{-3} + 10^{-3} + 3 \times 10^{-3} + 0.2)\lambda] = 0.0097$. For Cu(II) at cathode, $|i_m| = (1/2)(0.0097)(6) = 0.03$, $i_d = 6 - 0.03 = 5.97$.

Cu(II) now reaches the cathode by diffusion, and only 0.5% of the total flux is by migration. Thus the addition of an excess of indifferent or nonelectroactive ions (the *supporting electrolyte*) decreases the contribution of migration to mass transfer of the electroactive species and product molecules, and simplifies the mathematical treatment of electrochemical systems by elimination of the $\nabla\phi$ or $\partial\phi/\partial x$ term in the mass transport equations (e.g., equations 4.1.10 and 4.1.11). Moreover, the supporting electrolyte serves the important function of decreasing the cell resistance (and, in analytical applications, it may decrease or eliminate matrix effects).

The effect of supporting electrolyte on the limiting current can be illustrated by the results in Table 4.2.1 showing the limiting current for the reduction of Pb(II) at the dropping mercury electrode (polarography) with different amounts of KNO_3 added to the solution. The current at low concentrations of KNO_3 is appreciably higher than that at high concentrations because at low concentrations Pb(II) carries an appreciable fraction of the total current. Migration of the positively charged Pb(II) species to the negative electrode occurs. At high KNO_3 concentrations, K^+ migration replaces that of the Pb(II) species, and the observed current is essentially a pure diffusion current.

4.3 DIFFUSION

The mathematical equations for diffusion were presented in a phenomenological way in Sections 1.4 and 4.1. Because we will be concerned with mass transfer by diffusion in so many electroanalytical techniques, it is appropriate to take a closer look at the phenomenon of diffusion and the mathematical models describing it (12–14).

Table 4.2.1

Effect of Added KNO_3 on Limiting Current for Reduction of Lead(II)[a]

Added KNO_3, M	Limiting Current, μA
0	17.6
0.0001	16.2
0.0002	15.0
0.0005	13.4
0.001	12.0
0.005	9.8
0.1	8.45
1.0	8.45

[a] The concentration of $PbCl_2$ was 0.95×10^{-3} M. [From J. J. Lingane and I. M. Kolthoff, *J. Am. Chem. Soc.*, **61**, 1045 (1939)].

4.3.1 A Microscopic View—Discontinuous Source Model

Diffusion, which normally leads to the homogenization of a mixture, occurs by a "random walk" process. A simple picture of this process can be obtained by considering a one-dimensional random walk. Consider a molecule constrained to a one-dimensional path, and, buffeted by solvent molecules undergoing Brownian motion, moving in steps of length, l, with one step being made per unit time, τ. We can ask, "Where will the molecule be after a time, t?" We can only answer this by giving the probability of the molecule being at different locations at t.† Alternately we can envision a large number of molecules concentrated in a line at $t = 0$, and ask what the distribution of molecules will be at time t.

The distribution follows the usual probability distribution (Figure 4.3.2). Thus at

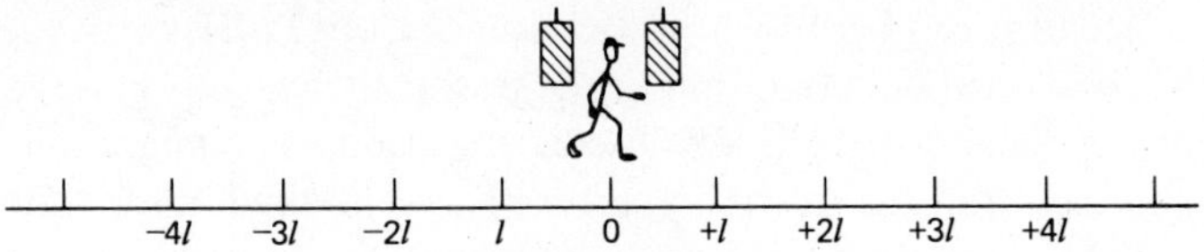

Figure 4.3.1

The one-dimensional random-walk or "drunken sailor" problem.

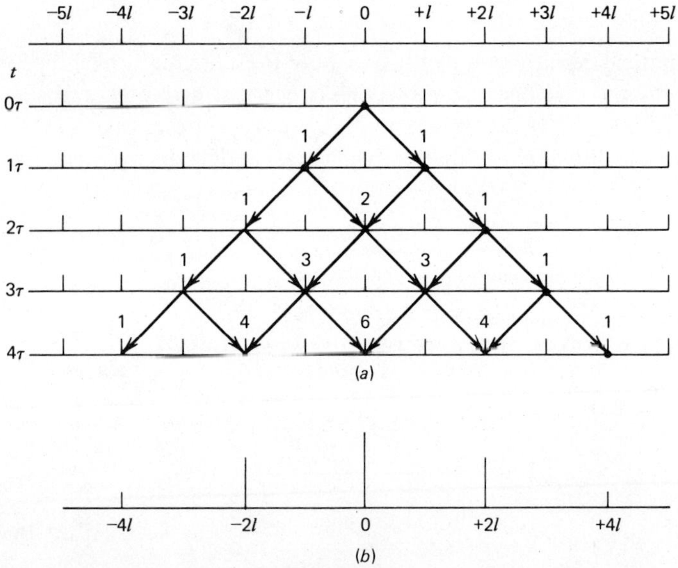

Figure 4.3.2

(a) Probability distribution for one-dimensional random walk problem for zero to four time units. (b) Bar graph showing distribution at $t = 4\tau$. At this time, probability of being at $x = 0$ is 6/16, at $x = \pm 2l$ is 4/16, and at $x = \pm 4l$ is 1/16.

† This is sometimes called the "drunken sailor" problem, where we envision a very drunk sailor emerging from a bar and staggering randomly left and right (with a stagger step size l, one step every τ seconds). What is the probability that the sailor will get down the street a certain distance after a certain time t? (See Figure 4.3.1.)

time τ, it is equally likely that the molecule is at $+l$ and $-l$; and at time 2τ the relative probabilities of being at $+2l$, 0, and $-2l$, are 1, 2, and 1, respectively, etc. The general expression for the probability, $P(m, r)$, that the molecule is at a given location after m time units ($m = t/\tau$) is given by the binomial coefficient

$$P(m, r) = \frac{m!}{r!\,(m - r)!}\left(\frac{1}{2}\right)^m \tag{4.3.1}$$

where the set of locations is defined by $x = (-m + 2r)l$ with $r = 0, 1, \ldots m$. The mean (or average) square displacement of the molecule, $\overline{\Delta^2}$, can be calculated by summing the squares of the displacements and dividing by the total number of possibilities (2^m). The squares of the displacements are used, just as when one obtains the standard deviation in statistics, because movement is possible in both the positive and negative directions, and the sum of the displacements is always zero. This procedure is shown in Table 4.3.1. In general $\overline{\Delta^2}$ is given by

$$\overline{\Delta^2} = ml^2 = \frac{t}{\tau}\,l^2 = 2Dt \tag{4.3.2}$$

where D, the diffusion coefficient, $l^2/2\tau$, is a constant of the experiment related to the step size and step frequency.† It has dimensions of length²/time, or, in electrochemical studies, usually cm²/sec. The root-mean-square displacement at time t is thus

$$\boxed{\bar{\Delta} = \sqrt{2Dt}} \tag{4.3.3}$$

This equation provides a handy rule of thumb for estimating the thickness of a diffusion layer (e.g., how far product molecules have moved, on the average, from an electrode in a certain time). A typical value of D for aqueous solutions is 5×10^{-6} cm²/sec, so that a diffusion layer thickness of 10^{-4} cm is built up in 1 msec, 10^{-3} cm in 0.1 sec, and 10^{-2} cm in 10 sec.

Table 4.3.1

Distributions for a Random Walk Process[a]

t	n[b]	Δ[c]	$\Sigma\,\Delta^2$	$\overline{\Delta^2} = \frac{1}{n}\Sigma\,\Delta^2$
0τ	$1(=2^0)$	0	0	0
1τ	$2(=2^1)$	$\pm l(1)$	$2l^2$	l^2
2τ	$4(=2^2)$	$0(2), \pm 2l(1)$	$8l^2$	$2l^2$
3τ	$8(=2^3)$	$\pm l(3), \pm 3l(1)$	$24l^2$	$3l^2$
4τ	$16(=2^4)$	$0(6), \pm 2l(4), \pm 4l(1)$	$64l^2$	$4l^2$
$m\tau$	2^m		$mnl^2(=m2^ml^2)$	ml^2

[a] l = step size, $1/\tau$ = step frequency, $t = m\tau$ = time interval.
[b] n = total number of possibilities.
[c] Δ = possible positions; relative probabilities are parenthesized.

† This was derived by Einstein in another way in 1905. D is sometimes given as $fl^2/2$, where f is the number of displacements per unit time ($= 1/\tau$).

As $m \to \infty$, a continuous form of equation 4.3.1 can be given. For example, for N_0 molecules located at the origin at $t = 0$, a Gaussian curve will describe the distribution at some later time t. The number of molecules $N(x, t)$ in a segment Δx wide centered on position x is (15)

$$\frac{N(x, t)}{N_0} = \frac{\Delta x}{2\sqrt{\pi D t}} \exp\left(\frac{-x^2}{4Dt}\right) \tag{4.3.4}$$

4.3.2 Fick's Laws of Diffusion

Fick's laws are differential equations describing the flux of a substance and its concentration as functions of time and position. Consider the case of linear (one-dimensional) diffusion. The flux of a substance O at a given location x at a time t, written as $J_O(x, t)$, is the net mass transfer rate of O in units of amount per unit time per unit area (e.g., mol sec^{-1} cm^{-2}). Thus $J_O(x, t)$ represents the number of moles of O that pass a given location per second per cm^2 of area normal to the axis of diffusion. *Fick's first law* states that the flux is proportional to the concentration gradient, $\partial C_O/\partial x$:

$$\boxed{-J_O(x, t) = D_O \frac{\partial C_O(x, t)}{\partial x}} \tag{4.3.5}$$

This equation can be derived from the microscopic model as follows. Consider the location x, and assume $N_O(x)$ molecules are to left of x and $N_O(x + \Delta x)$ molecules are to the right at time t (Figure 4.3.3). During the time increment Δt, by the random walk process, half of these molecules move Δx in either direction, so that the net flux through an area A at x is given by the difference between the number of molecules moving from left to right and those moving right to left:

$$J_O(x, t) = \frac{1}{A} \frac{\dfrac{N_O(x)}{2} - \dfrac{N_O(x + \Delta x)}{2}}{\Delta t} \tag{4.3.6}$$

By multiplying by $\Delta x^2/\Delta x^2$ and noting that the concentration of O is given by $C_O = N_O/A\,\Delta x$, we derive from equation 4.3.6

$$-J_O(x, t) = \frac{\Delta x^2}{2\,\Delta t} \frac{C_O(x + \Delta x) - C_O(x)}{\Delta x} \tag{4.3.7}$$

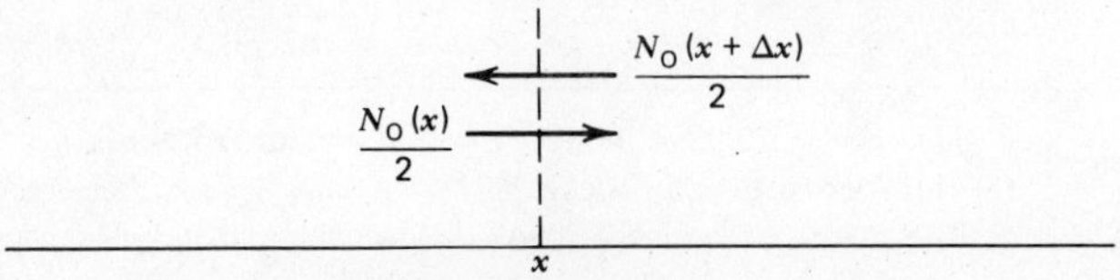

Figure 4.3.3
Fluxes at point x in solution.

From the definition of the diffusion coefficient, (4.3.2), $D_O = \Delta x^2/2\,\Delta t$, and allowing Δx and Δt to approach zero, we obtain (4.3.5).

Fick's second law pertains to the change in concentration of O with time:

$$\frac{\partial C_O(x, t)}{\partial t} = D_O\left(\frac{\partial^2 C_O(x, t)}{\partial x^2}\right) \tag{4.3.8}$$

This equation is derived from the first law as follows. The change in concentration at a location x is given by the differences in flux into and flux out of an element of width dx (Figure 4.3.4).

$$\frac{\partial C_O(x, t)}{\partial t} = \frac{J(x, t) - J(x + dx, t)}{dx} \tag{4.3.9}$$

Note that J/dx has dimensions of (mol $\sec^{-1}$ cm^{-2})/cm or change in concentration per unit time, as required. The flux at $x + dx$ can be given in terms of that at x by the general equation

$$J(x + dx, t) = J(x, t) + \frac{\partial J(x, t)}{\partial x}\,dx \tag{4.3.10}$$

However, from equation 4.3.5 we obtain

$$-\frac{\partial J(x, t)}{\partial x} = \frac{\partial}{\partial x} D_O \frac{\partial C_O(x, t)}{\partial x} \tag{4.3.11}$$

Combination of equations 4.3.9 to 4.3.11 yields

$$\frac{\partial C_O(x, t)}{\partial t} = \left(\frac{\partial}{\partial x}\right)\left[D_O\left(\frac{\partial C_O(x, t)}{\partial x}\right)\right] \tag{4.3.12}$$

When D_O is not a function of x, (4.3.8) obtains. We will have many occasions in future chapters to solve (4.3.8) under a variety of boundary conditions. Solutions of this equation yield concentration profiles, $C_O(x, t)$, while the flux at the electrode surface will give the current.

The general formulation of Fick's second law, for any geometry, is

$$\frac{\partial C_O}{\partial t} = D_O \nabla^2 C_O \tag{4.3.13}$$

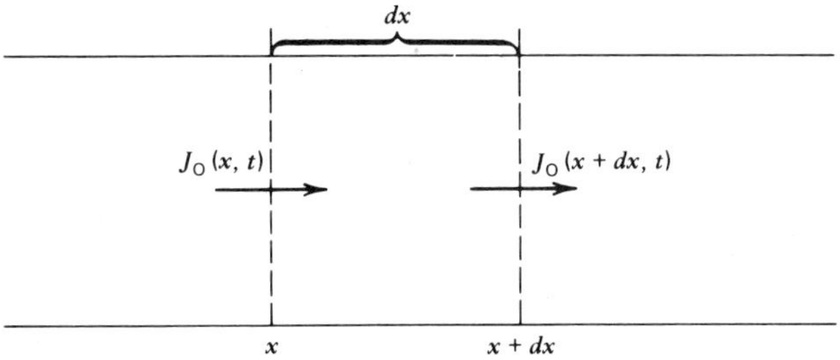

Figure 4.3.4

Fluxes into and out of an element at x.

Table 4.3.2
Forms of the Laplacian Operator for Different Geometries

Type	Variables	∇^2	Example
Linear	x	$\partial^2/\partial x^2$	Planar disk electrode
Rectangular	x, y, z	$\partial^2/\partial x^2 + \partial^2/\partial y^2 + \partial^2/\partial z^2$	Cube-shaped electrode
Spherical	r	$\partial^2/\partial r^2 + (2/r)(\partial/\partial r)$	Hanging drop electrode
Cylindrical (axial)	r	$\partial^2/\partial r^2 + (1/r)(\partial/\partial r)$	Wire electrode

where ∇^2 is the Laplacian operator. Forms of ∇^2 for different geometries are given in Table 4.3.2. Thus for problems involving a planar electrode, the linear diffusion equation (4.3.8) is appropriate (Figure 4.3.5*a*). For problems involving a spherical electrode, such as the hanging mercury drop electrode, the spherical form of the diffusion equation must be employed (Figure 4.3.5*b*).

$$\boxed{\frac{\partial C_O(r, t)}{\partial t} = D_O\left(\frac{\partial^2 C_O(r, t)}{\partial r^2} + \frac{2}{r}\frac{\partial C_O(r, t)}{\partial r}\right)} \qquad (4.3.14)$$

The difference between the linear and spherical equations arises because spherical diffusion takes place through an increasing area as r increases.

If O represents the electroactive substance that undergoes the electrode reaction

$$O + ne \rightarrow R \qquad (4.3.15)$$

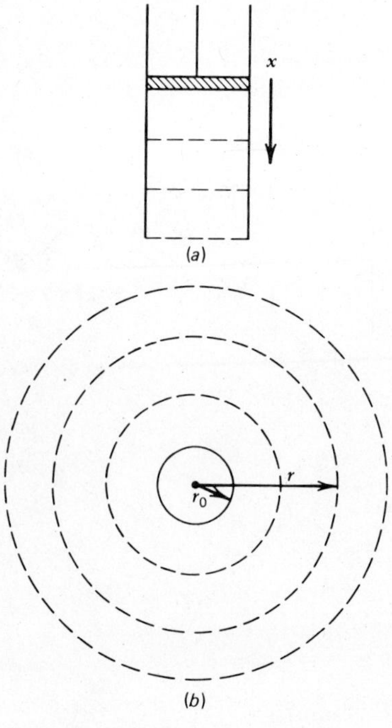

Figure 4.3.5
Types of diffusion occurring at different electrodes.
(*a*) Linear diffusion to a planar electrode.
(*b*) Spherical diffusion to a hanging drop electrode.

(and assuming no other electrode reactions occur), the current density, i/A, is related to the flux of O at the electrode surface ($x = 0$), $J_O(0, t)$, by the equation

$$-J_O(0, t) = \frac{i}{nFA} = D_O\left[\frac{\partial C_O(x, t)}{\partial x}\right]_{x=0} \tag{4.3.16}$$

because the total number of electrons transferred at the electrode in a unit time must be proportional to the quantity of O reaching the electrode in that time period. If several electroactive species exist in the solution, then the current is related to the sum of the fluxes of all these at the electrode surface. Thus for q reducible species,

$$\frac{i}{FA} = \sum_{k=1}^{q} n_k J_k(0, t) = \sum_{k=1}^{q} n_k D_k\left[\frac{\partial C_k(x, t)}{\partial x}\right]_{x=0} \tag{4.3.17}$$

4.3.3 Boundary Conditions in Electrochemical Problems

In solving the mass-transfer part of an electrochemical problem, a diffusion equation (or in general a mass-transfer equation) is written for each species in solution (O, R, . . .). The solution of these equations, that is, the discovery of an equation for the computation of C_O, C_R, . . . as functions of x and t, requires that an initial condition (values at $t = 0$) and two boundary conditions (values at certain values of x) be given. Typical initial and boundary conditions include the following.

(a) Initial Conditions. These are usually of the form

$$C_O(x, 0) = f(x) \tag{4.3.18}$$

For example, if O is uniformly distributed throughout the solution at a bulk concentration C_O^* at the start of the experiment, the initial condition is

$$C_O(x, 0) = C_O^* \qquad \text{(for all } x\text{)} \tag{4.3.19}$$

If R is initially absent from the solution, then

$$C_R(x, 0) = 0 \qquad \text{(for all } x\text{)} \tag{4.3.20}$$

(b) Semi-Infinite Boundary Conditions. Usually the electrolysis cell is so large compared to the length of the diffusion path in the time of an experiment that effects at the walls of the cell are not felt at the electrode. Thus one can assume that at large distances from the electrode ($x \to \infty$) the concentration reaches a constant value, for example, the same as the initial one, so that

$$\lim_{x\to\infty} C_O(x, t) = C_O^* \qquad \text{(at all } t\text{)} \tag{4.3.21}$$

$$\lim_{x\to\infty} C_R(x, t) = 0 \qquad \text{(at all } t\text{)} \tag{4.3.22}$$

For thin-layer electrochemical cells, where the cell wall at a distance, l, from the electrode is of the order of the diffusion path length, boundary conditions at $x = l$ replace those for $x \to \infty$.

(c) Electrode Surface Boundary Conditions. The additional boundary conditions usually relate to the concentrations or concentration gradients at the electrode surface.

For example,

$$C_O(0, t) = f(E) \tag{4.3.23}$$

$$\frac{C_O(0, t)}{C_R(0, t)} = f(E) \tag{4.3.24}$$

where $f(E)$ is some function of the electrode potential that is obtained from the general current-potential characteristic or one of its special cases (e.g., the Nernst equation). If the current is the controlled quantity, the flux equation yields the boundary condition; for example,

$$-J_O(0, t) = \frac{i}{nFA} = D_O\left[\frac{\partial C_O(x, t)}{\partial x}\right]_{x=0} = f(t) \tag{4.3.25}$$

The conservation of matter in an electrode reaction can also be used. For example, when O is converted to R at the electrode (equation 4.3.15), and both O and R are soluble in the solution phase, then for each O that undergoes electron transfer at the electrode, an R must be produced. Hence

$$D_O\left[\frac{\partial C_O(x, t)}{\partial x}\right]_{x=0} + D_R\left[\frac{\partial C_R(x, t)}{\partial x}\right]_{x=0} = 0 \tag{4.3.26}$$

since $J_O(0, t) = -J_R(0, t)$.

In the chapters that follow we will examine the solution of the diffusion equations under a variety of conditions. The mathematical methods for attacking these problems are discussed briefly in Appendix A.

4.4 REFERENCES

1. M. Planck, *Ann. Physik.*, **39**, 161; **40**, 561 (1890).
2. J. Newman, *Electroanal. Chem.*, **6**, 187 (1973).
3. J. Newman, *Adv. Electrochem. Electrochem. Eng.*, **5**, 87 (1967).
4. C. W. Tobias, M. Eisenberg, and C. R. Wilke, *J. Electrochem. Soc.*, **99**, 359C (1952).
5. W. Vielstich, *Z. Elektrochem.*, **57**, 646 (1953).
6. N. Ibl, *Chemie-Ingenieur-Technik*, **35**, 353 (1963).
7. G. Charlot, J. Badoz-Lambling, and B. Tremillion, "Electrochemical Reactions," Elsevier, Amsterdam, 1962, pp. 18–21, 27–28.
8. I. M. Kolthoff and J. J. Lingane, "Polarography," 2nd ed., Interscience, New York, 1952, Vol. 1, Chap. 7.
9. K. Vetter, "Electrochemical Kinetics," Academic Press, New York, 1967.
10. J. Koryta, J. Dvorak, and V. Bohackova, "Electrochemistry," Methuen & Co., Ltd., London, 1970, pp. 88–112.
11. J. Coursier, as quoted in Reference 7.
12. W. Jost, *Angew. Chem.* (*Intl. Ed.*), **3**, 713 (1964).
13. W. J. Moore, "Physical Chemistry," 3rd ed., Prentice Hall, New York, 1962, pp. 232–4.
14. R. B. Bird, W. E. Stewart, and E. N. Lightfoot, "Transport Phenomena," Wiley, New York, 1960.

15. See, for example, L. B. Anderson and C. N. Reilley, *J. Chem. Educ.*, **44**, 9 (1967).

4.5 PROBLEMS

4.1 Consider the electrolysis of a 0.10 M NaOH solution at platinum electrodes, where the reactions are:

$$\text{(anode)} \quad 2OH^- \rightarrow \tfrac{1}{2}O_2 + H_2O + 2e$$

$$\text{(cathode)} \quad 2H_2O + 2e \rightarrow H_2 + 2OH^-$$

Show the balance sheet for the system operating at steady state. Assume 20e passed in the external circuit and use the λ_0 values in Table 2.3.2 to estimate transference numbers.

4.2 Consider the electrolysis of a solution containing 10^{-3} M $Fe(ClO_4)_3$ and 10^{-3} M $Fe(ClO_4)_2$ (assume both salts are completely dissociated) at platinum electrodes:

$$\text{(anode)} \quad Fe^{2+} \rightarrow Fe^{3+} + e$$

$$\text{(cathode)} \quad Fe^{3+} + e \rightarrow Fe^{2+}$$

Assume the λ values for Fe^{3+}, Fe^{2+}, and ClO_4^- are equal and that 10e are passed in the external circuit. Show the balance sheet for the steady-state operation of this system.

4.3 For a given electrochemical system to be described by equations involving semi-infinite boundary conditions, the cell wall must be at least about five "diffusion layer thicknesses" away from the electrode. For a substance with $D = 10^{-5}$ cm²/sec, what distance between the working electrode and the cell wall is required for a 100-sec experiment?

4.4 The mobility, u_j, is related to the diffusion coefficient, D_j, by the equation (see equation 4.1.15)

$$u_j = \frac{|z_j|FD_j}{RT}$$

(a) From the mobility data in Table 2.3.2, estimate the diffusion coefficients of H^+, I^-, and Li^+ at 25 °C.

(b) Write the equation for the estimation of D from the λ value.

4.5 Using the procedure of Section 4.3.2, derive Fick's second law for spherical diffusion (equation 4.3.14). [*Hint*: Because of the different areas through which diffusion occurs at r and at $r + dr$, it is more convenient to obtain the change of concentration in dr by considering the number of moles diffusing per second (rather than the flux).]

chapter 5 Controlled Potential Microelectrode Techniques—Potential Step Methods

The next two chapters are concerned with methods in which the working electrode potential is forced to adhere to a known program. It may be constant or it may vary with time in a predetermined manner, and the current is measured as a function of time or potential. In the methods covered in the next five chapters, the electrode is a microelectrode, and the solution volume is large enough that the passage of current does not alter the bulk concentrations of electroactive species. Such circumstances are known as *small A/V conditions*. In this chapter, we will consider systems in which the mass transport of electroactive species occurs only by diffusion. Also, we will restrict our view for the moment to methods involving perturbations of the working electrode by step-functional changes in its potential. This family of techniques is probably the largest single group, and it contains some of the most powerful experimental approaches available to electrochemistry.

5.1 OVERVIEW OF STEP EXPERIMENTS

5.1.1 Types of Techniques

Figure 5.1.1 is a picture of the basic experimental system. An instrument known as a *potentiostat* has control of the voltage across the working electrode–counter electrode pair, and it adjusts this voltage in order to maintain the potential difference between the working and reference electrodes (which it senses through a high-impedance feedback loop) in accord with the program supplied by a function generator. One can view the potentiostat alternatively as an active element whose job is to force through the working electrode whatever current is required to achieve the desired potential at any time. Since the current and the potential are related functionally, that current is

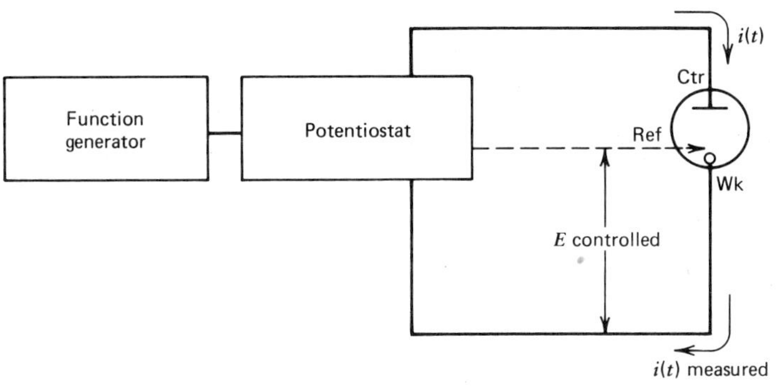

Figure 5.1.1
Schematic experimental arrangement for controlled potential experiments.

unique. Thus the potentiostat's response (the current) actually is the experimental observable. For an introduction to the design of such apparatus, see Chapter 13.

Figure 5.1.2*a* is a diagram of the waveform applied in a basic potential step experiment. Let us consider its effect on the interface between a solid electrode and an unstirred solution containing an electroactive species. As an example, take anthracene in deoxygenated DMF. We know that there generally is a potential region where faradaic processes do not occur; let E_1 be in this region. On the other hand, we can also find a potential at which the kinetics for reduction of anthracene become so rapid that no anthracene can coexist with the electrode, and its surface concentration goes

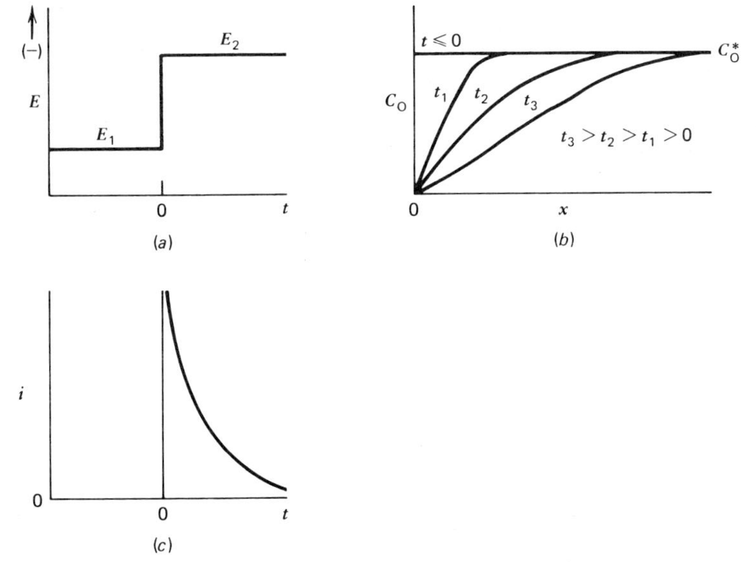

Figure 5.1.2
(*a*) Waveform for a step experiment in which species O is electroinactive at E_1 but is reduced at a diffusion-limited rate at E_2. (*b*) Concentration profiles for various times into the experiment. (*c*) Current flow versus time.

nearly to zero. Consider E_2 to be in this "mass transfer limited" region. What is the response of the system to the step perturbation?

First, the electrode must instantly reduce the anthracene nearby to the stable anion radical:

$$\text{An} + e \rightarrow \text{An}^{\overline{\cdot}} \tag{5.1.1}$$

This act requires a very large current, but subsequently current is required only because the reduction has created a concentration gradient that produces, in turn, a net flux of anthracene to the electrode surface. Since this material cannot coexist with the electrode at E_2, it must be eliminated by reduction. The flux of anthracene, hence the current as well, is proportional to the concentration gradient at the electrode surface. Note, however, that the continued anthracene flux causes the zone of anthracene depletion to thicken; thus the slope of the concentration profile at the surface declines with time, and so does the current. Both of these effects are depicted in Figures 5.1.2*b* and 5.1.2*c*. This kind of experiment is called *chronoamperometry*, because current is recorded as a function of time.

Suppose we now consider a series of step experiments in the anthracene solution discussed above. Between each experiment the solution is stirred, so that the initial conditions are always the same. Similarly, the initial potential (before the step) is chosen to be at a constant value where no faradaic processes occur. The change from experiment to experiment is in the step potential, as depicted in Figure 5.1.3*a*. Suppose further that experiment 1 involves a step to a potential at which anthracene is not yet electroactive; that experiments 2 and 3 involve potentials where anthracene is reduced, but not so effectively that its surface concentration is zero; and that 4 and 5 have step

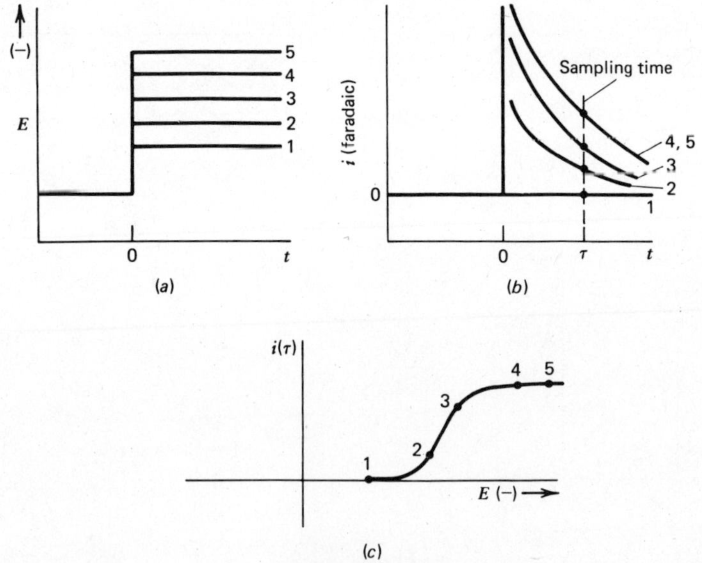

Figure 5.1.3

Sampled-current voltammetry. (*a*) Step waveforms applied in a series of experiments. (*b*) Current-time curves observed in response to the steps. (*c*) Sampled-current voltammogram.

potentials in the mass-transfer-limited region. Obviously experiment 1 yields no faradaic current, and experiments 4 and 5 yield the same current we obtained in the chronoamperometric case above. In both 4 and 5, the surface concentration is zero; hence anthracene arrives as fast as diffusion can bring it, and the current is limited by this factor. Once the electrode potential becomes so extreme that this condition applies, the potential no longer influences the electrolytic current. In experiments 2 and 3 the story is different because the reduction process is not so dominant that some anthracene cannot coexist with the electrode. Still, its concentration is less than the bulk value, so anthracene does diffuse to the surface where it must be eliminated by reduction. Since the difference between the bulk and surface concentrations is smaller than in the mass-transfer-limited case, less material arrives at the surface per unit time, and the currents (for corresponding times) are smaller than in experiments 4 and 5. Nonetheless, the depletion effect still applies, which means that the current still decays with time.

Now suppose we sample the current at some fixed time τ into each of these step experiments; then we can plot the sampled current $i(\tau)$ *vs.* the potential to which the step takes place. As shown in Figures 5.1.3*b* and 5.1.3*c*, the current-potential curve has a waveshape much like that encountered in earlier considerations of steady-state voltammetry (Chapter 1) under convective conditions. This kind of experiment is called *sampled-current voltammetry*, and we will see that it really is the basis for *dc polarography* (voltammetry at the dropping mercury electrode) as well as the newer pulse polarographic methods.

Now consider the effect of the potential program displayed in Figure 5.1.4*a*. The forward step, that is, the transition from E_1 to E_2 at $t = 0$, is exactly the experiment that we have just discussed. For a period τ, it causes a buildup of the reduction product (e.g., anthracene anion radical) in the region near the electrode.

However, in the second phase of the experiment, after $t = \tau$, the potential returns to E_1, where only the oxidized form (e.g., anthracene) is stable at the electrode. The anion radical cannot coexist there; hence a large anodic current flows as it begins to reoxidize, then the current declines in magnitude (Figure 5.1.4*b*) as the depletion effect sets in.

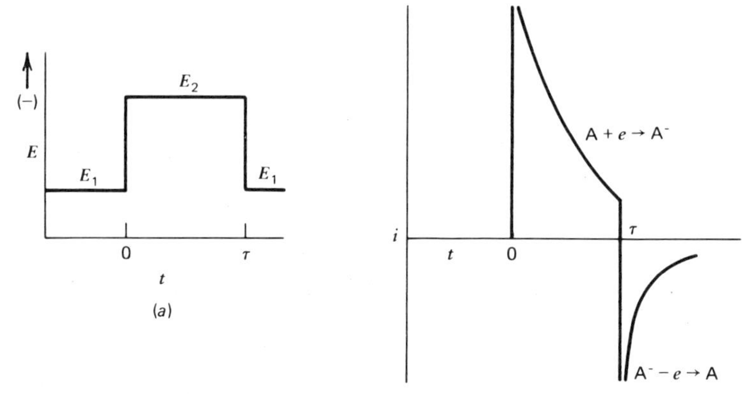

Figure 5.1.4
Double potential step chronoamperometry. (*a*) Typical waveform. (*b*) Current response.

This experiment, called *double potential step chronoamperometry*, is our first example of a *reversal technique*. These methods comprise a large and growing class of approaches that all feature an initial generation of an electrolytic product (in the forward phase), then a reversal of electrolysis so that the first product is examined electrolytically in a direct fashion. These methods make up a powerful arsenal for studies of complex electrode reactions, and we will have much to say about them.

5.1.2 Detection

The usual observables in controlled potential experiments are currents as functions of time or potential. In some experiments, it is useful to record the integral of the current versus time. Since this integral is the amount of charge passed, these methods are *coulometric* approaches. The most prominent examples are *chronocoulometry* and *double potential step chronocoulometry*, which are the integral analogs of the corresponding chronoamperometric approaches. Figure 5.1.5 is a display of the coulometric response to the double-step program of Figure 5.1.4*a*.

One can easily see the linkage, through the integral, between Figures 5.1.4*b* and 5.1.5. The charge that is injected by reduction in the forward step is withdrawn by oxidation in the reversal.

Occasionally it is useful to record the derivative of the current versus time or potential. Nevertheless, these *derivative techniques* are infrequently used because they tend to enhance noise on the signal (see Chapter 13).

Several more sophisticated detection modes involving convolution (or semi-integration), semidifferentiation, or other transformations of the current function also find useful applications. Since they tend to rest on fairly subtle mathematics, we will defer discussions of them for now (see Section 6.7).

5.1.3 Applicable Current-Potential Characteristics

With a qualitative understanding of the experiments described in the preceding section, we saw that it was possible to develop a rationale for the general shapes of the current functions that arose. However, we are interested in obtaining quantitative information

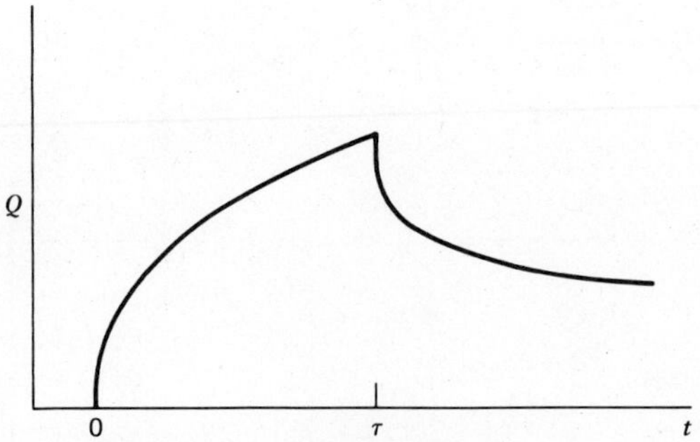

Figure 5.1.5
Response curve for double potential step chronocoulometry. Step waveform is similar to that in Figure 5.1.4*a*.

about electrode processes from these current-time or current-potential curves, and doing so requires the creation of theory that can predict, quantitatively, the response functions in terms of the experimental parameters of time, potential, concentration, mass transfer coefficients, kinetic parameters, and so on. In general, a controlled potential experiment can be treated by invoking the current-potential characteristic:

$$i = nFAk^0[C_O(0, t)\, e^{-\alpha nf(E - E^{0'})} - C_R(0, t)\, e^{(1-\alpha)nf(E - E^{0'})}] \tag{5.1.2}$$

in conjunction with Fick's laws, which can give the time-dependent surface concentrations $C_O(0, t)$ and $C_R(0, t)$. This approach is nearly always difficult, and it sometimes fails to yield closed-form solutions. One is then forced to numerical solutions or to approximate solutions.

The usual alternative in science is to design experiments so that simpler mathematics can be used. Several special cases are easily identified:

(a) Large-Amplitude Potential Step. If the potential is stepped to the mass-transfer-controlled region, the concentration of the electroactive species is nearly zero at the electrode surface and the current is totally controlled by mass transfer. Electrode kinetics no longer influence the current, hence the general *i-E* characteristic is not needed at all. For this case, i is independent of E.

(b) Small-Amplitude Potential Changes. If a perturbation in potential is small in size and both redox forms of a couple are present (so that an equilibrium potential exists), then current and potential are linked by the linearized i-η relation (Section 3.5.3).

$$i = \frac{-nFi^0}{RT}\eta = \frac{-\eta}{R_{ct}} \tag{5.1.3}$$

(c) Reversible (Nernstian) Electrode Process. For very rapid electrode kinetics, we have seen that the general *i-E* relation (5.1.2) collapses to a relation of the Nernst form (Section 3.5.5):

$$E = E^{0'} + \frac{RT}{nF}\ln\frac{C_O(0, t)}{C_R(0, t)} \tag{5.1.4}$$

Again the kinetic parameters k^0 and α are not involved, and mathematical treatments are nearly always greatly simplified.

(d) Totally Irreversible Electrode Process. When the electrode kinetics are very sluggish (k^0 is very small), the anodic and cathodic terms of (5.1.2) are never simultaneously significant. That is, when an appreciable net cathodic current is flowing, the second term in (5.1.2) has a negligibly small effect, and vice versa. In order to observe the net current, the forward process has to be so strongly activated (by application of an overpotential) that the back reaction is totally inhibited. In such cases, observations are always made in the "Tafel region," hence one of the terms in (5.1.2) can be neglected (see also Section 3.5.3).

Unfortunately, electrode processes are not always very facile or very sluggish, and we sometimes must consider the whole *i-E* characteristic. In these *quasi-reversible* (or *quasi-nernstian* cases), we recognize that the net current involves appreciable activated components from the forward and reverse charge transfers.

5.2 POTENTIAL STEP UNDER DIFFUSION CONTROL

5.2.1 A Planar Microelectrode

Above, we considered an experiment involving an instantaneous change in potential from a value where no electrolysis occurs to a value in the mass-transfer-controlled region for reduction of anthracene, and we were able to grasp qualitatively the current-time response. Here we will develop a quantitative treatment of such an experiment. A planar electrode (e.g., a platinum disk) and an unstirred solution are presumed. In place of the anthracene example, we can consider the general reaction O + $ne \rightarrow$ R. Regardless of whether the kinetics of this process are basically facile or sluggish, they can be activated by a sufficiently negative potential (unless the solvent or supporting electrolyte is reduced first)† so that the surface concentration of O becomes effectively zero. This condition will then hold at any more extreme potential. We will consider our instantaneous step to terminate in this region.

(a) Solution of the Diffusion Equation. The calculation of the diffusion-limited current, i_d, and the concentration profile $C_O(x, t)$ involves the solution of the linear diffusion equation:

$$\frac{\partial C_O(x, t)}{\partial t} = D_O \frac{\partial^2 C_O(x, t)}{\partial x^2} \tag{5.2.1}$$

under the boundary conditions:

$$C_O(x, 0) = C_O^* \tag{5.2.2}$$

$$\lim_{x \to \infty} C_O(x, t) = C_O^* \tag{5.2.3}$$

$$C_O(0, t) = 0 \qquad (\text{for } t > 0) \tag{5.2.4}$$

The *initial condition* (5.2.2) merely expresses the homogeneity of the solution before the experiment starts at $t = 0$, and the *semi-infinite condition* (5.2.3) is an assertion that regions sufficiently distant from the electrode are unperturbed by the experiment. The third condition (5.2.4) expresses the surface condition after the potential transition, and it really embodies the particular experiment we have at hand.

Section A.1.6 contains a demonstration showing that after Laplace transformation of (5.2.1), application of conditions (5.2.2) and (5.2.3) yields

$$\bar{C}_O(x, s) = \frac{C_O^*}{s} + A(s)\, e^{-\sqrt{s/D_O}\,x} \tag{5.2.5}$$

By applying the surface condition (5.2.4), the function $A(s)$ can be evaluated, and then

† The large currents arising from reduction of these major constituents usually make electrochemical studies at potentials much more negative than their onset rather impractical. The potential at which they begin to interfere seriously in a given experiment is called the *cathodic background limit*. Likewise, the *anodic background limit* is set by oxidation of major components.

$\bar{C}_O(x, s)$ can be inverted to obtain the concentration profile for species O. Transforming (5.2.4) gives

$$\bar{C}_O(0, s) = 0 \tag{5.2.6}$$

which implies that

$$\bar{C}_O(x, s) = \frac{C_O^*}{s} - \frac{C_O^*}{s} e^{-\sqrt{s/D_O}x} \tag{5.2.7}$$

In Chapter 4, we saw that the flux at the electrode surface is proportional to the current; specifically,

$$-J_O(0, t) = \frac{i(t)}{nFA} = D_O\left[\frac{\partial C_O(x, t)}{\partial x}\right]_{x=0} \tag{5.2.8}$$

which is transformed to

$$\frac{\bar{i}(s)}{nFA} = D_O\left[\frac{\partial \bar{C}_O(x, s)}{\partial x}\right]_{x=0} \tag{5.2.9}$$

The derivative in (5.2.9) can be evaluated from (5.2.7). Substitution yields

$$\bar{i}(s) = \frac{nFAD_O^{1/2}C_O^*}{s^{1/2}} \tag{5.2.10}$$

and inversion produces the current-time response

$$\boxed{i(t) = i_d(t) = \frac{nFAD_O^{1/2}C_O^*}{\pi^{1/2}t^{1/2}}} \tag{5.2.11}$$

which is known as the *Cottrell equation* (1). Its validity was verified in detail by the classic experiments of Kolthoff and Laitinen, who left no parameter freely adjustable (2, 3). Note that the effect of depleting the electroactive species near the surface is characterized by an inverse $t^{1/2}$ function. We will encounter this kind of time dependence frequently in other kinds of experiments.

In practical measurements of the $i - t$ behavior under "Cottrell conditions" one must be aware of instrumental and experimental limitations:

1. *Potentiostatic limitations.* Equation 5.2.11 predicts very high currents at short times, but the actual maximum current may depend on the current and voltage output characteristics of the potentiostat (see Chapter 13).

2. *Limitations in the recording device.* During the initial part of the current transient, the vertical amplifier of the oscilloscope (or other recording device) may be overdriven, and some time may be required for the amplifier to recover, so that accurate readings can be displayed.

3. *Limitations imposed by* R_u *and* C_d. As shown in Section 1.2.4, a nonfaradaic current will also flow during a potential step. This current will decay exponentially with a time constant R_uC_d (where R_u is the uncompensated resistance and C_d is the double-layer capacitance). Within this time period, some contribution of charging current to the total measured current will exist.

4. *Limitations due to convection.* At longer times the buildup of density gradients and the existence of stray vibrations will cause convective disruption of the diffusion layer, and will usually result in currents larger than those predicted by the Cottrell equation. The time for the onset of convective interference depends on the orientation of the electrode, the existence of a glass mantle around the electrode, and so on (2, 3). In general, diffusion measurements for times longer than about 300 sec are difficult, and even measurements longer than 20 sec may show some convective effects. The available "time window" for Cottrell measurements depends on the exact experimental arrangement; under the best of conditions it lies approximately between 20 μsec and 200 sec.

(b) Concentration Profile. Inversion of (5.2.7) yields

$$C_O(x, t) = C_O^*\left\{1 - \operatorname{erfc}\left[\frac{x}{2(D_O t)^{1/2}}\right]\right\} \tag{5.2.12}$$

or

$$\boxed{C_O(x, t) = C_O^* \operatorname{erf}\left[\frac{x}{2(D_O t)^{1/2}}\right]} \tag{5.2.13}$$

Figure 5.2.1 comprises several plots from (5.2.13) for various values of time. The depletion of O near the electrode is easily seen, as is the time-dependent falloff in the concentration gradient at the electrode surface, which leads to the monotonically decreasing i_d function of (5.2.11).

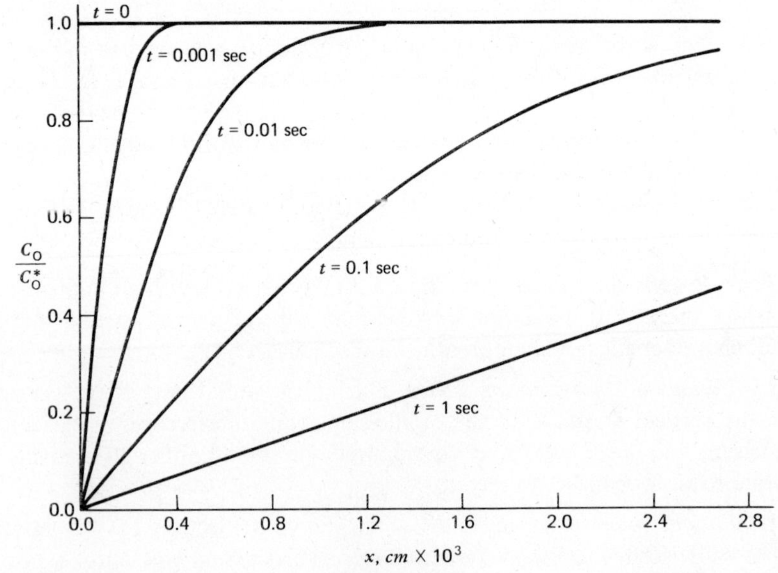

Figure 5.2.1
Concentration profiles for several times after the start of a Cottrell experiment. $D_O = 1 \times 10^{-5}$ cm²/sec.

5.2.2 Semi-Infinite Spherical Diffusion

If the electrode in the step experiment is spherical rather than planar (e.g., a hanging mercury drop), one must consider a spherical diffusion field, and Fick's second law becomes

$$\frac{\partial C_O(r, t)}{\partial t} = D_O\left\{\frac{\partial^2 C_O(r, t)}{\partial r^2} + \frac{2}{r}\frac{\partial C_O(r, t)}{\partial r}\right\} \tag{5.2.14}$$

where r is the radial distance from the electrode center. The boundary conditions are then

$$C_O(r, 0) = C_O^* \qquad (r > r_0) \tag{5.2.15}$$

$$\lim_{r \to \infty} C_O(r, t) = C_O^* \tag{5.2.16}$$

$$C_O(r_0, t) = 0 \qquad (t > 0) \tag{5.2.17}$$

where r_0 is the radius of the electrode.

(a) Solution of the Diffusion Equation. The substitution, $v(r, t) = rC_O(r, t)$, converts (5.2.14) into an equation having the same form as the linear problem. The details are left to the reader (Problem 5.1). The resulting diffusion current is

$$\boxed{i_d(t) = nFAD_OC_O^*\left[\frac{1}{(\pi D_O t)^{1/2}} + \frac{1}{r_0}\right]} \tag{5.2.18}$$

which can be written

$$i_d(\text{spherical}) = i_d(\text{linear}) + \frac{nFAD_OC_O^*}{r_0} \tag{5.2.19}$$

Thus the diffusion current for the spherical case is just that for the linear situation plus a constant term. For a planar electrode,

$$\lim_{t \to \infty} i_d = 0 \tag{5.2.20}$$

but in the spherical case,

$$\lim_{t \to \infty} i_d = \frac{nFAD_OC_O^*}{r_0} \tag{5.2.21}$$

The reason for this curious nonzero limit is that one converges on a situation in which the growth of the depletion region fails to affect the concentration gradients at the surface because the diffusion field is able to draw material from a continually larger area at its outer limit. In actual experiments, convection caused by density gradients or vibration becomes important at longer times and contributes to mass transfer; thus the diffusive steady state is rarely reached.

(b) Concentration Profile. The distribution of the electroactive species near the electrode also can be obtained from the solution to the diffusion equation, and it turns out to be

$$C_O(r, t) = C_O^*\left[1 - \frac{r_0}{r}\operatorname{erfc}\left(\frac{r - r_0}{2(D_O t)^{1/2}}\right)\right] \tag{5.2.22}$$

Since $r - r_0$ is the distance from the electrode surface, this profile strongly resembles that for the linear case (5.2.12). The difference is the factor r_0/r and, if the diffusion layer is thin compared to the electrode's radius, the linear and spherical cases are indistinguishable. The situation is directly analogous to our experience in living on a spherical planet. The zone of our activities above the earth's surface is small compared to its radius of curvature; hence we usually cannot distinguish it from a rough plane.

(c) Applicability of the Linear Approximation. These ideas indicate that linear diffusion will adequately describe mass transport to a sphere, provided the sphere's radius is large enough and the time domain of interest is small enough. More precisely, the linear treatment is adequate as long as the second (constant) term of (5.2.18) is small compared to the Cottrell term. For accuracy within $a\%$,

$$\frac{nFAD_O C_O^*}{r_0} \leq \frac{a}{100} \cdot \frac{nFAD_O^{1/2} C_O^*}{(\pi t)^{1/2}} \tag{5.2.23}$$

or

$$\frac{\pi^{1/2} D_O^{1/2} t^{1/2}}{r_0} \leq \frac{a}{100} \tag{5.2.24}$$

With $a = 10$ and $D_O = 10^{-5}$ cm^2/sec, $t^{1/2}/r_0 \leq 18$ sec$^{1/2}$/cm. A typical mercury drop might be 0.1 cm in radius, and thus the linear treatment holds within 10% for about 3 sec.

5.3 LIMITING CURRENTS AT THE DROPPING MERCURY ELECTRODE

An instrument of enormous importance to electroanalytical chemistry, both historically and currently, is the *dropping mercury electrode*, which was invented by Heyrovsky (4) for measurements of surface tension (see also Section 12.2.1). Several excellent discussions of the construction and operation of the electrode are available (5–9). Figure 5.3.1 is a drawing of a typical device. A capillary with an internal diameter of $\sim 5 \times 10^{-3}$ cm is fed by a head of mercury perhaps 20 to 100 cm high. Mercury issues through the capillary to form a nearly spherical drop, which grows until its weight can no longer be supported by the surface tension. A mature drop typically has a diameter on the order of 0.1 cm. If the electrolysis occurs during the drop's growth, the current obviously has a time dependence that reflects both the expansion of the spherical electrode and the depletion effects of electrolysis. Upon falling, the drop stirs the solution and largely (but not completely) erases the depletion effects, so that each drop is born into fresh solution. If the potential does not change during the lifetime of a drop (2–6 sec), the experiment really is indistinguishable from a step experiment in which the potential transition coincides with the birth of a new drop. Each drop's lifetime is itself a new experiment.

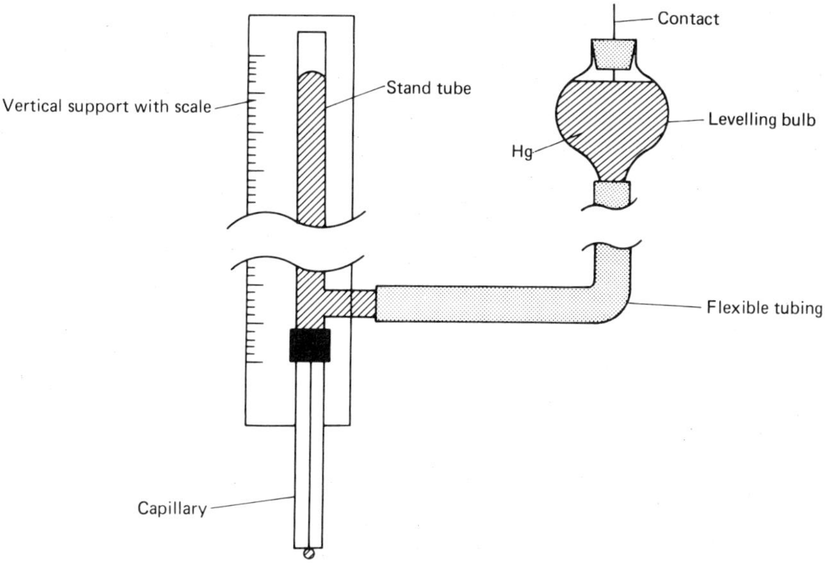

Figure 5.3.1
A dropping mercury electrode.

5.3.1 The Ilkovic Equation

Let us consider here the current that flows during a single lifetime when the DME is held at a potential in the mass-transfer-controlled region for electrolysis. That is, we seek the diffusion-limited current. The problem has been solved rigorously by Koutecky (10, 11), but the mathematics require consideration of the relative convective movement between the electrode and the solution during drop growth. This leads to rather complicated mathematics that give little intuitive feel for the effects one must consider. The treatment we will follow, originally due to Lingane and Loveridge (12), makes no pretense to rigor. It really is only an outline to the problem, but it highlights the differences between a DME and a stationary electrode (6, 12–15).

The typical values of drop lifetime and drop diameter at maturity ensure that linear diffusion holds at a DME to a good approximation. Thus, we begin by invoking the Cottrell relation (5.2.11), while remembering that for the moment we are considering only electrolysis at potentials on the diffusion-limited portion of the polarographic response curve. Notice, however, that since the drop area is a function of time, we must determine an expression for $A(t)$. If the rate of mercury flow from the DME capillary (mass/time) is m and the density of mercury is d_{Hg}, then the weight of the drop at time t is

$$mt = \tfrac{4}{3}\pi r_0{}^3 d_{\mathrm{Hg}} \tag{5.3.1}$$

The drop's radius and its area are then given by

$$r_0 = \left(\frac{3mt}{4\pi d_{\mathrm{Hg}}}\right)^{1/3} \tag{5.3.2}$$

$$A = 4\pi\left(\frac{3mt}{4\pi d_{\text{Hg}}}\right)^{2/3} \tag{5.3.3}$$

Substitution into the Cottrell relation gives

$$i_d = \left[4\pi^{1/2}F\left(\frac{3}{4\pi d_{\text{Hg}}}\right)^{2/3}\right]nD_O^{1/2}C_O^*m^{2/3}t^{1/6} \tag{5.3.4}$$

In addition to the effect of the changing area, which progressively enlarges the diffusion field, there is a second consideration that we might call the "stretching (or convective) effect." That is, at any time t, expansion of the drop causes the existing diffusion layer to stretch over a still larger sphere. This has the effect of making the layer thinner than it otherwise would be, and the concentration gradient at the electrode surface is enhanced, so that larger currents flow. It turns out that the result is the same as if the effective diffusion coefficient were $(7/3)D_O$; hence (5.3.4) requires multiplication by $(7/3)^{1/2}$:

$$i_d = \left[4\left(\frac{7\pi}{3}\right)^{1/2}F\left(\frac{3}{4\pi d_{\text{Hg}}}\right)^{2/3}\right]nD_O^{1/2}C_O^*m^{2/3}t^{1/6} \tag{5.3.5}$$

Evaluating the constant in brackets, we have

$$\boxed{i_d = 708nD_O^{1/2}C_O^*m^{2/3}t^{1/6}} \tag{5.3.6}$$

where i_d is in amperes, D_O in cm^2/sec, C_O^* in mol/cm^3, m in mg/sec, and t in seconds. Alternatively, i_d is taken in μA, and C_O^* in mM units.

Ilkovic was first to derive (5.3.6); hence this famous relation bears his name (13–17). His approach was much more exact than ours has been, as was that of MacGillavry and Rideal (18), who provided an alternative derivation a few years afterward. Actually the Lingane-Loveridge approach is not independent of these more rigorous treatments, for they arrived at the $(7/3)^{1/2}$ stretching coefficient merely by comparing the bracketed factor in (5.3.4) with the factor 708 given by Ilkovic and by MacGillavry and Rideal. We should note also that all three treatments are based on linear diffusion. Figure 5.3.2 is an illustration of the current-time curves for several drops as predicted by the Ilkovic equation. One of the things we notice is that the effects of drop expansion (increasing area and stretching of the diffusion layer) more than counteract depletion of the electroactive substance near the electrode. The current is a monotonically *increasing* function of time, in direct contrast to the Cottrell behavior we found at a stationary planar electrode. Two important consequences of the increasing current-time function are that the current is greatest and its rate of change is lowest just at the end of the drop's life. As we will see, these aspects are helpful for applications of the DME in sampled-current voltammetric experiments.

Polarograms are usually recorded on strip-chart or *x-y* recorders as the potential of the DME is scanned linearly (but slowly) with time. The oscillations in current arising from the growth and fall of the individual drops are ordinarily quite apparent. A typical case is shown in Figure 5.3.3. Recorders usually are not fast enough to follow the rapidly changing currents accompanying drop fall and the early stages of growth in the succeeding drop, but they have little difficulty accurately portraying the current

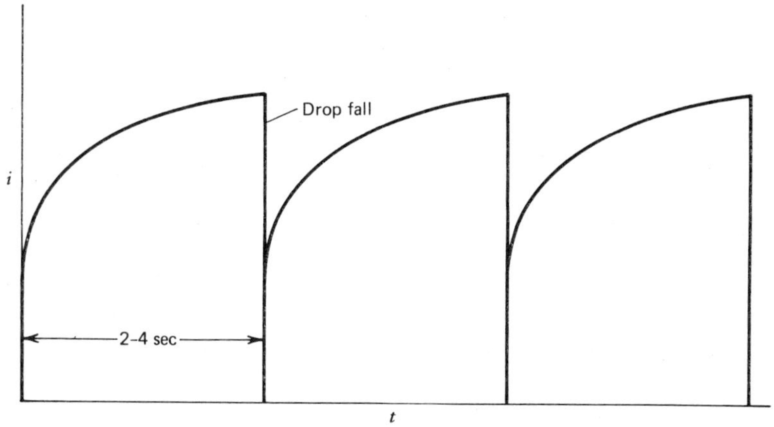

Figure 5.3.2
Current growth during three successive drops of a DME.

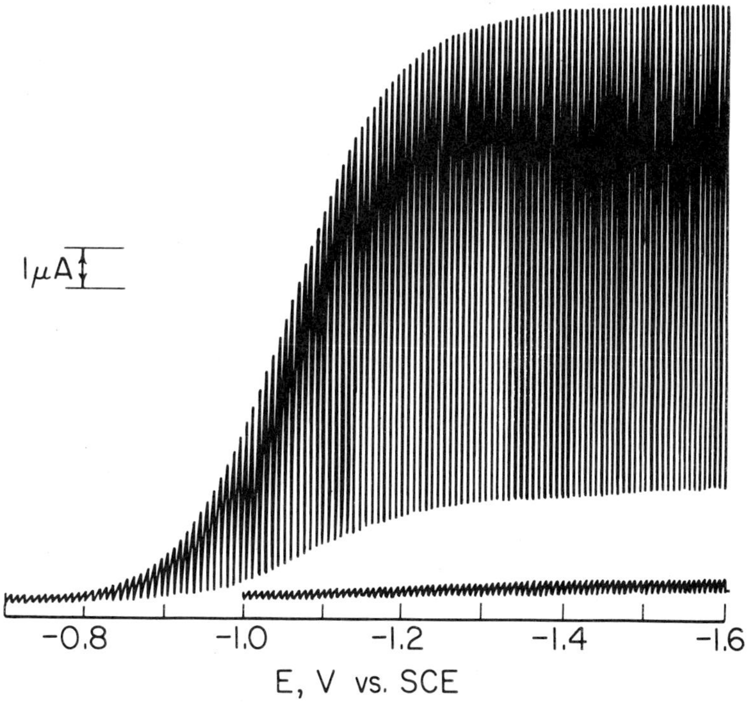

Figure 5.3.3
Polarogram for 1 mM CrO_4^{2-} in deaerated 0.1 M NaOH. The lower curve is the residual current observed in the absence of CrO_4^{2-}.

late in a drop's life. Thus the current-time profile will usually only approximate the true function as shown in Figure 5.3.4. The most easily measured current is that which flows just before drop fall, and within the linear approximation it is given by

$$(i_d)_{\max} = 708nD_O^{1/2}C_O^*m^{2/3}t_{\max}^{1/6} \quad (5.3.7)$$

where $t_{\max}$ is the lifetime of a drop (usually called the *drop time* and often symbolized merely as t).

Much of the older polarographic literature involves measurements of the *average* current flowing during a drop's lifetime. This practice grew up when recording was usually carried out by light beam galvanometers and photographic plates, because the damped galvanometers responded to average current in a way depicted in Figure 5.3.4. From the Ilkovic equation, one can calculate the average current $\bar{i}_d$ as

$$\bar{i}_d = \frac{\int_0^{t_{\max}} i_d\,dt}{\int_0^{t_{\max}} dt} \quad (5.3.8)$$

which yields

$$\bar{i}_d = 607nD_O^{1/2}C_O^*m^{2/3}t_{\max}^{1/6} \quad (5.3.9)$$

Thus the average current is merely six-sevenths of the maximum current.

5.3.2 Extensions to the Ilkovic Equation

Several workers have attempted to take electrode sphericity into account in their derivations of the current-time function at the DME (10–15, 19). Lingane and Loveridge (12) and Strehlow and von Stackelberg (19) first approached the problem

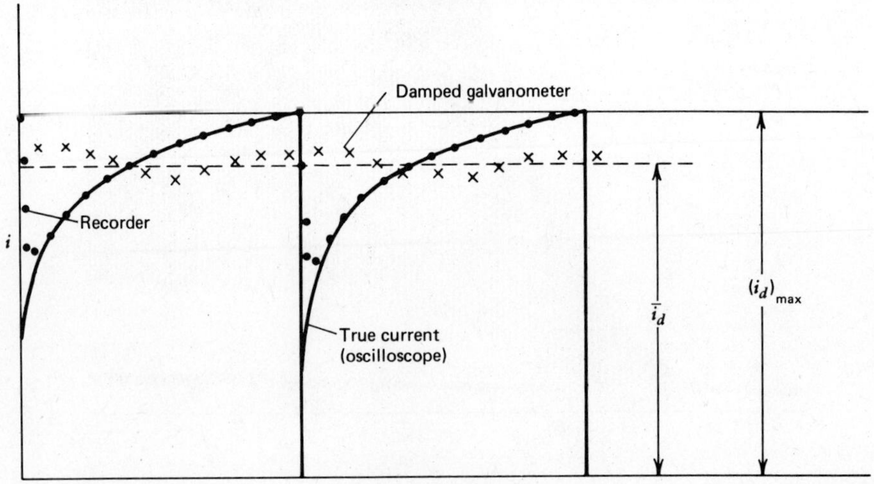

Figure 5.3.4
Current-time curves during individual drops at a DME as observed with three different recording devices.

in a way resembling our outlined derivation of the Ilkovic equation above. The effect of sphericity was introduced by starting with the equation for diffusion-limited electrolysis at a stationary sphere:

$$i_d = \frac{nFAD_O^{1/2}C_O^*}{\pi^{1/2}t^{1/2}} + \frac{nFAD_OC_O^*}{r_0} \tag{5.3.10}$$

Both groups arrived at a relation of the same form for the average current at a DME:

$$\bar{i}_d = 607nD_O^{1/2}C_O^*m^{2/3}t_{\max}^{1/6}\left(1 + \frac{KD_O^{1/2}t_{\max}^{1/6}}{m^{1/3}}\right) \tag{5.3.11}$$

where K has been evaluated at 17–39 by several investigators using different approaches to the area expansion and stretching effects on the second term of (5.3.10).

Koutecky (10, 11) has solved the problem in a stricter manner and has given

$$\boxed{\bar{i}_d = 607nD_O^{1/2}C_O^*m^{2/3}t_{\max}^{1/6}\left[1 + 34.7\frac{D_O^{1/2}t_{\max}^{1/6}}{m^{1/3}} + 100\left(\frac{D_O^{1/2}t_{\max}^{1/6}}{m^{1/3}}\right)^2\right]} \tag{5.3.12}$$

The third term of this relation is not significant in practical terms, but the second amounts to 5 to 10% of the first for typical diffusion coefficients and capillary characteristics.

Very extensive tests of the Ilkovic and Koutecky relations have been carried out, and some interesting discrepancies have been uncovered (14, 15, 20–22). For example, the current early in a drop's life is typically much lower than the predicted value; yet the actual currents are higher than the theoretical values later in the growth cycle. The early differences probably arise because the electrode and the diffusion field cannot possibly resemble the postulated spherical shapes until some appreciable time has lapsed. The capillary tip is bound to restrict the diffusion process and cause lower currents. The positive errors late in the period may indicate effects of convection or time dependence in m. Another interesting effect is that currents flowing for the first drop after imposition of the potential are 10 to 20% larger than those at succeeding drops, which differ little from each other. This behavior has been ascribed to a failure of drop fall to renew completely the solution near the electrode. Given these observations, it is perhaps surprising that the Koutecky equation gives an almost exact description of *average* currents flowing at successive drops (after the first drop) in solutions containing a small amount of *maximum suppressor*.† However, later currents in practice are somewhat larger than the predicted values, and the ratio of average to maximum current is closer to 0.80 than to the theoretical figures near 0.85.

† Sometimes polarographic current-potential curves show peaks, called *polarographic maxima*, which can greatly exceed the limiting currents due to diffusion. These excess currents arise from convection around the growing mercury drop. The convection, in turn, apparently comes about (a) because differences in current density at different points on the drop (e.g., at the shielded top versus the accessible bottom) cause variations in surface tension across the interface, or (b) because the inflow of mercury causes disturbances of the surface. Surfactants, such as gelatin or Triton X-100, have been found experimentally to suppress these maxima and are routinely added in small quantities to test solutions when the maxima themselves are not of primary interest. The literature on this subject should be consulted for more detailed information (23–25).

For many purposes the Ilkovic relation is an adequate description of diffusion-limited currents at the DME. Its functional form is almost wholly accurate, and one only occasionally must recall that the constants 607 and 708 are low by several percent.

5.3.3 Polarographic Analysis

The dropping mercury electrode possesses many characteristics that give polarographic techniques a favored position among electroanalytical methods (6, 9, 26, 27). The dropping action is very reproducible, and the surface is continuously renewed. These factors make high-precision measurements possible (8), and the latter carries the additional advantage that the electrode is not permanently modified by electrode reactions (e.g., many metal ion reductions) that deposit material in or on the electrode. The current-time curves at the DME, which feature minimal rates of current change in the region of maximum current (and maximum current at the end of drop life), are especially well adapted to sampled-current voltammetry. That technique, in turn, is convenient for quantitative measurements because current plateaus are obtained in the mass-transfer-limited region of each wave, and hence flat (or at least linear) baselines apply to each of several successive waves. The repeated dropping and stirring action makes it possible, in effect, to carry out a succession of step experiments with merely a constant potential or a slowly varying ramp applied to the electrode, and sampling can be carried out just by observing the locus of maximum currents on the current-potential curve (the *polarogram*). More sophisticated methods feature electronic sampling for reasons that will be discussed in Section 5.8.

Another important advantage to the DME is the very high overpotentials for hydrogen discharge on mercury surfaces. In many media, this process is the cathodic background reaction; thus the high overpotential means that the background limit is pushed to more negative potentials, and it becomes possible to observe electrode reactions that are thermodynamically less favorable. An example is the reduction of the sodium ion to sodium amalgam in basic aqueous media, which is observable as a clean wave well before the background limit is reached. The exothermicity and the vigor of sodium's reaction with water testifies to the much lower *energy changes* involved in hydrogen reduction compared to sodium reduction. It is the sluggish *kinetics* of the former that makes possible observations of higher-energy processes.

Actually, this particular case is aided by another nice feature of the DME, that is, its ability to form amalgams. Since sodium amalgam is spontaneously formed:

$$\text{Na} + \text{Hg} \rightarrow \text{Na(Hg)} \qquad \Delta G^0 < 0 \tag{5.3.13}$$

the amalgam is at a lower position relative to metallic sodium on the free energy scale. The free energy of Na^+ reduction to Na(Hg) is therefore less than that to Na(c), and the standard potential is not so negative. This is a general feature of electrode processes that involve reductions to amalgams, and it, like the high hydrogen overpotentials, contributes to a widening of the range of processes that can be studied at the DME.

The chief disadvantage of polarographic methods lies in the inability of the DME to operate at potentials much more positive than 0.0 V *vs.* SCE. The anodic limit, which arises from the oxidation of mercury, is always near that potential, although it depends somewhat on the medium.

Quantitative polarographic analysis is based on the linear linkage between the diffusion current and bulk concentration of the electroactive species. In general, the most precise measurements of concentration are carried out by construction of a calibration curve with a set of standard solutions. In routine work, $\pm 1\%$ precision can be obtained (6, 9, 26, 27); however, Lingane (8) has shown that $\pm 0.1\%$ is possible with careful precautions. His work has demonstrated that most sources of imprecision involve temperature effects of some sort. Chief among these, of course, is the temperature dependence of mass transport itself, for the diffusion coefficient increases by 1 to 2% per degree. Even at the 1% precision level, thermostating of the cell is required. For more details on actual measurements, the literature should be consulted (6, 9, 26, 27).

Standard addition and internal standard methods are also used in concentration measurements. They are implemented in the obvious ways and usually are capable of precisions of a few percent.

Unique to polarographic analysis is the "absolute" method of evaluation of concentrations, which was advocated early by Lingane (28). A rearrangement of the Ilkovic equation [(5.3.7) or (5.3.9)], placing all the experimental variables (i_d, t_{max}, m, and C_O^*) on one side, gives

$$\bar{I} = \frac{\bar{i}_d}{m^{2/3} t_{max}^{1/6} C_O^*} = 607 n D_O^{1/2} \tag{5.3.14}$$

for average currents and

$$(I)_{max} = \frac{(i_d)_{max}}{m^{2/3} t_{max}^{1/6} C_O^*} = 708 n D_O^{1/2} \tag{5.3.15}$$

for maximum currents. Note that $\bar{I}$ and $(I)_{max}$, which are called *diffusion current constants*, are independent of the specific values of m, t, and C_O^* used in the measurement. Since they depend only on n and D_O, they are constants of the electroactive substance and the medium in much the same way the molar absorptivity, ε, is a constant of the system for optical measurements. Given $\bar{I}$ or $(I)_{max}$ for the system at hand, one can evaluate C_O^* simply by measuring i_d, t, and m. Since no standards are needed, this technique is especially useful when one-shot analyses are called for. The method is not wholly accurate, as the Koutecky equation indicates, because (5.3.14) and (5.3.15) are not entirely true. However, it is sufficiently useful that many workers have taken the trouble to report $\bar{I}$ and $(I)_{max}$ values, and large tabulations exist (29–33). For concentrations within the 0.5- to 10-mM range, one can expect accuracies within about 5%. The most important point to remember is that t_{max}, being a strong function of potential, must be determined for the potential at which i_d is taken. On the other hand, m varies little with potential (or even from day to day with constant mercury column height and with care), and therefore it needs checking only infrequently.

Diffusion current constants also find frequent use in characterizations of electrode processes, particularly as an indicator of the n value. Consider Table 5.3.1, which is a list of several $\bar{I}$ values for metal-ion reductions in aqueous media. Note that processes having the same n value possess similar diffusion current constants. In general, one-electron reactions have $\bar{I} \simeq 1.5$–2.0, two-electron reactions have $\bar{I} \simeq 3.0$–4.0, and three-electron processes have $\bar{I} \simeq 4.5$–6.0, whenever media with water-like viscosities

Table 5.3.1

Diffusion Current Constants for Selected Reductions[a]

Redox Couples	$\bar{I}$(μA-sec$^{1/2}$ mg$^{-2/3}$-mM^{-1})	Medium
U(VI), U(V)	1.54	0.1 M HCl
Fe(III), Fe(II)	1.46	3 M HCl
Cd^{2+}, Cd(Hg)	3.58	1 M HCl
Pb^{2+}, Pb(Hg)	3.86	1 M HCl
Sb^{3+}, Sb(Hg)	5.57	1 M HCl
Bi^{3+}, Bi(Hg)	5.23	1 M HCl

[a] Tabulated from Reference 30.

(~1 cP) are employed. This criterion can be extended to other media via *Walden's Rule* (15), which notes that for most substances the diffusion coefficients in two media, 1 and 2, are related to the viscosities η_1 and η_2 by

$$D_1\eta_1 \simeq D_2\eta_2 \tag{5.3.16}$$

Thus

$$\frac{\bar{I}_1}{\bar{I}_2} \simeq \left(\frac{\eta_2}{\eta_1}\right)^{1/2} \tag{5.3.17}$$

Of course all of this rests upon the fact that diffusion coefficients for most ions and small molecules do not vary much within a single medium. Exceptions include H^+ and OH^- in aqueous media, oxygen generally, and polymers and large biomolecules. It is also clear that $\bar{I}$ or $(I)_{max}$ can be used for estimates of D values, but the comments of Section 5.3.2 indicate the discretion that is called for in the matter.

Polarographic analysis is carried out most precisely in the range from 0.1 to 10 mM, and comments above about precision generally apply to this region. Useful work can sometimes be carried out below 10^{-5} M, but charging currents (see below) become a severe interference at low concentrations and effectively set the detection limit. Above 10 mM, electrode processes tend to cause such large alterations in solution composition near the electrode that density gradients arise, convection becomes a problem, and currents may be erratic.

5.3.4 Effect of Mercury Column Height (14, 15)

The mercury column height above the capillary tip governs the pressure driving mercury through the DME; thus it is a key determinant of m. In fact, m can be expressed by the Poiseuille equation, which describes the flow of a fluid under laminar conditions through any open linear tube of radius r_c and length l:

$$m = \frac{\pi r_c{}^4 d_{Hg}}{8\, l\eta_{Hg}} P \tag{5.3.18}$$

The driving pressure is P and the viscosity and the density of mercury are η_{Hg} and d_{Hg}, respectively.

To a first approximation, P is simply the pressure due to the mercury column; that is, $d_{Hg}gh$, where the height h is measured from the tip. Of course, g is the gravitational acceleration. In reality, two refinements to P are required because (a) the solution exerts a back pressure that is proportional to the depth of the tip's immersion, h_{soln}, and (b) the expansion of the mercury drop requires work against the surface tension, an effect that is manifested as a second back pressure term. The solution back pressure is given by $h_{soln}d_{soln}g$, and the effect of surface tension is $2\gamma/r_0$, where r_0 is the radius of the drop and γ is the surface tension. In equation 5.3.2 we saw that r_0 is proportional to $(mt)^{1/3}$; thus the second back pressure is time dependent. If one averages over the drop life, it amounts to $4.31\ \gamma d_{Hg}^{1/3}/m^{1/3}t_{max}^{1/3}$, where the variables are in the usual units. We can now express the effective pressure as

$$P_{eff} = h_{Hg}gd_{Hg} - h_{soln}gd_{soln} - \frac{4.31\gamma d_{Hg}^{1/3}}{m^{1/3}t_{max}^{1/3}} \tag{5.3.19}$$

It is customary to factor P_{eff} as follows:

$$P_{eff} = gd_{Hg}\left(h_{Hg} - h_{soln}\frac{d_{soln}}{d_{Hg}} - \frac{4.31\gamma}{d_{Hg}^{2/3}gm^{1/3}t_{max}^{1/3}}\right) \tag{5.3.20}$$

The term in parentheses has dimensions of length, and is known as the *corrected column height*, h_{corr}. Typically, h_{Hg} is 20 to 80 cm, the solution back pressure is ~0.1 cm, and the surface tension term is ~1.5 cm. For most work the second term is negligible, but the third is not. However, it is small enough that one can satisfactorily approximate it using an average γ for aqueous solutions of 400 dynes/cm and the density of mercury at room temperature, so that

$$\boxed{h_{corr} = h_{Hg} - h_{soln}\left(\frac{d_{soln}}{d_{Hg}}\right) - \frac{3.1}{(mt_{max})^{1/3}} \quad \text{(in cm)}} \tag{5.3.21}$$

From (5.3.18) and (5.3.20) one can see that m is proportional to h_{corr}. It is also true that mt_{max} is a constant—the maximum mass that the surface tension can support—and can be expressed as follows:

$$mt_{max}g = 2\pi r_c\gamma \tag{5.3.22}$$

Since $t_{max} = 2\pi r_c\gamma/mg$, it follows that the drop time is inversely proportional to h_{corr}.

The Ilkovic relations show that the diffusion-limited current is proportional to $m^{2/3}t^{1/6}$, which in turn is proportional to $h_{corr}^{2/3}\,h_{corr}^{-1/6} = h_{corr}^{1/2}$. This square-root dependence of the limiting current on corrected column height is characteristic of processes that are limited by the rate of diffusion to the DME, and it is often used as a diagnostic criterion to distinguish this case from other kinds of current limitation. For example, the current could be limited by the amount of space available on an electrode surface for adsorption of a faradaic product, or it might be limited by the rate of production of an electroreactant in a preceding homogeneous chemical reaction. Chapters 11 and 12 deal with such cases.

5.3.5 Residual Current

In the absence of an electroactive substance of interest, and between the anodic and cathodic background limits, the current that flows is called the *residual current* (6, 9, 15, 34). It is composed of a current caused by oxidation or reduction of traces of impurities (e.g., heavy metal ions or oxygen from the solvent or supporting electrolyte) and a current due to double-layer charging. The first of these is a faradaic residual current, and the second is denoted variously as the *charging*, *capacitive*, or *condenser current*. Even in highly purified solutions, the nonfaradaic component can make the residual current rather large.

An expression for the charging current, i_c, can be obtained as follows. The charge on the double-layer is given by

$$q = C_i A(E_z - E) \tag{5.3.23}$$

in terms of the integral capacitance C_i, the electrode area A, and the potential relative to the point of zero charge E_z (see Section 12.2.2). Then,

$$i_c = \frac{dq}{dt} = C_i(E_z - E)\frac{dA}{dt} \tag{5.3.24}$$

since C_i and E are both effectively constant during a drop's lifetime. From (5.3.3), dA/dt can be obtained, and one finds that

$$\boxed{i_c = 0.00567 C_i(E_z - E) m^{2/3} t^{-1/3}} \tag{5.3.25}$$

where i_c is in μA if C_i is given in μF/cm^2. Typically, C_i is 10 to 20 μF/cm^2. Several important conclusions can be drawn from (5.3.25):

1. The average charging current over a drop's lifetime is

$$\bar{i}_c = \frac{\int_0^{t_{max}} i_c\, dt}{\int_0^{t_{max}} dt} \tag{5.3.26}$$

which is

$$\boxed{\bar{i}_c = 0.0085 C_i(E_z - E) m^{2/3} t_{max}^{-1/3}} \tag{5.3.27}$$

This current typically is about the same magnitude as the average faradaic current for an electroactive substance present at 10^{-5} M levels; thus we understand why the limiting current-to-residual current ratio degrades badly in this concentration range. Charging current, more than any other factor, limits detection by normal polarography to concentrations above 5×10^{-6} M or so.

2. From the considerations of the previous section and either (5.3.25) or (5.3.27), one can see that charging current increases in direct proportion to the corrected mercury column height, h_{corr}. This dependence is characteristic of processes that are limited by the rate of increase of the electrode area, and it contrasts with the square-root function found earlier for diffusion-limited processes. It is interesting that the signal-to-background ratio in a diffusion current measurement (i.e., i_d/i_c) actually degrades with increased column height, even though the signal itself (i_d) increases.

3. Note that if C_i and t_{max} are not strongly varying functions of potential, i_c is linear with E. As shown in Figure 5.3.5, experimental residual current curves are

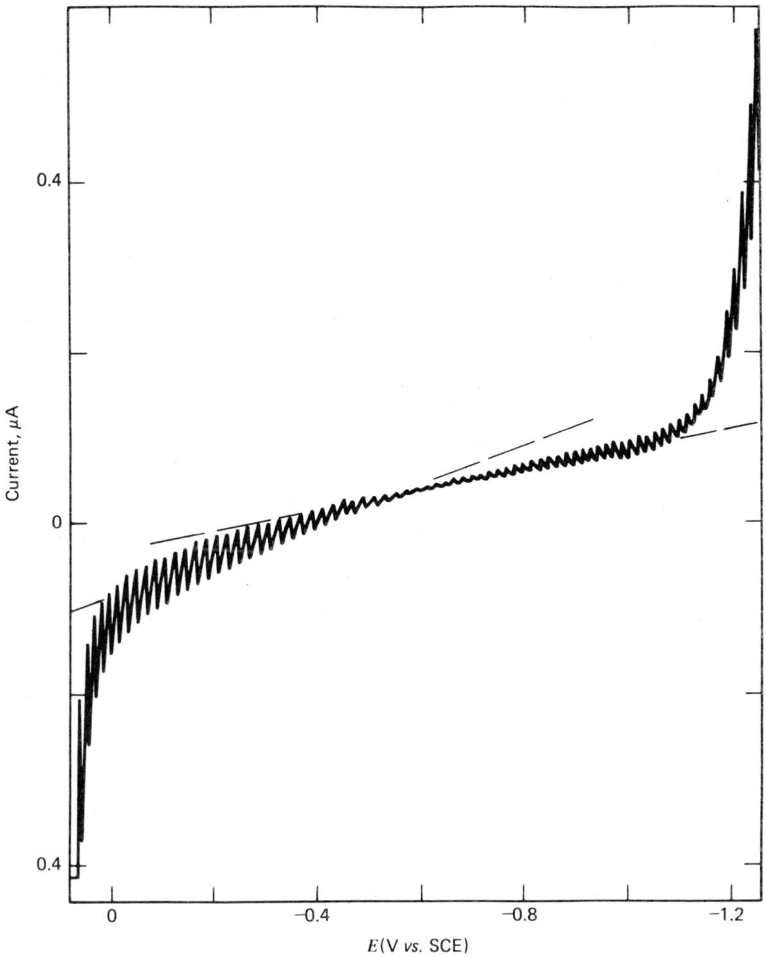

Figure 5.3.5

Residual current curve for 0.1 M HCl. (From L. Meites, "Polarographic Techniques," 2nd ed., Wiley-Interscience, New York, 1965, p. 101, with permission of John Wiley and Sons, Inc.)

fairly linear over wide ranges, and this evidence provides the justification for the common practice of measuring i_d for a polarographic wave by extrapolating the preceding residual current as shown in Figure 5.3.6. Note also that the capacitive current vanishes and changes sign at $E = E_z$.

4. Another important contrast between i_c and i_d is in their time dependences. As we have seen, i_d increases monotonically and reaches its maximum value at t_{max}. On the other hand, the charging current decreases monotonically [see (5.3.25)] as $t^{-1/3}$ because the rate of area increase slows as the drop ages. Thus, the charging current is at its minimal value at t_{max}. This contrast underlies some approaches to increasing polarographic sensitivity by discriminating against i_c in favor of i_d. We will discuss them in Section 5.8.

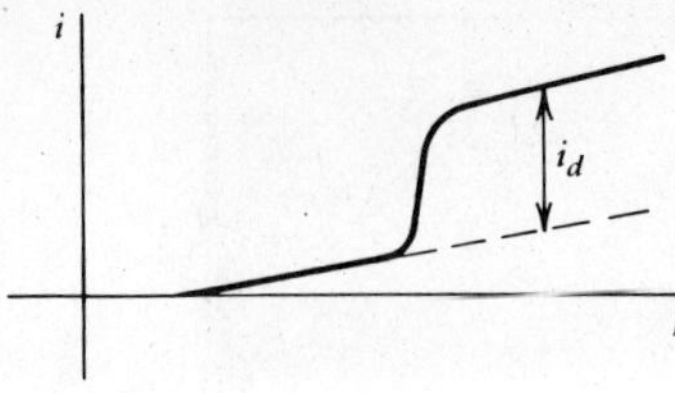

Figure 5.3.6
Method for obtaining i_d from a wave superimposed on a sloping baseline of residual current.

5.4 SAMPLED-CURRENT VOLTAMMETRY FOR A REVERSIBLE ELECTRODE REACTION. THE SHAPE OF THE WAVE

5.4.1 Potential Steps at a Planar Electrode (13)

Consider again the reaction $O + ne \rightleftharpoons R$, but this time let us treat potential steps of any magnitude. We begin each experiment at a potential at which no current flows; and at $t = 0$, we change E instantaneously to a value anywhere on the reduction wave. We assume here that charge transfer kinetics are very rapid, so that

$$E = E^{0'} + \frac{RT}{nF} \ln \frac{C_O(0, t)}{C_R(0, t)} \tag{5.4.1}$$

always.

The equations governing this case are†

$$\frac{\partial C_O(x, t)}{\partial t} = D_O \frac{\partial^2 C_O(x, t)}{\partial x^2} \qquad \frac{\partial C_R(x, t)}{\partial t} = D_R \frac{\partial^2 C_R(x, t)}{\partial x^2} \tag{5.4.2}$$

$$C_O(x, 0) = C_O^* \qquad C_R(x, 0) = 0 \tag{5.4.3}$$

$$\lim_{x \to \infty} C_O(x, t) = C_O^* \qquad \lim_{x \to \infty} C_R(x, t) = 0 \tag{5.4.4}$$

and the flux balance is

$$D_O \left(\frac{\partial C_O(x, t)}{\partial x} \right)_{x=0} + D_R \left(\frac{\partial C_R(x, t)}{\partial x} \right)_{x=0} = 0 \tag{5.4.5}$$

It is convenient to rewrite (5.4.1) as

$$\theta = \frac{C_O(0, t)}{C_R(0, t)} = \exp\left[\frac{nF}{RT}(E - E^{0'})\right] \tag{5.4.6}$$

In Section 5.2.1 we saw that application of the Laplace transform to (5.4.2) and consideration of conditions (5.4.3) and (5.4.4) would yield

$$\bar{C}_O(x, s) = \frac{C_O^*}{s} + A(s)\, e^{-\sqrt{s/D_O}x} \tag{5.4.7}$$

$$\bar{C}_R(x, s) = B(s)\, e^{-\sqrt{s/D_R}x} \tag{5.4.8}$$

† Clearly (5.4.3) implies that R is initially absent. The case for $C_R(x, 0) = C_R^*$ follows analogously, and is left as Problem 5.10.

Transformation of (5.4.5) gives

$$D_O\left(\frac{\partial \bar{C}_O(x, s)}{\partial x}\right)_{x=0} + D_R\left(\frac{\partial \bar{C}_R(x, s)}{\partial x}\right)_{x=0} = 0 \tag{5.4.9}$$

which can be simplified by evaluating the derivatives from (5.4.7) and (5.4.8):

$$-A(s)D_O^{1/2}s^{1/2} - B(s)D_R^{1/2}s^{1/2} = 0 \tag{5.4.10}$$

Thus, $B = -A(s)\xi$, where $\xi = (D_O/D_R)^{1/2}$. So far we have not invoked the Nernst relation (5.4.1); hence our results:

$$\bar{C}_O(x, s) = \frac{C_O^*}{s} + A(s)\, e^{-(s/D_O)^{1/2}x} \tag{5.4.11}$$

$$\bar{C}_R(x, s) = -A(s)\xi\, e^{-(s/D_R)^{1/2}x} \tag{5.4.12}$$

hold for any i-E *characteristic*. We will make use of this fact in Section 5.5.

We introduce the assumption of reversibility in order to evaluate $A(s)$. Transformation of (5.4.6) shows that $\bar{C}_O(0, s) = \theta\bar{C}_R(0, s)$; thus

$$\frac{C_O^*}{s} + A(s) = -\xi\theta A(s) \tag{5.4.13}$$

and therefore $A(s)$ equals $-C_O^*/s(1 + \xi\theta)$. The transformed profiles are then

$$\bar{C}_O(x, s) = \frac{C_O^*}{s} - \frac{C_O^*\, e^{-(s/D_O)^{1/2}x}}{s(1 + \xi\theta)} \tag{5.4.14}$$

$$\bar{C}_R(x, s) = \frac{\xi C_O^*\, e^{-(s/D_R)^{1/2}x}}{s(1 + \xi\theta)} \tag{5.4.15}$$

Equation 5.4.14 differs from (5.2.7) only in the factor, $1/(1 + \xi\theta)$, in the second term. Since $(1 + \xi\theta)$ is independent of x and t, the current can be obtained exactly as in the treatment of the Cottrell experiment by evaluating $\bar{i}(s)$ and then inverting:

$$\boxed{i(t) = \frac{nFAD_O^{1/2}C_O^*}{\pi^{1/2}t^{1/2}(1 + \xi\theta)}} \tag{5.4.16}$$

This relation is the general response function for a step experiment; and the Cottrell equation (5.2.11) is a special case for the diffusion-limited region, which implies a very negative $E - E^{0\prime}$, so that $\theta \to 0$. It is convenient to represent the Cottrell current as $i_d(t)$ and rewrite (5.4.16) as

$$\boxed{i(t) = \frac{i_d(t)}{1 + \xi\theta}} \tag{5.4.17}$$

Now we see that for a reversible couple, every current-time curve has the same shape; but its magnitude is scaled by $1/(1 + \xi\theta)$ according to the potential to which the step is made. Note that for very positive potentials (relative to $E^{0\prime}$), this scale factor is zero. Thus $i(t)$ has a value between zero and $i_d(t)$, depending on E, as shown in Figure 5.1.3.

5.4.2 The Shape of the Current-Potential Curve

In the sampled-current voltammetric experiment, our goal is to obtain an $i(\tau)$-E curve by (a) performing several step experiments with different final potentials E, (b) sampling the current response at a fixed time τ after the step, and (c) plotting $i(\tau)$ *vs.* E. Here we consider the shape of this curve for a reversible couple and the kinds of information one can obtain from it.

Equation 5.4.17 really answers the question for us. For a fixed sampling time τ,

$$i(\tau) = \frac{i_d(\tau)}{1 + \xi\theta} \tag{5.4.18}$$

which can be rewritten as

$$\xi\theta = \frac{i_d(\tau) - i(\tau)}{i(\tau)} \tag{5.4.19}$$

and expanded:

$$E = E^{0'} + \frac{RT}{nF} \ln \frac{D_R^{1/2}}{D_O^{1/2}} + \frac{RT}{nF} \ln \frac{i_d(\tau) - i(\tau)}{i(\tau)} \tag{5.4.20}$$

When $i(\tau) = \frac{1}{2}i_d(\tau)$, the current ratio becomes unity so that the third term vanishes. The potential for which this is so is $E_{1/2}$, the *half-wave potential*:

$$\boxed{E_{1/2} = E^{0'} + \frac{RT}{nF} \ln \frac{D_R^{1/2}}{D_O^{1/2}}} \tag{5.4.21}$$

and (5.4.20) is often written

$$\boxed{E = E_{1/2} + \frac{RT}{nF} \ln \frac{i_d(\tau) - i(\tau)}{i(\tau)}} \tag{5.4.22}$$

It is interesting to compare (5.4.20) and (5.4.22) with the wave shape equations derived in a naive way for steady-state voltammetry in Chapter 1. They are identical in form.

These relations predict a wave that rises from baseline to the diffusion-controlled limit over a fairly narrow potential region (~150 mV) centered on $E_{1/2}$. Since the ratio of diffusion coefficients in (5.4.21) is nearly unity in almost any case, $E_{1/2}$ is usually a very good approximation to $E^{0'}$ for a reversible couple.

Note also that E *vs.* $\log[(i_d - i)/i]$ for the voltammetric wave should be a linear plot with a slope of $2.303RT/nF$ or $59.1/n$ mV at 25 °C. These "wave slopes" are often computed for experimental data in order to test for reversibility.

A quicker test is that $|E_{3/4} - E_{1/4}| = 56.4/n$ mV at 25° [Tomeš (35) criterion of reversibility]. The potentials $E_{3/4}$ and $E_{1/4}$ are those for which $i = 3i_d/4$ and $i = i_d/4$, respectively.

5.4.3 Concentration Profiles

Taking the inverse transforms of (5.4.14) and (5.4.15) yields the concentration profiles:

$$C_O(x, t) = C_O^* - \frac{C_O^*}{1 + \xi\theta} \operatorname{erfc}\left[\frac{x}{2(D_O t)^{1/2}}\right] \tag{5.4.23}$$

$$C_R(x, t) = \frac{\xi C_O^*}{1 + \xi\theta} \operatorname{erfc}\left[\frac{x}{2(D_R t)^{1/2}}\right] \tag{5.4.24}$$

Some other convenient equations relating concentrations can also be written. Let us solve for $A(s)$ and $B(s)$ in (5.4.7) and (5.4.8) in terms of the transformed surface concentrations $\bar{C}_O(0, s)$ and $\bar{C}_R(0, s)$; then we substitute into (5.4.10):

$$D_O^{1/2}\left(\bar{C}_O(0, s) - \frac{C_O^*}{s}\right) + D_R^{1/2}\bar{C}_R(0, s) = 0 \tag{5.4.25}$$

or, using the inverse transform,

$$D_O^{1/2}C_O(0, t) + D_R^{1/2}C_R(0, t) = C_O^* D_O^{1/2} \tag{5.4.26}$$

The more general relation for R initially present is

$$D_O^{1/2}C_O(0, t) + D_R^{1/2}C_R(0, t) = C_O^* D_O^{1/2} + C_R^* D_R^{1/2} \tag{5.4.27}$$

For the special case when $D_O = D_R$,

$$C_O(0, t) + C_R(0, t) = C_O^* + C_R^* \tag{5.4.28}$$

Equations 5.4.26 to 5.4.28 were derived without reference to any particular electrochemical perturbation or *i-E* function, and they hold for virtually any electrochemical method.†

Returning now to the step experiments for which (5.4.23) and (5.4.24) apply, we see that the surface concentrations are

$$\boxed{C_O(0, t) = C_O^*\left(1 - \frac{1}{1 + \xi\theta}\right)} \tag{5.4.29}$$

$$\boxed{C_R(0, t) = C_O^*\left(\frac{\xi}{1 + \xi\theta}\right)} \tag{5.4.30}$$

Since (5.4.17) shows that $i(t)/i_d(t) = 1/(1 + \xi\theta)$,

$$C_O(0, t) = C_O^*\left[1 - \frac{i(t)}{i_d(t)}\right] \tag{5.4.31}$$

$$C_R(0, t) = \xi C_O^*\left[\frac{i(t)}{i_d(t)}\right] \tag{5.4.32}$$

---

† Note also that for the step experiments under discussion, (5.4.23) and (5.4.24) show that $C_O(x, t) + C_R(x, t) = C_O^*$ at any point along the profiles, when $D_O = D_R$.

and, evaluating i_d from the Cottrell relation, we have

$$i(t) = \frac{nFAD_O^{1/2}}{\pi^{1/2}t^{1/2}}[C_O^* - C_O(0, t)] \tag{5.4.33}$$

$$i(t) = \frac{nFAD_R^{1/2}}{\pi^{1/2}t^{1/2}} C_R(0, t) \tag{5.4.34}$$

Since these relations hold at any time along the current decay, for sampled voltammetry we can replace t by the sampling time τ.

It is important to recognize the direct analogy between these rigorous relations [(5.4.31) to (5.4.34)] and those *assumed* for the current-concentration relation in the naive approach to mass transport used in Chapter 1. The equations are of precisely the same form; one need only replace m_O with $D_O^{1/2}/\pi^{1/2}t^{1/2}$ and m_R with $D_R^{1/2}/\pi^{1/2}t^{1/2}$ to interchange them exactly. The two approaches to deriving the *i-E* curve can be compared further as follows:

Naive Approach

Nernstian behavior and $i = nFAm_O[C_O^* - C_O(0, t)]$, $i = nFAm_R[C_R(0, t) - C_R^*]$ were assumed $\xrightarrow[\text{math}]{\text{Simple}}$ *i-E* curve

Rigorous Approach

Nernstian behavior, diffusion equations, and boundary conditions were assumed $\xrightarrow[\text{math}]{\text{More complex}}$ *i-E* curve as before and $i = nFAm_O[C_O^* - C_O(0, t)]$, $i = nFAm_R[C_R(0, t) - C_R^*]$ also as before

The rigorous treatment has therefore justified the *i-C* linkages we used before, and it increases our confidence in the simpler approach as a means for treating other systems.

5.4.4 Wave Shapes and Surface Concentrations at the DME (13, 36, 37)

In Section 5.3 we saw that, to a first approximation, current flow at the DME can be treated as a linear diffusion problem. The time dependence of the area is taken into account directly in terms of $m^{2/3}t^{1/6}$, and a multiplicative factor, $(7/3)^{1/2}$, accounts for increased mass transport due to the "stretching" of the diffusion layer. These concepts apply equally well to $i(t)$, as expressed in (5.4.17); thus the wave shape found for sampled-current voltammetry (5.4.22) applies also to polarography, which is in essence a sampled-current voltammetric experiment if the rate of potential sweep is sufficiently slow that E is virtually constant during a drop's lifetime.

Similarly, the surface concentrations are described by (5.4.29) and (5.4.30); hence (5.4.31) and (5.4.32) are valid for the DME. In this case, however, the maximum diffusion current is given by (5.3.7), and the analogs to (5.4.33) and (5.4.34) are

$$(i)_{max} = 708nD_O^{1/2}m^{2/3}t_{max}^{1/6}[C_O^* - C_O(0, t)] \tag{5.4.35}$$

$$(i)_{max} = 708nD_O^{1/2}m^{2/3}t_{max}^{1/6}C_R(0, t) \tag{5.4.36}$$

For average currents, the factor 708 would be replaced by 607. Obviously these relations still follow the forms

$$(i)_{max} = nFAm_O[C_O^* - C_O(0, t)] \tag{5.4.37}$$

$$(i)_{max} = nFAm_R[C_R(0, t) - C_R^*] \tag{5.4.38}$$

where m_O is now $[(7/3)D_O/\pi t_{max}]^{1/2}$ and m_R is defined analogously. The points emphasized at the end of the previous section also apply to polarography.

5.4.5 Applications of Reversible i-E Curves (9, 36 39)

We have seen that reversible (nernstian) systems are always at equilibrium. The kinetics are so facile that the interface is solely governed by thermodynamic aspects. It is not surprising, then, that reversible systems can provide no kinetic information. Instead they are useful for obtaining thermodynamic properties, such as standard potentials, free energies of reaction, and various equilibrium constants, just as potentiometric measurements are.

As an example, we consider the reversible reduction of a complex ion. To treat this problem, we use the simplified approach to the derivation of the *i-E* curve, as justified in the preceding sections. Suppose the process is represented schematically as

$$MX_p + ne + Hg \rightleftharpoons M(Hg) + pX \tag{5.4.39}$$

where the charges on the metal M and the ligands X are omitted for simplicity. A real example might be

$$Zn(NH_3)_4^{2+} + 2e + Hg \rightleftharpoons Zn(Hg) + 4NH_3 \tag{5.4.40}$$

For $M + ne + Hg \rightleftharpoons M(Hg)$,

$$E = E_M^{0'} + \frac{RT}{nF} \ln \frac{C_M(0, t)}{C_{M(Hg)}(0, t)} \tag{5.4.41}$$

and for $MX_p \rightleftharpoons M + pX$

$$K_d = \frac{C_M C_X^p}{C_{MX_p}} \tag{5.4.42}$$

The presumption of reversibility implies that both of these processes, as well as (5.4.39), are all simultaneously at equilibrium. Substituting (5.4.42) into (5.4.41),

$$E = E_M^{0'} + \frac{RT}{nF} \ln K_d - \frac{pRT}{nF} \ln C_X(0, t) + \frac{RT}{nF} \ln \frac{C_{MX_p}(0, t)}{C_{M(Hg)}(0, t)} \tag{5.4.43}$$

Let us now add the assumptions (a) that initially $C_{M(Hg)} = 0$, $C_{MX_p} = C_{MX_p}^*$, and $C_X = C_X^*$ and (b) that $C_X^* \gg C_{MX_p}^*$ so that the electrode process has little effect on the value of C_X at the surface. That is, $C_X(0, t) \simeq C_X^*$. Then the following relations apply:

$$i(t) = nFAm_C[C_{MX_p}^* - C_{MX_p}(0, t)] \tag{5.4.44}$$

$$i(t) = nFAm_A C_{M(Hg)}(0, t) \tag{5.4.45}$$

$$i_d(t) = nFAm_C C_{MX_p}^* \tag{5.4.46}$$

or,

$$C_{MX_p}(0, t) = \frac{i_d(t) - i(t)}{nFAm_C} \quad (5.4.47)$$

$$C_{M(Hg)}(0, t) = \frac{i(t)}{nFAm_A} \quad (5.4.48)$$

Substituting into (5.4.43), we obtain

$$\boxed{E = E_{1/2}^C + \frac{RT}{nF} \ln\left[\frac{i_d(t) - i(t)}{i(t)}\right]} \quad (5.4.49)$$

with

$$\boxed{E_{1/2}^C = E_M^{0\prime} + \frac{RT}{nF} \ln K_d - \frac{pRT}{nF} \ln C_X^* + \frac{RT}{nF} \ln \frac{m_A}{m_C}} \quad (5.4.50)$$

It is clear now that the wave *shape* is the same as that for the simple redox couple $O + ne \rightleftharpoons R$, but the *location* of the wave depends on K_d and C_X^*, in addition to the formal potential of the metal/amalgam couple. For constant K_d, increased concentrations of the complexing agent shift the wave to more extreme potentials. In effect, complexation stabilizes the oxidized form of M and raises the free energy required for its reduction. The stronger the binding in the complex (i.e., the smaller K_d is), the larger is the shift from the free metal potential $E_M^{0\prime}$. Actually K_d can be evaluated from this displacement:

$$E_{1/2}^C - E_M^{0\prime} = \frac{RT}{nF} \ln K_d - \frac{RT}{nF} p \ln C_X^* + \frac{RT}{nF} \ln \frac{m_A}{m_C} \quad (5.4.51)$$

In a practical situation, $E_M^{0\prime}$ is usually estimated from the voltammetric half-wave potential for the metal in a solution free of X, so that

$$E_{1/2}^C - E_{1/2}^M = \frac{RT}{nF} \ln K_d - \frac{RT}{nF} p \ln C_X^* + \frac{RT}{nF} \ln \frac{m_M}{m_C} \quad (5.4.52)$$

Note also the m_M/m_C is $D_M^{1/2}/D_C^{1/2}$, where D_C and D_M are the diffusion coefficients for the complex and the free metal ion in the solution phase.

From a plot of $E_{1/2}^C$ *vs.* $\ln C_X^*$ one also has access to the stoichiometric number p. Equation 5.4.50 shows that such a plot should have a slope of $-pRT/nF$.

Treatments of this sort are easily worked out for other types of electrode reactions, such as

$$O + mH^+ + ne \rightleftharpoons R + H_2O$$

$$2O + ne \rightleftharpoons R$$

$$O + ne \rightleftharpoons R \text{ (insoluble)}$$

Details are available in references on polarography and voltammetry (9, 36–39).

5.5 SAMPLED-CURRENT VOLTAMMETRY FOR AN IRREVERSIBLE ELECTRODE REACTION. THE SHAPE OF THE WAVE.

In this section, we will treat the reaction $O + ne \rightarrow R$ using the general (quasi-reversible) *i-E* characteristic. The two special cases, the totally irreversible system and the linearized system, will then follow.

5.5.1 Linear Diffusion: Current-Time Behavior (40)

The treatment of semi-infinite linear diffusion with the current governed by both mass transfer and charge transfer kinetics follows generally that given in Section 5.4.1. The diffusion equations for O and R are needed, as are the initial conditions, the semi-infinite conditions, and the flux balance condition. As we noted before, these lead to

$$\bar{C}_{\mathrm{O}}(x, s) = \frac{C_{\mathrm{O}}^*}{s} + A(s)\, e^{-(s/D_{\mathrm{O}})^{1/2}x} \tag{5.5.1}$$

$$\bar{C}_{\mathrm{R}}(x, s) = -\xi A(s)\, e^{-(s/D_{\mathrm{R}})^{1/2}x} \tag{5.5.2}$$

where $\xi = (D_{\mathrm{O}}/D_{\mathrm{R}})^{1/2}$

For the quasi-reversible case, we can evaluate $A(s)$ by applying the condition:

$$\frac{i}{nFA} = D_{\mathrm{O}}\left(\frac{\partial C_{\mathrm{O}}(x, t)}{\partial x}\right)_{x=0} = k_f C_{\mathrm{O}}(0, t) - k_b C_{\mathrm{R}}(0, t) \tag{5.5.3}$$

where

$$k_f = k^0\, e^{-\alpha n_a f(E - E^{0\prime})}$$

and

$$k_b = k^0\, e^{(1-\alpha)n_a f(E - E^{0\prime})}$$

with $f = F/RT$.†

The transform of (5.5.3) is

$$D_{\mathrm{O}}\left(\frac{\partial \bar{C}_{\mathrm{O}}(x, s)}{\partial x}\right)_{x=0} = k_f \bar{C}_{\mathrm{O}}(0, s) - k_b \bar{C}_{\mathrm{R}}(0, s) \tag{5.5.4}$$

and, by substitution from (5.5.1) and (5.5.2),

$$A(s) = -\frac{k_f}{D_{\mathrm{O}}^{1/2}} \frac{C_{\mathrm{O}}^*}{s(H + s^{1/2})} \tag{5.5.5}$$

where

$$H = \frac{k_f}{D_{\mathrm{O}}^{1/2}} + \frac{k_b}{D_{\mathrm{R}}^{1/2}} \tag{5.5.6}$$

† Note the distinction here between n and n_a. The former is the number of electrons involved in the overall electrode process, whereas n_a is the number involved in the rate-determining step (see Section 3.6).

Then,

$$\bar{C}_O(x, s) = \frac{C_O^*}{s} - \frac{k_f C_O^* e^{-(s/D_O)^{1/2}x}}{D_O^{1/2} s(H + s^{1/2})} \tag{5.5.7}$$

From (5.5.3)

$$\bar{i}(s) = nFAD_O\left[\frac{\partial \bar{C}_O(x, s)}{\partial x}\right]_{x=0} = \frac{nFAk_f C_O^*}{s^{1/2}(H + s^{1/2})} \tag{5.5.8}$$

or, taking the inverse transform,

$$\boxed{i(t) = nFAk_f C_O^* \exp(H^2 t) \operatorname{erfc}(Ht^{1/2})} \tag{5.5.9}$$

For the case when R is initially present at C_R^*, equation 5.5.9 becomes

$$\boxed{i(t) = nFA(k_f C_O^* - k_b C_R^*) \exp(H^2 t) \operatorname{erfc}(Ht^{1/2})} \tag{5.5.10}$$

At a constant potential, k_f, k_b, and H are constants and, since $e^{x^2} \operatorname{erfc}(x) = 1$ for $x = 0$ and $\lim_{x \to \infty} e^{x^2} \operatorname{erfc}(x) = 0$, the current-time curve has the shape shown in Figure 5.5.1. Note that the sluggish kinetics limit the current at $t = 0$ to a finite value proportional to k_f (with R initially absent).

5.5.2 Alternate Expression in Terms of η

An alternate expression for (5.5.10) can be given by noting that

$$k_f C_O^* - k_b C_R^* = k^0[C_O^* e^{-\alpha nf(E - E^{0\prime})} - C_R^* e^{(1-\alpha)nf(E - E^{0\prime})}] \tag{5.5.11}$$

or, by substituting for k^0 in terms of i_0 by (3.5.11),

$$k_f C_O^* - k_b C_R^* = \frac{i_0}{nFA}[e^{-\alpha nf\eta} - e^{(1-\alpha)nf\eta}] \tag{5.5.12}$$

Therefore, (5.5.10) may be written

$$i = i_0[e^{-\alpha nf\eta} - e^{(1-\alpha)nf\eta}] \exp(H^2 t) \operatorname{erfc}(Ht^{1/2}) \tag{5.5.13}$$

By similar substitutions into the expression for H, one has

$$H = \frac{i_0}{nFA}\left[\frac{e^{-\alpha nf\eta}}{C_O^* D_O^{1/2}} + \frac{e^{(1-\alpha)nf\eta}}{C_R^* D_R^{1/2}}\right] \tag{5.5.14}$$

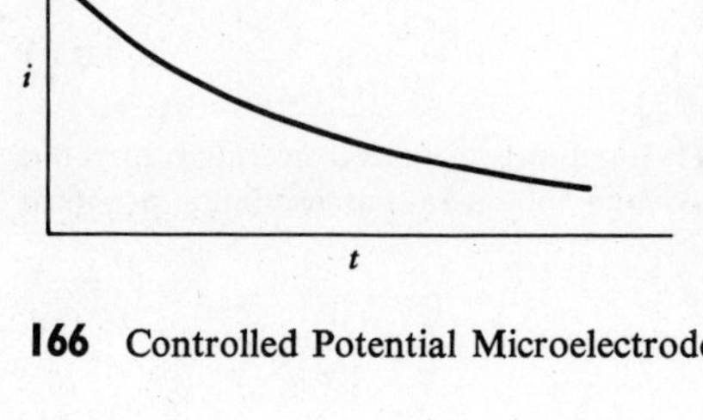

Figure 5.5.1
Current decay after the application of a step to a potential where species O is reduced with quasi-reversible kinetics.

Note that the form of (5.5.10) and (5.5.13) is

$$i = [i \text{ in the absence of mass transfer effects}] \cdot [f(H, t)]$$

where $f(H, t)$ accounts for the effects of mass transfer.

5.5.3 Linearized Current-Time Curve

For small values of $Ht^{1/2}$, the factor $\exp(H^2t)\,\mathrm{erfc}(Ht^{1/2})$ can be linearized. The expression

$$e^{x^2}\,\mathrm{erfc}(x) = \left(1 + x^2 + \frac{x^4}{2!} + \frac{x^6}{3!} + \cdots\right)\cdot\left[1 - \frac{2}{\pi^{1/2}}\left(x - \frac{x^3}{3} + \frac{x^5}{5\cdot 2!} - \cdots\right)\right] \tag{5.5.15}$$

applies generally, but at very small values of x (when higher powers can be neglected), it becomes

$$e^{x^2}\,\mathrm{erfc}(x) \simeq 1 - \frac{2x}{\pi^{1/2}} \tag{5.5.16}$$

Then, (5.5.9) becomes

$$i = nFAk_fC_O^*\left(1 - \frac{2Ht^{1/2}}{\pi^{1/2}}\right) \tag{5.5.17}$$

In a system for which R is initially absent we can apply a step to the potential region at the foot of the wave (where k_f, hence H, is still small), plot i *vs.* $t^{1/2}$, extrapolate the linear plot to $t = 0$, and obtain k_f from the intercept.

Likewise, (5.5.13) can be written

$$i = i_0[e^{-\alpha nf\eta} - e^{(1-\alpha)nf\eta}]\left(1 - \frac{2Ht^{1/2}}{\pi^{1/2}}\right) \tag{5.5.18}$$

This relation applies only to a system containing both O and R initially, so that E_{eq} is defined. Stepping from E_{eq} to another potential gives a step of magnitude η, and a plot of i *vs.* $t^{1/2}$ has as its intercept the kinetically controlled current free of mass transfer effects. A plot of $i_{t=0}$ *vs.* η then be used to obtain i_0.

For small values of η, the linearized i-η characteristic can be used, so that (5.5.13) becomes

$$i = -\frac{nFi_0\eta}{RT}\exp(H^2t)\,\mathrm{erfc}(Ht^{1/2}) \tag{5.5.19}$$

Then for small η and small $Ht^{1/2}$, one has a "completely linearized" form:

$$i = -\left(\frac{nFi_0}{RT}\right)(\eta)\left(1 - \frac{2Ht^{1/2}}{\pi^{1/2}}\right) \tag{5.5.20}$$

5.5.4 Totally Irreversible Reactions (40)

When one current component (e.g., the anodic component) is negligible, (5.5.9) becomes

$$i = nFAC_O^*k_f\exp\left(\frac{k_f^2t}{D_O}\right)\mathrm{erfc}\left[k_f\left(\frac{t}{D_O}\right)^{1/2}\right] \tag{5.5.21}$$

This may be rewritten

$$i = \left(\frac{nFAD_O^{1/2}C_O^*}{t^{1/2}}\right)\lambda \exp(\lambda^2)\,\mathrm{erfc}(\lambda) \tag{5.5.22}$$

where $\lambda = Ht^{1/2} = k_f t^{1/2}/D_O^{1/2}$. The first factor is quickly recognized as the Cottrell current i_d times $\pi^{1/2}$, so that

$$\boxed{\frac{i}{i_d} = F_1(\lambda) = \pi^{1/2}\lambda \exp(\lambda^2)\,\mathrm{erfc}(\lambda)} \tag{5.5.23}$$

This equation can be viewed in two ways. For a given step experiment, k_f and D_O are constant; hence λ is a dimensionless parameter proportional to time and (5.5.23) is clearly a dimensionless representation of $i(t)$, the current decay. On the other hand for sampled-current voltammetry, since the time is fixed at τ, λ becomes a function of the terminal step potential because of its proportionality to k_f. In this situation, (5.5.23) describes the shape of the $i(\tau)$-E curve. In fact, it is the shape of *every possible* totally irreversible wave (i.e., for any set of C_O^*, τ, k^0, $E^{0'}$, and α), because it is dimensionless.

A plot of this function like that in Figure 5.5.2 can be used conveniently to extract kinetic information from step experiments. For example, one can measure $i(\tau)/i_d(\tau)$ at some potential on the rising portion of a sampled-current wave and, by reference to Figure 5.5.2, λ for that potential is found. Since $\lambda = k_f\tau^{1/2}/D_O^{1/2}$, k_f is easily obtained if $D_O^{1/2}$ is known. If k_f is available for several potentials, k^0 and α are available from a plot of $\log k_f$ *vs.* E, which will be linear if the theoretical kinetic relation $k_f = k^0 \exp[-\alpha n_a f(E - E^{0'})]$ holds. A nonlinear plot may indicate more than one rate-determining step in the heterogeneous charge transfer mechanism, or it could reflect a potential dependence of α.

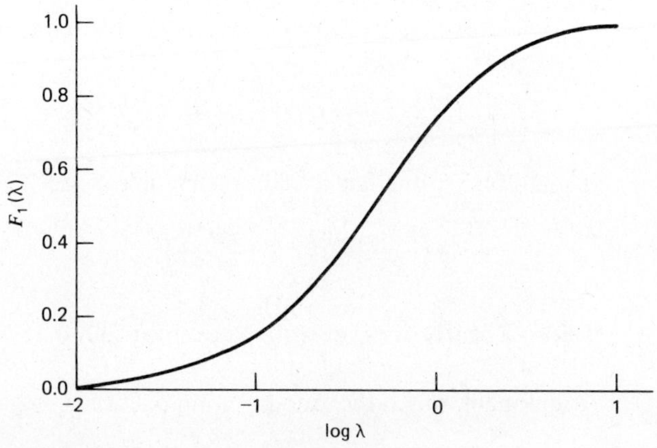

Figure 5.5.2
Wave shape function for sampled-current voltammetry of a totally irreversible system.

5.5.5 Irreversible Polarographic Waves (37, 40–42)

In the past, we have been able to obtain current-time curves for diffusion at the expanding sphere by taking the corresponding result for linear diffusion and accounting explicitly for the changing area with time and the stretching effect. The latter factor seemed to manifest itself as an effective diffusion coefficient of $(7/3)D$. The same tricks will almost work again here.

For a totally irreversible system under conditions of linear diffusion, equation 5.5.23 is the essential result. Since it deals with the ratio i/i_d, the changing area divides out and needs no further consideration. If we substitute as before for the diffusion coefficient,

$$\frac{i}{i_d} \cong \pi^{1/2}(\tfrac{3}{7})^{1/2}\lambda \exp[(\tfrac{3}{7})\lambda^2] \operatorname{erfc}[(\tfrac{3}{7})^{1/2}\lambda] \tag{5.5.24}$$

or

$$\frac{i}{i_d} \cong F_1\left(\frac{\chi}{2}\right) \tag{5.5.25}$$

if we define $\chi = 2(\frac{3}{7})^{1/2}\lambda = (\frac{12}{7})^{1/2}\lambda$.

Koutecky solved this problem more rigorously and expressed the result as (41–42)

$$\boxed{\frac{i}{i_d} = F_2(\chi)} \tag{5.5.26}$$

where $F_2(\chi)$ is a numeric function computed from a power series. Table 5.5.1 gives some representative values. A comparison between the *approximate* function $F_1(\chi)$ and the *exact* function $F_2(\chi)$ will reveal that $F_1(\chi)$ is consistently high by margins ranging up to about 10%. The discrepancy arises because the surface concentrations in this case are determined by an interplay of diffusion and faradaic reaction, rather than being fixed by the electrode potential almost alone, as they are in a reversible system.

A simplified equation for the treatment of totally irreversible polarographic waves has been proposed by Meites and Israel (37, 43). From the definition of χ and k_f,

$$\chi = (\tfrac{12}{7})^{1/2} \frac{k_f^0 t^{1/2}}{D_O^{1/2}} e^{-\alpha n_a f E} \tag{5.5.27}$$

where we recall that k_f^0 is the value of k_f at $E = 0$ on the potential scale in use. Its value is $k^0 e^{\alpha n_a f E^{0'}}$. From (5.5.27) one easily obtains

$$\ln(\tfrac{7}{12})^{1/2}\chi = \ln \frac{k_f^0 t^{1/2}}{D_O^{1/2}} - \alpha n_a f E \tag{5.5.28}$$

or

$$E = \frac{2.303RT}{\alpha n_a F} \log \frac{k_f^0 t^{1/2}}{D_O^{1/2}} - \frac{2.303RT}{\alpha n_a F} \log(\tfrac{7}{12})^{1/2}\chi \tag{5.5.29}$$

From Koutecky's values of $F(\chi)$, Meites and Israel found that the equation

$$\log(\tfrac{7}{12})^{1/2}\chi \cong -0.130 + 0.9163 \log \frac{i}{i_d - i} \tag{5.5.30}$$

Table 5.5.1

Shape Function of a Totally Irreversible Wave[a]

χ	i/i_d	χ	i/i_d
0.05	0.0428	1.4	0.5970
0.1	0.0828	1.6	0.6326
0.2	0.1551	1.8	0.6623
0.3	0.2189	2.0	0.6879
0.4	0.2749	2.5	0.7391
0.5	0.3245	3.0	0.773
0.6	0.3688	4.0	0.825
0.7	0.4086	5.0	0.8577
0.8	0.4440	10.0	0.9268
0.9	0.4761	20.0	0.9629
1.0	0.5050	50.0	0.9851
1.2	0.5552	∞	1

[a] Originally reported in References 41 and 42; also available in Reference 40.

is valid for $0.1 < (i/i_d) < 0.94$. Substitution into (5.5.29) therefore gives, at 25°,

$$E = \frac{0.059}{\alpha n_a} \log\left(1.349 \frac{k_f^0 t^{1/2}}{D_O^{1/2}}\right) + \frac{0.0542}{\alpha n_a} \log\left(\frac{i_d - i}{i}\right) \tag{5.5.31}$$

or

$$E = E_{1/2} + \frac{0.0542}{\alpha n_a} \log\left(\frac{i_d - i}{i}\right) \tag{5.5.32}$$

with

$$E_{1/2} = \left(\frac{0.059}{\alpha n_a}\right) \log\left(1.349 \frac{k_f^0 t^{1/2}}{D_O^{1/2}}\right)$$

Equations 5.5.31 and 5.5.32 involve an implicit assumption that t is the drop time, t_{max}.†

Obviously a plot of E *vs.* $\log[(i_d - i)/i]$ should be linear with a slope of $54.2/\alpha n_a$ mV. Alternatively, it is easily shown that $|E_{3/4} - E_{1/4}| = 51.7/\alpha n_a$ mV at 25°. Since $\alpha < 1$ generally, both the wave slope and the Tomeš criterion for a totally irreversible system will be significantly larger than for a reversible system with an equivalent n value. These figures of merit are not without ambiguity, however. Consider the predicted wave slope for $n_a = 2$ and $\alpha = 0.45$. One could diagnose the system as either a reversible one-electron or an irreversible multielectron process. The ambiguity can sometimes be removed by obtaining an independent estimate of n (e.g., from the wave height or from coulometry).

† Corrections for electrode sphericity are available. See the original literature for details (37, 44).

Figure 5.5.3 is a display of actual data reported by Meites and Israel (43) for the polarographic reduction of chromate. Note that the abscissa contains an explicit correction for the effect of the drop time on the position of the wave as predicted in (5.5.32), and that the data from several runs with different drop times then fall on a single line.

In general, irreversible waves are more drawn out than reversible ones, and their half-wave potentials are much more extreme than $E^{0'}$, which may be available independently from potentiometric measurements. Figure 5.5.4 is a display of the contrasting behavior expected for reversible and irreversible systems with the same n value and the same $E^{0'}$.

5.5.6 Information from Irreversible Waves (37, 40, 45, 46)

By this point, it is clear that irreversible systems can furnish only *kinetic* information. We can determine values of α, k_f, k_b, k^0, and n_a, but *thermodynamic* results such as $E^{0'}$ and free energies are not available. As a rule of thumb, a system with $k^0 > 2 \times 10^{-2}$ cm/sec appears reversible on a polarographic (2–4 sec) time scale. A heterogeneous charge transfer with $k^0 < 3 \times 10^{-5}$ cm/sec will behave in a totally irreversible manner, and one can evaluate the rate parameters as described above. Systems with k^0 between these limits are quasi-reversible, and some kinetic information can be obtained from

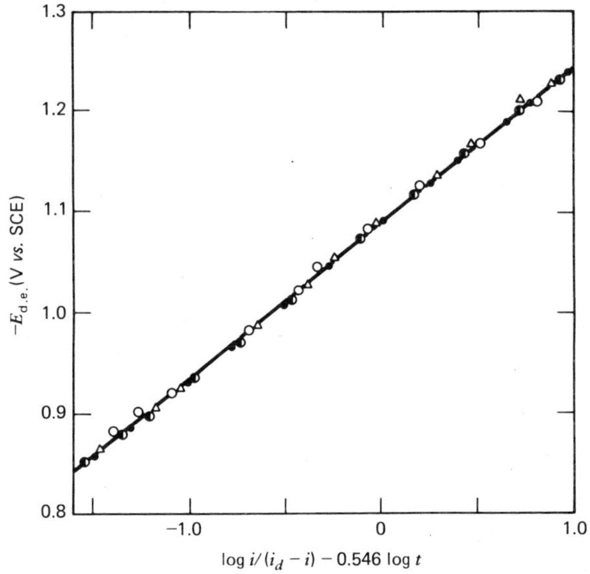

Figure 5.5.3

Wave slope plots for the reduction of 1.0 mM CrO_4^{2-} in 0.1 M NaOH. The different symbols refer to curves recorded with different drop times at −0.80 V *vs.* SCE: $t_{max} = 7.5$ sec (open circles), 5.5 sec (triangles), 4.1 sec (half-filled circles), and 3.4 sec (filled circles). See Figure 5.3.3 for an actual polarogram for this system. [Reprinted with permission from L. Meites and Y. Israel, *J. Am. Chem. Soc.*, **83**, 4903 (1961). Copyright 1961, American Chemical Society.]

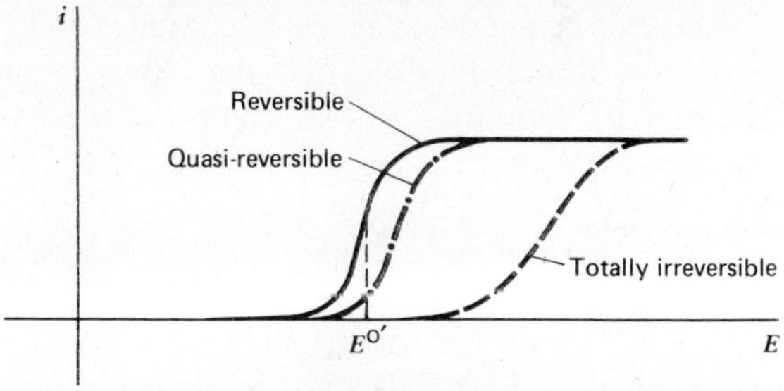

Figure 5.5.4
Wave shapes and positions expected for the reduction of species O to R under reversible, quasi-reversible, and totally irreversible kinetics. The formal potential for the couple is $E^{0'}$, as indicated.

them through the treatment prescribed by Randles (45, 46). Naturally, the precision of the kinetic information deteriorates as the reversible limit is approached.

In principle, any system can be made to appear irreversible if the experimental time scale is short enough. Note that λ in (5.5.23) depends on the product $k_f t$ (or, for sampled-current voltammetry, on $k_f \tau$). Suppose $\tau = 1$ sec and one is taking data well out onto the plateau of a wave that appears reversible. Of course, these conditions imply that k_f is large, hence λ is very large and $i \approx i_d$. [Since the working potential is well past $E^{0'}$, k_b is small compared to k_f and (5.5.23) would apply.] If k_f were infinite, this condition would always apply no matter how small τ were made. However, in a real system, τ can be reduced until λ at the working potential approaches unity. Then i/i_d is noticeably less than unity; hence one has forced the system to appear totally irreversible. What has been done, of course, is that the electrochemical perturbation has been made so rapid that the system can no longer move fast enough to keep up (i.e., maintain equilibrium).

Practical step experiments can be carried out on time scales on the order of 1 μsec if the potential change is not too large. For big excursions, double-layer charging becomes a greater interference and lengthens the minimum pulse width. See Chapter 13 for more details.

In general kinetic parameters extracted from totally irreversible polarographic waves should be viewed with caution. The treatment here involves a simple electron transfer reaction with no other perturbations. Such electron transfers, unless accompanied by major structural reorganization, are frequently quite rapid. The irreversibility observed in many cases can be attributed to homogeneous chemical reactions preceding or following the electron transfer step. For example, the reduction of nitrobenzene in aqueous solutions is irreversible, ultimately leading, depending on the pH, to phenylhydroxylamine (29, 47):

$$PhNO_2 + 4H^+ + 4e \rightarrow Ph\overset{H}{N}OH + H_2O \tag{5.5.33}$$

However, the first electron transfer step

$$PhNO_2 + e \rightarrow PhNO_2^{\cdot -} \tag{5.5.34}$$

is probably quite rapid, as found from measurements in nonaqueous solvents, such as DMF (47). The irreversibility observed in aqueous solutions probably arises because of the series of protonations and electron transfers following the first electron addition. If one fitted the observed polarographic curve of nitrobenzene using the totally irreversible electron transfer model, kinetic parameters for the electron transfer might be obtained, but they would be of no significance. Treatment of such complex systems requires a more complete elucidation of the electrode reaction mechanism as discussed in Chapter 11.

5.6 MULTICOMPONENT SYSTEMS AND MULTISTEP CHARGE TRANSFERS

Consider the case in which two reducible substances, O and O′, are present in the same solution, so that the consecutive electrode reactions $O + ne \rightarrow R$ and $O' + n'e \rightarrow R'$ can occur. Suppose the first process takes place at less extreme potentials than the second and that the second does not commence until the mass-transfer-limited region has been reached for the first. The reduction of species O can then be studied without interference from O′. In contrast, one must observe the current from O′ superimposed on that caused by the mass-transfer-limited flux of O.

In the potential region where both processes are limited by the rates of mass transfer [i.e., $C_O(0, t) = C_{O'}(0, t) = 0$], the total current is simply the sum of the individual diffusion currents. For a stationary planar electrode, one has

$$(i_d)_{\text{total}} = \frac{FA}{\pi^{1/2}t^{1/2}}(nD_O^{1/2}C_O^* + n'D_{O'}^{1/2}C_{O'}^*) \tag{5.6.1}$$

but for polarography, the approximation for maximum currents is

$$(i_d)_{\text{max,total}} = 708m^{2/3}t^{1/6}(nD_O^{1/2}C_O^* + n'D_{O'}^{1/2}C_{O'}^*) \tag{5.6.2}$$

Note that t in (5.6.1) is a variable time following the potential step, whereas in (5.6.2) it stands for the drop time.

In making measurements by sampled-current voltammetry, one would obtain traces like those in Figure 5.6.1. The diffusion current for the first wave can be subtracted

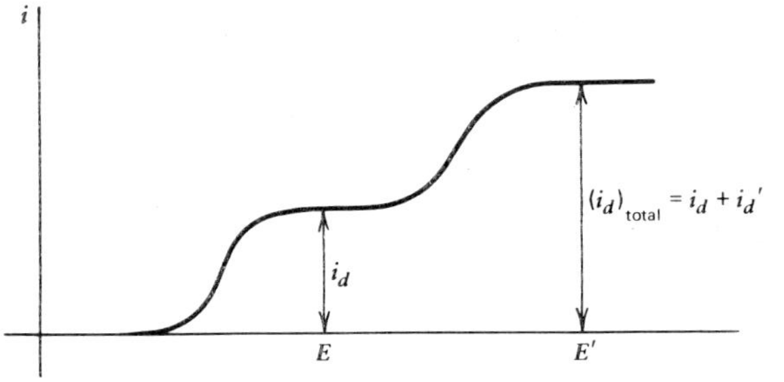

Figure 5.6.1
Sampled-current voltammogram for a two-component system.

from the total current of the composite wave to obtain the current attributable to O′ alone. That is,

$$i_d' = (i_d)_{\text{total}} - i_d \tag{5.6.3}$$

where i_d and i_d' are the current components due to O and O′, respectively.

If polarography is the technique employed (Figure 5.6.2), the drop times will generally be different for the two potentials used for the measurements of the single-component and composite currents. A correction must therefore be applied in the subtraction used to obtain i_d'. The measured currents can be expressed:

$$i_d(E) = 708nD_O^{1/2}C_O^* m^{2/3}t^{1/6} \tag{5.6.4}$$

$$(i_d)_{\text{total}}(E') = 708m^{2/3}t'^{1/6}(nD_O^{1/2}C_O^* + n'D_{O'}^{1/2}C_{O'}^*) \tag{5.6.5}$$

where the potential dependence has been explicitly declared for the currents. The drop times are t and t'. Inspection shows that

$$i_d'(E') = (i_d)_{\text{total}}(E') - \left(\frac{t'}{t}\right)^{1/6} i_d(E) \tag{5.6.6}$$

Similar considerations hold for a system in which a single species O is reduced in several steps, depending on potential, to more than one possible product. That is,

$$O + n_1e \rightarrow R_1 \tag{5.6.7}$$

$$R_1 + n_2e \rightarrow R_2 \tag{5.6.8}$$

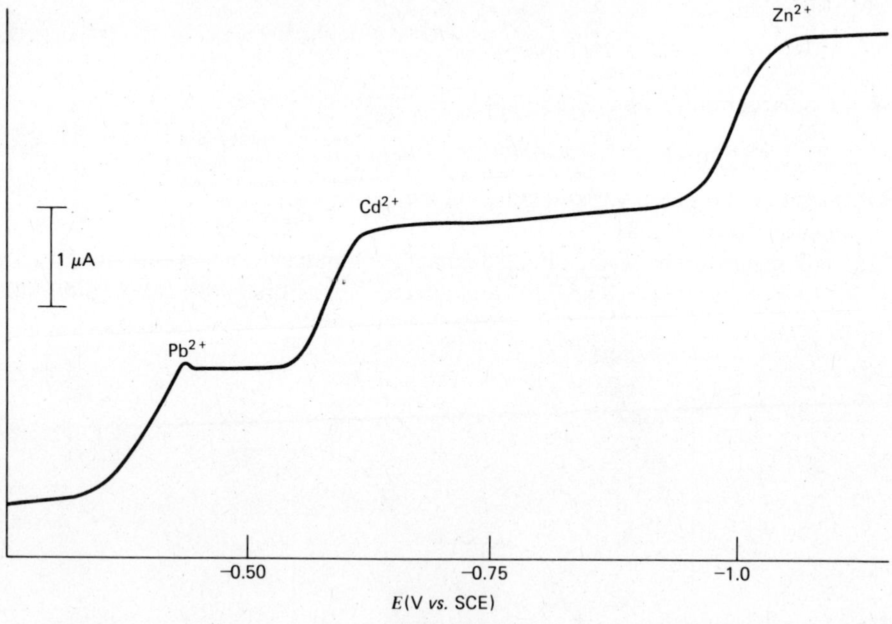

Figure 5.6.2

Polarogram of 0.2 mM Pb^{2+}, Cd^{2+}, and Zn^{2+} in 0.05 M KCl. The wave for Pb^{2+} shows a slight maximum of the type described in the note to Section 5.3.2. Only the envelope of the polarogram is shown here. Current oscillations are not depicted.

where the second step occurs at more extreme potentials than the first. A simple example is molecular oxygen, which is reduced in two steps in neutral solution, as shown in Figure 5.6.3. In the first, oxygen goes to hydrogen peroxide in a two-electron change, which is manifested at the DME by a wave near −0.1 V *vs.* SCE. A second two-electron step takes hydrogen peroxide to water. At potentials less extreme than about −0.5 V, the second step does not occur to any appreciable extent; hence one sees only a single wave corresponding to a diffusion-limited, two-electron process. At still more negative potentials, the second step occurs more readily, and beyond −1.2 V oxygen is reduced completely to water at the diffusion-limited rate.

For the general process [(5.6.7) to (5.6.8)] it is clear that at potentials for which the reduction of O to R_2 is diffusion controlled, the current following a potential step is simply

$$i_d = \frac{FAD_O^{1/2}C_O^*}{\pi^{1/2}t^{1/2}}(n_1 + n_2) \tag{5.6.9}$$

Equations for currents measured in sampled-current voltammetric experiments, including polarography, can be written analogously.

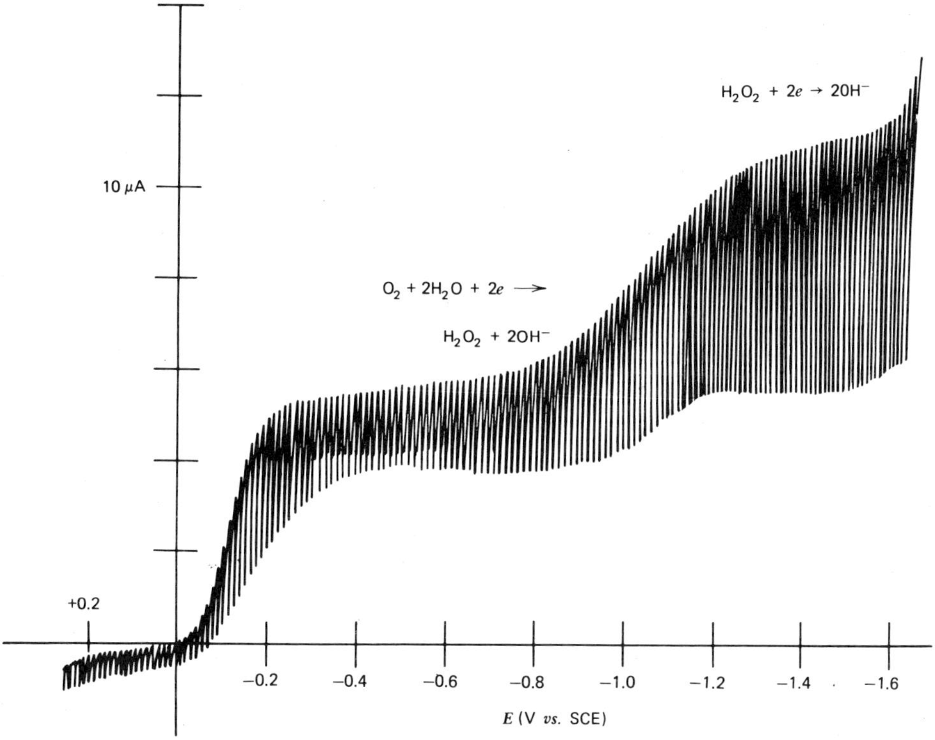

Figure 5.6.3
Polarogram of air-saturated 0.1 *M* KNO_3 with Triton X-100 added as a maximum suppressor.

5.7 CHRONOAMPEROMETRIC REVERSAL TECHNIQUES

After the application of an initial potential step, one might wish to apply an additional step, or even a complex sequence of steps. The most common arrangement is the double-step technique, in which the first step is used to generate some species of interest and the second is used to examine it. The latter step might be made to any potential within the working range, but it usually is used to reverse the effects of the initial step. An example is shown in Figure 5.7.1. Suppose an electrode is immersed in a solution of species O that is reversibly reduced at $E^{0'}$. If the initial potential E_i is much more positive than $E^{0'}$, no electrolysis occurs until, at $t = 0$, the potential is changed abruptly to E_f, which is far more negative than $E^{0'}$. Species R is generated electrolytically for a period τ, then the second step shifts the electrode to the comparatively positive value E_r. In many cases, E_r is equal to E_i. The reduced form R is then incompatible with the electrode, and it is reoxidized to O. This experiment, in common with most reversal techniques, is designed to provide a direct observation of R after its electrogeneration. That feature is exceedingly useful for evaluating R's participation in homogeneous chemical reactions on a time scale comparable to τ.

5.7.1 Potential Step Reversal: Nernstian Behavior (48, 49)

Consider the experiment described just above, under the added condition that semi-infinite linear diffusion applies. The potential for $t > 0$ can be written:

$$E(t) = E_f + S_\tau(t)(E_r - E_f) \tag{5.7.1}$$

The step function $S_\tau(t)$ is zero for $t < \tau$ and unity for $t \geq \tau$. Since the couple is reversible, a nernstian relation links $C_O(0, t)$ and $C_R(0, t)$ with E at all times.

(a) Approaches to the Problem. To obtain a quantitative description of this experiment, one might consider first the result of the forward step, then use the concentration profiles applicable at τ as initial conditions for the diffusion equation describing

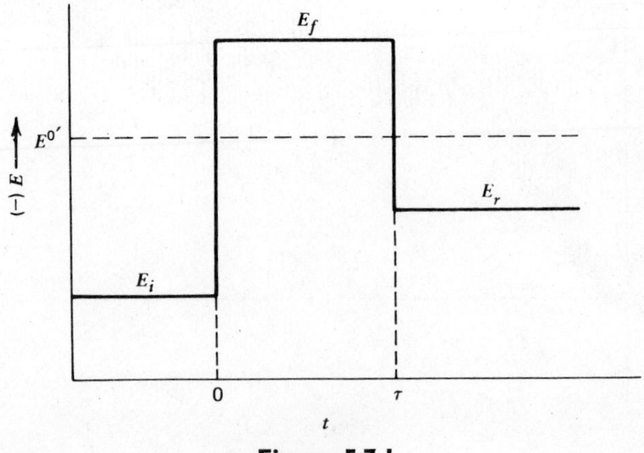

Figure 5.7.1
General waveform for a double potential step experiment.

events in the reversal step. In the present case, the effects of the forward step are well known (see Section 5.4.1), and this direct approach can be followed straightforwardly. Usually, however, reversal experiments present very complex concentration profiles to the theoretician attempting to describe the second phase, and it is often simpler to resort to methods based on the principle of superposition. We will introduce such a technique here as a means for solving the present problem.

The applied potential can be represented as the superposition of two signals—a constant component E_f for all $t > 0$ and a step component $E_r - E_f$ superimposed on the constant perturbation for $t > \tau$. Figure 5.7.2 is an expression of this idea, which is also embodied in equation 5.7.1. Similarly, the concentrations of O and R can be expressed as a superposition of two concentrations that may be regarded as responsive to the separate potential components:

$$C_O(x, t) = C_O{}^{I}(x, t) + S_\tau(t)C_O{}^{II}(x, t - \tau) \tag{5.7.2}$$

$$C_R(x, t) = C_R{}^{I}(x, t) + S_\tau(t)C_R{}^{II}(x, t - \tau) \tag{5.7.3}$$

Of course, the boundary conditions and initial conditions for this problem are most easily formulated in terms of the actual concentrations $C_O(x, t)$ and $C_R(x, t)$, and we write the initial situation as

$$C_O(x, 0) = C_O^* \qquad C_R(x, 0) = 0 \tag{5.7.4}$$

During the forward step we have

$$C_O(0, t) = C_O' \qquad C_R(0, t) = C_R' \tag{5.7.5}$$

and

$$C_O' = \theta' C_R' \tag{5.7.6}$$

where

$$\theta' = \exp[nf(E_f - E^{0\prime})] \tag{5.7.7}$$

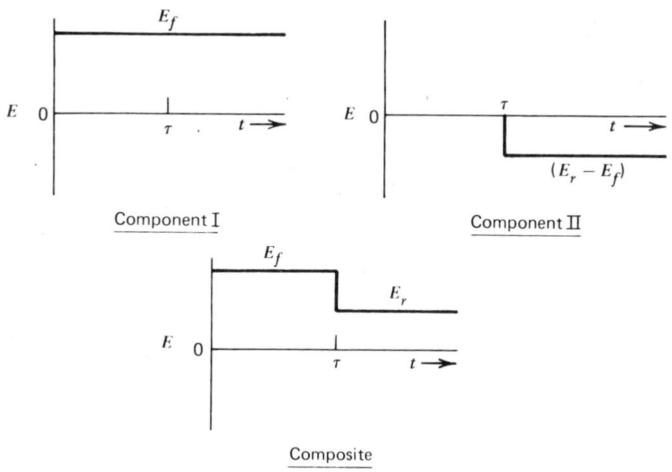

Figure 5.7.2
A double-step waveform as a superposition of two components.

The reversal step is defined by

$$C_O(0, t) = C_O'' \qquad C_R(0, t) = C_R'' \tag{5.7.8}$$

and

$$C_O'' = \theta'' C_R'' \tag{5.7.9}$$

where

$$\theta'' = \exp[nf(E_r - E^{0'})] \tag{5.7.10}$$

At all times, the semi-infinite conditions:

$$\lim_{x\to\infty} C_O(x, t) = C_O^* \qquad \lim_{x\to\infty} C_R(x, t) = 0 \tag{5.7.11}$$

and the flux balance:

$$J_O(0, t) = -J_R(0, t) \tag{5.7.12}$$

are applicable.

(b) Concentration Profiles. Note that all these conditions, as well as the diffusion equations for O and R, are linear. An important mathematical consequence is that the component concentrations $C_O{}^{\mathrm{I}}$, $C_O{}^{\mathrm{II}}$, $C_R{}^{\mathrm{I}}$, and $C_R{}^{\mathrm{II}}$ can all be carried through the problem separately. Each makes a separable contribution to every condition. We can therefore solve individually for each component, then we combine them through (5.7.2) and (5.7.3) to obtain the real concentration profiles.

Consider $C_O{}^{\mathrm{I}}(x, t)$ and $C_R{}^{\mathrm{I}}(x, t)$ first. We write the Fick's law relations as usual:

$$\frac{\partial C_O{}^{\mathrm{I}}(x, t)}{\partial t} = D_O \frac{\partial^2 C_O{}^{\mathrm{I}}(x, t)}{\partial x^2} \qquad \frac{\partial C_R{}^{\mathrm{I}}(x, t)}{\partial t} = D_R \frac{\partial^2 C_R{}^{\mathrm{I}}(x, t)}{\partial x^2} \tag{5.7.13}$$

Since $C_O{}^{\mathrm{II}}$ and $C_R{}^{\mathrm{II}}$ make no contribution at all before the second step, it is clear that $C_O{}^{\mathrm{I}}$ and $C_R{}^{\mathrm{I}}$ must satisfy all conditions applicable before $t = \tau$. Thus we find that

$$C_O{}^{\mathrm{I}}(x, 0) = C_O^* \qquad C_R{}^{\mathrm{I}}(x, 0) = 0 \tag{5.7.14}$$

$$C_O{}^{\mathrm{I}}(0, t) = C_O' \qquad C_R{}^{\mathrm{I}}(0, t) = C_R' \tag{5.7.15}$$

$$\lim_{x\to\infty} C_O{}^{\mathrm{I}}(x, t) = C_O^* \qquad \lim_{x\to\infty} C_R{}^{\mathrm{I}}(x, t) = 0 \tag{5.7.16}$$

Likewise, a flux balance can be written:

$$D_O\left(\frac{\partial C_O{}^{\mathrm{I}}}{\partial x}\right)_{x=0} = -D_R\left(\frac{\partial C_R{}^{\mathrm{I}}}{\partial x}\right)_{x=0} \tag{5.7.17}$$

These conditions, together with the nernstian relation (5.7.6), exactly restate the problem treated in Section 5.4.1. The results, left in transform space, are

$$\bar{C}_O{}^{\mathrm{I}}(x, s) = \frac{C_O^*}{s} - \frac{C_O^*}{s(1 + \xi\theta')} e^{-(s/D_O)^{1/2}x} \tag{5.7.18}$$

$$\bar{C}_R{}^{\mathrm{I}}(x, s) = \frac{\xi C_O^*}{s(1 + \xi\theta')} e^{-(s/D_R)^{1/2}x} \tag{5.7.19}$$

where $\xi = (D_O/D_R)^{1/2}$.

Now consider the components $S_\tau(t)C_O^{II}(x, t - \tau)$ and $S_\tau(t)C_R^{II}(x, t - \tau)$, which we carry symbolically as $F_O(x, t)$ and $F_R(x, t)$ for manipulative simplicity. They must, of course, satisfy the diffusion equations

$$\frac{\partial F_O(x, t)}{\partial t} = D_O \frac{\partial^2 F_O(x, t)}{\partial x^2} \qquad \frac{\partial F_R(x, t)}{\partial t} = D_R \frac{\partial^2 F_R(x, t)}{\partial x^2} \tag{5.7.20}$$

The presence of $S_\tau(t)$ shows immediately that

$$F_O(x, 0) = 0 \qquad F_R(x, 0) = 0 \tag{5.7.21}$$

and, from comparisons between (5.7.8) and (5.7.15), we have

$$F_O(0, t) = S_\tau(t)[C_O'' - C_O'] \qquad F_R(0, t) = S_\tau(t)[C_R'' - C_R'] \tag{5.7.22}$$

Also, the differences between (5.7.11) and (5.7.16) show that

$$\lim_{x \to \infty} F_O(x, t) = \lim_{x \to \infty} F_R(x, t) = 0 \tag{5.7.23}$$

The flux balance (5.7.12) can be written

$$D_O\left[\frac{\partial C_O^{I}(x, t)}{\partial x} + \frac{\partial F_O(x, t)}{\partial x}\right]_{x=0} = -D_R\left[\frac{\partial C_R^{I}(x, t)}{\partial x} + \frac{\partial F_R(x, t)}{\partial x}\right]_{x=0} \tag{5.7.24}$$

yet we have already established (5.7.17); therefore, it must be true that

$$D_O\left(\frac{\partial F_O(x, t)}{\partial x}\right)_{x=0} = -D_R\left(\frac{\partial F_R(x, t)}{\partial x}\right)_{x=0} \tag{5.7.25}$$

Finally, there are the nernstian relations (5.7.6) and (5.7.9). These conditions permit the evaluation of $F_O(x, t)$ and $F_R(x, t)$.

We can see in Section A.1.6 that transformation of (5.7.20) and application of the initial conditions (5.7.21) and the semi-infinite conditions (5.7.23) yield

$$\bar{F}_O(x, s) = A(s)\, e^{-(s/D_O)^{1/2}x} \tag{5.7.26}$$

$$\bar{F}_R(x, s) = B(s)\, e^{-(s/D_R)^{1/2}x} \tag{5.7.27}$$

By transforming (5.7.22) and comparing with (5.7.26) and (5.7.27), we find

$$\bar{F}_O(0, s) = A(s) = \frac{e^{-\tau s}}{s}(C_O'' - C_O') \tag{5.7.28}$$

$$\bar{F}_R(0, s) = B(s) = \frac{e^{-\tau s}}{s}(C_R'' - C_R') \tag{5.7.29}$$

and, according to the nernstian conditions, (5.7.28) can be rewritten as

$$\bar{F}_O(0, s) = A(s) = \frac{e^{-\tau s}}{s}(\theta'' C_R'' - \theta' C_R') \tag{5.7.30}$$

Application of the flux condition to (5.7.26) and (5.7.27), followed by substitution from (5.7.29) and (5.7.30), then gives

$$s^{1/2} D_O^{1/2} \frac{e^{-\tau s}}{s}(\theta'' C_R'' - \theta' C_R'') = -s^{1/2} D_R^{1/2} \frac{e^{-\tau s}}{s}(C_R'' - C_R') \tag{5.7.31}$$

which can be rearranged to

$$C_R'' = \frac{1 + \xi\theta'}{1 + \xi\theta''} C_R' \tag{5.7.32}$$

From (5.7.15) and (5.7.19) we find that

$$C_R' = \frac{\xi C_O^*}{1 + \xi\theta'} \tag{5.7.33}$$

hence

$$C_R'' = \frac{\xi C_O^*}{1 + \xi\theta''} \tag{5.7.34}$$

Thus

$$\bar{F}_O(x, s) = \frac{\xi C_O^* e^{-\tau s}}{s} \left(\frac{\theta''}{1 + \xi\theta''} - \frac{\theta'}{1 + \xi\theta'} \right) e^{-(s/D_O)^{1/2}x} \tag{5.7.35}$$

$$\bar{F}_R(x, s) = \frac{\xi C_O^* e^{-\tau s}}{s} \left(\frac{1}{1 + \xi\theta''} - \frac{1}{1 + \xi\theta'} \right) e^{-(s/D_R)^{1/2}x} \tag{5.7.36}$$

From (5.7.2) and (5.7.3) it is clear that

$$\overline{C_O}(x, s) = \overline{C_O^I}(x, s) + \bar{F}_O(x, s) \tag{5.7.37}$$

$$\overline{C_R}(x, s) = \overline{C_R^I}(x, s) + \bar{F}_R(x, s) \tag{5.7.38}$$

hence we obtain the concentration profile upon reverse transformation of (5.7.18) (5.7.19), (5.7.35), and (5.7.36):

$$\frac{C_O(x, t)}{C_O^*} = 1 - \left(\frac{1}{1 + \xi\theta'}\right) \text{erfc}\left[\frac{x}{2(D_O t)^{1/2}}\right] + S_\tau(t)\left(\frac{\xi\theta''}{1 + \xi\theta''} - \frac{\xi\theta'}{1 + \xi\theta'}\right) \text{erfc}\left\{\frac{x}{2[D_O(t - \tau)]^{1/2}}\right\} \tag{5.7.39}$$

$$\frac{C_R(x, t)}{C_O^*} = \left(\frac{\xi}{1 + \xi\theta'}\right) \text{erfc}\left[\frac{x}{2(D_R t)^{1/2}}\right] + S_\tau(t)\left(\frac{\xi}{1 + \xi\theta''} - \frac{\xi}{1 + \xi\theta'}\right) \text{erfc}\left\{\frac{x}{2[D_R(t - \tau)]^{1/2}}\right\} \tag{5.7.40}$$

(c) Current-Time Curves. Since the experiment from $0 < t \leq \tau$ is identical to that treated in Section 5.4.1, the current is given by (5.4.16), which is restated for the present context as

$$i_f(t) = \frac{nFAD_O^{1/2}C_O^*}{\pi^{1/2}t^{1/2}(1 + \xi\theta')} \tag{5.7.41}$$

The current during the reversal step is

$$-i_r(t) = nFAD_R\left[\frac{\partial C_R(x, t)}{\partial x}\right]_{x=0} \tag{5.7.42}$$

which is transformed to

$$-\bar{i}_r(s) = nFAD_R\left[\frac{\partial \bar{C}_R(x, s)}{\partial x}\right]_{x=0} \tag{5.7.43}$$

From (5.7.19), (5.7.36), and (5.7.38), we obtain the factor $(\partial \bar{C}_R(x, s)/\partial x)_{x=0}$. Substitution then gives

$$-\bar{i}_r(s) = nFAD_R^{1/2}\xi C_O^*\left[\frac{e^{-\tau s}}{s^{1/2}}\left(\frac{1}{1+\xi\theta'} - \frac{1}{1+\xi\theta''}\right) - \frac{1}{s^{1/2}(1+\xi\theta')}\right] \tag{5.7.44}$$

and upon inversion we have

$$-i_r(t) = \frac{nFAD_O^{1/2}C_O^*}{\pi^{1/2}}\left\{\left(\frac{1}{1+\xi\theta'} - \frac{1}{1+\xi\theta''}\right)\left[\frac{1}{(t-\tau)^{1/2}}\right] - \frac{1}{(1+\xi\theta')t^{1/2}}\right\} \tag{5.7.45}$$

A special case of interest involves stepping in the forward phase to a potential on the diffusion plateau of the reduction wave ($\theta' \approx 0$, $C_O' \approx 0$), then reversing to a potential on the diffusion plateau for reoxidation ($\theta'' \rightarrow \infty$, $C_R'' \approx 0$). In that instance, (5.7.45) simplifies to the result first obtained by Kambara (48):

$$\boxed{-i_r(t) = \frac{nFAD_O^{1/2}C_O^*}{\pi^{1/2}}\left[\frac{1}{(t-\tau)^{1/2}} - \frac{1}{t^{1/2}}\right]} \tag{5.7.46}$$

Note that this relation could also have been derived under the conditions $C_O' = 0$ and $C_R'' = 0$ without requiring nernstian behavior. It therefore holds also for irreversible systems, provided large enough potential steps are employed.

(d) Interpretation of Data. Figure 5.7.3 shows the kind of current response predicted by (5.7.45) and (5.7.46). In comparing a real experiment to the prediction, it is inconvenient to deal with absolute currents because they are proportional to $AD_O^{1/2}$, which is often difficult to ascertain. Instead, the reversal current $-i_r$ is usually divided by some particular value of the forward current. If t_r and t_f are the times at which the current measurements are made, then

$$\frac{-i_r}{i_f} = \left(\frac{t_f}{t_r - \tau}\right)^{1/2} - \left(\frac{t_f}{t_r}\right)^{1/2} \tag{5.7.47}$$

for the purely diffusion-limited case described by (5.7.46). If t_r and t_f values are selected in pairs so that $t_r - \tau = t_f$ always, then

$$\boxed{\frac{-i_r}{i_f} = 1 - \left(1 - \frac{\tau}{t_r}\right)^{1/2}} \tag{5.7.48}$$

When one calculates these ratios for several different values of t_r, they ought to fall on

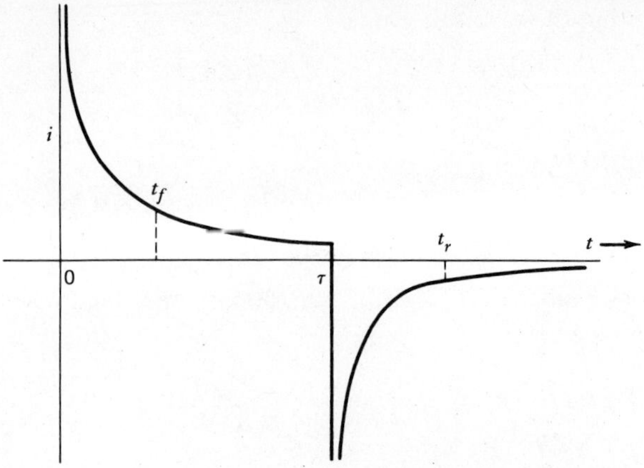

Figure 5.7.3
Current response in double-step chronoamperometry.

the working curve shown in Figure 5.7.4. A convenient quick reference for a stable system is that $-i_r(2\tau)/i_f(\tau) = 0.293$.

Deviations from the working curves indicate kinetic complications in the electrode reaction. For example, if species R decays to an electroinactive species, then i_r will be smaller than that predicted by (5.7.46) and the current ratio $-i_r/i_f$ will deviate negatively from that given in Figure 5.7.4. Chapter 11 covers in more detail the ways in which these experiments can be used to diagnose and quantify complex electrode processes.

5.7.2 Potential Step Reversal: Quasi-Reversible Behavior (50, 51)

The treatment for a quasi-reversible electrode reaction without homogeneous kinetics follows in a manner similar to that used in Section 5.7.1. The superposition principle

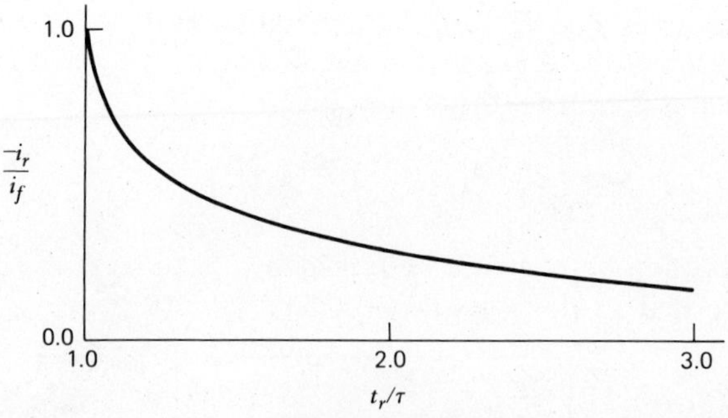

Figure 5.7.4
Working curve for $-i_r(t_r)/i_f(t_f)$ for $t_r = \tau + t_f$. The system is $O \pm ne \rightleftharpoons R$, with both O and R being stable on the time scale of observation.

is again assumed to be true, and the boundary conditions [(5.7.4) to (5.7.12)] are applied, except that (5.7.6) and (5.7.9) are replaced by

$$D_O\left(\frac{\partial C_O}{\partial x}\right)_{x=0} = k_f' C_O' - k_b' C_R' \qquad (t \leq \tau) \tag{5.7.49}$$

and

$$D_O\left(\frac{\partial C_O}{\partial x}\right)_{x=0} = k_f'' C_O'' - k_b'' C_R'' \qquad (t > \tau) \tag{5.7.50}$$

where

$$k_f' = k^0 e^{-\alpha nf(E_f - E^{0\prime})} \qquad k_f'' = k^0 e^{-\alpha nf(E_r - E^{0\prime})} \tag{5.7.51}$$

and

$$k_b' = k^0 e^{(1-\alpha)nf(E_f - E^{0\prime})} \qquad k_b'' = k^0 e^{(1-\alpha)nf(E_r - E^{0\prime})} \tag{5.7.52}$$

The development follows that of Section 5.5, and the final result is

$$\begin{aligned} -i_r = nFAC_O^*\{k_f' \exp(H'^2 t) \operatorname{erfc}(H' t^{1/2}) \\ + S_\tau(t)[k_f'' \exp[H''^2(t-\tau)] \operatorname{erfc}[H''(t-\tau)^{1/2}] \\ - k_f' \exp[H'^2(t-\tau)] \operatorname{erfc}[H'(t-\tau)^{1/2}]]\} \end{aligned} \tag{5.7.53}$$

where $C_R^* = 0$ and

$$H' = \frac{k_f'}{D_O^{1/2}} + \frac{k_b'}{D_R^{1/2}} \qquad H'' = \frac{k_f''}{D_O^{1/2}} + \frac{k_b''}{D_R^{1/2}} \tag{5.7.54}$$

Corrigan and Evans (51) have recently treated this problem by digital simulation (Appendix B). They have reported extensive theoretical results and have made comparisons with experimental data.

5.8 PULSE POLAROGRAPHIC METHODS

We have already noted that the limits of detection by conventional polarography are usually set by the magnitude of the charging current resulting from the continuous expansion of the mercury drop. These limits, which typically lie in the 10^{-5} M range, are comparable to those encountered in the competing methods of molecular and atomic absorption spectroscopy. Of course, lower limits are always useful, hence much effort has gone into the development of voltammetric methods with improved sensitivities. Three of them—tast polarography (9, 52–54), pulse polarography (9, 55–58), and differential pulse polarography (9, 55–58)—will be discussed here. As the names imply, these techniques were developed for use with the DME. All rely heavily on reducing the charging current's relative contribution to the measured current, and all are directed toward analytical measurements of concentration, so that sensitivity and resolution are the key features of performance. On the other hand, they are rarely used for diagnosis of electrode processes; hence detailed knowledge of wave shapes is not usually important.

5.8.1 Tast Polarography (9, 52–54)

In describing current flow at the DME under the usual polarographic conditions, we noted that the limiting faradaic current increases monotonically during the life of the drop and is described approximately by the Ilkovic equation (Section 5.3):

$$i_d = 708nD_O^{1/2}C_O^*m^{2/3}t^{1/6} \tag{5.8.1}$$

where t is time measured from the birth of the drop. In contrast, the charging current decreases steadily (Section 5.3) according to

$$i_c = 0.00567C_i(E_z - E)m^{2/3}t^{-1/3} \tag{5.8.2}$$

This contrast is illustrated in Figure 5.8.1. Clearly one can optimize the ratio of faradaic to charging current—and thus the sensitivity—by sampling the current at the instant just before drop fall.

Tast polarography features precisely this scheme, which is described in detail in Figure 5.8.2. At a fixed time τ after the birth of a drop, the current is sampled electronically, and this sample is presented to a recorder as a constant readout, until it is replaced at the sampling time during the next drop. The potentiostat is active at all times, and the potential is changed linearly with time or in small steps, as in conventional polarography. The record of the experiment is therefore a trace of the sampled currents versus potential, which is equivalent to time. Figure 5.8.3*b* shows an example for 10^{-5} M Cd^{2+} in 0.01 M HCl. Usually the drop time is enforced at some fixed value by dislodging each drop mechanically just after the current sample is taken. This

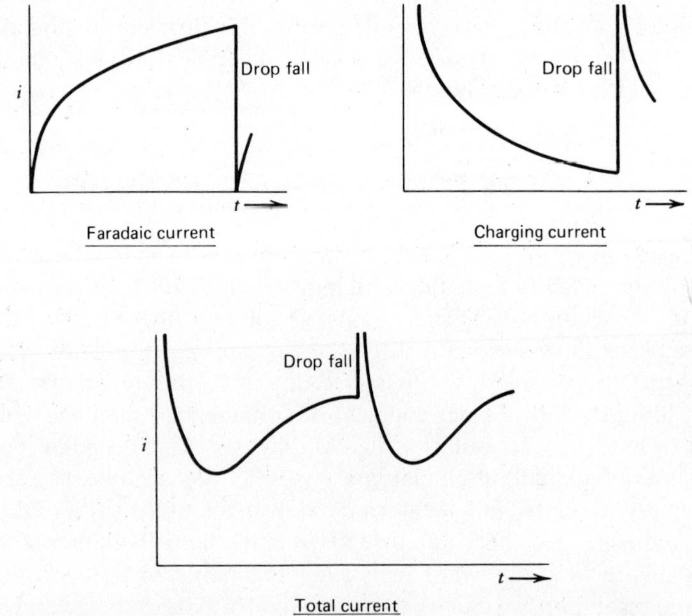

Figure 5.8.1
Superposition of capacitive and faradaic currents at a DME.

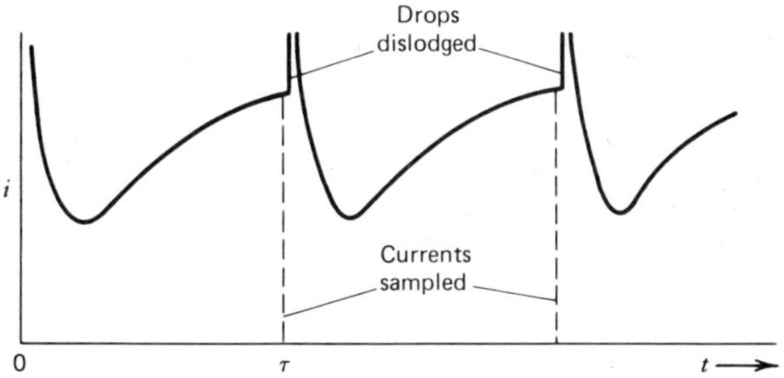

Figure 5.8.2
Sampling scheme for tast polarography.

procedure allows an even drop time over the entire potential range, and the pulse applied to the drop knocker, usually a solenoid linked to the DME, serves as a convenient time marker for the measurement of the next sampling interval τ and the next drop lifetime. Figure 5.8.4 is a diagram of the experimental arrangement.

Since the potentiostat is always active and the potential program is identical to that used in conventional polarography, the actual current flow at the electrode is the same as that observed in a conventional experiment featuring controlled drop times. The difference is that the recorder is fed only signals proportional to the sampled currents. The faradaic component to the limiting sampled current must be

$$i_d(\tau) = 708nD_O^{1/2}C_O^*m^{2/3}\tau^{1/6} \tag{5.8.3}$$

whereas the charging component is always

$$i_c(\tau) = 0.00567C_i(E_z - E)m^{2/3}\tau^{-1/3} \tag{5.8.4}$$

Of at least equal importance to this optimized faradaic-to-charging current ratio is the fact that the tast method eliminates from the recorded polarogram the sharp nonfaradaic spikes appearing at each drop fall. These spikes really represent a kind of

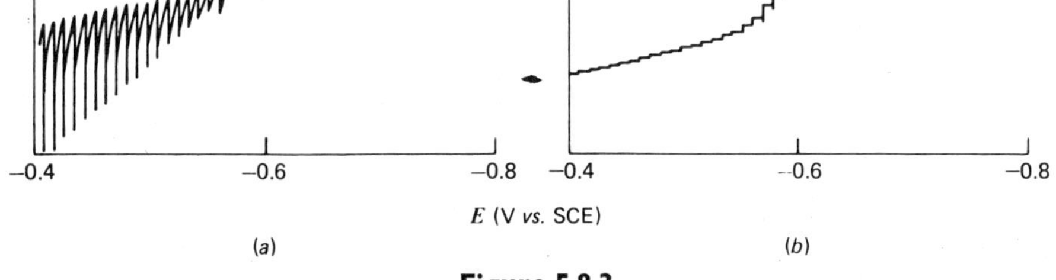

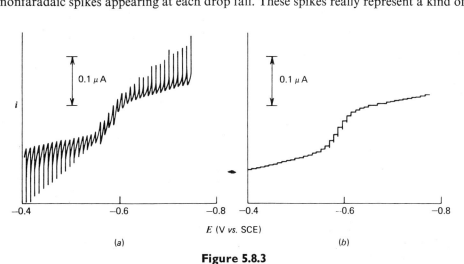

Figure 5.8.3
Polarograms for 10^{-5} M Cd^{2+} in 0.01 M HCl. (*a*) Conventional dc mode. (*b*) Tast mode.

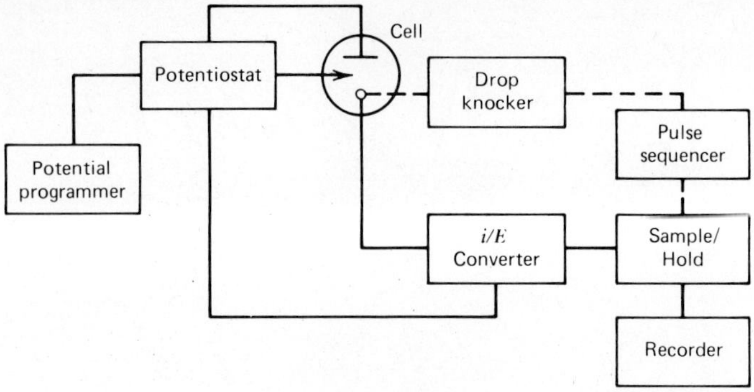

Figure 5.8.4
Schematic experimental arrangement for tast polarography.

noise level that interferes with the quantitative measurement of faradaic currents arising from solutes at low concentrations in conventional polarography. This point is illustrated quite effectively in Figure 5.8.3. The improvements in this method yield detection limits near 10^{-6} M, which are somewhat lower than those of conventional polarography. Note, however, that since tast measurements are only sampled-current presentations of conventional polarographic currents, all conclusions about the shapes of waves and all diagnostics developed for conventional measurements of maximum currents apply to the tast technique.

5.8.2 Normal Pulse Polarography (9, 55–58)

Since tast measurements actually record the current only during a very small time period late in a drop's life, all the faradaic current flow that occurs before the sampling period serves no useful purpose. Actually it even works to the detriment of sensitivity because it depletes the region near the electrode of the substance being measured and necessarily reduces its flux to the surface at the time of actual measurement. Normal pulse polarography is designed to eliminate this effect by blocking electrolysis prior to the measurement period. Figure 5.8.5 is an outline of the way in which this goal is achieved.

The electrode is held at a base potential, E_b, at which negligible electrolysis occurs, for most of the life of each mercury drop from the DME. After a fixed waiting period τ', measured from the birth of the drop, the potential is changed abruptly to value E for a period about 50 msec in duration. The potential pulse is ended by a return to the base value E_b. The current is sampled at a time τ near the end of the pulse, and a signal proportional to this sampled value is presented as a constant output to a recording device until the sample taken in the next drop lifetime replaces it. The drop is dislodged just after the pulse ends, then the whole cycle is repeated with successive drops, except that the step potential is made a few millivolts more extreme with each additional cycle. The output is a plot of sampled current versus step potential E, and it takes the form shown in Figure 5.8.6*a*. A block diagram of the apparatus is shown in Figure 5.8.7.

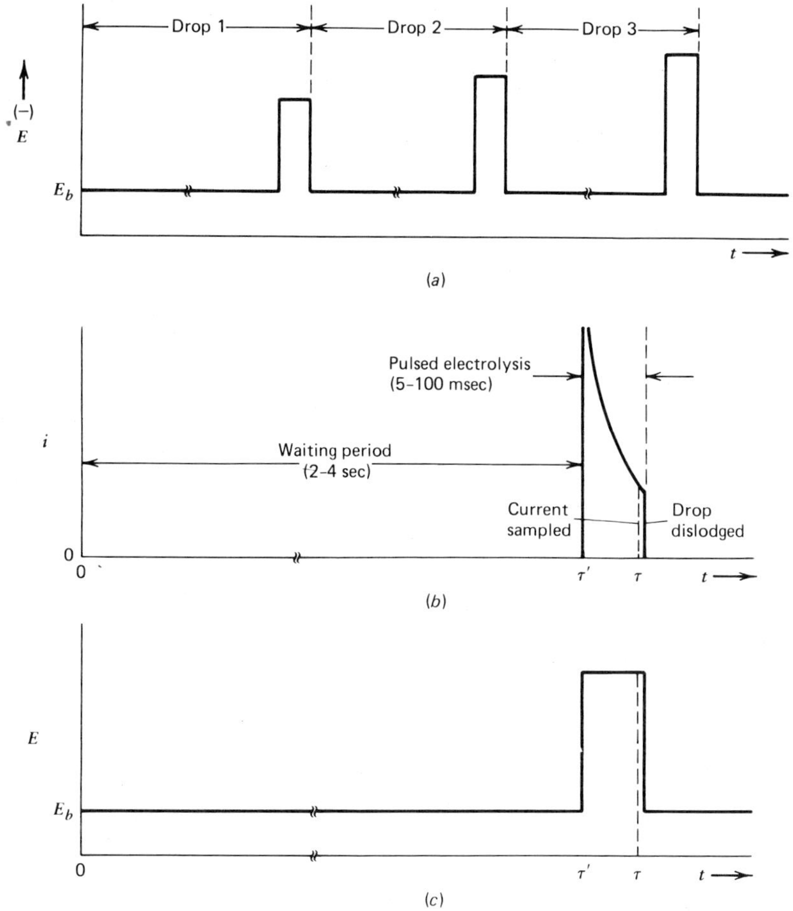

Figure 5.8.5

Sampling scheme for normal pulse polarography. (*a*) Potential program. (*b*) and (*c*) Current and potential during a single drop's lifetime.

This experiment, first performed by Barker and Gardner (55), is immediately recognizable as a sampled-current voltammetric measurement exactly on the model described in Sections 5.1, 5.2, 5.4, and 5.5. Since electrolysis during the waiting time is negligible, the initially uniform character of the concentration distribution in solution is preserved until the pulse is applied. Even though the electrode is approximately spherical, it acts as a planar surface during the short time of the actual electrolysis (Section 5.2.2), and therefore the sampled faradaic current on the plateau is

$$i_d = \frac{nFAD_O^{1/2}C_O^*}{\pi^{1/2}(\tau - \tau')^{1/2}} \tag{5.8.5}$$

where $(\tau - \tau')$ is time measured from the pulse rise.

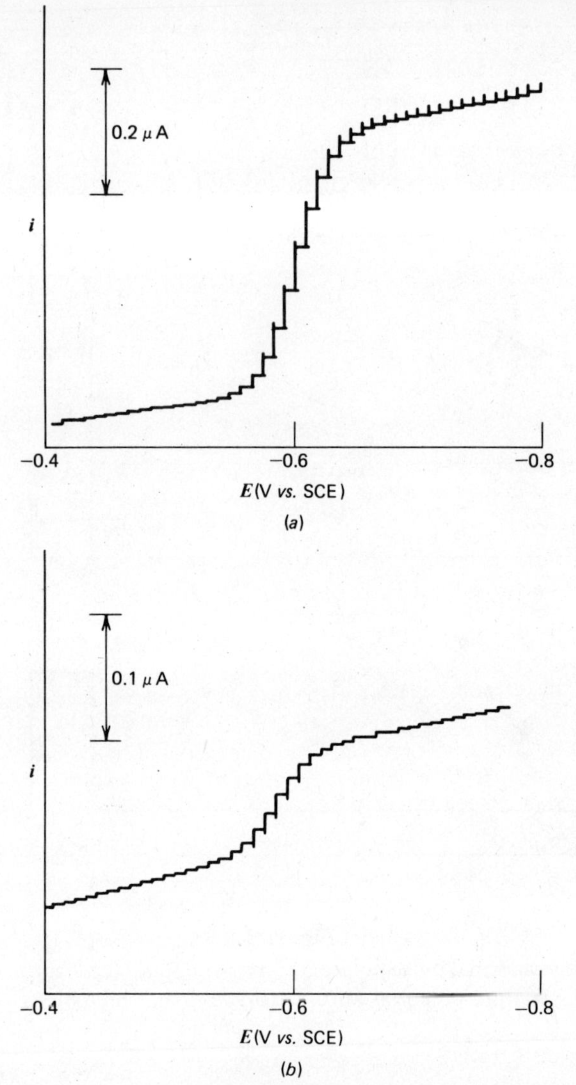

Figure 5.8.6
Polarograms for 10^{-5} M Cd^{2+} in 0.01 M HCl. (*a*) Normal pulse mode. (*b*) Tast mode.

In comparing this current to that measured in the tast experiment, it is useful to recall (Section 5.3.1) that (5.8.3) can be rewritten as

$$(i_d)_{\text{tast}} = \frac{nFA(7/3)^{1/2} D_O^{1/2} C_O^*}{\pi^{1/2} \tau^{1/2}} \tag{5.8.6}$$

thus (56)

$$\frac{(i_d)_{\text{pulse}}}{(i_d)_{\text{tast}}} = \left(\frac{3}{7}\right)^{1/2} \left(\frac{\tau}{\tau - \tau'}\right)^{1/2} \tag{5.8.7}$$

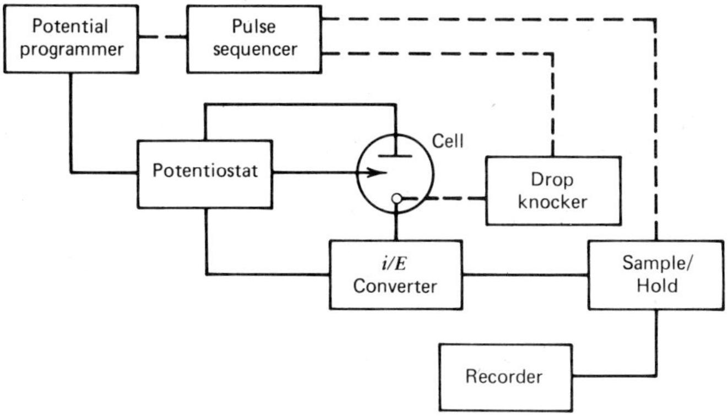

Figure 5.8.7
Schematic experimental arrangement for normal pulse polarography.

for experiments in which the current-sampling times for both methods are equal to τ. For the typical values of $\tau = 4$ sec, and $(\tau - \tau') = 50$ msec, this ratio is about 6; thus the expected increase in faradaic current is substantial. Figure 5.8.6 is a comparison of results obtained by normal pulse and tast polarography for a solution of 10^{-5} M Cd^{2+} in 0.01 M HCl. The larger sampled currents obtained with the pulse method are obvious.

At the time of measurement by normal pulse polarography, since dE/dt is virtually zero, the charging current contributing to the sampled total current comes almost completely from the continuous expansion of the electrode's area under potentiostatic conditions at potential E. It therefore is identical to that contributing to tast measurements at the same potential, provided that τ and m are the same for the two types of measurement. Equation 5.8.4 quantifies the nonfaradaic component. It is now clear that pulse polarography preserves entirely the sensitivity improvements achieved in tast polarography by discrimination against the charging current. In addition, the pulse method gains enhanced sensitivity through the increased faradaic currents, by comparison to those observed in tast or conventional polarography. Sensitivity limits are usually between 10^{-6} and 10^{-7} M (56–58).

In recent years, normal pulse polarography has been used very widely as an analytical tool for the measurement of low-level concentrations of heavy metals and organics, particularly in environmental samples (9, 57, 58). Section 5.8.4 deals specifically with its application to practical analysis.

Since normal pulse polarography is usually viewed as an analytical (rather than diagnostic) tool, the shapes of waves are not ordinarily examined in much detail. Nonetheless, the theory for the analysis does exist, since we have already noted that the method is essentially a sampled-current voltammetric method. The characteristic time scale of ~50 msec is, of course, much shorter than the ~3-sec time scale of conventional polarography. It is, therefore, possible for a chemical system to behave reversibly in a conventional polarographic experiment and quasi-reversibly or irreversibly in the normal pulse mode. Many systems that show moderately sluggish electrode kinetics behave in just this way. Notice also that the reverse behavior can be seen, too. If a system shows fast electrode kinetics, but the product of the electrode reaction decays on a 1-sec time scale, then the normal pulse experiment will show reversibility, because little product decay will occur during the measurement; yet the conventional polarogram will show the kinds of distortion that are characteristic of homogeneous reactions following charge transfer (see Chapter 11).

5.8.3 Differential Pulse Polarography (9, 55–58)

Sensitivities even better than those of normal pulse polarography can be obtained with the small-amplitude pulse scheme shown in Figures 5.8.8 and 5.8.9. The approach resembles normal pulse polarography, but several major differences are evident: (a) The base potential applied during most of a drop's lifetime is not constant from drop to drop, but instead is changed steadily in small increments. (b) The pulse height is only 10 to 100 mV and is maintained at a constant level with respect to the base potential. (c) Two current samples are taken during each drop's lifetime. One is at time τ', immediately before the pulse, and the second is at time τ, late in the pulse and just before the drop is dislodged. (d) The record of the experiment is a plot of the current difference, $i(\tau) - i(\tau')$, versus the base potential. Obviously the name of the method is derived from its reliance on this differential current measurement. The pulse width (~50 msec) and the waiting period for drop growth (0.5 to 4 sec) are both similar to the analogous periods in the normal pulse method.

Figure 5.8.10 is a block diagram of the experimental system and Figure 5.8.11*a* is an actual polarogram for 10^{-6} M Cd^{2+} in 0.01 M HCl. For comparison, the normal pulse response from the same system is given in Figure 5.8.11*b*.

Note that the differential measurement gives a peaked output, rather than the wavelike response to which we have grown accustomed. The underlying reason is easily understood qualitatively. Early in the experiment, when the base potential is much more positive than $E^{0'}$ for Cd^{2+}, no faradaic current flows during the time before the pulse, and the change in potential manifested in the pulse is too small to stimulate the faradaic process. Thus $i(\tau) - i(\tau')$ is virtually zero, at least for the faradaic component. Late in the experiment, when the base potential is in the diffusion-limited-current region, Cd^{2+} is reduced during the waiting period at the maximum possible rate. The pulse cannot increase the rate further, and hence the difference $i(\tau) - i(\tau')$ is again

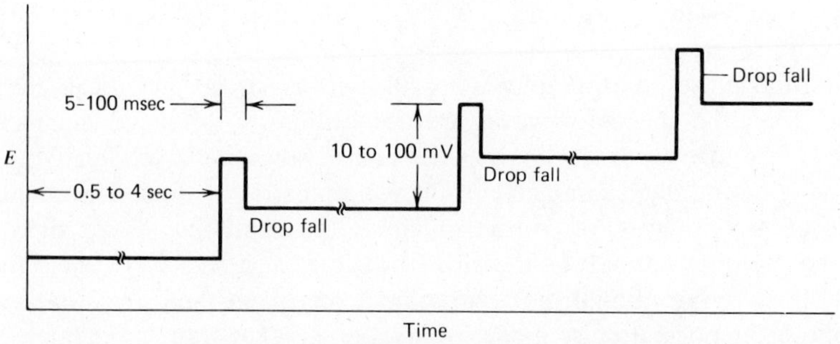

Figure 5.8.8
Potential program for several drops in a differential pulse polarographic experiment.

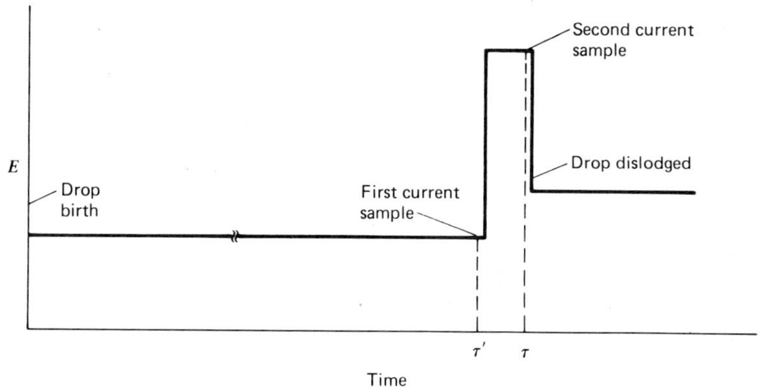

Figure 5.8.9
Events for a single drop of a differential pulse polarographic experiment.

small. Only in the region of $E^{0'}$ (for this reversible system) is an appreciable faradaic difference current observed. There the base potential is such that Cd^{2+} is reduced during the waiting period at some rate less than the maximum, since the surface concentration $C_O(0, t)$ is not zero. Application of the pulse drives $C_O(0, t)$ to a lower value; hence the flux of O to the surface and the faradaic current are both enhanced. Only in potential regions where a small potential difference can make a sizable difference in current flow does the differential pulse technique show a response.

The shape of the response function and the height of the peak can be treated quantitatively in a straightforward manner. Note that the events during each drop's lifetime actually comprise a double-step experiment. From the birth of the drop at $t = 0$ until the application of the pulse at $t = \tau'$, the base potential E is enforced. At later times, the potential is $E + \Delta E$, where ΔE is the pulse height. Each drop is born into a solution of the bulk composition, but generally electrolysis occurs during the

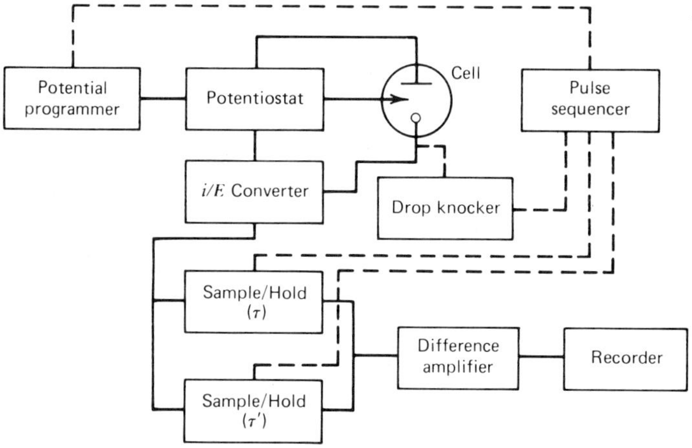

Figure 5.8.10
Schematic experimental arrangement for differential pulse polarography.

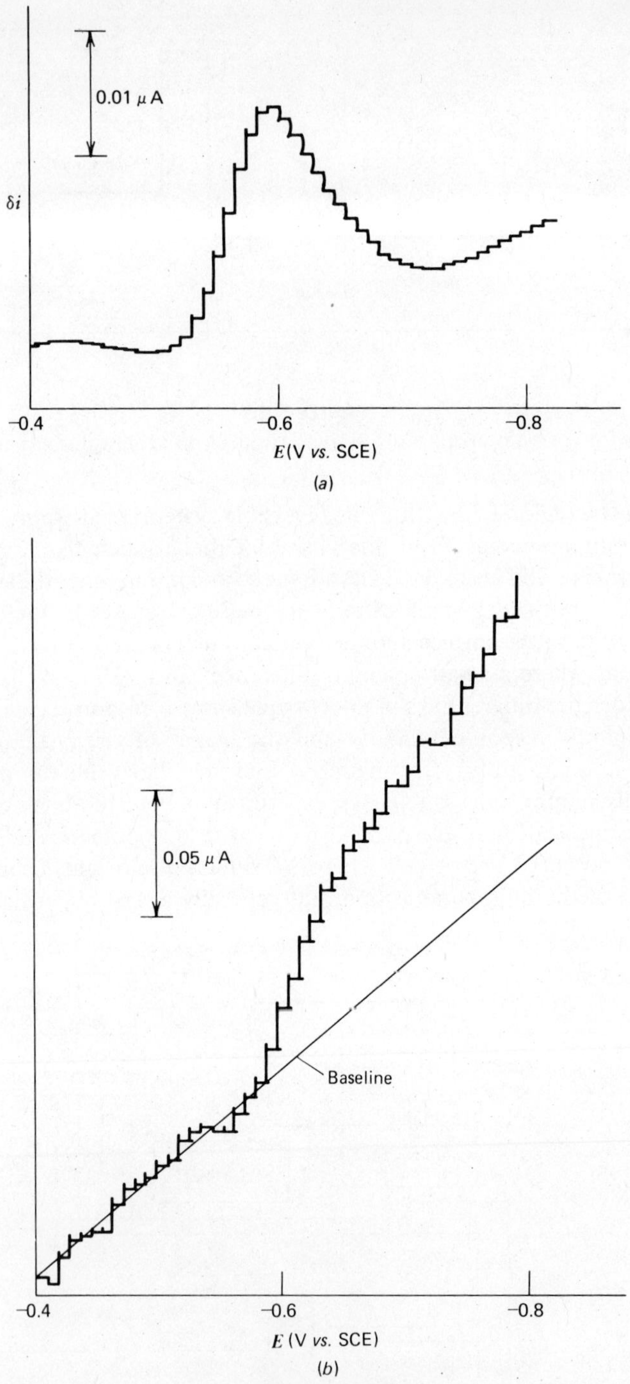

Figure 5.8.11

Polarograms for 10^{-6} *M* Cd^{2+} in 0.01 *M* HCl. (*a*) Differential pulse mode, $\Delta E = -50$ mV. (*b*) Normal pulse mode.

period before τ', and the pulse operates on the concentration profiles that it creates. This situation is analogous to that considered in Section 5.7, and it can be treated by the techniques developed there. Even so, we will not take that approach, because dealing with the expanding sphere is cumbersome, and the essential simplicity of the problem is obscured.

Instead, we begin by noting that the preelectrolysis period τ' is perhaps 60 times longer than the pulse duration $\tau - \tau'$. Thus the preelectrolysis establishes a very thick diffusion layer, and the pulse is able to modify only a small part of it. In fact, the whole experiment can be approximated by assuming that the pulse cannot distinguish the actual finite concentration profiles appearing at its start from a semi-infinite homogeneous solution with bulk concentrations equal to the values of $C_O(0, t)$ and $C_R(0, t)$ enforced by potential E. The role of the preelectrolysis is to set up "apparent bulk concentrations" that vary during successive drops from pure O to pure R (or vice versa), as the scan is made. For a given drop, the differential faradaic current is taken as the current that would flow at time $\tau - \tau'$ after a potential step from E to $E + \Delta E$.

Now let us restrict our consideration to a nernstian system in which R is initially absent. The results from Section 5.4 show that the surface concentrations during preelectrolysis at potential E are

$$C_O(0, t) = C_O^*\left(\frac{\xi\theta}{1 + \xi\theta}\right) \tag{5.8.8}$$

$$C_R(0, t) = C_O^*\left(\frac{\xi}{1 + \xi\theta}\right) \tag{5.8.9}$$

where $\theta = \exp[(nF/RT)(E - E^{0'})]$. We regard these values as the apparent bulk concentrations $(C_O^*)_{app}$ and $(C_R^*)_{app}$ for the pulse. Since the system is nernstian, they are in equilibrium with potential E. The problem is now simply to find the faradaic current flow after a step from equilibrium to $E + \Delta E$ in a homogeneous medium of bulk concentrations $(C_O^*)_{app}$ and $(C_R^*)_{app}$.

Through the approach of Section 5.4 (see also Problem 5.10), that current is straightforwardly found to be

$$i = \frac{nFAD_O^{1/2}}{\pi^{1/2}t^{1/2}} \cdot \frac{[(C_O^*)_{app} - \theta'(C_R^*)_{app}]}{(1 + \xi\theta')} \tag{5.8.10}$$

where $\theta' = \exp[(nF/RT)(E + \Delta E - E^{0'})]$. Substitution by (5.8.8) and (5.8.9) gives

$$i = \frac{nFAD_O^{1/2}C_O^*}{\pi^{1/2}t^{1/2}} \cdot \frac{(\xi\theta - \xi\theta')}{(1 + \xi\theta)(1 + \xi\theta')} \tag{5.8.11}$$

It is convenient here to introduce (56) the parameters P_A and σ, where

$$P_A = \xi \exp\left[\frac{nF}{RT}\left(E + \frac{\Delta E}{2} - E^{0'}\right)\right] \tag{5.8.12}$$

and

$$\sigma = \exp\left(\frac{nF}{RT}\frac{\Delta E}{2}\right) \tag{5.8.13}$$

In this notation, $\xi\theta = P_A/\sigma$ and $\xi\theta' = P_A\sigma$; thus

$$i = \frac{nFAD_O^{1/2}C_O^*}{\pi^{1/2}t^{1/2}}\left[\frac{P_A(1-\sigma^2)}{(\sigma+P_A)(1+P_A\sigma)}\right] \tag{5.8.14}$$

and we take the differential faradaic current $\delta i = i(\tau) - i(\tau')$ as

$$\delta i = \frac{nFAD_O^{1/2}C_O^*}{\pi^{1/2}(\tau-\tau')^{1/2}}\left[\frac{P_A(1-\sigma^2)}{(\sigma+P_A)(1+P_A\sigma)}\right] \tag{5.8.15}$$

The bracketed factor describes δi as a function of potential. When E is far more positive than $E^{0'}$, P_A is large and δi is virtually zero. When E is much more negative then $E^{0'}$, P_A approaches zero, and so does δi. Through the derivative $d(\delta i)/dP_A$, one can easily show (56) that δi is maximized at $P_A = 1$, which implies

$$E_{\max} = E^{0'} + \frac{RT}{nF}\ln\left(\frac{D_R}{D_O}\right)^{1/2} - \frac{\Delta E}{2} = E_{1/2} - \frac{\Delta E}{2} \tag{5.8.16}$$

Since ΔE is small, the potential of maximum current lies close to the polarographic half-wave potential $E_{1/2}$.

The height of the peaked waveform is then

$$(\delta i)_{\max} = \frac{nFAD_O^{1/2}C_O^*}{\pi^{1/2}(\tau-\tau')^{1/2}}\cdot\left(\frac{1-\sigma}{1+\sigma}\right) \tag{5.8.17}$$

The quotient $(1-\sigma)/(1+\sigma)$ decreases monotonically with diminishing $|\Delta E|$, and it reaches zero for $\Delta E = 0$. When ΔE is negative δi is positive (or cathodic), and vice versa. The quotient's maximum magnitude, which applies at large pulse amplitudes, is unity. In that limit, $(\delta i)_{\max}$ is equal to the faradaic current sampled on top of the normal pulse polarographic wave obtained under the same timing conditions. As (5.8.5) notes, that current is $nFAD^{1/2}C_O^*/\pi^{1/2}(\tau-\tau')^{1/2}$. Under usual conditions, ΔE is not large enough to realize this greatest possible $(\delta i)_{\max}$. Table 5.8.1 shows the

Table 5.8.1

Effect of Pulse Amplitude on Peak Height[a]

	$(1-\sigma)/(1+\sigma)$		
ΔE, mV	$n = 1$	$n = 2$	$n = 3$
−10	0.0971	0.193	0.285
−50	0.453	0.750	0.899
−100	0.750	0.960	0.995
−150	0.899	0.995	—
−200	0.960	—	—

[a] From E. P. Parry and R. A. Osteryoung, *Anal. Chem.*, **37**, 1634 (1965).

influence of $|\Delta E|$ over $(1 - \sigma)/(1 + \sigma)$, which is also the ratio of the peak height to the limiting value. For analysis, a typical ΔE is 50 mV, which gives a peak current from 45% to 90% as large as the limiting value, depending on n.

One refrains from increasing ΔE much past 100 mV, because resolution is degraded. The width of the peak at half height, $W_{1/2}$, increases as the pulse height grows larger, because differential behavior can be seen over a greater range of base potential. The precise form of $W_{1/2}$ as a function of ΔE is complicated and is of no real use. However, it is of interest to note the limiting width as ΔE approaches zero. By simple, but tedious algebra, that turns out to be (56)

$$W_{1/2} = 3.52RT/nF \tag{5.8.18}$$

At 25°, the limiting widths for $n = 1, 2,$ and 3 are 90.4, 45.2, and 30.1 mV, respectively. Real peaks are wider, especially if the pulse height is comparable to or larger than the limiting width.

Since the faradaic current measured in differential pulse polarography is never larger than the faradaic wave height found in the corresponding normal pulse experiment, the sensitivity gain in the differential method obviously does not come from enhanced faradaic response. Instead, the improvement comes from a reduced charging current contribution. Since both current samples $i(\tau)$ and $i(\tau')$ are taken under potentiostatic conditions, the charging currents appear only because dA/dt is never zero at the DME. From (5.3.25), we express these contributions as

$$i_c(\tau) = 0.00567C_i(E_z - E - \Delta E)m^{2/3}\tau^{-1/3} \tag{5.8.19}$$

$$i_c(\tau') = 0.00567C_i(E_z - E)m^{2/3}\tau'^{-1/3} \tag{5.8.20}$$

thus the contribution to the differential current is

$$\delta i_c = i_c(\tau) - i_c(\tau') = 0.00567C_i m^{2/3}\tau^{-1/3}\left[(E_z - E - \Delta E) - \left(\frac{\tau}{\tau'}\right)^{1/3}(E_z - E)\right] \tag{5.8.21}$$

where C_i has been taken as constant over the range from E to $E + \Delta E$. For the usual operating conditions, $(\tau/\tau')^{1/3}$ is very close to unity; hence the bracketed factor is approximately $-\Delta E$:

$$\boxed{\delta i_c \simeq -0.00567C_i\,\Delta E m^{2/3}\tau^{-1/3}} \tag{5.8.22}$$

For a negative scan, δi_c is positive, and vice versa. A comparison between (5.8.4) and (5.8.22) shows that the capacitive contribution to differential pulse measurements differs from that in tast and normal pulse polarography by the factor $\Delta E/(E_z - E)$. Over most regions of polarographic operation ΔE is smaller than $E_z - E$ by an order of magnitude or more. Note also that the capacitive background in differential pulse polarography is flat, insofar as C_i is constant over a potential range. In contrast, normal pulse and tast measurements feature a sloping background because of the dependence on $(E_z - E)$. This difference is apparent in Figure 5.8.11, and the greater ease in evaluating the differential faradaic response is obvious.†

The improvements manifested in the differential method yield sensitivities that are often an order of magnitude better than those of normal pulse polarography. Detection limits as low as 10^{-8} M can be achieved, but doing so requires close attention to selection of the medium. See Section 5.8.4 for more details.

We will not treat the application of differential measurements to irreversible systems. Instead we will note only that the response waveform in any differential scan approaches the derivative of the normal pulse polarographic response as $|\Delta E|$ tends toward zero. This fact is easily demonstrated for the reversible system (see Problem 5.17). Thus one can expect to see a differential response for an irreversible system, but the peak will be shifted from $E^{0'}$ toward more extreme potentials by an activation overpotential, and the peak width will be larger than for a reversible system, because the rising portion of an irreversible wave extends over a larger potential range. Since the maximum slope on the rising portion is smaller than in the corresponding reversible case, $(\delta i)_{max}$ will be smaller than the value predicted by (5.8.17).

The time scale of the differential pulse experiment is usually exactly the same as for normal pulse polarography, hence a given system ordinarily shows the same degree of reversibility toward either approach. However, the degree of reversibility toward the pulse methods may differ from that shown toward conventional polarography for reasons discussed in Section 5.8.2.

5.8.4 Analysis by Pulse Polarographic Methods (9, 57, 58)

The normal pulse and differential pulse techniques are among the most sensitive means for the direct evaluation of concentrations, and they find very wide use for trace analysis. When they can be applied, they are often far more sensitive than molecular or atomic absorption spectroscopy or most chromatographic approaches. In addition, they can provide information about the chemical form in which an analyte appears. Oxidation states can be defined, complexation can often be detected, and acid-base chemistry can be characterized. This information is frequently overlooked in competing methods. The weaknesses of pulse analysis are those common to most electroanalytical techniques. Resolution of very complex systems is difficult, and analysis time can be fairly long, particularly if deaeration is required.

Pulse measurements are sufficiently sensitive that one must pay special attention to impurity levels in solvents and supporting electrolytes. Contamination from the latter can be reduced by lowering their concentrations from the usual 0.1 to 1 M range to 0.01 M or even 0.001 M. The lower limit is fixed by the maximum cell resistance that

† Princeton Applied Research Corporation has introduced an automated device, called the static mercury drop electrode (SMDE), which extrudes a drop from a capillary very quickly following an electronic signal. It then holds the drop without change until a drop knocker dislodges it at the behest of another electronic signal. This device can serve as an HMDE or, in a repetitive mode, as a replacement for the DME. In the latter role, it retains all of the important advantages of the DME, and it has the added feature that $dA/dt = 0$ at the time of measurement, so that charging current does not contribute significantly to any step experiment. It is likely that devices such as these will see increasingly wide usage in analytical applications.

can be tolerated, if it is not set first by chemical considerations, such as the role of the supporting electrolyte in complexation or pH determination. In most analyses, aqueous media are used, both for convenience and for compatibility with the chemistry of sample preparation; however, other solvents can provide superior working ranges and may merit consideration for new applications. Note in this regard that the working range for any medium is much narrower for trace analysis by differential pulse polarography than it is for conventional polarography, simply because the residual faradaic background becomes intolerably high at less extreme potentials. This point is clear from the data in Figure 5.8.12.

Concentrations can be evaluated by comparisons with external standards, by standard addition, or by the internal standard technique. Since subtle interferences often apply to analyses at low concentrations, standard addition is probably the method of choice for such cases.

Under some circumstances, the pulse techniques can produce distorted views of a sample's composition. Note that a fundamental assumption underlying analysis by normal pulse polarography is that the solution's composition near the working electrode at the start of each pulse is the same as that of the bulk. This assumption can hold only if negligible electrolysis occurs at the working electrode during the waiting period before τ'. Therefore, the base potential must be either the equilibrium value itself or in a range over which the electrode behaves as an IPE (59). Otherwise, electrolysis modifies the solution's composition near the electrode before the pulse begins.

Figure 5.8.13 is a display of conventional and pulse polarograms for 1 m*M* Fe^{3+} and 10^{-4} *M* Cd^{2+} in 0.1 *M* HCl. Since $E^{0'}$ for Fe^{3+}/Fe^{2+} is more positive than the anodic limit of the DME, no wave is seen for that couple, but instead the diffusion-limited reduction current for Fe^{3+} is recorded from the positive end of the working range. In the conventional polarogram (Figure 5.8.13*a*), the limiting current for Fe^{3+} is about

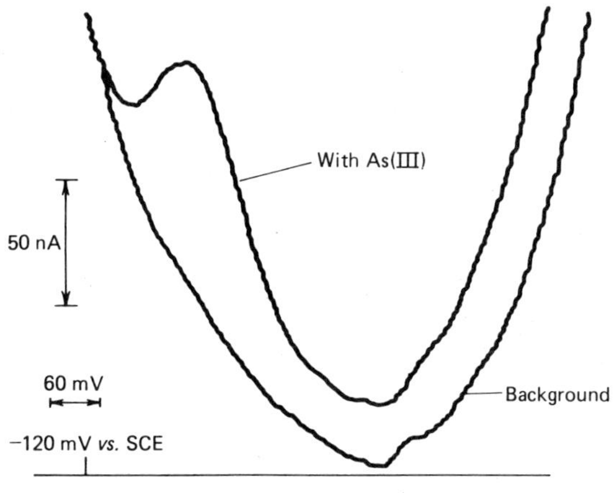

Figure 5.8.12

Differential pulse polarogram for 4.84×10^{-7} *M* As(III) in 1 *M* HCl containing 0.001% Triton X-100. $t_{max} = 2$ sec, $\Delta E = -100$ mV. [From J. G. Osteryoung and R. A. Osteryoung, *Amer. Lab*, **4** (7), 8 (1972), with permission.]

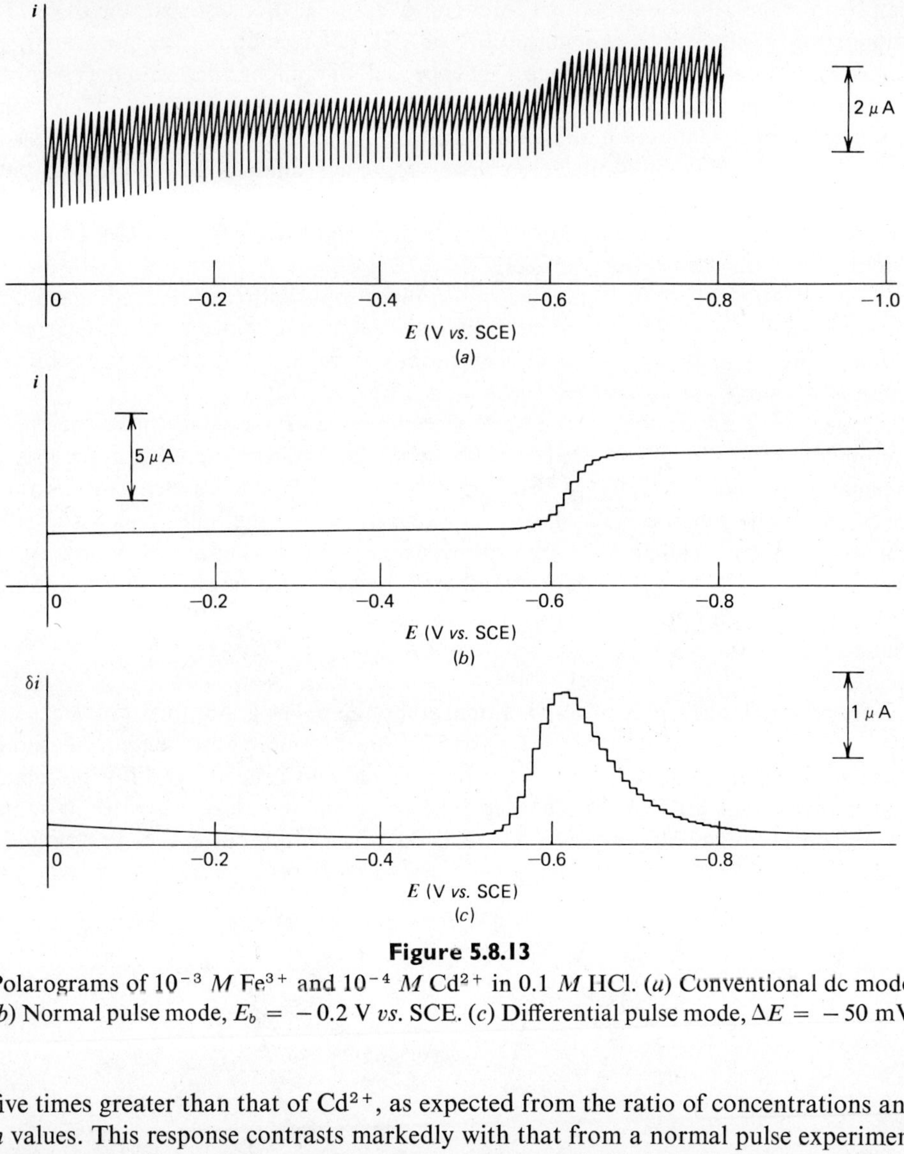

Figure 5.8.13

Polarograms of 10^{-3} *M* Fe^{3+} and 10^{-4} *M* Cd^{2+} in 0.1 *M* HCl. (*a*) Conventional dc mode. (*b*) Normal pulse mode, $E_b = -0.2$ V *vs.* SCE. (*c*) Differential pulse mode, $\Delta E = -50$ mV.

five times greater than that of Cd^{2+}, as expected from the ratio of concentrations and *n* values. This response contrasts markedly with that from a normal pulse experiment (Figure 5.8.13*b*) in which E_b was -0.2 V *vs.* SCE. The limiting current for Fe^{3+} is actually smaller than that for Cd^{2+}, because electrolysis at the base potential depletes the diffusion layer of Fe^{3+} before the pulse has a chance to measure it faithfully. Since the wave height for Cd^{2+} is unaffected, this effect would be alarming only if an accurate picture of the concentration ratio $C^*_{Fe^{3+}}/C^*_{Cd^{2+}}$ were desired. In fact it can actually be useful for suppressing the response of a concentrated interferent.

Differential pulse polarography also produces an ambiguous record for this kind of situation, as shown in Figure 5.8.13*c*. A peak is seen only for the Cd^{2+} reduction, essentially because the trace covers potentials only on the negative side of the Fe^{3+} wave. We note again that the differential pulse polarogram approximates the derivative of the normal pulse record; hence distinct peaks will not be seen in the former unless distinct waves appear in the latter.

Aside from this type of problem, the differential pulse method is particularly well-suited to the analysis of multicomponent systems because its readout format usually allows the separation of signals from individual components along a common baseline. This point is illustrated in Figure 5.8.14. Note also from that figure that this method is applicable to a variety of analytes encompassing considerably more than heavy metal species.

Although pulse techniques were developed specifically for the DME, they can be employed with other kinds of electrodes. As important examples, one can cite differential pulse anodic stripping at a hanging mercury drop or at a thin mercury film on a rotating substrate. See Section 10.8 for details. Applications to stationary electrodes are hampered by the requirement that each interval between pulses be actively utilized to erase the electrolytic effects of the previous pulse. In a reversible system, the reversal of pulse electrolysis that occurs during a waiting period tends to accomplish this end, but without reversibility, one finds cumulative depletion effects that may severely distort the voltammetric record (59).

5.9 CHRONOCOULOMETRY

Till now, this chapter has exclusively concerned either current-time transients stimulated by potential steps or voltammograms constructed by sampling those curves. An alternative, and very useful, mode for recording the electrochemical response is to

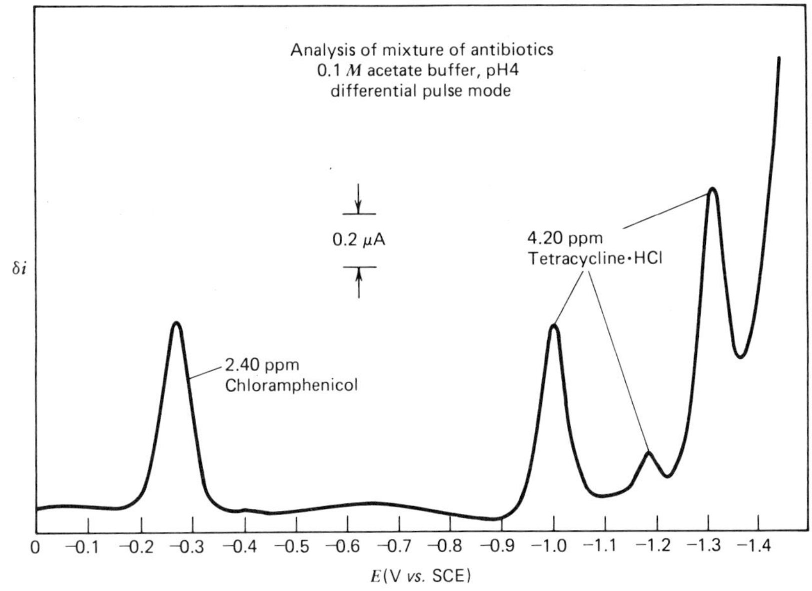

Figure 5.8.14

Differential pulse polarogram for a mixture of tetracycline and chloramphenicol. $\Delta E = -25$ mV. [From Application Note AN-111, EG & G Princeton Applied Research, Princeton, N.J., with permission.]

integrate the current, so that the charge passed as a function of time, $Q(t)$, is obtained. This *chronocoulometric* mode was popularized by Anson (60) and co-workers and is widely employed in place of chronoamperometry because it offers important experimental advantages: (a) The measured signal often grows with time, and hence the later parts of the transient, which are most accessible experimentally and are least distorted by nonideal potential rise, offer better signal-to-noise ratios than the early time results. The opposite is true for chronoamperometry. (b) The act of integration smooths random noise on the current transients, and hence the chronocoulometric records are inherently cleaner. (c) Contributions to $Q(t)$ from double-layer charging and from electrode reactions of adsorbed species can be distinguished from those due to diffusing electroreactants. An analogous separation of the components of a current transient is not generally feasible. This latter advantage of chronocoulometry is especially valuable for the study of surface processes.

5.9.1 Large-Amplitude Potential Step (60)

The simplest chronocoulometric experiment is the same as the Cottrell case discussed in Section 5.2.1. One begins with a quiescent, homogeneous solution of species O, in which a planar working electrode is held at some potential E_i where insignificant electrolysis takes place. At $t = 0$, the potential is shifted to E_f, which is sufficiently negative to enforce a diffusion-limited current. The Cottrell equation (5.2.11) describes the chronoamperometric response, and its integral from $t = 0$ gives the cumulative charge passed in reducing the diffusing reactant:

$$Q_d = \frac{2nFAD_O^{1/2}C_O^*t^{1/2}}{\pi^{1/2}} \tag{5.9.1}$$

As shown in Figure 5.9.1, Q_d rises with time, and a plot of its value *vs.* $t^{1/2}$ is linear. Given C_O^* and A, the slope of this plot is useful for evaluating D_O.

Equation 5.9.1 shows that the diffusional component to the charge is zero at $t = 0$, yet a plot of the total charge Q *vs.* $t^{1/2}$ generally does not pass through the origin, because additional components arise from double-layer charging and from the electroreduction of any O molecules that might be adsorbed at E_i. The charges devoted to these processes are utilized very quickly compared to the slow accumulation of the diffusional component; hence they may be accounted by two additive terms:

$$\boxed{Q = \frac{2nFAD_O^{1/2}C_O^*t^{1/2}}{\pi^{1/2}} + Q_{dl} + nFA\Gamma_O} \tag{5.9.2}$$

where Q_{dl} is the capacitive charge and $nFA\Gamma_O$ quantifies the faradaic component given to the reduction of Γ_O mol/cm^2 of adsorbed O.

The intercept of Q *vs.* $t^{1/2}$ is therefore $Q_{dl} + nFA\Gamma_O$. Separating these two interfacial components is of interest if one wishes to evaluate the *surface excess* Γ_O. However, doing so reliably usually requires other experiments, such as those described in the next section. An approximate value of $nFA\Gamma_O$ can be had by comparing the intercept of the Q-$t^{1/2}$ plot obtained for a solution containing O, with the "instantaneous" charge passed in the same experiment performed with supporting electrolyte only. The

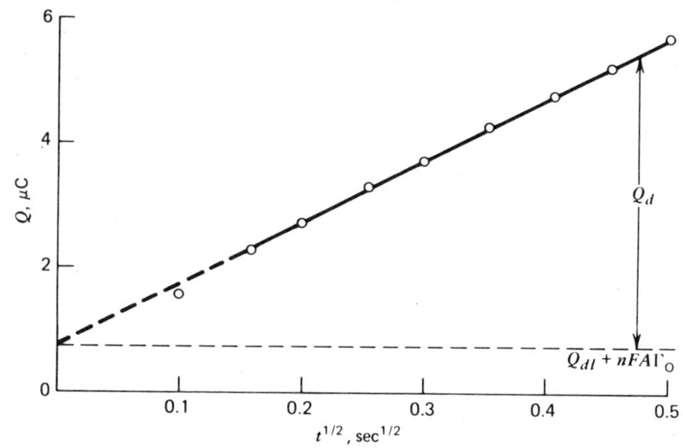

Figure 5.9.1
Linear plot of chronocoulometric response at a planar platinum disk. System is 0.95 m*M* 1,4-dicyanobenzene (DCB) in benzonitrile containing 0.1 *M* tetra-*n*-butylammonium fluoborate. Initial potential: 0.0 V *vs.* Pt QRE. Step potential: −1.892 V *vs.* Pt QRE. $T = 25°$, $A = 0.018$ cm^2. $E^{0\prime}$ for DCB + $e \rightleftharpoons$ DCB$^{\overline{\cdot}}$ is −1.63 V *vs.* QRE. The actual chronocoulometric trace is the part of Figure 5.9.2 corresponding to $t < 250$ msec. [Courtesy of R. S. Glass.]

latter quantity is Q_{dl} for the background solution, and it may approximate Q_{dl} for the complete system. Note, however, that these two capacitive components will not be the same if O is adsorbed, because adsorption influences the interfacial capacitance (see Chapter 12).

5.9.2 Reversal Experiments Under Diffusion Control (60–62)

Chronocoulometric reversal experiments are nearly always designed with step magnitudes that are large enough to ensure that any electroreactant diffuses to the electrode at its maximum rate. A typical experiment begins exactly like the one described just above. At $t = 0$, the potential is shifted from E_i to E_f, where O is reduced under diffusion-limited conditions. That potential is enforced for a fixed period τ, then the electrode is returned to E_i, where R is reconverted to O, again at the limiting rate. This sequence is a special case of the general reversal experiment considered in Section 5.7.1, and we have already found that the chronoamperometric response for $t > \tau$ is

$$i_r = \frac{-nFAD_O^{1/2}C_O^*}{\pi^{1/2}}\left[\frac{1}{(t-\tau)^{1/2}} - \frac{1}{t^{1/2}}\right] \tag{5.9.3}$$

Before τ, the experiment is clearly the same as that treated above; hence the cumulative charge devoted to the diffusional component after τ is

$$Q_d(t > \tau) = \frac{2nFAD_O^{1/2}C_O^*\tau^{1/2}}{\pi^{1/2}} + \int_\tau^t i_r\,dt \tag{5.9.4}$$

or

$$Q_d(t > \tau) = \frac{2nFAD_O^{1/2}C_O^*}{\pi^{1/2}}[t^{1/2} - (t-\tau)^{1/2}] \tag{5.9.5}$$

This function declines with increasing t, because the second step actually withdraws the charge injected in the forward step. The overall experimental record would resemble the curve of Figure 5.9.2, and one could expect a linear plot of $Q(t > \tau)$ *vs.* $[t^{1/2} - (t - \tau)^{1/2}]$. Note that there is no capacitive component to $Q(t > \tau)$, because the net potential change is zero. Although Q_{dl} was injected with the rise of the forward step, it was withdrawn upon reversal.

Now consider explicitly the quantity of charge removed in the reversal, $Q_r(t > \tau)$, which experimentally is the difference $Q(\tau) - Q(t > \tau)$, as depicted in Figure 5.9.2.

$$Q_r(t > \tau) = Q_{dl} + \frac{2nFAD_O^{1/2}C_O^*}{\pi^{1/2}}[\tau^{1/2} + (t - \tau)^{1/2} - t^{1/2}] \tag{5.9.6}$$

The bracketed factor is sometimes denoted as θ. For simplicity, we consider the case in which R is not adsorbed. A plot of Q_r *vs.* θ should be linear and possess the same slope magnitude seen in the other chronocoulometric plots (see Figure 5.9.3). Its intercept is Q_{dl}.

The pair of graphs depicting $Q(t < \tau)$ *vs.* $t^{1/2}$ and $Q_r(t > \tau)$ *vs.* θ is extremely useful for quantifying electrode reactions of adsorbed species. In the case we have considered, the difference between their intercepts is simply $nFA\Gamma_O$. This difference cancels Q_{dl} and leaves only a net faradaic charge devoted to adsorbate, which in general is $nFA(\Gamma_O - \Gamma_R)$. For details of interpretation concerning the various possible situations, the original literature should be consulted (60–62). (See also Section 12.5.6.)

We note in passing that (5.9.3), (5.9.5), and (5.9.6) are all based on the assumption that the concentration profiles at the start of the second step are exactly those that would be produced by an uncomplicated Cottrell experiment. In other words, we have regarded those profiles as being unperturbed by the additions or subtractions of diffusing material that are implied by adsorption and desorption. This assumption obviously cannot hold strictly. Christie et al. have avoided it in their rigorous treat-

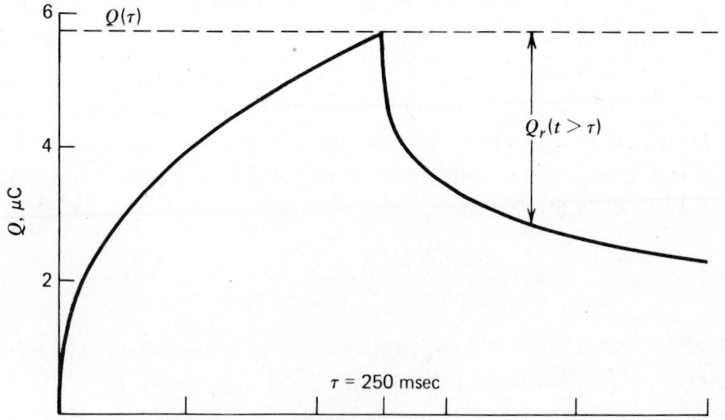

Figure 5.9.2
Chronocoulometric response for a double-step experiment performed on the system of Figure 5.9.1. The reversal step was made to 0.0 V *vs.* QRE. (Courtesy of R. S. Glass.)

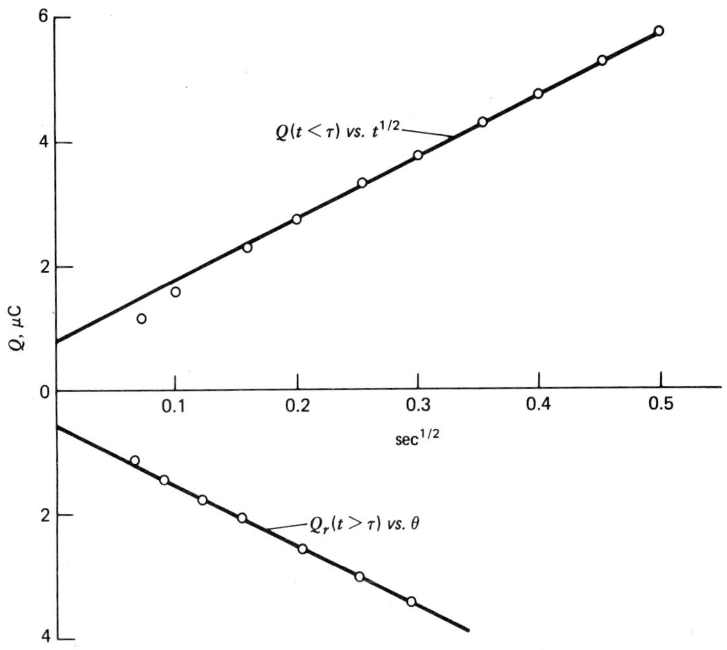

Figure 5.9.3

Linear chronocoulometric plots for data from the trace shown in Figure 5.9.2. For $Q(t < \tau)$ *vs.* $t^{1/2}$, the slope is 9.89 μC/sec$^{1/2}$ and the intercept is 0.79 μC. For $Q_r(t > \tau)$ *vs.* θ, the slope is 9.45 μC/sec$^{1/2}$ and the intercept is 0.66 μC. (Courtesy of R. S. Glass.)

ments, and they have shown how conventional chronocoulometric data can be corrected for such effects (61).

Reversal chronocoulometry is also extremely useful for characterizing the homogeneous chemistry of O and R. The diffusive faradaic component $Q_d(t)$ is especially sensitive to solution-phase reactions (62, 63), and it can be conveniently separated from the overall charge $Q(t)$ as described above.

If both O and R are stable, and are not adsorbed, then $Q_d(t)$ is fully described by (5.9.1) and (5.9.5). Let us consider the result of dividing $Q_d(t)$ by the Cottrell charge passed in the forward step, that is, $Q_d(\tau)$. This charge ratio takes a particularly simple form:

$$\frac{Q_d(t \leq \tau)}{Q_d(\tau)} = \left(\frac{t}{\tau}\right)^{1/2} \tag{5.9.7}$$

$$\boxed{\frac{Q_d(t > \tau)}{Q_d(\tau)} = \left(\frac{t}{\tau}\right)^{1/2} - \left[\left(\frac{t}{\tau}\right) - 1\right]^{1/2}} \tag{5.9.8}$$

which is independent of the specific experimental parameters n, C_O^*, D_O, and A. For a given value of t/τ, the charge ratio is even independent of τ. Equations 5.9.7 and 5.9.8, which are plotted in Figure 5.9.4, clearly describe the essential *shape* of the chronocoulometric response for a stable system. If the experimental results for any

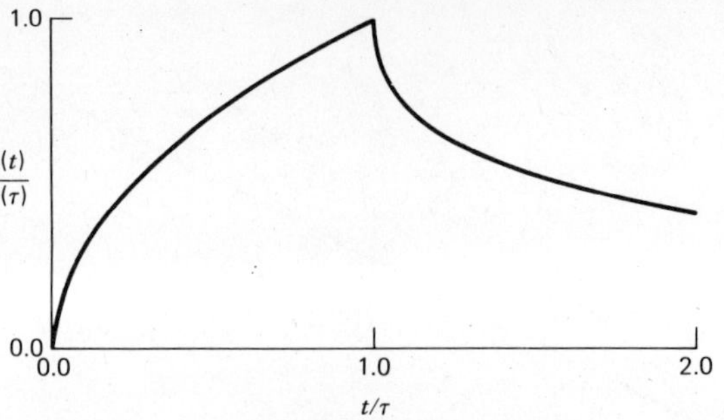

Figure 5.9.4
Dimensionless response curve for a completely stable system $O + ne \rightleftharpoons R$ subjected to reversal chronocoulometry with electrolysis at the diffusion-limited rate in each phase. This plot describes faradaic charge only.

real system do not adhere to this shape function, then chemical complications are indicated.

For example, consider the nernstian O/R couple in which R rapidly decays to electroinactive X. In the forward step O is reduced at the diffusion-controlled rate and (5.9.7) is obeyed. However, (5.9.8) is not followed, because species R cannot be fully reoxidized. The ratio $Q_d(t > \tau)/Q_d(\tau)$ falls less rapidly than for a stable system, and in the limit of completely effective conversion of R to X, no reoxidation is seen at all. Then $Q_d(t > \tau)/Q_d(\tau) = 1$ for all $t > \tau$.

Various other kinds of departure from (5.9.7) and (5.9.8) can be observed. See Chapter 11 for a discussion concerning the diagnosis of prominent homogeneous reaction mechanisms.

For a quick examination of chemical stability, one can conveniently evaluate the charge ratio $Q_d(2\tau)/Q_d(\tau)$ or, alternatively, the ratio $[Q_d(\tau) - Q_d(2\tau)]/Q_d(\tau)$. Equation 5.9.8 shows that these ratios for a stable system are 0.414 and 0.586, respectively.

5.9.3 Effects of Heterogeneous Kinetics (64, 65)

In the foregoing discussion, we have examined only situations in which electroreactants arrived at the electrode at the diffusion-limited rate. At the extreme potentials required to enforce that condition, the heterogeneous rate parameters are experimentally inaccessible. On the other hand, if one wished to evaluate those parameters, it would be useful to obtain a chronocoulometric response governed wholly or partially by the interfacial charge transfer kinetics. That goal can be reached by using a step potential that is insufficiently extreme to enforce diffusion-controlled electrolysis throughout the experimental time domain. In other words, steps must be made to potentials in the rising portion of the sampled-current voltammogram corresponding to the time scale of interest, and that time scale must be sufficiently short that electrode kinetics govern current flow for a significant period.

The usual experiment involves a step at $t = 0$ from an initial potential where electrolysis does not occur, to potential E, where it does. Let us consider the special case in which species O is initially present at concentration C_O^* and species R is initially absent. In Section 5.5.1, we found that the current transient for quasi-reversible electrode kinetics was given by (5.5.9). Integration from $t = 0$ provides the chronocoulometric response:

$$Q(t) = \frac{nFAk_fC_O^*}{H^2}\left[\exp(H^2t)\,\mathrm{erfc}(Ht^{1/2}) + \frac{2Ht^{1/2}}{\pi^{1/2}} - 1\right] \tag{5.9.9}$$

where $H = (k_f/D_O^{1/2}) + (k_b/D_R^{1/2})$. For $Ht^{1/2} > 5$, the first term in the brackets is negligible compared to the others; hence (5.9.9) takes the limiting form:

$$\boxed{Q(t) = nFAk_fC_O^*\left(\frac{2t^{1/2}}{H\pi^{1/2}} - \frac{1}{H^2}\right)} \tag{5.9.10}$$

A plot of this faradaic charge *vs.* $t^{1/2}$ should therefore be linear and display a negative intercept on the Q-axis and a positive intercept on the $t^{1/2}$-axis. The latter involves a shorter extrapolation, as shown in Figure 5.9.5, hence it can be evaluated more precisely. Designating it as $t_i^{1/2}$, we find H by the relation:

$$H = \frac{\pi^{1/2}}{2t_i^{1/2}} \tag{5.9.11}$$

With H in hand, k_f is found from the linear slope, $2nFAk_fC_O^*/(H\pi^{1/2})$.

Note that when E is very negative, H approaches $k_f/D_O^{1/2}$, and the slope approaches the Cottrell slope, $2nFAD_O^{1/2}C_O^*/\pi^{1/2}$. Moreover, H is large, so that the intercepts approach the origin. This limiting case is clearly that treated in Section 5.9.1.

Equations 5.9.9 and 5.9.10 do not include contributions from adsorbed species or double-layer charging. For accurate application of this treatment, those terms must be compensated or be rendered negligibly small compared to the diffusive component to

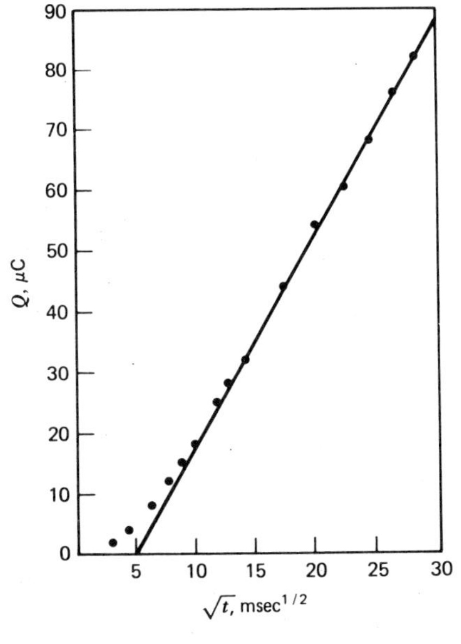

Figure 5.9.5
Chronocoulometric response for 10 mM Cd^{2+} in 1 M Na_2SO_4. The working electrode was an HMDE with $A = 2.30 \times 10^{-2}$ cm^2. The initial potential was -0.470 V *vs.* SCE, and the step potential was -0.620 V. The slope of the plot is 3.52 μC/msec$^{1/2}$ and $t_i^{1/2} = 5.1$ msec$^{1/2}$. [From J. H. Christie, G. Lauer, and R. A. Osteryoung, *J. Electroanal. Chem.*, 7, 60 (1964), with permission.]

the charge. Since compensation is difficult, workers using this technique have tended to employ high concentrations of electroactive species in an effort to deemphasize the interfacial components (65).

The advantage of chronocoulometric, as opposed to chronoamperometric, evaluation of rate parameters is that extrapolation is made from a later time domain; therefore, it is more easily applied to fast reactions. In effect, the act of integration preserves the kinetic limitations on current flow, which are manifested at very early times, for accurate examination over a much later period. Observing the early current flow directly is difficult to do in an accurate way because it has a high rate of change and is subject to distortions due, for example, to slow recovery from an overload of the current-monitoring amplifiers.

5.10 REFERENCES

1. F. G. Cottrell, *Z. Physik. Chem.*, **42**, 385 (1902).
2. H. A. Laitinen and I. M. Kolthoff, *J. Am. Chem. Soc.*, **61**, 3344 (1939).
3. H. A. Laitinen, *Trans. Electrochem. Soc.*, **82**, 289 (1942).
4. J. Heyrovsky, *Chem. Listy*, **16**, 256 (1922).
5. I. M. Kolthoff and J. J. Lingane, "Polarography," 2nd ed., Wiley-Interscience, New York, 1952, Chap. 17.
6. J. J. Lingane, "Electroanalytical Chemistry," 2nd ed., Wiley-Interscience, New York, 1958, Chap. 11.
7. L. Meites, "Polarographic Techniques," 2nd ed., Wiley-Interscience, New York, 1958, Chap. 2.
8. J. J. Lingane, *Anal. Chim. Acta*, **44**, 411 (1969).
9. A. Bond, "Modern Polarographic Methods in Analytical Chemistry," Marcel Dekker, New York, 1980.
10. J. Koutecky, *Czech. Cas. Fys.*, **2**, 50 (1953).
11. J. Koutecky and M. v. Stackelberg in "Progress in Polarography," Vol. 1, P. Zuman and I. M. Kolthoff, Eds., Wiley-Interscience, New York, 1962.
12. J. J. Lingane and B. A. Loveridge, *J. Am. Chem. Soc.*, **72**, 438 (1950).
13. P. Delahay, "New Instrumental Methods in Electrochemistry," Wiley-Interscience, New York, 1954, Chap. 3.
14. I. M. Kolthoff and J. J. Lingane, *op. cit.*, Chap. 4.
15. L. Meites, *op. cit.*, Chap. 3.
16. D. Ilkovic, *Collect. Czech. Chem. Commun.*, **6**, 498 (1934).
17. D. Ilkovic, *J. Chim. Phys.*, **35**, 129 (1938).
18. D. MacGillavry and E. K. Rideal, *Rec. Trav. Chim.*, **56**, 1013 (1937).
19. H. Strehlow and M. V. Stackelberg, *Z. Elektrochem.*, **54**, 51 (1950).
20. J. M. Markowitz and P. J. Elving, *Chem. Rev.*, **58**, 1047 (1958).
21. T. E. Cummings and P. J. Elving, *Anal. Chem.*, **50**, 480 (1978).
22. T. E. Cummings and P. J. Elving, *Anal. Chem.*, **50**, 1980 (1978).
23. I. M. Kolthoff and J. J. Lingane, *op. cit.*, Chap. 10.
24. L. Meites, *op. cit.*, Chap. 6.

25. H. H. Bauer, *Electroanal. Chem.*, **8**, 170 (1975).
26. I. M. Kolthoff and J. J. Lingane, *op. cit.*, Chap. 18.
27. L. Meites, *op. cit.*, Chaps. 5 and 7.
28. J. J. Lingane, *Ind. Eng. Chem., Anal. Ed.*, **15**, 588 (1943).
29. I. M. Kolthoff and J. J. Lingane, *op. cit.*, Vol. 2.
30. L. Meites, *op. cit.*, Appendices B and C.
31. L. Meites, Ed., "Handbook of Analytical Chemistry," McGraw-Hill, New York, 1963, pp. 4-43 to 5-103.
32. A. J. Bard and H. Lund, Eds., "Encyclopedia of the Electrochemistry of the Elements," Marcel Dekker, New York, 1973–1980.
33. L. Meites and P. Zuman, "Electrochemical Data," Wiley, New York, 1974.
34. I. M. Kolthoff and J. J. Lingane, *op. cit.*, Chap. 9.
35. J. Tomeš, *Collect. Czech. Chem. Commun.*, **9**, 12, 81, 150 (1937).
36. I. M. Kolthoff and J. J. Lingane, *op. cit.*, Chap. 11.
37. L. Meites, *op. cit.*, Chap. 4.
38. I. M. Kolthoff and J. J. Lingane, *op. cit.*, Chaps. 12 and 14.
39. L. Meites, *op. cit.*, Chap. 5.
40. P. Delahay, *op. cit.*, Chap. 4.
41. J. Koutecky, *Chem. Listy*, **47**, 323 (1953); *Collect. Czech. Chem. Commun.*, **18**, 597 (1953).
42. J. Weber and J. Koutecky, *ibid.*, **20**, 980 (1955).
43. L. Meites and Y. Israel, *J. Electroanal. Chem.*, **8**, 99 (1964).
44. J. Koutecky and J. Čižek, *Collect. Czech. Chem. Commun.*, **21**, 836 (1956).
45. J. E. B. Randles, *Can. J. Chem.*, **37**, 238 (1959).
46. J. E. B. Randles in "Progress in Polarography," I. M. Kolthoff and P. Zuman, Eds., Wiley-Interscience, 1962, Chap. 6.
47. C. K. Mann and K. K. Barnes, "Electrochemical Reactions in Nonaqueous Solvents," Marcel Dekker, New York, 1970, Chap. 11.
48. T. Kambara, *Bull. Chem. Soc. Japan*, **27**, 523 (1954).
49. D. D. Macdonald, "Transient Techniques in Electrochemistry," Plenum, New York, 1977.
50. W. M. Smit and M. D. Wijnen, *Rec. Trav. Chim.*, **79**, 5 (1960) present a treatment of a special case of this problem.
51. D. A. Corrigan and D. H. Evans, *J. Electroanal. Chem.*, in press.
52. E. Wahlin, *Radiometer Polarog.*, **1**, 113 (1952).
53. E. Wahlin and A. Bresle, *Acta Chem. Scand.*, **10**, 935 (1956).
54. L. Meites, *op. cit.*, Chap. 10.
55. G. C. Barker and A. W. Gardner, *Z. Anal. Chem.*, **173**, 79 (1960).
56. E. P. Parry and R. A. Osteryoung, *Anal. Chem.*, **37**, 1634 (1964).
57. J. G. Osteryoung and R. A. Osteryoung, *Amer. Lab.*, **4** (7), 8 (1972).
58. J. B. Flato, *Anal. Chem.*, **44** (11), 75A (1972).
59. J. L. Morris, Jr., and L. R. Faulkner, *Anal. Chem.*, **49**, 489 (1977).
60. F. C. Anson, *Anal. Chem.*, **38**, 54 (1966).
61. J. H. Christie, R. A. Osteryoung, and F. C. Anson, *J. Electroanal. Chem.*, **13**, 236 (1967).
62. J. H. Christie, *J. Electroanal. Chem.*, **13**, 79 (1967).

63. M. K. Hanafey, R. L. Scott, T. H. Ridgway, and C. N. Reilley, *Anal. Chem.*, **50**, 116 (1978).
64. J. H. Christie, G. Lauer, R. A. Osteryoung, and F. C. Anson, *Anal. Chem.*, **35**, 1979 (1963).
65. J. H. Christie, G. Lauer, and R. A. Osteryoung, *J. Electroanal. Chem.*, **7**, 60 (1964).

5.11 PROBLEMS

5.1 Fick's law for diffusion to a spherical electrode of radius r_0 is written

$$\frac{\partial C(r, t)}{\partial t} = D\left[\frac{\partial^2 C(r, t)}{\partial r^2} + \frac{2}{r}\frac{\partial C(r, t)}{\partial r}\right]$$

Solve this expression for $C(r, t)$ with the conditions

$$C(r, 0) = C^*, \qquad C(r_0, t) = 0 \qquad (t > 0), \quad \text{and} \quad \lim_{r \to \infty} C(r, t) = C^*$$

Show that the current i follows the expression

$$i = nFADC^*\left[\frac{1}{r_0} + \frac{1}{(\pi Dt)^{1/2}}\right]$$

[*Hint*: By making the substitution $v(r, t) = rC(r, t)$ in Fick's equation and in the boundary conditions, the problem becomes essentially the same as that for linear diffusion.]

5.2 Calculate the current for diffusion-controlled electrolysis at (a) a planar electrode and (b) a spherical electrode (see Problem 5.1) under the conditions:

$$n = 1, \qquad C^* = 1.00\ \text{m}M, \qquad A = 0.02\ \text{cm}^2, \qquad D = 10^{-5}\ \text{cm}^2/\text{sec}$$

at $t = 0.1, 0.5, 1, 2, 3, 5, 10$ sec, and as $t \to \infty$. Plot both i *vs.* t curves on the same graph. How long can the electrolysis proceed before the current at the spherical electrode exceeds that at the planar electrode by 10%?

5.3 The following measurements were made on a reversible polarographic wave at 25 °C. The process could be written $O + ne \rightleftharpoons R$.

E(volts *vs.* SCE)	$\bar{i}$(μA)
−0.395	0.48
−0.406	0.97
−0.415	1.46
−0.422	1.94
−0.431	2.43
−0.445	2.92

$\bar{i}_d = 3.24\ \mu\text{A}$

Calculate:

(a) The number of electrons involved in the electrode reaction.
(b) The formal potential (*vs.* NHE) of the couple involved in the electrode reaction, assuming $D_O = D_R$.

5.4 Derive equation 5.7.53.

5.5 Derive the polarographic current-potential curve for the reduction of a simple metal ion to a metal that is insoluble in mercury (or that plates out on the electrode). Assume the electrode reaction is

$$M^{n+} + ne \rightleftharpoons M\ (\text{solid})$$

that the electrode reaction is reversible, and that the activity of solid M is constant and equal to 1. How does $E_{1/2}$ vary with i_d? With the concentration of M^{n+}?

5.6 The following measurements were made at 25 °C on the reversible wave for the reduction of a metallic complex ion to metal amalgam:

Concentration of Ligand Salt NaX (*M*)	$E_{1/2}$ (volts *vs.* SCE)
0.10	−0.448
0.50	−0.531
1.00	−0.566

(a) Calculate the number of ligands X^- associated with the metal M^{2+} in the complex ($n = 2$).

(b) Calculate the instability constant of the complex, if $E_{1/2}$ for the reversible reduction of the simple metal ion is +0.081 V *vs.* SCE. Assume that the *D* values for the complex ion and the metal atom are equal, and that all activity coefficients are unity.

5.7 (a) Reductions of many organic substances involve the hydrogen ion. Derive the polarographic equation for the reversible reaction

$$O + pH^+ + ne \rightleftharpoons R$$

where both O and R are soluble substances, and only O is initially present in solution at a concentration C_O^*.

(b) What experimental procedure would be useful for determining p?

5.8 Derive the equations for $i(t)$ and $C_O(x, t)$ during a step to a constant potential for a totally irreversible process:

$$O + ne \xrightarrow{k_f} R$$

by solving Fick's equation, with the following conditions:

(1) $C_O(x, 0) = C_O^*$ (2) $\lim_{x \to \infty} C_O(x, t) = C_O^*$

(3) $D_O\left(\frac{\partial C_O(x, t)}{\partial x}\right)_{x=0} = k_f C_O(0, t)$

Show that $i = nFAC_O^* k_f \exp(H^2 t)\, \text{erfc}(H t^{1/2})$ where $H = k_f / D_O^{1/2}$.

5.9 The following data were obtained for an apparently totally irreversible polarographic wave:

E (volts *vs.* SCE)	$i_{t=\tau}$ (μA)
−0.419	0.31
−0.451	0.62
−0.491	1.24
−0.515	1.86
−0.561	2.48
−0.593	2.79
−0.680	3.10
−0.720	3.10

The overall reaction is known to be $O + 2e \rightarrow R$; and $m = 1.26$ mg/sec, $\tau = 3.53$ sec (constant at all potentials), and $C_O^* = 0.88$ mM.

(a) Use Table 5.5.1 to determine k_f at each potential. From these values, determine αn_a and k_f^0 (i.e., the value of k_f at $E = 0.0$ V *vs.* NHE).

(b) Use the treatment of Meites and Israel to obtain the same information.

(c) It is possible to calculate the concentration $C_O(0, t)$ for any i. [*Hint*: See boundary condition (3) in Problem 5.8 above.] Calculate $C_O(0, t)$ for each $i_{t=\tau}$. Plot $C_O(0, t)$ as a function of potential. On the same graph plot the $i_{t=\tau}$ values listed above.

5.10 Consider the reversible system

$$O + ne \rightleftharpoons R$$

in which both O and R are present initially. Derive the current-time curve for a step experiment in which the initial potential is the equilibrium potential and the final potential is any arbitrary value E. Derive the shape of the current-potential curve that would be recorded in a sampled-current experiment performed in the manner described here. What is the value of $E_{1/2}$? Does it depend on concentration?

5.11 Derive the Tomeš criterion for a reversible polarographic wave.

5.12 Consider a material A, which can be reduced at a dropping mercury electrode to form B. A 1-mM solution of A in acetonitrile shows a wave with $E_{1/2}$ at −1.90 V *vs.* SCE. The wave has a log slope of 60.5 mV at 25° and it gives $(I)_{max} = 2.15$ in the usual units. When dibenzo-15-crown-5 is added to the solution, the polarographic behavior changes.

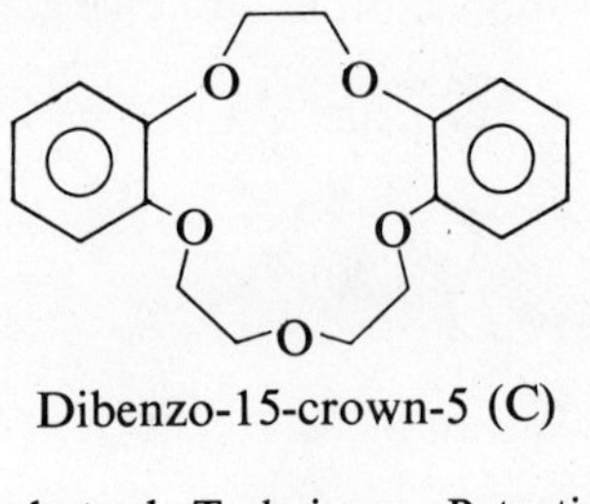

Dibenzo-15-crown-5 (C)

The following observations were made:

Concentration of C, M	$E_{1/2}$, V	Wave Slope, mV	$(I)_{max}$
10^{-3}	-2.15	60.3	2.03
10^{-2}	-2.21	59.8	2.02
10^{-1}	-2.27	59.8	2.04

What interpretation can be made of these results? Can any thermodynamic data be derived from the data? Can you suggest an identity for species A?

5.13 Derive the shape of the sampled-current voltammogram that would be recorded at a stationary platinum microelectrode immersed in a solution containing only I^-. The couple:

$$I_3^- + 2e \rightleftharpoons 3I^-$$

is reversible. What is the half-wave potential? Does it depend on the bulk concentration of I^-? Is this situation directly comparable to the case $O + ne \rightleftharpoons R$?

5.14 A polarogram of molecular oxygen in air-saturated 0.1 M KNO_3 is like that of Figure 5.6.3. The concentration of O_2 is about 0.25 mM. At $E = -0.4$ V *vs.* SCE, $(i_d)_{max} = 3.9$ μA, $t_{max} = 3.8$ sec, and $m = 1.85$ mg/sec. At $E = -1.7$ V *vs.* SCE, $(i_d)_{max} = 6.5$ μA, $t_{max} = 3.0$ sec, and $m = 1.85$ mg/sec. Calculate $(I)_{max}$ at each potential. Is the ratio of the two values what you expect? Explain any discrepancy in chemical terms. Calculate the diffusion coefficient for O_2 using the more appropriate constant. Defend your choice of the two.

5.15 Consider an analysis for the toxic ion Tl(I) in waste water which also contains Pb(II) and Zn(II) in 10 to 100-fold excesses. Outline any obstacles that would impede a polarographic determination and suggest means for circumventing them without implementing separation techniques. For 0.1 M KCl:

$$E_{1/2}(Tl^+/Tl) = -0.46 \text{ V } vs. \text{ SCE}$$

$$E_{1/2}(Pb^{2+}/Pb) = -0.40 \text{ V } vs. \text{ SCE}$$

$$E_{1/2}(Zn^{2+}/Zn) = -0.995 \text{ V } vs. \text{ SCE}$$

5.16 Sketch the normal pulse voltammogram expected for a substance undergoing an irreversible electrode reaction at a gold disk (e.g., $O_2 \rightarrow H_2O_2$ in 1 M KCl). Assume that the reduced form is not initially present and that no electrolysis occurs at the base potential in the starting solution. Explain the shape of the trace. (Note that the *location* of the trace is not of interest here.) Do the same for the case in which the electrode reaction is reversible. How would the curves differ if the disk were rotated during the recording of the polarograms? Do your curves say anything about applications of normal pulse voltammetry to potentials more positive than 0.0 V *vs.* SCE?

5.17 (a) Show that the derivative of a reversible sampled-current voltammetric wave is

$$\frac{di}{dE} = \frac{n^2F^2AC_O^*}{RT\pi^{1/2}\tau^{1/2}} \frac{\xi\theta}{(1 + \xi\theta)^2}$$

(b) Show that (5.8.15) approaches this form for $\delta i/\Delta E \approx di/dE$ as $\Delta E \to 0$.

5.18 Calculate k_f for the reduction of Cd^{2+} to the amalgam from the data in Figure 5.9.5.

5.19 Devise a chronocoulometric experiment for measuring the diffusion coefficient of Tl in mercury.

5.20 Consider the data in Figures 5.9.1 to 5.9.3. Calculate the diffusion coefficient of DCB. How well do the slopes of the two lines in Figure 5.9.3 bear out the expectations for a completely stable, reversible system? These data are typical for a solid planar electrode in nonaqueous media. Offer at least two possible explanations for the slight inequalities in the magnitudes of the slopes and intercepts in Figure 5.9.3.

chapter 6
Controlled Potential Microelectrode Techniques—Potential Sweep Methods

6.1 INTRODUCTION

The complete electrochemical behavior of a system can be obtained through a series of steps to different potentials with recording of the current-time curves, as described in Sections 5.4 and 5.5, to yield a three-dimensional *i-t-E* surface (Figure 6.1.1*a*). However, the accumulation and analysis of these data can be tedious especially when a stationary electrode is used. Also, it is not easy to recognize the presence of different species (i.e., to observe waves) from the recorded *i-t* curves alone, and potential steps that are very closely spaced (e.g., 1 mV apart) are needed for the derivation of well-resolved *i-E* curves. More information can be gained in a single experiment by sweeping the potential with time and recording the *i-E* curve directly. This amounts, in a qualitative way, to traversing the three-dimensional *i-t-E* realm (Figure 6.1.1*b*). Usually the potential is varied linearly with time (i.e., the applied signal is a voltage ramp) with sweep rates v ranging from 0.04 V/sec (1 V traversed in 25 sec) to about 1000 V/sec. In this experiment, it is customary to record the current as a function of potential, which is obviously equivalent to recording current versus time. The technical name for the method is *linear potential sweep chronoamperometry* but most workers refer to it as *linear sweep voltammetry* (LSV).†

A typical LSV response curve for a system like the anthracene example considered in Section 5.1 is shown in Figure 6.1.2*b*. If the scan is begun at a potential well positive of $E^{0'}$ for the reduction, only nonfaradaic currents flow for awhile. When the electrode potential reaches the vicinity of $E^{0'}$ the reduction begins and current starts to flow.

† This method is also called *stationary electrode polarography*; however, we will adhere to the recommended practice of reserving the term *polarography* for voltammetric measurements at the DME.

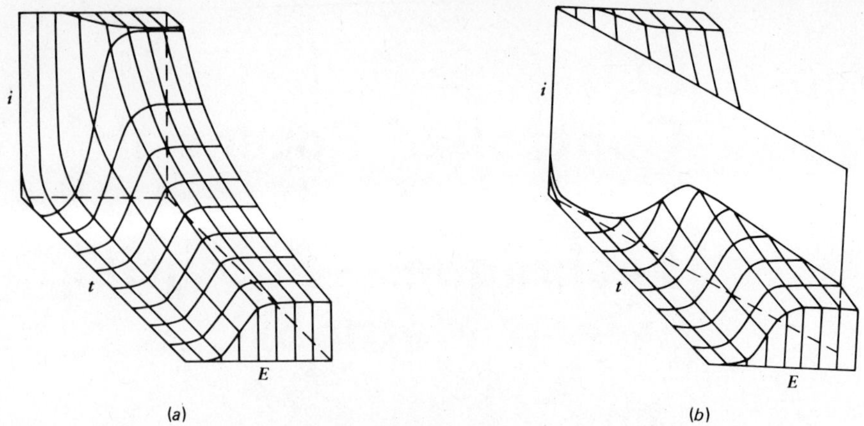

Figure 6.1.1

(*a*) Representation of a portion of the *i-t-E* surface for a nernstian reaction. Potential axis is in units of $60/n$ mV. (*b*) Linear potential sweep across this surface. [Reprinted with permission from W. H. Reinmuth, *Anal. Chem.*, **32**, 1509 (1960). Copyright 1960, American Chemical Society.]

As the potential continues to grow more negative, the surface concentration of anthracene must drop; hence the flux to the surface (and the current) increase. As the potential moves past $E^{0'}$, the surface concentration drops to near zero, mass transfer of anthracene to the surface reaches a maximum rate, and then it declines as the depletion effect sets in. The observation is therefore a peaked current-potential curve like that depicted.

At this point, the concentration profiles near the electrode are like those shown in Figure 6.1.2*c*. Let us consider what happens if we reverse the potential scan (see

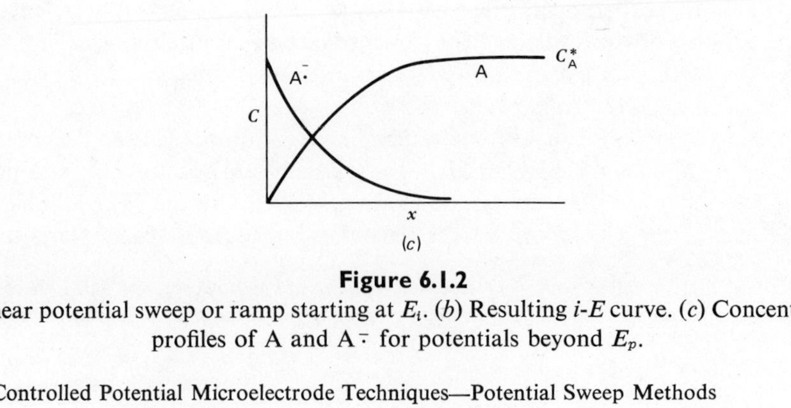

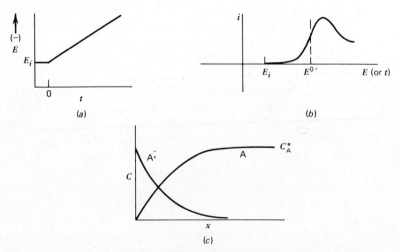

Figure 6.1.2

(*a*) Linear potential sweep or ramp starting at E_i. (*b*) Resulting *i-E* curve. (*c*) Concentration profiles of A and $A^{\cdot-}$ for potentials beyond E_p.

Figure 6.1.3). Suddenly the potential is sweeping in a positive direction, and in the electrode's vicinity there is a large concentration of the oxidizable anion radical of anthracene. As the potential approaches, then passes, $E^{0'}$, the electrochemical balance at the surface grows more and more favorable toward the anthracene neutral species. Thus the anion becomes reoxidized and an anodic current flows. This reversal current has a shape much like that of the forward peak for essentially the same reasons.

This experiment, which is called *cyclic voltammetry*, is a reversal technique and is the potential-scan equivalent of double potential step chronoamperometry, discussed in Section 5.7. Cyclic voltammetry has become a very popular technique for initial electrochemical studies of new systems and has proven very useful in obtaining information about fairly complicated electrode reactions. These will be discussed in more detail in Chapter 11.

6.2 NERNSTIAN (REVERSIBLE) SYSTEMS

6.2.1 Solution of the Boundary Value Problem

We consider again the reaction $O + ne \rightleftharpoons R$, assuming semi-infinite linear diffusion and a solution initially containing only species O, with the electrode held initially at a potential E_i, where no electrode reaction occurs. These initial conditions are identical to those in Section 5.4. The potential is swept linearly at v V/sec so that the potential at any time is

$$E(t) = E_i - vt \tag{6.2.1}$$

With the assumption that the rate of electron transfer is so rapid at the electrode surface that species O and R immediately adjust to the ratio dictated by the Nernst equation, the equations of Section 5.4, that is, (5.4.2) to (5.4.6), still apply. However, (5.4.6) can be written more clearly as

$$\frac{C_O(0, t)}{C_R(0, t)} = f(t) = \exp\left[\frac{nF}{RT}(E_i - vt - E^{0'})\right] \tag{6.2.2}$$

to show that this ratio is now a function of time. The difference is significant, since the Laplace transformation of (6.2.2) cannot be obtained as it could be in deriving

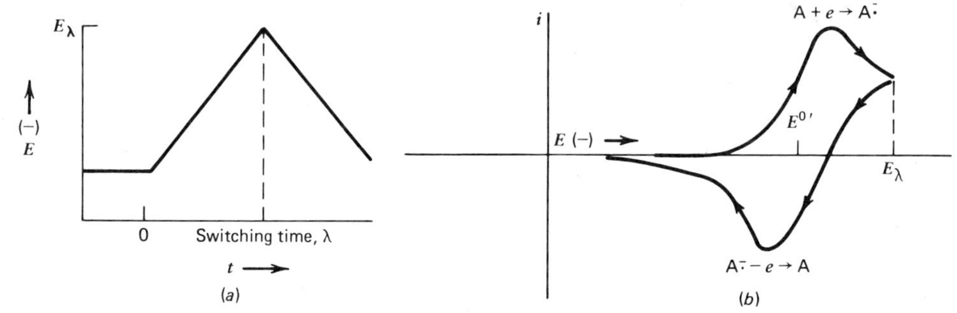

Figure 6.1.3
(*a*) Cyclic potential sweep. (*b*) Resulting cyclic voltammogram.

(5.4.13).† This inability to use the Laplace transform procedure greatly complicates the mathematics in this case. The problem was first considered by Randles (1) and Sevcik (2); the treatment and notation here follow the later work of Nicholson and Shain (3). The boundary condition (6.2.2) can be written

$$\frac{C_O(0, t)}{C_R(0, t)} = \theta\, e^{-\sigma t} = \theta S(t) \tag{6.2.3}$$

where $S(t) = e^{-\sigma t}$, $\theta = \exp[(nF/RT)(E_i - E^{0'})]$, and $\sigma = (nF/RT)v$. As before (see Section 5.4) application of the Laplace transform to the diffusion equations and the boundary conditions leads to [see (5.4.7)]

$$\bar{C}_O(x, s) = \frac{C_O^*}{s} + A(s) \exp\left[-\left(\frac{s}{D_O}\right)^{1/2} x\right] \tag{6.2.4}$$

We note that the transform of the current is given by [see (5.2.9)]

$$\bar{i}(s) = nFAD_O\left[\frac{\partial \bar{C}_O(x, s)}{\partial x}\right]_{x=0} \tag{6.2.5}$$

Combining this with (6.2.4) and inverting, by making use of the convolution theorem (see Appendix A), we obtain‡

$$\boxed{C_O(0, t) = C_O^* - [nFA(\pi D_O)^{1/2}]^{-1} \int_0^t i(\tau)(t - \tau)^{-1/2}\, d\tau} \tag{6.2.6}$$

By letting

$$f(\tau) = \frac{i(\tau)}{nFA} \tag{6.2.7}$$

(6.2.6) can be written

$$C_O(0, t) = C_O^* - (\pi D_O)^{-1/2} \int_0^t f(\tau)(t - \tau)^{-1/2}\, d\tau \tag{6.2.8}$$

Similarly from (5.4.12) an expression for $C_R(0, t)$ can be obtained (assuming R is initially absent):

$$C_R(0, t) = (\pi D_R)^{-1/2} \int_0^t f(\tau)(t - \tau)^{-1/2}\, d\tau \tag{6.2.9}$$

The derivation of (6.2.8) and (6.2.9) employed only the linear diffusion equations, initial conditions, semi-infinite conditions, and the flux balance. No assumption related to electrode kinetics or technique has been made, so that (6.2.8) and (6.2.9) are

---

† The Laplace transform of $C_O(0, t) = \theta C_R(0, t)$ is $\bar{C}_O(0, s) = \theta \bar{C}_R(0, s)$ only when θ is not a function of time; it is only under this condition that θ can be removed from the Laplace integral.
‡ This derivation is left as an exercise for the reader (see Problem 6.1). Equation 6.2.6 is often a useful starting point in other electrochemical treatments involving semi-infinite linear diffusion. τ in the integral is just a dummy variable that is lost when the definite integral is evaluated.

completely general. From these equations and the boundary condition for LSV, (6.2.3), we obtain

$$\int_0^t f(\tau)(t-\tau)^{-1/2}\,d\tau = \frac{C_O^*}{[\theta S(t)(\pi D_R)^{-1/2} + (\pi D_O)^{-1/2}]} \tag{6.2.10}$$

$$\int_0^t i(\tau)(t-\tau)^{-1/2}\,d\tau = \frac{nFA\pi^{1/2}D_O^{1/2}C_O^*}{[\theta S(t)\xi + 1]} \tag{6.2.11}$$

where, as before, $\xi = (D_O/D_R)^{1/2}$. The solution of this integral equation would be of the form $i(t) = (\text{constant})g(t)$ [where $g(t)$ is some function of time], and would thus yield the desired current-time curve or, since potential is linearly related to time, the current-potential equation. A closed-form solution of (6.2.11) cannot be obtained, and a numerical method must be employed. Before solving (6.2.11) numerically, it is convenient (a) to change from $i(t)$ to $i(E)$, since that is the way in which the data are usually considered, and (b) to put the equation in a dimensionless form so that a single numerical solution will give results that will be useful under any experimental conditions. This is accomplished by using the following substitution:

$$\sigma t = \frac{nF}{RT}vt = \left(\frac{nF}{RT}\right)(E_i - E) \tag{6.2.12}$$

Let $f(\tau) = g(\sigma\tau)$. With $z = \sigma\tau$, so that $\tau = z/\sigma$, $d\tau = dz/\sigma$, $z = 0$ at $\tau = 0$, and $z = \sigma t$ at $\tau = t$, we obtain

$$\int_0^t f(\tau)(t-\tau)^{-1/2}\,d\tau = \int_0^{\sigma t} g(z)\left(t - \frac{z}{\sigma}\right)^{-1/2}\frac{dz}{\sigma} \tag{6.2.13}$$

so that (6.2.11) can be written

$$\int_0^{\sigma t} g(z)(\sigma t - z)^{-1/2}\sigma^{-1/2}\,dz = \frac{C_O^*(\pi D_O)^{1/2}}{1 + \xi\theta S(\sigma t)} \tag{6.2.14}$$

or finally, dividing by $C_O^*(\pi D_O)^{1/2}$, we obtain

$$\int_0^{\sigma t} \frac{\chi(z)\,dz}{(\sigma t - z)^{1/2}} = \frac{1}{1 + \xi\theta S(\sigma t)} \tag{6.2.15}$$

where

$$\chi(z) = \frac{g(z)}{C_O^*(\pi D_O\sigma)^{1/2}} = \frac{i(\sigma t)}{nFAC_O^*(\pi D_O\sigma)^{1/2}} \tag{6.2.16}$$

Note that (6.2.15) is the desired equation in terms of the dimensionless variables $\chi(z)$, ξ, θ, $S(\sigma t)$ and σt. The current can be obtained from (6.2.16):

$$\boxed{i = nFAC_O^*(\pi D_O\sigma)^{1/2}\chi(\sigma t)} \tag{6.2.17}$$

Thus at any value of $S(\sigma t)$, which is a function of E, $\chi(\sigma t)$ can be obtained by solution of (6.2.15) and, from it, by (6.2.17), the current is available. Note that $\chi(\sigma t)$ at any given point is a pure number, so that (6.2.17) gives the functional relationship between the current at any point on the LSV curve and the variables. Specifically, i is proportional to C_O^* and $v^{1/2}$. The solution of (6.2.15) has been carried out numerically by computer [Nicholson and Shain (3)], by a series solution [Sevcik (2), Reinmuth (4)] and analytically, in terms of an integral that must be evaluated numerically [Matsuda and Ayabe (5), Gokhshtein (6)]. The general result of solving (6.2.15) is a table of values of $\chi(\sigma t)$ as a function of σt or $n(E - E_{1/2})$ (see Table 6.2.1 and Figure 6.2.1).

6.2.2 Peak Current and Potential

The function $\pi^{1/2}\chi(\sigma t)$, and hence the current, reaches a maximum (for 25°) at $n(E_p - E_{1/2}) = -28.50$ mV at $\pi^{1/2}\chi_p(\sigma t) = 0.4463$. From (6.2.17) the *peak current* i_p is

$$i_p = 0.4463nFAC_O^*\left(\frac{nF}{RT}\right)^{1/2} v^{1/2} D_O^{1/2} \tag{6.2.18}$$

or, at 25 °C, for A in cm², D_O in cm²/sec, C_O^* in mol/cm³, and v in V/sec, i_p in amperes is

$$\boxed{i_p = (2.69 \times 10^5)n^{3/2}AD_O^{1/2}v^{1/2}C_O^*} \tag{6.2.19}$$

$$E_p - E_{1/2} = E_p - E^{0'} + \left(\frac{RT}{nF}\right)\ln\xi = -1.109\left(\frac{RT}{nF}\right) \tag{6.2.20}$$

Table 6.2.1

Current Functions $\sqrt{\pi}\chi(\sigma t)$ for Reversible Charge Transfer (3)[a]

$(E - E_{1/2})n$[b] mV	$\sqrt{\pi}\chi(\sigma t)$	$\phi(\sigma t)$	$(E - E_{1/2})n$[b] mV	$\sqrt{\pi}\chi(\sigma t)$	$\phi(\sigma t)$
120	0.009	0.008	−5	0.400	0.548
100	0.020	0.019	−10	0.418	0.596
80	0.042	0.041	−15	0.432	0.641
60	0.084	0.087	−20	0.441	0.685
50	0.117	0.124	−25	0.445	0.725
45	0.138	0.146	−28.50	0.4463	0.7516
40	0.160	0.173	−30	0.446	0.763
35	0.185	0.208	−35	0.443	0.796
30	0.211	0.236	−40	0.438	0.826
25	0.240	0.273	−50	0.421	0.875
20	0.269	0.314	−60	0.399	0.912
15	0.298	0.357	−80	0.353	0.957
10	0.328	0.403	−100	0.312	0.980
5	0.355	0.451	−120	0.280	0.991
0	0.380	0.499	−150	0.245	0.997

[a] To calculate the current:
1. $i = i(\text{plane}) + i(\text{spherical correction})$.
2. $i = nFA\sqrt{\sigma D_O}C_O^*\sqrt{\pi}\chi(\sigma t) + nFAD_OC_O^*(1/r_O)\phi(\sigma t)$.
3. $i = 602\, n^{3/2}A\sqrt{D_O v}C_O^*[\sqrt{\pi}\chi(\sigma t) + 0.160(\sqrt{D_O}/r_O\sqrt{nv})\phi(\sigma t)]$ amperes at 25°. Units for step 3 are: A, cm²; D_O, cm²/sec; v, V/sec; C_O^*, moles/liter; r_O, cm.

[b] $E_{1/2} = E^{0'} + (RT/nF)\ln(D_R/D_O)^{1/2}$

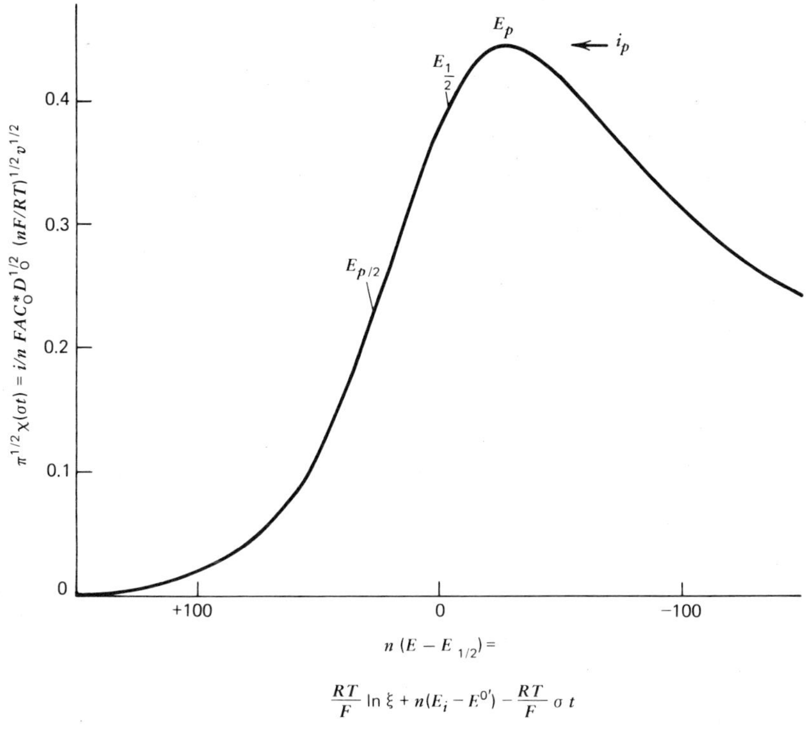

Figure 6.2.1
Linear potential sweep voltammogram in terms of dimensionless current function.

where $-1.109(RT/nF) = -28.5/n$ mV at 25 °C. Because the peak is somewhat broad, so that the peak potential may be difficult to determine, it is sometimes more convenient to report the potential at $\frac{1}{2}i_p$, called the *half-peak potential*, $E_{p/2}$, which is

$$E_{p/2} = E_{1/2} + 1.09\frac{RT}{nF} = E_{1/2} + 28.0/n \text{ mV at } 25\text{ °C} \tag{6.2.21}$$

Note that the polarographic $E_{1/2}$ value is located just about midway between E_p and $E_{p/2}$, and that a convenient diagnostic for a nernstian wave is

$$|E_p - E_{p/2}| = 2.2\frac{RT}{nF} = 56.5/n \text{ mV at } 25\text{ °C} \tag{6.2.22}$$

Thus for a reversible wave, E_p is independent of scan rate, and i_p (as well as the current at any other point on the wave) is proportional to $v^{1/2}$. [The latter property indicates diffusion control and is analogous to the variation of i_d with $h^{1/2}$ in polarography (see Section 5.3.4)]. A convenient constant in LSV is $i_p/v^{1/2}C_O^*$ (sometimes called *the current function*), which depends on $n^{3/2}$ and $D_O^{1/2}$. This constant can be used to estimate n for an electrode reaction, if a value of D_O can be estimated, for example, from the LSV of a compound of similar size or structure, which undergoes an electrode reaction with a known n value.

6.2.3 Spherical Corrections

For LSV with a spherical electrode (e.g., a hanging mercury drop), a similar treatment can be presented (4); the resulting current is

$$i = i(\text{plane}) + \frac{nFAD_O C_O^* \phi(\sigma t)}{r_0} \tag{6.2.23}$$

where r_0 is the radius of the electrode and $\phi(\sigma t)$ is a tabulated function (see Table 6.2.1). The peak current in amperes is given by (6.2.24):

$$i_p = i_p(\text{plane}) + (0.725 \times 10^5)\frac{nAD_O C_O^*}{r_0} \tag{6.2.24}$$

where r_0 is in centimeters and the units for the other variables are as in (6.2.19). Note that for large values of v the i_p (plane) term becomes much larger than the spherical correction term, and the electrode can be considered planar under these conditions.

6.2.4 Effect of C_d and R_u

For a potential step experiment at a stationary, constant area electrode, the charging current dies away after a time equivalent to a few time constants ($R_u C_d$) (see Section 1.2.4). For a potential sweep experiment, since the potential is continuously changing, a charging current, i_c, always flows (see equation 1.2.15):

$$|i_c| = AC_d v \tag{6.2.25}$$

and the faradaic current must always be measured from a baseline of charging current (Figure 6.2.2). Note that while i_p varies with $v^{1/2}$, i_c varies with v, so that i_c becomes relatively more important at faster scan rates. From (6.2.19) and (6.2.25)

$$\frac{|i_c|}{i_p} = \frac{C_d v^{1/2}(10^{-5})}{2.69 n^{3/2} D_O^{1/2} C_O^*} \tag{6.2.26}$$

or for $D_O = 10^{-5}$ cm^2/sec and $C_d = 20\ \mu$F/cm^2,

$$\frac{|i_c|}{i_p} \approx \frac{(2.4 \times 10^{-8})v^{1/2}}{n^{3/2} C_O^*} \tag{6.2.27}$$

Thus at high v and low C_O^* values, severe distortion of the LSV wave occurs. This effect often sets the limit of the maximum useful scan rate.

If R_u is large enough that $i_p R_u$ is appreciable compared to the accuracy of the measurement (e.g., a few millivolts), then the potential of the working electrode will not be the value desired from the potentiostat (i.e., $E_i \pm vt$), but will be $E + iR_u$. Thus the sweep will not be truly linear and the condition given by (6.2.1) does not hold.

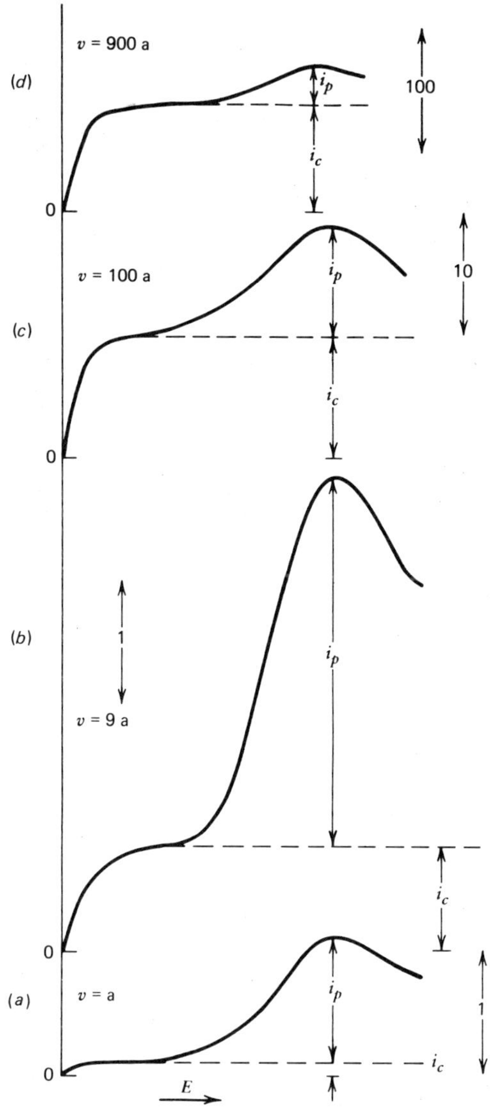

Figure 6.2.2
Effect of double-layer charging at different sweep rates on linear potential sweep voltammogram. Curves are plotted with the assumption that C_d is independent of E. The magnitude of the charging current, i_c, and the faradaic peak current, i_p, is shown. Note that the current scale in (*c*) is 10× and in (*d*) is 100× that in (*a*) and (*b*).

The practical effect of R_u is to flatten the wave and shift the reduction peak toward more negative potentials. Since the current increases with $v^{1/2}$, the larger the scan rate, the more E_p will be shifted, so that appreciable R_u causes E_p to be a function of v. It moves systematically in a negative direction with increasing v (for a reduction), even for a nernstian electrode reaction.

6.3 TOTALLY IRREVERSIBLE SYSTEMS

6.3.1 The Boundary Value Problem

For a totally irreversible reaction ($O + ne \rightarrow R$) the nernstian boundary condition, (6.2.2), is replaced by (see Section 5.5)

$$\frac{i}{nFA} = D_O\left[\frac{\partial C_O(x, t)}{\partial x}\right]_{x=0} = k_f(t)C_O(0, t) \qquad (6.3.1)$$

where

$$k_f(t) = k^0 \exp\{-\alpha n_a f[E(t) - E^{0\prime}]\} \qquad (6.3.2)$$

Introducing $E(t)$ from (6.2.1) into (6.3.2) yields

$$k_f(t)C_O(0, t) = k_{fi}C_O(0, t)\, e^{bt} \qquad (6.3.3)$$

where

$$k_{fi} = k^0 \exp[-\alpha n_a f(E_i - E^{0\prime})] \qquad (6.3.4)$$

$$b = \alpha n_a fv \qquad f = \frac{F}{RT} \qquad (6.3.5)$$

The solution follows in an analogous manner to that described in Section 6.2.1 (3, 5) and again requires a numerical solution of an integral equation. The current is given by

$$i = nFAC_O^*(\pi D_O b)^{1/2}\chi(bt) \qquad (6.3.6)$$

$$\boxed{i = nFAC_O^* D_O^{1/2}v^{1/2}\left(\frac{\alpha n_a F}{RT}\right)^{1/2}\pi^{1/2}\chi(bt)} \qquad (6.3.7)$$

where $\chi(bt)$ is a tabulated function [different from $\chi(\sigma t)$, Table 6.3.1]. Again i at any point on the wave varies with $v^{1/2}$ and C_O^*.

6.3.2 Peak Current and Potential

The function $\chi(bt)$ goes through a maximum at $\pi^{1/2}\chi(bt) = 0.4958$. Introduction of this value into (6.3.7) yields the following for the peak current (in amperes):

$$\boxed{i_p = (2.99 \times 10^5)n(\alpha n_a)^{1/2}AC_O^* D_O^{1/2}v^{1/2}} \qquad (6.3.8)$$

where the units are the same as for (6.2.19). This value occurs when

$$\alpha n_a(E_p - E^{0\prime}) + \left(\frac{RT}{F}\right)\ln\left[\frac{(\pi D_O b)^{1/2}}{k^0}\right] = -5.34 \text{ mV} \qquad (6.3.9)$$

Table 6.3.1
Current Functions $\sqrt{\pi}\chi(bt)$ for Irreversible Charge Transfer (3)[a,b]

Potential, mV	$\sqrt{\pi}\chi(bt)$	$\phi(bt)$	Potential, mV	$\sqrt{\pi}\chi(bt)$	$\phi(bt)$
160	0.003		15	0.437	0.323
140	0.008		10	0.462	0.396
120	0.016		5	0.480	0.482
110	0.024		0	0.492	0.600
100	0.035		−5	0.496	0.685
90	0.050		−5.34	0.4958	0.694
80	0.073	0.004	−10	0.493	0.755
70	0.104	0.010	−15	0.485	0.823
60	0.145	0.021	−20	0.472	0.895
50	0.199	0.042	−25	0.457	0.952
40	0.264	0.083	−30	0.441	0.992
35	0.300	0.115	−35	0.423	1.00
30	0.337	0.154	−40	0.406	
25	0.372	0.199	−50	0.374	
20	0.406	0.253	−70	0.323	

[a] The potential scale is $(E - E^{0\prime})\alpha n_a + (RT/F)\ln\sqrt{\pi D_O b}/k^0$.
[b] To calculate the current:
1. $i = i(\text{plane}) + i(\text{spherical correction})$.
2. $i = nFA\sqrt{bD_O}C_O^*\sqrt{\pi}\chi(bt) + nFAD_OC_O^*(1/r_0)\phi(bt)$.
3. $i = 602\, n(\alpha n_a)^{1/2}A\sqrt{D_O v}C_O^*[\sqrt{\pi}\chi(bt)] + 0.160(\sqrt{D_O}/r_0\sqrt{\alpha n_a v})\phi(bt)$ (at 25°).

Units for step 3 are the same as in Table 6.2.1.

or (in millivolts)

$$E_p = E^{0\prime} - \frac{RT}{\alpha n_a F}\left[0.780 + \ln\left(\frac{D_O^{1/2}}{k^0}\right) + \ln\left(\frac{\alpha n_a F v}{RT}\right)^{1/2}\right] \tag{6.3.10}$$

$$|E_p - E_{p/2}| = \frac{1.857RT}{\alpha n_a F} = \frac{47.7}{\alpha n_a}\text{ mV at } 25° \tag{6.3.11}$$

Thus, for a totally irreversible wave, i_p is also proportional to C_O^* and $v^{1/2}$, but E_p is a function of scan rate, shifting (for a reduction) in a negative direction by an amount $1.15RT/\alpha n_a F$ (or $30/\alpha n_a$ mV at 25°) for each tenfold increase in v. Note that in this case E_p occurs beyond $E^{0\prime}$ by an activation overpotential related to k^0. An alternate expression for i_p in terms of E_p can be obtained by combining (6.3.10) with (6.3.7), so that the result contains the value of $\chi(bt)$ at the peak. After rearrangement and evaluation of the constants, the following equation is obtained (3, 6):

$$i_p = 0.227nFAC_O^*k^0\exp\left[-\left(\frac{\alpha n_a F}{RT}\right)(E_p - E^{0\prime})\right] \tag{6.3.12}$$

A plot of $\ln(i_p)$ *vs.* $E_p - E^{0'}$ (assuming $E^{0'}$ could be obtained) determined at different scan rates should have a slope of $-\alpha n_a f$ and an intercept proportional to k^0.

For spherical electrodes, a procedure analogous to that employed at planar electrodes has been proposed. The spherical correction factor, $\phi(bt)$, which can be employed in the equation

$$i = i(\text{plane}) + \frac{nFAD_O C_O^* \phi(bt)}{r_0} \tag{6.3.13}$$

is tabulated in Table 6.3.1.

6.4 QUASI-REVERSIBLE SYSTEMS

The treatment of these systems was first described by Matsuda and Ayabe (5), who coined the term *quasi-reversible* for reactions that show electron transfer kinetic limitations where the reverse reaction has to be considered. The boundary condition for this case is [from (5.5.3)]

$$D_O\left(\frac{\partial C_O(x, t)}{\partial x}\right)_{x=0} = k^0 e^{-\alpha nf[E(t)-E^{0'}]}\left\{C_O(0, t) - C_R(0, t)\, e^{nf[E(t)-E^{0'}]}\right\} \tag{6.4.1}$$

The shape of the peak and the various peak parameters were shown to be functions of α and the parameter Λ, defined as

$$\boxed{\Lambda = k^0 \Big/ \left[D_O^{1-\alpha} D_R^{\alpha}\left(\frac{nF}{RT}\right) v\right]^{1/2}} \tag{6.4.2}$$

or, for $D_O = D_R = D$,

$$\Lambda = k^0 / D^{1/2}\left(\frac{nF}{RT}\right)^{1/2} v^{1/2} \tag{6.4.3}$$

The current is given by

$$i = nFAC_O^* D_O^{1/2}\left(\frac{nF}{RT}\right)^{1/2} \Psi(E) v^{1/2} \tag{6.4.4}$$

where $\Psi(E)$ is shown in Figure 6.4.1. Note that when $\Lambda \gtrsim 10$, the behavior approaches that of a reversible system. The i_p, E_p, and $E_{p/2}$ values depend on Λ and α. The appropriate expression for the peak current is

$$i_p = i_p(\text{rev}) K(\Lambda, \alpha) \tag{6.4.5}$$

where $i_p(\text{rev})$ is the value for the reversible i_p value (equation 6.2.18), and $K(\Lambda, \alpha)$ is shown in Figure 6.4.2. Note that for a quasi-reversible reaction, i_p is not proportional to $v^{1/2}$. The peak potential is

$$E_p - E_{1/2} = -\Xi(\Lambda, \alpha)\left(\frac{RT}{nF}\right) = -26\Xi(\Lambda, \alpha)/n \text{ mV at } 25° \tag{6.4.6}$$

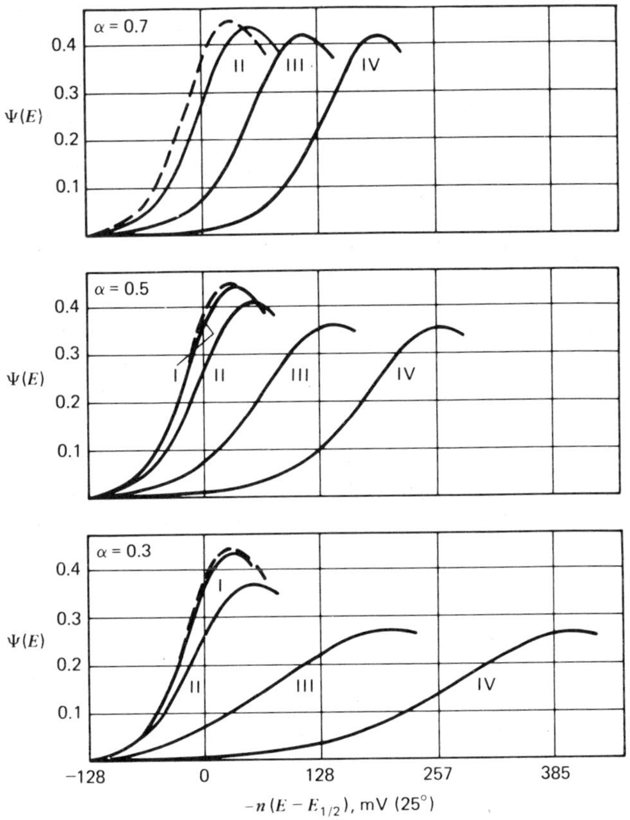

Figure 6.4.1

Variation of quasi-reversible current function, $\Psi(E)$, for different values of α (0.7, 0.5, 0.3, as indicated) and the following values of Λ: I, $\Lambda = 10$; II, $\Lambda = 1$; III, $\Lambda = 0.1$; IV, $\Lambda = 10^{-2}$. Dashed curve is for a reversible reaction.

$$\Psi(E) = i/nFAC_O^* D_O^{1/2}(nF/RT)^{1/2}v^{1/2}$$
$$\Lambda = k^0/D^{1/2}(nF/RT)^{1/2}v^{1/2} \qquad (\text{for } D_O = D_R = D)$$

[From H. Matsuda and Y. Ayabe, *Z. Elektrochem.*, **59**, 494 (1955), with permission.]

where $\Xi(\Lambda, \alpha)$ is shown in Figure 6.4.3. For the half-peak potential, we have

$$E_{p/2} - E_p = \Delta(\Lambda, \alpha)\left(\frac{RT}{nF}\right) = 26\Delta(\Lambda, \alpha)/n \text{ mV at } 25° \qquad (6.4.7)$$

where $\Delta(\Lambda, \alpha)$ is given in Figure 6.4.4. These parameters attain limiting values characteristic of reversible or totally irreversible processes as Λ varies. For example, consider $\Delta(\Lambda, \alpha)$. For $\Lambda \geq 10$, $\Delta(\Lambda, \alpha) \approx 2.2$, yielding the $E_p - E_{p/2}$ value characteristic of a reversible wave (6.2.22). For $\Lambda < 10^{-2}$ and $\alpha = 0.5$, $\Delta(\Lambda, \alpha) \approx 3.7$, yielding the totally irreversible characteristic, (6.3.11). Thus a system may show nernstian, quasi-reversible, or totally irreversible behavior, depending on Λ, or experimentally, on the scan rate employed. The appearance of kinetic effects depends on the time window of the experiment, which is essentially the time needed to traverse the LSV wave. At small

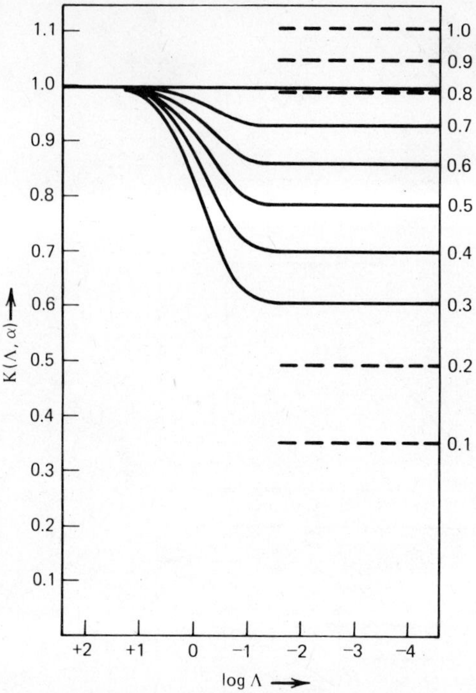

Figure 6.4.2
Variation of $K(\Lambda, \alpha)$ with Λ for different values of α. Dashed lines show functions for totally irreversible reaction.

$$K(\Lambda, \alpha) = i_p/i_p(\text{rev})$$

[From H. Matsuda and Y. Ayabe, *Z. Electrochem.*, **59**, 494 (1955), with permission.]

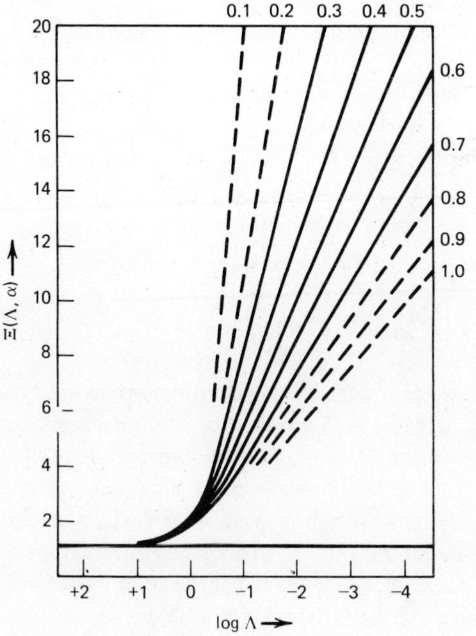

Figure 6.4.3
Variation of $\Xi(\Lambda, \alpha)$ with Λ for different values of α. Dashed lines show functions for totally irreversible reaction.

$$\Xi(\Lambda, \alpha) = -(E_p - E_{1/2})\frac{nF}{RT}$$

[From H. Matsuda and Y. Ayabe, *Z. Elektrochem.*, **59**, 494 (1955), with permission.]

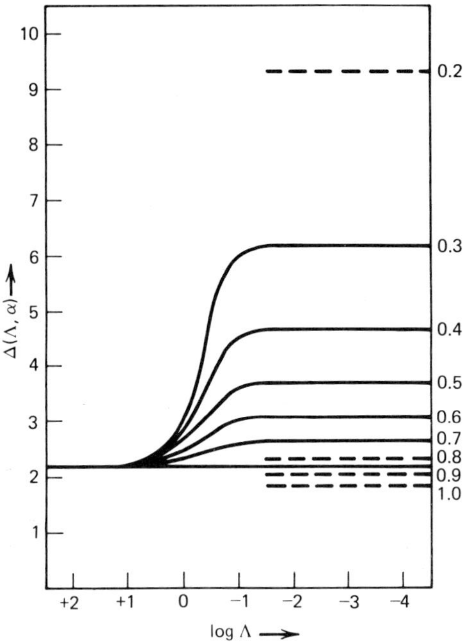

Figure 6.4.4
Variation of $\Delta(\Lambda, \alpha)$ with Λ and α. Dashed lines show values for totally irreversible reactions.

$$\Delta(\Lambda, \alpha) = (E_{p/2} - E_p)\frac{nF}{RT}$$

[From H. Matsuda and Y. Ayabe, *Z. Elektrochem.*, **59**, 494 (1955), with permission.]

v (or long times), systems may yield reversible waves, while at large v (or short times), irreversible behavior is observed. Matsuda and Ayabe suggest the following zone boundaries:†

Reversible (nernstian)	$\Lambda \geq 15$; $k^0 \geq 0.3v^{1/2}$ cm/sec
Quasi-reversible	$15 \geq \Lambda \geq 10^{-2(1+\alpha)}$; $0.3v^{1/2} \geq k^0 \geq 2 \times 10^{-5}v^{1/2}$ cm/sec
Totally irreversible	$\Lambda \leq 10^{-2(1+\alpha)}$; $k^0 \leq 2 \times 10^{-5}v^{1/2}$ cm/sec

6.5 REVERSAL TECHNIQUES

Reversal techniques with linear scan voltammetry are carried out by reversing the direction of the scan at a certain time, $t = \lambda$ (or at the switching potential, E_λ). Thus the potential is given at any time by

$$(0 < t \leq \lambda) \qquad E = E_i - vt \tag{6.5.1}$$

$$(t > \lambda) \qquad E = E_i - 2v\lambda + vt \tag{6.5.2}$$

While it is possible to use a different scan rate (v') on reversal (7), this is not usually done, and only the case of a symmetrical triangular wave is considered here. The theoretical treatment follows that of Section 6.2, except that (6.5.2) is used in the concentration-potential equation rather than (6.2.1) for $t > \lambda$.

† The $k°$ values are based on $n = 1$, $\alpha = 0.5$, and $T = 25°$, with v in V/sec, and D in cm²/sec; $\Lambda \approx k°/(39Dv)^{1/2}$. Values are for $D = 10^{-5}$ cm²/sec.

6.5.1 Nernstian Systems

Application of (6.5.2) into the equation for a nernstian system (5.4.6) yields (6.2.3), where $S(t)$ is now given by

$$(t > \lambda) \qquad S(t) = e^{\sigma t - 2\sigma\lambda} \tag{6.5.3}$$

The derivation then proceeds as in Section 6.2. The shape of the curve on reversal depends on the switching potential, E_λ, or how far beyond the cathodic peak the scan is allowed to proceed before reversal. However, if E_λ is at least $35/n$ mV past the cathodic peak, the reversal peaks all have the same general shapes, basically consisting of a curve shaped like the forward *i-E* curve plotted in the opposite direction on the current axis, with the decaying current of the cathodic wave used as a baseline. Typical *i-t* curves for different switching potentials are shown in Figure 6.5.1. This type of presentation would result if the curves were recorded on a strip-chart recorder. The more usual *i-E* presentation, with curves recorded on an X-Y recorder, is shown in Figure 6.5.2. The measured parameters of interest on these *i-E* curves (*cyclic voltammograms*) are i_{pa}/i_{pc}, the ratio of peak currents, and $E_{pa} - E_{pc}$, the separation of peak potentials. For a nernstian wave with stable product, the ratio $i_{pa}/i_{pc} = 1$ regardless of scan rate, E_λ (for $E_\lambda > 35/n$ mV past E_{pc}), and diffusion coefficients, when i_{pa} is measured from the decaying cathodic current as a baseline (see Figures 6.5.1 and 6.5.2). This baseline can be determined by the methods described in Section 6.6. Note that if the cathodic sweep is stopped and the current is allowed to decay to zero (Figure 6.5.2, curve 4), the resulting anodic *i-E* curve is identical in shape to the cathodic one, but is plotted in the opposite direction on both the *i* and *E* axes. This is so because allowing the cathodic current to decay to zero results in the diffusion

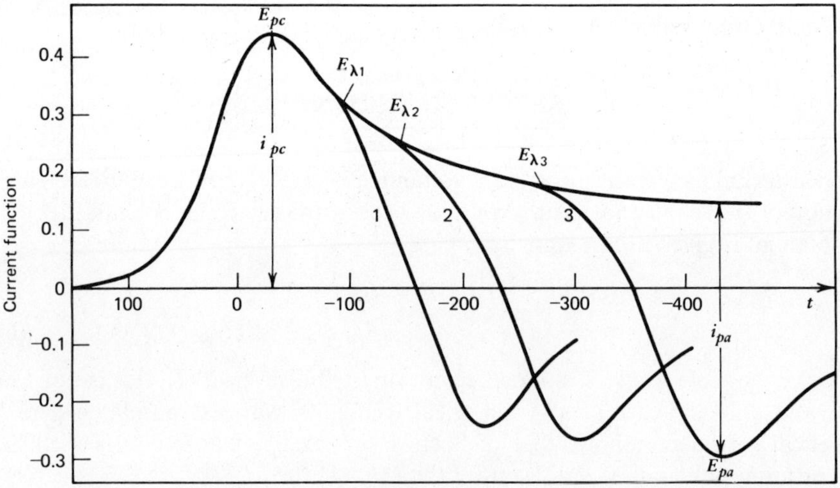

Figure 6.5.1
Cyclic voltammograms for reversal at different E_λ values with presentations as they appear on a strip-chart recorder (*i-t* curves).

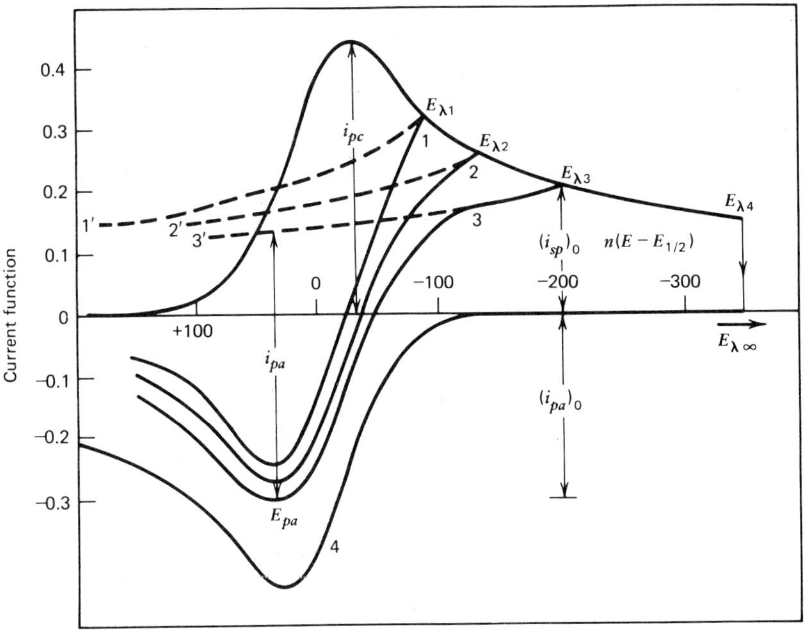

Figure 6.5.2

Cyclic voltammograms under the same conditions as in Figure 6.5.1 with presentations as they appear on X-Y recorder (*i-E* curves). E_λ of (1) $E_{1/2} - 90/n$; (2) $E_{1/2} - 130/n$; (3) $E_{1/2} - 200/n$ mV; (4) for potential held at $E_{\lambda 4}$ until the cathodic current decays to zero. [This curve results from reflection of the cathodic *i-E* curve through the *E* axis and then through the $n(E - E_{1/2}) = 0$ line. The curves in (1), (2), and (3) result by addition of this curve to the decaying current of the cathodic *i-E* curve.]

layer being depleted of O and populated with R at a concentration near C_O^*, so that the anodic scan is virtually the same as that which would result from an initial anodic scan in a solution containing only R. Deviation of the ratio i_{pa}/i_{pc} from unity is indicative of kinetic or other complications in the electrode process. Nicholson (9) suggested that if the actual baseline for measuring i_{pa} cannot be determined, the ratio can be calculated from (a) the uncorrected anodic peak current $(i_{pa})_0$ with respect to the zero current baseline (see Figure 6.5.2, curve 3) and (b) the current at E_λ, $(i_{sp})_0$, by the expression:

$$\frac{i_{pa}}{i_{pc}} = \frac{(i_{pa})_0}{i_{pc}} + \frac{0.485(i_{sp})_0}{i_{pc}} + 0.086 \tag{6.5.4}$$

The difference between E_{pa} and E_{pc} (ΔE_p) is a useful diagnostic test of a nernstian reaction. Although ΔE_p is slightly a function of E_λ, it is always close to $2.3RT/nF$ (or $59/n$ mV at 25°). Actual values at 25° as a function of E_λ are shown in Table 6.5.1. For repeated cycling the cathodic peak current decreases and the anodic one increases until a steady-state pattern is attained. At steady state $\Delta E_p = 58/n$ mV at 25° (5).

Table 6.5.1
Separation of Peak Potentials for a Nernstian Wave as a Function of E_λ[a]

$n(E_{pc} - E_\lambda)$ (mV)	$n(E_{pa} - E_{pc})$ (mV)
71.5	60.5
121.5	59.2
171.5	58.3
271.5	57.8
∞	57.0

[a] Adapted from Reference 3.

6.5.2 Quasi-Reversible Reactions

By using the potential program given by (6.5.2) and (6.5.3) in the equations for linear scan voltammetry in Section 6.4, the *i-E* curves for quasi-reversible electrode reactions can be derived. In this case the wave shape and ΔE_p are functions of v, k_0, α, and E_λ. As before, however, if E_λ is at least $90/n$ mV beyond the cathodic peak, the effect of E_λ is small. In this case the curves are functions of the dimensionless parameters α and either Λ (see equation 6.4.2) or an equivalent parameter ψ defined by†

$$\psi = \Lambda\pi^{-1/2} = \frac{\left(\dfrac{D_O}{D_R}\right)^{\alpha/2} k^0}{[D_O \pi v (nF/RT)]^{1/2}} \tag{6.5.5}$$

Typical results are shown in Figure 6.5.3. However, for $0.3 < \alpha < 0.7$ the ΔE_p values are nearly independent of α and depend only on ψ; typical results are given in Table 6.5.2 (9). This method is very useful in estimating k_0 for quasi-reversible reactions by determining the variation of ΔE_p with v, and from this variation, ψ. It is closely related to the determination of the electron transfer kinetics by the shift of E_p with v as described in Section 6.4.

With both of these approaches one must be sure that the uncompensated resistance, R_u, is sufficiently small that the voltage drops caused by the effect (of the order of $i_p R_u$) are negligible compared to the ΔE_p attributable to kinetic effects. In fact Nicholson (9) has shown that since the effect of uncompensated resistance on the ΔE_p-v behavior is almost the same as that of ψ, resistive effects cannot be distinguished simply from the ΔE_p-v plot. The effect of R_u is most important when the i_p values are large and for large k^0 values (where ΔE_p for the kinetic effect differs only slightly from the reversible value). It is especially difficult not to have a few ohms of uncompensated resistance in nonaqueous solvents (such as acetonitrile or tetrahydrofuran), even with positive-feedback circuitry (Chapter 13), and some reported studies made under these conditions have suffered from this problem.

† Note that ψ in (6.5.5) is not the same as $\Psi(E)$ in (6.4.4).

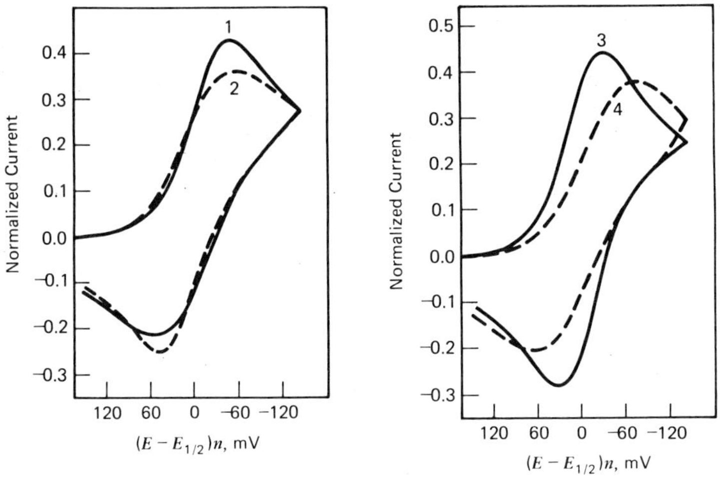

Figure 6.5.3

Theoretical cyclic voltammograms showing effect of ψ and α on curve shape. Curve 1: —— $\psi = 0.5$, $\alpha = 0.7$. Curve 2: ···· $\psi = 0.5$, $\alpha = 0.3$. Curve 3: —— $\psi = 7.0$, $\alpha = 0.5$. Curve 4: ···· $\psi = 0.25$, $\alpha = 0.5$. [Reprinted with permission from R. S. Nicholson, *Anal. Chem.*, **37**, 1351 (1965). Copyright 1965, American Chemical Society.]

Table 6.5.2

Variation of Peak Potential Separation (ΔE_p) with Kinetic Parameter ψ (9)[a]

ψ	$n(E_{pa} - E_{pc})$[b] mV
20	61
7	63
6	64
5	65
4	66
3	68
2	72
1	84
0.75	92
0.50	105
0.35	121
0.25	141
0.10	212

[a] For $E_\lambda = E_p - 112.5/n$ and $\alpha = 0.5$. ψ is defined in equation 6.5.5.
[b] $T = 25$ °C.

6.6 MULTICOMPONENT SYSTEMS AND MULTISTEP CHARGE TRANSFERS

The situation for the consecutive reduction of two substances O and O′ in a potential scan experiment is more complicated than that of the potential step (or sampled-current voltammetric) experiment treated in Section 5.6 (10, 11). As before we consider that the reactions $O + ne \rightarrow R$ and $O' + n'e \rightarrow R'$ occur. If the diffusion of O and O′ occurs independently, the fluxes are additive and the *i-E* (or *i-t*) curve for the mixture is the sum of the individual *i-E* curves of O and O′ (Figure 6.6.1). Note, however, that the measurement of i'_p must be made using the decaying current of the first wave as the baseline. Usually this baseline in a mixture is obtained by assuming that the current past the peak potential follows that for the large-amplitude potential step and decays as $t^{-1/2}$. A better fit based on an equation with two adjustable parameters has been suggested by Polcyn and Shain (11); since this fitting procedure depends on the reversibility of the reactions and is a little messy, it is not used very frequently. An experimental approach to obtaining the baseline was suggested by Reinmuth (unpublished). Since the concentration of O at the electrode falls essentially to zero at potentials just beyond E_p, the current beyond E_p is independent of potential. Thus if the voltammogram of a single-component system is recorded on a strip-chart (rather than on an X-Y) recorder and the potential scan is held at about $60/n$ mV beyond E_p while the chart continues to move, the current-time curve that results will be the same as that obtained with the potential sweep continuing (until a new wave or background reduction occurs). For a two-component system this technique allows establishing the baseline for the second wave by halting the scan somewhere before the foot of the second wave and recording the *i-t* curve, and then repeating the experiment (after stirring the solution and allowing it to come to rest to reestablish the initial

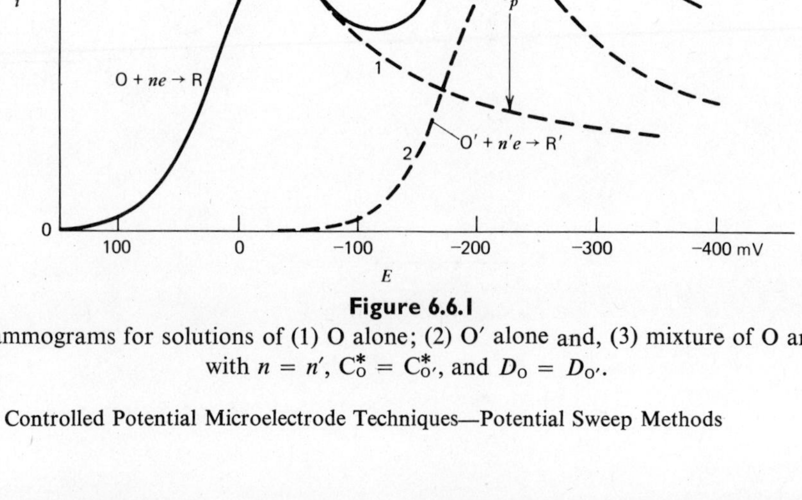

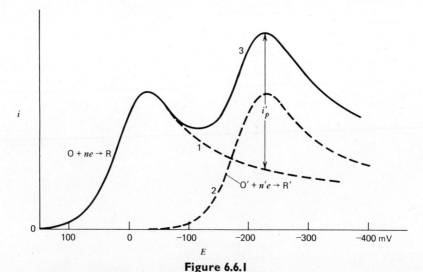

Figure 6.6.1

Voltammograms for solutions of (1) O alone; (2) O′ alone and, (3) mixture of O and O′, with $n = n'$, $C_O^* = C_{O'}^*$, and $D_O = D_{O'}$.

conditions). The second run is made at the same rate and continues beyond the second peak (Figure 6.6.2). An alternate experimental approach involves stopping the sweep beyond E_p as before and allowing the current to decay to a small value (so that O is depleted in the vicinity of the electrode or the concentration gradient of O is essentially zero near the electrode). Then one continues the scan and measures i_p' from the potential axis as a baseline (Figure 6.6.3). The application of this method requires convection-free conditions (quiet, vibration-free solutions, and shielded electrodes oriented to prevent density gradients, etc; see Section 7.3.6), because the waiting time for the current decay has to be appreciable (approximately 20 to 50 times the time needed to traverse a peak).

For the stepwise reduction of a substance O, that is, $O + n_1e \rightarrow R_1$ (E_1^0), $R_1 + n_2e \rightarrow R_2$ (E_2^0), the situation is similar but more complicated. If E_1^0 and E_2^0 are well separated, with $E_1^0 > E_2^0$ (i.e., O reduces before R_1), then two separate waves are observed. The first wave corresponds to reduction of O to R_1 in an n_1-electron reaction, with R_1 diffusing into the solution as the wave is traversed. At the second wave, O continues to reduce at the electrode in an $(n_1 + n_2)$-electron reaction and R_1 diffuses

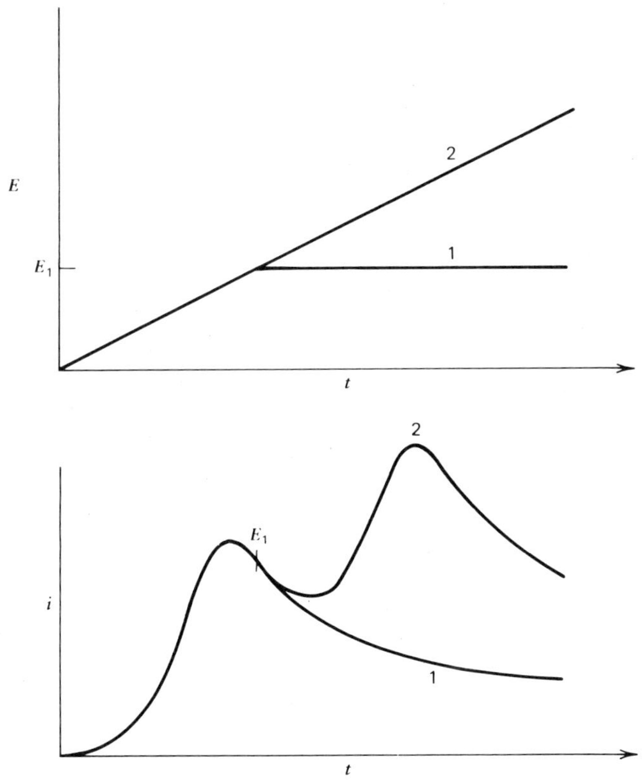

Figure 6.6.2
Method for obtaining baseline for measurement of i_p' of second wave. *Upper curves*: potential programs. *Lower curves*: resulting voltammograms with curve 1 potential stopped at E_1, curve 2 potential scan continued. System as in Figure 6.6.1.

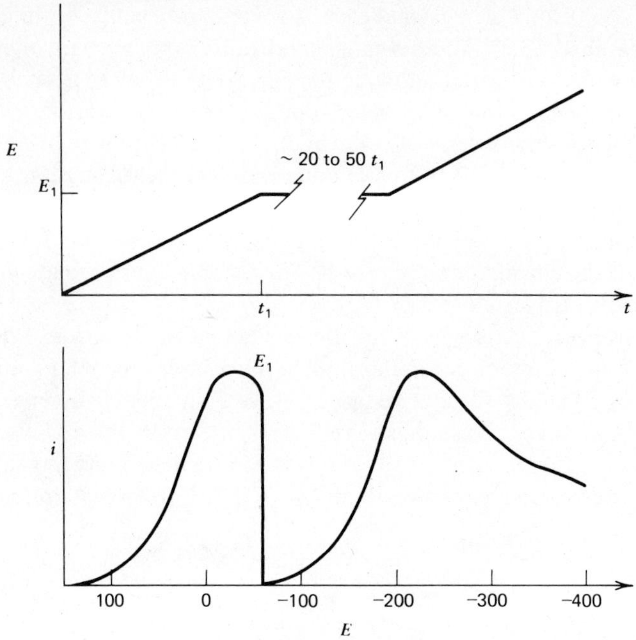

Figure 6.6.3
Method of allowing current of first wave to decay before scanning second wave. *Upper curve*: potential program. *Lower curve*: resulting voltammogram. System as in Figure 6.6.1.

back toward the electrode to be reduced in an n_2-electron reaction. The voltammogram for this case resembles that of Figure 6.6.1. In general the nature of the *i-E* curve depends on $\Delta E^0 (= E_2^0 - E_1^0)$, the reversibility of each step, and n_1 and n_2. Calculated cyclic voltammograms for different values of ΔE^0 in a system with two one-electron steps are shown in Figure 6.6.4. When ΔE^0 is between 0 and -100 mV, the individual waves are merged into a broad wave whose E_p is independent of scan rate. When $\Delta E^0 = 0$, a single peak with a peak current intermediate between those of single-step 1*e* and 2*e* reactions is found, and $E_p - E_{p/2} = 21$ mV. For $\Delta E^0 > 180$ mV (i.e., the second step is easier than the first), a single wave characteristic of a direct 2*e* reduction ($O + 2e \rightarrow R_2$) is observed (i.e., $\Delta E_p = 2.3\ RT/2F$). A particular case of interest occurs when $\Delta E^0 = -(2RT/F) \ln 2 = -35.6$ mV (25 °C). This ΔE^0 occurs when there is no interaction between the reducible groups on O, and the additional difficulty in adding the second electron arises purely from statistical factors (12). Under these conditions the observed wave has all of the characteristics of a one-electron transfer even though it is actually the result of two merged one-electron transfers. This same concept can be extended to the reduction of molecules containing *k* equivalent, noninteracting, reducible centers (e.g., reducible polymers). For this case the ΔE°'s between the first and kth electron transfers is given by

$$E_k^0 - E_1^0 = -\left(\frac{2RT}{F}\right) \ln k \qquad (6.6.1)$$

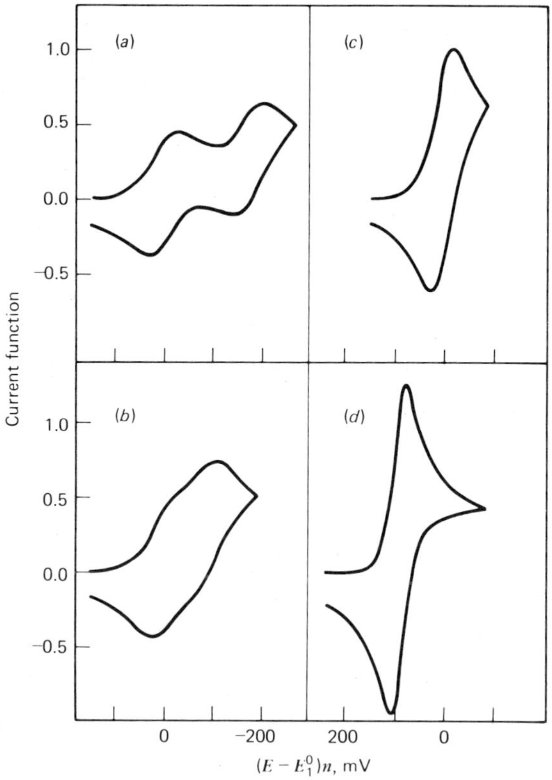

Figure 6.6.4

Cyclic voltammograms for a reversible two-step system. Current function is analogous to $\chi(z)$ defined in (6.2.16). $n_2/n_1 = 1.0$. (*a*) $\Delta E^0 = -180$ mV. (*b*) $\Delta E^0 = -90$ mV. (*c*) $\Delta E^0 = 0$ mV. (*d*) $\Delta E^0 = 180$ mV. [Reprinted with permission from D. S. Polcyn and I. Shain, *Anal. Chem.*, **38**, 370 (1966). Copyright 1966, American Chemical Society.]

and again the reduction wave, now involving k merged waves, appears like a single one-electron wave (13). From these considerations it is clear that for two one-electron transfer reactions, ΔE^0 values more positive than $-(2RT/F)\ln 2$ represent positive interactions (i.e., the second electron transfer is assisted by the first), while ΔE^0 values more negative than $-(2RT/F)\ln 2$ represent negative interactions. A fuller discussion of the theoretical treatment for several cases, including those involving irreversible electron transfers, is given by Polcyn and Shain (11).

Linear sweep and cyclic voltammetric methods have been employed for numerous basic studies of electrochemical systems and for analytical purposes. For example, the technique can be used for in vivo monitoring of substances in the kidney or brain (14); a typical example that employed a miniature carbon paste electrode to study ascorbic acid in a rat brain is illustrated in Figure 6.6.5. These techniques are especially powerful tools in the study of electrode reaction mechanisms (Chapter 11) and of adsorbed species (Chapter 12).

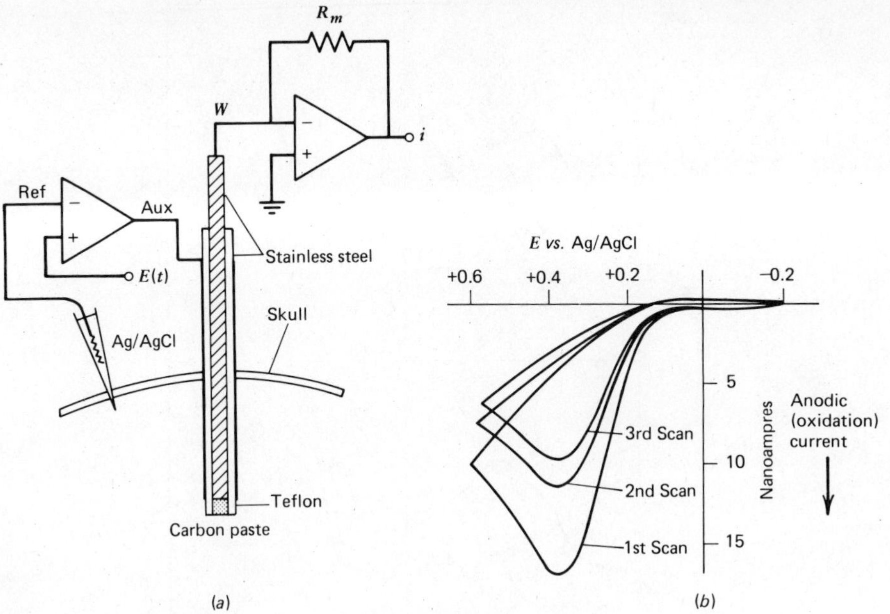

Figure 6.6.5

Application of cyclic voltammetry to in vivo analysis in brain tissue. (*a*) Carbon paste working electrode, stainless steel auxiliary electrode (18-gauge cannula), Ag/AgCl reference electrode, and other apparatus for voltammetric measurements. (*b*) Cyclic voltammogram for ascorbic acid oxidation at C-paste electrode positioned in the caudate nucleus of an anesthetized rat. [From P. T. Kissinger, J. B. Hart, and R. N. Adams, *Brain Res.*, **55**, 20 (1973), with permission.]

6.7 CONVOLUTIVE OR SEMI-INTEGRAL TECHNIQUES

6.7.1 Principles and Definitions

By proper treatment of the linear potential sweep data, the voltammetric *i-E* (or *i-t*) curves can be transformed into forms, closely resembling the steady-state voltammetric curves, which are frequently more convenient for further data processing. This transformation makes use of the convolution principle (A.1.21) and has been facilitated by the availability of digital computers for the processing and acquisition of data. The solution of the diffusion equation for semi-infinite linear diffusion conditions, and for species O initially present at a concentration C_O^*, yields for any electrochemical technique, the expression (see equations 6.2.4 to 6.2.6)

$$C_O(0, t) = C_O^* - \frac{1}{nFAD_O^{1/2}}\left[\frac{1}{\pi^{1/2}}\int_0^t \frac{i(u)}{(t-u)^{1/2}}\,du\right] \tag{6.7.1}$$

If the term in brackets, which represents a particular (convolutive) transformation of the experimental $i(t)$ data, is defined as $I(t)$, then equation 6.7.1 becomes (15)

$$C_O(0, t) = C_O^* - \frac{I(t)}{nFAD_O^{1/2}} \tag{6.7.2}$$

where

$$\boxed{I(t) = \frac{1}{\pi^{1/2}} \int_0^t \frac{i(u)}{(t-u)^{1/2}}\, du} \tag{6.7.3}$$

Following the generalized definition of the Riemann-Liouville operators, this integral can be considered as the *semi-integral* of $i(t)$, generated by the operator $d^{-1/2}/dt^{-1/2}$, so that (16, 17)

$$\frac{d^{-1/2}}{dt^{-1/2}} i(t) = m(t) = I(t) \tag{6.7.4}$$

Both $m(t)$ and $I(t)$, which represent the integral in equation 6.7.3, have been used in discussing this transformation technique; clearly the convolutive (15) and semi-integral (16, 17) approaches are equivalent. Thus the transformed current data can be used directly, by (6.7.2), to obtain $C_O(0, t)$. Under conditions where $C_O(0, t) = 0$ (i.e., under purely diffusion-controlled conditions), $I(t)$ reaches its limiting or maximum value, I_l [or, in semi-integral notation, $m(t)_{max}$] where

$$I_l = nFAD_O^{1/2}C_O^* \tag{6.7.5}$$

or

$$C_O(0, t) = \frac{[I_l - I(t)]}{nFAD_O^{1/2}} \tag{6.7.6}$$

Note the similarity between this expression for the transformed current and that for the steady-state concentration in terms of the actual current, (1.4.11). Similarly for species R, assumed absent initially, the expression that results is (see equation 6.2.9)

$$C_R(0, t) = \frac{I(t)}{nFAD_R^{1/2}} \tag{6.7.7}$$

Let us stress that these equations hold for any form of signal excitation in any electrochemical technique applied under the above conditions (semi-infinite diffusion, absence of migration, convection, etc.), and no assumptions have been made concerning the reversibility of the charge transfer reaction or even the form of the dependence of $C_O(0, t)$ and $C_R(0, t)$ on E. Thus, with the application of any excitation signal that eventually drives $C_O(0, t)$ to zero, the transformed current $I(t)$ will attain a limiting value, I_l, that can be used to determine C_O^* by equation 6.7.5 (17). If the electron transfer reaction is nernstian, the application of equations 6.7.6 and 6.7.7 immediately yields

$$E = E_{1/2} + \frac{RT}{nF} \ln \frac{I_l - I(t)}{I(t)} \tag{6.7.8}$$

where $E_{1/2} = E^{0'} + (RT/2nF)\ln(D_R/D_O)$. Note that this expression is identical in form to those for the steady-state or sampled-current i-E curves (equations 1.4.16 and

5.4.22). Transformation of a linear potential sweep *i-E* response thus converts the peaked *i-E* curve to an S-shaped one resembling a polarogram (Figure 6.7.1).

6.7.2 Transformation of the Current—Evaluation of I(t)

Although analog circuits that approximate $I(t)$ have been proposed (18), the function is usually evaluated by a numerical integration technique on a computer. Several different algorithms have been proposed for the evaluation (19, 21). The *i-t* data are usually divided into N equally spaced time intervals between $t = 0$ and $t = t_f$, indexed by j; then $I(t)$ becomes $I(k\Delta t)$ (where k varies between 0 and N, representing $t = 0$ and $t = t_f$; $\Delta t = t_f/N$) (Figure 6.7.2). One convenient algorithm, which follows directly from the definition of $I(t)$, is (19)

$$I(t) = I(k\Delta t) = \frac{1}{\pi^{1/2}} \sum_{j=1}^{j=k} \frac{i(j\Delta t - \frac{1}{2}\Delta t)\,\Delta t}{\sqrt{k\Delta t - j\Delta t + \frac{1}{2}\,\Delta t}} \tag{6.7.9}$$

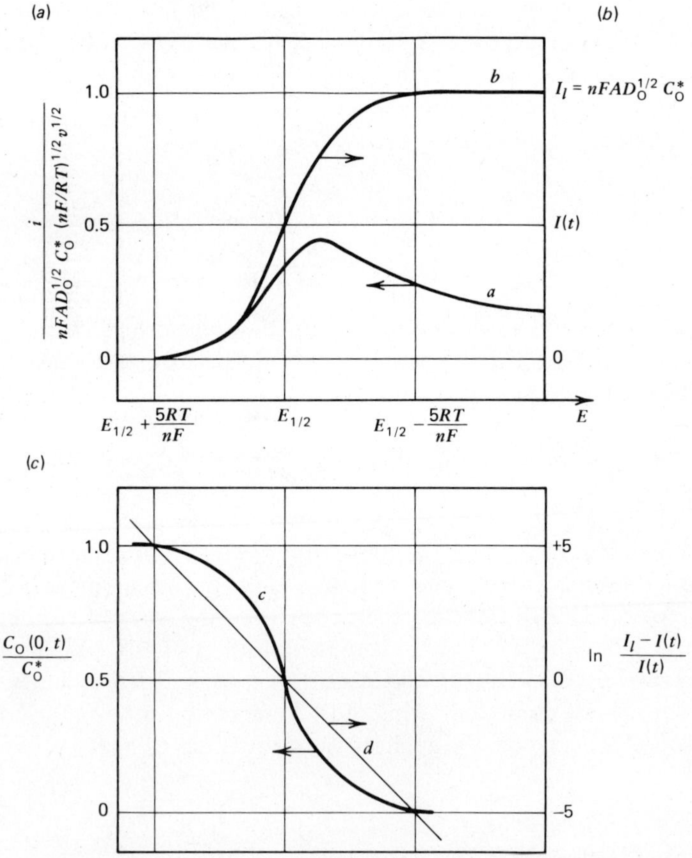

Figure 6.7.1

Variation of i, I, $C_O(0, t)/C_O^*$, and $\ln[(I_l - I)/I]$ with E. [Adapted from J. C. Imbeaux and J. M. Saveant, *J. Electroanal. Chem.*, **44**, 169 (1973), with permission.]

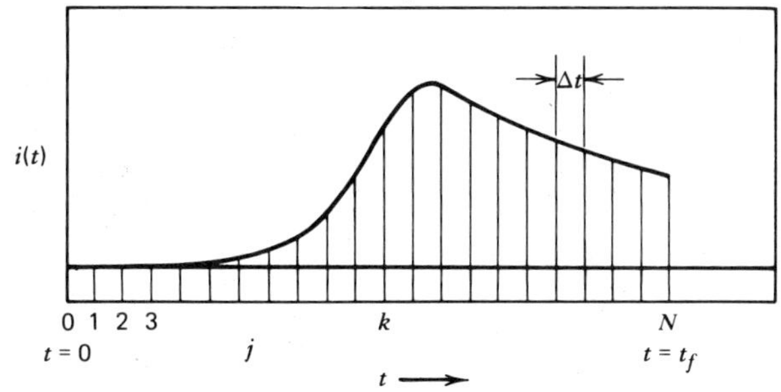

Figure 6.7.2
Division of experimental $i(t)$ *vs.* t [or *vs.* $E(t)$] curve for digital evaluation of $I(t)$.

which is obtained by using $t = k\Delta t$ and $u = j\Delta t$, and measuring i at the midpoint of each interval. This can be simplified to

$$I(k\Delta t) = \frac{1}{\pi^{1/2}} \sum_{j=1}^{j=k} \frac{i(j\Delta t - \frac{1}{2}\Delta t)\Delta t^{1/2}}{\sqrt{k - j + \frac{1}{2}}} \tag{6.7.10}$$

Another algorithm, which is especially convenient for digital computer processing, is

$$I(k\Delta t) = \frac{1}{\pi^{1/2}} \sum_{j=1}^{j=k} \frac{\Gamma(k - j + \frac{1}{2})}{(k - j)!} \Delta t^{1/2} i(j\Delta t) \tag{6.7.11}$$

where $\Gamma(x)$ is the Gamma function of x, where $\Gamma(\frac{1}{2}) = \pi^{1/2}$, $\Gamma(3/2) = 1/2\pi^{1/2}$, $\Gamma(5/2) = \frac{3}{2}\frac{1}{2}\pi^{1/2}$, etc. Other algorithms based on standard methods of numerical evaluation of definite integrals also have been used (15–17, 20, 21).

6.7.3 Irreversible and Quasi-Reversible Reactions

The convolutive form for a totally irreversible reaction follows directly from the i-E expression, equation 3.3.14, with no back reaction:

$$i = nFAk^0 C_O(0, t)\, e^{-\alpha nf(E - E^{0'})} \tag{6.7.12}$$

and the expression for $C_O(0, t)$, equation 6.7.6. Thus (15)

$$i(t) = k^0 D_O^{-1/2}[I_l - I(t)]\, e^{-\alpha nf(E - E^{0'})} \tag{6.7.13}$$

or

$$E = E^{0'} - \frac{RT}{\alpha nF} \ln \frac{D_O^{1/2}}{k^0} - \frac{RT}{\alpha nF} \ln \frac{i(t)}{I_l - I(t)} \tag{6.7.14}$$

For a quasi-reversible reaction the full equation 3.3.14 is employed, along with equations 6.7.6 and 6.7.7, to yield

$$i(t) = k^0\{D_O^{-1/2}[I_l - I(t)]\, e^{-\alpha nf(E - E^{0'})} - D_R^{-1/2} I(t)\, e^{(1-\alpha)nf(E - E^{0'})}\} \tag{6.7.15}$$

$$i(t) = k^0 D_O^{-1/2}\, e^{-\alpha nf(E - E^{0'})}\{I_l - I(t)[1 + \xi\, e^{nf(E - E^{0'})}]\} \tag{6.7.16}$$

$$E = E^{0'} + \frac{RT}{\alpha nF} \ln \frac{k^0}{D_O^{1/2}} + \frac{RT}{\alpha nF} \ln \frac{\{I_l - I(t)[1 + \xi\, e^{(nF/RT)(E - E^{0'})}]\}}{i(t)} \tag{6.7.17}$$

where $\xi = (D_O/D_R)^{1/2}$.

In deriving (6.7.14) and (6.7.17), we assumed that the Butler-Volmer expression for the rate constant for electron transfer applied [as given in (3.3.12) and (3.3.13), which produced the i-E characteristic of (3.3.14)]. Indeed this assumption (or the adoption of some other model) is necessary a priori before equations can be derived for a particular electrochemical method. However, with the convolutive approach, this assumption is not needed and the rate law can be written in the general form (22):

$$i(t) = nFAk_f(E)[C_O(0, t) - C_R(0, t)\, e^{(nF/RT)(E - E^{0'})}] \tag{6.7.18}$$

where $k_f(E)$ is the potential-dependent rate constant of the forward reaction. Thus, with (6.7.6) and (6.7.7),

$$\ln k_f(E) = \ln D_O^{1/2} - \ln\left\{\frac{I_l - I(t)[1 + \xi\, e^{(nF/RT)(E - E^{0'})}]}{i(t)}\right\} \tag{6.7.19}$$

Analysis of experimental linear potential sweep experiments according to (6.7.19) or the equivalent expression for a totally irreversible reduction $\{\xi \exp[(nF/RT) \times (E - E^{0'})] \ll 1\}$,

$$\ln k_f(E) = \ln D_O^{1/2} - \ln \frac{[I_l - I(t)]}{i(t)} \tag{6.7.20}$$

yields $\ln k_f(E)$ as a function of E at different v. If pure Butler-Volmer kinetics apply, then a plot of $\ln k_f(E)$ *vs.* E should be linear with a slope $\alpha nF/RT$. In an analysis of experimental data for the electroreduction of tertnitrobutane in aprotic solvents, Saveant and Tessier (22) noted significant deviations from linearity (after necessary corrections for double-layer effects were carried out), demonstrating that the α value was potential dependent, as is indeed predicted by other theories of electron transfer reactions.

6.7.4 Applications

The convolution technique offers a number of advantages in the treatment of the linear sweep data (and perhaps also in other electrochemical techniques). For a reversible reaction in a cyclic voltammetric experiment, the curves of $I(t)$ *vs.* E for the forward and backward scans superimpose, with $I(t)$ returning to zero at sufficiently positive potentials [where $C_R(0, t) = 0$]. This behavior has been verified experimentally (15, 21, 23) (Figure 6.7.3*a*). For a quasi-reversible reaction, however, the forward and backward $I(t)$ curves do not coincide (Figure 6.7.3*b*). One can regard this effect as a consequence of the shifting of the E_{pc} and E_{pa} values from their reversible values. The procedure used to obtain the reversible $E_{1/2}$ value for such a system is shown in Figure 6.7.3*b* (22).

Correction for the uncompensated resistance, R_u, is also much more straightforward for $I(t)$-E curves than for the $i(t)$-E curve, since R_u affects the linearity of the potential sweep at the electrode (see Section 6.2.4). For the $I(t)$-E curves, correction is accomplished simply by replacing the applied potential E by $E' = E + iR_u$ (15, 21, 23).

A procedure for correcting the $I(t)$ curves for charging current has also been

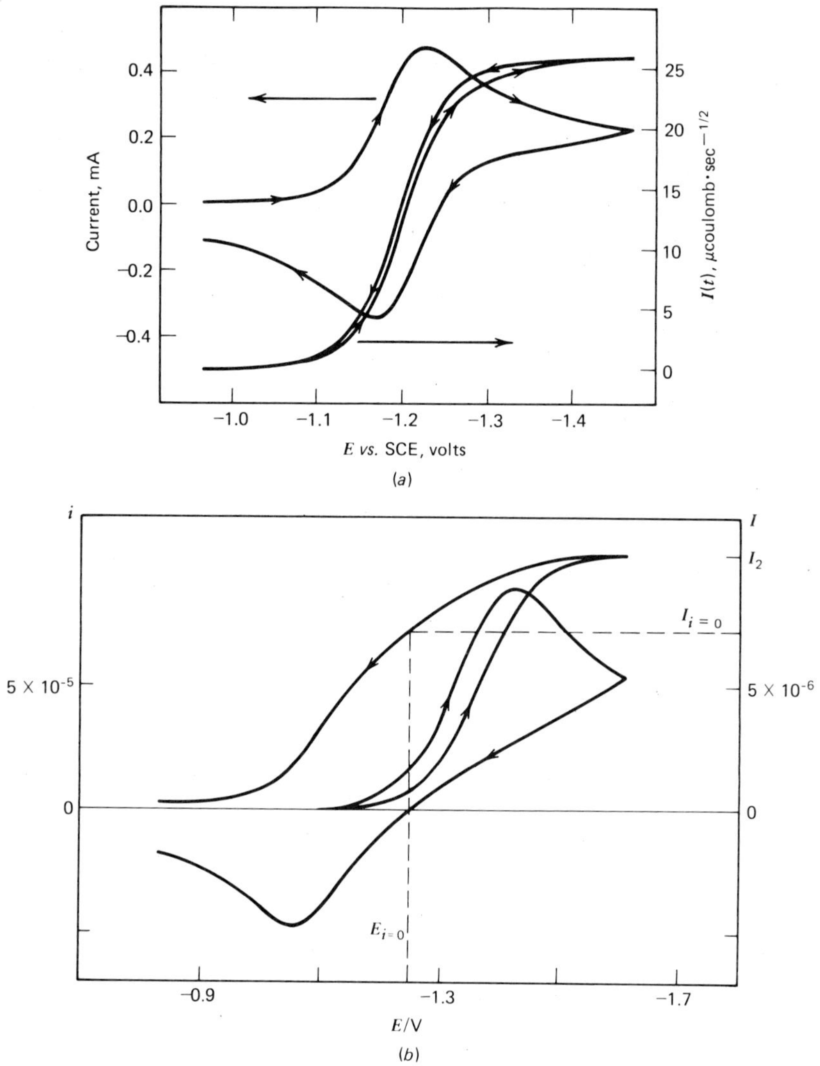

Figure 6.7.3

Experimental cyclic voltammogram and convolution of (*a*) 1.84 m*M* *p*-nitrotoluene in acetonitrile containing 0.2 *M* TEAP at HMDE, $v = 50$ V/sec. [Reprinted with permission from P. E. Whitson, H. W. Vanden Born, and D. H. Evans, *Anal. Chem.*, **45**, 1298 (1975). Copyright 1975, American Chemical Society.] (*b*) 0.5 m*M* tert-nitrobutane in DMF containing 0.1 *M* TBAI, $v = 17.9$ V/sec. $E_{1/2}$ determined for quasi-reversible system from

$$E_{1/2} = E^0 - (RT/F)\ln[(I_l - I_{i=0})/I_{i=0}]$$

[From J. M. Saveant and D. Tessier, *J. Electroanal. Chem.*, **65**, 57 (1975), with permission.]

presented (23). The charging current for a blank experiment in the absence of electroactive compound, i_c^b, is

$$i_c^b = \frac{-C_d\, dE'}{dt} \tag{6.7.21}$$

where C_d is the potential-dependent capacitance (μF) and E' is the potential corrected for R_u:

$$E' = E + i_c^b R_u = E_i - vt + i_c^b R_u \tag{6.7.22}$$

Thus

$$C_d(E') = \frac{i_c^b}{v - R_u(di_c^b/dt)} \tag{6.7.23}$$

Thus if R_u is known, the measured values of i_c^b and di_c^b/dt allow calculation of $C_d(E')$. When substance O is introduced into solution, the total current, i_t, is

$$i_t = i_f + i_c \tag{6.7.24}$$

where i_c may differ from i_c^b because of the presence of O. However, it is still true that

$$i_c = \frac{-C_d\, dE'}{dt} \tag{6.7.25}$$

$$E' = E - vt + i_t R_u \tag{6.7.26}$$

and finally

$$i_f = i_t - C_d v + C_d R_u \left(\frac{di_t}{dt}\right) \tag{6.7.27}$$

where C_d is assumed to be the same function of potential as determined in the blank experiment. Correction of i_t for C_d can be accomplished by (6.7.27) to obtain i_f on a computer before the convolution is performed.

Convolution methods simplify to some extent the treatment of data for electrode processes with coupled chemical reactions (15) and may be useful in analytical applications (17). The manual digitizing of experimental *i-E* curves is, of course, tedious, and even the posttreatment of digitally acquired data on a separate computer is not convenient. However, the advent of low-cost minicomputers and microprocessors has led to the development of instruments where data acquisition and treatment can be accomplished at the time the electrochemical experiment is carried out. The widespread use of these instruments will undoubtedly encourage the application of convolution, as well as other transform techniques.

6.8 REFERENCES

1. J. E. B. Randles, *Trans. Faraday Soc.*, **44**, 327 (1948).
2. A. Sevcik, *Collect. Czech. Chem. Commun.*, **13**, 349 (1948).
3. R. S. Nicholson and I. Shain, *Anal. Chem.*, **36**, 706 (1964).

4. W. H. Reinmuth, *J. Am. Chem. Soc.*, **79**, 6358 (1957).
5. H. Matsuda and Y. Ayabe, *Z. Elektrochem.*, **59**, 494 (1955).
6. Y. P. Gokhshtein, *Dokl. Akad. Nauk SSSR*, **126**, 598 (1959).
7. J. M. Savéant, *Electrochim. Acta*, **12**, 999 (1967).
8. W. M. Schwarz and I. Shain, *J. Phys. Chem.*, **70**, 845 (1966).
9. R. S. Nicholson, *Anal. Chem.*, **37**, 1351 (1965).
10. Y. P. Gokhshtein and A. Y. Gokhshtein, in "Advances in Polarography," I. S. Longmuir, Ed., Vol. 2, p. 465, Pergamon Press, New York, 1960; *Dokl. Akad. Nauk SSSR*, **128**, 985 (1959).
11. D. S. Polcyn and I. Shain, *Anal. Chem.*, **38**, 370 (1966).
12. F. Ammar and J. M. Savéant, *J. Electroanal. Chem.*, **47**, 215 (1973).
13. J. B. Flanagan, S. Margel, A. J. Bard, and F. C. Anson, *J. Am. Chem. Soc.*, **100**, 4248 (1978).
14. P. T. Kissinger, J. B. Hart, and R. N. Adams, *Brain Res.*, **55**, 20 (1973).
15. J. C. Imbeaux and J. M. Savéant, *J. Electroanal. Chem.*, **44**, 1969 (1973).
16. K. B. Oldham and J. Spanier, *ibid.*, **26**, 331 (1970).
17. K. B. Oldham, *Anal. Chem.*, **44**, 196 (1972).
18. K. B. Oldham, *ibid.*, **45**, 39 (1973).
19. R. J. Lawson and J. T. Maloy, *ibid.*, **46**, 559 (1974).
20. J. H. Carney and H. C. Miller, *ibid.*, **45**, 2175 (1973).
21. P. E. Whitson, H. W. Vanden Born, and D. H. Evans, *ibid.*, **45**, 1298 (1975).
22. J. M. Savéant and D. Tessier, *J. Electroanal. Chem.*, **65**, 57 (1975).
23. L. Nadjo, J. M. Savéant, and D. Tessier, *ibid.*, **52**, 403 (1974).

6.9 PROBLEMS

6.1 Derive (6.2.6) from (6.2.4) and (6.2.5).

6.2 From the data in Table 6.3.1 plot the linear potential sweep voltammograms, that is, $\sqrt{\pi}\chi(bt)$ *vs.* potential for several values of k^0 with $\alpha = 0.5$, $T = 25$ °C, $n_a = 1$, $v = 100$ mV/sec, and $D_O = 10^{-5}$ cm^2/sec. Compare these results with those for a nernstian reaction shown in Fig. 6.2.1.

6.3 T. R. Mueller and R. N. Adams (see R. N. Adams, "Electrochemistry at Solid Electrodes," Marcel Dekker, New York, 1969, p. 128) suggested that by measurement of $i_p/v^{1/2}$ for a nernstian linear potential sweep voltammetric curve, and by carrying out a potential step experiment in the same solution at the same electrode to obtain the limiting value of $it^{1/2}$, the n value of an electrode reaction can be determined without the need to know A, C_O^*, or D_O. Demonstrate that this is the case. Why would this method be unsuitable for irreversible reactions?

6.4 The oxidation of *o*-dianisidine (*o*-DIA) occurs in a nernstian $2e$ reaction. For a 2.27 mM solution of (*o*-DIA) in 2 M H_2SO_4 at a carbon paste electrode of area 2.73 mm^2 with a scan rate of 0.500 V/min, $i_p = 8.19$ μA. Calculate the D value for *o*-DIA. What i_p is expected for $v = 100$ mV/sec? What i_p will be obtained for $v = 50$ mV/sec and 8.2 mM *o*-DIA?

6.5 Figure 6.9.1 shows a cyclic voltammogram taken for a solution containing benzophenone (BP) and tri-p-tolylamine (TPTA), both at 1 mM in acetonitrile. Benzophenone can be reduced inside the working range of acetonitrile and TPTA can be oxidized. However, benzophenone cannot be oxidized, and TPTA cannot be reduced. The scan shown here begins at 0.0 V *vs.* QRE and first moves toward positive potentials. Account for the shape of the voltammogram.

(a) Assign the voltammetric features between +0.5 and 1.0 V and between −1.5 and −2.0 V to appropriate electrode reactions. Comment on the heterogeneous and homogeneous kinetics pertaining to these electrode reactions.

(b) Why does the falloff in current appear between 0.7 and 1.0 V *vs.* QRE?

(c) What constitutes the anodic and cathodic currents seen at −1.0 V *vs.* QRE?

6.6 K. M. Kadish, L. A. Bottomley, and J. S. Cheng presented results of a study of the interactions between Fe(II) phthalocyanine (FePc) and various nitrogen bases, such as imidazole (Im).

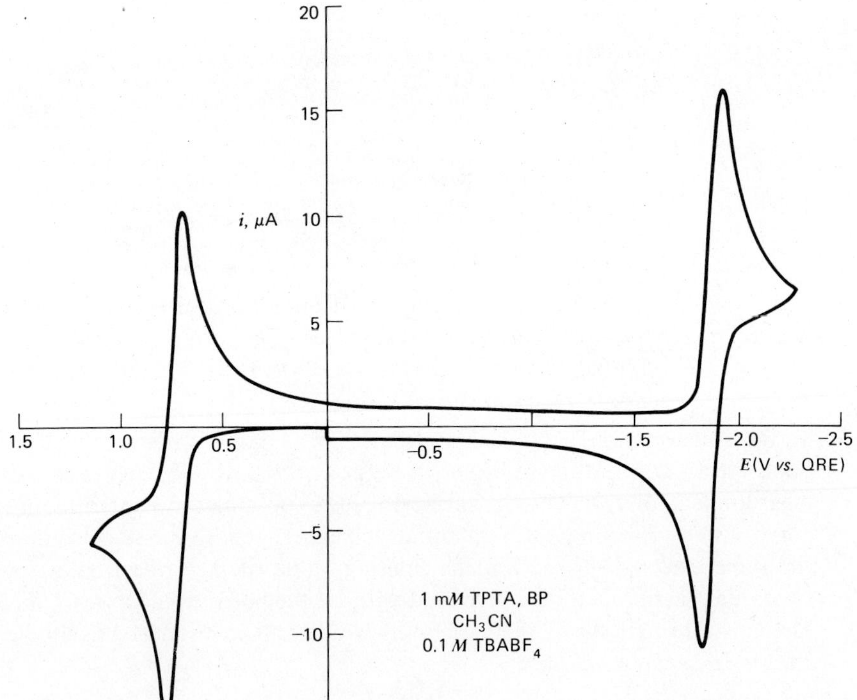

Figure 6.9.1

Cyclic voltammogram of benzophenone and tri-p-tolylamine in acetonitrile. (From P. R. Michael, PhD Thesis, University of Illinois at Urbana-Champaign, 1977, with permission.)

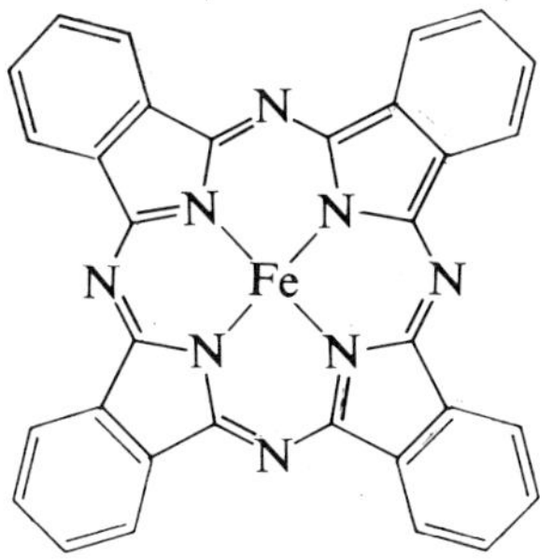

Iron phthalocyanine, FePc

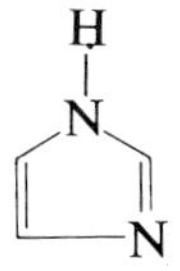

Im

The work was carried out in dimethylsulfoxide (DMSO) containing 0.1 *M* tetraethylammonium perchlorate (TEAP). Some results are shown in Figure 6.9.2. In (*a*), couples I and II both show peak potentials and current functions that are invariant with scan rate. Interpret the voltammetric properties of the system before and after addition of imidazole.

6.7 Consider the electrochemical reduction of molecular oxygen in an aprotic solvent such as pyridine or acetonitrile. In general, a cyclic voltammogram like that in Figure 6.9.3 is obtained. The polarogram (at a DME) gives a

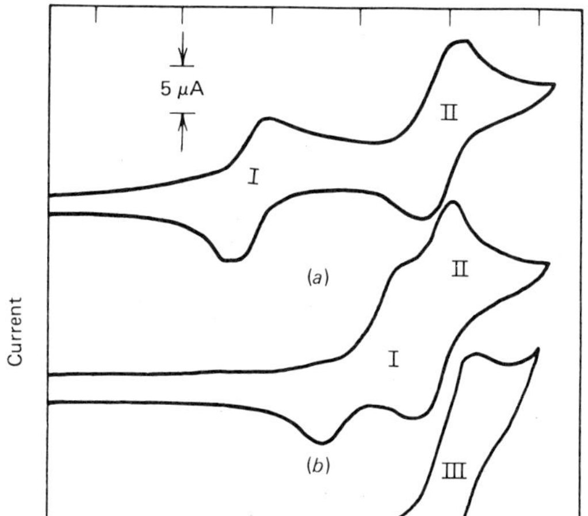

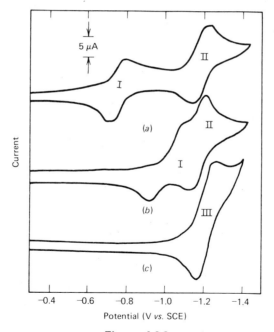

Figure 6.9.2

Cyclic voltammograms of 1.18 m*M* FePc in Me_2SO/imidazole mixtures containing 0.1 *M* TEAP. Scan rate 0.100 V/sec. Imidazole concentrations: (*a*) 0.00 *M*; (*b*) 0.01 *M*; (*c*) 0.95 *M*. [Reprinted with permission from K. M. Kadish, L. A. Bottomley, and J. S. Cheng, *J. Am. Chem. Soc.*, **100**, 2731 (1978). Copyright 1978, American Chemical Society.]

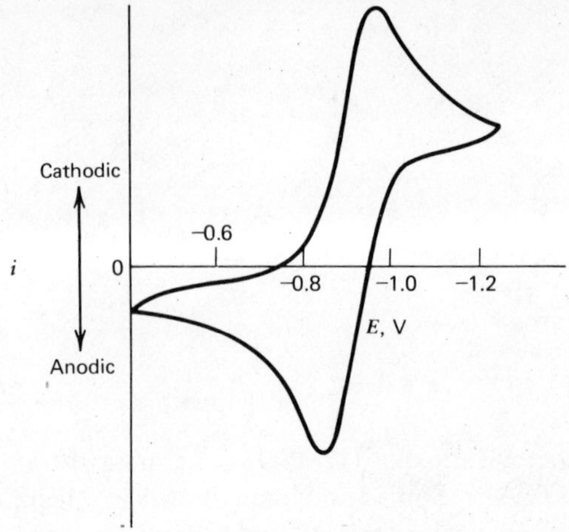

Figure 6.9.3
Cyclic voltammogram at a stationary drop electrode of oxygen in pyridine—0.2 *M* TBAP. Frequency 0.1 cps. [From M. E. Peover and B. S. White, *Electrochim. Acta*, **11**, 1061 (1966), with permission of Pergamon Press, Ltd.]

linear plot of E *vs.* $\log[(i_d - i)/i]$ with a slope of 63 mV. The reduction product at −1.0 V *vs.* SCE gives an ESR signal. If methanol is added in small quantities, the cyclic voltammogram shifts toward positive potentials, the forward peak rises in magnitude, and the reverse peak disappears. These trends continue with increasing methanol concentration until a limit is reached with reduction near −0.4 *V vs.* SCE. The polarogram under these limiting conditions is approximately twice as high as it was in methanol-free solution, and the wave slope is 78 mV.

(a) Identify the reduction product in methanol-free solution.
(b) Identify the reduction product under limiting conditions in methanol-containing solution.
(c) Comment on the charge transfer kinetics in methanol-free solution.
(d) Explain the voltammetric and polarographic responses.

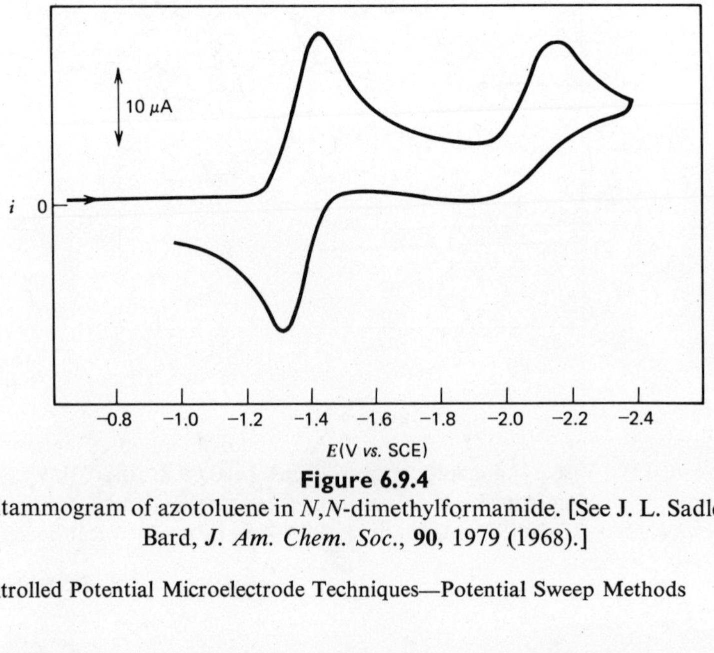

Figure 6.9.4
Cyclic voltammogram of azotoluene in *N*,*N*-dimethylformamide. [See J. L. Sadler and A. J. Bard, *J. Am. Chem. Soc.*, **90**, 1979 (1968).]

6.8 The cyclic voltammetry of azotoluene was studied under the following conditions:

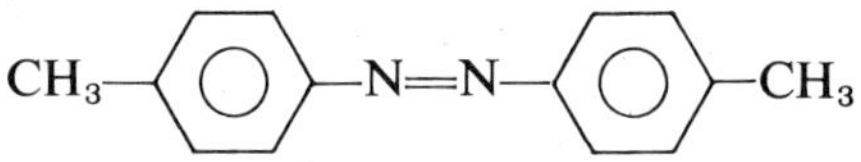

Azotoluene

Solution: *N*,*N*-dimethylformamide containing 0.10 *M* tetra-*n*-butylammonium perchlorate as supporting electrolyte and 0.68 m*M* in azotoluene; *working electrode:* planar platinum disk, 1.54 mm^2; *reference electrode:* SCE; *temperature:* 25 °C.

A typical cyclic voltammogram is shown in Figure 6.9.4 and the data obtained are as follows:

	First Wave[a]				Second Wave		
Scan Rate (mV/sec)	i_{pc} (μA)	i_{pa} (μA)	$-E_{pc}$ (V *vs.* SCE)	$-E_{pa}$ (V *vs.* SCE)	i_{pc} (μA)	$-E_{pc}$ (V *vs.* SCE)	$-E_{p/2}$ (V *vs.* SCE)
430	8.0	8.0	1.42	1.36	7.0	2.10	2.00
298	6.7	6.7	1.42	1.36	6.5	2.09	2.00
203	5.2	5.2	1.42	1.36	4.7	2.08	2.00
91	3.4	3.4	1.42	1.36	3.0	2.07	1.99
73	3.0	2.9	1.42	1.36	2.8	2.06	1.98

[a] For scan reversed 100 mV past E_{pc}.

Coulometry shows that the first reduction step involves one electron. Work up this set of data and discuss what information is obtained about the reversibility of the reactions, stability of products, diffusion coefficients, etc. (This is a set of actual data, so don't expect numbers to necessarily *exactly* conform to theoretical treatments.)

6.9 R. W. Johnson described the electrochemical behavior of 1,3,5-tri-*tert*-butylpentalene (I).

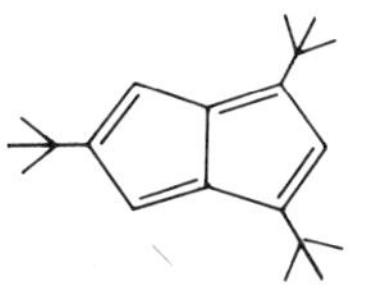

(I)

Solutions of I in CH_3CN with 0.1 *M* tetra-*n*-butylammonium perchlorate (TBAP) were subjected to polarographic and cyclic voltammetric examination. The results were as follows:

Polarography. One wave at $E_{1/2} = -1.46$ V *vs.* SCE. Wave slope of 59 mV.

Cyclic voltammetry. Illustrated in Figure 6.9.5; the scan starts at 0.0 V *vs.* SCE and moves first in a positive direction.

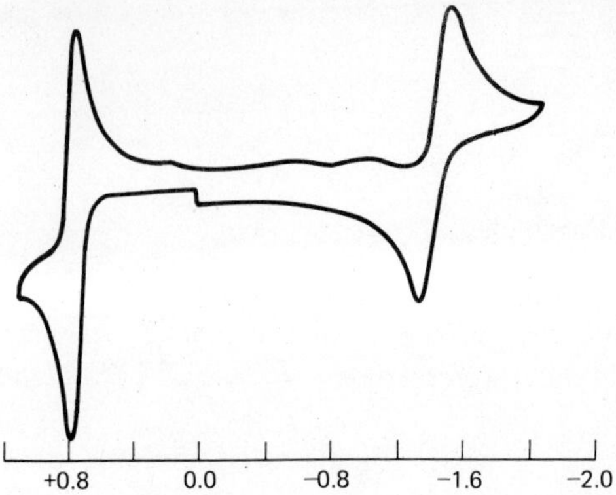

Figure 6.9.5
Cyclic voltammogram of I in CH_3CN with 0.1 M TBAP at a Pt electrode *vs.* SCE and a scan rate of 500 mV/sec. [Reprinted with permission from R. W. Johnson, *J. Am. Chem. Soc.*, **99**, 1461 (1977). Copyright 1977, American Chemical Society.]

In addition, bulk electrolysis at $+1.0$ V produced a green solution giving a well-resolved ESR spectrum, and bulk electrolysis at -1.6 V gave a magenta solution that also produced a well-resolved ESR spectrum. Both bulk transformations were carried out in CH_2Cl_2.

(a) Describe the chemistry of the system.

(b) Account for the shape of the cyclic voltammetric curve. Identify all peaks.

(c) Interpret the polarogram and relate the cyclic voltammogram to it.

(d) What would you expect the diffusion current constant for the polarogram to be (in conventional units)? Take $D_I = 2 \times 10^{-5}$ cm²/sec.

(e) Make sketches showing the expected variations with v of forward peak current and ΔE_p for the couple responsible for the green solution. Do the same for the couple responsible for the magenta solution.

chapter 7 Controlled Current Microelectrode Techniques

7.1 INTRODUCTION

We discussed in Chapters 5 and 6 methods in which the potential of an electrode was controlled (or was the independent variable), while the current (the dependent variable) was determined as a function of time. In this chapter we consider the opposite case, where the current is controlled (frequently held constant), and the potential becomes the dependent variable, which is determined as a function of time. The other conditions assumed in Chapters 5 and 6, such as small ratio of electrode area to solution volume and semi-infinite diffusion, are also assumed here. The experiment is carried out by applying the controlled current between the working and auxiliary electrodes with a current source (called a *galvanostat*) and determining the potential between the working electrode and a reference electrode (usually with a recorder or oscilloscope, often by a voltage follower) (Figure 7.1.1). These techniques are generally called *chronopotentiometric* techniques, because E is determined as a function of time, or *galvanostatic* techniques, because a small constant current is applied to the working electrode.

7.1.1 Comparison with Controlled Potential Methods

Since the general aspects of controlled potential and controlled current experiments are so similar, we might consider the basic differences between the two types of experiments and the relative advantages of each. The instrumentation for controlled current experiments is simpler than the potentiostats required in controlled potential ones, since no feedback from the reference electrode to the control device is required. Although electronic and operational amplifier constant current sources are easily constructed and frequently used, a simple circuit employing a high-voltage power supply (e.g., a 400-V power supply or several 90-V batteries) and a large resistor

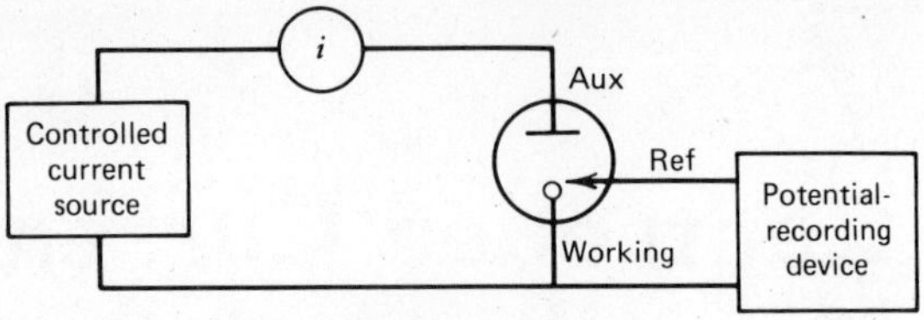

Figure 7.1.1
Simplified block diagram of apparatus for chronopotentiometric measurements.

will often be adequate. The mathematical treatment is also different. In controlled current experiments the surface boundary condition is based on the known current or fluxes (i.e., the concentration gradients) at the electrode surface, while in controlled potential methods, the concentrations (as functions of E) at $x = 0$ provide the boundary conditions. Usually the mathematics involved in solving the diffusion equations in controlled current problems are much simpler, and closed-form analytical solutions are usually obtained.

A fundamental disadvantage of controlled current techniques is that double-layer charging effects are frequently larger and occur throughout the experiment in such a way that straightforward correction for them is not simple. Treating data from multicomponent systems and stepwise reactions is more complicated in controlled current methods, and the waves observed in E-t transients are usually less well-defined than those of potential sweep i-E curves.

7.1.2 Classification and Qualitative Description

The different types of constant current techniques that have been employed are illustrated in Figure 7.1.2. In Figure 7.1.2*a*, *constant current chronopotentiometry*, the constant current i applied to the electrode causes the electroactive species [e.g., the anthracene (A) used as an example in Section 5.1.1] to be reduced at a constant rate to product (e.g., $A^{\overline{\cdot}}$). The potential of the electrode moves to values characteristic of the couple and varies with time as the $A/A^{\overline{\cdot}}$ concentration ratio changes at the electrode surface. The process can be regarded as a titration of the A in the vicinity of the electrode by the continuous flux of electrons, resulting in a curve like that obtained for a potentiometric titration (E as a function of titrant added, $i \cdot t$). Eventually, after the concentration of A drops to zero at the electrode surface, the flux of A to the surface is insufficient to accept all of the electrons being forced across the electrode-solution interface. The potential of the electrode will then rapidly change toward more negative values until a new, second reduction process can start. The time after application of the constant current for this potential transition to occur is called τ, the *transition time*. This time is related to the concentration and the diffusion coefficient and is the chronopotentiometric analog to the peak or limiting current in controlled potential experiments. The shape of the E-t curve is governed by the reversibility, or the heterogeneous rate constant, of the electrode reaction.

Rather than a constant current, an applied current that varies as a known function of time (e.g., $i = \beta t$, a current ramp) can be employed (Figure 7.1.2*b*). Although this technique, called *programmed current chronopotentiometry*, can be treated theoretically

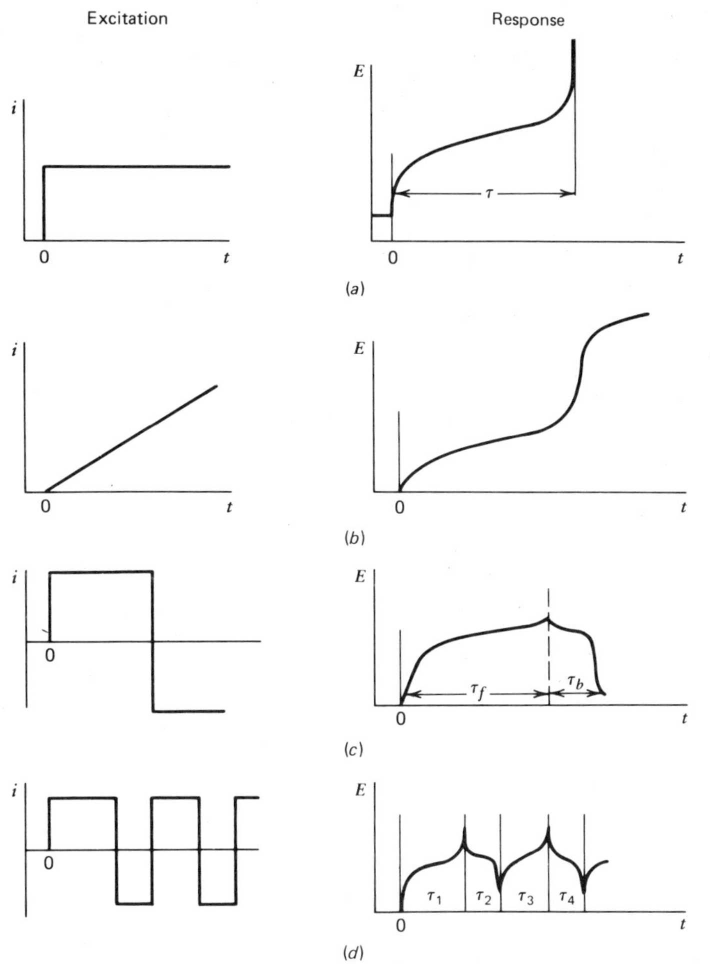

Figure 7.1.2

Different types of controlled current techniques. (*a*) Constant current chronopotentiometry. (*b*) Chronopotentiometry with linearly increasing current. (*c*) Current reversal chronopotentiometry. (*d*) Cyclic chronopotentiometry.

with little difficulty, it has been employed only infrequently. The current can also be reversed after some time (*current reversal chronopotentiometry*) (Figure 7.1.2*c*). For example, if in the case considered above, the current is suddenly changed to an anodic current of equal magnitude at, or before, the transition time, the $A^{\overline{\cdot}}$ formed during the forward step will start oxidizing. The potential will move in a positive direction as the $A/A^{\overline{\cdot}}$ concentration ratio increases and, when the $A^{\overline{\cdot}}$ concentration falls to zero at the electrode surface, a potential transition toward positive potentials occurs and a reverse transition time can be measured. In an extension of this technique the current can be continuously reversed at each transition, resulting in *cyclic chronopotentiometry* (Figure 7.1.2*d*). Finally, as in the treatment of controlled potential data, the derivatives of the *E-t* curves can be obtained or differential methods can be employed.

7.2 GENERAL THEORY OF CONTROLLED CURRENT METHODS

7.2.1 Mathematics of Semi-Infinite Linear Diffusion

We again consider the simple electron transfer reaction, O + $ne \rightarrow$ R. A planar working electrode and an unstirred solution are assumed, with only species O initially present at a concentration C_O^*. These conditions are the same as those in Section 5.2, so that the diffusion equations and general boundary conditions, (5.2.1) to (5.2.3), apply:

$$\frac{\partial C_O(x, t)}{\partial t} = D_O\left[\frac{\partial^2 C_O(x, t)}{\partial x^2}\right] \tag{7.2.1}$$

$$\frac{\partial C_R(x, t)}{\partial t} = D_R\left[\frac{\partial^2 C_R(x, t)}{\partial x^2}\right] \tag{7.2.2}$$

$$\left.\begin{array}{c}\text{At } t = 0 \text{ (for all } x)\\ \text{and}\\ \text{as } x \rightarrow \infty \text{ (for all } t)\end{array}\right\} C_O(x, t) = C_O^* \qquad C_R(x, t) = 0 \tag{7.2.3}$$

$$D_O\left[\frac{\partial C_O(x, t)}{\partial x}\right]_{x=0} + D_R\left[\frac{\partial C_R(x, t)}{\partial x}\right]_{x=0} = 0 \tag{7.2.4}$$

Since the applied current $i(t)$ is presumed known, the flux at the electrode surface is also known at any time, by the equation [see (5.2.8)]:

$$D_O\left[\frac{\partial C_O(x, t)}{\partial x}\right]_{x=0} = \frac{i(t)}{nFA} \tag{7.2.5}$$

Note that this boundary condition involving the concentration *gradient* allows the diffusion problem to be solved without reference to the rate of the electron transfer reaction, in contrast with the concentration-potential boundary conditions required for controlled potential methods. Although in many controlled current experiments the applied current is constant, the more general case for any arbitrarily applied current, $i(t)$, can be solved readily and includes the constant current case, as well as reversal experiments and several others of interest.

As before, application of the Laplace transform method to (7.2.1) and (7.2.3) yields

$$\bar{C}_O(x, s) = \frac{C_O^*}{s} + B(s) \exp\left[-\left(\frac{s}{D_O}\right)^{1/2} x\right] \tag{7.2.6}$$

The transform of (7.2.5) is

$$D_O\left[\frac{\partial \bar{C}_O(x, s)}{\partial x}\right]_{x=0} = \frac{\bar{i}(s)}{nFA} \tag{7.2.7}$$

The combination of (7.2.6) and (7.2.7) with elimination of the integration constant $B(s)$ finally yields

$$\bar{C}_O(x, s) = \frac{C_O^*}{s} - \left[\frac{\bar{i}(s)}{nFAD_O^{1/2}s^{1/2}}\right] \exp\left[-\left(\frac{s}{D_O}\right)^{1/2} x\right] \tag{7.2.8}$$

By substitution of the known function, $\bar{i}(s)$, and employing the inverse transform, $C_O(x, t)$ can be obtained. Similarly, the following expression for $\bar{C}_R(x, s)$ can be derived:

$$\bar{C}_R(x, s) = \left[\frac{\bar{i}(s)}{nFAD_R^{1/2}s^{1/2}}\right] \exp\left[-\left(\frac{s}{D_R}\right)^{1/2} x\right] \tag{7.2.9}$$

Note that direct inversion of (7.2.8) and (7.2.9) using the convolution property leads to (6.2.6) and (6.2.9). These integral forms are also convenient for solving controlled current problems.

7.2.2 Constant Current Electrolysis—The Sand Equation

If $i(t) = i$ (constant), then $\bar{i}(s) = i/s$ and (7.2.8) becomes

$$\bar{C}_O(x, s) = \frac{C_O^*}{s} - \left[\frac{i}{nFAD_O^{1/2}s^{3/2}}\right] \exp\left[-\left(\frac{s}{D_O}\right)^{1/2} x\right] \tag{7.2.10}$$

The inverse transform of this equation yields the expression for $C_O(x, t)$:

$$C_O(x, t) = C_O^* - \frac{i}{nFAD_O}\left\{2\left(\frac{D_O t}{\pi}\right)^{1/2} \exp\left(-\frac{x^2}{4D_O t}\right) - x \operatorname{erfc}\left[\frac{x}{2(D_O t)^{1/2}}\right]\right\} \tag{7.2.11}$$

Typical concentration profiles at various times during a constant current electrolysis are given in Figure 7.2.1. Note that $[\partial C_O(x, t)/\partial x]_{x=0}$ is constant at all times after the onset of electrolysis and $C_O(0, t)$ decreases continually. An expression for $C_O(0, t)$ can be obtained by setting $x = 0$ in (7.2.11), or directly by inverse transform of (7.2.8) with $x = 0$:

$$\bar{C}_O(0, s) = \frac{C_O^*}{s} - \frac{i}{nFAD_O^{1/2}s^{3/2}} \tag{7.2.12}$$

to yield

$$C_O(0, t) = C_O^* - \frac{2it^{1/2}}{nFAD_O^{1/2}\pi^{1/2}} \tag{7.2.13}$$

At a certain characteristic time τ, called the *transition time*, $C_O(0, t)$ drops to zero. At this point (7.2.13) becomes

$$\boxed{\frac{i\tau^{1/2}}{C_O^*} = \frac{nFAD_O^{1/2}\pi^{1/2}}{2} = 85.5nD_O^{1/2}A\,\frac{\text{mA-sec}^{1/2}}{\text{mM}} \quad (\text{with } A \text{ in cm}^2)} \tag{7.2.14}$$

This equation, known as the *Sand equation*, was first derived by H. J. S. Sand (1). As discussed in Section 7.1.2, the flux of O to the electrode surface beyond the transition time is not large enough to satisfy the applied current, and the potential jumps to a value where another electrode process can occur (Figure 7.2.2). The actual shape of the *E-t* curve is discussed in the next sections. The measured value of τ at known i (or better, the values of $i\tau^{1/2}$ obtained at various currents) can be used to determine C_O^* or D_O. A lack of constancy of the *transition time constant*, $i\tau^{1/2}/C_O^*$, with i or C_O^* indicates complications to the electrode reaction from coupled homogeneous chemical

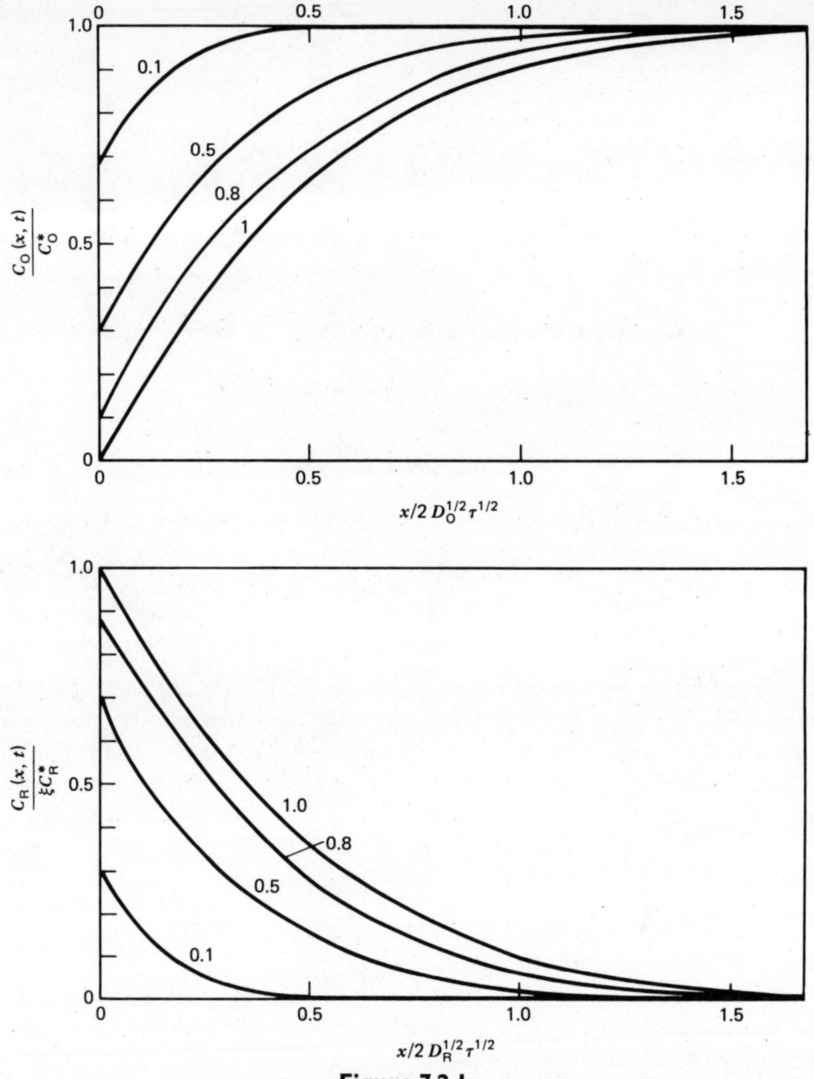

Figure 7.2.1

Concentration profiles of O and R (given in dimensionless form) at various values of t/τ indicated on the curves.

reactions, adsorption, or measurement artifacts (double-layer charging or the onset of convection, see Section 7.3.6). Note that (7.2.11) can be written in a convenient form with dimensionless groupings $C_O(x, t)/C_O^*$, t/τ, and $\lambda_O = x/2(D_O t)^{1/2}$ for $(0 \leq t \leq \tau)$:

$$\frac{C_O(x, t)}{C_O^*} = 1 - \left(\frac{t}{\tau}\right)^{1/2} [\exp(-\lambda_O{}^2) - \pi^{1/2}\lambda_O \operatorname{erfc}(\lambda_O)] \quad (7.2.15)$$

$$\boxed{\frac{C_O(0, t)}{C_O^*} = 1 - \left(\frac{t}{\tau}\right)^{1/2}} \quad (7.2.16)$$

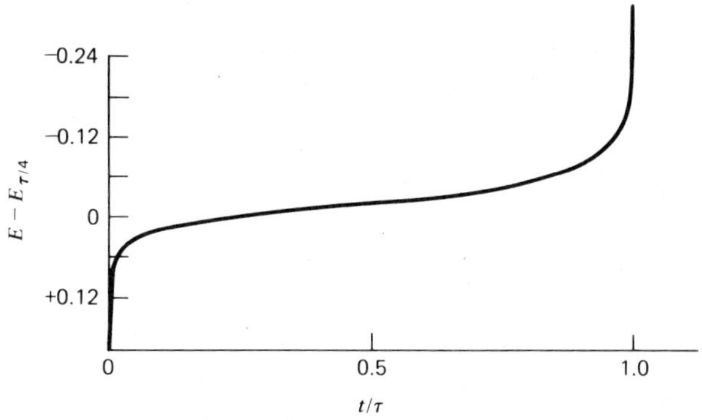

Figure 7.2.2
Theoretical chronopotentiogram for a nernstian electrode process.

In a similar way, the following equations hold for $C_R(x, t)$ when $(0 \leq t \leq \tau)$:

$$\frac{C_R(x, t)}{C_O^*} = \xi\left(\frac{t}{\tau}\right)^{1/2} [\exp(-\lambda_R{}^2) - \pi^{1/2}\lambda_R \operatorname{erfc}(\lambda_R)] \tag{7.2.17}$$

where

$$\lambda_R = \frac{x}{2(D_R t)^{1/2}} \text{ and } \xi = (D_O/D_R)^{1/2},$$

$$\boxed{C_R(0, t) = \frac{2it^{1/2}}{nFA\pi^{1/2}D_R^{1/2}} = \xi\left(\frac{t}{\tau}\right)^{1/2} C_O^*} \tag{7.2.18}$$

See Figure 7.2.1.

7.2.3 Programmed Current Chronopotentiometry

It is possible to use currents that are programmed to vary with time in a special way, rather than remaining constant (2). For example, a current that increases linearly with time could be used:

$$i(t) = \beta t \tag{7.2.19}$$

The treatment follows that for a constant current electrolysis. In this case the transform is

$$\bar{i}(s) = \frac{\beta}{s^2} \tag{7.2.20}$$

so that (7.2.9) becomes, at $x = 0$,

$$\bar{C}_O(0, s) = \frac{C_O^*}{s} - \frac{\beta}{nFAD_O^{1/2}s^{5/2}} \tag{7.2.21}$$

$$C_O(0, t) = C_O^* - \frac{2\beta t^{3/2}}{nFAD_O^{1/2}\Gamma(5/2)} \tag{7.2.22}$$

where $\Gamma(5/2)$ is the Gamma function, equal with this argument to 1.33. This same treatment can be employed with any power function of time.

A particularly interesting applied current is one varying with the square root of time:

$$i(t) = \beta t^{1/2} \tag{7.2.23}$$

$$\bar{i}(s) = \frac{\beta\pi^{1/2}}{2s^{3/2}} \tag{7.2.24}$$

$$\bar{C}_O(0, s) = \frac{C_O^*}{s} - \frac{\beta\pi^{1/2}}{2nFAD_O^{1/2}s^2} \tag{7.2.25}$$

$$C_O(0, t) = C_O^* - \frac{\beta\pi^{1/2}t}{2nFAD_O^{1/2}} \tag{7.2.26}$$

Again, defining the transition time τ as that time when $C_O(0, t) = 0$, an expression equivalent to the constant current Sand equation, but with τ (rather than $\tau^{1/2}$) proportional to C_O^* and β, results:

$$\frac{\beta\tau}{C_O^*} = 2nFA\pi^{-1/2}D_O^{1/2} \tag{7.2.27}$$

Because this current excitation function is fairly difficult to generate, the technique has not been used very much. Nevertheless, it would be especially advantageous for stepwise electron transfer reactions and multicomponent systems (see Section 7.5).

7.3 POTENTIAL-TIME CURVES IN CONSTANT CURRENT ELECTROLYSIS

7.3.1 Reversible (Nernstian) Waves

For rapid electron transfer, the Nernst equation applies. Substitution of the expressions for $C_O(0, t)$ and $C_R(0, t)$ (equations 7.2.16 and 7.2.18) into it yields (3)

$$\boxed{E = E_{\tau/4} + \frac{RT}{nF}\ln\frac{\tau^{1/2} - t^{1/2}}{t^{1/2}}} \tag{7.3.1}$$

where $E_{\tau/4}$, the *quarter-wave potential*, is

$$E_{\tau/4} = E^{0'} - \frac{RT}{2nF}\ln\frac{D_O}{D_R} \tag{7.3.2}$$

so that $E_{\tau/4}$ is the chronopotentiometric equivalent of the voltammetric $E_{1/2}$ value (Figure 7.2.2). The test for reversibility of an *E-t* curve is linearity of an E *vs.* log $[(\tau^{1/2} - t^{1/2})/t^{1/2}]$ plot with a slope of $0.059/n$ V, or a value of $|E_{\tau/4} - E_{3\tau/4}| = 47.9/n$ mV (at 25 °C).

7.3.2 Totally Irreversible Waves

For a totally irreversible cathodic reaction, i is related to E by either of the following equations (4):

$$\frac{i}{nFA} = k_f^{0} C_O(0, t) \exp\left[\frac{-\alpha n_a FE}{RT}\right] \tag{7.3.3}$$

$$\frac{i}{nFA} = k^{0} C_O(0, t) \exp\left[\frac{-\alpha n_a F(E - E^{0'})}{RT}\right] \tag{7.3.4}$$

When the expression for $C_O(0, t)$, (7.2.16), is substituted into these equations, the following expressions result:

$$E = \frac{RT}{\alpha n_a F} \ln\left(\frac{nFAC_O^* k_f^{0}}{i}\right) + \frac{RT}{\alpha n_a F} \ln\left[1 - \left(\frac{t}{\tau}\right)^{1/2}\right] \tag{7.3.5a}$$

$$E = E^{0'} + \frac{RT}{\alpha n_a F} \ln\left(\frac{nFAC_O^* k^{0}}{i}\right) + \frac{RT}{\alpha n_a F} \ln\left[1 - \left(\frac{t}{\tau}\right)^{1/2}\right] \tag{7.3.5b}$$

Equivalent expressions can be obtained by using the Sand equation and substituting for $\tau^{1/2}$:

$$E = \left(\frac{RT}{\alpha n_a F}\right) \ln\left[\frac{2k_f^{0}}{(\pi D_O)^{1/2}}\right] + \left(\frac{RT}{\alpha n_a F}\right) \ln(\tau^{1/2} - t^{1/2}) \tag{7.3.6a}$$

$$\boxed{E = E^{0'} + \left(\frac{RT}{\alpha n_a F}\right) \ln\left[\frac{2k^{0}}{(\pi D_O)^{1/2}}\right] + \left(\frac{RT}{\alpha n_a F}\right) \ln(\tau^{1/2} - t^{1/2})} \tag{7.3.6b}$$

Thus, for a totally irreversible reduction wave, the whole E-t wave shifts toward more negative potentials with increasing current, with a tenfold increase in current causing a shift of $2.3RT/\alpha n_a F$ (or $59/\alpha n_a$ mV at 25 °C). Note that uncompensated resistance between the reference and working electrode will also cause the E-t curve to shift with increasing i. For a totally irreversible wave, $|E_{\tau/4} - E_{3\tau/4}| = 33.8/\alpha n_a$ mV at 25 °C.

7.3.3 Quasi-Reversible Waves

The general equation for the E-t curve results from combining the general current-potential-concentration characteristic, equation 3.5.27, with the equations for $C_O(0, t)$, (7.2.13), and $C_R(0, t)$, (7.2.18), (including, so that a starting equilibrium potential can be defined, an initial concentration of R of C_R^*) (5, 6). The result, in terms of η, is

$$\boxed{\frac{i}{i_0} = \left[1 - \frac{2i}{nFAC_O^*}\left(\frac{t}{\pi D_O}\right)^{1/2}\right] e^{-\alpha n f \eta} - \left[1 + \frac{2i}{nFAC_R^*}\left(\frac{t}{\pi D_R}\right)^{1/2}\right] e^{(1-\alpha) n f \eta}} \tag{7.3.7}$$

Alternate forms can be written, for example,

$$j = k_f\left[nFC_O^* - 2j\left(\frac{t}{\pi D_O}\right)^{1/2}\right] - k_b\left[nFC_R^* + 2j\left(\frac{t}{\pi D_R}\right)^{1/2}\right] \tag{7.3.8a}$$

or, when $C_R^* = 0$,

$$j = nFk_f C_O^* - \frac{2jt^{1/2}}{\pi^{1/2}}\left(\frac{k_f}{D_O^{1/2}} + \frac{k_b}{D_R^{1/2}}\right) \tag{7.3.8b}$$

where k_f and k_b are defined in (3.2.9) and (3.2.10), and j is the current density.

Usually the study of the kinetics of quasi-reversible electrode reactions by constant current techniques (generally called the *galvanostatic* or *current step* method) involves the use of such small current perturbations that the potential change is small. When both O and R are initially present, the linearized current-potential-concentration characteristic, equation 3.5.33, can then be employed; and combination of that with (7.2.13) and (7.2.18) [or linearization of (7.3.7)] yields

$$\boxed{-\eta = \frac{RT}{nF} i\left[\frac{2t^{1/2}}{nFA\pi^{1/2}}\left(\frac{1}{C_O^* D_O^{1/2}} + \frac{1}{C_R^* D_R^{1/2}}\right) + \frac{1}{i_0}\right]} \tag{7.3.9}$$

Thus a plot of η *vs.* $t^{1/2}$, for small values of η, will be linear, and i_0 can be obtained from the intercept. This method is the constant current analog of the potentiostatic or potential step method discussed in Section 5.5.3.

7.3.4 Effect of Double-Layer Capacity

Because the potential is changing during the application of the current step, there is always a nonfaradaic current that contributes to charging of the double-layer capacitance. It is given by

$$i_c = -AC_d(d\eta/dt) = -AC_d(dE/dt) \tag{7.3.10}$$

Thus, of the total applied constant current, i, only a portion, i_f, goes to the faradaic reaction:

$$i_f = i - i_c \tag{7.3.11}$$

Since dE/dt is a function of time, i_c and i_f also vary with time, even when i is constant. This can be treated as a case of programmed current chronopotentiometry, if an explicit form of dE/dt or $d\eta/dt$ is known. Alternatively, instrumental or derivative methods can be used (see Section 7.6). For the simplest case, where the general η-t expression can be linearized, the following equation holds (7):

$$-\eta = \frac{RT}{nF} i\left[\frac{2t^{1/2}}{\pi^{1/2}} N - \frac{RT}{nF} AC_d N^2 + \frac{1}{i_0}\right] \tag{7.3.12}$$

where

$$N = \frac{1}{nFA}\left(\frac{1}{C_O^* D_O^{1/2}} + \frac{1}{C_R^* D_R^{1/2}}\right) \tag{7.3.13}$$

Thus the intercept of the η-$t^{1/2}$ plot allows the determination of $1/i_0$ only when this term is appreciable compared to $(RT/nF)AC_dN^2$ (7). To overcome this problem for fast electron transfer reactions, where i_0 is large, the galvanostatic double-pulse method has been proposed.

7.3.5 The Galvanostatic Double-Pulse (GDP) Method

As seen from the above discussion, the single-pulse galvanostatic method cannot be used for fast electron transfer reactions (i.e., those with large i_0), because during the initial moments following the application of the current step, the current is primarily nonfaradaic and contributes mostly to charging C_d. Gerischer and Krause (8) developed the double-pulse method in which two constant current pulses are applied to the electrode (Figure 7.3.1). The first large pulse (i_1; $0 < t < t_1$) mainly serves to charge the double layer to a potential that corresponds to the value at the instant of application of a second, smaller pulse of current i_2. The basic idea is to use the short first pulse (typically 0.5 to 1 μsec long) to drive the system to an overpotential that exactly supports the second current. Then η will not change abruptly when i_2 is applied, and there is no significant effect from charging the double layer during the second phase. A block diagram of apparatus for this technique is shown in Figure 7.3.2.

Some faradaic current does flow during the first pulse, and its effects must be taken into account. The theory of the GDP method was extended along that line by Matsuda, Oka, and Delahay (9). Valuable instrumental advances were later made by Aoyagui and co-workers (10–12).

When the ratio of pulse heights (i_1/i_2) is adjusted properly, and this is determined by trial and error, the E-t curve following the cessation of the first pulse is horizontal (Figure 7.3.3). Under these conditions, the overpotential is given by (9)

$$-\eta = \frac{RT}{nF}\frac{i_2}{i_0}\left[1 + \frac{4Ni_0}{3\pi^{1/2}}t_1^{1/2} + \left(1 - \frac{9\pi}{32}\right)\left(\frac{4Ni_0}{3\pi^{1/2}}\right)^2 t_1 + \ldots\right] \qquad (7.3.14)$$

or, at sufficiently small values of t_1,

$$-\eta \simeq \frac{RT}{nF} i_2\left(\frac{1}{i_0} + \frac{4N}{3\pi^{1/2}} t_1^{1/2}\right) \qquad (7.3.15)$$

Thus one can carry out a series of experiments with different pulse widths t_1, and plot the value of η at the onset of i_2 *vs.* $t_1^{1/2}$ to obtain the exchange current from the intercept. Note the similarity between (7.3.9) and (7.3.15). The $t^{1/2}$ term in each case accounts for electrolytic modification of the surface concentrations. In the situation yielding (7.3.9), it was induced by current i; but in this case it is first induced by i_1 and then supported by i_2.

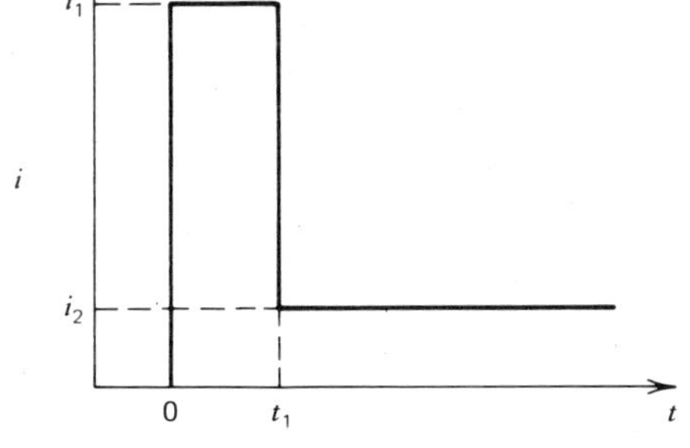

Figure 7.3.1
Excitation waveform for the galvanostatic double-pulse method.

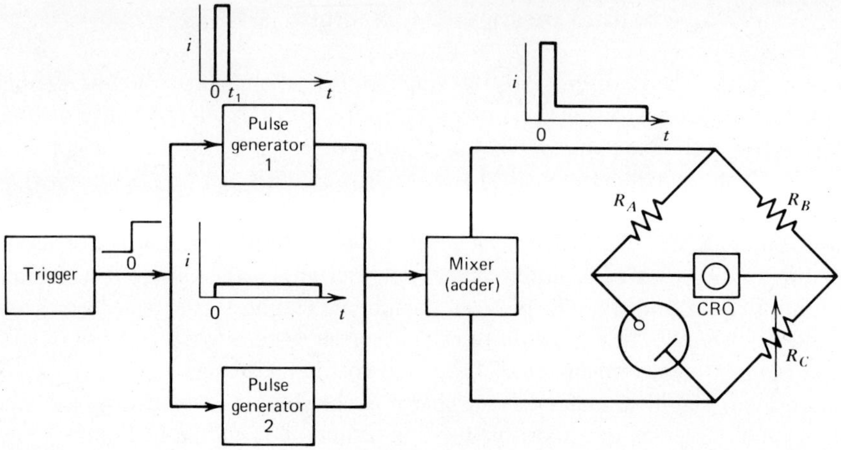

Figure 7.3.2

Block diagram for galvanostatic double-pulse method. A two-electrode cell arrangement (i.e., with a common counter-reference electrode) in a bridge circuit for compensation of the cell ohmic resistance, R_S, is shown. The bridge is adjusted with $R_A = R_B$, $R_C = R_S$, $R_A \gg R_S$ so that the pulse generators produce an essentially constant current through the cell. Provision is sometimes made in double-pulse circuits, however, for a three-electrode cell and potentiostatic control of the working electrode before the application of the galvanostatic pulse. For details of such apparatus, see References 10 to 12.

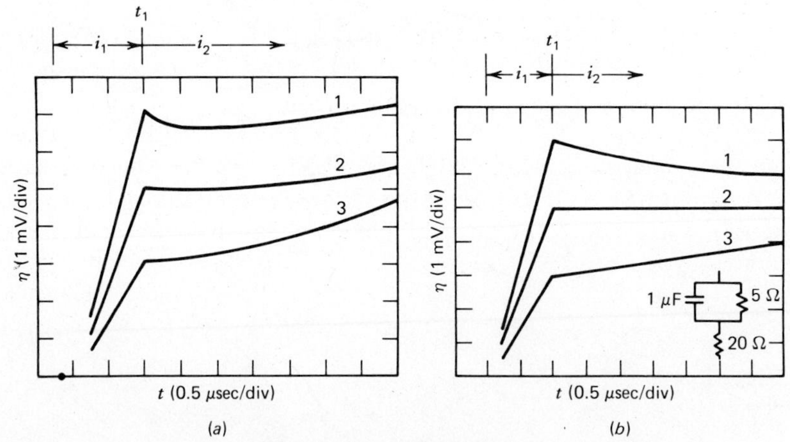

Figure 7.3.3

(*a*) Overpotential-time traces for galvanostatic double-pulse method for reduction of 0.25 m*M* Hg_2^{2+} in 1 *M* $HClO_4$ at hanging mercury drop electrode. Ratio i_2/i_1 was (1) 7.8, (2) 5.3, (3) 3.2; (*b*) shows the desired response for utilization of (7.3.15). (*b*) Voltage-time traces for constant-current double pulses applied to equivalent circuit shown. i_1 was (1) 7.6, (2) 5.5, (3) 3.3 mA; and i_2 = 1 mA. [From M. Kogoma, T. Nakayama, and S. Aoyagui, *J. Electroanal. Chem.*, **34**, 123 (1972), with permission.]

The differential double-layer capacitance can also be obtained from these data by the equation:

$$C_d = \lim_{t_1 \to 0} \frac{nFt_1 i_0}{RTA}\left(\frac{i_1}{i_2}\right)\left(1 - \frac{4Ni_0 t_1^{1/2}}{3\pi^{1/2}}\right)^{-1} \tag{7.3.16}$$

This relation rests on the idea that the total charge in the first step, $i_1 t_1$, is nonfaradaic in the limit of very short t_1.

The GDP method does not require knowledge of the diffusion coefficients for reactant and products, or of the C_d value, for the calculation of i_0. Measurements using improved instrumentation by Aoyagi and co-workers suggest that rate constants of very rapid electrode reactions (~1 cm/sec) can be determined using this technique.

7.3.6 Effect of C_d on Transition Time

As discussed in Section 7.3.4, the presence of a finite double-layer capacity results in a charging current contribution proportional to dE/dt (equation 7.3.10) and causes i_f to differ from the total applied current i. This effect, which is largest immediately after application of the current and near the transition (where dE/dt is relatively large), affects the overall shape of the E-t curve and makes measurement of τ difficult and inaccurate. A number of authors have examined this problem and proposed techniques for measuring τ from distorted E-t curves or correcting τ values obtained in the presence of significant double-layer effects. In the simplest approach, it is assumed that i_c is constant for $0 < t < \tau$; this is clearly not so, of course, since dE/dt and C_d (a function of E) change throughout the E-t curve (13, 14). This approximation leads to the equations:

$$i = i_f + i_c \tag{7.3.17}$$

$$\frac{i\tau^{1/2}}{C_O^*} = \frac{i_f\tau^{1/2}}{C_O^*} + \frac{i_c\tau}{C_O^*\tau^{1/2}} \tag{7.3.18}$$

where $i_f\tau^{1/2}/C_O^*$ is the "true" chronopotentiometric constant, A, equal to $nFD_O^{1/2}\pi^{1/2}/2$. In the last term, $i_c\tau$ is the total number of coulombs needed to charge an average double-layer capacitance from the initial potential to the potential at which τ is measured (ΔE), so that $i_c\tau \approx (C_d)_{avg}\Delta E$, and is represented by a correction factor B. Thus the final equation is

$$i\tau^{1/2}/C_O^* = A + B/C_O^*\tau^{1/2} \tag{7.3.19}$$

Plots based on (7.3.19) can thus be used to extract A and B from the observed data (e.g., a plot of $i\tau$ *vs.* $\tau^{1/2}$ yields a slope AC_O^* and an intercept B). Note that an equation of this form can also be used to correct for formation of an oxide film (e.g., on a platinum electrode during an electrochemical oxidation) and for electrolysis of adsorbed material in addition to diffusing species. Under these conditions, (7.3.17) becomes (14):

$$i = i_f + i_c + i_{ox} + i_{ads} \tag{7.3.20}$$

where i_{ox} is the current going to formation (or reduction) of the oxide film and i_{ads} is the current required for the adsorbed material. A similar treatment again yields (7.3.19), where B is now an overall correction factor including $Q_{ox} = i_{ox}\tau$ and $Q_{ads} = nF\Gamma$, where Γ is the number of moles of adsorbed species per square centimeter (see Section 12.5.7). Although these approximations are very rough, treatments of actual experimental data by (7.3.19) yield fairly good results, even at rather low concentrations and short transition times, where these surface effects are most important (15).

A more rigorous approach involves only the assumption that C_d is independent of E (16–18). In this case the diffusion equation (7.2.1), boundary conditions (7.2.3), and (7.3.11), written in the form

$$i = nFAD_O\left(\frac{\partial C_O}{\partial x}\right)_{x=0} + AC_d\left(\frac{dE}{dt}\right) \tag{7.3.21}$$

as well as the appropriate *i-E-C* equation (i.e., for a reversible, totally irreversible, or quasi-reversible reaction), must be solved simultaneously. The resulting nonlinear integral equation must be solved numerically or by digital simulation techniques. Typical results showing the effect of different relative contributions of double-layer charging on i_f (at constant i) and on the *E-t* curves of a nernstian reaction are shown in Figures 7.3.4 and 7.3.5. The charging contribution is represented by the dimensionless parameter, K, where

$$K = \left(\frac{RT}{nF}\right)\frac{C_d}{nFC_O^*(\pi D_O\tau)^{1/2}} \tag{7.3.22}$$

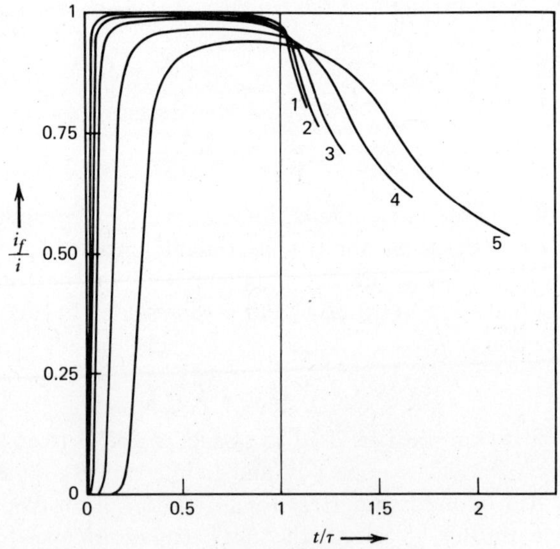

Figure 7.3.4
Variation of fraction of total current contributing to faradaic process (i_f/i) with time for various values of K for a nernstian electrode process. K values of (1) 5×10^{-4}; (2) 10^{-3}; (3) 2×10^{-3}; (4) 5×10^{-3}; (5) 0.01. [From W. T. de Vries, *J. Electroanal. Chem.*, **17**, 31 (1968), with permission.]

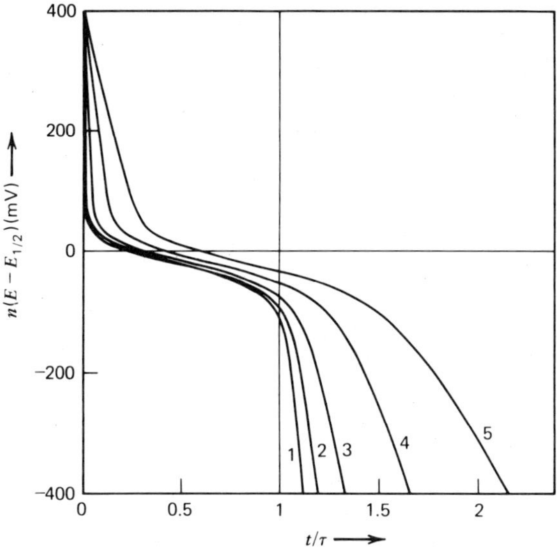

Figure 7.3.5
Effect of double-layer capacitance on chronopotentiograms for nernstian electrode reaction. Values of K as in Figure 7.3.4. [From W. T. de Vries, *J. Electroanal. Chem.*, **17**, 31 (1968), with permission.]

Unfortunately these treatments depend on the reversibility of the electron transfer reaction and lead to rather complex expressions that are not particularly attractive for treating real experimental data.

The effect of double-layer charging is clearly most important at small τ values (see equation 7.3.22), and problems with distorted E-t curves and measurements of corrected τ values have discouraged the use of controlled current methods as opposed to controlled potential ones. Problems arise in both controlled current and potential methods at long experimental times because of the onset of convection and nonlinearity of diffusion. Convective effects, caused by motion of the solution with respect to the electrode, can arise by accidental vibrations transmitted to the cell (e.g., by hood fans, vacuum pumps, passing traffic) or as a result of density gradients building up at the electrode surface because of differences in density between reactants and products (so-called "natural convection"). Convective effects can be minimized by using electrodes with glass mantles (shielded electrodes; Figure 7.3.6) and orienting the electrode horizontally so that the denser species is always below the less dense one (19, 20). Vertically oriented electrodes (e.g., foils or wires) usually suffer from convection effects even at not very long times (e.g., 60 to 80 sec). The shielded electrode also has the virtue of constraining diffusion to lines normal to the electrode surface so that true linear diffusion conditions are approached. An unshielded electrode, such as a platinum disk imbedded in glass can show appreciable "sphericity" effects when the diffusion layer thickness is not negligible with respect to the electrode dimensions; that is, material can diffuse to the unshielded electrode from the sides. This effect causes increases in the transition time (or anomalously large currents in

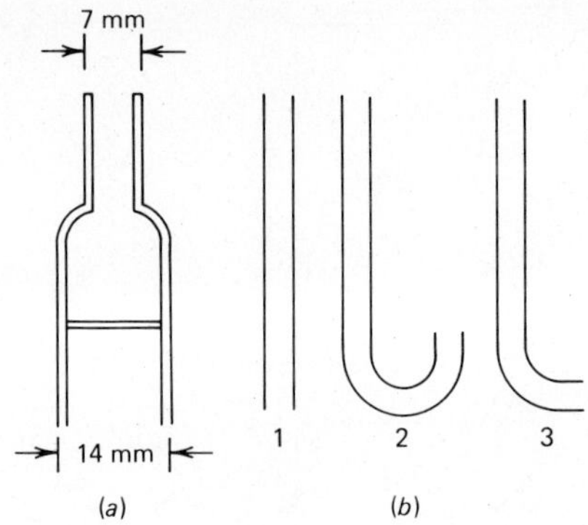

Figure 7.3.6
Shielded electrode for maintaining linear diffusion and suppressing convection. (*a*) Shielded electrode. (*b*) Tubes to which shielded electrode is attached to provide: (1) horizontal electrode, diffusion upward; (2) horizontal electrode, diffusion downward; (3) vertical electrode. [Reprinted with permission from A. J. Bard, *Anal. Chem.*, **33**, 11 (1961). Copyright 1961, American Chemical Society.]

controlled potential methods). With properly oriented shielded electrodes, however, linear diffusion conditions can be maintained for as long as 300 sec.

7.4 REVERSAL TECHNIQUES

7.4.1 Response Function Principle

A useful technique for treating reversal methods in chronopotentiometry (and other techniques in electrochemistry) is based on the *response function principle* (2, 21). This method, which is based on analogous procedures used to treat electrical circuits, considers the system's response to a perturbation or excitation signal, as applied in Laplace transform space. Thus one can write the general response function equation (2)

$$\bar{R}(s) = \bar{\Psi}(s)\bar{S}(s) \tag{7.4.1}$$

where $\bar{\Psi}(s)$ is the excitation function transform, $\bar{R}(s)$ is the response transform, which describes how the system responds to the excitation, and $\bar{S}(s)$ is the system transform, which connects the excitation and the response. For example for current excitation we can write, from (7.2.8) at $x = 0$,

$$\bar{C}_O(0, s) = C_O^*/s - [nFAD_O^{1/2}s^{1/2}]^{-1}\bar{i}(s) \tag{7.4.2}$$

or

$$C_O^* - \bar{C}_O(0, s) = [nFAD_O^{1/2}s^{1/2}]^{-1}\bar{i}(s) \tag{7.4.3}$$

In this case $\bar{\Psi}(s) = \bar{i}(s)$ (the transform of the applied current perturbation), $\bar{R}(s) = \bar{C}_O^* - \bar{C}_O(0, s)$, the transform of the concentration response to the perturbation, and $\bar{S}(s) = [nFAD_O^{1/2}s^{1/2}]^{-1}$, which is characteristic of the nature of the system under excitation (semi-infinite linear diffusion). For controlled current problems involving different systems (e.g., spherical or cylindrical diffusion, first-order kinetic complications†) other system transforms would be employed. We have illustrated how this

† This approach, and transform methods in general, are useful only for linear problems; hence second-order reactions or nonlinear complications cannot be treated by this technique.

equation could be employed for constant and programmed current methods, using appropriate $\bar{i}(s)$ functions. We now extend its use to reversal techniques.

7.4.2 Current Reversal (22, 23)

Consider a solution where only O is present initially at a concentration C_O^*, semi-infinite linear diffusion conditions prevail, and a constant cathodic current i is applied for a time t_1 (where $t_1 \leq \tau_1$, τ_1 = forward transition time). At t_1 the current is reversed, that is, the direction of the current is changed from cathodic to anodic, so that R formed during the forward step is oxidized to O, and the time τ_2 at which C_R at the electrode surface drops to zero (measured from t_1) is noted. At τ_2 the potential shows a rapid change toward positive values. We desire an expression for τ_2. This is most easily accomplished using the "zero shift theorem" of Section A.1.7. Since for $0 < t \leq t_1$, $i(t) = i$, and for $t_1 < t \leq \tau_2$, $i(t) = -i$, the expression for the current, using step function notation, is

$$i(t) = i + S_{t_1}(t)(-2i) \tag{7.4.4}$$

so that the transform is given by

$$\bar{i}(s) = \frac{i}{s} - \frac{(2e^{-t_1 s})i}{s} = \left(\frac{i}{s}\right)(1 - 2\,e^{-t_1 s}) \tag{7.4.5}$$

Introducing this into (7.4.2), we obtain

$$\frac{C_O^*}{s} - \bar{C}_O(0, s) = \left(\frac{i}{s}\right)(1 - 2\,e^{-t_1 s})(nFAD_O^{1/2}s^{1/2})^{-1} \tag{7.4.6}$$

The analogous expression for $\bar{C}_R(0, s)$ [see (7.2.9)] is

$$\bar{C}_R(0, s) = \left(\frac{i}{s}\right)(1 - 2\,e^{-t_1 s})(nFAD_R^{1/2}s^{1/2})^{-1} \tag{7.4.7}$$

The inverse transform of (7.4.7) yields an expression for $C_R(0, t)$ at any time (recall $S_{t_1}(t) = 0$ for $t \leq t_1$; $S_{t_1}(t) = 1$ for $t > t_1$):

$$C_R(0, t) = \frac{2i}{nFA\pi^{1/2}D_R^{1/2}}\,[t^{1/2} - 2S_{t_1}(t)(t - t_1)^{1/2}] \tag{7.4.8}$$

At $t = t_1 + \tau_2$, $C_R(0, t) = 0$, so that

$$(t_1 + \tau_2)^{1/2} = 2\tau_2^{1/2}$$

$$\boxed{\tau_2 = t_1/3} \tag{7.4.9}$$

Thus the reverse transition time is always (for stable R) 1/3 that of the forward time up to and including τ_1, independent of D_O, D_R, C_O^* and the rate of the electron transfer (assuming it is sufficiently rapid to show a reverse transition, i.e., is not totally irreversible) (Figure 7.4.1). The factor of 1/3 means that of the total amount of R generated during the forward step (equal to it_1/nF mol), only one-third returns to the electrode

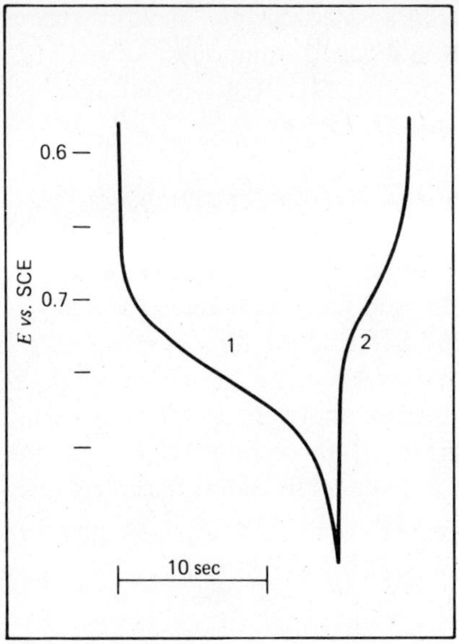

Figure 7.4.1
Typical experimental chronopotentiogram with current reversal. Oxidation of diphenylpicrylhydrazyl (DPPH) followed by reduction of the stable radical cation, DPPH$^{+\cdot}$. Solution was acetonitrile containing 1.04 m*M* DPPH and 0.1 *M* $NaClO_4$. The current was 100 μA and a shielded platinum electrode of area 1.2 cm^2 was employed. [Reprinted with permission from E. Solon and A. J. Bard, *J. Am. Chem. Soc.*, **86**, 1926 (1964). Copyright 1964, American Chemical Society.]

during the backward step up to τ_2, with a large amount diffusing into the bulk solution. At first thought one might wonder at the independence of τ_2/t_1 on D_R. If D_R is very large, then a larger amount might be expected to diffuse away. However, a large D_R also implies that a larger amount will diffuse back during the reverse step, and the mathematics demonstrates that this exactly compensates for the diffusion away from the electrode. Thus the τ_2/t_1 ratio of 1/3 in chronopotentiometry is the analog of $i_r(2\tau)/i_f(\tau) = 0.293$ in potential step reversal (Section 5.7.1) and $i_{pc}/i_{pa} = 1.00$ in cyclic voltammetry (Section 6.5.1). Expressions can also be written for the potential-time behavior by combining the appropriate *i-E-C* expression with the expressions for $C_O(0, t)$ and $C_R(0, t)$. For example, for a nernstian wave, $E_{0.215\tau_2} = E_{\tau_1/4}$. For a quasi-reversible system the separation between $E_{\tau_1/4}$ and $E_{0.215\tau_2}$ can be used to determine k^0 (24).

7.4.3 Cyclic Chronopotentiometry (21, 25)

The technique of current reversal can be continued with successive reversals at each transition. The theoretical treatment involves an extension of the single reversal technique, so that

$$i = i + S_{\tau_1}(-2i) + S_{\tau_2}(2i) + \dots \qquad (7.4.10)$$

A typical cyclic chronopotentiogram (Figure 7.4.2) shows a series of reversal transitions of different relative lengths, a_n $(= \tau_n/\tau_1)$. For stable soluble oxidized and reduced forms the sequence is $a_1 = 1.000$, $a_2 = 0.333$, $a_3 = 0.588$, $a_4 = 0.355$, $a_5 = 0.546$, $a_6 = 0.366$, etc. For the special case where the reduced form is insoluble and precipitates on the electrode (e.g., the electrodeposition of silver on a platinum electrode), the transition time for the first reversal equals τ_1, since the reduced form does not diffuse into the bulk solution but remains on the electrode. The τ_3 value is larger than

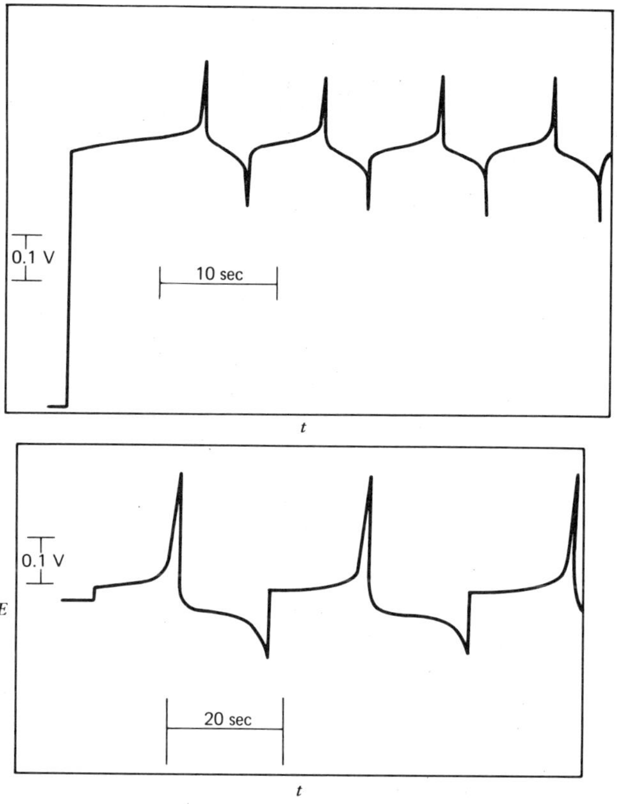

Figure 7.4.2

Typical experimental cyclic chronopotentiograms. *Upper curve*: reduction of Cd(II). Solution contained 1.71 m*M* Cd(II) and 0.1 *M* KCl. Current density was 0.370 mA/cm^2 at mercury pool electrode. *Lower curve*: reduction of Ag(I). Solution contained 9.8 m*M* Ag(I) and 0.2 *M* KNO_3. Current density was 0.865 mA/cm^2 at platinum disk electrode. [Reprinted with permission from H. B. Herman and A. J. Bard, *Anal. Chem.*, **35**, 1121 (1963). Copyright 1963, American Chemical Society.]

τ_1, however, since during production of O (e.g., Ag^+ ion) during the reversal step, additional O diffuses to the electrode surface. The sequence of a_n values for this case is: $a_1 = a_2 = 1.000$, $a_3 = a_4 = 1.174$, $a_5 = a_6 = 1.263$, etc.

7.5 MULTICOMPONENT SYSTEMS AND MULTISTEP REACTIONS (23, 26–28)

Consider a solution containing two reducible substances, O_1 and O_2, at concentrations C_1^* and C_2^*, respectively, where the reduction $O_1 + n_1e \rightarrow R_1$ occurs first, and then, at more negative potentials, $O_2 + n_2e \rightarrow R_2$. Again, assuming semi-infinite linear diffusion, the following response-function equations can be written:

$$n_1FAD_1^{1/2}\left[\frac{C_1^*}{s} - \bar{C}_1(0, s)\right] = \frac{\bar{i}_1(s)}{s^{1/2}} \tag{7.5.1}$$

$$n_2FAD_2^{1/2}\left[\frac{C_2^*}{s} - \bar{C}_2(0, s)\right] = \frac{\bar{i}_2(s)}{s^{1/2}} \tag{7.5.2}$$

where $\bar{i}_1(s)$ and $\bar{i}_2(s)$ are the transforms of the individual currents [$i_1(t)$ and $i_2(t)$] involved in the reduction of O_1 and O_2, respectively. Since the total applied current, $i(t)$, is $i_1(t) + i_2(t)$, then $\bar{i}(s) = \bar{i}_1(s) + \bar{i}_2(s)$. Then, from (7.5.1) and (7.5.2),

$$n_1D_1^{1/2}\left[\frac{C_1^*}{s} - \bar{C}_1(0, s)\right] + n_2D_2^{1/2}\left[\frac{C_2^*}{s} - \bar{C}_2(0, s)\right] = \frac{\bar{i}(s)}{FAs^{1/2}} \tag{7.5.3}$$

This equation is true at all times. For the times when the potential is not sufficiently negative for O_2 reduction to occur (i.e., when $t \leq \tau_1$), $\bar{C}_2(0, s) = C_2^*/s$, $\bar{i}_2(s) = 0$, and (7.5.3) becomes identical to the simple equation (7.4.2), so that up to τ_1 the behavior is unaffected by the presence of O_2. For $t > \tau_1$, $\bar{C}_1(0, s) = 0$, so that (7.5.3) becomes

$$\frac{n_1D_1^{1/2}C_1^*}{s} + n_2D_2^{1/2}\left[\frac{C_2^*}{s} - \bar{C}_2(0, s)\right] = \frac{\bar{i}(s)}{FAs^{1/2}} \tag{7.5.4}$$

The second transition time ($t = \tau_1 + \tau_2$) occurs when the concentration of O_2 drops to zero at the electrode surface, that is, $\bar{C}_2(0, s) = 0$. Thus, for a constant current, $\bar{i}(s) = i/s$, and (7.5.4) for this time becomes

$$\frac{n_1D_1^{1/2}C_1^*}{s} + \frac{n_2D_2^{1/2}C_2^*}{s} = \frac{i}{FAs^{3/2}} \tag{7.5.5}$$

Inversion yields

$$\boxed{(n_1D_1^{1/2}C_1^* + n_2D_2^{1/2}C_2^*)\left(\frac{FA\pi^{1/2}}{2}\right) = i(\tau_1 + \tau_2)^{1/2}} \tag{7.5.6}$$

For example, for the special case $n_1D_1^{1/2}C_1^* = n_2D_2^{1/2}C_2^*$, $\tau_2 = 3\tau_1$. Thus, while in controlled potential voltammetric methods two substances at equal concentration with equal diffusion coefficients show two waves of equal height, in chronopotentiometry unequal transition times arise. The long second transition results from the continued diffusion of O_1 to the electrode after τ_1, so that only a fraction of the applied current is available for reduction of O_2 (Figure 7.5.1).

Similar reasoning shows that for a stepwise process:

$$O + n_1e \rightarrow R_1 \tag{7.5.7}$$

$$R_1 + n_2e \rightarrow R_2 \tag{7.5.8}$$

the transition time ratio is given by

$$\frac{\tau_2}{\tau_1} = \frac{2n_2}{n_1} + \left(\frac{n_2}{n_1}\right)^2 \tag{7.5.9}$$

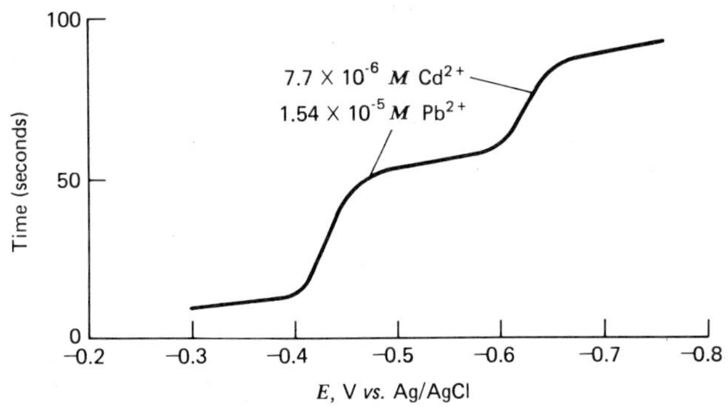

Figure 7.5.1

Consecutive reduction of Pb(II) and Cd(II) at a mercury pool electrode. Note that if *E-t* curve is plotted in this manner, it resembles a voltammogram. [Reprinted with permission from C. N. Reilley, G. W. Everett, and R. H. Johns, *Anal. Chem.*, **27**, 483 (1955). Copyright 1955, American Chemical Society.]

Thus for $n_2 = n_1$, $\tau_2 = 3\tau_1$ (Figure 7.5.2). By use of the response-function principle one can show (see Problem 7.4) that if a current of the form $i(t) = \beta t^{1/2}$ is used, then equal transition times result when $n_2 = n_1$.

7.6 DERIVATIVE METHODS

By rather straightforward instrumental approaches the derivative of the chronopotentiogram, that is, a curve of dE/dt *vs.* t, can be obtained. The theoretical form of the derivative curve can readily be found by differentiation of the appropriate *E-t* expression. Thus for a nernstian process, from (7.3.1),

$$\frac{dE}{dt} = -\left(\frac{RT}{2nF}\right)\left[\frac{\tau^{1/2}}{t(\tau^{1/2} - t^{1/2})}\right] \tag{7.6.1}$$

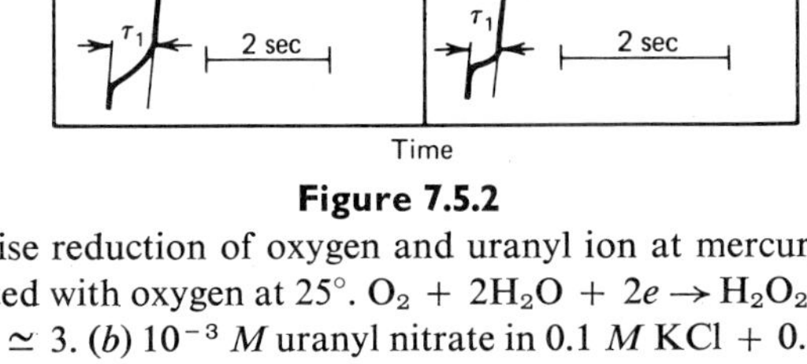

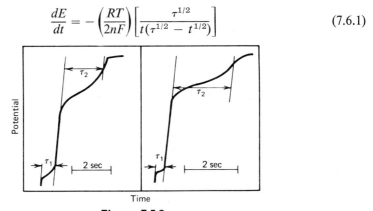

Figure 7.5.2

E-t curves for stepwise reduction of oxygen and uranyl ion at mercury electrode. (*a*) 1 *M* LiCl solution saturated with oxygen at 25°. $O_2 + 2H_2O + 2e \rightarrow H_2O_2 + 2OH^-$; $H_2O_2 + 2e^- \rightarrow 2OH^-$; $\tau_2/\tau_1 \simeq 3$. (*b*) 10^{-3} *M* uranyl nitrate in 0.1 *M* KCl + 0.01 *M* HCl. U(VI) + $e \rightarrow$ U(V); U(V) + $2e \rightarrow$ U(III); $\tau_2/\tau_1 \simeq 8$. [Reprinted with permission from T. Berzins and P. Delahay, *J. Am. Chem. Soc.*, **75**, 4205 (1953). Copyright, 1953, American Chemical Society.]

While finding τ from the maximum of the derivative curve is possible, its determination by this approach suffers the same problems as the direct measurement approach. As an alternative, Peters and Burden (29) recommended that the minimum in the derivative curve, which for a nernstian process occurs at $t = 4\tau/9$, be evaluated. Since

$$(dE/dt)_{\text{min,rev}} = -(27/8)\left(\frac{RT}{nF\tau}\right) \quad (7.6.2)$$

$$= -0.08664/n\tau \text{ V sec}^{-1} \text{ at } 25\ ^\circ\text{C} \quad (7.6.3)$$

τ can be obtained by a direct evaluation of $(dE/dt)_{\text{min}}$. Typical experimental results are shown in Figure 7.6.1.

For a totally irreversible reaction, (see equation 7.3.6),

$$\frac{dE}{dt} = -\left(\frac{RT}{2\alpha n_a F}\right)[t^{1/2}(\tau^{1/2} - t^{1/2})]^{-1} \quad (7.6.4)$$

and $(dE/dt)_{\text{min}}$ occurs at $t = \tau/4$ with the value

$$\left(\frac{dE}{dt}\right)_{\text{min,irrev}} = -\frac{2RT}{\alpha n_a F\tau} \quad (7.6.5)$$

Determination of τ by this approach is freer from problems of double-layer charging, because it is evaluated at a position in the curve before the transition time region where an appreciable charging current contribution exists. However, the large charging current contribution at the start of the chronopotentiogram still contributes. Even so, by using τ values evaluated from $(dE/dt)_{\text{min}}$ and applying the double-layer correction approach embodied in equation 7.3.19, extensions of chronopotentiometric measurements to low concentrations (e.g., 10^{-6} M Cd^{2+}) and very short times (e.g., τ values in the μsec region) are possible (15, 30, 31). This derivative approach does suffer from a need for knowledge about the degree of reversibility of the electrode reaction and, if it is irreversible, knowing the α value.

An alternate instrumental approach involves using the generated (dE/dt) value to produce a feedback signal representing i_c. This is then added to i applied to the cell to yield a more constant i_f (32). This approach requires a fair amount of instrumental complexity and is only partially successful, since a known and assumed constant value of C_d must be used to obtain i_c. Applications of this method, and chronopotentiometric techniques in general, to analytical or mechanistic problems have been rather sparse compared to the use of controlled potential techniques (27, 33, 34).

7.7 CHARGE STEP (COULOSTATIC) METHODS

7.7.1 Principles

In the *charge step* (or *coulostatic impulse*) *method* a very short-duration (e.g., 0.1 to 1 μsec) current pulse is applied to the cell, and the variation of the electrode potential with time after the pulse (i.e., at open circuit) is recorded. The current-pulse length is chosen to be sufficiently short that it only causes charging of the electrical double

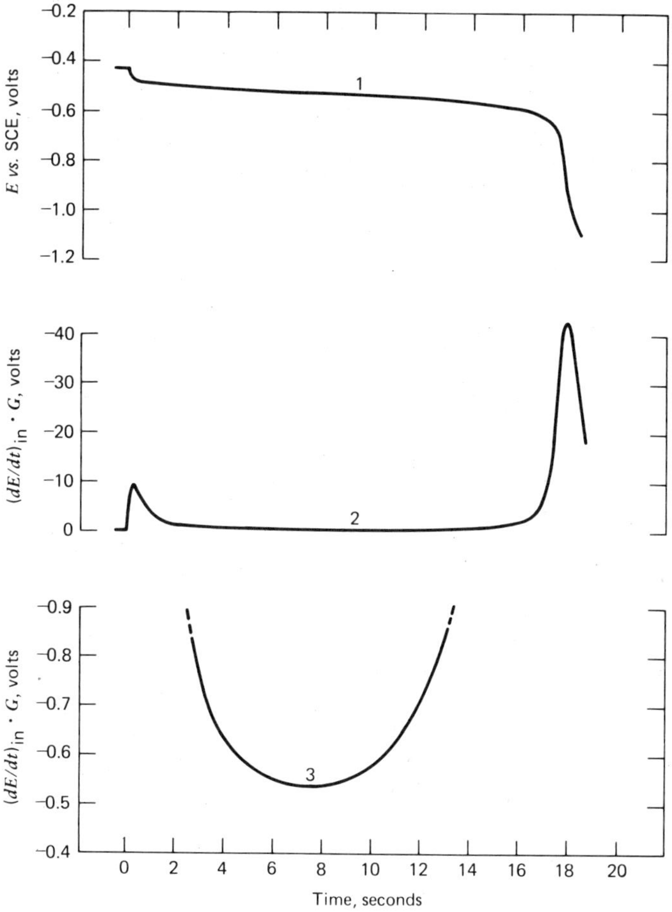

Figure 7.6.1
Experimental conventional and derivative chronopotentiograms. Reduction of 6.28 m*M* Tl(I) in 0.1 *M* KNO_3 at mercury pool cathode (1.38 cm^2 area) at $i = 0.503$ mA. Curve 1: Conventional *E-t* curve. Curve 2: Derivative curve, $(dE/dt)_{in} \cdot G$, represents the value of dE/dt (mV/sec) multiplied by gain in circuit, G (0.1095 V mV^{-1} sec). Curve 3: As in curve 2, with portion near region of minimum enlarged. [Reprinted with permission from D. G. Peters and S. L. Burden, *Anal. Chem.*, **38**, 530 (1966). Copyright 1966, American Chemical Society.]

layer, so that even a very fast charge transfer reaction does not proceed to an appreciable extent during this time. The pulse then serves only to inject a charge increment, Δq, and, in fact, under these conditions the method of charge injection or the actual shape of the injecting pulse (the coulostatic impulse) is unimportant. For example, the charge can be injected by discharging a small capacitor across the electrochemical cell (Figure 7.7.1) or with a pulse generator connected to the cell by a capacitor or switching diodes. For the circuit in Figure 7.7.1, when the relay is in position *A*, the

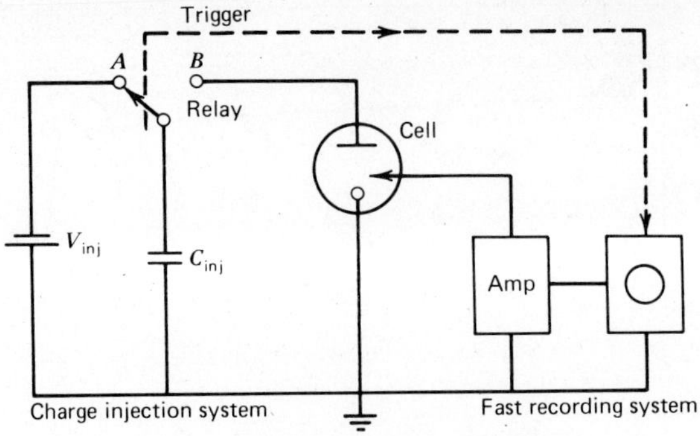

Figure 7.7.1
Schematic circuit for charge step or coulostatic pulse method. In practice the cell may be held initially at a potential E_{eq} by means of a potentiostat that is disconnected immediately before the charge injection.

capacitor, C_{inj}, is charged by the voltage source, V_{inj}, until the capacitor is charged by an amount

$$\Delta q = C_{inj} V_{inj} \tag{7.7.1}$$

For example, for $V_{inj} = 10$ V and $C_{inj} = 10^{-9}$ F, $\Delta q = 0.01\ \mu$C. When the relay switches to position B the charge is delivered to the electrochemical cell. Because the double-layer capacitance, C_d, is much larger than C_{inj}, essentially all of the charge will flow into the cell. The time required for this charge injection will depend on the cell resistance, R_Ω (Figure 7.7.2), with the time constant for injection being essentially

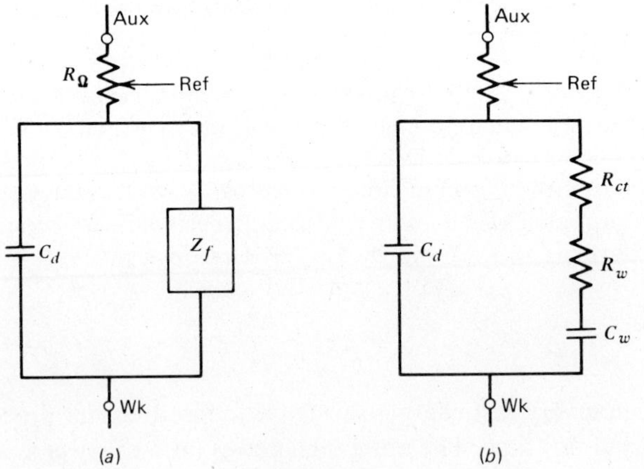

Figure 7.7.2
Equivalent circuit of cell with (*a*) R_Ω, the solution resistance, C_d, the double-layer capacitance, and Z_f, the faradaic impedance. The components of Z_f shown in (*b*) are the charge transfer resistance R_{ct} and the components of the Warburg impedance, R_w and C_w, determined by the rate of diffusion (see Section 9.1.3).

$C_{\text{inj}}R_{\Omega}$ (see Problem 7.6). This injected charge causes the potential of the electrode to deviate from its original value E_{eq} to a value $E(t = 0)$, where

$$E(t = 0) - E_{eq} = \eta(t = 0) = \frac{-\Delta q}{C_d} \tag{7.7.2}$$

The charge on C_d now discharges through the faradaic impedance, and the open-circuit potential moves back toward E_{eq} as $\eta(t)$ decreases to zero. In this case, since the total external current i is zero, then from (7.3.10) and (7.3.11),

$$i_f = -i_c = C_d\left(\frac{d\eta}{dt}\right) \tag{7.7.3}$$

or

$$\eta(t) = \eta(t = 0) + \frac{1}{C_d}\int_0^t i_f\, dt \tag{7.7.4}$$

Solution of (7.7.4) with the appropriate expression for i_f yields the desired expression for the variation of E (or η) with t. Note that if no faradaic reaction is possible at $E(t = 0)$ (i.e., at an ideally polarized electrode), C_d remains charged and the potential will not decay [i.e., for $i_f = 0$, $E = E_{eq} + \eta = E_{eq} + \eta(t = 0)$ at all t]. We will now examine the E-t behavior following a coulostatic impulse for several cases of interest. Details of the theoretical treatments have been given by Delahay (35, 36) and Reinmuth (37, 38) and their co-workers, who first described the application of this technique.

7.7.2 Small-Signal Analysis

When the potential excursion is sufficiently small, that is, when $\eta(t = 0) \ll RT/nF$, the linearized i-E characteristic (3.5.33) can be used in (7.7.4). Let us first consider the case of a totally irreversible reaction with negligible concentration polarization, so that the i-η relation is simply

$$-\eta = \left(\frac{RT}{nFi_0}\right)i \tag{7.7.5}$$

Then, from (7.7.4),

$$\eta(t) = \eta(t = 0) - \frac{i_0 nF}{RTC_d}\int_0^t \eta(t)\, dt \tag{7.7.6}$$

This equation can be solved readily by the Laplace transform method (Problem 7.7) to yield

$$\boxed{\eta(t) = \eta(t = 0)\exp\left(\frac{-t}{\tau_c}\right)} \tag{7.7.7}$$

$$\boxed{\tau_c = \frac{RTC_d}{nFi_0}} \tag{7.7.8}$$

Thus under these conditions the potential relaxes exponentially toward E_{eq} with a time constant τ_c, governed by the rate of the charge transfer reaction (Figure 7.7.3). This result can also be obtained from the equivalent circuit in Figure 7.7.2*b*, by noting that R_w and C_w are negligible, and that C_d then discharges through the charge transfer resistance R_{ct}, given by (3.5.13), with a time constant $C_d R_{ct}$. When (7.7.7) holds, a plot of $\ln|\eta|$ *vs.* t is linear with an intercept $|\eta(t = 0)|$ [which can be used to determine C_d by (7.7.2)] and a slope $-1/\tau_c$, which yields the exchange current density.

On the other hand, when R_{ct} is negligible compared to the mass transfer terms, the following expression applies:

$$\eta(t) = \eta(t = 0) \exp\left(\frac{t}{\tau_D}\right) \operatorname{erfc}\left[\left(\frac{t}{\tau_D}\right)^{1/2}\right] \tag{7.7.9}$$

$$\tau_D^{1/2} = \frac{RTC_d}{n^2F^2}\left(\frac{1}{C_O^* D_O^{1/2}} + \frac{1}{C_R^* D_R^{1/2}}\right) \tag{7.7.10}$$

The general small-signal expression, where both charge and mass transfer terms are significant, is (38)

$$\boxed{\eta(t) = \frac{\eta(t = 0)}{\gamma - \beta}\left[\gamma \exp(\beta^2 t) \operatorname{erfc}(\beta t^{1/2}) - \beta \exp(\gamma^2 t) \operatorname{erfc}(\gamma t^{1/2})\right]} \tag{7.7.11}$$

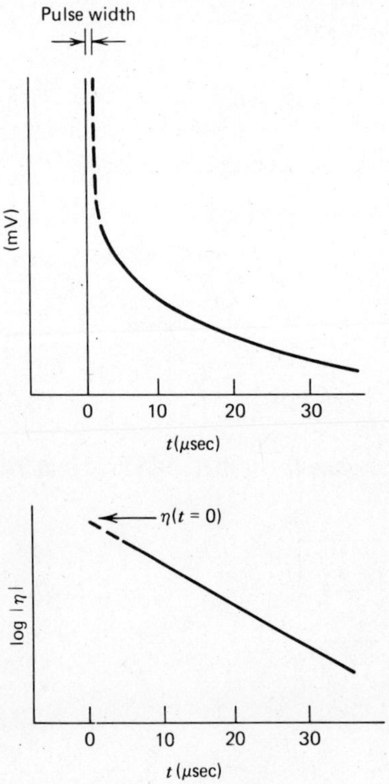

Figure 7.7.3
Typical charge step overpotential-time relaxation curves for totally irreversible reaction.

$$\beta, \gamma = \frac{\tau_D^{1/2}}{2\tau_c} \pm \frac{[(\tau_D/4\tau_c) - 1]^{1/2}}{\tau_C^{1/2}} \tag{7.7.12}$$

(where the + is associated with β and the − with γ). Note that $\beta + \gamma = \tau_D^{1/2}/\tau_c$ and $\beta\gamma = 1/\tau_c$.

Clearly the analysis of experimental data for the determination of i_0 is easiest when (7.7.7) applies; this requires that $\tau_c \gg \tau_D$. Detailed discussions of the analysis of coulostatic data and relaxation curves have appeared (39, 40).

7.7.3 Large Steps—Coulostatic Analysis

Consider the application of a charge step sufficiently large that the potential changes from E_{eq} to a value, $E(t = 0)$, corresponding to the diffusion plateau of the voltammetric wave. We will assume that the double-layer capacity, C_d, is independent of potential in this region. The faradaic current that flows under these conditions at a planar electrode is given by (5.2.11). Introduction of this expression into (7.7.4) yields

$$E(t) = E(t = 0) + \left(\frac{nFAD_O^{1/2}C_O^*}{\pi^{1/2}C_d}\right)\int_0^t t^{-1/2}\, dt \tag{7.7.13}$$

$$\Delta E = E(t) - E(t = 0) = \frac{2nFAD_O^{1/2}C_O^* t^{1/2}}{\pi^{1/2}C_d} \tag{7.7.14}$$

The sign of ΔE is positive, since the electrode relaxes from a more negative initial potential toward more positive values. A plot of ΔE *vs.* $t^{1/2}$ is linear with a zero intercept and a slope $2nFAD_O^{1/2}C_O^*/\pi^{1/2}C_d$, which is proportional to the solution concentration (Figure 7.7.4). This method has been suggested for the determination of small concentrations of electroactive materials (41, 42), but it has not been widely applied,

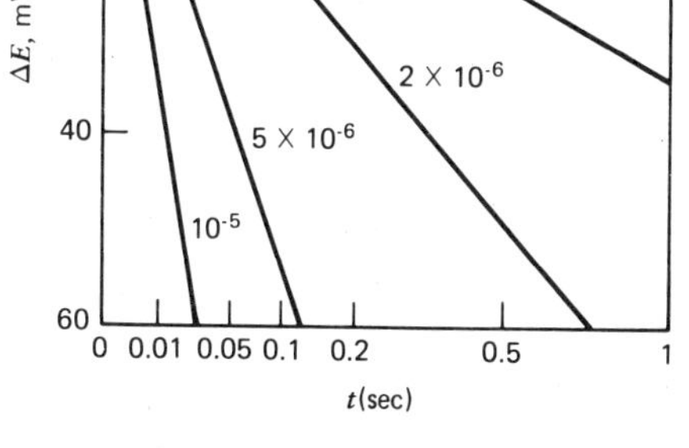

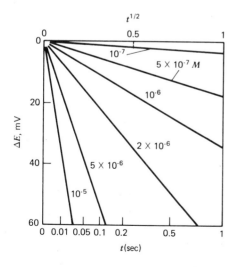

Figure 7.7.4
ΔE *vs.* t curves for plane electrode for several values of C_O^*, with $n = 2$, $D = 10^{-5}$ cm²/sec, and $C_d = 20\ \mu$F/cm². [Reprinted with permission from P. Delahay, *Anal. Chem.*, **34**, 1267 (1962). Copyright 1962, American Chemical Society.]

probably because it requires recording of the E-t curve and is less readily automated than, for example, pulse polarography. The technique can also be extended to cover large potential excursions to the rising portion of the voltammetric wave. For example, for a reversible process, the appropriate equation for the faradaic current would be (5.4.16), as long as measurements were made at small ΔE values so that E remained close to $E(t = 0)$ and the surface concentrations did not change appreciably during the measurement. A plot of the slope of the ΔE *vs.* $t^{1/2}$ curve at any potential *vs.* E then yields a charge-step voltammogram that resembles an ordinary voltammogram (43). Because the method requires the determination of a slope at each data point, incremental changes in Δq to cause changes in potential, and renewal of initial conditions before each step, it requires digital computer control and data acquisition.

7.7.4 Application of Charge-Step Methods

The charge-step or coulostatic methods described here have some advantages in the study of electrode reactions. Since the measurement is made at open circuit with no net external current flow, the ohmic drop is not of importance and measurements in highly resistive media can be made. Moreover, because the relaxation occurs by discharge of the double-layer capacitance, the usual competition between faradaic and charging current is replaced by an equality of i_c and i_f, and C_d no longer interferes in the measurement. However, some fundamental limitations still exist in charge-step measurements. High values of R_Ω increase the time required to deliver charge to the cell. Also, the high voltage, V_{inj}, appears across the cell at the instant of charge injection; this overloads the measuring amplifier, which must be adjusted to a high sensitivity to determine the small changes in ΔE and an amplifier capable of rapid recovery from overloads is required. Parasitic oscillations caused by stray capacitance and unmatched impedances generally occur following the pulse. Thus, in general, measurements cannot be made for a time interval of about 0.5 μsec following the pulse. The technique appears to be useful for the determination of i_0 values up to about 0.1 A/cm^2 or k^0 values up to 0.4 cm/sec. A discussion of the experimental conditions and a review of applications of charge-step methods has appeared (40).

7.8 REFERENCES

1. H. J. S. Sand, *Phil. Mag.*, **1**, 45 (1901).
2. R. W. Murray and C. N. Reilley, *J. Electroanal. Chem.*, **3**, 64, 182 (1962).
3. Z. Karaoglanoff, *Z. Elektrochem.*, **12**, 5 (1906).
4. P. Delahay and T. Berzins, *J. Am. Chem. Soc.*, **75**, 2486 (1953).
5. L. B. Anderson and D. J. Macero, *Anal. Chem.*, **37**, 322 (1965).
6. Y. Okinaka, S. Toshima, and H. Okaniwa, *Talanta*, **11**, 203 (1964).
7. P. Delahay and T. Berzins, *J. Chem. Phys.*, **23**, 972 (1955); *J. Am. Chem. Soc.*, **77**, 6448 (1955); *Z. Elektrochem.*, **59**, 792 (1955).
8. H. Gerischer and M. Krause, *Z. Physik. Chem.*, *N.F.*, **10**, 264 (1957); **14**, 184 (1958).
9. H. Matsuda, S. Oka, and P. Delahay, *J. Am. Chem. Soc.*, **81**, 5077 (1959).

10. M. Kogoma, T. Nakayama, and S. Aoyagi, *J. Electroanal. Chem.*, **34,** 123 (1972).
11. T. Rohko, M. Kogoma, and S. Aoyagi, *J. Electroanal. Chem.*, **38**, 45 (1972).
12. M. Kogoma, Y. Kanzaki, and S. Aoyagi, *Chem. Instr.*, **7**, 193 (1976).
13. J. J. Lingane, *J. Electroanal. Chem.*, **1**, 379 (1960).
14. A. J. Bard, *Anal. Chem.*, **35**, 340 (1963).
15. P. E. Sturrock, G. Privett, and A. R. Tarpley, *J. Electroanal. Chem.*, **14**, 303 (1967).
16. W. T. deVries, *J. Electroanal. Chem.*, **17**, 31 (1968).
17. R. S. Rodgers and L. Meites, *J. Electroanal. Chem.*, **16**, 1 (1968).
18. M. L. Olmstead and R. S. Nicholson, *J. Phys. Chem.*, **72**, 1650 (1968).
19. A. J. Bard, *Anal. Chem.*, **33**, 11 (1961).
20. H. A. Laitinen and I. M. Kolthoff, *J. Am. Chem. Soc.*, **61**, 3344 (1939).
21. H. B. Herman and A. J. Bard, *Anal. Chem.*, **35**, 1121 (1963).
22. A. C. Testa and W. H. Reinmuth, *Anal. Chem.*, **33**, 1320, 1324 (1961).
23. T. Berzins and P. Delahay, *J. Am. Chem. Soc.*, **75**, 4205 (1953).
24. F. H. Beyerlein and R. S. Nicholson, *Anal. Chem.*, **40**, 286 (1968).
25. H. B. Herman and A. J. Bard, *Anal. Chem.*, **35**, 1121 (1963).
26. P. Delahay and G. Mamantov, *Anal. Chem.*, **27**, 478 (1955).
27. C. N. Reilley, G. W. Everett, and R. H. Johns, *Anal. Chem.*, **27**, 483 (1955).
28. H. B. Herman and A. J. Bard, *Anal. Chem.*, **36**, 971 (1964).
29. D. G. Peters and S. L. Burden, *Anal. Chem.*, **38**, 530 (1966).
30. P. E. Sturrock, J. L. Hughey, B. Vaudreuil, G. O'Brien, and R. H. Gibson, *J. Electrochem. Soc.*, **122**, 1195 (1975).
31. P. E. Sturrock and R. H. Gibson, *J. Electrochem. Soc.*, **123**, 629 (1976).
32. W. D. Shults, F. E. Haga, T. R. Mueller, and H. C. Jones, *Anal. Chem.*, **37**, 1415 (1965).
33. L. Gierst and A. Juliard, *Proc. Intern. Comm. Electrochem. Thermodynam. and Kinet.*, 2nd meeting, 1950, pp. 117, 279.
34. D. G. Davis, *Electroanal. Chem.*, **1**, 157 (1966).
35. P. Delahay, *J. Phys. Chem.*, **66**, 2204 (1962); *Anal. Chem.*, **34**, 1161 (1962).
36. P. Delahay and A. Aramata, *J. Phys. Chem.*, **66**, 2208 (1962).
37. W. H. Reinmuth and C. E. Wilson, *Anal. Chem.*, **34**, 1159 (1962).
38. W. H. Reinmuth, *Anal. Chem.*, **34**, 1272 (1962).
39. J. M. Kudirka, P. H. Daum, and C. G. Enke, *Anal. Chem.*, **44**, 309 (1972).
40. H. P. van Leeuwen, *Electrochim. Acta*, **23**, 207 (1978).
41. P. Delahay, *Anal. Chem.*, **34**, 1267 (1962).
42. P. Delahay and Y. Ide, *Anal. Chem.*, **34**, 1580 (1962).
43. J. M. Kudirka, R. Abel, and C. G. Enke, *Anal. Chem.*, **44**, 425 (1972).

7.9 PROBLEMS

7.1 Derive equation 7.3.12 under the assumptions given in the text.

7.2 For current reversal chronopotentiometry involving the forward reduction of a species O under conditions of semi-infinite linear diffusion, the reverse transition time can be made equal to forward electrolysis time by a proper choice of currents during the forward (reduction) reaction, i_f, and the reverse (oxidation) reaction, i_r. Find the ratio i_f/i_r that will yield $\tau_r = t_f$.

7.3 An analyst determines a mixture of lead and cadmium at a mercury pool cathode by chronopotentiometry. In the cell used in the determination, a 1.00 mM solution of Pb^{2+} at a current of 273 mA yielded $\tau = 25.9$ sec and $E_{\tau/4} = -0.38$ V *vs.* SCE. A 0.69 mM solution of Cd^{2+} with a current of 136 mA gave $\tau = 42.0$ sec and $E_{\tau/4} = -0.56$ V *vs.* SCE.

An unknown mixture of Pb^{2+} and Cd^{2+} reduced at a current of 56.5 mA produced a double wave, with $\tau_1 = 7.08$ sec and $\tau_2 = 7.00$ sec. Calculate the concentrations of Pb^{2+} and Cd^{2+} in the mixture. Neglect double-layer and other background effects.

7.4 Show that if programmed current chronopotentiometry with $i(t) = \beta t^{1/2}$ is used, then for the stepwise reduction of a substance with $n_1 = n_2$, $\tau_1 = \tau_2$.

7.5 Examine the results in Figure 7.4.1. Estimate the transition times and work up the data to yield information about the electrode reaction.

7.6 Consider the circuit in Figure 7.9.1, which is characteristic of that used for the injection of charge in a coulostatic impulse experiment. C_{inj} is initially charged completely with a 10-V battery. At equilibrium, after the switch is closed, how much charge will reside on C_d and on C_{inj}? About how long will it take to charge C_d?

7.7 Solve (7.7.6) by the Laplace transform method to yield (7.7.7).

7.8 Derive the equation for the large-step coulostatic response in the diffusion-limiting region, analogous to (7.7.14), for a spherical indicator electrode.

7.9 Consider a 1 mM solution of cadmium in 0.1 M HCl, which is being examined coulostatically at a hanging mercury drop 0.05 cm^2 in area. The formal potential for the $Cd^{2+}/Cd(Hg)$ couple is -0.61 V *vs.* SCE. Suppose the electrode is initially at rest at -0.4 V *vs.* SCE, then a sufficient charge is applied to shift instantaneously its potential to -1.0 *V vs.* SCE. Assume the differential and integral double-layer capacitances to be 10 $\mu F/cm^2$. How much charge is required for the initial potential excursion? How long would it take for the potential to fall back to -0.9 V after the charge injection? Take $D = 10^{-5}$ cm^2/sec.

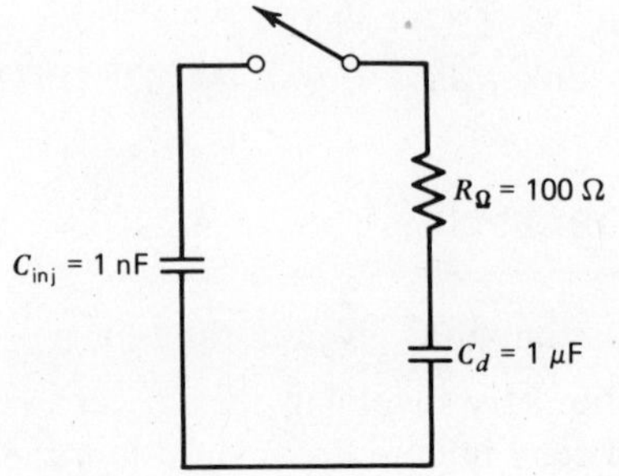

Figure 7.9.1

7.10 Barker et al. [*Faraday Disc. Chem. Soc.*, **56**, 41 (1974)] have performed experiments in which 15-nsec pulses from a frequency-doubled ruby laser have been used to illuminate a mercury pool working electrode. This causes ejection of electrons from the electrode. The electrons seem to travel about 50 Å before becoming solvated and available for reaction.

When electrons are emitted into a solution of N_2O in water containing 1 *M* KCl, the following reaction occurs:

$$e_{aq} + N_2O + H_2O \rightarrow OH\cdot + N_2 + OH^-$$

The hydroxyl radicals are easily reduced at the electrode at potentials more negative than -1.0 V *vs.* SCE.

The response of the illuminated working electrode to the flash is followed coulostatically. Curves like those shown in Figure 7.9.2 can be obtained. Explain their shapes. ΔE is measured with respect to the initial potential.

7.11 Barker's technique (see Problem 7.10) can also be used to create hydrogen atoms and study their electrochemistry. The reaction producing them in acid media is

$$H_3O^+ + e_{aq} \rightarrow H\cdot + H_2O$$

Investigators studying the hydrogen discharge reaction have often suggested that $H\cdot$ is an intermediate and that hydrogen gas is produced by reducing it further in a fast heterogeneous process:

$$(H\cdot)_{free} + e + H_3O^+ \rightarrow H_2 + H_2O \qquad \text{(a)}$$

or

$$(H\cdot)_{free} \xrightarrow{k} (H\cdot)_{ads}$$
$$(H\cdot)_{ads} + e + H_3O^+ \rightarrow H_2 + H_2O \qquad \text{(b)}$$

Whether $H\cdot$ is free or adsorbed has been debated. Barker addressed the question by comparing, in effect, the rate of $H\cdot$ electroreduction to the rate of its homogeneous reaction with ethanol (leading to electroinactive products). He found [*Ber. Bunsenges. Phys. Chem.*, **75**, 728 (1971)] that the fraction of $H\cdot$ undergoing electroreduction was independent of potential from -0.9 V to -1.3 V *vs.* SCE. What do his observations tell us about the choice between (a) and (b)?

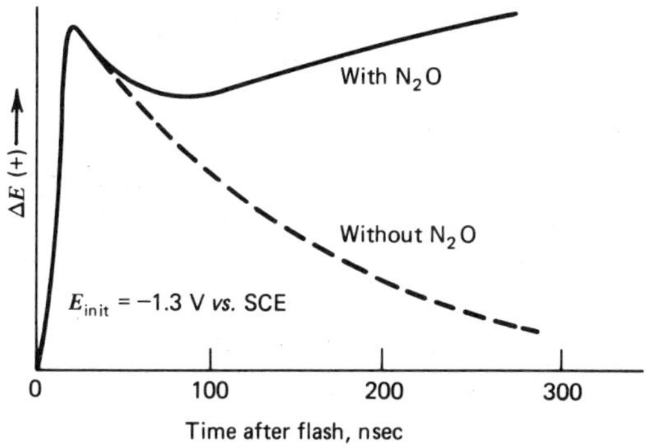

Figure 7.9.2

chapter 8

Methods Involving Forced Convection—Hydrodynamic Methods

8.1 INTRODUCTION

There are many electrochemical techniques in which the electrode moves with respect to the solution. These involve systems where the electrode is in motion (e.g., rotating disks, rotating wires, streaming mercury electrodes, rotating mercury electrodes, vibrating electrodes) or ones where there is forced solution flow past a stationary electrode (conical, tubular, screen, and packed-bed electrodes in fluid streams, bubbling electrodes). Indeed the dropping mercury electrode is actually such a system, but was treated by the approximate method described in Chapter 5. Methods involving convective mass transport of reactants and products are sometimes called *hydrodynamic* methods; for example, the techniques involving measurement of limiting currents or *i-E* curves are called *hydrodynamic amperometry* and *voltammetry*, respectively.

The advantage of hydrodynamic methods is that a steady state is attained rather quickly and measurements can be made with high precision (e.g., with digital voltmeters), often without the need for recorders or oscilloscopes. In addition, at steady state, double-layer charging does not enter the measurement. Also, the rates of mass transfer at the electrode surface in these methods are much larger than the rates of diffusion, so that the relative contribution of the effect of mass transfer to electron transfer kinetics is smaller. Although it might first appear that the valuable time variable is lost in steady-state convective methods, this is not so, because time enters the experiment as the rotation rate of the electrode or the solution velocity with respect to the electrode. Dual-electrode techniques can be employed to provide the same kind of information that reversal methods do in stationary electrode techniques. These methods are also of interest in the continuous monitoring of flowing liquids and in treatments of large-scale reactors such as those employed for electrosynthesis (see Section 10.6).

Construction of hydrodynamic electrodes that provide known and reproducible mass transfer conditions is more difficult than for stationary electrodes. The theoretical treatments involved in these methods are also much more difficult and involve solving a hydrodynamic problem (i.e., determination of solution flow velocity profiles as functions of rotation rates, solution viscosities, and densities) before the electrochemical one can be tackled. Rarely can closed-form or exact solutions be obtained. Even though the number of possible electrode configurations and flow patterns possible in these methods is limited by the imagination and resources of the experimenter, the most convenient and widely used system involves the rotating disk electrode. This electrode is amenable to rigorous theoretical treatment and is easy to construct with a variety of electrode materials. Most of what follows deals with it and its variations.

8.2 THEORETICAL TREATMENT OF CONVECTIVE SYSTEMS

The simplest treatments of convective systems are based on a diffusion layer approach. In this model, it is assumed that convection maintains the concentrations of all species uniform and equal to the bulk values up to a certain distance from the electrode, δ. Within the layer, $0 \leq x \leq \delta$, no solution movement occurs and mass transfer takes place by diffusion. Thus the convection problem is converted to a diffusional one in which the adjustable parameter δ is introduced. This is basically the approach that was used in Chapter 1 to deal with the steady-state mass transport problem. However, this approach does not yield equations that show how currents are related to flow rates, rotation rates, solution viscosity, and electrode dimensions. Neither can it be employed for dual-electrode techniques or for predicting relative mass transfer rates of different substances. A more rigorous approach begins with the convective-diffusion equation and the velocity profiles in the solution. They are solved either analytically or, more frequently, numerically; in most cases only the steady-state solution is desired.

8.2.1 The Convective-Diffusion Equation

The general equation for the flux of species j, $\mathbf{J}_j$, is (equation 4.1.9)

$$\mathbf{J}_j = -D_j \nabla C_j - \frac{z_j F}{RT} D_j C_j \nabla \phi + C_j \mathbf{v} \tag{8.2.1}$$

where on the right-hand side, the first term represents diffusion, the second, migration, and the last, convection. For solutions containing an excess of supporting electrolyte, the ionic migration term can be neglected; we will assume this to be the case for most of this chapter (see, however, Section 8.3.5). The velocity vector, $\mathbf{v}$, represents the motion of the solution and, in rectilinear coordinates, is given by

$$\mathbf{v}(x, y, z) = \mathbf{i}v_x + \mathbf{j}v_y + \mathbf{k}v_z \tag{8.2.2}$$

where $\mathbf{i}$, $\mathbf{j}$, and $\mathbf{k}$ are unit vectors, and v_x, v_y, and v_z are the magnitudes of the solution velocities in the x, y, and z directions, respectively, at point (x, y, z). Similarly, in rectilinear coordinates,

$$\nabla C_j = \mathbf{grad}\ C_j = \mathbf{i}\frac{\partial C_j}{\partial x} + \mathbf{j}\frac{\partial C_j}{\partial y} + \mathbf{k}\frac{\partial C_j}{\partial z} \tag{8.2.3}$$

The variation of C_j with time is given by

$$\frac{\partial C_j}{\partial t} = -\nabla \cdot \mathbf{J}_j = \mathbf{div}\ \mathbf{J}_j \tag{8.2.4}$$

By combining (8.2.1) and (8.2.4), assuming that migration is absent and that D_j is not a function of x, y, and z, we obtain the general convective-diffusion equation:

$$\frac{\partial C_j}{\partial t} = D_j \nabla^2 C_j - \mathbf{v} \cdot \nabla C_j \tag{8.2.5}$$

The forms for the Laplacian operator, ∇^2, are given in Table 4.3.2. For example, for one-dimensional diffusion and convection, (8.2.5) is

$$\frac{\partial C_j}{\partial t} = D_j \frac{\partial^2 C_j}{\partial y^2} - v_y \frac{\partial C_j}{\partial y} \tag{8.2.6}$$

Note that in the absence of convection (i.e., $\mathbf{v} = 0$ or $v_y = 0$), (8.2.5) and (8.2.6) are reduced to the diffusion equations. Before the convective-diffusion equation can be solved for the concentration profiles, $C_j(x, y, z)$, and subsequently for the currents from the concentration gradients at the electrode surface, expressions for the velocity profile, $\mathbf{v}(x, y, z)$, must be obtained in terms of x, y, z, rotation rate, and so on.

8.2.2 Determination of the Velocity Profile

Although it is beyond the scope of this chapter to treat hydrodynamics in any depth, a brief discussion of some of the concepts, terms, and equations is included to provide some feeling for the approach and the results that follow. For an incompressible fluid (i.e., a fluid whose density is constant in time and space), the velocity profile is obtained by solution of the *continuity equation*, (8.2.7), and the *Navier-Stokes equation*, (8.2.8), with the appropriate boundary conditions. The continuity equation:

$$\nabla \cdot \mathbf{v} = \mathbf{div}\ \mathbf{v} = 0 \tag{8.2.7}$$

is a statement of incompressibility, whereas the equation:

$$d_s \frac{d\mathbf{v}}{dt} = -\nabla P + \eta_s \nabla^2 \mathbf{v} + \mathbf{f} \tag{8.2.8}$$

represents Newton's first law ($\mathbf{F} = m\mathbf{a}$) for a fluid; the left side represents $m\mathbf{a}$ (per unit volume; d_s is the density), and the right side represents the forces on a volume element (P is the pressure; η_s the viscosity; and $\mathbf{f}$ the force/volume exerted on an element of the fluid by gravity). The term $\eta_s \nabla^2 \mathbf{v}$ represents frictional forces. This equation is usually written in the form

$$\frac{d\mathbf{v}}{dt} = \frac{-1}{d_s} \nabla P + \nu \nabla^2 \mathbf{v} + \frac{\mathbf{f}}{d_s} \tag{8.2.9}$$

where $\nu = \eta_s/d_s$ is called the *kinematic viscosity* and has units of cm^2/sec; for water and dilute aqueous solutions near 20 °C, ν is about 0.01 cm^2/sec. The term **f** represents the effect of natural convection arising from the buildup of density gradients in the solution.

Two different types of fluid flow are usually considered in hydrodynamic problems (Figure 8.2.1). When the flow is smooth and steady and occurs as if separate layers (laminae) of the fluid have steady and characteristic velocities, the flow is said to be *laminar*. For example, the flow of water through a smooth pipe will be laminar, with the flow velocity zero right at the walls (because of friction between the fluid and the wall) and with some maximum value in the middle of the pipe. When the flow involves unsteady and chaotic motion, in which only on the average is there a net flow in a particular direction, it is termed *turbulent* flow. This type of flow might result from a barrier being placed in a pipe to obstruct the flow stream.

The solution of the hydrodynamic equations requires modelling the system and writing the equations in the appropriate coordinate system (linear, cylindrical, etc.), specifying the boundary conditions, and usually, numerical solution. In electrochemical problems, only the steady-state velocity profile is of interest, and therefore (8.2.9) is solved for $d\mathbf{v}/dt = 0$. Often the equations are rewritten in terms of dimensionless groups of variables. One that occurs in many hydrodynamic problems is the *Reynolds number*, Re. This number is formed by choosing a characteristic velocity, v_{ch} (cm/sec), and a characteristic length, l(cm), in the particular problem, and converting all velocities and distances in the problem to dimensionless ones by dividing by v_{ch} or l. The dimensionless parameter (Re) that arises in applying this procedure to (8.2.9) is given by

$$\mathrm{Re} = v_{ch}l/\nu \tag{8.2.10}$$

It is proportional to fluid velocity, so that high Reynolds numbers imply high flow or electrode rotation rates. For flow rates below a level characterized by a certain critical Reynolds number, Re_{cr}, the flow remains laminar. When $\mathrm{Re} > \mathrm{Re}_{cr}$ the flow regime becomes turbulent.

General treatments of the formulation and solution of hydrodynamic problems, especially as they relate to problems in electrochemistry, are available (1–5).

8.3 ROTATING DISK ELECTRODE

One of the few convective electrode systems for which the hydrodynamic equations and the convective-diffusion equation have been solved rigorously for the steady state is the rotating disk electrode (RDE). This electrode is rather simple to construct and consists of a disk of the electrode material imbedded in a rod of an insulating material. For example, a commonly used form involves a platinum wire sealed in glass tubing with the sealed end ground smooth and perpendicular to the rod axis. More frequently the metal is imbedded into Teflon, epoxy resin, or another plastic (Figure 8.3.1). Although the literature suggests that the shape of the insulating mantle is critical and that exact alignment of the disk is important (6), in practice these factors are usually not troublesome except perhaps at high rotation rates where turbulence and vortex

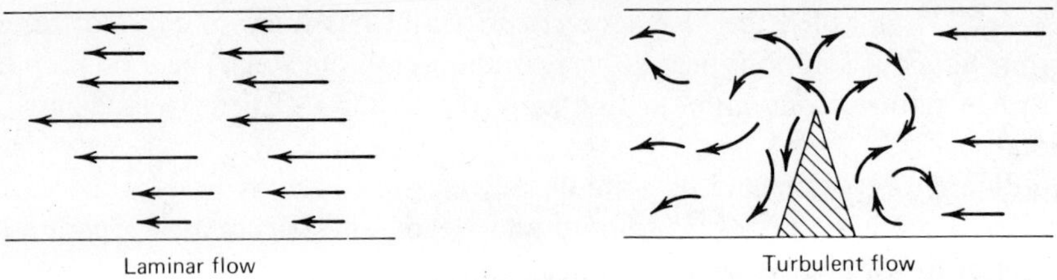

Figure 8.2.1
Types of fluid flow. Arrows represent instantaneous local fluid velocities.

formation may occur. It is more important that there is no leakage of the solution between the electrode material and the insulator. The rod is attached to a motor directly by a chuck or by a flexible rotating shaft or pulley arrangement and rotated at a certain frequency, f (revolutions per second). The parameter of interest is the angular velocity, $\omega(\sec^{-1})$, where $\omega = 2\pi f$. The electrical connection is made to the electrode by means of a brush contact; the noise level observed in the current at the RDE depends on this contact, and carbon-silver (Graphalloy) materials are frequently used. Details of the construction and application of RDEs are given in excellent reviews by Riddiford (6) and Adams (7); RDEs are also available commercially.

8.3.1 The Velocity Profile at the RDE

The velocity profile, **v**, of a fluid near a rotating disk was obtained by von Karman and Cochran (1) by solving the hydrodynamic equations under steady-state conditions. Qualitatively, the spinning disk drags the fluid at its surface along with it and, because of centrifugal force, flings the solution outwards from the center in a radial direction.

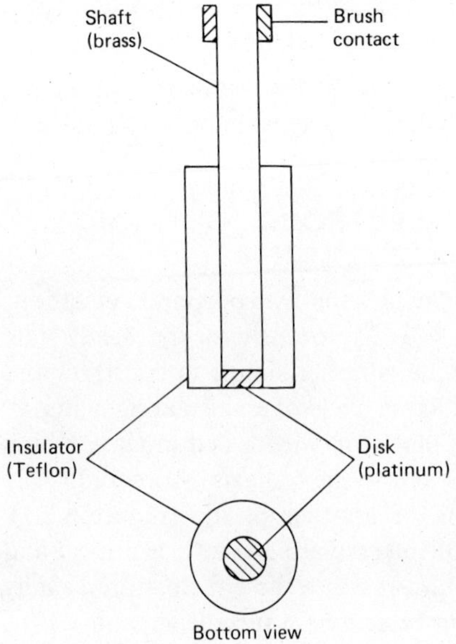

Figure 8.3.1
Rotating disk electrode.

The fluid at the disk surface is replenished by a flow normal to the surface. Because of the symmetry of the system, it is convenient to write the hydrodynamic equations in terms of the cylindrical coordinates y, r, and ϕ (Figure 8.3.2). For cylindrical coordinates,

$$\mathbf{v} = \boldsymbol{\mu}_1 v_r + \boldsymbol{\mu}_2 v_y + \boldsymbol{\mu}_3 v_\phi \tag{8.3.1}$$

$$\nabla = \boldsymbol{\mu}_1(\partial/\partial r) + \boldsymbol{\mu}_2(\partial/\partial y) + (\boldsymbol{\mu}_3/r)(\partial/\partial\phi) \tag{8.3.2}$$

where $\boldsymbol{\mu}_1$, $\boldsymbol{\mu}_2$, and $\boldsymbol{\mu}_3$ are unit vectors in the directions of positive changes of r, y, and ϕ at a given point. In contrast to the usual cartesian vectors $\mathbf{i}$, $\mathbf{j}$, and $\mathbf{k}$, the vectors $\boldsymbol{\mu}_1$ and $\boldsymbol{\mu}_3$ have directions that depend on the position of the point; thus the divergence and the Laplacian take on more complex forms. In particular,

$$\nabla \cdot \mathbf{v} = \frac{1}{r^2}\left[\frac{\partial}{\partial r}(v_r r^2) + \frac{\partial}{\partial y}(v_y r^2) + \frac{\partial}{\partial \phi} v_\phi\right] \tag{8.3.3}$$

$$\nabla^2 = \frac{1}{r}\left[\frac{\partial}{\partial r}\left(r\frac{\partial}{\partial r}\right) + \frac{\partial}{\partial y}\left(r\frac{\partial}{\partial y}\right) + \frac{\partial}{\partial \phi}\left(\frac{1}{r}\frac{\partial}{\partial \phi}\right)\right] \tag{8.3.4}$$

It is assumed that gravitational effects are absent ($\mathbf{f} = 0$) and that there are no special flow effects at the edge of the disk. At the disk surface ($y = 0$), $v_r = 0$, $v_y = 0$, and $v_\phi = \omega r$. This implies that the solution is dragged along at the surface of the disk at the angular velocity ω. In the bulk solution ($y \to \infty$), $v_r = 0$, $v_\phi = 0$, and $v_y = -U_0$. Thus, far from the disk, there is no flow in the r and ϕ directions, but the solution flows at a limiting velocity, U_0, toward the disk with U_0 determined by the solution of the problem. The treatment of this problem by von Karman and Cochran yielded values of the velocities in the form of infinite series in terms of the dimensionless variable γ, where

$$\gamma = \left(\frac{\omega}{\nu}\right)^{1/2} y \tag{8.3.5}$$

$$v_r = r\omega F(\gamma) = r\omega\left(a\gamma - \frac{\gamma^2}{2} - \tfrac{1}{3}b\gamma^3 + \cdots\right) \tag{8.3.6}$$

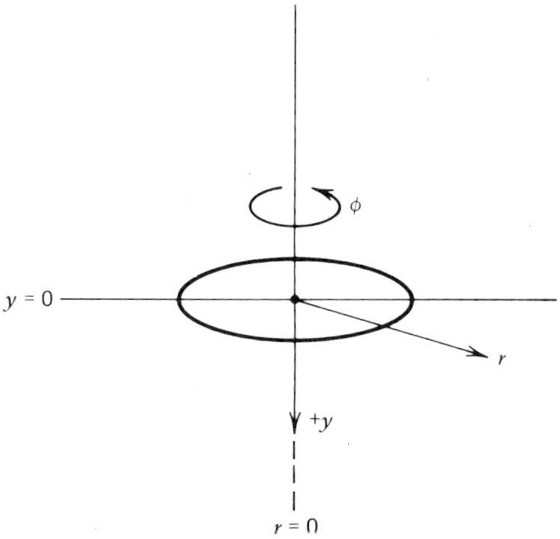

Figure 8.3.2
Cylindrical polar coordinates for rotating disk.

$$v_\phi = r\omega G(\gamma) = r\omega(1 + b\gamma + \tfrac{1}{3}a\gamma^3 + \cdots) \tag{8.3.7}$$

$$v_y = (\omega\nu)^{1/2}H(\gamma) = (\omega\nu)^{1/2}\left(-a\gamma^2 + \frac{\gamma^3}{3} + \frac{b\gamma^4}{6} + \cdots\right) \tag{8.3.8}$$

$$a = 0.51023 \qquad b = -0.6159$$

For the rotating disk electrode as employed in electrochemical studies, the important velocities are v_r and v_y (Figure 8.3.3). Near the surface of the rotating disk, $y \rightarrow 0$ (or $\gamma \rightarrow 0$), and these velocities are given by:

$$v_y = (\omega\nu)^{1/2}(-a\gamma^2) = -0.51\omega^{3/2}\nu^{-1/2}y^2 \tag{8.3.9}$$

$$v_r = r\omega(a\gamma) = 0.51\omega^{3/2}\nu^{-1/2}ry \tag{8.3.10}$$

A vector representation of the flow velocities is shown in Figure 8.3.4. The limiting velocity in the y direction, U_0, is

$$U_0 = \lim_{y\rightarrow\infty} v_y = -0.88447(\omega\nu)^{1/2} \tag{8.3.11}$$

At $\gamma = (\omega/\nu)^{1/2}y = 3.6$, $v_y \simeq 0.8U_0$. The corresponding distance, $y_h = 3.6(\nu/\omega)^{1/2}$, is called the *hydrodynamic* (or sometimes the *momentum* or *Prandtl*) *boundary layer thickness* and roughly represents the thickness of the layer of liquid dragged by the rotating disk. For water ($\nu \simeq 0.01$ cm^2/sec) at ω values of 100 and 10^4 sec^{-1}, y_h is 0.036 and 3.6×10^{-3} cm, respectively.

8.3.2 Solution of the Convective-Diffusion Equation

Once the velocity profile has been determined, the convective-diffusion equation, (8.2.5), for the rotating disk electrode, written in convenient coordinates and with appropriate boundary conditions, can be solved. Let us first consider the steady-state limiting current. When ω is fixed and a steady velocity profile has been attained, a potential step in the limiting current region [i.e., where $C_O(y = 0) \approx 0$] will cause the appearance of a current transient similar to that observed in the absence of convection.

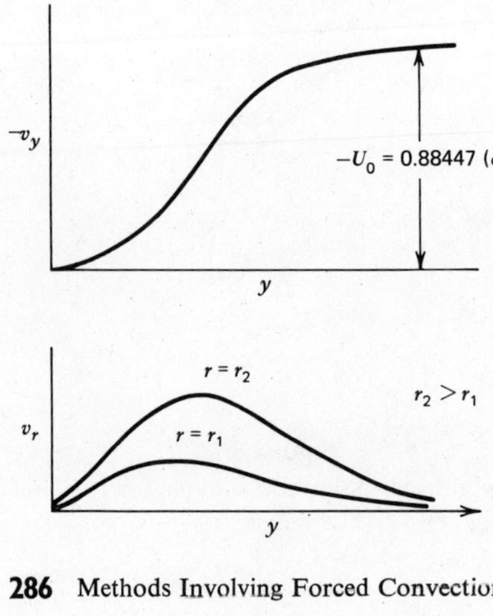

Figure 8.3.3
Variation of normal (v_y) and radial (v_r) fluid velocities as functions of y and r.

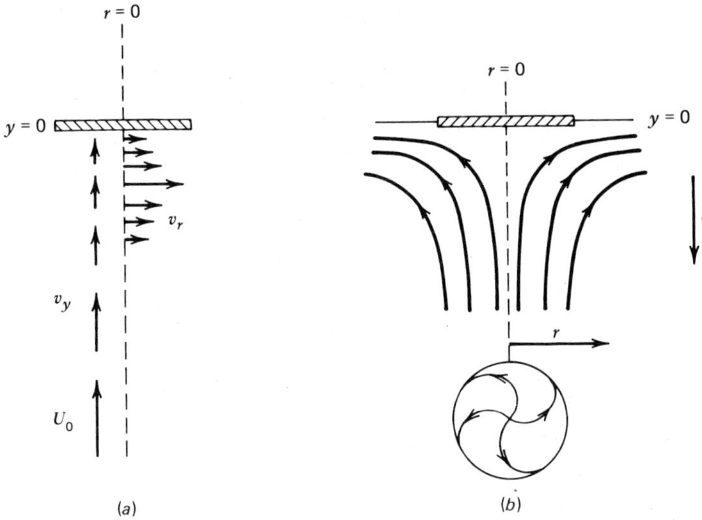

Figure 8.3.4
(*a*) Vector representation of fluid velocities near disk. (*b*) Schematic resultant streamlines (or flows).

However, in contrast to the transient that appears in an unstirred solution at a planar electrode, which decays toward zero, the current at the RDE decays to a steady-state value. Under these conditions the concentrations near the electrode are no longer functions of time, $\partial C_O/\partial t = 0$, and the steady-state convective-diffusion equation, written in terms of cylindrical coordinates, becomes

$$v_r\left(\frac{\partial C_O}{\partial r}\right) + \frac{v_\phi}{r}\left(\frac{\partial C_O}{\partial \phi}\right) + v_y\left(\frac{\partial C_O}{\partial y}\right) = D_O\left[\frac{\partial^2 C_O}{\partial y^2} + \frac{\partial^2 C_O}{\partial r^2} + \frac{1}{r}\frac{\partial C_O}{\partial r} + \frac{1}{r^2}\left(\frac{\partial^2 C_O}{\partial \phi^2}\right)\right] \quad (8.3.12)$$

For the limiting current condition, at $y = 0$, $C_O = 0$, and $\lim_{y\to\infty} C_O = C_O^*$. For reasons of symmetry C_O is not a function of ϕ; therefore, $\partial C_O/\partial\phi = (\partial^2 C_O/\partial\phi^2) = 0$. Also, v_y does not depend on r (8.3.8), and at $y = 0$, $(\partial C_O/\partial r) = 0$. Thus, across the face of the disk electrode, that is, $0 \le r \le r_1$, where r_1 is the disk radius, $(\partial C_O/\partial r) = 0$ for all y. This leads to a considerable simplification in (8.3.12), yielding

$$v_y\left(\frac{\partial C_O}{\partial y}\right) = D_O\frac{\partial^2 C_O}{\partial y^2} \quad (8.3.13)$$

or, by substitution of the value of v_y from (8.3.9) and rearrangement,

$$\frac{\partial^2 C_O}{\partial y^2} = \frac{-y^2}{B}\frac{\partial C_O}{\partial y} \quad (8.3.14)$$

where $B = D_O\omega^{-3/2}\nu^{1/2}/0.51$. This can be solved directly by integration. To make the job easier, let $X = \partial C_O/\partial y$, so that $\partial X/\partial y = \partial^2 C_O/\partial y^2$. At $y = 0$, $X = X_0 =$

$(\partial C_O/\partial y)_{y=0}$. Then (8.3.14) becomes

$$\frac{\partial X}{\partial y} = \left(\frac{-y^2}{B}\right) X \tag{8.3.15}$$

$$\int_{X_0}^{X} \left(\frac{dX}{X}\right) = \left(\frac{-1}{B}\right) \int_0^y y^2\, dy \tag{8.3.16}$$

$$\frac{X}{X_0} = \exp\left(\frac{-y^3}{3B}\right) \tag{8.3.17}$$

$$\frac{\partial C_O}{\partial y} = \left(\frac{\partial C_O}{\partial y}\right)_{y=0} \exp\left(\frac{-y^3}{3B}\right) \tag{8.3.18}$$

Integrating once more, we find that

$$\int_0^{C_O^*} dC_O = \left(\frac{\partial C_O}{\partial y}\right)_{y=0} \int_0^{\infty} \exp\left(\frac{-y^3}{3B}\right) dy \tag{8.3.19}$$

The definite integral on the right side is obtained by making the substitution $z = y^3/3B$, and is $(3B)^{1/3}\Gamma(4/3)$ or $0.8934(3B)^{1/3}$. Thus

$$C_O^* = \left(\frac{\partial C_O}{\partial y}\right)_{y=0} 0.8934\left(\frac{3D_O\omega^{-3/2}\nu^{1/2}}{0.51}\right)^{1/3} \tag{8.3.20}$$

The current, as before, is the flux at the electrode surface, that is,

$$i = nFAD_O\left(\frac{\partial C_O}{\partial y}\right)_{y=0} \tag{8.3.21}$$

where, under the limiting current conditions [integration of the left side of (8.3.19) from $C_O = 0$], $i = i_{l,c}$. From (8.3.20) and (8.3.21) we obtain the *Levich equation*:

$$\boxed{i_{l,c} = 0.620nFAD_O^{2/3}\omega^{1/2}\nu^{-1/6}C_O^*} \tag{8.3.22}$$

This equation thus applies to the totally mass-transfer-limited condition at the RDE and predicts that $i_{l,c}$ is proportional to C_O^* and $\omega^{1/2}$. One can define the *Levich constant*, $i_{l,c}/\omega^{1/2}C_O^*$, which is the RDE analog of the diffusion current constant or current function in voltammetry or the transition time constant in chronopotentiometry. Recall that the simple steady-state diffusion layer model yielded (equation 1.4.9)

$$i_{l,c} = nFAm_OC_O^* = nFA\left(\frac{D_O}{\delta_O}\right)C_O^* \tag{8.3.23}$$

Thus for the RDE

$$\boxed{m_O = \frac{D_O}{\delta_O} = 0.620D_O^{2/3}\omega^{1/2}\nu^{-1/6}} \tag{8.3.24}$$

$$\boxed{\delta_O = 1.61D_O^{1/3}\omega^{-1/2}\nu^{1/6}} \tag{8.3.25}$$

The concepts and results of the diffusion-layer model can often be used in RDE problems and, when needed, the appropriate value of δ_O can be substituted to yield the final equation.†

While the Levich equation (8.3.22) suffices for many purposes, improved forms based on derivations utilizing more terms in the velocity expression are available (8).

8.3.3 The Concentration Profile

The concentration profile at the limiting current condition can be obtained from (8.3.19) by integrating between 0 and $C_O(y)$; thus

$$\int_0^{C_O(y)} dC_O = C_O(y) = \left(\frac{\partial C_O}{\partial y}\right)_{y=0} \int_0^y \exp\left(\frac{-y^3}{3B}\right) dy \qquad (8.3.26)$$

From (8.3.20) we have

$$\left(\frac{\partial C_O}{\partial y}\right)_{y=0} = \frac{C_O^*}{0.8934}(3B)^{1/3} \qquad (8.3.27)$$

This can be put in a more convenient form by letting $u^3 = y^3/3B$; $dy = du(3B)^{1/3}$. Then (8.3.27) becomes

$$C_O(y) = \left(\frac{C_O^*}{0.8934}\right) \int_0^Y \exp(-u^3)\, du \qquad (8.3.28)$$

where $Y = y/(3B)^{1/3}$. The concentration profile for C_O under these conditions is shown in Figure 8.3.5.

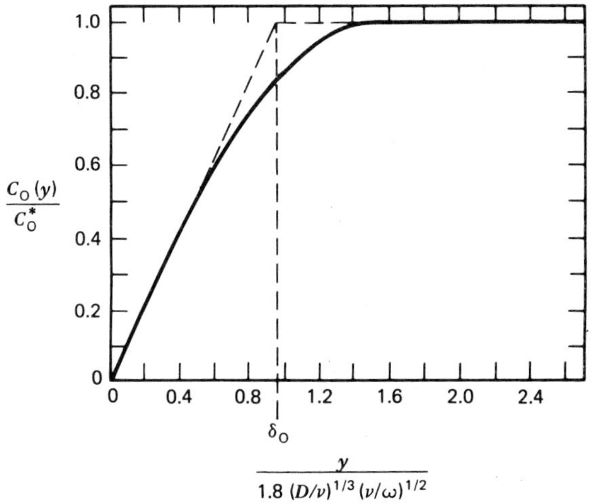

Figure 8.3.5
Concentration profile of species O given in terms of dimensionless coordinates.

† From the expression for the hydrodynamic boundary layer, y_h, and (8.3.25) we obtain $y_h/\delta_O \sim 2(\nu/D)^{1/3}$. For H_2O, $\nu = 0.01$ cm²/sec, $D_O \approx 10^{-5}$ cm²/sec, so that $\delta_O \approx 0.05\, y_h$. The dimensionless ratio (ν/D) occurs frequently in hydrodynamic problems and is called the *Schmidt number*, Sc.

8.3.4 General Current-Potential Curves at the RDE

For nonlimiting current conditions, again only a change in integration limits in (8.3.19) is required. In general, at $y = 0$, $C_O = C_O(y = 0)$ and $(\partial C_O/\partial y)_{y=0}$ is given by an analog to (8.3.20), which yields

$$C_O^* - C_O(y = 0) = \left(\frac{\partial C_O}{\partial y}\right)_{y=0} \int_0^\infty \exp\left(\frac{-y^3}{3B}\right) dy \tag{8.3.29}$$

Thus,

$$i = 0.620nFAD_O^{2/3}\omega^{1/2}\nu^{-1/6}[C_O^* - C_O(y = 0)] \tag{8.3.30}$$

or, from (8.3.22),

$$i = i_{l,c}\left[\frac{C_O^* - C_O(y = 0)}{C_O^*}\right] \tag{8.3.31a}$$

Alternately, (8.3.30) can be written in terms of δ_O as defined in (8.3.25) to yield

$$\boxed{i = \frac{nFAD_O[C_O^* - C_O(y = 0)]}{\delta_O} = nFAm_O[C_O^* - C_O(y = 0)]} \tag{8.3.31b}$$

Note that this equation is identical to that derived by the steady-state approximations in Section 1.4.

The current-potential curves at the RDE can be derived from (8.3.30) and the equivalent expression for the reduced form (assuming the simple reaction $O + ne \rightleftharpoons R$):

$$i = i_{l,a}\left[\frac{C_R^* - C_R(y = 0)}{C_R^*}\right] \tag{8.3.32}$$

$$i_{l,a} = -0.620nFAD_R^{2/3}\omega^{1/2}\nu^{-1/6}C_R^* \tag{8.3.33}$$

Thus, for a nernstian reaction, combination of the Nernst equation for the O, R couple with the equations for the various currents and limiting currents yields the familiar voltammetric wave equation:

$$\boxed{E = E_{1/2} + \frac{RT}{nF} \ln \frac{(i_{l,c} - i)}{(i - i_{l,a})}} \tag{8.3.34}$$

where

$$\boxed{E_{1/2} = E^{0'} + \frac{RT}{nF} \ln \left(\frac{D_R}{D_O}\right)^{2/3}} \tag{8.3.35}$$

Note that the shape of the wave for a totally reversible reaction is independent of ω. Thus, since i_l varies as $\omega^{1/2}$, i at any potential should vary as $\omega^{1/2}$. A deviation of a plot of i vs. $\omega^{1/2}$ from a straight line that intersects the origin suggests some kinetic step involved in the electron transfer reaction. For example, for a totally irreversible reaction [see (3.2.5)], the disk current is

$$i_D = nFAk_f(E)C_O(y = 0) \tag{8.3.36}$$

where $k_f(E) = k^0 \exp[-\alpha n_a F(E - E^{0'})/RT]$. From (8.3.31)

$$i = nFAk_f(E)C_O^*\left[1 - \left(\frac{i}{i_{l,c}}\right)\right] \tag{8.3.37}$$

or, with rearrangement and defining

$$i_K = nFAk_f(E)C_O^* \tag{8.3.38}$$

(where i_K represents the current in the absence of any mass transfer effects, e.g., at the foot of the wave of the irreversible electron transfer),

$$\boxed{\frac{1}{i} = \frac{1}{i_K} + \frac{1}{i_{l,c}} = \frac{1}{i_K} + \frac{1}{0.620nFAC_O^* D_O^{2/3}\nu^{-1/6}\omega^{1/2}}} \tag{8.3.39}$$

Clearly $i/\omega^{1/2}C$ is a constant only when i_K [or $k(E)$] is very large. When this is not the case a plot of i *vs.* $\omega^{1/2}$ will be curved and tend toward the limit $i = i_K$ as $\omega^{1/2} \rightarrow \infty$ (Figure 8.3.6). A plot of $1/i$ *vs.* $1/\omega^{1/2}$ should be linear and can be extrapolated to $\omega^{-1/2} = 0$ to yield $1/i_K$. Determination of i_K at different values of E then allows determination of the kinetic parameters k^0 and α (Figure 8.3.7). A typical application of this procedure is illustrated in Figure 8.3.8, which shows such plots for the reduction of O_2 to HO_2^- at a gold electrode in alkaline solution.

The general *i-E* equation (i.e., for a quasi-reversible reaction) can be derived in a similar manner. Thus the *i-η* equation (3.5.10) can be written

$$\frac{i}{i_0} = \left[\frac{C_O(y = 0)}{C_O^*}\right]b^{-\alpha} - \left[\frac{C_R(y = 0)}{C_R^*}\right]b^{1-\alpha} \tag{8.3.40}$$

where $b = \exp(nF\eta/RT)$. This equation combined with (8.3.31) and (8.3.32) yields

$$\frac{1}{i} = \frac{b^\alpha}{1 - b}\left(\frac{1}{i_0} + \frac{b^{-\alpha}}{i_{l,c}} - \frac{b^{1-\alpha}}{i_{l,a}}\right) \tag{8.3.41}$$

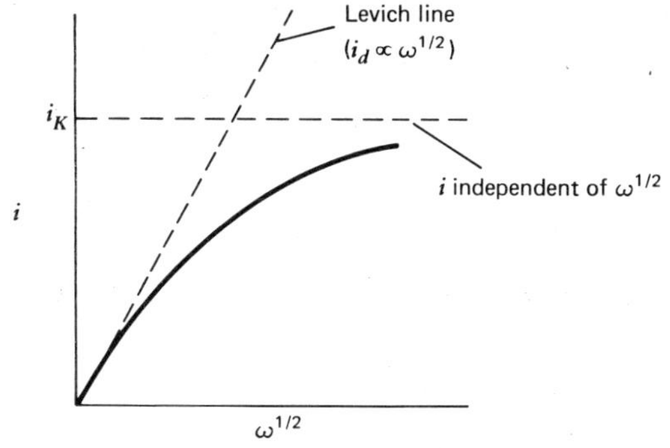

Figure 8.3.6
Variation of i with $\omega^{1/2}$ at RDE for electrode reaction with slow kinetics (at constant E_D).

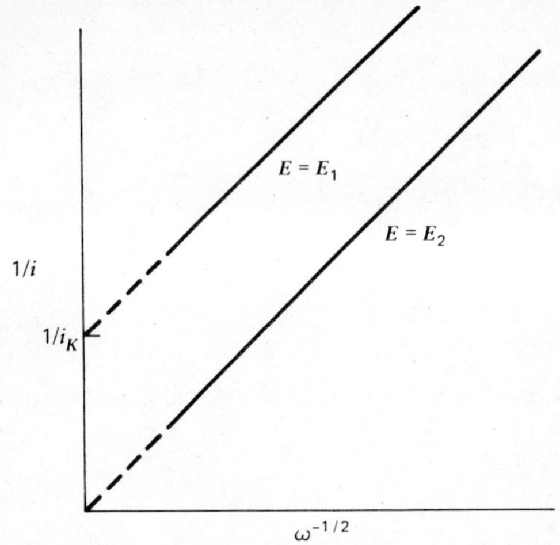

Figure 8.3.7
Variation of $1/i$ with $\omega^{-1/2}$ at potential E_1, where rate of electron transfer is sufficiently slow to act as a limiting factor, and at E_2 where electron transfer is rapid, for example, in limiting current region of curve. In both cases the slope of the lines is $(0.620\, nFAC_O^* D_O^{2/3} \nu^{-1/6})^{-1}$.

$$\boxed{\frac{1}{i} = \frac{b^\alpha}{1-b}\left[\frac{1}{i_0} + \frac{1}{0.62nFA\nu^{-1/6}\omega^{1/2}}\left(\frac{b^{-\alpha}}{D_O^{2/3}C_O^*} + \frac{b^{1-\alpha}}{D_R^{2/3}C_R^*}\right)\right]} \tag{8.3.42}$$

Thus $1/i$ *vs.* $\omega^{-1/2}$ at a given value of η is predicted to be linear for this case as well, and the intercept of this plot allows the determination of kinetic parameters.

Some alternate forms of (8.3.39) and (8.3.42) are sometimes given in the literature and are listed here for convenience. If the more general kinetic relation (3.2.8) is used in the derivation, then the equation for $1/i$ at the disk becomes

$$\frac{1}{i} = \frac{1}{nFA(k_f C_O^* - k_b C_R^*)}\left[1 + \frac{D_O^{-2/3}k_f + D_R^{-2/3}k_b}{0.62\nu^{-1/6}\omega^{1/2}}\right] \tag{8.3.43}$$

If the reverse (e.g., anodic) reaction can be ignored, then (8.3.43) yields

$$i = \frac{nFAk_f C_O^*}{1 + k_f/0.62\nu^{-1/6}D_O^{2/3}\omega^{1/2}} = \frac{nFAk_f C_O^*}{1 + k_f \delta_O/D_O} \tag{8.3.44}$$

where δ_O is as defined in (8.3.25). This equation is useful in defining the conditions for kinetic or mass transfer control at the RDE. When $k_f \delta_O/D_O \ll 1$, the current is completely under kinetic (or activation) control. When $k_f \delta_O/D_O \gg 1$, the mass-transfer-controlled equation results. Thus if the RDE is to be used for kinetic measurements, $k_f \delta_O/D_O$ should be small, say less than 0.1; that is, $k_f \leq 0.1 D_O/\delta_O$. Applications of RDE techniques to electrochemical problems have been reviewed (6–9).

8.3.5 Current Distribution at the RDE

In the preceding derivations, we assumed that the resistance of the solution was very small; under these conditions the current density is uniform across the disk, that is, it is independent of the radial distance. Although this is frequently the case, the actual current distribution will depend on the solution resistance as well as the mass and

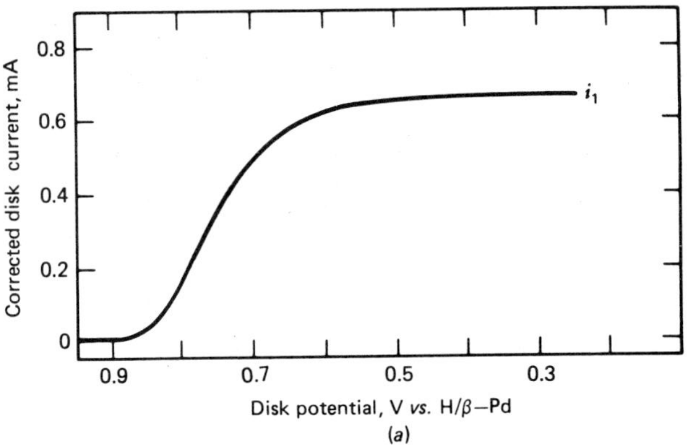

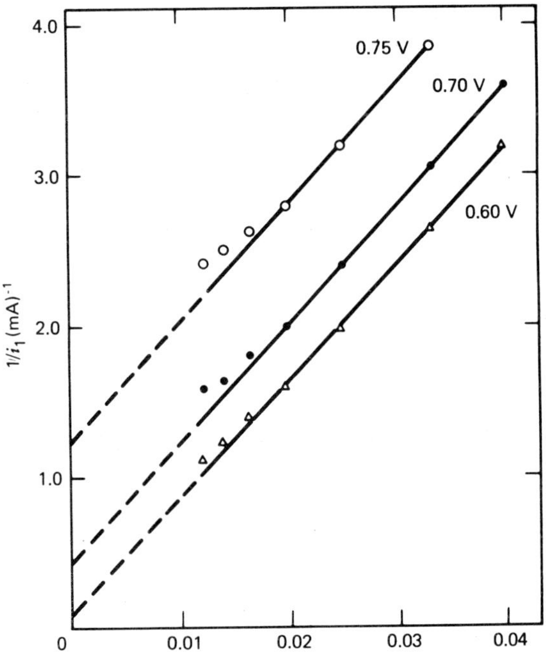

Figure 8.3.8

(*a*) i_D *vs.* E plot at 2500 rpm and (*b*) plot of $1/i_1$ *vs.* $\omega^{-1/2}$ for the reduction of O_2 to HO_2^- at a gold electrode in O_2-saturated (1.0 mM) 0.1 M NaOH at an RDE (A = 0.196 cm^2). The potential was swept at 1 V/min and T = 26 °C. (i_1 represents the corrected current attributable to O_2 reduction.) [From R. W. Zurilla, R. K. Sen, and E. Yeager, *J. Electrochem. Soc.*, **125**, 1103 (1978). Reprinted by permission of the publisher, The Electrochemical Society, Inc.]

charge transfer parameters of the electrode reaction. This topic has been treated by Newman (10) and discussed by Albery and Hitchman (11).

Consider first the *primary current distribution*, which represents the distribution when the surface overpotentials (activation and concentration) are neglected and the electrode is taken as an equipotential surface. Under such conditions for a disk electrode of radius r_1 embedded in a large insulating plane with a counter electrode at infinity, the potential distribution is as shown in Figure 8.3.9. The current flows in a direction perpendicular to the equipotential surfaces, and the current density is not uniform across the disk surface, being much larger at the edges ($r = r_1$) than at the center ($r = 0$). This arises because the ionic flux at the edges occurs from the sides as well as from a direction normal to the disk. The total current flowing to the disk under total resistive control is (4, 10)

$$i = 4\kappa r_1(\Delta E) \tag{8.3.45}$$

where κ is the specific conductivity of the bulk solution, and ΔE is the potential

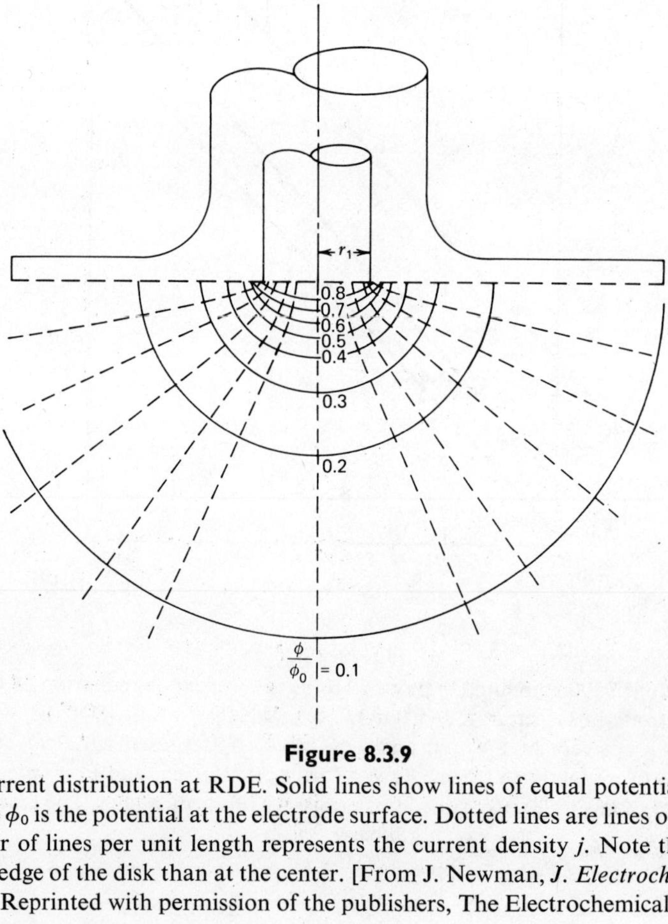

Figure 8.3.9

Primary current distribution at RDE. Solid lines show lines of equal potential at values of ϕ/ϕ_0, where ϕ_0 is the potential at the electrode surface. Dotted lines are lines of current flow. The number of lines per unit length represents the current density j. Note that j is higher toward the edge of the disk than at the center. [From J. Newman, *J. Electrochem. Soc.*, **113**, 501 (1966). Reprinted with permission of the publishers, The Electrochemical Society, Inc.]

difference in solution between the disk and counter electrode. Thus the overall resistance, R_Ω, is

$$R_\Omega = 1/4\kappa r_1 \tag{8.3.46}$$

When electrode kinetics and mass transfer effects are included, the current distribution (now called the *secondary current distribution*) is more nearly uniform than the primary one. Albery and Hitchman (11) have shown that the current distribution can be considered in terms of the dimensionless parameter ρ, given by

$$\rho = \frac{R_\Omega}{R_E} \tag{8.3.47}$$

where R_E is the electrode resistance due to both charge transfer and concentration polarization. The secondary current distribution as a function of ρ is shown in Figure 8.3.10. Note that as $\rho \to \infty$ (i.e., high solution resistance and small R_E), the current distribution approaches the primary one. Conversely, for small values of ρ (highly conductive solutions and large R_E) a fairly uniform current distribution is obtained. To avoid a nonuniform distribution, the conditions must be such that $\rho < 0.1$ (11). By taking

$$R_E + R_\Omega = \frac{dE}{di} \tag{8.3.48}$$

(where dE/di is the slope of the current-potential curve at a given value of E) and combining with (8.3.46) and (8.3.47), we obtain the condition for a uniform distribution (11):

$$\frac{di}{dE} < 0.36 r_1 \kappa \tag{8.3.49}$$

A plot of the values of di/dE that satisfy this condition at different values of r_1 and κ, taken from Albery and Hitchman, is shown in Figure 8.3.11. Note that at the limiting current, di/dE approaches zero, so that a uniform current distribution is always obtained in this circumstance.

8.3.6 Applicable Range of ω for RDE

The equations derived for the RDE will not apply at very small or very large values of ω. When ω is small, the hydrodynamic boundary layer $[y_h \simeq 3(\nu/\omega)^{1/2}]$ becomes large, and when it approaches the disk radius r_1, the approximations break down. Thus the lower limit for ω is obtained from the condition $r_1 > 3(\nu/\omega)^{1/2}$; that is, $\omega > 10\nu/r_1^2$. For $\nu = 0.01$ cm^2/sec and $r_1 = 0.1$ cm, ω should be larger than 10 sec^{-1}. Another problem occurs when recording *i-E* curves at the RDE at low values of ω. The derivation involved an assumed steady-state concentration at the electrode surface (i.e., $\partial C_O/\partial t = 0$). Thus the rate at which the electrode potential is scanned (V/sec) must be small with respect to ω to allow the steady-state concentrations to be achieved. If the scan rate is too large for a given ω, the *i-E* curves will not have the S-shape predicted by, for example, (8.3.34), but will instead show a peak, as in linear scan voltammetry at a stationary electrode. The question of transients and the time response at the RDE is dealt with further in Section 8.5.1.

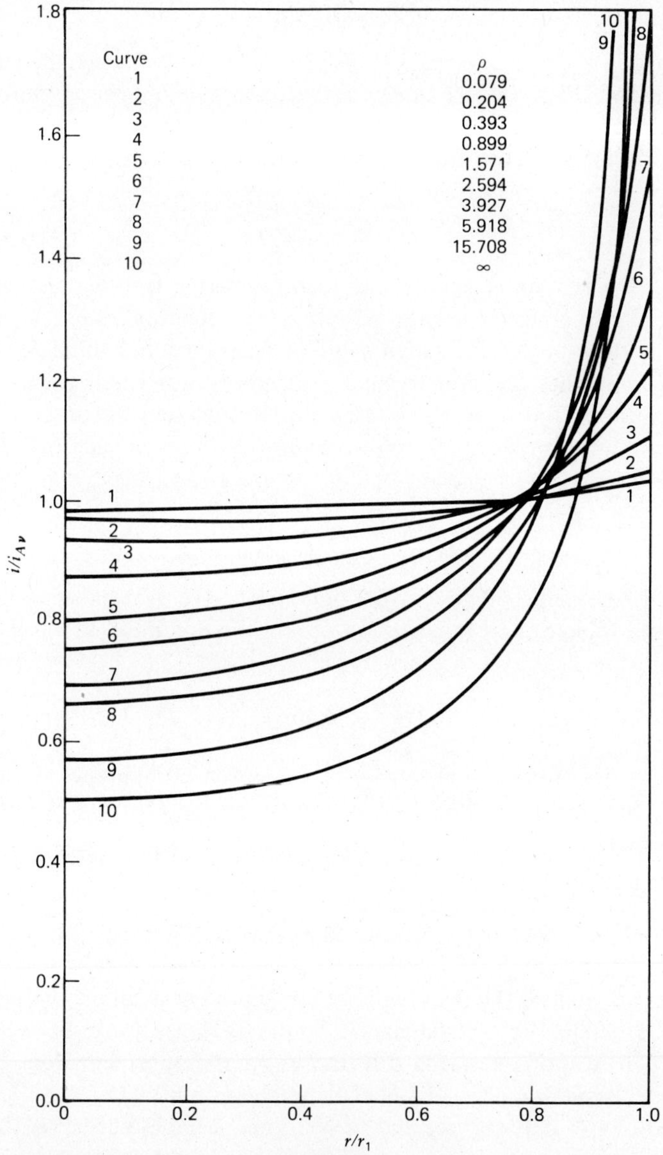

Figure 8.3.10

Secondary current distribution at an RDE. [From J. Newman, *J. Electrochem. Soc.*, **113**, 1235 (1966) as modified by W. J. Albery and M. L. Hitchman, "Ring-Disc Electrodes," Clarendon Press, Oxford, 1971, Chapter 4, with permission of the publishers, The Electrochemical Society, Inc., and Oxford University Press.]

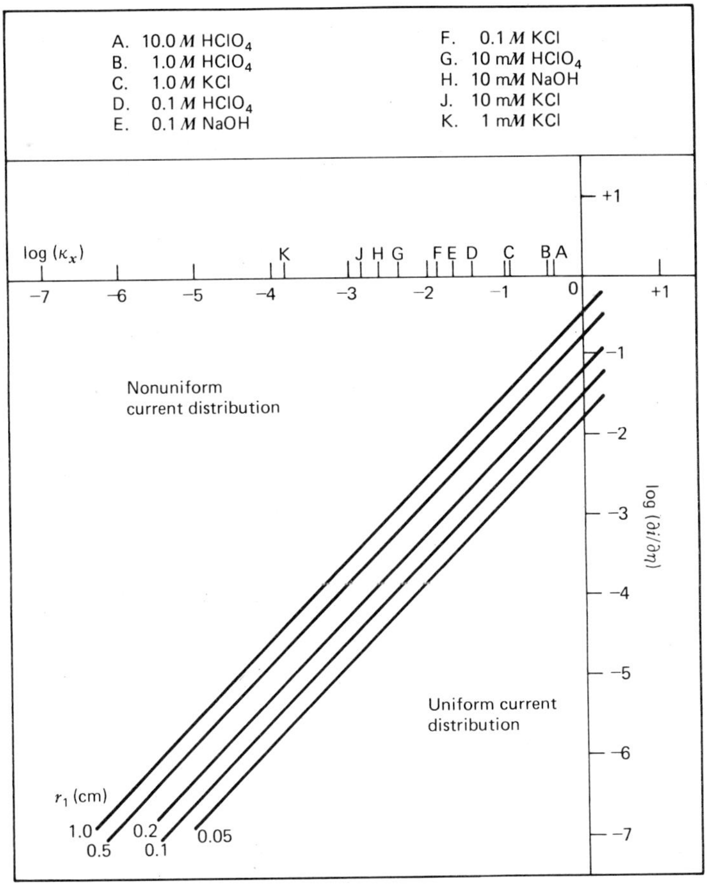

Figure 8.3.11

Diagnostic plot for the uniformity of current distribution at an RDE. Some typical background electrolytes at 25 °C in aqueous solution marked on log (κ_∞) scale in the figure. *Note.* $di/d\eta$ is in units of Ω^{-1}; κ_∞ is bulk electrolyte conductivity in Ω^{-1} cm^{-1}. [From W. J. Albery and M. L. Hitchman, "Ring-Disc Electrodes," Clarendon Press, Oxford, 1971, Chapter 4, with permission.]

The upper limit for ω is governed by the onset of turbulent flow. This occurs at the RDE at a Reynolds number, Re_{Cr}, above about 2×10^5 (6, 9). In this system, v_{ch} is the velocity of the edge of the disk ωr_1, and the characteristic distance l is r_1 itself. Thus, from (8.2.10),

$$\text{Re} = \frac{v_{\text{ch}} l}{\nu} = \frac{\omega r_1^2}{\nu} \tag{8.3.50}$$

the condition for nonturbulent flow is $\omega < 2 \times 10^5 \nu / r_1^2$. For the r_1 and ν values assumed, ω should be less than 2×10^5 sec^{-1}. The transition to turbulent flow can occur at much lower values of ω when the surface of the disk is not perfectly polished, when there are small bends or eccentricities in the RDE shaft, or when the cell walls are too close to the electrode surface. Also, at very high rotation rates, excessive splashing or vortex formation around the electrode occurs. In practice the maximum rotation rates are frequently set at 10,000 rpm or at $\omega \approx 1000\ \text{sec}^{-1}$. Thus in most RDE studies the ranges of ω and f are given by: $10\ \text{sec}^{-1} < \omega < 1000\ \text{sec}^{-1}$ or $100\ \text{rpm} < f < 10{,}000\ \text{rpm}$.

8.4 ROTATING RING AND RING-DISK ELECTRODES

Reversal techniques are obviously not available with the RDE, since the product of the electrode reaction, R, is continuously swept away from the surface of the disk. Thus at the RDE, reversal of the direction of the potential sweep, under conditions that the scan rate is sufficiently slow compared to ω (i.e., when no peak is seen during the forward scan), will just retrace the i-E curve of the forward scan. Information equivalent to that obtained by reversal techniques at a stationary electrode is obtained by the addition of an independent ring electrode surrounding the disk (Figure 8.4.1). By

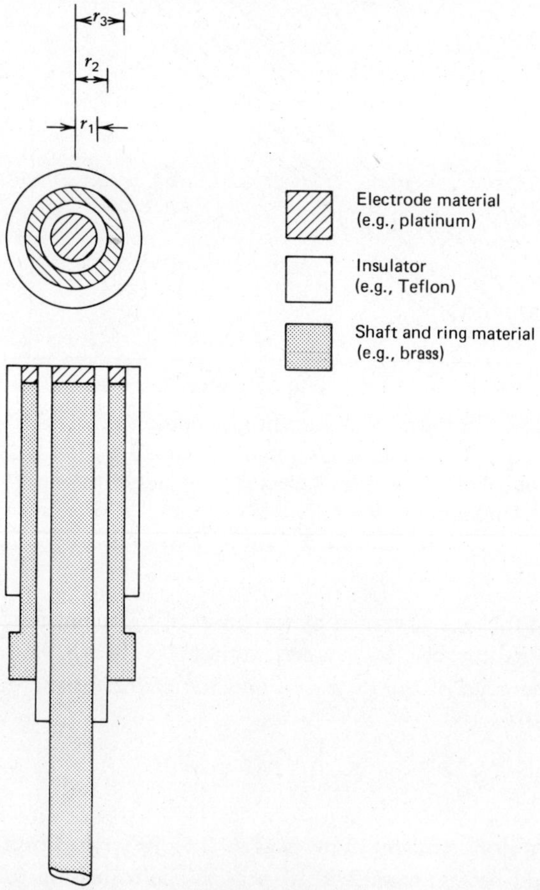

Figure 8.4.1
Ring-disk electrode.

measuring the current at the ring electrode with the potential maintained at a given value, some knowledge about what is occurring at the disk electrode surface can be obtained. For example, if the potential of the ring is held at a value at the foot of the $O + ne \rightarrow R$ wave, product R formed at the disk will be swept over to the ring by the radial flow streams where it will be oxidized back to O (or "collected"). The ring alone can also be used as an electrode (the rotating ring electrode); for example, when the disk is disconnected. The mass transfer to a ring electrode is larger than that to a disk at a given ω, because flow of fresh solution to it occurs radially from the area inside the ring, as well as normally from the bulk solution. The theoretical treatment of ring electrodes is more complicated than that of the RDE, since the radial mass transfer term must be included in the convective-diffusion equation. While the mathematics sometimes become difficult in these problems, the results are quite easy to understand and to apply. The problem will only be outlined here; details of the mathematical treatment are given in References 6 and 11.

8.4.1 Rotating Ring Electrode

Consider a ring electrode with an inner radius r_2 and outer radius r_3 [$A_r = \pi(r_3^2 - r_2^2)$]. This can be the RRDE illustrated in Figure 8.4.1 with the disk electrode at open circuit. When this electrode is rotated at an angular velocity, ω, the solution flow velocity profile is that discussed in Section 8.3.1. The steady-state, convective-diffusion equation that must be solved in this case is

$$v_r\left(\frac{\partial C_O}{\partial r}\right) + v_y\left(\frac{\partial C_O}{\partial y}\right) = D_O\left(\frac{\partial^2 C_O}{\partial y^2}\right) \tag{8.4.1}$$

This is obtained from (8.3.12). As with the RDE, symmetry considerations require that the concentrations be independent of ϕ, so that the derivatives in ϕ vanish. In addition, the mass transfer by diffusion in the radial direction, represented by the terms $D_O[(\partial^2 C_O/\partial r^2) + (1/r)(\partial C_O/\partial r)]$, is, at usual flow rates, small compared to convection in the radial direction, $(v_r\, \partial C_O/\partial r)$, so that these terms are neglected. The boundary conditions for the limiting ring current are:

$$C_O = C_O^* \qquad \text{for} \qquad y \rightarrow \infty$$

$$C_O = 0 \text{ at } y = 0 \qquad \text{for} \qquad r_2 \leq r < r_3$$

$$\frac{\partial C_O}{\partial y} = 0 \text{ at } y = 0 \qquad \text{for} \qquad r < r_2$$

When the values of v_r and v_y are introduced [see (8.3.9) and (8.3.10)], we obtain

$$(B'ry)\left(\frac{\partial C_O}{\partial r}\right) - B'y^2\left(\frac{\partial C_O}{\partial y}\right) = D_O\left(\frac{\partial^2 C_O}{\partial y^2}\right) \tag{8.4.2}$$

$$r\left(\frac{\partial C_O}{\partial r}\right) - y\left(\frac{\partial C_O}{\partial y}\right) = \left(\frac{D_O}{B'}\right)\frac{1}{y}\left(\frac{\partial^2 C_O}{\partial y^2}\right) \tag{8.4.3}$$

where $B' = 0.51\omega^{3/2}\nu^{-1/2}$. The current at the ring electrode is given by†

$$i_R = nFD_O 2\pi \int_{r_2}^{r_3} \left(\frac{\partial C_O}{\partial y}\right)_{y=0} r\, dr \qquad (8.4.4)$$

The solution to these equations yields (12) the limiting ring current:

$$\boxed{i_{R,l,c} = 0.620nF\pi(r_3^3 - r_2^3)^{2/3} D_O^{2/3}\nu^{-1/6}\omega^{1/2}C_O^*} \qquad (8.4.5)$$

or, in general,

$$i_R = i_{R,l,c}\{[C_O^* - C_O(y=0)]/C_O^*\} \qquad (8.4.6)$$

This can be written in terms of the disk current (8.3.30), which would be observed under identical conditions for a disk of radius r_1 to yield

$$i_R = i_D \frac{(r_3^3 - r_2^3)^{2/3}}{r_1^2} \qquad (8.4.7)$$

or

$$\boxed{\frac{i_R}{i_D} = \beta^{2/3} = \left(\frac{r_3^3}{r_1^3} - \frac{r_2^3}{r_1^3}\right)^{2/3}} \qquad (8.4.8)$$

Notice that for given reaction conditions (C_O^* and ω) a ring electrode will produce a larger current than a disk electrode of the same area. Thus the analytical sensitivity of a ring electrode (i.e., the current caused by a mass-transfer-controlled reaction of an electroactive species divided by the residual current) is better than that of a disk electrode, and this is especially true of a thin ring electrode. However, constructing a rotating ring electrode is usually more difficult than an RDE.

8.4.2 The Rotating Ring-Disk Electrode

In the rotating ring-disk electrode (RRDE), the current-potential characteristics of the disk electrode are unaffected by the presence of the ring, and the properties of the disk are as described in Section 8.3. (In fact, if in use the disk current changes upon variation of the ring potential or current, a defective RRDE or undesirable coupling of the ring and disk through solution uncompensated resistance is suggested). Since RRDE experiments involve the examination of two potentials (that of the disk, E_D, and that of the ring, E_R) and two currents (disk, i_D, and ring, i_R), the representation

† The area of an infinitesimal section of ring of thickness δr at a radius r is $\pi(r + \delta r)^2 - \pi r^2 \simeq 2\pi r\delta r$, assuming the term $\pi(\delta r)^2$ is negligible. The current through this section is

$$\frac{(i_R)_{\delta r}}{nFA} = \frac{(i_R)_{\delta r}}{nF2\pi r\delta r} = D_O\left(\frac{\partial C_O}{\partial y}\right)_{y=0}$$

The total ring current is the summation of $(i_R)_{\delta r}$:

$$i_R = \sum_{r=r_2}^{r_3} (i_R)_{\delta r} = nFD_O 2\pi \sum_{r=r_2}^{r_3} (\partial C_O/\partial y)_{y=0} r\, \delta r$$

which yields, as $\delta r \to 0$, eq. (8.4.4).

of the results involves more dimensions than that of experiments involving a single working electrode. RRDE experiments are usually carried out with a *bipotentiostat* (Section 13.4.4), which allows separate adjustment of E_D and E_R (Figure 8.4.2*a*). However, since most RRDE measurements involve steady-state conditions, it is possible to use an ordinary potentiostat to control the ring circuit and a simple floating power supply in the disk circuit (Figure 8.4.2*b*). Several different types of experiments are possible at the RRDE; *collection* experiments, where the disk-generated species is observed at the ring, and *shielding* experiments, where the flow of bulk electroactive species to the ring is perturbed because of the disk reaction, are the most frequent.

(a) Collection Experiments. Consider the experiment in which the disk is held at a potential E_D where the reaction $O + ne \rightarrow R$ occurs and produces a cathodic current i_D and the ring is maintained at a sufficiently positive potential, E_R, so that any R that reaches the ring is oxidized in the reaction $R \rightarrow O + ne$, with the concentration of R at the ring surface essentially zero. We are interested in the magnitude of the ring current i_R under these conditions, that is, how much of the disk-generated R is collected at the ring. We again must solve the steady-state ring convective-diffusion equation (8.4.3), this time for species R:

$$r\left(\frac{\partial C_{\mathrm{R}}}{\partial r}\right) - y\left(\frac{\partial C_{\mathrm{R}}}{\partial y}\right) = \left(\frac{D_{\mathrm{R}}}{B'}\right)\frac{1}{y}\left(\frac{\partial^2 C_{\mathrm{R}}}{\partial y^2}\right) \tag{8.4.9}$$

The boundary conditions are

1. At the disk ($0 \leq r < r_1$) the flux of R is related to that of O by the usual conservation equation:

$$D_{\mathrm{R}}\left(\frac{\partial C_{\mathrm{R}}}{\partial y}\right)_{y=0} = -D_{\mathrm{O}}\left(\frac{\partial C_{\mathrm{O}}}{\partial y}\right)_{y=0} \tag{8.4.10}$$

Thus, from the results in Section 8.3.2,

$$\left(\frac{\partial C_{\mathrm{R}}}{\partial y}\right)_{y=0} = \frac{-i_D}{nFAD_{\mathrm{R}}} = \frac{-i_D}{\pi r_1{}^2 nFD_{\mathrm{R}}} \tag{8.4.11}$$

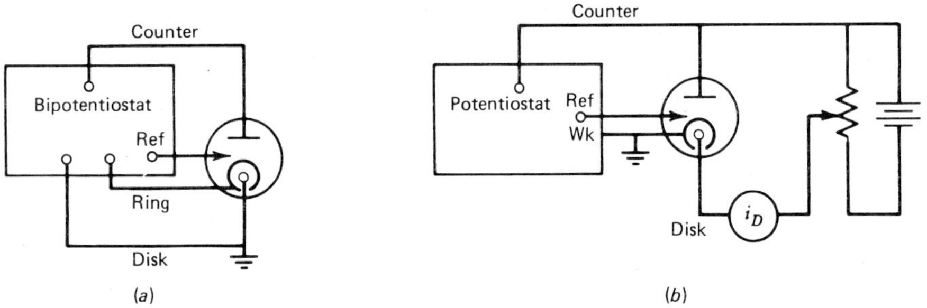

Figure 8.4.2
Block diagram of RRDE apparatus. (*a*) Bipotentiostat. (*b*) Ordinary (three-electrode) potentiostat and voltage divider.

2. In the insulating gap region ($r_1 \le r < r_2$) no current flows, so that

$$\left(\frac{\partial C_R}{\partial y}\right)_{y=0} = 0 \tag{8.4.12}$$

3. At the ring ($r_2 \le r < r_3$), under limiting current conditions,

$$C_R(y = 0) = 0 \tag{8.4.13}$$

We assume R is initially absent from the bulk solution ($\lim_{y \to \infty} C_R = 0$) and the bulk concentration of O is C_O^*. As in (8.4.4) the ring current is given by

$$i_R = nFD_R 2\pi \int_{r_2}^{r_3} \left(\frac{\partial C_R}{\partial y}\right)_{y=0} r\, dr \tag{8.4.14}$$

The mathematics of the problem involve solving the problem in terms of dimensionless variables in the various zones using the Laplace transform method to obtain solutions in terms of Airy functions (13, 14). The result is that the ring current is related to the disk current by a quantity N, *the collection efficiency*; this can be calculated from the electrode geometry, since it depends only on r_1, r_2, and r_3 and is *independent* of ω, C_O^*, D_O, D_R, etc:

$$\boxed{N = \frac{-i_R}{i_D}} \tag{8.4.15}$$

N is calculated by using the rather messy equation:

$$N = 1 - F(\alpha/\beta) + \beta^{2/3}[1 - F(\alpha)] - (1 + \alpha + \beta)^{2/3}\{1 - F[(\alpha/\beta)(1 + \alpha + \beta)]\} \tag{8.4.16}$$

where $\alpha = (r_2/r_1)^3 - 1$, β is given by (8.4.8), and the F values are defined by

$$F(\theta) = \left(\frac{\sqrt{3}}{4\pi}\right) \ln\left\{\frac{(1 + \theta^{1/3})^3}{1 + \theta}\right\} + \frac{3}{2\pi} \arctan\left(\frac{2\theta^{1/3} - 1}{3^{1/2}}\right) + \frac{1}{4} \tag{8.4.17}$$

The function $F(\theta)$ and values of N for different ratios r_2/r_1 and r_3/r_2 are tabulated in Reference 13. N can also be determined experimentally for a given electrode, by measuring $-i_R/i_D$ for a system where R is stable; once N is determined, it is a constant known value for that RRDE. For example, for an RRDE with $r_1 = 0.187$ cm, $r_2 = 0.200$ cm, and $r_3 = 0.332$ cm, $N = 0.555$, that is, 55.5% of the product generated at the disk is collected at the ring. Qualitatively N becomes larger as the gap thickness ($r_2 - r_1$) decreases and as the ring size ($r_3 - r_2$) increases. The concentration profiles of R in the vicinity of the RRDE surface are shown in Figure 8.4.3. Typically, in a collection experiment, plots of i_D and i_R as functions of E_D (at a constant E_R) are plotted (Figure 8.4.4*a*). Stability of the product is assured if N is independent of i_D and ω. If R decomposes at a rate sufficiently high that some is lost in its passage from disk to ring, the collection efficiency will be smaller than the N determined for that electrode and will be a function of ω, i_D, or C_O^*. Information about the rate and mechanism of decay of R can thus be obtained from RRDE collection experiments (see Chapter 11). Information about the reversibility of the electrode reaction can be

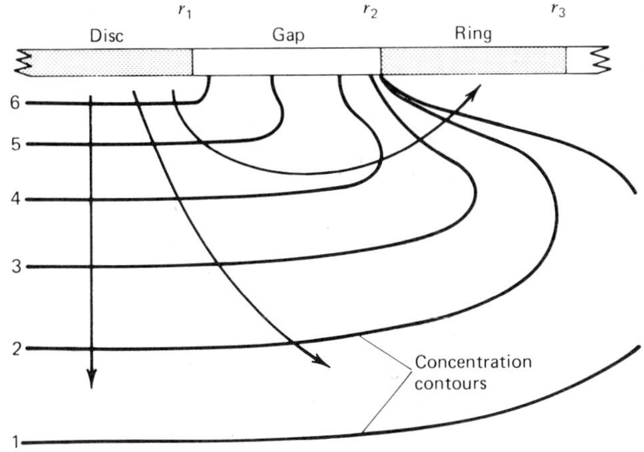

Figure 8.4.3

Schematic concentration profiles of species R at RRDE. Concentrations increase in proceeding from 1 to 6. Note for $0 < r < r_1$, $\partial C_R/\partial r = 0$; gap ($r_1 < r < r_2$), $y = 0$, $(\partial C_R/\partial y)_{y=0} = 0$ and at ring surface ($r_2 < r < r_3$), $y = 0$, $C_R(y = 0) = 0$. [From W. J. Albery and M. L. Hitchman, "Ring-Disc Electrodes," Clarendon Press, Oxford, 1971, Chapter 3, with permission.]

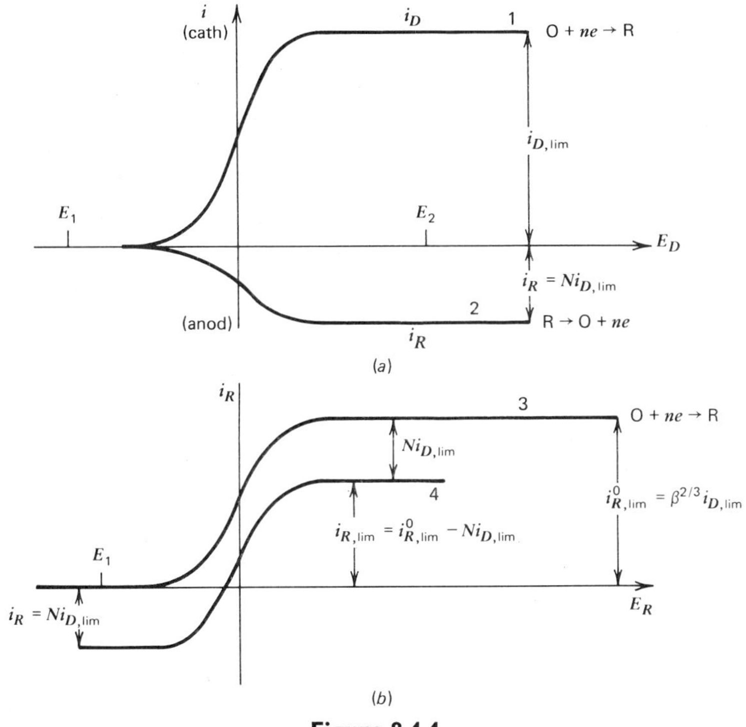

Figure 8.4.4

(*a*) Disk voltammogram. (1), i_D *vs.* E_D and (2) i_R *vs.* E_D with $E_R = E_1$. (*b*) Ring voltammograms. (3) i_R *vs.* E_R, $i_D = 0$ ($E_D = E_1$) and (4) i_R *vs.* E_R, $i_D = i_{D,l,c}$ ($E_D = E_2$).

obtained by plotting the ring voltammogram (i_R *vs.* E_R) at a constant value of E_D, and comparing the $E_{1/2}$ with that of the disk voltammogram (Figure 8.4.4*b*).

(b) Shielding Experiments. The current at the ring electrode for the reduction of O to R when the disk is at open circuit is given by (8.4.5) to (8.4.8). The limiting current at the ring with $i_D = 0$, denoted $i^0_{R,l}$, compared to the limiting current at the disk electrode, $i_{D,l}$, is given by (8.4.8), which is rewritten as

$$i^0_{R,l} = \beta^{2/3} i_{D,l} \tag{8.4.18}$$

If the disk current is changed to a finite value i_D, the flux of O to the ring will be decreased. The extent of this decrease will be the same as the flux of stable product R to the ring in a collection experiment—Ni_D. Hence the limiting ring current for any i_D, $i_{R,l}$ is given by

$$\boxed{i_{R,l} = i^0_{R,l} - Ni_D} \tag{8.4.19}$$

(This equation holds for any value of i_D, including $i_D = 0$ and $i_D = i_{D,l}$.) Using (8.4.18), we have for the special case $i_D = i_{D,l}$,

$$\boxed{i_{R,l} = i^0_{R,l}(1 - N\beta^{-2/3})} \tag{8.4.20}$$

Thus, when the disk current is at its limiting value, the ring current is decreased by the factor $(1 - N\beta^{-2/3})$. This factor, always less than unity, is called the *shielding factor*. These relations are easier to understand when the complete $i - E$ curves are considered (Figure 8.4.4*b*). One sees that the effect of switching i_D from 0 to $i_{D,l}$ is to shift the entire ring voltammogram (i_R *vs.* E_R), which is assumed to be reversible, by the amount $Ni_{D,l}$.

8.5 TRANSIENTS AT THE RDE AND RRDE

Although a major advantage of rotating disk electrode techniques, compared to stationary electrode methods, is the ability to make measurements at steady state without the need of considering the time of electrolysis, the observation of current transients at the disk or ring following a potential step can sometimes be of use in understanding an electrochemical system. For example the adsorption of a component, A, on the disk electrode can be studied by noticing the transient shielding of the ring current for the electrolysis of A upon stepping the disk potential to a value where A is adsorbed.

8.5.1 Transients at the RDE

The solution of the non-steady-state problem at the RDE requires solution of the usual disk convective-diffusion equation (8.3.14), but with inclusion of the $\partial C/\partial t$ term, that is,

$$\frac{\partial C_O}{\partial t} = D_O\left(\frac{\partial^2 C_O}{\partial y^2}\right) - B'y^2\left(\frac{\partial C_O}{\partial y}\right) \tag{8.5.1}$$

$B' = 0.51\omega^{3/2}\nu^{-1/2}$. This has been accomplished by approximation methods (15, 16) and by digital simulation (17). Thus, for a potential step to the limiting current region of the *i-E* curve, the instantaneous value of i_l, denoted $i_l(t)$, is given approximately by (15)

$$R(t) = \frac{i_l(t)}{i_l(ss)} = 1 + 2\sum_{m=1}^{\infty} \exp\left(\frac{-m^2\pi^2 D_O t}{\delta_O^2}\right) \tag{8.5.2}$$

where $i_l(ss)$ is the value of i_l as $t \to \infty$, and δ_O is given in (8.3.25). An implicit approximate equation for $R(t)$ obtained by the "method of moments" has also been proposed (16):

$$\frac{D_O t}{\delta_O^2} = \frac{1}{6}\left(\frac{1.8049}{1.6116}\right)^2\left\{\frac{1}{2}\ln\left(\frac{1 - R(t)^3}{[1 - R(t)]^3}\right) + \sqrt{3}\left[\frac{\pi}{6} - \arctan\left(\frac{2R(t) + 1}{\sqrt{3}}\right)\right]\right\} \tag{8.5.3}$$

Both of these results are in good agreement with the digital simulation (17); a typical disk transient is shown in Figure 8.5.1. At short times, when the diffusion layer thickness is much thinner than δ_O, the potential step transient follows that for a stationary electrode [equation 5.2.11]. The time required for the current to attain its steady-state value can be obtained from the curve in Figure 8.5.1. The current is within 1% of $i_l(ss)$ at a time τ when

$$\omega\tau(D/\nu)^{1/3}(0.51)^{2/3} \geq 1.3 \tag{8.5.4}$$

or, taking $(D/\nu)^{1/3} \simeq 0.1$, when $\omega\tau \gtrsim 20$. Thus for $\omega = 100$ sec (or a rotation rate of about 1000 rpm), $\tau \simeq 0.2$ sec.

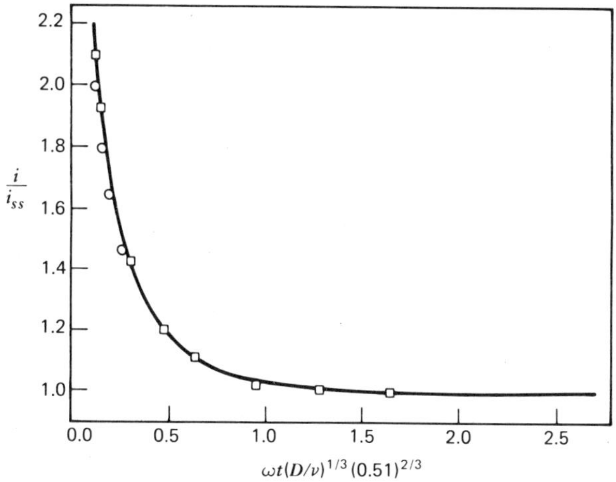

Figure 8.5.1

Simulated disk current transient for potential step at disk: ○ theoretical work of Bruckenstein and Prager; □ theoretical work of Siver. [From K. B. Prater and A. J. Bard, *J. Electrochem. Soc.*, **117**, 207 (1970), with permission of the publisher, The Electrochemical Society, Inc.]

8.5.2 Transients at the RRDE

Consider the experiment in which the ring of the RRDE is maintained at a potential where oxidation of species R to O can occur, and the disk is at open circuit or at a potential where no R is produced. If R is now generated at the disk by a potential step to an appropriate value or by a constant current step, a certain time will be required for R to transit the gap from the outside of the disk to the inside edge of the ring (the *transit time*, t'). An additional time will be required until the disk current attains its steady-state value. The rigorous solution for the ring current transient $[i_R(t)]$ involves solving the non-steady-state form of (8.4.9):

$$\frac{\partial C_R}{\partial t} = D_R\left(\frac{\partial^2 C_R}{\partial y^2}\right) + B'y^2\left(\frac{\partial C_R}{\partial y}\right) - B'ry\left(\frac{\partial C_R}{\partial r}\right) \tag{8.5.5}$$

This rather difficult problem is discussed by Albery and Hitchman (18), and several approaches to the solution and approximate equations are given. These may also be obtained by the digital simulation method (see Appendix B.5) (17), and typical simulated ring current transients for both a current step [to $i_{D,l}(ss)$] and a potential step to the limiting current region at the disk electrode are shown in Figure 8.5.2. Note that the ring current rises more rapidly when a potential step is applied. This effect can be attributed to the large instantaneous current that flows at the disk electrode when the potential is stepped (Figure 8.5.1).

An approximate value for the transit time can be obtained by the method suggested by Bruckenstein and Feldman (19). The radial velocity near the electrode surface is given by (8.3.10), which can be written

$$v_r = \frac{dr}{dt} = 0.51\omega^{3/2}\nu^{-1/2}ry \tag{8.5.6}$$

A molecule of R generated at the edge of the disk ($r = r_1$) must diffuse normal to the disk to reach the ring, since v_r at $y = 0$ is zero. It is then swept in a radial direction

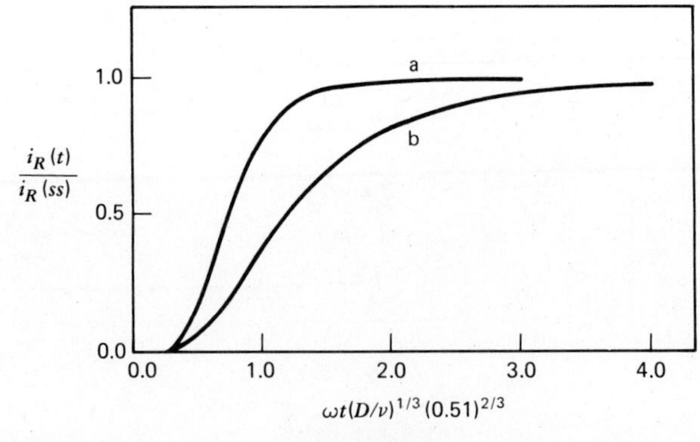

Figure 8.5.2

Simulated ring current transients. Curve a: Potential step at the disk. Curve b: Current step at the disk. [From K. B. Prater and A. J. Bard, *J. Electrochem. Soc.*, **117**, 207 (1970), with permission of the publisher, The Electrochemical Society, Inc.]

and then moves by diffusion and convection in the y direction to reach the inner edge of the ring. This path can be described by some average trajectory and some time-dependent distance y from the electrode surface. Integration of (8.5.6) yields

$$\ln\left(\frac{r_2}{r_1}\right) = 0.51\omega^{3/2}\nu^{-1/2}\int_0^{t'} y\,dt \tag{8.5.7}$$

If one uses the approximation $y \simeq \sqrt{Dt}$, substitutes this value in (8.5.7), and carries out the integration, the result is†

$$\omega t' = 3.58(\nu/D)^{1/3}\left[\log\left(\frac{r_2}{r_1}\right)\right]^{2/3} \tag{8.5.8}$$

With $(D/\nu)^{1/3} = 0.1$, $\omega = 100\ \text{sec}^{-1}$, and an electrode with $r_2/r_1 = 1.07$ (which represents a rather narrow gap), the transit time from (8.5.8) is about 30 msec. Simulated ring transients for electrodes of different geometries are shown in Figure 8.5.3; the points on the curves represent the t' values calculated from (8.5.8). This t' more closely represents the time required for the ring current to attain about 2% of the steady value.

A study of ring current transients can be useful in determining adsorption of a disk-generated intermediate at the disk, since this will cause a delay in the appearance of this species at the ring (20). The transients can also be used in a qualitative way to study electrode processes. Consider a "shielding transient" experiment described by Bruckenstein and Miller (21) concerning the reduction of oxygen at a copper ring–platinum disk electrode. Oxygen is reduced more readily on Pt than on Cu. If the

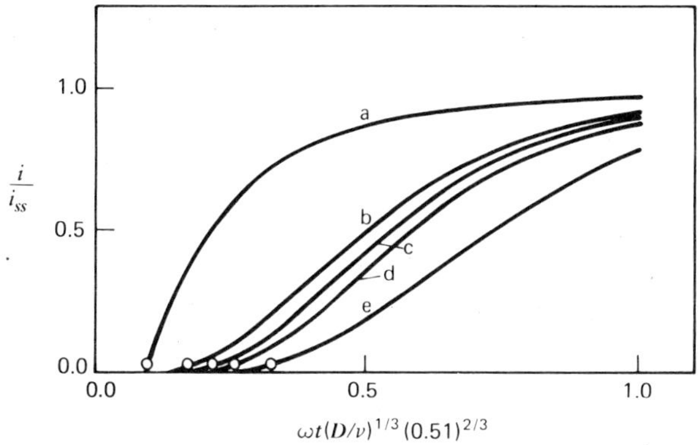

Figure 8.5.3

Simulated ring transients for electrodes of different geometries (r_1, r_2, r_3). Curve a: 2.00, 2.04, 2.08. Curve b: 1.00, 1.05, 1.07. Curve c: 1.00, 1.07, 1.48. Curve d: 1.00, 1.09, 1.52. Curve e: 0.83, 0.94, 1.59. Points show transit times, t', as calculated from (8.5.8). [From K. B. Prater and A. J. Bard, *J. Electrochem. Soc.*, **117**, 207 (1970), with permission of the publisher, The Electrochemical Society, Inc.]

† Different choices of y (e.g., $2\sqrt{Dt}$) or a different method of carrying out the integration will result in an alternate value of the constant. For example, Bruckenstein and Feldman (19) considered the case in which the species diffuses outward for $t < t'/2$, then back toward the surface for $t'/2 < t \leq t'$. Then the constant in (8.5.8) is 4.51.

electrode is immersed in a solution containing 0.2 M H_2SO_4 and 2×10^{-5} M Cu(II) and saturated with oxygen, with E_D at 1.00 V *vs.* SCE, no reaction occurs at the disk. If E_R is held at -0.25 V *vs.* SCE, a cathodic ring current due to the reaction $Cu(II) + 2e \rightarrow Cu$ of about 11 μA flows (Figure 8.5.4). Oxygen is not reduced at the copper ring at this potential (even though the reversible potential for the reaction $O_2 + 4H^+ + 4e \rightarrow H_2O$ is much more positive) because the reaction rate is very slow. If E_D is now stepped to 0.0 V, a cathodic disk current of about 700 μA flows. This current represents reduction of oxygen at the platinum disk and also plating of copper by reduction of the Cu(II). (This plating of copper on the platinum substrate occurs at potentials more positive than those where copper would deposit on bulk copper and is called *underpotential deposition.*) The reduction of Cu(II) at the disk shields the ring, so that a drop in i_R is observed. As copper deposits on the disk, however, the reduction of oxygen is hindered, and the disk current falls. After about a monolayer of copper has deposited on the disk, further underpotential deposition is no longer possible, Cu(II) reduction at the disk ceases, and the ring current returns to its unshielded value. This rather simple experiment demonstrates quite clearly the "poisoning" of the oxygen reduction process by copper. Incidentally, Cu(II) is a rather common impurity in distilled water and mineral acids, and this experiment demonstrates that underpotential deposition of a monolayer of copper from solutions containing as little as 1 ppm Cu can drastically affect the behavior of a platinum electrode. Adsorption of small amounts of other impurities (i.e., organic molecules) can also have an effect on solid electrode behavior. Thus electrochemical experiments often require making great efforts and taking precautions to maintain solution purity.

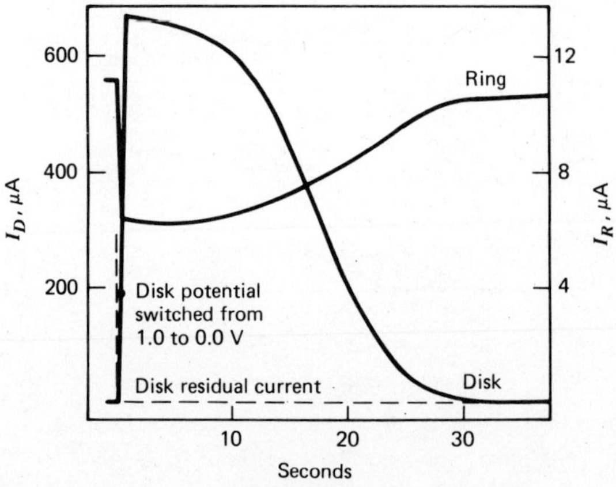

Figure 8.5.4

Time dependence of the oxygen-reduction current at the disk and the Cu(II) reduction current at the ring of a platinum ring-disk electrode. Solution: 0.2 M H_2SO_4 and 2×10^{-5} M Cu(II) with air saturation. Rotation speed 2500 rpm. Disk potential held at 0.00 V *vs.* SCE for $t > 0$; ring potential $= -0.25$ V *vs.* SCE at all t. Disk area 0.458 cm^2, $\beta^{2/3} = 0.36$, collection efficiency 0.183. [Reprinted with permission from S. Bruckenstein and B. Miller, *Accounts Chem. Res.*, **10**, 54 (1977). Copyright 1977, American Chemical Society.]

8.6 HYDRODYNAMICALLY MODULATED RDE

In all of the methods discussed so far in this chapter, we assumed that the rotation rate of the electrode was constant and at its steady-state value of ω when measurements of currents were carried out. However, it can also be useful to measure currents under conditions when the ω value is changing with time.

The simplest case involves a variation of ω as a function of time (e.g., $\omega \propto t^2$) and an automatic plotting of i_D *vs.* $\omega^{1/2}$. These "automated Levich plots" can be of value compared to manual (and possibly more precise) point-by-point measurements, when the electrode surface is changing with time (e.g., during an electrodeposition, or with impurity or product adsorption) and a rapid scan is needed. References to this technique and related methods are given in Reference 22.

A very useful technique involving changes of ω with time features the sinusoidal variation of $\omega^{1/2}$ [called *sinusoidal hydrodynamic modulation* (22)]. Consider a RDE whose rotation rate is varied sinusoidally about a fixed center speed, $\omega_0^{1/2}$, at a frequency σ so that the instantaneous value of $\omega^{1/2}$ is given by

$$\omega^{1/2} = \omega_0^{1/2} + \Delta\omega^{1/2} \sin(\sigma t) \tag{8.6.1}$$

For example, if $\omega_0^{1/2} = 19.4 \text{ sec}^{-1/2}$ (rotation rate, 3600 rpm), $\Delta\omega^{1/2} = 1.94 \text{ sec}^{-1/2}$ (36 rpm), and the modulation frequency is 3 Hz [$\sigma = 2\pi(3)$], then the rotation rate varies between 3636 and 3564 rpm three times per second (Figure 8.6.1). $\Delta\omega$ is always

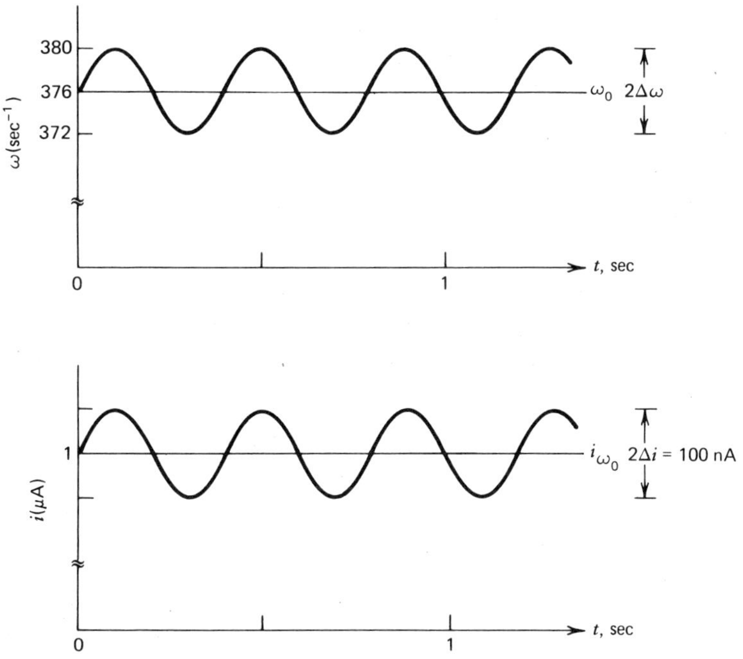

Figure 8.6.1
Representation of $\Delta\omega$ and Δi for a sinusoidally hydrodynamically modulated RDE.

smaller than ω_0, and is usually only about 1% of ω_0. The usable modulation frequency depends on the inertia and response of the motor-electrode system and is usually 3 to 6 Hz. If the system follows the Levich equation (8.3.22), the current is given by

$$i(t) = A'[\omega_0^{1/2} + \Delta\omega^{1/2} \sin(\sigma t)] \tag{8.6.2}$$

where $A' = 0.620nFAD_O^{2/3}\nu^{-1/6}C_O^* = i_{\omega_0}/\omega_0^{1/2}$. Thus

$$i(t) = i_{\omega_0}\left[1 + \left(\frac{\Delta\omega}{\omega_0}\right)^{1/2} \sin \sigma t\right] \tag{8.6.3}$$

or the amplitude Δi of the modulated current is†

$$\boxed{\Delta i = (\Delta\omega/\omega_0)^{1/2} i_{\omega_0}} \tag{8.6.4}$$

This varying component of the disk current is most conveniently recorded after filtering and passage through a lock-in amplifier or full-wave rectifier (Figure 8.6.2). Although the Δi value is much smaller in magnitude than i_{ω_0}, it has the very important advantage of being free from any contributions that arise from factors that do not depend on the mass transfer rate. Thus Δi is essentially free from double-layer charging components and processes associated with oxidation and reduction of the electrode or of adsorbed species, and is relatively insensitive to the anodic and cathodic background currents. Thus sinusoidal hydrodynamic modulation is a useful technique for

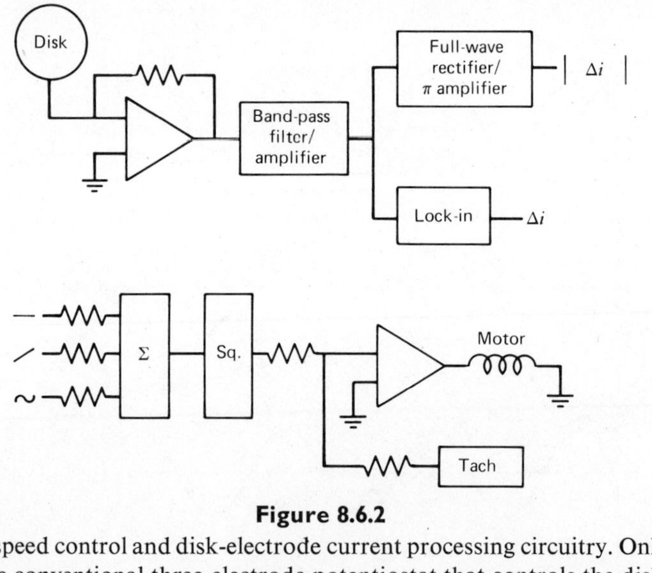

Figure 8.6.2
Schematic of speed control and disk-electrode current processing circuitry. Only the current follower of the conventional three-electrode potentiostat that controls the disk electrode is shown. [Reprinted with permission from B. Miller and S. Bruckenstein, *Anal. Chem.*, **46**, 2026 (1974), Copyright 1974, American Chemical Society.]

† This equation was derived for limiting current conditions. It is also valid for currents in the rising portion of a reversible wave. Note that it will not be valid if the rate of change in ω is comparable to the rate of hydrodynamic relaxation. Thus $\Delta\omega/\omega$ and σ are usually kept small.

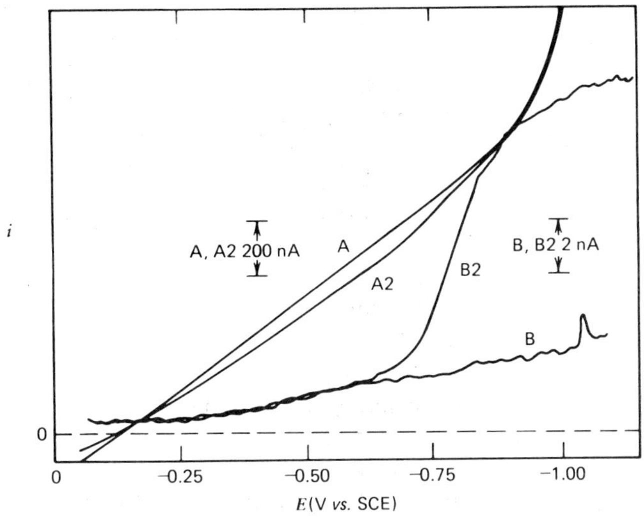

Figure 8.6.3

Controlled potential cathodic scans of Tl(I) at amalgamated gold disk. All A traces are RDE curves, all B traces are HMRDE curves. (A, B) 0.01 M $HClO_4$; (A2, B2) 2.0×10^{-7} M Tl^+ in 0.01 M $HClO_4$. Current sensitivities indicated by markers, zero current for all curves is the dashed line. For all curves, $\omega_0^{1/2} = 60$ rpm$^{1/2}$. B, $\Delta\omega^{1/2} = 6$ rpm$^{1/2}$, $\sigma/2\pi =$ 3 Hz, averaging time constant is 3 sec, and scan rate is 2 mV/sec. [Reprinted with permission from B. Miller and S. Bruckenstein, *Anal. Chem.*, **46**, 2026 (1974). Copyright 1974, American Chemical Society.]

the determination of very low (submicromolar) concentrations with the RDE, for studies in the presence of surface complications, and for measurements near the background limits of solvent or supporting electrolyte oxidation or reduction. For example, a study of the reduction of 0.2 μM Tl(I) at a RDE with an amalgamated gold disk (which is frequently used when studies at a mercury surface are desired) is shown in Figure 8.6.3 (23). Note that while the faradaic current for the Tl(I) reduction cannot be distinguished from the residual current on the i_D-E scan, a clear reduction wave is found by measuring Δi. It is also possible to obtain kinetic information from hydrodynamic modulation experiments, by employing σ values where deviations from Levich behavior occur (24).

8.7 REFERENCES

1. V. G. Levich, "Physicochemical Hydrodynamics," Prentice-Hall, Englewood Cliffs, N.J., 1962.
2. R. B. Bird, W. E. Stewart, and E. N. Lightfoot, "Transport Phenomena," Wiley, New York, 1960.
3. J. N. Agar, *Disc. Faraday Soc.*, **1**, 26 (1947).
4. J. Newman, *Electroanal. Chem.*, **6**, 187 (1973).

5. J. S. Newman, "Electrochemical Systems," Prentice-Hall, Englewood Cliffs, N.J., 1973.
6. A. C. Riddiford, *Adv. Electrochem. Electrochem. Eng.*, **4**, 47 (1966).
7. R. N. Adams, "Electrochemistry at Solid Electrodes," Marcel Dekker, New York, 1969, pp. 67–114.
8. J. S. Newman, *J. Phys. Chem.*, **70**, 1327 (1966).
9. V. Yu. Filanovsky and Yu. V. Pleskov, *Prog. in Surf. Membrane Sci.*, **10**, 27 (1976).
10. J. Newman, *J. Electrochem. Soc.*, **113**, 501, 1235 (1966).
11. W. J. Albery and M. L. Hitchman, "Ring-Disc Electrodes," Clarendon Press, Oxford, 1971, Chapter 4.
12. V. G. Levich, *op. cit.*, p. 107.
13. W. J. Albery and M. Hitchman, *op. cit.*, Chapter 3.
14. W. J. Albery and S. Bruckenstein, *Trans. Faraday Soc.*, **62**, 1920 (1966).
15. Yu G. Siver, *Russ. J. Phys. Chem.*, **33**, 533 (1959).
16. S. Bruckenstein and S. Prager, *Anal. Chem.*, **39**, 1161 (1967).
17. K. B. Prater and A. J. Bard, *J. Electrochem. Soc.*, **117**, 207 (1970).
18. W. J. Albery and M. Hitchman, *op. cit.*, Chapter 10.
19. S. Bruckenstein and G. A. Feldman, *J. Electroanal. Chem.*, **9**, 395 (1965).
20. S. Bruckenstein and D. T. Napp, *J. Am. Chem. Soc.*, **90**, 6303 (1968).
21. S. Bruckenstein and B. Miller, *Accounts Chem. Res.*, **10**, 54 (1977).
22. S. Bruckenstein and B. Miller, *J. Electrochem. Soc.*, **117**, 1032 (1970).
23. B. Miller and S. Bruckenstein, *Anal. Chem.*, **46**, 2026 (1974).
24. K. Tokuda, S. Bruckenstein, and B. Miller, *J. Electrochem. Soc.*, **122**, 1316 (1975).

8.8 PROBLEMS

8.1 Consider an RDE with a disk radius r_1 of 0.20 cm, immersed in an aqueous solution of a substance A ($C_A^* = 10^{-2}\ M$, $D_A = 5 \times 10^{-6}$ cm²/sec), and rotated at 1000 rpm. A is reduced in a 1e reaction. $\nu \approx 0.01$ cm²/sec. Calculate: v_r and v_y at a distance 0.1 cm normal to the disk surface at the edge of the disk; v_r and v_y at the electrode surface; U_0; $i_{l,c}$; m_A; δ_A; and the Levich constant.

8.2 What dimensions (r_2 and r_3) can a rotating ring electrode have to produce the same limiting current as an RDE with $r_1 = 0.20$ cm? (Note that many possible combinations of r_2 and r_3 are suitable.) What is the area of the ring electrode?

8.3 From the data in Figure 8.3.8, calculate the diffusion coefficient for O_2 in 0.1 M NaOH, and k_f for the reduction of oxygen at 0.75 V (assumed totally irreversible). Take $\nu = 0.01$ cm²/sec.

8.4 Current-potential curves at an RRDE electrode for a solution of 5 mM $CuCl_2$ in 0.5 M KCl showing (1) i_D *vs.* E_D and (2) i_R *vs.* E_R are given in Figure 8.8.1. $N = 0.53$.

(a) Analyze these data to determine D, β, and any other possible information about the electrode reaction at the first reduction step [Cu(II) + e → Cu(I)].

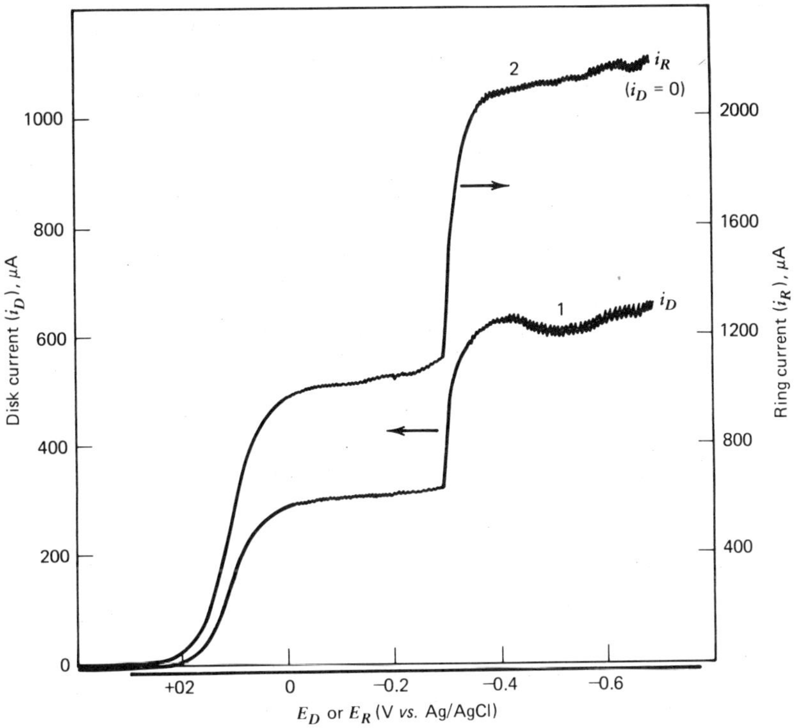

Figure 8.8.1

Voltammograms at RRDE showing: curve 1, i_D *vs.* E_D and curve 2, i_R *vs.* E_R (at $i_D = 0$) for a solution containing 5 mM $CuCl_2$ and 0.5 M KCl. $\omega = 201$ sec^{-1}; disk area = 0.0962 cm^2; $\nu = 0.011$ cm^2/sec.

(b) If the ring voltammogram was obtained with $E_D = -0.10$ V, what value of $i_{R,l,c}$ (at $E_R = -0.10$ V) is expected?
(c) If E_R is held at $+0.4$ V and $E_D = -0.10$ V, what value of i_R is expected?
(d) What process occurs at the second wave? Account for this wave shape.
(e) Assume the ring is held at $+0.4$ V. Sketch the expected plot of i_R *vs.* E_D as E_D is scanned from $+0.4$ to -0.6 V.

8.5 The ring voltammograms for an RRDE in 5 mM $K_3Fe(CN)_6$ and 0.1 M KCl are shown in Figure 8.8.2. From the information in these curves, calculate N for this electrode and D [for $Fe(CN)_6^{3-}$]. What is the predicted slope of a curve of i_D *vs.* $\omega^{1/2}$? What are the predicted values of the limiting disk ($i_{D,l,c}$) and ring currents ($i_{R,l,c}$ with $i_D = 0$ and $i_D = i_{D,l,c}$) at 5000 rpm? Assume $\nu = 0.01$ cm^2/sec.

8.6 Experiments are performed at an RRDE with the following dimensions: $r_1 = 0.20$ cm, $r_2 = 0.22$ cm, $r_3 = 0.32$ cm. A disk voltammogram (i_D *vs.* E_D) is to be recorded for a rotation rate of 2000 rpm. What maximum potential sweep should be employed to prevent non-steady-state effects from occurring? What is the transit time with this electrode?

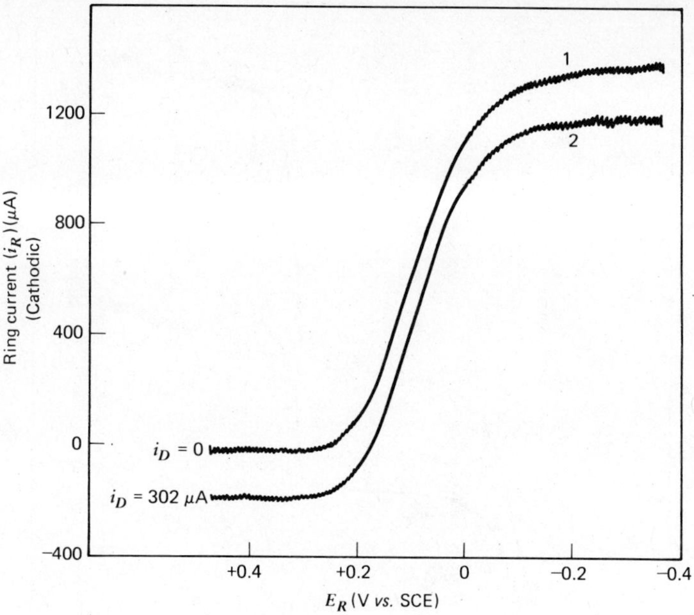

Figure 8.8.2

Ring voltammograms (i_R *vs.* E_R) with (1) $i_D = 0$ and (2) i_D at its limiting value, 302 μA for an RRDE electrode with $r_2 = 0.188$ cm and $r_3 = 0.325$ cm, rotated at 48.6 r/sec for a solution of 5.0 mM $K_3Fe(CN)_6$ and 0.10 M KCl.

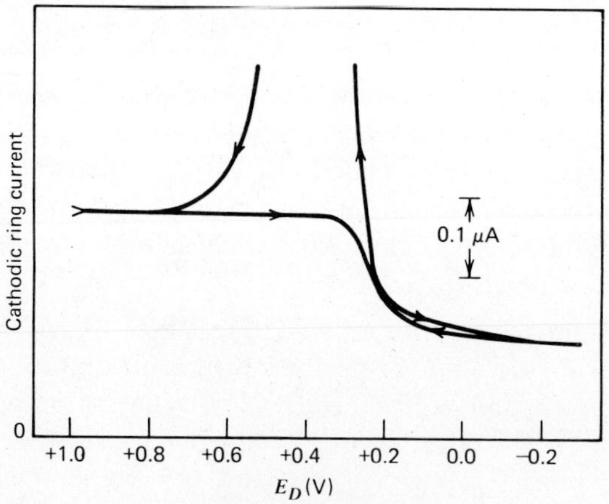

Figure 8.8.3

$i_{R,l,c}$ *vs.* E_D for reduction of Bi(III) to Bi(0) for a solution containing 4.86×10^{-7} M Bi(III) and 0.1 M HNO_3. The ring electrode was held at -0.25 V and the disk potential scanned from $+1.0$ V at 200 mV/sec. For this electrode the slope of $i_{D,l,c}$ *vs.* $C^*_{\mathrm{Bi(III)}}$ is 0.934 μA/μM. [Reprinted with permission from S. Bruckenstein and P. R. Gifford, *Anal. Chem.*, **51**, 250 (1979). Copyright 1979, American Chemical Society.]

8.7 The diffusion coefficient can be obtained from a limiting current measurement at an RDE and a transient measurement (e.g., a potential step measurement) at the same electrode (at $\omega = 0$) under identical conditions without requiring a knowledge of electrode area, n, or C^*. Explain how this is accomplished and discuss the possible errors in this procedure.

8.8 S. Bruckenstein and P. R. Gifford [*Anal. Chem.*, **51**, 250 (1979)] proposed that ring electrode shielding measurements at the RRDE could be employed for the analysis of micromolar solutions by using the equation

$$\Delta i_{R,l} = 0.62 n F \pi r_1^2 D^{2/3} \nu^{-1/6} \omega^{1/2} N C^*$$

where $\Delta i_{R,l}$ represents the change in the limiting ring current for $i_D = 0$ and $i_D = i_{D,l}$. (a) Derive this equation. (b) A plot of $i_{R,l}$ *vs.* E_D is given in Figure 8.8.3 for the reduction of Bi(III) to Bi(0) in 0.1 M HNO_3. Mass-transfer-limited reduction of Bi(III) occurs at -0.25 V. From the data on the curve during the forward scan (E_D varied from $+1.0$ to -0.2 V), calculate N for this RRDE. Explain the large ring current transient observed during the reverse scan (E_D from -0.2 to $+1.0$ V).

chapter 9
Techniques Based on Concepts of Impedance

9.1 INTRODUCTION

In previous chapters we have discussed ways of studying electrode reactions through large perturbations on the system. By imposing potential sweeps, potential steps, or current steps, we generally drive the electrode to a condition far from equilibrium, and we observe the response, which is usually a transient signal. Another approach is to perturb the cell with an alternating signal of small magnitude and observe the way in which the system follows the perturbation at steady state. Many advantages accrue to these techniques. Among the most important are (a) an experimental ability to make high-precision measurements because the response may be indefinitely steady and can therefore be averaged over a long term, and (b) an ability to treat the response theoretically by linearized (or otherwise simplified) current-potential characteristics. Since one usually works close to equilibrium, one often does not require detailed knowledge about the behavior of the *i-E* response curve over great ranges of overpotential. This advantage leads to important simplifications in treating kinetics and diffusion.

In deriving the theory below, we will rely frequently on analogies between the electrochemical cell and networks of resistors and capacitors that are thought to behave like the cell. This feature may seem at times to disembody the interpretation from the chemical system, so let us emphasize beforehand that the ideas and the mathematics used in the interpretation are basically simple. We will do our best to tie them to the chemistry at every possible point, and we hope readers will avoid letting the details of interpretation obscure their view of the great power and beauty of these methods.

9.1.1 Types of Techniques (1–7)

The prototypical experiment is the *faradaic impedance* measurement, in which one fills the cell with a solution containing both forms of a redox couple, so that the working electrode is poised. For example, one might use 1 mM Eu^{2+} and 1 mM Eu^{3+}

in 1 M $NaClO_4$. A mercury drop of fixed area might be employed as the working electrode, and it might be paired with a nonpolarizable reference such as an SCE, which would act also as the counter electrode. The cell is inserted as the unknown impedance into an impedance bridge, and the bridge is balanced by adjusting R and C in the opposite arm of the bridge, as shown in Figure 9.1.1.

This operation determines the values of R and C that, in series, behave as the cell does at the measurement frequency. The job of theory is to interpret those equivalent resistance and capacitance values in terms of interfacial phenomena. The mean potential of the working electrode (the dc potential) is simply the equilibrium potential determined by the ratio of oxidized and reduced forms of the couple. Measurements are made at other potentials by preparing additional solutions with different concentration ratios. This method is capable of very high precision and is frequently used for the evaluation of heterogeneous charge transfer parameters and for studies of double-layer structure.

A variation on the faradaic impedance method is *ac voltammetry* (or, with a DME, *ac polarography*). In these experiments, a three-electrode cell is used in the conventional manner, and the potential program imposed on the working electrode is a dc mean value, E_{dc}, which is scanned slowly with time, plus a sinusoidal component, E_{ac}, of perhaps 5-mV, peak-to-peak amplitude. The measured responses are the magnitude of the ac component of the current at the frequency of E_{ac} and its phase angle with respect to E_{ac}.† A typical experimental arrangement is shown schematically in Figure 9.1.2. As we will see presently, this measurement is equivalent to determining

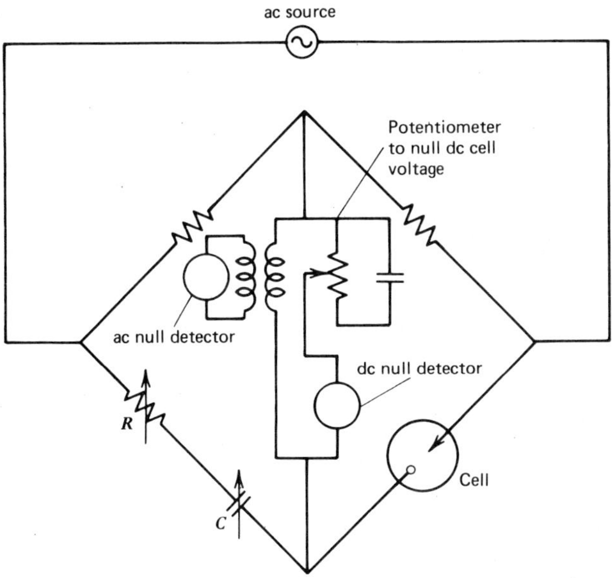

Figure 9.1.1
A bridge circuit for measurements of electrochemical cell impedances.

† Alternatively, one could measure the current components in phase with E_{ac} and 90° out of phase with E_{ac}. They provide equivalent information.

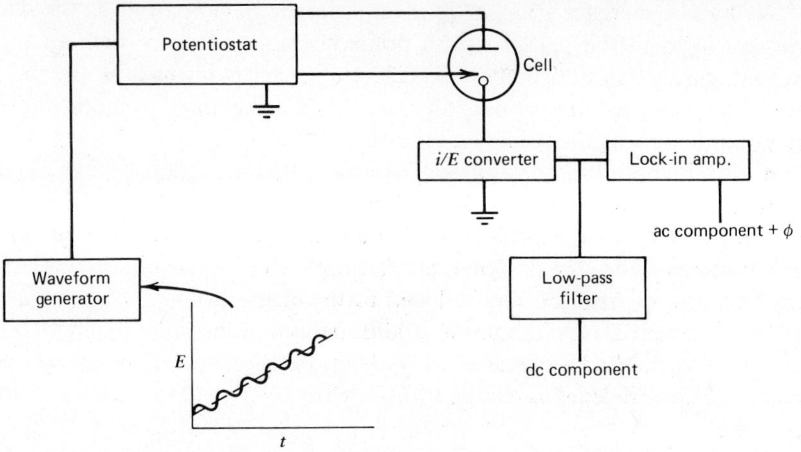

Figure 9.1.2
Schematic diagram of apparatus for an ac voltammetric experiment.

the faradaic impedance. The role of the dc potential is to set the mean surface concentrations of O and R. In general, this potential differs from the true equilibrium value; hence $C_O(0, t)$ and $C_R(0, t)$ differ from C_O^* and C_R^*, and a diffusion layer exists. Note, however, that since E_{dc} is effectively steady, this layer soon becomes so thick that its dimensions greatly exceed those of the diffusion zone affected by the rapid perturbations from E_{ac}. Thus, the mean surface concentrations $C_O(0, t)$ and $C_R(0, t)$ look like bulk concentrations to the ac part of the experiment. This same effect is exploited in differential pulse polarography (see Section 5.8.3). One usually starts with a solution containing only one redox form, for examples Eu^{3+}, and obtains continuous plots of the ac current amplitude and the phase angle *vs.* E_{dc}. In effect, these plots represent the faradaic impedances at continuous ratios of $C_O(0, t)$ and $C_R(0, t)$, all recorded without changing the solution. The amplitude plot is also useful for analytical measurements of concentration.

Faradaic impedance and conventional ac polarographic measurements involve excitation signals E_{ac} of very low amplitude, and they essentially depend on the fact that current-overpotential relations are virtually linear at low overpotentials. In a linear system, excitation at frequency ω provides a current also of frequency ω (and only of frequency ω). On the other hand, a nonlinear *i-E* relation gives a distorted response that is not purely sinusoidal. Even so, it is periodic and can be represented as a superposition (Fourier synthesis) of signals at frequencies ω, 2ω, 3ω, . . . , etc. The current-overpotential function for an electrode reaction is nonlinear over moderate ranges of overpotential, and the effects of this nonlinearity can be observed and put to use. For example, consider *second* (and higher) *harmonic ac voltammetry*, which is nearly the same as the first harmonic ac experiment described above. It differs in that one detects an ac component current at 2ω, 3ω, . . . , etc., instead of the component at the excitation frequency ω. *Faradaic rectification* features excitation with a purely sinusoidal source and measurement of the dc component of current flow. *Intermodulation voltammetry* depends on the mixing properties of a nonlinear characteristic.

One excites with two superimposed signals at frequencies ω_1 and ω_2 and observes the current at the combination frequencies (sidebands or beat frequencies) $\omega_1 + \omega_2$ or $\omega_1 - \omega_2$. A big advantage common to all techniques based on nonlinearity is comparative freedom from charging currents. The double-layer capacitance is generally much more linear than faradaic processes are, hence charging currents are largely restricted to the excitation frequencies.

9.1.2 Review of ac Circuits

A purely sinusoidal voltage can be expressed as

$$e = E \sin \omega t \tag{9.1.1}$$

where ω is the angular frequency, which is 2π times the conventional frequency in hertz. It is convenient to think of this voltage as a rotating vector (or *phasor*) quantity like that pictured in Figure 9.1.3. Its length is the amplitude E and its frequency of rotation is ω. The observed voltage e is the component of the phasor projected on some particular axis (e.g., that at 0°) at any time.

One frequently wishes to consider the relationship between two related sinusoidal signals, such as the current i and the voltage e. Each is then represented as a separate phasor, $\dot{I}$ or $\dot{E}$, rotating at the same frequency. As shown in Figure 9.1.4, they generally will not be in phase, and their phasors will be separated by a *phase angle*, ϕ. One of the phasors, usually $\dot{E}$, is taken as a reference signal, and ϕ is measured with respect to it. In the figure, the current lags the voltage. It can be expressed generally as

$$i = I \sin(\omega t + \phi) \tag{9.1.2}$$

where ϕ is a signed quantity, which is negative in this case.

The relationship between two phasors at the same frequency remains constant as they rotate, and hence the phase angle is constant. Consequently, we can usually drop the references to rotation in the phasor diagrams and study the relationships between phasors simply by plotting them as vectors having a common origin and separated by the appropriate angles.

Let us apply these concepts to the analysis of some simple circuits. Consider first a pure resistance R across which a sinusoidal voltage, $e = E \sin \omega t$, is applied. Since

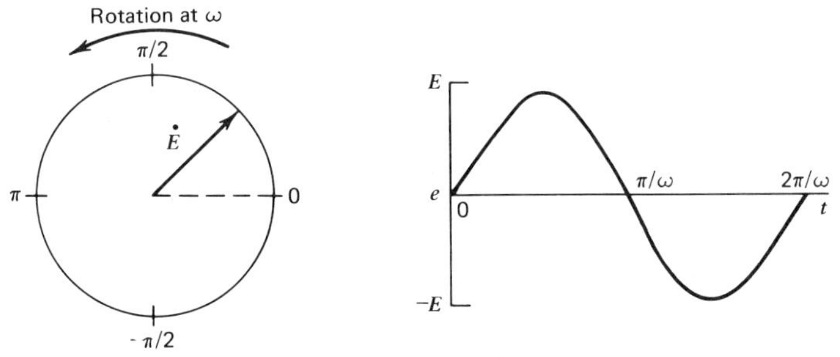

Figure 9.1.3
Phasor diagram for an alternating voltage $e = E \sin \omega t$.

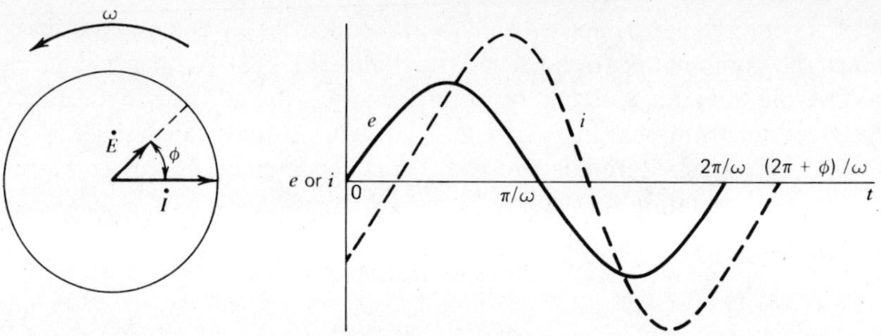

Figure 9.1.4
Phasor diagram showing the relationship between alternating current and voltage signals at frequency ω.

Ohm's law always holds, the current is $(E/R)\sin\omega t$ or, in phasor notation,

$$\dot{I} = \frac{\dot{E}}{R} \tag{9.1.3}$$

$$\dot{E} = \dot{I}R \tag{9.1.4}$$

The phase angle is zero, and the vector diagram is that of Figure 9.1.5.

Suppose we now substitute a pure capacitance C for the resistor. The fundamental relation of interest is then $q = Ce$, or $i = C(de/dt)$; thus

$$i = \omega CE \cos\omega t \tag{9.1.5}$$

$$i = \frac{E}{X_C}\sin\left(\omega t + \frac{\pi}{2}\right) \tag{9.1.6}$$

where X_C is the *capacitive reactance*, $1/\omega C$.

The phase angle is $\pi/2$, and the current leads the voltage, as shown in Figure 9.1.6. Since the vector diagram has now expanded to a plane, it is convenient to represent phasors in terms of complex notation. Components along the ordinate are assigned as imaginary and are multiplied by $j = \sqrt{-1}$. Components along the abscissa are real. Introducing complex notation here is only a bookkeeping measure to help keep the vector components straight. We handle them mathematically as "real" or "imaginary," but both types are real in the sense of being measurable by phase angle. In circuit analysis, it turns out to be advantageous to plot the current phasor along the abscissa

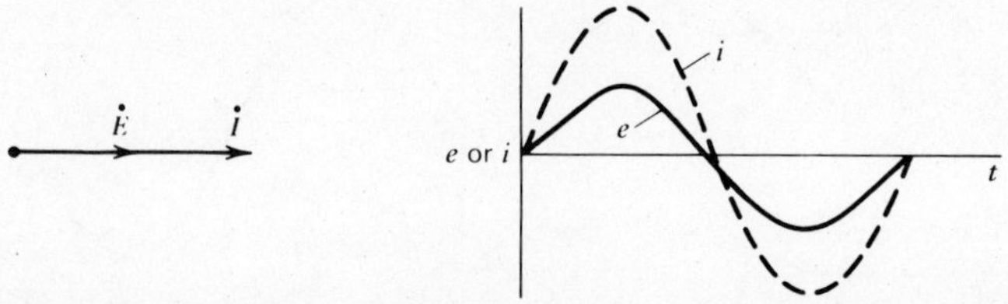

Figure 9.1.5
Relationship between the voltage across a resistor and current through the resistor.

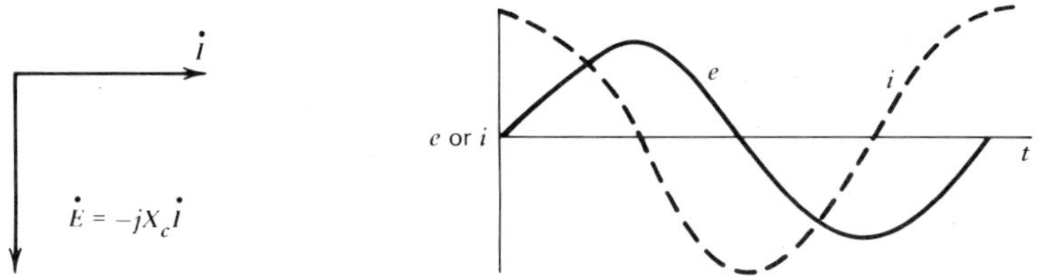

Figure 9.1.6
Relationship between an alternating voltage across a capacitor and the alternating current through the capacitor.

as shown in Figure 9.1.6, even though the current's phase angle is measured with respect to the voltage. If that is done, it is clear that

$$\dot{E} = -jX_C\dot{I} \tag{9.1.7}$$

Of course, this relation must hold regardless of where $\dot{I}$ is plotted with respect to the abscissa, since only the relation between $\dot{E}$ and $\dot{I}$ is significant. A comparison of equations 9.1.4 and 9.1.7 shows that X_C must carry dimensions of resistance but, unlike R, its magnitude falls with increasing frequency.

Now consider a resistance R and a capacitance C in series. A voltage $\dot{E}$ is applied across them, and at all times it must equal the sum of the individual voltage drops across the resistor and the capacitor; thus

$$\dot{E} = \dot{E}_R + \dot{E}_C \tag{9.1.8}$$

$$\dot{E} = \dot{I}(R - jX_C) \tag{9.1.9}$$

$$\dot{E} = \dot{I}\mathbf{Z} \tag{9.1.10}$$

Thus we find that the voltage is linked to the current through a vector $\mathbf{Z} = R - jX_C$ called the *impedance*. Figure 9.1.7 is a display of the relationships between these various quantities. The magnitude of $\mathbf{Z}$ is $(R^2 + X_C{}^2)^{1/2}$ and the phase angle is given by

$$\tan\phi = \frac{X_C}{R} = \frac{1}{\omega RC} \tag{9.1.11}$$

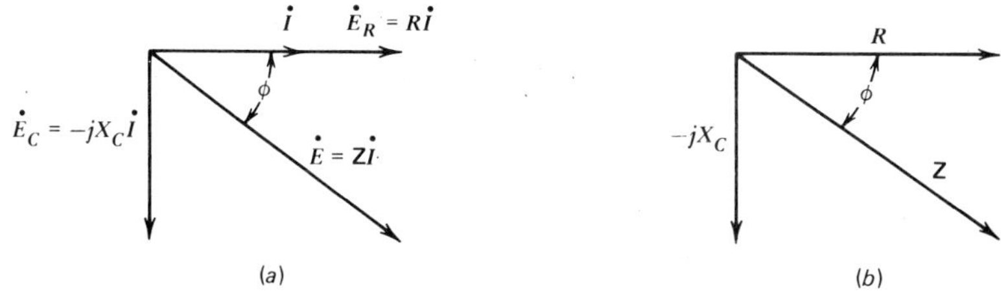

Figure 9.1.7
(*a*) Phasor diagram showing the relationship between the current and the voltages in a series RC network. $\dot{E}$ is the voltage across the whole network, and $\dot{E}_R$ and $\dot{E}_C$ are its components across the resistance and the capacitance. (*b*) An impedance vector diagram derived from the phasor diagram in (*a*).

The impedance is a kind of generalized resistance, and equation 9.1.10 is a generalized version of Ohm's law. It embodies both (9.1.4) and (9.1.7) as special cases. The phase angle expresses the balance between capacitive and resistive components in the series circuit. For a pure resistance, $\phi = 0$; for a pure capacitance, $\phi = \pi/2$; and for mixtures, intermediate phase angles are observed.

More complex circuits can be analyzed by combining impedances according to rules analogous to those applicable to resistors. For impedances in series, the overall impedance is the sum of the individual values (expressed as complex vectors). For impedances in parallel, the inverse of the overall impedance is the sum of the reciprocals of the individual vectors. Figure 9.1.8 shows a simple application.

Sometimes it is advantageous to analyze ac circuits in terms of the *admittance*, **Y**, which is the inverse impedance $1/\mathbf{Z}$, and therefore represents a kind of conductance. Our generalized Ohm's law (9.1.10) can then be rewritten as $\dot{I} = \dot{E}\mathbf{Y}$. These concepts are especially useful in the analysis of parallel circuits, because the overall admittance of parallel elements is simply the sum of the individual admittances.

Later we will be interested in the vector relationship between **Z** and **Y**. If **Z** is written in its polar form (Section A.5):

$$\mathbf{Z} = Z\, e^{j\phi} \tag{9.1.12}$$

then the admittance is

$$\mathbf{Y} = \frac{1}{Z}\, e^{-j\phi} \tag{9.1.13}$$

Here we see that **Y** is a vector with magnitude $1/Z$ and a phase angle equal to that of **Z**, but opposite in sign. Figure 9.1.9 is a picture of the arrangement.

9.1.3 Equivalent Circuit of a Cell (1, 4, 8, 9)

In a general sense, an electrochemical cell is simply an impedance to a small sinusoidal excitation; hence we ought to be able to represent its performance by an *equivalent circuit* of resistors and capacitors that pass current with the same amplitude and phase angle that the real cell does under a given excitation. A typical circuit is shown in Figure 9.1.10*a*. The parallel elements are introduced because the total current through the working interface is the sum of distinct contributions from the faradaic process i_f and double-layer charging i_c. The double-layer capacitance closely resembles a pure capacitance; hence it is represented in the equivalent circuit by the element C_d. The faradaic process must be considered as a general impedance Z_f. Of course, all the

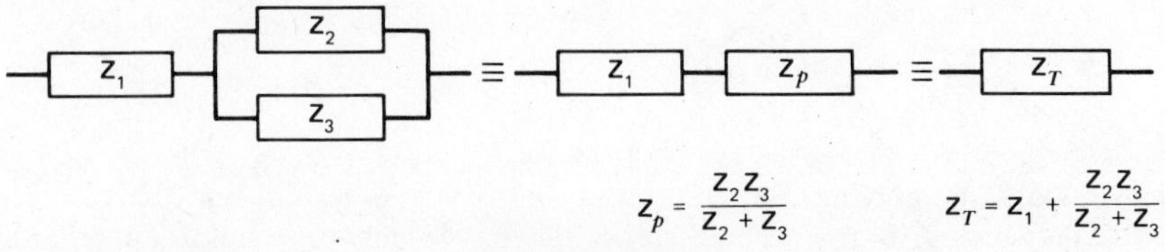

Figure 9.1.8
Calculation of a total impedance from component impedances.

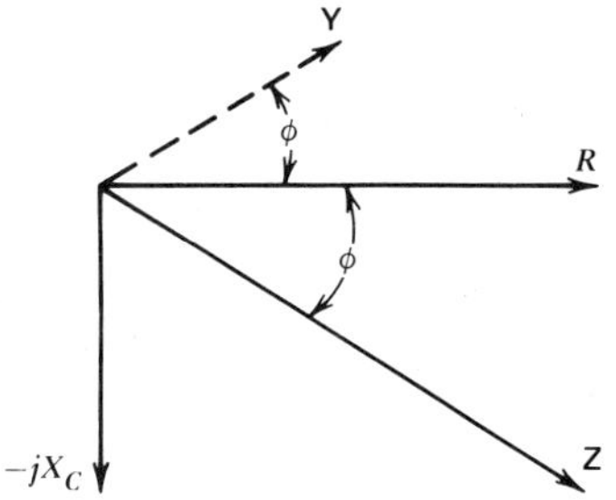

Figure 9.1.9
Relationship between the impedance **Z** and the admittance **Y**.

current must pass through the uncompensated solution resistance, and therefore R_Ω is inserted as a series element to represent this effect in the equivalent circuit.†

The faradaic impedance has been considered in the literature in various ways. Figure 9.1.10*b* shows two equivalences that have been made. The simplest representation is to take it as a series resistance-capacitance combination comprising the *series resistance* R_s and the *pseudocapacity* C_s.‡ An alternative is to separate a pure resistance R_{ct}, the *charge transfer resistance* (Sections 1.3 and 3.5.3), from another general impedance, Z_w, the *Warburg impedance*, which represents a kind of resistance to mass transfer. In contrast to R_Ω and C_d, which are nearly ideal circuit elements, the components of the faradaic impedance are not ideal, because they change with frequency ω. A given equivalent circuit represents cell performance at a given frequency, but not at other frequencies. In fact, a chief objective of a faradaic impedance experiment is to discover the frequency dependencies of R_s and C_s. Theory is then applied to transform these functions into chemical information.

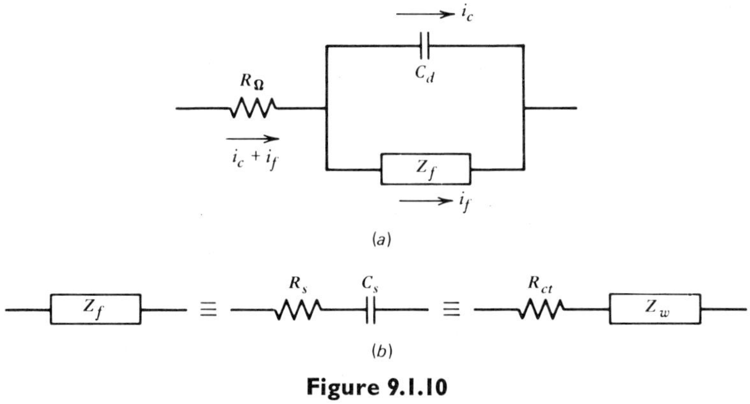

Figure 9.1.10
(*a*) Equivalent circuit of an electrochemical cell. (*b*) Subdivision of Z_f into R_s and C_s or into R_{ct} and Z_w.

† In the faradaic impedance measurements described above (Section 9.1.1), the impedance determined by the bridge is a whole-cell impedance and includes contributions from the counter electrode's interface. Processes there are not usually of interest; hence the impedance at that interface is intentionally reduced to insignificance by employing a counter electrode of large area.

‡ In some treatments, R_s is called the *polarization resistance*. However, that name is applied to other variables in electrochemistry, so we avoid it here.

The circuits considered here are based on the simplest electrode processes. Many others have been devised in order to account for more complex situations, for example, those involving adsorption of electroreactants, multistep charge transfer, or homogeneous chemistry. For specific information, the original or review literature should be consulted (1, 4, 8–10).

9.2 INTERPRETATION OF THE FARADAIC IMPEDANCE

9.2.1 Characteristics of the Equivalent Circuit

The measurement of the cell characteristics in a bridge yields values of R_B and C_B that in series are equivalent to the whole cell impedance, including the contributions from R_Ω and C_d, which may not be of interest to studies of the faradaic process. In general, R_Ω and C_d can be evaluated individually from the frequency dependencies of R_B and C_B, or from separate experiments in the absence of the electroactive couple, if O and R do not affect R_Ω and C_d appreciably. Techniques for making such determinations are considered in detail in Section 9.5. For now, let us assume that the faradaic impedance, expressed as the series combination R_s and C_s, is measurable (see Figure 9.1.10).

Now consider the behavior of this impedance as a sinusoidal current is forced through it. The total voltage drop is

$$E = iR_s + \frac{q}{C_s} \tag{9.2.1}$$

hence

$$\frac{dE}{dt} = R_s \frac{di}{dt} + \frac{i}{C_s} \tag{9.2.2}$$

If the current is

$$i = I \sin \omega t \tag{9.2.3}$$

then

$$\frac{dE}{dt} = (R_s I \omega) \cos \omega t + \left(\frac{I}{C_s}\right) \sin \omega t \tag{9.2.4}$$

This equation is the link we will use to identify R_s and C_s in electrochemical terms. We will find that the response of the electrode process to the current stimulus (9.2.3) will also give dE/dt having the form of (9.2.4). That is, sine and cosine terms will appear; thus R_s and C_s can be identified by equating the coefficients of those terms in the electrical and chemical equations.

9.2.2 Properties of the Chemical System (1, 4, 8, 9)

For our standard system, $O + ne \rightleftharpoons R$, with both O and R soluble, we can write

$$E = E[i, C_O(0, t), C_R(0, t)] \tag{9.2.5}$$

hence,

$$\frac{dE}{dt} = \left(\frac{\partial E}{\partial i}\right)\frac{di}{dt} + \left[\frac{\partial E}{\partial C_O(0, t)}\right]\frac{dC_O(0, t)}{dt} + \left[\frac{\partial E}{\partial C_R(0, t)}\right]\frac{dC_R(0, t)}{dt} \tag{9.2.6}$$

or

$$\frac{dE}{dt} = R_{ct}\frac{di}{dt} + \beta_O\frac{dC_O(0, t)}{dt} + \beta_R\frac{dC_R(0, t)}{dt} \tag{9.2.7}$$

where

$$R_{ct} = \left(\frac{\partial E}{\partial i}\right)_{C_O(0,t),C_R(0,t)} \tag{9.2.8}$$

$$\beta_O = \left[\frac{\partial E}{\partial C_O(0, t)}\right]_{i,C_R(0,t)} \tag{9.2.9}$$

$$\beta_R = \left[\frac{\partial E}{\partial C_R(0, t)}\right]_{i,C_O(0,t)} \tag{9.2.10}$$

Obtaining an expression for dE/dt depends on our ability to evaluate the six factors on the right of (9.2.7). The three parameters R_{ct}, β_O, and β_R depend specifically on the kinetic properties of the electrode reaction. Special cases will be considered later. The remaining three factors can be evaluated generally for current flow according to (9.2.3). One of them is trivial:

$$\frac{di}{dt} = I\omega \cos \omega t \tag{9.2.11}$$

The others are evaluated by considering mass transfer.†

Assuming semi-infinite linear diffusion with initial conditions $C_O(x, 0) = C_O^*$ and $C_R(x, 0) = C_R^*$, we can write from our experience in Section 7.2.1 that

$$\bar{C}_O(0, s) = \frac{C_O^*}{s} + \frac{\bar{i}(s)}{nFAD_O^{1/2}s^{1/2}} \tag{9.2.12}$$

$$\bar{C}_R(0, s) = \frac{C_R^*}{s} - \frac{\bar{i}(s)}{nFAD_R^{1/2}s^{1/2}} \tag{9.2.13}$$

† Note that the equivalent impedance was analyzed just above in terms of current as it is usually defined for circuit analysis. That is, a positive change in E causes a positive change in i. On the other hand, the electrochemical current convention followed elsewhere in this book denotes cathodic currents as positive; hence a *negative* change in E causes a positive change in i. If we adhere to this convention now, confusion will reign when we try to make comparisons between the electrical equivalents and the chemical systems. We must have a common basis for the current. Since the interpretations of the measurements are closely linked to electronic circuit analysis, it is advantageous to adopt the electronic convention. *For this chapter, then, we take an anodic current as positive.*

This expedient will turn out not to cause much trouble, because we never really follow the instantaneous sign of the current in ac experiments. Instead, we measure the amplitude of the sinusoidal component and its phase angle with respect to the sinusoidal potential. Of course the phase angle would depend on our choice of current convention, but it is advantageous even here to take the electronic custom, because the electronic devices used to measure phase angle are based on it.

Inversion by convolution gives

$$C_O(0, t) = C_O^* + \frac{1}{nFAD_O^{1/2}\pi^{1/2}} \int_0^t \frac{i(t-u)}{u^{1/2}}\, du \tag{9.2.14}$$

$$C_R(0, t) = C_R^* - \frac{1}{nFAD_R^{1/2}\pi^{1/2}} \int_0^t \frac{i(t-u)}{u^{1/2}}\, du \tag{9.2.15}$$

From (9.2.3), we can substitute for $i(t - u)$, and hence the problem becomes one of evaluating the integral common to both of these relations.

We begin with the trigonometric identity:

$$\sin \omega(t-u) = \sin \omega t \cos \omega u - \cos \omega t \sin \omega u \tag{9.2.16}$$

which implies that

$$\int_0^t \frac{I \sin \omega(t-u)}{u^{1/2}}\, du = I \sin \omega t \int_0^t \frac{\cos \omega u}{u^{1/2}}\, du - I \cos \omega t \int_0^t \frac{\sin \omega u}{u^{1/2}}\, du \tag{9.2.17}$$

Now let us consider the range of times in which we are interested. Before the current is turned on, the surface concentrations are C_O^* and C_R^*, and after a few cycles we can expect them to reach a *steady state* in which they cycle repeatedly through constant patterns. We can be sure of this point because no net electrolysis takes place in any full cycle of current flow. Our interest is not in the transition from initial conditions to steady state, but in the steady state itself. The two integrals on the right side of (9.2.17) embody the transition period. Because $u^{1/2}$ appears in their denominators, the integrands are appreciable only at short times. After a few cycles, each integral must reach a constant value characteristic of the steady state. We can obtain it by letting the integration limits go to infinity:

$$\int_{\substack{\text{Steady}\\ \text{state}}} \frac{I \sin \omega(t-u)}{u^{1/2}}\, du = I \sin \omega t \int_0^\infty \frac{\cos \omega u}{u^{1/2}}\, du - I \cos \omega t \int_0^\infty \frac{\sin \omega u}{u^{1/2}}\, du \tag{9.2.18}$$

It is easy to show that both integrals on the right side of (9.2.18) are equal to $(\pi/2\omega)^{1/2}$; hence we have by substitution into (9.2.14) and (9.2.15)

$$C_O(0, t) = C_O^* + \frac{I}{nFA(2D_O\omega)^{1/2}} (\sin \omega t - \cos \omega t) \tag{9.2.19}$$

$$C_R(0, t) = C_R^* - \frac{I}{nFA(2D_R\omega)^{1/2}} (\sin \omega t - \cos \omega t) \tag{9.2.20}$$

Now we can evaluate the derivatives of the surface concentrations as required above:†

$$\frac{dC_O(0, t)}{dt} = \frac{I}{nFA}\left(\frac{\omega}{2D_O}\right)^{1/2} (\sin \omega t + \cos \omega t) \tag{9.2.21}$$

$$\frac{dC_R(0, t)}{dt} = -\frac{I}{nFA}\left(\frac{\omega}{2D_R}\right)^{1/2} (\sin \omega t + \cos \omega t) \tag{9.2.22}$$

† In general, we ought to consider the current as $i = i_{dc} + I \sin \omega t$, where i_{dc} is steady or varies only slowly with time. However, we are interested now in derivatives of surface concentrations, and they will be dominated by the higher-frequency ac signal. Relations (9.2.21) and (9.2.22) will still apply to a very high approximation. This is a mathematical manifestation of the way in which the ac part of the experiment can usually be uncoupled from the dc part.

9.2.3 Identification of R_s and C_s

By substitution of (9.2.11), (9.2.21), and (9.2.22) into (9.2.7), we obtain

$$\frac{dE}{dt} = \left(R_{ct} + \frac{\sigma}{\omega^{1/2}}\right) I\omega \cos \omega t + I\sigma\omega^{1/2} \sin \omega t \tag{9.2.23}$$

where

$$\sigma = \frac{1}{nFA\sqrt{2}} \left(\frac{\beta_{\mathrm{O}}}{D_{\mathrm{O}}^{1/2}} - \frac{\beta_{\mathrm{R}}}{D_{\mathrm{R}}^{1/2}}\right) \tag{9.2.24}$$

Now we can identify R_s and C_s by comparison with (9.2.4):

$$\boxed{R_s = R_{ct} + \frac{\sigma}{\omega^{1/2}}} \tag{9.2.25}$$

$$\boxed{C_s = \frac{1}{\sigma\omega^{1/2}}} \tag{9.2.26}$$

The complete evaluation of R_s and C_s depends on finding relations for R_{ct}, β_{O}, and β_{R}. We will see below that R_{ct} is primarily determined by the heterogeneous charge transfer kinetics, and we have already observed above that the terms $\sigma/\omega^{1/2}$ and $1/\sigma\omega^{1/2}$ come from mass transfer effects. Recognition of this situation has led to a division of the faradaic impedance into the *charge transfer resistance* R_{ct} and the *Warburg impedance* Z_w as shown in Figure 9.1.10*b*. Equations 9.2.25 and 9.2.26 demonstrate that this latter impedance can be regarded as a frequency-dependent resistance $R_w = \sigma/\omega^{1/2}$ in series with the pseudo-capacitance $C_w = C_s = 1/\sigma\omega^{1/2}$.

9.3 KINETIC PARAMETERS FROM IMPEDANCE MEASUREMENTS (1, 4, 6, 8–10)

From the description of the faradaic impedance experiment given in Section 9.1, it is clear that measurements are made with the working electrode's mean potential at equilibrium. Since the amplitude of the sinusoidal perturbation is small, we can use the linearized i-η characteristic to describe the electrical response to the departure from equilibrium. In terms of the electronic current convention,

$$\eta = \frac{RT}{nF}\left[\frac{C_{\mathrm{O}}(0, t)}{C_{\mathrm{O}}^*} - \frac{C_{\mathrm{R}}(0, t)}{C_{\mathrm{R}}^*} + \frac{i}{i_0}\right] \tag{9.3.1}$$

hence

$$\boxed{R_{ct} = \frac{RT}{nFi_0}} \tag{9.3.2}$$

$$\beta_{\mathrm{O}} = \frac{RT}{nFC_{\mathrm{O}}^*} \tag{9.3.3}$$

$$\beta_{\mathrm{R}} = \frac{-RT}{nFC_{\mathrm{R}}^*} \tag{9.3.4}$$

Now we see that

$$R_s - \frac{1}{\omega C_s} = R_{ct} = \frac{RT}{nFi_0} \tag{9.3.5}$$

so that the exchange current, and therefore k^0, can be evaluated easily when R_s and C_s are known. The bridge method allows a precise definition of these electrical equivalents; thus it can yield kinetic data of very high quality.

Equation 9.3.5 shows that one can, in principle, evaluate i_0 from data taken at a single frequency. However, doing so is not really wise, because one has no experimental assurance that the equivalent circuit actually mirrors the performance of the system. The best way to check for agreement is to examine the frequency dependence of the impedance. For example, (9.2.25) and (9.2.26) predicts that R_s and $1/\omega C_s$ should both be linear with $\omega^{-1/2}$ and should have a common slope, σ, which is quantitatively predictable from the constants of the experiment; that is,

$$\boxed{\sigma = \frac{RT}{n^2F^2A\sqrt{2}}\left(\frac{1}{D_O^{1/2}C_O^*} + \frac{1}{D_R^{1/2}C_R^*}\right)} \tag{9.3.6}$$

Figure 9.3.1 is a display of these relationships.

The plot of R_s should have an intercept R_{ct}, from which i_0 can be evaluated. Extrapolation to the intercept is equivalent to estimating the system's performance at infinite frequency. The Warburg impedance drops out at high frequencies, because the time scale is so short that diffusion cannot manifest itself as a factor influencing the current. Since the surface concentrations never change significantly from the mean values [see (9.2.19) and (9.2.20)], charge transfer kinetics alone dictate the current.

If the linear behavior typified in Figure 9.3.1 is not observed, then the electrode process is not as simple as we assume here and a more complex situation must be considered. The availability of this kind of check for internal consistency is an extremely important asset of the impedance technique. See Section 9.5 for more details.

Let us now consider the general impedance properties of a reversible system, which is an important limiting case. When charge transfer kinetics are very facile, $i_0 \to \infty$; hence $R_{ct} \to 0$. Thus $R_s \to \sigma/\omega^{1/2}$. The corresponding impedance plot is shown in Figure 9.3.2*a*. Since the resistance and the capacitive reactance are exactly equal, the

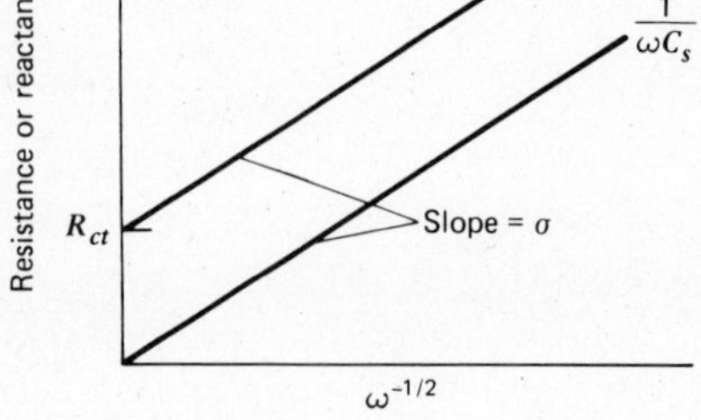

Figure 9.3.1
Dependence of R_s and $1/\omega C_s$ on frequency.

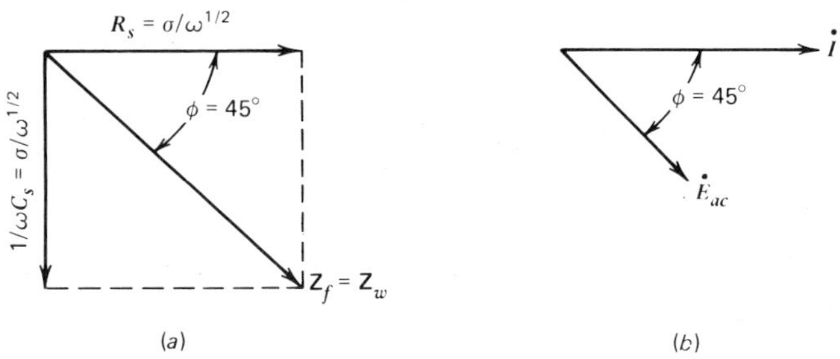

Figure 9.3.2
(*a*) Vector diagram showing the components of the faradaic impedance for a reversible system. (*b*) Phase relationship between ac current and the ac component of potential.

faradaic impedance is

$$Z_f = \left(\frac{2}{\omega}\right)^{1/2} \sigma \tag{9.3.7}$$

which is the Warburg impedance alone.

Since this is a mass transfer impedance that applies to any electrode reaction, it is a minimum impedance. If kinetics are observable, another factor, R_{ct}, contributes and Z_f must be greater, as Figure 9.3.3*a* depicts. Thus the amplitude of the sinusoidal current flowing in response to a given excitation signal $\dot{E}_{ac}$ is maximal for reversible systems and decreases correspondingly for more sluggish kinetics. If the heterogeneous redox process is very immobile, R_{ct} and Z_f are so large that there is only a very small ac component to the current, and the limit of detection sets the lower bound on rate constants that can be measured by this approach. See Section 9.5 for more details about quantitative working ranges.

It is interesting also to examine the effect of concentration, which is manifested through σ. In general, higher concentrations reduce the mass transfer impedance, as we would expect intuitively. Of greater interest, however, is the effect of the concentration ratio C_O^*/C_R^*. One does change it experimentally in order to vary the equilibrium potential for a series of impedance measurements. Both large and small ratios

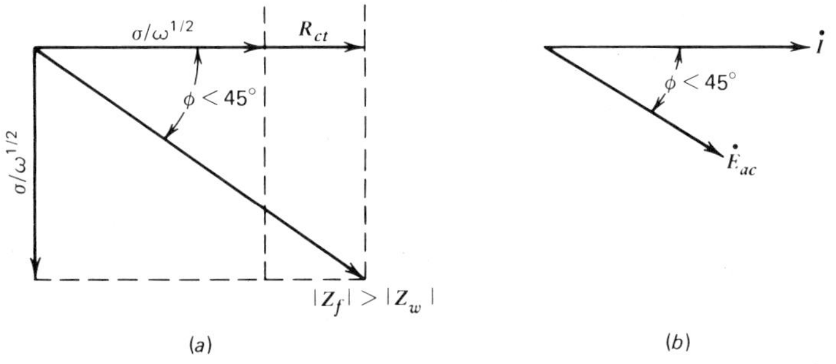

Figure 9.3.3
(*a*) Vector diagram showing the effect of R_{ct} on the impedance. (*b*) Phase relationship between $\dot{I}$ and $\dot{E}_{ac}$ for a system with significant R_{ct}.

imply that one of the concentrations is small; hence σ and Z_f must be large. Current response to $\dot{E}_{ac}$ is not very great, because the supply of one reagent is insufficient to permit a high reaction rate for the cyclic, reversible electrode process that causes the ac current. Large rates can be achieved only when both electroreactants are present at comparable concentrations; hence we expect Z_f to be minimal near $E^{0'}$. Impedance measurements are most easily made in that potential region, and they gradually become more difficult as one departs from it either positively or negatively. This effect presages the shape of the ac voltammetric response, which will be derived in the next section.

A final point of interest is the phase angle between the current phasor $\dot{I}_{ac}$ and the potential $\dot{E}_{ac}$. Since $\dot{I}_{ac}$ lies along R_s and $\dot{E}_{ac}$ lies along $\mathbf{Z}_f$, the phase angle is readily calculated as

$$\phi = \tan^{-1}\frac{1}{\omega R_s C_s} = \tan^{-1}\frac{\sigma/\omega^{1/2}}{R_{ct} + \sigma/\omega^{1/2}} \tag{9.3.8}$$

For the reversible case, $R_{ct} = 0$; hence $\phi = 45°$ or $\pi/4$. A quasi-reversible system shows $R_{ct} > 0$; hence $\phi < \pi/4$. However, ϕ must always be greater than zero, unless $R_{ct} \rightarrow \infty$; but then the reaction would be so sluggish that little alternating current would flow anyway in a conventional impedance measurement. This sensitivity of ϕ to kinetics suggests that R_{ct} might be extracted from the phase angle. It can be, and often it is, by ac voltammetric experiments. Before we proceed to a discussion of them, let us note that since $0 \leq \phi \leq 45°$, there is always a component of i_{ac} that is in phase (0°) with $\dot{E}_{ac}$, and it can be measured with a phase-sensitive detector (e.g., a lock-in amplifier) referenced to $\dot{E}_{ac}$. This feature is extremely useful as a basis for discriminating against charging current in ac voltammetry.

9.4 AC VOLTAMMETRY

We noted in Section 9.1 that ac voltammetry is basically a faradaic impedance technique in which the mean potential E_{dc} is imposed potentiostatically at arbitrary values that usually differ from the equilibrium value. Ordinarily, it is varied systematically (e.g., linearly) on a long time scale compared to that of the superimposed ac variation $\dot{E}_{ac}$ (10 Hz to 100 kHz). The output is a plot of the magnitude of the ac component of the current *vs.* E_{dc}. The phase angle between the alternating current and E_{ac} is also of interest.

Treatments of this kind of problem are greatly simplified by uncoupling the long-term diffusion due to E_{dc} from the rapid diffusional fluctuations due to $\dot{E}_{ac}$. We do that by recognizing that E_{dc} sets up mean surface concentrations that look like bulk values to the ac perturbation because of the difference in time scale. In the previous section, we defined the faradaic impedance in terms of bulk concentrations; thus the current response in ac voltammetry as a function of E_{dc} is readily obtained by substituting the surface concentrations imposed by E_{dc} directly into these impedance relations. Since this strategy is simple and intuitive, we will pursue it. More rigorous treatments are available in the literature for the interested reader (2, 3, 5). The results are the same by either approach.

The mean surface concentrations enforced by E_{dc} depend on many factors: (a) the way in which E_{dc} is varied; (b) whether or not there is periodic renewal of the diffusion layer; (c) the applicable current-potential characteristic; and (d) homogeneous or heterogeneous chemical complications associated with the overall electrode reaction. For example, one could vary E_{dc} in a sequential potentiostatic manner with periodic renewal of the diffusion layer, as in sampled-current voltammetry. This is the technique that is actually used in *ac polarography*, which features a DME and effectively constant E_{dc} during the lifetime of each drop. Alternatively one could use a stationary electrode and a fairly fast sweep without renewal of the diffusion layer. Both techniques are useful, and both are considered below. The effects of different kinds of charge transfer kinetics will also be examined here, but the effects of homogeneous complications are deferred to Chapter 11. Throughout the discussion, one should keep in mind that the chief strength of ac voltammetry is the access it gives to exceptionally precise quantitative information about electrode processes. Diagnostic aspects certainly exist, but they are more subtle than with other methods.

9.4.1 ac Polarography in a Reversible System

Let us consider the ac response at a dropping mercury electrode immersed in a solution containing initially only species O of the nernstian couple $O + ne \rightleftharpoons R$. The dc potential starts at a value considerably more positive than $E^{0\prime}$ and is scanned slowly in a negative direction. During the lifetime of a single drop, E_{dc} is effectively constant, and hence the dc part of the experiment is conventional polarography and is treated as a series of individual step experiments (see Sections 5.3 and 5.4).

Since the charge transfer resistance is completely negligible, (9.3.6) always applies where

$$\sigma = \frac{RT}{n^2F^2A\sqrt{2}}\left[\frac{1}{D_O^{1/2}C_O(0, t)_m} + \frac{1}{D_R^{1/2}C_R(0, t)_m}\right] \tag{9.4.1}$$

and the mean concentrations $C_O(0, t)_m$ and $C_R(0, t)_m$ are determined by the nernstian relation:

$$\frac{C_O(0, t)_m}{C_R(0, t)_m} = \theta_m = \exp\left[\frac{nF}{RT}(E_{dc} - E^{0\prime})\right] \tag{9.4.2}$$

The arguments that led to (5.4.29) and (5.4.30) apply equally to the dc part of this experiment; hence we write

$$C_O(0, t)_m = C_O^*\left(\frac{\xi\theta_m}{1 + \xi\theta_m}\right) \tag{9.4.3}$$

$$C_R(0, t)_m = C_O^*\left(\frac{\xi}{1 + \xi\theta_m}\right) \tag{9.4.4}$$

where ξ is $D_O^{1/2}/D_R^{1/2}$, as before. Thus the faradaic impedance is obtained by substitution into (9.4.1) and then into (9.3.7):

$$Z_f = \frac{RT}{n^2F^2A\omega^{1/2}D_O^{1/2}C_O^*}\left(\frac{1}{\xi\theta_m} + 2 + \xi\theta_m\right) \tag{9.4.5}$$

Let us note now that $\xi\theta_m$ can be written

$$\xi\theta_m = e^a \tag{9.4.6}$$

where

$$a = \frac{nF}{RT}(E_{\text{dc}} - E_{1/2}) \tag{9.4.7}$$

and $E_{1/2}$ is the reversible half-wave potential defined in (5.4.21):

$$E_{1/2} = E^{0'} + \frac{RT}{nF} \ln \frac{D_{\text{R}}^{1/2}}{D_{\text{O}}^{1/2}} \tag{9.4.8}$$

By substitution from (9.4.6), we find that the term in parentheses (9.4.5) is $e^{-a} + 2 + e^a$, which is also $4 \cosh^2(a/2)$. Thus we have

$$Z_f = \frac{4RT}{n^2F^2A\omega^{1/2}D_{\text{O}}^{1/2}C_{\text{O}}^*} \cosh^2\left(\frac{a}{2}\right) \tag{9.4.9}$$

In Section 9.3 we saw that the faradaic current for a reversible system leads $\dot{E}_{\text{ac}}$ by exactly 45°. If $\dot{E}_{\text{ac}}$ is $\Delta E \sin \omega t$, then

$$i_{\text{ac}} = \frac{\Delta E}{Z_f} \sin\left(\omega t + \frac{\pi}{4}\right) \tag{9.4.10}$$

and the amplitude of this current, which is the chief observable, is simply

$$\boxed{I = \frac{\Delta E}{Z_f} = \frac{n^2F^2A\omega^{1/2}D_{\text{O}}^{1/2}C_{\text{O}}^* \, \Delta E}{4RT \cosh^2(a/2)}} \tag{9.4.11}$$

Figure 9.4.1 is a display of the ac polarogram defined by this equation. The bell shape derives from the factor $\cosh^{-2}(a/2)$, and it reflects the potential dependence of

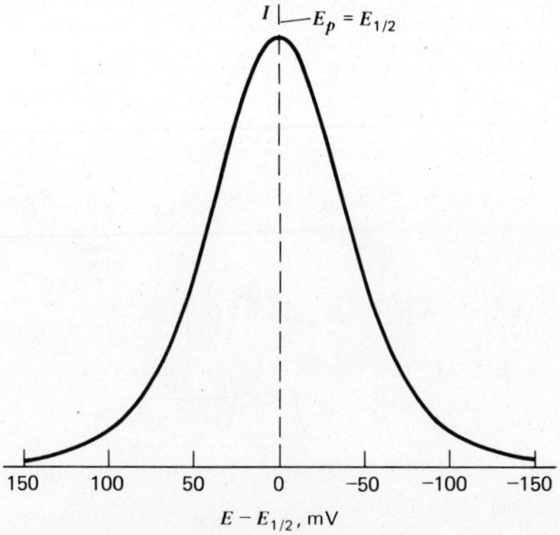

Figure 9.4.1
Shape of a reversible ac voltammetric peak for $n = 1$.

the impedance Z_f. The maximum in the current occurs at $a/2 = 0$ or at $E_{dc} = E_{1/2}$, which is near $E^{0'}$. As one moves away from that potential, either positively or negatively, the impedance rises sharply and the current falls off. The physical basis for this behavior was outlined in Section 9.3. In effect, the current is controlled by the limiting reagent, that is, the smaller of the two surface concentrations. At potentials far from $E^{0'}$, where only small amounts of one reagent can exist at the surface, only small currents can flow.

The peak current at $E_{dc} = E_{1/2}$ comes easily from (9.4.11). Since cosh(0) = 1,

$$I_p = \frac{n^2F^2A\omega^{1/2}D_O^{1/2}C_O^* \Delta E}{4RT} \tag{9.4.12}$$

From this relation and (9.4.11) one can show straightforwardly that the shape of the ac polarogram adheres to

$$E_{dc} = E_{1/2} + \frac{2RT}{nF}\ln\left[\left(\frac{I_p}{I}\right)^{1/2} - \left(\frac{I_p - I}{I}\right)^{1/2}\right] \tag{9.4.13}$$

(See Problem 9.1.)

We have not yet recognized the effect of drop growth on the polarogram, because everything we have noted thus far is based on a stationary planar electrode. The use of linear diffusion relations for the dc part of the experiments has already been justified (see Section 5.3.1), and that justification is even more valid for the ac part because of its shorter time scale. Thus the peculiarities of the expanding sphere are felt only in the changing area A with time, and that factor is directly accountable by substitution of (5.3.3) into (9.4.11). Since A grows as $t^{2/3}$ as the drop ages, the current also shows the same dependence. Thus we can expect the current to oscillate as successive drops grow and fall. Maxima should be observed at the end of each drop's life. Experimental results in Figure 9.4.2 bear out these expectations. Measurements carried out on the envelope of the ac polarogram can be treated by all relations derived above, provided that A is defined as the area just before drop fall.

A number of important properties of the reversible ac polarogram can be deduced from (9.4.11) to (9.4.13). Among them are the direct proportionalities between I_p and n^2, $\omega^{1/2}$, and C_O^*. There is also a proportionality to ΔE; however, this relation is a limited one, because the linearized i-E characteristic underlying the derivation of Z_f becomes invalid if ΔE is too large. For linearity within a few percent, ΔE must be less than about $10/n$ mV. Not surprisingly, the width of the peak at half height also depends on ΔE if large values are used. If it is kept below $10/n$ mV, there is a constant width of $90.4/n$ mV at 25°. At larger ΔE the peaks are broader (see also Section 5.8.3).

9.4.2 ac Polarographic Response to Quasi-Reversible and Irreversible Systems

When heterogeneous kinetics become sluggish enough to be visible, one requires a more elaborate theory to predict the ac polarographic response. The general case, in which k^0 can adopt any value, is very complex. The reader is referred to the literature for

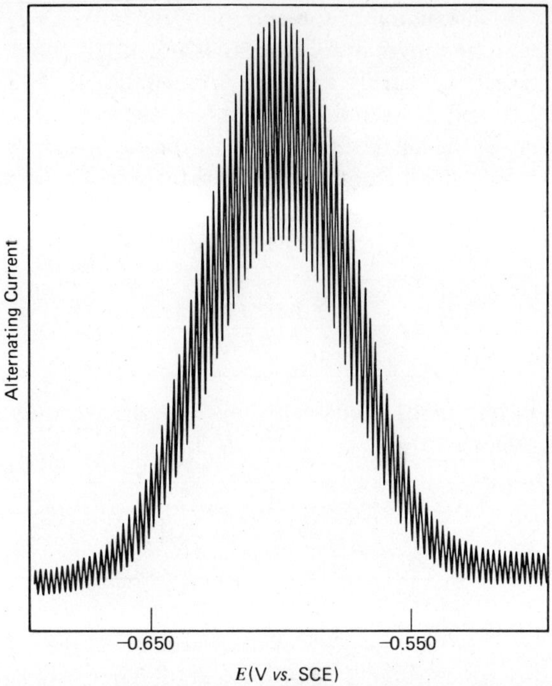

Figure 9.4.2
An ac polarogram for 3×10^{-3} M Cd^{2+} in 1.0 M Na_2SO_4. $\Delta E = 5$ mV, $\omega/2\pi = 320$ Hz. [Reprinted with permission from D. E. Smith, *Anal. Chem.*, **35**, 1811 (1963). Copyright 1963, American Chemical Society.]

complete discussions of it (2, 3, 5). Here we examine an important special case in detail, and it will provide us with a good intuitive understanding of the kinetic effects of interest.

That special case is the situation in which the dc process is effectively nernstian, whereas the ac process is not. This situation is frequently seen in real systems because the time scales of the two aspects can differ so greatly. That is, k^0 can be sufficiently large that the mean surface concentrations are kept in the ratio dictated in E_{dc} through the Nernst equation (9.4.2), even though that rate constant is not large enough to assure a negligible charge transfer resistance to the much faster ac perturbation.

The faradaic impedance in this situation involves both R_{ct} and σ, and can be written from (9.2.25) and (9.2.26) and the laws for obtaining impedances from reactances:

$$Z_f = \left[\left(R_{ct} + \frac{\sigma}{\omega^{1/2}}\right)^2 + \left(\frac{\sigma}{\omega^{1/2}}\right)^2\right]^{1/2} \tag{9.4.14}$$

The parameters R_{ct} and σ can both be defined through the assumption of dc reversibility. That premise allows us to use the same mean surface concentrations as in the previous section. They are defined by (9.4.3) and (9.4.4), and thus we can develop σ

by substitution into (9.4.1). Rearrangements equivalent to those used in obtaining (9.4.9) then yield

$$\sigma = \frac{4RT}{\sqrt{2}n^2F^2AD_O^{1/2}C_O^*}\cosh^2\left(\frac{a}{2}\right) \tag{9.4.15}$$

The charge transfer resistance R_{ct} is given by (9.3.2) in terms of the exchange current i_0. Normally we speak of i_0 as an equilibrium property defined by bulk concentrations of O and R according to (3.5.6). However, for the ac process, since the mean surface concentrations act like bulk values, we can recognize an effective exchange current for ac perturbation that would be given by

$$(i_0)_{\text{eff}} = nFAk^0[C_O(0, t)_m]^{(1-\alpha)}[C_R(0, t)_m]^{\alpha} \tag{9.4.16}$$

By changing the mean surface concentrations, E_{dc} controls $(i_0)_{\text{eff}}$ and, therefore, R_{ct}. A more explicit expression of this dependence is obtained by substitution from (9.4.3), (9.4.4), and (9.4.6), as above:

$$(i_0)_{\text{eff}} = nFAk^0C_O^*\xi^{\alpha}\left(\frac{e^{\beta a}}{1 + e^a}\right) \tag{9.4.17}$$

where $\beta = (1 - \alpha)$. Since $R_{ct} = RT/nF(i_0)_{\text{eff}}$, we have

$$R_{ct} = \frac{RT}{n^2F^2Ak^0C_O^*\xi^{\alpha}}\left(\frac{1 + e^a}{e^{\beta a}}\right) \tag{9.4.18}$$

Now that R_{ct} and σ are available, we can write Z_f as a function of E_{dc} by substitution into (9.4.14). That operation is straightforward, but it will yield a rather messy expression. Perhaps more instructive is to examine limiting behavior for high and low frequencies, which can be discerned from (9.4.14).

At very low frequencies, R_{ct} is small compared to $\sigma/\omega^{1/2}$; hence the system looks reversible. This is not surprising; after all we are bringing the time domain of the ac process toward that of the dc perturbation, which evokes a reversible response. Everything we found in the previous section about reversible ac response should also apply to the quasi-reversible system at the low-frequency limit.

As the frequency is elevated, R_{ct} becomes appreciable in comparison to $\sigma/\omega^{1/2}$; hence reversibility is vitiated. The ac time domain has become shortened enough to strain the heterogeneous kinetics. Finally, at the high-frequency limit, R_{ct} greatly exceeds $\sigma/\omega^{1/2}$, and Z_f approaches R_{ct} itself. The amplitude of the alternating current is then

$$\boxed{I = \frac{\Delta E}{R_{ct}} = \frac{n^2F^2Ak^0C_O^*\Delta E\xi^{\alpha}}{RT}\left(\frac{e^{\beta a}}{1 + e^a}\right)} \tag{9.4.19}$$

This equation describes the shape of the ac polarogram. In general, the response as a function of dc potential is bell-shaped, much as in the reversible situation. This point is seen by noting the behavior of the factor in parentheses as a becomes large either positively or negatively. However, positive deviations from $E_{1/2}$ do not evoke the same response as negative deviations; that is, the response is not symmetric and the bell shape is skewed.

The peak is easily found by differentiating (9.4.19) with respect to a. The maximum is reached when $e^a = \beta/\alpha$, or

$$E_{\mathrm{dc}} = E_{1/2} + \frac{RT}{nF}\ln\frac{\beta}{\alpha} \tag{9.4.20}$$

The peak current amplitude is therefore

$$I_p = \frac{n^2F^2Ak^0C_O^*\,\Delta E\xi^\alpha}{RT}\,\beta^\beta\alpha^\alpha \tag{9.4.21}$$

These equations, together with those describing the reversible, low-frequency limit, give a good picture of the behavior of the system as ω changes. The peak current is at first linear with $\omega^{1/2}$, but with increasing frequency that dependence is reduced until, at the high-frequency limit, I_p becomes independent of ω. It is easy to see from preceding developments that the frequency dependence reflects the mass transport effects manifested in the Warburg impedance. The lack of a frequency dependence in (9.4.19) and (9.4.21) comes about because the current is totally controlled at high ω by heterogeneous kinetics. Mass transfer plays no role. Not surprisingly, then, I_p is proportional to k^0 at high ω, and it is totally insensitive to k^0 at low ω. The proportionalities between I_p and ΔE and C_O^* hold at all frequencies.

Note also that since kinetic control of I at high frequencies implies a faradaic impedance that greatly exceeds the Warburg impedance at those frequencies, the currents must be much smaller than that for a truly reversible system, which shows only the Warburg impedance at any frequency. The general reduction in ac response in quasi-reversible systems is illustrated in Figure 9.4.3. The k^0 values for all curves shown there are sufficiently great that the assumption of dc reversibility holds. It is easy to see the trend in responses with decreasing k^0; hence one recognizes that there will be a rather small ac response if k^0 falls below 10^{-4} to 10^{-5} cm/sec. Systems showing totally irreversible dc polarograms can be almost invisible to the ac experiment. This fact is extremely useful for analytical work (see Section 9.7).†

The position of the peak is also of interest. Relations derived above show that there is a slight shift with increasing frequency. At low ω, the peak comes at $E_{\mathrm{dc}} = E_{1/2}$, just as for a reversible system at any frequency. As ω becomes greater the peak potential deviates from this value until it reaches the limiting position defined by (9.4.20). Since α and β are generally comparable, we can expect the extent of this shift, $(RT/nF)\ln(\beta/\alpha)$, to be quite small. In other words, the peak potential for an ac polarogram always is near the formal potential for the couple, provided dc reversibility applies.

† The totally irreversible case does yield an ac current, contrary to the impression one might gain from this line of argument. The current arises from the simple modulation of the dc wave (11, 12). Since the shape of that wave is independent of k^0 (Section 5.5), the ac peak height is also independent of k^0. The peak lies near the half-wave potential of the dc wave; hence it is shifted substantially from $E^{0'}$ by an amount related to the size of k^0.

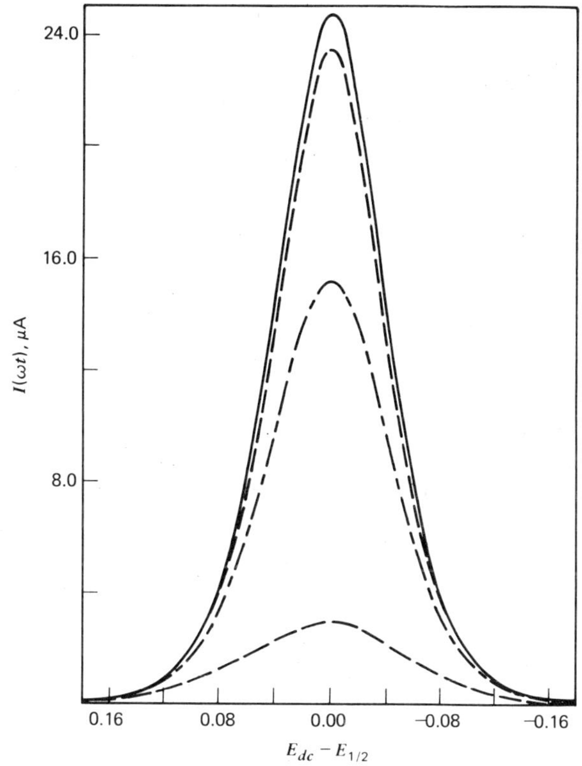

Figure 9.4.3

Calculated ac polarograms for quasi-reversible systems. Curves (from the top) are the $k^0 \to \infty$, $k^0 = 1$, $k^0 = 0.1$, and $k^0 = 0.01$ cm/sec. Other parameters are as follows: $\omega = 2500$ sec^{-1}, $\alpha = 0.500$, $D = 9 \times 10^{-6}$ cm^2/sec, $A = 0.035$ cm^2, $C_O^* = 1.00 \times 10^{-3}$ M, $T = 298$ K, $\Delta E = 5.00$ mV, and $n = 1$. The curves show the faradaic current at t_{max} [From D. E. Smith, *Electroanal. Chem.*, **1**, 1 (1966), courtesy of Marcel Dekker, Inc.]

The phase angle of i_{ac} with respect to $\dot{E}_{ac}$ is of great interest as a source of kinetic information. This point was suggested in Section 9.3, and it is rooted in equation 9.3.8. We can rewrite that relation as

$$\cot \phi = 1 + \frac{R_{ct}\omega^{1/2}}{\sigma} \tag{9.4.22}$$

Substitution from (9.4.15) and (9.4.18) and rearrangement gives

$$\boxed{\cot \phi = 1 + \frac{(2D_O{}^{\beta} D_R{}^{\alpha}\omega)^{1/2}}{k^0}\left[\frac{1}{e^{\beta a}(1 + e^{-a})}\right]} \tag{9.4.23}$$

The bracketed factor shows that $\cot \phi$ depends on the dc potential. Large positive and negative values of a force $\cot \phi$ to unity, and hence there must be a maximum in this parameter near the peak of the polarogram. The precise position is easily found by differentiation, and one ascertains that $e^{-a} = \beta/\alpha$ at that point. Thus,

$$\boxed{E_{dc} = E_{1/2} + \frac{RT}{nF}\ln\frac{\alpha}{\beta}} \tag{9.4.24}$$

This maximum point is independent of nearly all experimental variables, for example, ΔE, C_O^*, and, most notably, A and ω. The difference between $E_{1/2}$ and the potential of maximum cot ϕ provides convenient access to the transfer coefficient α.

Actual cot ϕ data are shown in Figure 9.4.4 for $TiCl_4$ in oxalic acid solution (13). The electrode reaction is the one-electron reduction of Ti(IV) to Ti(III). Note that the potential of maximum cot ϕ is independent of frequency, as predicted above.

Plots of cot ϕ *vs.* $\omega^{1/2}$ yield k^0, once α is known from the position of $[\cot\phi]_{max}$ and the diffusion coefficients are known from other measurements. This point is easily seen from (9.4.23), which holds for any value of E_{dc}. In practice, these plots are usually made for special values of E_{dc} that give simplified forms of the linear relation.

A very convenient procedure is to choose $E_{dc} = E_{1/2}$, for then $a = 0$, and we have

$$[\cot\phi]_{E_{1/2}} = 1 + \left(\frac{D_O{}^{\beta} D_R{}^{\alpha}}{2}\right)^{1/2} \frac{\omega^{1/2}}{k^0} \tag{9.4.25}$$

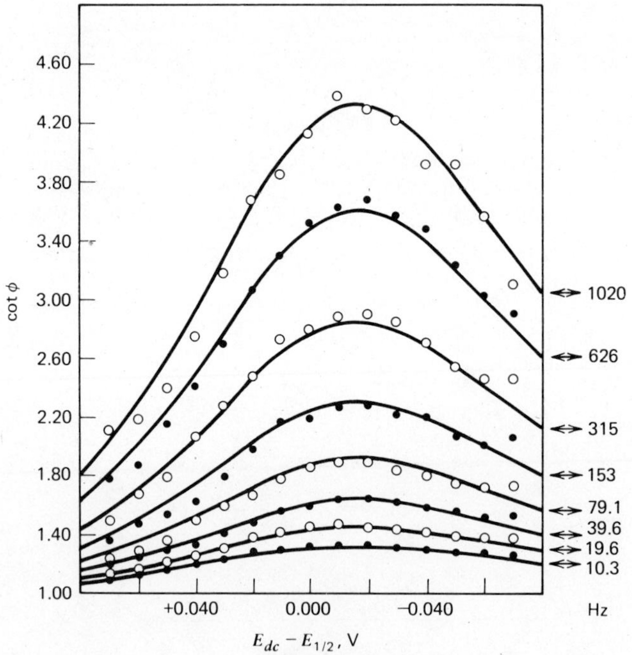

Figure 9.4.4
Dependence of the phase angle on E_{dc}. The system is 3.36 m*M* $TiCl_4$ in 0.200 *M* $H_2C_2O_4$. $\Delta E = 5.00$ mV, $T = 25$ °C. Points are experimental; curves are predicted from experimental parameters by (9.4.23). [Reprinted with permission from D. E. Smith, *Anal. Chem.*, **35**, 610 (1963). Copyright 1963, American Chemical Society.]

If one can take $D_O = D_R = D$, then $D_O{}^{\alpha} D_R{}^{\beta} = D$, and the slope of this particular plot becomes independent of α. Figure 9.4.5 is an example in which the data from Figure 9.4.4 at $E_{dc} = E_{1/2} = -0.290$ V *vs.* SCE have been plotted *vs.* $\omega^{1/2}$.

Another simplified version of (9.4.23) can be obtained for the potential of maximum cot ϕ. By substituting $e^{-a} = \beta/\alpha$, we obtain

$$[\cot \phi]_{\max} = 1 + \frac{(2D_O{}^{\beta} D_R{}^{\alpha})^{1/2}}{k^0\left[\left(\frac{\alpha}{\beta}\right)^{-\alpha} + \left(\frac{\alpha}{\beta}\right)^{\beta}\right]}\,\omega^{1/2} \tag{9.4.26}$$

The product of the diffusion coefficients can usually be simplified as above, but α still must be known for an evaluation of k^0 because of the bracketed factor.

Quantitative information about heterogeneous charge transfer kinetics obtained from ac polarographic data nearly always comes from the behavior of cot ϕ with potential and frequency, rather than from the heights, shapes, or positions of the polarograms. One reason for this favor toward cot ϕ is that many experimental variables do not have to be controlled closely or even be known. Among them are C_O^*, ΔE, and A. Freedom from knowing A can be a significant advantage. However, the most important reason for evaluating kinetics through cot ϕ is that relations (9.4.23) to (9.4.26) hold for any quasi-reversible or irreversible system. We have derived them for the situation in which dc reversibility applies; however, they hold regardless of that condition. Demonstrations of this point are available in the literature (3, 5). Their unconditional validity is a big asset, for it frees the experimenter from having to achieve special limiting conditions.

As in the previous section, we have assumed semi-infinite linear diffusion to a planar electrode throughout the mathematical discussion here. With a reversible dc process,

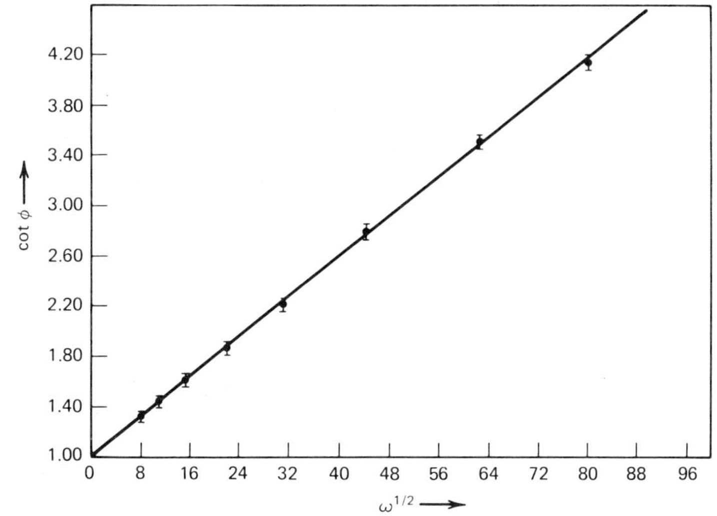

Figure 9.4.5

Plot of cot ϕ *vs.* $\omega^{1/2}$ for 3.36 m*M* $TiCl_4$ in 0.200 m*M* $H_2C_2O_4$. $E_{dc} = E_{1/2} = -0.290$ V *vs.* SCE. $\Delta E = 5.00$ mV, $T = 25$ °C. [Reprinted with permission from D. E. Smith, *Anal. Chem.*, **35**, 610 (1963). Copyright 1963, American Chemical Society.]

the effects of sphericity and drop growth at the DME are exactly as discussed in Section 9.4.1. In general, the sphericity has a negligible impact and drop growth can be accommodated by using an explicit expression for A as a function of time. If dc reversibility does not apply, these factors influence the ac response in more complex ways (3, 5, 11). The reader is referred to the literature for details.

9.4.3 Linear Sweep ac Voltammetry at Stationary Electrodes (14, 15)

The previous two sections have dealt generally with ac voltammetry as recorded by the application of E_{dc} in successive steps and with a renewal of the diffusion layer between each step. The dropping mercury electrode permits the most straightforward application of that technique, but other electrodes can be used if there is a means for stirring the solution between steps so that the diffusion layer is renewed. On the other hand, this requirement for periodic renewal is rather inconvenient when one wishes to use stationary electrodes, such as metal or carbon disks, or a hanging mercury drop. Then one prefers to apply E_{dc} as a ramp and to renew the diffusion layer only between scans. In this section, we will examine the expected ac voltammograms for reversible and quasi-reversible systems when E_{dc} is imposed as a linear sweep and we will compare them with the results obtained above for effectively constant E_{dc}.

The strategy is exactly that used before. The time domains associated with variations in E_{dc} and $\dot{E}_{ac}$ are assumed to differ greatly, so that the diffusional aspects of the two parts of the experiment can be uncoupled. This assumption will hold as long as the scan rate v is not too large compared to the ac frequency (16). More precisely, $dE_{dc}/dt = v$ should be much smaller than the amplitude of $d\dot{E}_{ac}/dt$, which is $\Delta E\omega$. Then, we can take the mean surface concentrations enforced by E_{dc} as effective bulk values for the ac perturbation, just as we did earlier. The current amplitude and phase angle then follow easily from the impedance properties.

(a) Reversible Systems. Let us consider a completely nernstian system $O + ne \rightleftharpoons R$ in which R is initially absent. The starting potential for the linear sweep is rather positive with respect to $E^{0'}$, and the scan direction is negative. Semi-infinite linear diffusion is assumed. The mean surface concentrations $C_O(0, t)_m$ and $C_R(0, t)_m$ are exactly those obtained in the analogous linear sweep experiment without superposed ac excitation, and they adhere always to the nernstian relation (9.4.2).

The arguments leading to equation 5.4.26 show that it applies without reference to the kinetic properties of the electrode reaction or the nature of the excitation waveform. For the present purpose, we can rewrite it as

$$D_O^{1/2}C_O(0, t)_m + D_R^{1/2}C_R(0, t)_m = C_O^* D_O^{1/2} \tag{9.4.27}$$

Substitution from (9.4.2) then reveal that the mean surface concentrations are exactly as given in (9.4.3) and (9.4.4). In other words, those relations, which were derived earlier expressly for step excitation, have been shown here to apply regardless of the manner by which E_{dc} is attained.†

† Equation 9.4.27 is based on semi-infinite linear diffusion; hence this conclusion applies strictly only to planar electrodes. Work at an HMDE can be affected by sphericity (15).

This conclusion is very important because it implies that all relations and all qualitative conclusions presented in Section 9.4.1 also hold for linear sweep ac voltammetry of reversible systems at a stationary electrode.

(b) Quasi-Reversible Systems. An important special case of quasi-reversibility is the situation in which the electrode kinetics are sufficiently facile to maintain a reversible dc response, but not facile enough to show a negligible charge transfer resistance R_{ct} to the ac perturbation.

If the dc process is nernstian, (9.4.2) and (9.4.27) hold, and the mean surface concentrations are given by (9.4.3) and (9.4.4), which are the same relations used in the treatment of Section 9.4.2. Thus, all of the equations and qualitative conclusions reached there for quasi-reversible ac polarograms also apply to the corresponding linear sweep ac voltammograms.

These precise parallels between linear sweep voltammetry and ac polarography no longer persist when there is a lack of dc reversibility. Treating such a case is more complex than the situations we have examined above because the mean surface concentrations are affected by the concentration profiles throughout the diffusion layer, and the surface values applicable at any potential generally depend on the waveform used to attain that potential (14, 15).

Linear sweep ac voltammetry is receiving attention because it allows precise, rapid kinetic measurements at solid electrodes. It can therefore be used to characterize these electrodes themselves, which may be of considerable interest, or it can be applied to the study of many electrode reactions that will not yield to the DME. In this group are electrode processes operating outside the working range of mercury and processes taking place in controlled environments where the DME may be inconvenient.

9.4.4 Cyclic ac Voltammetry (14, 15)

Cyclic ac voltammetry is a simple extension of the linear sweep technique; one simply adds the reversal scan in E_{dc}. This technique is attractive because it retains the best features of two powerful, complementary methodologies. Conventional cyclic voltammetry is especially informative about the qualitative aspects of an electrode process. However, the response waveforms lend themselves poorly to quantitative evaluations of parameters. Cyclic ac voltammetry retains the diagnostic utility of conventional cyclic measurements, but it does so with an improved response function that permits quantitative evaluations as precise as those obtainable with the usual ac approaches. This technique is still in the development stage, but it may prove useful.

Treatments of cyclic ac voltammetry follow the familiar pattern. The ac and dc time scales are independently variable, but are assumed to differ markedly. Then a treatment of the dc aspect yields mean surface concentrations, which are used to calculate faradaic impedances that define the ac response by amplitude and phase angle. The electrode is assumed to be stationary and the solution is regarded as quiescent for the duration of the dc cycle.

(a) Reversible Systems. Cyclic ac voltammograms for completely nernstian systems are easy to predict on the basis of results from the previous section. The mean surface concentrations $C_O(0, t)_m$ and $C_R(0, t)_m$ adhere to (9.4.3) and (9.4.4) unconditionally; hence at any potential they are the same for both the forward and reverse scans. The cyclic ac voltammogram should therefore show superimposed forward and reverse traces of ac current amplitude *vs.* E_{dc}. We expect a peak-shaped voltammogram that adheres in every way to the conclusions reached in Section 9.4.1 about the general ac voltammetric response to a reversible system at a planar electrode.

Figure 9.4.6 contrasts the responses from the ac and dc versions of cyclic voltammetry for the purely nernstian case. Kinetic reversibility is shown in the dc experiment by a peak separation near $60/n$ mV, regardless of scan rate. In the ac experiment, it is shown by identical forward and reverse peak potentials and by peak widths of $90/n$ mV, again regardless of scan rate. Chemical stability of the reduced form is demonstrated in the dc experiment by a peak current ratio $|i_{p,r}/i_{p,f}|$ of unity. Given charge transfer reversibility, the same thing is shown by the ratio of peak ac current amplitudes $I_{p,r}/I_{p,f}$. The advantage to the ac experiment is that the reversal response has an obvious baseline for quantitative measurements, whereas the baseline for reversal currents in the dc response is more difficult to fix.

(b) Quasi-Reversible Systems. It continues to be helpful to consider two separate cases of quasi-reversibility. In both, a significant polarization resistance is manifested in the ac response, but in one instance the dc process appears reversible, whereas more generally it is not.

When dc reversibility obtains, a theoretical description is straightforward, because the mean surface concentrations still adhere to (9.4.3) and (9.4.4), regardless of the manner in which the dc potential determining them was reached. Thus the forward and the reverse traces again overlap precisely. The shape of the peak and its position adhere to the relations derived in Section 9.4.2, where this kinetic case was considered in detail.

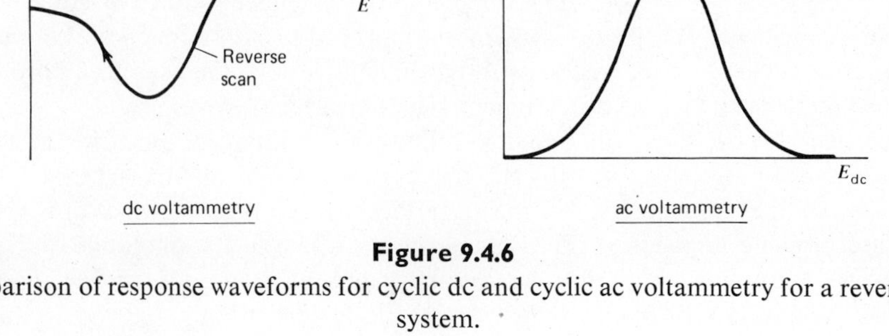

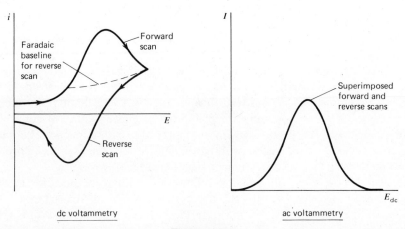

Figure 9.4.6
Comparison of response waveforms for cyclic dc and cyclic ac voltammetry for a reversible system.

If dc reversibility does not hold, then the situation becomes quite complex. The mean surface concentrations at a given dc potential tend to depend on the way in which that potential is reached. In general, the surface concentrations at any E_{dc} will differ for forward and reverse scans, and therefore we can expect the corresponding traces to differ in the voltammogram. In the dc cyclic voltammogram, increasingly sluggish electron transfer causes greater splitting of the forward and reverse peaks, because larger activation overpotentials are needed to motivate charge transfer. This splitting also manifests the fact that the surface concentrations undergo the transition from nearly pure O to virtually pure R in different potential regions for the two scan directions. Since the ac voltammogram shows a response only in the potential regions where such transition takes place, we can expect the cyclic ac voltammogram to show split peaks that are largely aligned with the forward and reverse dc voltammetric peaks. The standard potential $E^{0'}$ will lie between them. Some traces are shown in Figure 9.4.7.

Evidently there is a *crossover potential*, E_{co}, where both scans yield the same response. This potential can be rigorously shown to lie at

$$E_{co} = E_{1/2} + \frac{RT}{nF} \ln \frac{\alpha}{1 - \alpha} \tag{9.4.28}$$

regardless of the details of dc polarization (14). It serves as a convenient source for the evaluation of α. The amplitude and the phase angle of the ac response at E_{co} are totally independent of the dc process. In this respect E_{co} is a unique potential, and it may be the most convenient point for evaluating k^0 by a plot of cot ϕ *vs.* $\omega^{1/2}$. One can also derive k^0 from the separation of forward and reverse peaks in the ac voltammogram (14, 15).

Figure 9.4.8 is a display of an actual cyclic ac voltammogram for ferric acetylacetonate, $Fe(acac)_3$, in acetone containing 0.1 *M* tetraethylammonium perchlorate. Since this system is very nearly reversible to the dc process, the peak splitting is quite small, but easily detectable. The convenience of the waveform for quantitative work is also readily apparent.

(c) Homogeneous Chemical Complications. Conventional cyclic voltammetry's greatest utility is in the diagnosis of electrode reactions involving chemical complications, and the ac variant is also useful in meeting this kind of problem. The ratio $I_{p,r}/I_{p,f}$ is a sensitive indicator of product stability, just as the dc voltammetric ratio $|i_{p,r}/i_{p,f}|$ is. However, the ac ratio is easier to measure precisely, and it lends itself well to quantitative evaluation of homogeneous rate constants.

Actual results (17) for a complicated case involving two interrelated couples are shown in Figure 9.4.9. The species of interest are the complexes $Mo(CO)_2(DPE)_2$, where DPE is diphenylphosphinoethane. These complexes exist in *cis* and *trans* forms which are oxidized at different potentials. Moreover, the oxidized *cis* form (*cis*$^+$) homogeneously converts to the oxidized *trans* form (*trans*$^+$). That is,

$$trans - e \rightleftharpoons trans^+ \tag{9.4.29}$$

$$cis - e \rightleftharpoons cis^+ \tag{9.4.30}$$

$$cis^+ \xrightarrow{k} trans^+ \tag{9.4.31}$$

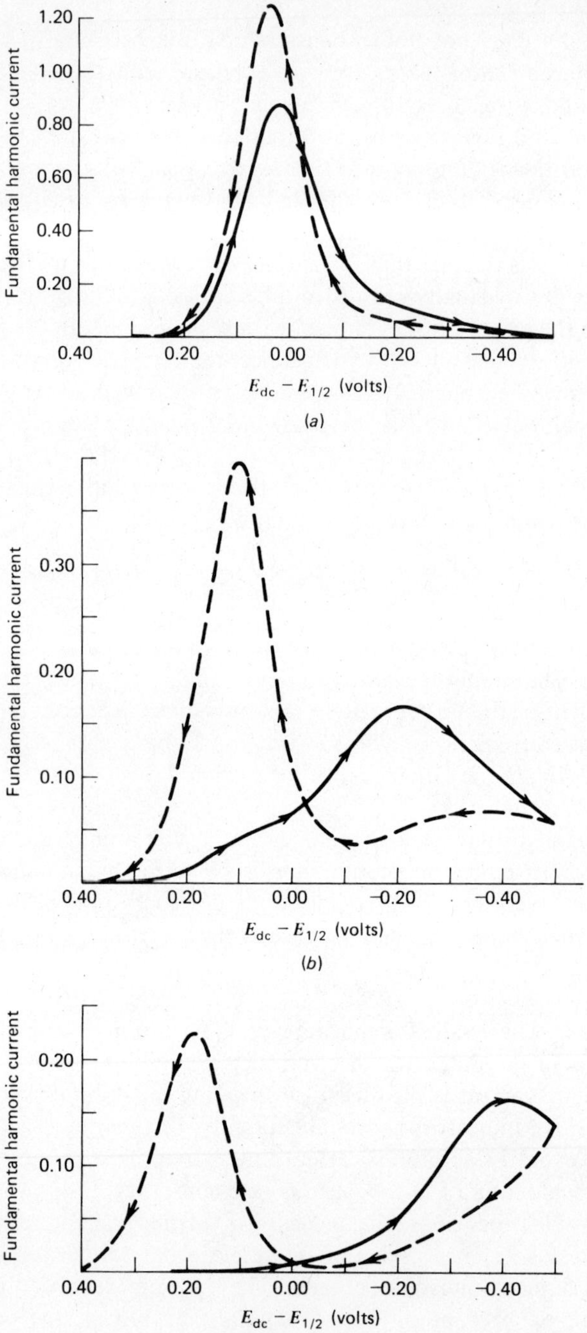

Figure 9.4.7

Predicted cyclic ac voltammograms for systems with nonnernstian dc behavior. (*a*) $k^0 = 4.4 \times 10^{-3}$ cm/sec. (*b*) 4.4×10^{-4} cm/sec, (*c*) 4.4×10^{-5} cm/sec. For $\omega/2\pi = 400$ Hz, $n = 1$, $T = 298$ K, $A = 0.30\ \text{cm}^2$, $C_O^* = 1.00$ mM, $D_O = D_R = 1.00 \times 10^{-5}\ \text{cm}^2/\text{sec}$, $v = 50$ mV/sec, $\Delta E = 5.00$ mV, and $\alpha = 0.5$. Ac amplitude given as the normalized function $RTI/n^2F^2A(2\omega D_O)^{1/2}C_O^*\Delta E$. [Reprinted with permission from A. M. Bond et al., *Anal. Chem.*, **48**, 872 (1976). Copyright 1976, American Chemical Society.]

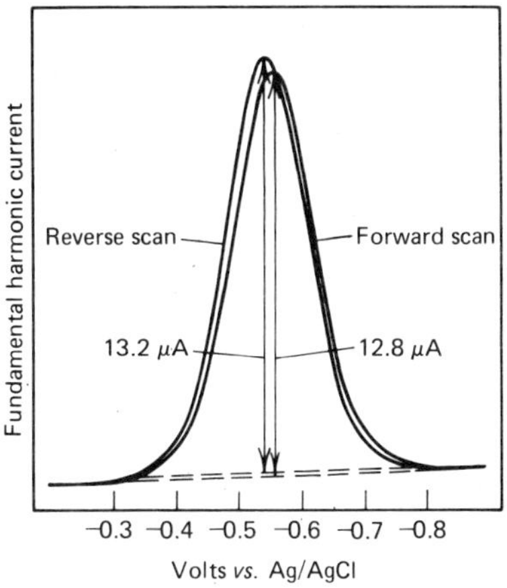

Figure 9.4.8
Cyclic ac voltammogram for 1.0 m*M* tris(acetylacetonate) Fe(III) in acetone containing 0.1 *M* tetraethylammonium perchlorate. Working electrode was a platinum disk. $T = 25°$, $\Delta E = 5$ mV, $v = 100$ mV/sec, $\omega/2\pi = 400$ Hz. [Reprinted with permission from A. M. Bond et al., *Anal. Chem.*, **48**, 872 (1976). Copyright 1976, American Chemical Society.]

The voltammograms of Figure 9.4.9 were obtained with a solution initially containing only *cis*-$Mo(CO)_2(DPE)_2$. Close study of these curves shows that the diagnostic utility of the dc voltammogram is preserved in the ac traces. In fact, it may be a bit more obvious from the ac curves that cis^+ does not decay completely during the experiment.

9.5 EFFECTS OF SOLUTION RESISTANCE AND DOUBLE-LAYER CAPACITANCE

Since Section 9.2, we have concentrated on the components of the faradaic impedance R_s and C_s; and we have assumed that they can be extracted readily from direct measurements of total impedance, which also includes the solution resistance R_Ω and the double-layer capacitance C_d. In this section, we will consider the methods for separating the faradaic impedance from those perturbations.

At a given frequency, the equivalent circuit of the cell can be detailed as in Figure 9.1.10, but we measure its impedance as the resistance value R_B and the capacitance value C_B in series. There is a set of such values for each combination of E_{dc} and ω. If one performs a parallel set of experiments in the absence of electroactive species, one obtains direct measures of R_Ω and C_d, since the faradaic path is inactive. Then the problem becomes one of graphically or analytically subtracting them from the R_B and C_B values obtained for the complete cell at the corresponding frequency and dc potential. This technique is satisfactory if the electroactive species does not alter R_Ω or C_d.

A second approach involves a study of the way in which the total impedance $\mathbf{Z} = R_B - j/\omega C_B$ varies with frequency. From this variation, one can extract R_Ω, C_d, R_s, and C_s directly. This method circumvents the need for separate measurements without the electroactive species, and it eliminates the need to assume that the electroreagents have no effect on the nonfaradaic impedance. However, one can evaluate parameters reliably only insofar as the equivalent circuit is accurately understood.

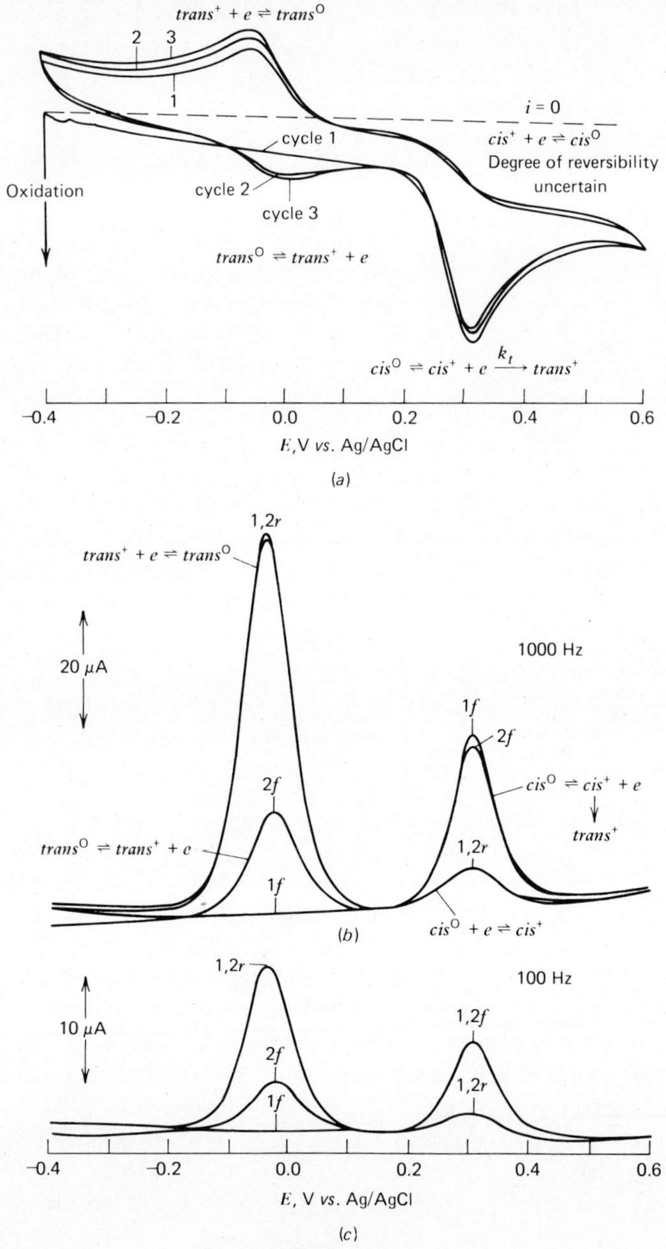

Figure 9.4.9

(*a*) Cyclic dc voltammograms of *cis*-$Mo(CO)_2(DPE)_2$ in acetone containing 0.1 *M* tetraethylammonium perchlorate. Solution was saturated with the molybdenum species. $v = 100$ mV/sec. (*b, c*) Cyclic ac voltammograms at the same platinum wire electrode. $v = 100$ mV/sec, $\Delta E = 5$ mV. *Notation*: *nf*, scan number *n* in forward direction; *nr*, scan number *n* in reverse direction. Anodic currents are plotted downward. [From A. M. Bond, *J. Electroanal. Chem.*, **50**, 285 (1974), with permission.]

9.5.1 Graphical Method

A graphical correction method based on impedance and admittance vectors was employed by Randles and by Delahay (1, 8, 18). This method is useful now in giving a mental picture of the operations involved, but the operations themselves are more conveniently carried out numerically. There are three steps, which we will discuss with reference to the frames of Figure 9.5.1.

(*a*) Since R_Ω corresponds to a series element in the equivalent cell, it can be subtracted directly from R_B. The new impedance vector $\mathbf{Z}'_B$ then represents the remaining combination of C_d, R_s, and C_s.

(*b*) Since the remaining elements are in parallel, it is convenient to deal with them as admittances. The admittance vector $\mathbf{Y}'_B$ is constructed at angle $-\phi'_B$ and with length $1/|\mathbf{Z}'_B|$. Now the admittance of the double-layer capacitance ωC_d can be subtracted. The remaining vector is the faradaic admittance $\mathbf{Y}_f$.

(*c*) The faradaic admittance is converted into the corresponding impedance vector $\mathbf{Z}_f$, which is then resolved into the components R_s and $1/\omega C_s$.

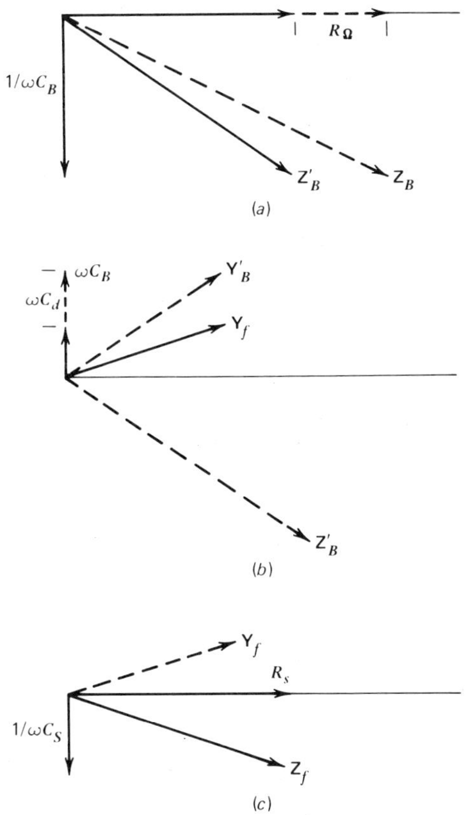

Figure 9.5.1
Extraction of R_s and C_s from R_B and C_B. See the text for a discussion of steps (*a*) to (*c*).

9.5.2 Analytical Method

Correction for the effects of R_Ω and C_d can also be effected by straightforward methods of circuit analysis through complex algebra. The technique is actually an analytical version of the graphical method that we have just discussed. A rather simple approach, described by Damaskin (19), is based on conversions between equivalent series and parallel RC networks.

Suppose we have the series RC circuit shown in Figure 9.5.2*a*, and we want to find the unique parallel circuit with the same impedance vector (Figure 9.5.2*b*). For the parallel network, we have

$$\frac{1}{\mathbf{Z}} = \frac{1}{R_p} + j\omega C_p \tag{9.5.1}$$

and, for the series combination,

$$\frac{1}{\mathbf{Z}} = \frac{\omega C}{\omega RC - j} = \frac{\omega C(\omega RC + j)}{(\omega RC)^2 + 1} \tag{9.5.2}$$

which can be simplified by defining $W \equiv (\omega RC)^2$:

$$\frac{1}{\mathbf{Z}} = \frac{W/R}{W + 1} + \frac{j\omega C}{W + 1} \tag{9.5.3}$$

By equating the real and imaginary components of (9.5.1) and (9.5.3), we obtain the useful conversion formulas:

$$\boxed{R_p = R\left(\frac{W + 1}{W}\right) \qquad C_p = \frac{C}{W + 1}} \tag{9.5.4}$$

Now suppose the problem is reversed, and we have a parallel circuit (Figure 9.5.2*b*) whose series equivalent (Figure 9.5.2*a*) is desired. The conversion formulas turn out to be

$$\boxed{R = \frac{R_p}{W_p + 1} \qquad C = \left(\frac{W_p + 1}{W_p}\right)C_p} \tag{9.5.5}$$

where $W_p = (\omega R_p C_p)^2$. The derivation is left to Problem 9.2.

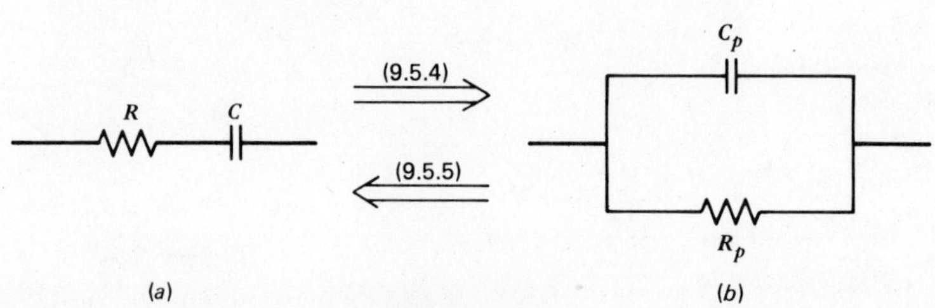

Figure 9.5.2
Interconversion of series and parallel equivalent circuits.

With (9.5.4) and (9.5.5) in hand, we are ready to face the problem of converting the measured electrochemical impedance parameters R_B and C_B into the faradaic parameters R_s and C_s. The plan of attack is laid out in Figure 9.5.3. The operations in each step are equivalent to those in the corresponding steps of the graphical procedure:

(*a*) Since R_Ω is a series element, it can be subtracted directly from R_B. The remaining series combination contains only the faradaic parameters and the double-layer capacitance.

$$R'_B = R_B - R_\Omega \tag{9.5.6}$$

(*b*) This combination is transformed to a parallel combination of R_p and C_p; thus

$$R_p = R'_B\left(\frac{1 + W}{W}\right) \qquad C_p = \frac{C_B}{1 + W} \tag{9.5.7}$$

where $W = (\omega R'_B C_B)^2$. The capacitive element C_p reflects a parallel combination of C_d and a faradaic parallel capacitance C'_p. We can subtract C_d from C_p to give only the faradaic element

$$C'_p = C_p - C_d \tag{9.5.8}$$

(*c*) Now we have a purely faradaic impedance expressed in a parallel equivalent. It is readily rewritten in terms of the usual series combination by (9.5.5):

$$R_s = \frac{R_p}{1 + W_p} \qquad C_s = \frac{C'_p(1 + W_p)}{W_p} \tag{9.5.9}$$

where $W_p = (\omega R_p C'_p)^2$.

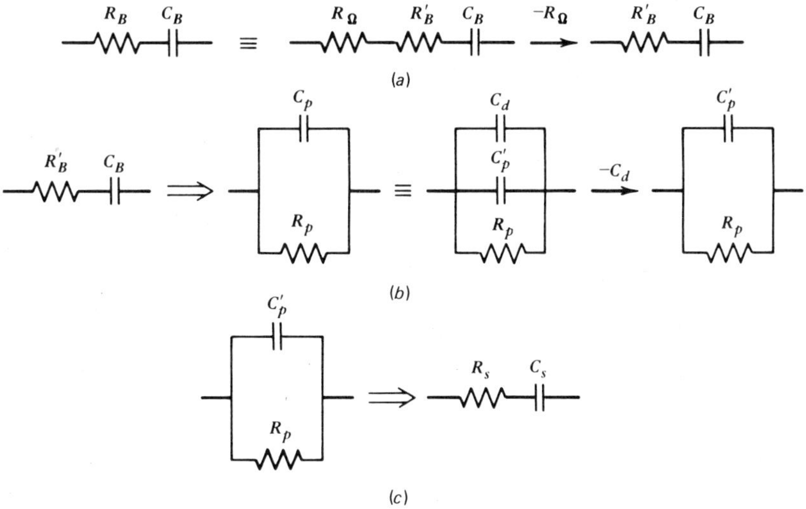

Figure 9.5.3
Extraction of R_s and C_s from R_B and C_B by the analytical method. See the text for an explanation of steps (*a*) to (*c*).

9.5.3 Variation of Total Impedance (4, 10)

An alternative approach, based on electrical engineering techniques, has been developed by Sluyters and co-workers (4). It deals with the variation of total impedance in the complex plane. An introduction is given here.

The measured total impedance of the cell, **Z**, is expressed as the series combination of R_B and C_B. These two elements provide the real and imaginary components of **Z**, i.e., $Z_{\mathrm{Re}} = R_B$ and $Z_{\mathrm{Im}} = 1/\omega C_B$. The electrochemical system is described theoretically in terms of a more complex equivalent circuit such as that in Figure 9.1.10. Its impedance is readily written down according to the methods of Section 9.1.2. The real part, which must equal the measured Z_{Re}, is

$$Z_{\mathrm{Re}} = R_B = R_\Omega + \frac{R_s}{A^2 + B^2} \tag{9.5.10}$$

where $A = (C_d/C_s) + 1$ and $B = \omega R_s C_d$. Similarly,

$$Z_{\mathrm{Im}} = \frac{1}{\omega C_B} = \frac{B^2/\omega C_d + A/\omega C_s}{A^2 + B^2} \tag{9.5.11}$$

Substitution for R_s and C_s by (9.2.25) and (9.2.26) provides

$$Z_{\mathrm{Re}} = R_\Omega + \frac{R_{ct} + \sigma\omega^{-1/2}}{(C_d\sigma\omega^{1/2} + 1)^2 + \omega^2 C_d^{\,2}(R_{ct} + \sigma\omega^{-1/2})^2} \tag{9.5.12}$$

$$Z_{\mathrm{Im}} = \frac{\omega C_d(R_{ct} + \sigma\omega^{-1/2})^2 + \sigma\omega^{-1/2}(\omega^{1/2}C_d\sigma + 1)}{(C_d\sigma\omega^{1/2} + 1)^2 + \omega^2 C_d^{\,2}(R_{ct} + \sigma\omega^{-1/2})^2} \tag{9.5.13}$$

Chemical information can be extracted by plotting Z_{Im} *vs.* Z_{Re} for different ω. For simplicity let us consider the limiting behavior at high and low ω.

(a) Low-Frequency Limit. As $\omega \to 0$, the functions (9.5.12) and (9.5.13) approach the limiting forms:

$$Z_{\mathrm{Re}} = R_\Omega + R_{ct} + \sigma\omega^{-1/2} \tag{9.5.14}$$

$$Z_{\mathrm{Im}} = \sigma\omega^{-1/2} + 2\sigma^2 C_d \tag{9.5.15}$$

Elimination of ω between these two gives

$$\boxed{Z_{\mathrm{Im}} = Z_{\mathrm{Re}} - R_\Omega - R_{ct} + 2\sigma^2 C_d} \tag{9.5.16}$$

Thus, the plot of Z_{Im} *vs.* Z_{Re} should be linear and have unit slope, as shown in Figure 9.5.4. The extrapolated line intersects the real axis at $R_\Omega + R_{ct} - 2\sigma^2 C_d$. One can see from (9.5.14) and (9.5.15) that the frequency dependence in this regime comes only from Warburg impedance terms; thus the linear correlation of Z_{Re} and Z_{Im} is characteristic of a diffusion-controlled electrode process. As the frequency rises, the charge transfer resistance R_{ct} and the double-layer capacitance become more important elements, and we can expect a departure from (9.5.16).

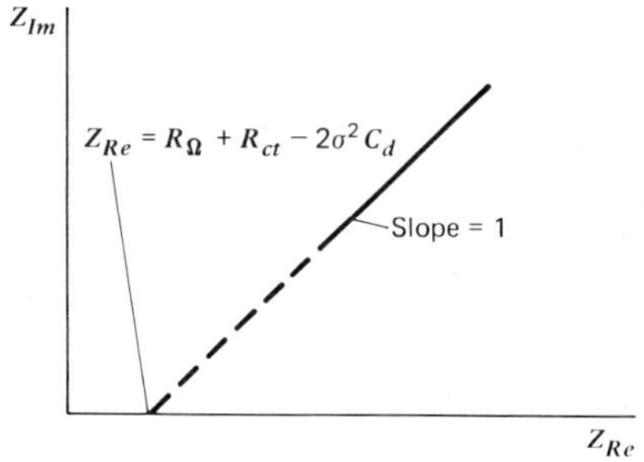

Figure 9.5.4
Impedance plane plot for low frequencies.

(b) High-Frequency Limit. At very high frequencies, the Warburg impedance becomes unimportant in relation to R_{ct}, and the equivalent circuit converges to that of Figure 9.5.5. The impedance is

$$\mathbf{Z} = R_\Omega - j\left(\frac{R_{ct}}{R_{ct}C_d\omega - j}\right) \tag{9.5.17}$$

which has the components

$$Z_{\text{Re}} = R_\Omega + \frac{R_{ct}}{1 + \omega^2 C_d{}^2 R_{ct}{}^2} \tag{9.5.18}$$

$$Z_{\text{Im}} = \frac{\omega C_d R_{ct}{}^2}{1 + \omega^2 C_d{}^2 R_{ct}{}^2} \tag{9.5.19}$$

Elimination of ω from this pair of equations yields

$$\boxed{\left(Z_{\text{Re}} - R_\Omega - \frac{R_{ct}}{2}\right)^2 + Z_{\text{Im}}{}^2 = \left(\frac{R_{ct}}{2}\right)^2} \tag{9.5.20}$$

Hence Z_{Im} *vs.* Z_{Re} should give a circular plot centered at $Z_{\text{Re}} = R_\Omega + R_{ct}/2$ and $Z_{\text{Im}} = 0$ and having a radius of $R_{ct}/2$. Figure 9.5.6 depicts the result.

The general features of the plot are readily grasped intuitively. The imaginary component to the impedance in the circuit of Figure 9.5.5 comes solely from C_d. Its contribution falls to zero at high frequencies because it offers no impedance. All of the current is charging current and the only impedance it sees is the ohmic resistance. As the frequency drops, the finite impedance of C_d manifests itself as a significant Z_{Im}. At very low frequencies, the capacitance C_d offers a high impedance, and hence current flow passes mostly through R_{ct} and R_Ω. Thus the imaginary impedance component falls off again. In general, we can expect to see a departure from this plot in this lower-frequency regime, because the Warburg impedance will become important.

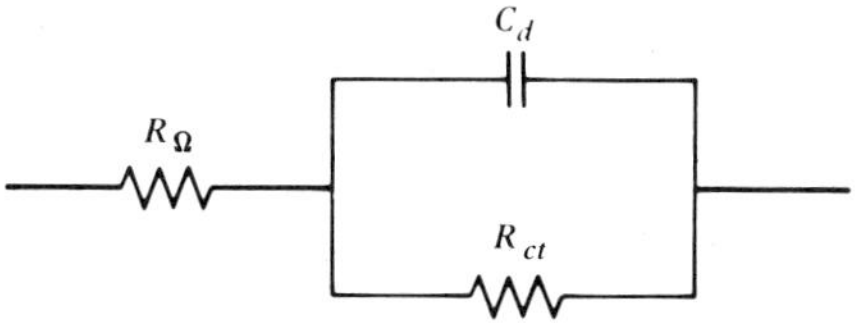

Figure 9.5.5
Equivalent circuit for a system in which the Warburg impedance is unimportant.

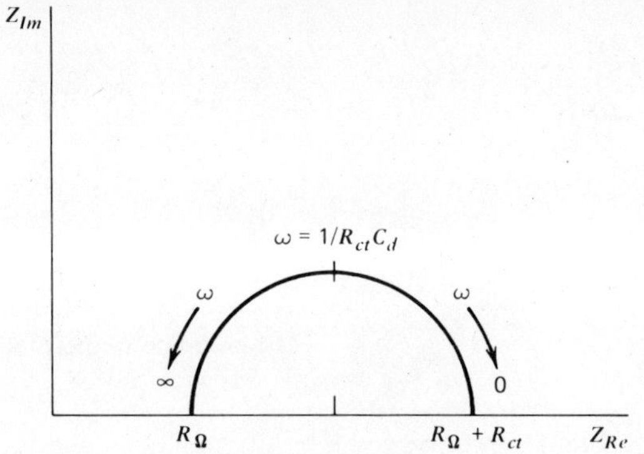

Figure 9.5.6
Impedance plane plot for the equivalent circuit of Figure 9.5.5.

(c) Application to Real Systems. An actual plot of impedance in the complex plane will combine the features of our two limiting cases as in Figure 9.5.7. However, both regions may not be well defined for any given system. The determining feature is the charge transfer resistance R_{ct} and its relation to the Warburg impedance, which is controlled by σ. If the chemical system is kinetically rather sluggish, it will show a large R_{ct}, and may display only a very limited-frequency region where mass transfer is a significant factor. This case is shown in Figure 9.5.8*a*. At the other extreme, R_{ct} might be inconsequentially small by comparison to the ohmic resistance and the Warburg impedance over nearly the whole available range of σ. Then the system is so kinetically facile that mass transfer always plays a role, and the semicircular region is not well defined. An example is shown in Figure 9.5.8*b*.

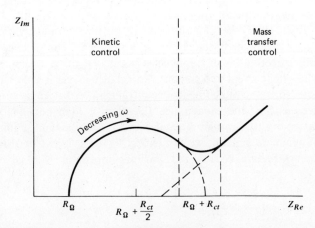

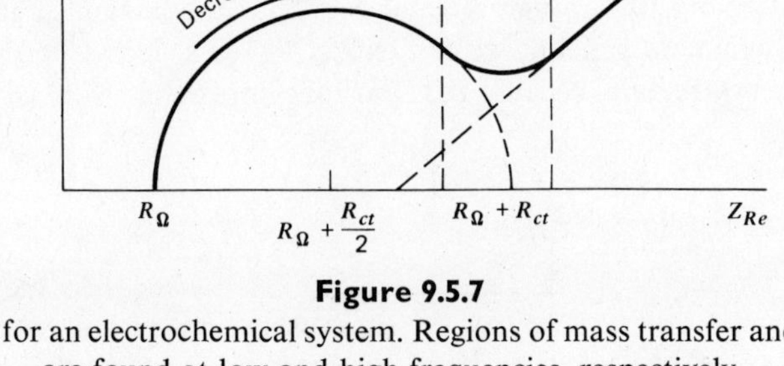

Figure 9.5.7
Impedance plot for an electrochemical system. Regions of mass transfer and kinetic control are found at low and high frequencies, respectively.

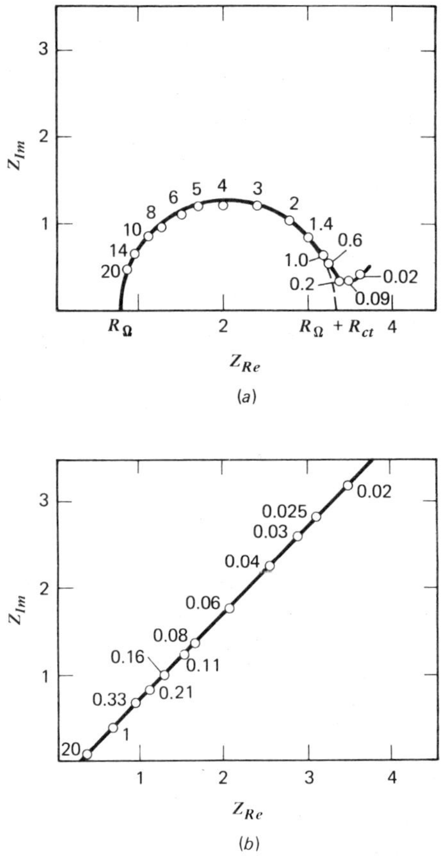

Figure 9.5.8
Impedance plane plots for actual chemical systems. Numbers by points are frequencies in kilohertz. (*a*) For the electrode reaction $Zn^{2+} + 2e \rightleftharpoons Zn(Hg)$. $C^*_{Zn^{2+}} = C^*_{Zn(Hg)} = 8 \times 10^{-3}$ *M*. Electrolyte was 1 *M* $NaClO_4$ plus 10^{-3} *M* $HClO_4$. (*b*) For the electrode reaction $Hg_2{}^{2+} + 2e \rightleftharpoons Hg$ in 1 *M* $HClO_4$. $C^{2+}_{Hg_2} = 2 \times 10^{-3}$ *M*. [From J. H. Sluyters and J. J. C. Oomen, *Rec. Trav. Chim.*, **79**, 1101 (1960), with permission.]

9.5.4 Matched Cells

In ac polarographic measurements, one can often employ a dual cell configuration in which one cell contains the electroactive species of interest and the other holds only the supporting electrolyte. If the DMEs are synchronized and matched in *m* values, and if both are controlled at the same potential, then their difference current represents the faradaic current of interest at every point in time.† The difference current is readily recorded directly by electronic means; hence this technique is essentially a method for compensating charging current in real time. It is most useful in kinetic studies involving sufficiently high concentrations of electroactive substances that the faradaic currents are relatively large. Of course, the fundamental assumption behind this scheme is that the electroactive substance of interest does not alter the interfacial capacitance.

† Note that the requirement that the two electrodes be at the same potential implies that the solution resistance is either negligible or is almost fully compensated. In ac measurements, uncompensated resistance also causes phase shifts of faradaic and nonfaradaic currents; hence the difference current may not even be fully faradaic if the solution resistance is appreciable (20).

9.5.5 Limits to Measurable k^0 by the Faradaic Impedance Method (1–6)

The foregoing paragraphs highlight the limitations in interpreting impedance data, and they lead naturally to the idea that k^0 must fall in some fairly well-defined range in order to be reliably measured by an impedance method. We can define the range semiquantitatively.

(a) Upper Limit. The parameter R_{ct} must make a significant contribution to R_s; hence $R_{ct} \gtrsim \sigma/\omega^{1/2}$. Substituting from (9.3.2), (9.3.6), and (3.5.7) and assuming $D_O = D_R$ and $C_O^* = C_R^*$, we obtain the condition that $k^0 \lesssim (D\omega/2)^{1/2}$. The highest practical values of ω are on the order of 2×10^5 sec^{-1}, and with $D \sim 10^{-5}$ cm^2/sec, we have $k^0 \lesssim 1$ cm/sec†. In addition, there are requirements that $C_s \gtrsim C_d$ and $R_s \gtrsim R_\Omega$.

(b) Lower Limit. For very large R_{ct}, the Warburg impedance is negligible, and the equivalent circuit of Figure 9.5.5 can be applied. The problem here is that R_{ct} cannot be so large that all the current takes the path through C_d. That is, $R_{ct} \lesssim 1/C_d\omega$ or $k^0 \gtrsim RTC_d\omega/n^2F^2C^*A$. If we choose the most favorable conditions of $C^* = 10^{-2}$ M and $\omega = 2\pi \times 10$ Hz, then at $T = 298$ K, $C_d/A = 20$ μF/cm^2, and $n = 1$, we obtain $k^0 \gtrsim 2 \times 10^{-5}$ cm/sec.

9.6 HIGHER HARMONICS (3, 5, 7)

To this point, we have found that excitation of an electrochemical system by a signal $\dot{E}_{ac} = \Delta E \sin \omega t$ produces a sinusoidal current response at the same frequency. That result rests on the fact that only the linearized current-potential relation has been used. The remaining terms in the Taylor expansion were dropped. If we include them, we find that the current response is not purely sinusoidal, but instead comprises a whole series of sinusoidal signals at ω, 2ω, 3ω, . . ., which are summed together. The current component of 2ω is the *second harmonic* response, while that at 3ω is the *third harmonic*, etc.‡ These higher harmonics arise from *curvature* in the *i-E* relation.

Each harmonic is individually detectable by circuits employing tuned amplifiers or lock-in amplifiers. The most common arrangement is a variant of ac voltammetry, as might be implemented according to Figure 9.6.1. The cell is excited exactly as in ac voltammetry, but the lock-in amplifier is tuned to 2ω and detects only that current contribution. The result is a trace of $I(2\omega)$ *vs.* E_{dc}.

An exact treatment of higher harmonic response is straightforward, but it is rather lengthy, so we will leave it to the specialized literature. Instead we will follow an intuitive approach that will reveal most of the distinctive features of second-harmonic

† The reductions and oxidations of aromatic species to anion and cation radicals in aprotic solvents are generally among the fastest known heterogeneous charge transfer reactions. The values of k^0 can exceed 1 cm/sec. See References 21 and 22 for recent measurements by impedance methods.
‡ This nomenclature differs from that used in electrical engineering, where the signal at ω is the *fundamental* and that at 2ω is the *first harmonic*. We will adhere to the usual electrochemical usage.

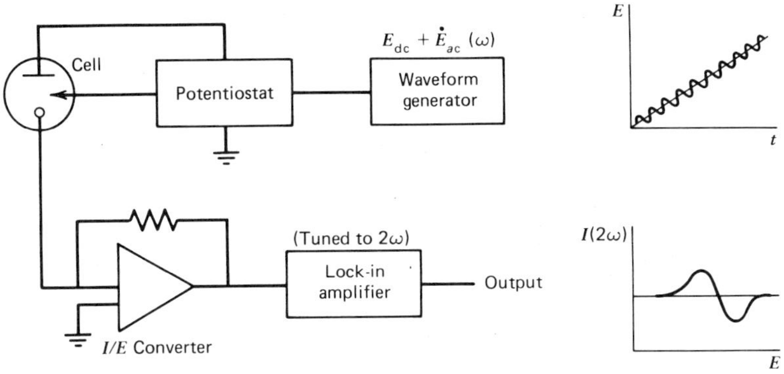

Figure 9.6.1
Block diagram of apparatus for recording a second-harmonic ac voltammogram.

ac voltammetry. For simplicity, we consider only a reversible system in which R is initially absent.

The mean surface concentrations $C_O(0, t)_m$ and $C_R(0, t)_m$ are set by the value of E_{dc} and are given by (9.4.3) and (9.4.4). Figure 9.6.2 is a graphical display of $C_R(0, t)_m$. The ac response is determined by the way in which $\dot{E}_{ac}$ causes small perturbations in the surface concentrations about those mean values. The fundamental (or first harmonic) component is controlled essentially by the linear elements of variation, which are the slopes $\partial C_O(0, t)_m/\partial E$ and $\partial C_R(0, t)_m/\partial E$. The higher harmonics reflect curvature; hence they are sensitive to the second and higher derivatives. Comprehension of this point allows us to predict the general shape of the second harmonic response.

Consider potentials E_1, E_3, and E_5 in Figure 9.6.2. Since they have the common feature that the curvature in $C_O(0, t)_m$ and $C_R(0, t)_m$ is zero, there is no second harmonic current. Of course, E_1 and E_5 lie at extreme values where there is also no fundamental response; but E_3 lies at the inflection point $E = E_{1/2}$, where the fundamental response is greatest. The potentials E_2 and E_4 are at points of maximum curvature, hence they should be the potentials of peak second-harmonic current. If we detect only the magnitude $I(2\omega)$, then we can expect a double-peaked voltammogram like that of Figure 9.6.3*a*.

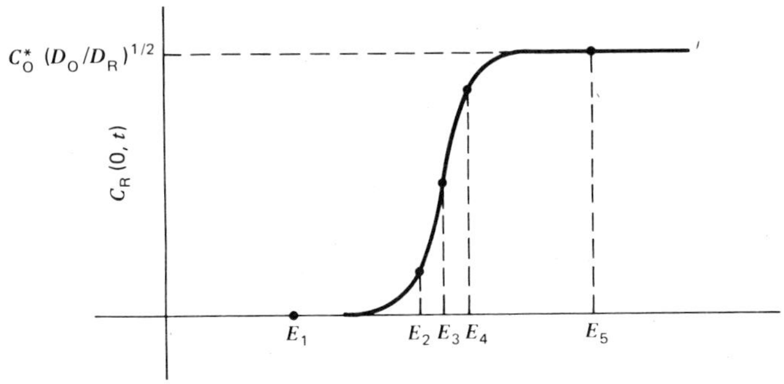

Figure 9.6.2
Dependence of the surface concentration of species R on the electrode potential.

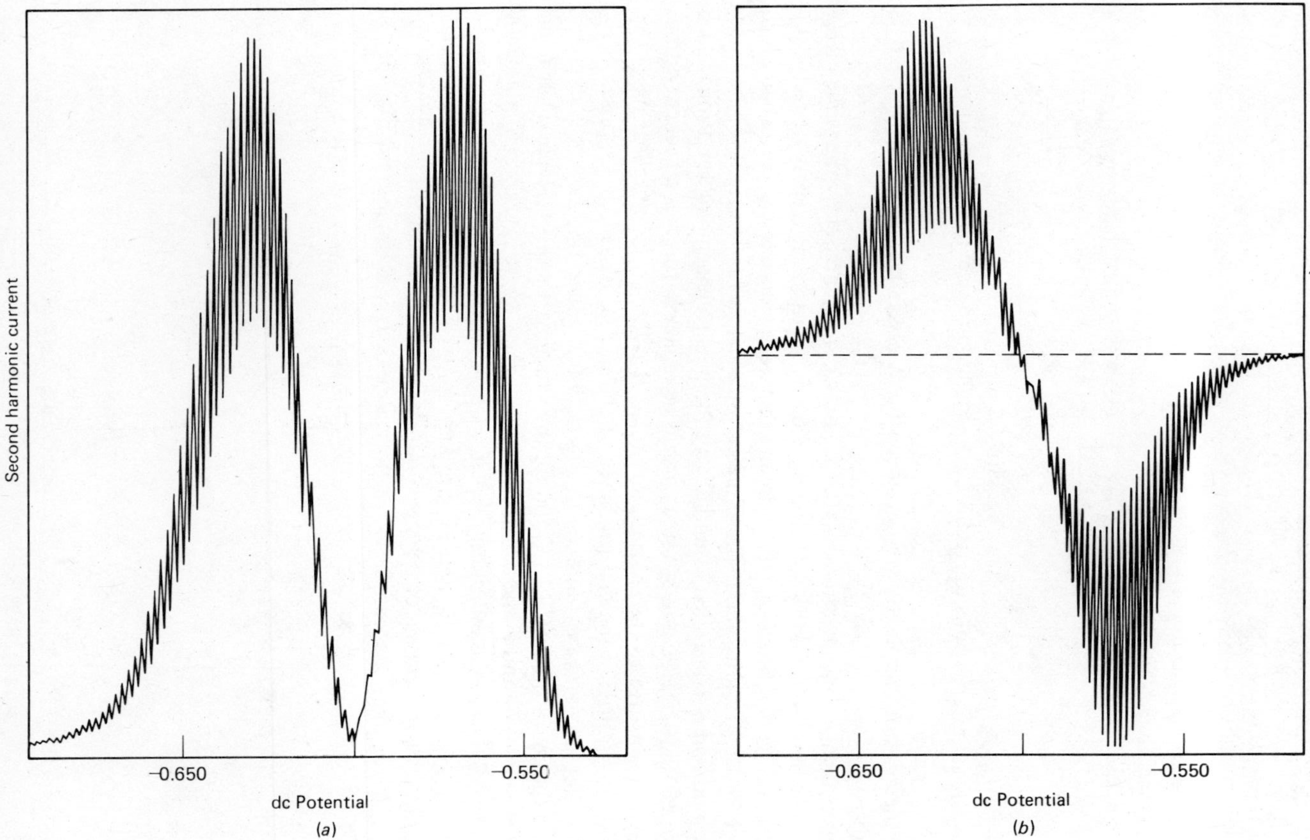

Figure 9.6.3

Second-harmonic ac polarograms for 3 mM Cd^{2+} in 1.0 M Na_2SO_4. $\omega/2\pi$ = 80 Hz, ΔE = 5 mV. (*a*) Total ac amplitude *vs.* E_{dc} (*vs.* SCE). (*b*) Phase-selective polarogram showing the ac amplitude at 0° with respect to $\dot{E}_{ac}$. [Reprinted with permission from D. E. Smith, *Anal. Chem.*, **35**, 1811 (1963). Copyright 1963, American Chemical Society.]

Let us note, though, that the curvature at E_2 is opposite to that at E_4. This difference implies that the second harmonic component undergoes a 180° phase shift when E_{dc} passes through the null point at $E_{1/2}$. Phase-sensitive detection of $I(2\omega)$ at a fixed phase angle will therefore produce a sign inversion at $E_{1/2}$. Figure 9.6.3*b* is an example.

The exact solution of this problem is

$$i(2\omega) = \frac{n^3F^3AC_O^*(2\omega D_O)^{1/2}\,\Delta E^2 \sinh(a/2)}{16R^2T^2\cosh^3(a/2)} \sin\left(2\omega t - \frac{\pi}{4}\right) \tag{9.6.1}$$

where a is defined in (9.4.7). This equation embodies the proportionalities of C_O^*, $\omega^{1/2}$, and $D_O^{1/2}$ that we have come to expect of diffusion-controlled processes. The phase angle of 45° has the same origin. Note, however, that $i(2\omega)$ is proportional to ΔE^2. This dependence manifests the greater importance of nonlinear effects for perturbations of larger magnitude. The two peak potentials are located at $E_{dc} = E_{1/2} \pm 34/n$ mV at 25°.

Second-harmonic techniques are useful for analytical purposes and for the quantitative evaluation of kinetic parameters (3, 5, 7, 23). Applications in both areas are attractive with respect to measurements at the fundamental frequency, because the double-layer capacitance is a rather linear element and contributes very small second-harmonic currents. Harmonics above the second have been examined briefly, but have not been applied to any great extent.

9.7 CHEMICAL ANALYSIS BY AC VOLTAMMETRY

Both fundamental and second-harmonic ac voltammetry are attractive as analytical techniques because they offer rather good sensitivities. Detection limits for the polarographic variants can reach the order of 10^{-7} *M*. Such performance is possible because both methods have ready means for discrimination against capacitive currents (5, 7).

In the fundamental mode, one employs phase-sensitive detection to measure the current component lying in phase with the excitation signal $\dot{E}_{ac}$. We noted in Section 9.3 that there is generally a faradaic contribution to this current, and Figure 9.7.1 reinforces the idea pictorially. In contrast, the charging current is ideally 90° out of phase with $\dot{E}_{ac}$, since it passes through a purely capacitive element. It therefore has no projection in phase with $\dot{E}_{ac}$. Thus we expect the current in phase to be purely faradaic,

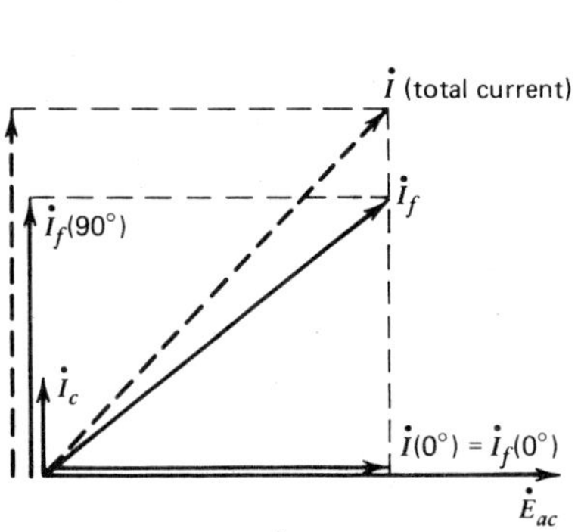

Figure 9.7.1
Phasor diagram showing the relationships between the faradaic ($\dot{I}_f$) and capacitive ($\dot{I}_c$) components to the total current ($\dot{I}$). Note that $\dot{I}_f$ has a component along $\dot{E}_{ac}$, whereas $\dot{I}_c$ does not.

whereas the current at 90° (*the quadrature current*) contains a second faradaic component plus the nonfaradaic contributions. By taking the current in phase as the analytical signal, we discriminate effectively against the capacitive interference.

The limitation to this scheme is imposed partially by the uncompensated resistance R_Ω. Since the charging current passes through R_Ω and C_d in series, this current does not lead $\dot{E}_{ac}$ by exactly 90°, but instead by some smaller angle. Then the current in phase with $\dot{E}_{ac}$ must contain a nonfaradaic element, which becomes significant to the measurement as the analyte concentration drops.

Second-harmonic ac voltammetry gains its freedom from nonfaradaic interference from the relative linearity of the double-layer capacitance as a circuit element. There is consequently only a very small second-harmonic capacitive current, although it too can become important at low analyte concentrations.

The shapes of the voltammograms generated in ac measurements are convenient for analysis. Detection of the fundamental current produces a peak whose height is readily measured and is linear with concentration. Phase-selective second-harmonic voltammetry gives the second-derivative waveform of Figure 9.6.3*b*. The peak-to-peak amplitude is linear with concentration and can be read with high precision. It is also relatively unaffected by the background signal (23).

Analytical measurements are usually carried out at excitation frequencies ranging from 10 Hz to 1 kHz, although Bond has noted that frequencies in the upper part of this range allow fuller exploitation of certain aspects of selectivity that are unique to ac methods (24). Their basis is the discrimination against the much smaller response from irreversible systems.

For example, one can effect a significant saving in analysis time by working directly with aerated solutions. Since the reduction of oxygen in most aqueous solutions is irreversible, it does not interfere with determinations made by ac voltammetry. Also, one can often control the medium in order to introduce selectivity toward certain analytes. Transition metals are especially susceptible to such manipulation because their electrode kinetics are often strongly affected by coordination. Thus, one can enhance their ac responses or mask them from the voltammogram by intelligent choice of electrolyte composition. Since many supporting electrolytes show irreversible reductions, there is considerable freedom to manipulate composition without introducing serious interferences.

9.8 USES OF THE FOURIER TRANSFORMATION IN THE ANALYSIS OF DATA

Applications of the Fourier transformation to spectroscopy have become widespread and are familiar to most chemists. They are attractive because they allow one to interpret experiments in which several different excitation signals are applied to a chemical system at the same time. The responses to those signals are superimposed on each other, but the Fourier transformation provides a means for resolving them. This capacity for simultaneous measurement is sometimes called the *multiplex advantage* of transform methods, and it is of great importance to applications in electrochemistry.

Research in this area includes some of the most promising new directions in electrochemical instrumentation and methods design; therefore, we will try in this section to sample its flavor and understand its distinctive strategies (see also Section A.6).

By now it should be clear that fully characterizing an electrochemical process by impedance methods is a tedious operation, because one requires information at a set of frequencies ranging over 2 to 3 decades and at a set of potentials ranging over $E^{0'} \pm 100$ mV. For example, the data in Figure 9.4.4 alone required eight ac polarograms, each scanned with the tuned circuitry set to a different frequency and each having the in-phase and quadrature currents separately recorded. Not only does the operation require time and patience but also there is the danger that the surface properties of the system will change during the procedure.

One can employ an alternative in which excitation signals of all desired frequencies are brought to bear at once (25–28). The idea is outlined in Figure 9.8.1, which shows that the excitation signal $\dot{E}_{ac}$ is actually a noise waveform, rather than a pure sinusoid. As before, $\dot{E}_{ac}$ is superimposed on a virtually constant level E_{dc}. Of course $\dot{E}_{ac}$ will stimulate a current flow showing related "noisy" variations. During a brief period, lasting perhaps 100 msec, late in the life of a DME drop, the output of the follower and the output of the i/E converter are digitized simultaneously and stored in a computer's memory.

Fourier transformation of these two transients gives the distribution of harmonics that comprise the signals. One therefore knows the amplitude of excitation and the corresponding amplitude and phase angle for current flow at each frequency in the Fourier distributions. In other words, one has the faradaic impedance as a function of ω for the potential E_{dc}. All of this has been obtained on a single drop with a 100-msec period of data acquisition; hence it is feasible to repeat the whole procedure on subsequent drops, so that more precise ensemble-averaged results are obtained. Changing E_{dc} for each complete set of measurements then provides the potential distribution, that is, $\dot{E}(E_{dc}, \omega)$.

In practice, it is desirable to have a special kind of excitation noise. Smith and his co-workers (26, 28) have demonstrated that the best choice is an odd-harmonic, phase-varying pseudorandom white noise of the type displayed in Figure 9.8.2. This noise is the superposition of signals at several frequencies (15 in the example), all of which are odd harmonics of the lowest frequency. The choice of odd harmonics ensures that second-harmonic components will not appear in the currents measured for the 15 fundamental frequencies. The amplitudes of the 15 excitation frequencies are equal ("white" noise) so that each carries equal weight; and their phase angles are randomized, so that the total excitation signal does not show large swings in amplitude.

Despite these demands, it is easy to generate such special noise by inverting the scheme for signal analysis. The method is sketched in Figure 9.8.2. One starts with the amplitude and phase-angle arrays, which have been tailored in the computer according to specifications. These are transformed into the complex plane; then the fast inverse Fourier transform is invoked, so that one obtains a digital representation of the time-domain noise signal. Feeding these numbers sequentially to a D/A converter at the desired rate yields an analog signal, which is filtered and passed on to the potentiostat's input. Repeated passage through the D/A conversion and filtering steps

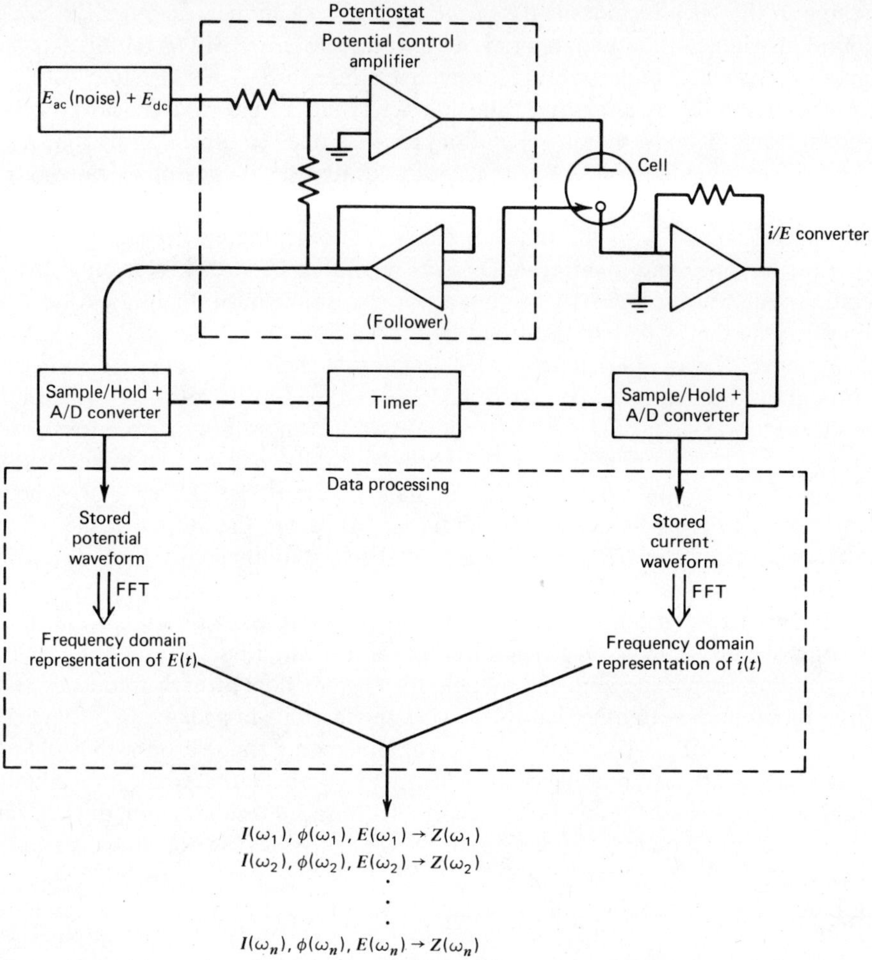

Figure 9.8.1
Schematic diagram showing apparatus and data-processing steps used in on-line Fourier analysis of ac voltammetric data. The steps in the large dashed box are carried out in data-processing equipment, such as a computer. The Fourier transformations are usually carried out by the fast Fourier transform (FFT) algorithm (see Section A.6).

yields a repetitive excitation waveform, which is applied continuously until a single measurement pass is completed. A new waveform with different randomized phase angles is generated for the next pass, and so on.

The quality of the results from these experiments is illustrated in Figure 9.8.3 by data for the $Cr(CN)_6^{3-}/Cr(CN)_6^{4-}$ couple. The run represents an average of 64 measurement passes, each taken on one DME drop and each requiring $\sim$2 sec for acquisition and reduction of the data. Compare the range of cot ϕ and its precision in this figure with that of the good manual data in Figure 9.4.5.

The ability of the Fourier transformation to dissect a complex waveform into its components can also be used to obtain higher harmonics (27, 28). In this case, one

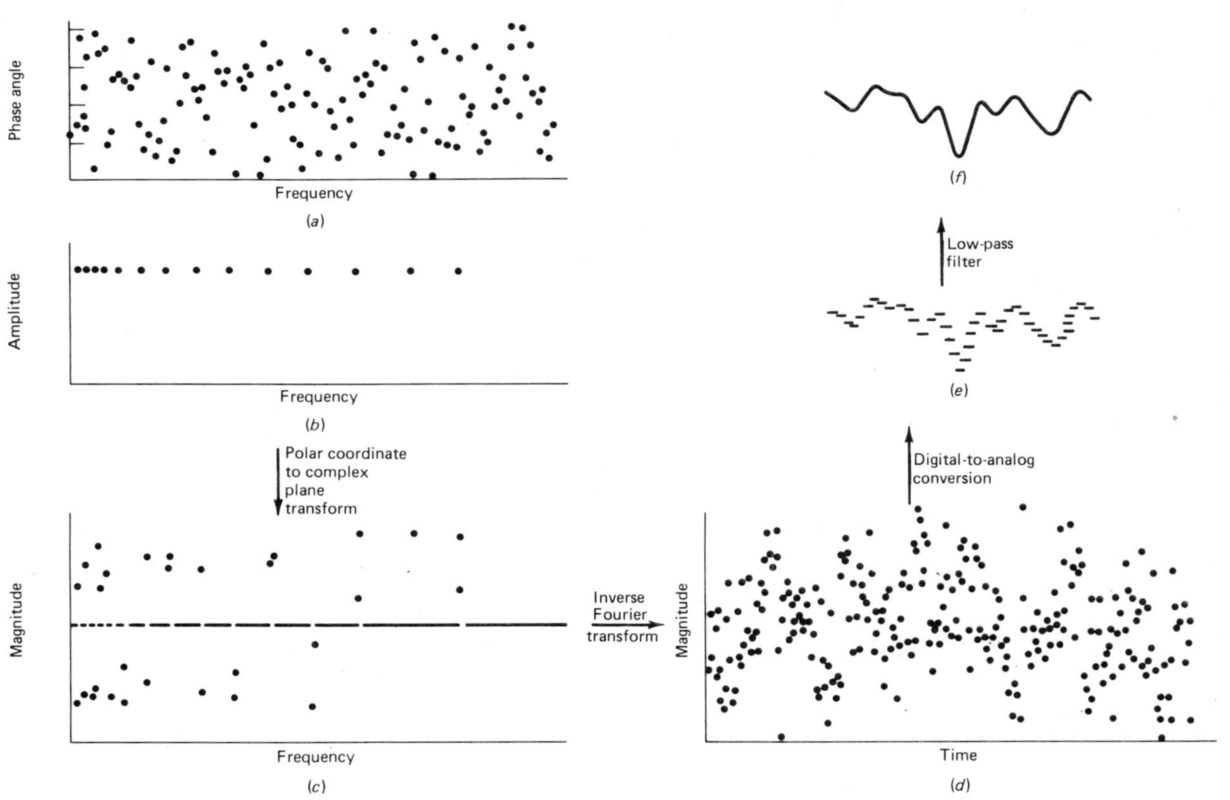

Figure 9.8.2

Procedure for generating a complex excitation waveform. (*b*) shows the chosen amplitudes for the various frequencies and (*a*) shows randomized phase angles. In (*c*) there is a complex plane representation of the arrays in (*a*) and (*b*). (*d*) is the time domain representation, which is subjected to digital/analog conversion to produce (*e*); and in turn, low-pass filtering yields (*f*). Only a small part of the waveform period is shown in (*e*) and (*f*). [From S. C. Creason et al., *J. Electroanal. Chem.*, **47**, 9 (1973), with permission.]

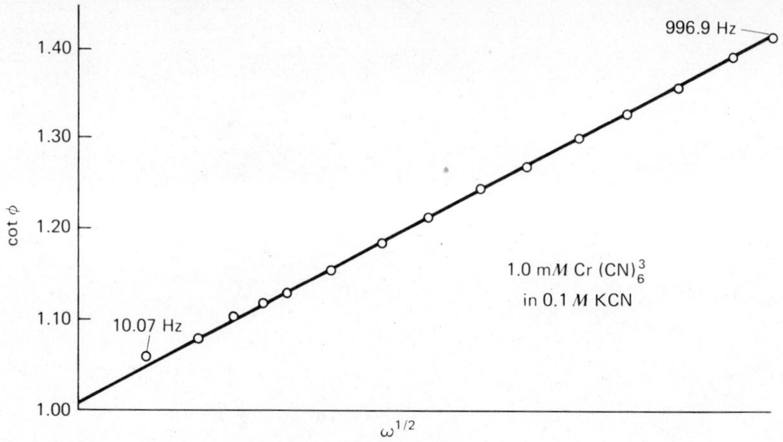

Figure 9.8.3
Cot ϕ *vs.* $\omega^{1/2}$ for the chromicyanide system. Data were obtained with a phase-varying, 15-component, odd-harmonic complex waveform. [From S. C. Creason et al., *J. Electroanal. Chem.*, **47**, 9 (1973), with permission.]

might excite with a pure sinusoid at frequency ω and examine the transformed current waveform. It will provide the dc current, the current and phase angle at the fundamental frequency ω, as well as amplitudes and phase angles of the higher harmonics. Repeating measurement passes at various values of E_{dc} allows one to trace out all of the corresponding polarograms from data obtained on a single run.

In addition to its applications as an integral component of the measurement process, the Fourier transformation can be extremely useful for various signal-conditioning operations, such as smoothing, convolution, and correlation. Smith et al. have discussed the intriguing possibilities in this area at some length, and the interested reader is urged to pursue their discussions (28, 29).

9.9 ANALYSIS OF DATA IN THE LAPLACE PLANE

There are many instances in electrochemistry when we find it very difficult to obtain an explicit relationship between current, potential, and time. Either the system itself is intrinsically complex (e.g., a quasi-reversible charge transfer involving adsorbed and diffusing reactant species) or the experimental conditions are less than ideal (e.g., step experiments carried out on a time domain sufficiently short that the risetime of the potentiostat is not negligible). It is usually true in these and other cases that much simpler relationships exist in the Laplace domain between the perturbations and the observables. Thus it can be useful to transform the data and carry out the analysis in transform space (30–33).

As an example, consider the case of potential steps applied to a system containing electroactive species O, which is reduced quasi-reversibly. In Section 5.5.1, we treated this case conventionally and found that the current-time function was

$$i = nFAk_f C_O^* \exp(H^2 t)\,\mathrm{erfc}(Ht^{1/2}) \tag{9.9.1}$$

where $H = k_f/D_O^{1/2} + k_b/D_R^{1/2}$. The complex time dependence embodied in (9.9.1) is difficult to handle in the analysis of real data; hence various attacks based on linearizations or extrapolations were devised. On the other hand we could use all of the data without introducing such approximations by considering the transformed current:

$$\bar{\imath}(s) = \frac{nFAk_f C_O^*}{s^{1/2}(H + s^{1/2})} \tag{9.9.2}$$

We might plot, for example, the function $1/\bar{\imath}(s)s^{1/2}$ *vs.* $s^{1/2}$. The slope and intercept of the resulting linear function would provide k_f and H. In doing this, we have elected to analyze the system in the s domain, rather than the time domain.

To implement such a plan, we must be able to obtain the function $\bar{\imath}(s)$ from the measured curve $i(t)$. That can be done by considering the definition of the Laplace transformation (see Section A.1):

$$\bar{\imath}(s) = \int_0^\infty i(t)\, e^{-st}\, dt \tag{9.9.3}$$

In a practical situation, $i(t)$ is usually a collection of data points. Thus, $\bar{\imath}(s)$ is calculated for a given value of s by multiplying each point by e^{-st} and then performing a numeric integration of the resulting curve. Algorithms for carrying out this task on a computer have been published (33). The whole process is repeated for each desired s value, and the final result is a new collection of data points describing $\bar{\imath}(s)$, just as the original data described $i(t)$. Since s has dimensions of frequency, $\bar{\imath}(s)$ is sometimes called the representation of the current in the *frequency domain.*

Many applications of this strategy are based on extensions of the concepts of impedance that have been developed earlier in this chapter (32–34). However, the excitation waveform is usually an impulse in potential, and a transient current is measured. One records *both* $E(t)$ and $i(t)$ as observed functions. Then both are subjected to transformations, and comparisons are made in the frequency domain between $\bar{E}(s)$ and $\bar{\imath}(s)$. Ratios of the form $\bar{\imath}(s)/\bar{E}(s)$ are *transient impedances,* which can be interpreted in terms of equivalent circuits in exactly the fashion we have come to understand. The advantages of this approach are (a) that the analysis of data is often simpler in the frequency domain, (b) that the multiplex advantage applies, and (c) the waveform $E(t)$ does not have to be ideal or even precisely predictable. The last point is especially useful in high-frequency regions, where potentiostat response is far from ideal. Laplace domain analyses have been carried out for frequency components above 10 MHz.

In general, it is useful to regard s as a complex number $s = \sigma + j\omega$ in these analyses (32, 33). Then one can calculate real-axis and imaginary-axis frequency domain representations of a function. For example, the *real axis transform* of $E(t)$ is

$$\boxed{\bar{E}(\sigma) = \int_0^\infty E(t)\, e^{-\sigma t}\, dt} \tag{9.9.4}$$

and the *imaginary axis transform* is

$$\boxed{\bar{E}(j\omega) = \int_0^\infty E(t)\, e^{-j\omega t}\, dt = \int_0^\infty E(t) \cos \omega t\, dt - j\int_0^\infty E(t) \sin \omega t\, dt} \tag{9.9.5}$$

Note that the real axis transform of any function is strictly real, but the imaginary axis transform is complex. It has both real and imaginary components. Since one can transform experimental pötential and current transients in this way, one can calculate a *real-axis transient impedance*, $Z(\sigma) = \bar{i}(\sigma)/\bar{E}(\sigma)$, and an *imaginary-axis transient impedance*, $Z(j\omega) = \bar{i}(j\omega)/\bar{E}(j\omega)$. Since $Z(j\omega)$ is complex, we can break it into real and imaginary components $Z(j\omega)_{\mathrm{Re}}$ and $Z(j\omega)_{\mathrm{Im}}$. One can easily show that $Z(j\omega)$ is the same as the conventional impedance (comprising resistances and reactances) that we have already discussed. These various functions are especially useful for the analysis of electrical response in terms of equivalent circuits. In general, all of the various transform functions contain the same chemical information; however, one of them may be more readily applied to data analysis. Since this treatment involves a complex s domain, it is often called *Laplace plane analysis.*

Consider as an example, a double-layer capacitance C_d in series with an uncompensated solution resistance R_u. The overall system obeys

$$E(t) = i(t)R_u + \frac{1}{C_d}\int_0^t i(t)\, dt \tag{9.9.6}$$

or, in the frequency domain,

$$\bar{E}(s) = \bar{i}(s)\left(R_u + \frac{1}{C_d s}\right) \tag{9.9.7}$$

Thus the various impedances are

$$Z(\sigma) = R_u + \frac{1}{C_d\sigma} \tag{9.9.8}$$

$$Z(j\omega)_{\mathrm{Re}} = R_u \qquad Z(j\omega)_{\mathrm{Im}} = \frac{-1}{C_d\omega} \tag{9.9.9}$$

and the phase angle ϕ, defined by the real and imaginary components of $Z(j\omega)$, is given by

$$\tan \phi = \frac{1}{\omega R_u C_d} \tag{9.9.10}$$

Thus we have four simple frequency domain relationships that allow the evaluation of R_u and C_d. It is probably most convenient to use $Z(\sigma)$ for that purpose, but the availability of the other functions is useful for cross-checking the validity of one's equivalent circuit as a model for any given chemical system.

An interesting recent application of Laplace plane analysis comes from the work by Pilla and Margules on ionic transport through biological membranes (34). Their experimental arrangement involved the use of the membrane as a separator between two solutions containing separate electrodes. A small voltage pulse was applied across the membrane and the transient current was measured. Transformation of the voltage and current functions allowed the calculation of the impedances described above.

The equivalent circuits used in the analysis are shown in Figure 9.9.1. The elements

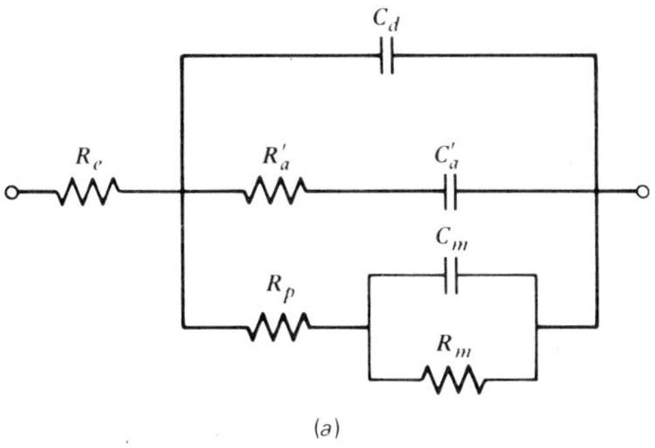

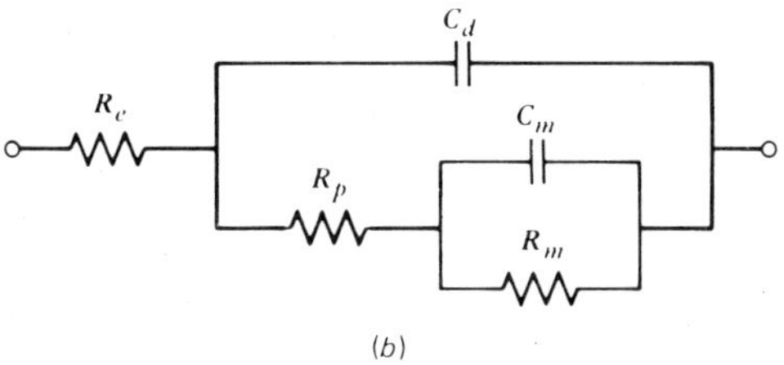

Figure 9.9.1

Equivalent circuits used to analyze the transient behavior of the toad urinary bladder membrane. R_e represents electrolyte resistance and C_d is the dielectric capacitance of the membrane. The branches involving R_p, C_m, and R_m are used to account for the transfer of charge across the membrane boundaries. They are analogous to R_{ct} and Z_w in electrode reactions. In circuit *a*, R'_a and C'_a model the effects of adsorption. [From A. A. Pilla and G. S. Margules, *J. Electrochem. Soc.*, **124**, 1697 (1977), reprinted by permission of the publisher, The Electrochemical Society, Inc.]

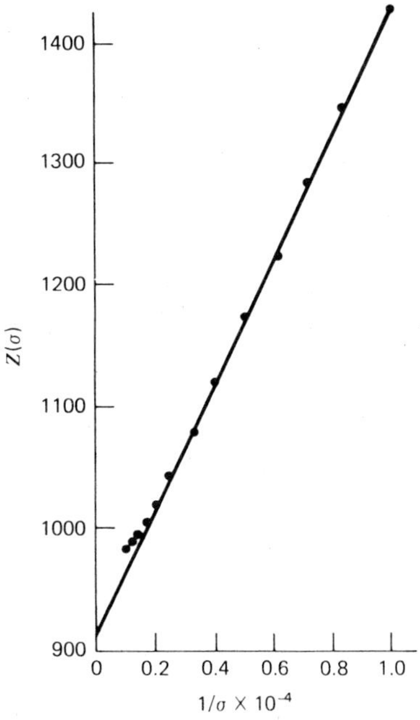

Figure 9.9.2

High-frequency plot of $Z(\sigma)$ *vs.* $1/\sigma$ for the toad urinary bladder membrane. [From A. A. Pilla and G. S. Margules, *J. Electrochem. Soc.*, **124**, 1697 (1977), reprinted by permission of the publisher, The Electrochemical Society, Inc.]

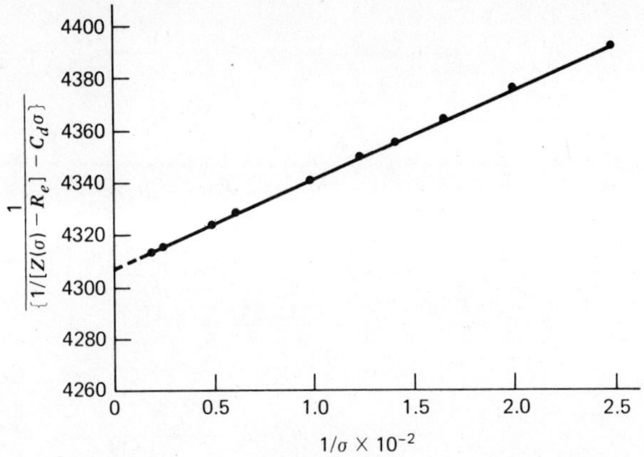

Figure 9.9.3

Intermediate-frequency plot of real axis impedance data for the toad urinary bladder membrane. [From A. A. Pilla and G. S. Margules, *J. Electrochem. Soc.*, **124**, 1697 (1977), reprinted by permission of the publisher, The Electrochemical Society, Inc.]

correspond to solution resistance, interfacial capacitances, and impedances associated with the transport of ions through the membrane and across the boundary between the solution and the membrane.

Figure 9.9.2 is an illustration of the behavior of $Z(\sigma)$ at high frequencies in an actual system (34). At these frequencies, both of the circuits in Figure 9.9.1 look essentially like a series combination of C_d and R_e, because the impedance of C_d would be much lower than the impedances of the parallel arms involving resistances. Thus $Z(\sigma)$ should adhere to (9.9.8), where R_u is the same as R_e for this example. The intercept and the slope of Figure 9.9.2 therefore allow a determination of R_e and C_d.

Data at lower frequencies contain information about the arms parallel to C_d, but extracting it requires correction for the effects of R_e and C_d. This is accomplished in Figure 9.9.3. The basis for that analysis is left as an exercise in Problem 9.10.

9.10 REFERENCES

1. P. Delahay, "New Instrumental Methods in Electrochemistry," Wiley-Interscience, New York, 1954, Chap. 7.
2. B. Breyer and H. H. Bauer, "Alternating Current Polarography and Tensammetry," Vol. 13 in the series "Chemical Analysis," P. J. Elving and I. M. Kolthoff, Eds., Wiley-Interscience, New York, 1963.
3. D. E. Smith, *Electroanal. Chem.*, **1**, 1 (1966).
4. M. Sluyters-Rehbach and J. H. Sluyters, *ibid.*, **4**, 1 (1970).
5. D. E. Smith, *Crit. Rev. Anal. Chem.*, **2**, 247 (1971).
6. D. D. Macdonald, "Transient Techniques in Electrochemistry," Plenum, New York, 1977.
7. A. M. Bond, "Modern Polarographic Methods in Analytical Chemistry," Marcel Dekker, New York, 1980.

8. J. E. B. Randles, *Disc. Faraday Soc.*, **1**, 11 (1947); D. C. Grahame, *J. Electrochem. Soc.*, **99**, C370 (1952).
9. L. Pospisil and R. de Levie, *J. Electroanal. Chem.*, **22**, 227 (1969); H. Moreira and R. de Levie, *ibid.*, **29**, 353 (1971); **35**, 103 (1972).
10. R. D. Armstrong, M. F. Bell, and A. A. Metcalfe, in "Electrochemistry" (A Specialist Periodical Report), Vol. 6, H. R. Thirsk, Senior Reporter, The Chemical Society, London, 1978.
11. B. Timmer, M. Sluyters-Rehbach, and J. H. Sluyters, *J. Electroanal. Chem.*, **14**, 169, 181 (1967).
12. D. E. Smith and T. G. McCord, *Anal. Chem.*, **40**, 474 (1968).
13. D. E. Smith, *ibid.*, **35**, 610 (1963).
14. A. M. Bond, R. J. O'Halloran, I. Ruzic, and D. E. Smith, *ibid.*, **48**, 872 (1976).
15. A. M. Bond, R. J. O'Halloran, I. Ruzic, and D. E. Smith, *ibid.*, **50**, 216 (1978).
16. W. L. Underkofler and I. Shain, *ibid.*, **37**, 218 (1965).
17. A. M. Bond, *J. Electroanal. Chem.*, **50**, 285 (1974).
18. P. Delahay and T. J. Adams, *J. Am. Chem. Soc.*, **75**, 5740 (1952).
19. B. B. Damaskin, "The Principles of Current Methods for the Study of Electrochemical Reactions," McGraw-Hill, New York, 1967, Chap. 3.
20. E. R. Brown, T. G. McCord, D. E. Smith, and D. D. DeFord, *Anal. Chem.*, **38**, 1119 (1966).
21. H. Kojima and A. J. Bard, *J. Electroanal. Chem.*, **63**, 117 (1975).
22. H. Kojima and A. J. Bard, *J. Am. Chem. Soc.*, **97**, 6317 (1975).
23. H. Blutstein, A. M. Bond, and A. Norris, *Anal. Chem.*, **46**, 1754 (1974).
24. A. M. Bond, *ibid.*, **45**, 2026 (1973).
25. H. Kojima and S. Fujiwara, *Bull. Chem. Soc. Japan*, **44**, 2158 (1971).
26. S. C. Creason, J. W. Hayes, and D. E. Smith, *J. Electroanal. Chem.*, **47**, 9 (1973).
27. D. E. Glover and D. E. Smith, *Anal. Chem.*, **45**, 1869 (1973).
28. D. E. Smith, *ibid.*, **48**, 221A, 517A (1976).
29. J. W. Hayes, D. E. Glover, D. E. Smith, and M. W. Overton, *ibid.*, **45**, 277 (1973).
30. M. D. Wijnen, *Rec. Trav. Chim.*, **79**, 1203 (1960).
31. E. Levart and E. Poirier d'Ange d'Orsay, *J. Electroanal. Chem.*, **19**, 335 (1968).
32. A. A. Pilla, *J. Electrochem. Soc.*, **117**, 467 (1970).
33. A. A. Pilla in "Computers in Chemistry and Instrumentation," Vol. 2, "Electrochemistry," J. S. Mattson, H. B. Mark, Jr., and H. C. MacDonald, Eds., Marcel Dekker, New York, 1972.
34. A. A. Pilla and G. S. Margules, *J. Electrochem. Soc.*, **124**, 1697 (1977).

9.11 PROBLEMS

9.1 Derive equation 9.4.13, describing the shape of a reversible polarographic wave, from equation 9.4.11.

9.2 Derive formulas (9.5.5) for converting a parallel resistance-capacitance network to a series equivalent.

9.3 The faradaic impedance is sometimes represented as a resistance and a capacitance in parallel rather than in series. Find the expressions for the components of the parallel representation of this impedance in terms of R_{ct}, β_O, β_R, and ω. (*Hint.* Use known expressions for series elements and equations for series-to-parallel circuit conversion.)

9.4 The faradaic impedance method is employed to study the reaction

$$O + e \rightleftharpoons R$$

by imposing a very small sinusoidal signal (5 mV) to the cell, and measuring the equivalent series resistance R_B and capacitance C_B of the cell. The following data are obtained for $C_O^* = C_R^* = 1.00$ mM, $T = 25$ °C, and $A = 1$ cm^2:

Frequency ($\omega/2\pi$) (Hz)	R_B (Ω)	C_B (μF)
49	146.1	290.8
100	121.6	158.6
400	63.3	41.4
900	30.2	25.6

In a separate experiment under exactly the same conditions, but in the absence of the electroactive species, the cell resistance R_Ω is found to be 10 Ω, and the double-layer capacity of the electrode C_d is found to be 20.0 μF.

(a) From these data calculate at each frequency R_s and C_s and the phase angle ϕ between the components of the faradaic impedance.

(b) Calculate i_0 and k^0 for the reaction, and estimate D (assuming $D_O = D_R$).

9.5 Derive (9.5.14) and (9.5.15) from (9.5.12) and (9.5.13).

9.6 Derive (9.5.20) from (9.5.18) and (9.5.19).

9.7 From the data in Figures 9.4.4 and 9.4.5, evaluate α and k^0 for the reduction of Ti(IV) to Ti(III) at a DME in oxalic acid solution. From other experiments, we know that $n = 1$ and $D_O = 6.6 \times 10^{-6}$ cm^2/sec. Assume $D_O = D_R$.

9.8 The reduction of nitrobenzene to its radical anion in *N,N*-dimethylformamide is reported to occur with $k^0 = 2.2 \pm 0.3$ cm/sec (see Reference 22). The value of D_O is given as 1.02×10^{-5} cm^2/sec at $22 \pm 2°$ C where k^0 was evaluated. The transfer coefficient α is 0.70. Calculate the phase angles expected for $\omega/2\pi = 10$, 100, 1000, and 10,000 Hz. Draw the corresponding plot of $\cot \phi$ *vs.* $\omega^{1/2}$ for $E = E_{1/2}$. Describe a means for obtaining $\cot \phi$ from the in-phase and quadrature phase-selective polarograms, and comment on the frequency range where it might be experimentally feasible to obtain $\cot \phi$ values sufficiently precise to allow a determination of k^0 for the system at hand.

9.9 Devise and justify an equivalent circuit for a system in which O and R are bound to the surface of the electrode as the result of a synthetic modification.

Follow the steps in Section 9.2 and 9.3 to evaluate the expected frequency dependence of the *faradaic* impedance for the case in which the electrode reaction is nernstian. What phase angle is expected?

9.10 Derive the equation underlying Figure 9.9.3. First show how the quantity plotted on the ordinate can be regarded as $Z(\sigma)$ corrected for the effects of R_e and C_d. Then derive the expression for $Z(\sigma)$ for the remaining elements of circuit B in Figure 9.3.1. Consider the behavior at low and intermediate frequencies with respect to Figure 9.3.3.

9.11 Plot the amplitude and phase arrays [analogous to segments (*a*) and (*b*) of Figure 9.8.2] for generating a complex waveform having components at 100, 200, 300, . . . Hz, all with phase angles equal to $\pi/2$. Let these arrays have 128 elements, with the zeroth element representing the dc level and the 127th element representing $\omega/2\pi = 1270$ Hz. What disadvantages would the waveform resulting from your arrays have with respect to that generated in Figure 9.8.2? Would your waveform have any advantages?

chapter 10 Bulk Electrolysis Methods

The methods described in Chapters 5 to 9 generally employ conditions featuring a small ratio of electrode area (A) to solution volume (V). These allow the experiments to be carried out over fairly long time periods without appreciable changes of the concentrations of the reactant and the products in the bulk solution, and they allow the semi-infinite boundary condition (e.g., $C_O(x, t) = C_O^*$ as $x \rightarrow \infty$) to be maintained over repeated trials. For example, consider a 5×10^{-3} M solution of O with $V = 100$ cm^3, $A = 0.1$ cm^2 and assume that during 1 hour of experimentation an average current of about 100 μA (i.e., current density, j, of 1 mA-cm^{-2}) flows. During this time period only 0.36 C of electricity will be passed, and the bulk concentration of electroactive species will have decreased by less than 1%. There are circumstances, however, where one desires to alter the composition of the bulk solution appreciably by electrolysis; these include analytical measurements (e.g., electrogravimetric or coulometric methods), techniques for removal or separation of solution components, and electrosynthetic methods. These *bulk* (or *exhaustive*) electrolytic methods are characterized by large A/V conditions and as effective mass transfer conditions as possible. Thus, if all of the conditions of the previous example hold, except that the electrode area is 100 cm^2 (so that the total current, assuming the same $j = 1$ mA-cm^{-2}, is 0.1 A), the electroactive species can be completely electrolyzed in less than 10 minutes (assuming $n = 1$ and 100% current efficiency). Although bulk electrolytic methods are generally characterized by large currents and time scales of experiments of the order of minutes and hours, the basic principles governing electrode reactions described in the previous chapters still apply.

10.1 CLASSIFICATION OF TECHNIQUES

The methods can be classified by the parameter controlled (E or i) and by the quantities actually measured or the process carried out. Thus in *controlled potential* techniques the potential of the working electrode is maintained constant with respect to a reference electrode. Since the potential of the working electrode is the basic variable that controls the degree of completion of an electrolytic process in most cases, controlled potential techniques are usually the most desirable for bulk electrolysis. However, these methods require potentiostats with large output current and voltage capabilities (but usually much less demanding response-time requirements compared to those needed for small-electrode transient experiments), as well as stable reference electrodes, carefully placed to minimize uncompensated resistance effects. Placement of the auxiliary electrode to provide a fairly uniform current distribution across the surface of the working electrode is usually desirable, and the auxiliary electrode is often placed in a separate compartment isolated from the working-electrode compartment by a sintered-glass disk, ion exchange membrane, or other separator. A related technique in which bulk electrolysis occurs without the use of an external power supply is implemented by choosing a counter electrode that makes the entire cell a galvanic cell, so that some degree of potential control occurs during discharge. This technique, called *internal electrolysis*, is now rarely used.

In *controlled current* techniques, the current passing through the cell is held constant (or sometimes programmed to change with time or in response to some indicator electrode signal). Although this technique frequently involves simpler instrumentation than controlled-potential methods, it requires either a special set of chemical conditions in the cell or specific detection methods to signal completion of the electrolysis and ensure 100% current efficiency. For preparative electrolysis (or *electrosynthesis*), constant current methods can sometimes be used as long as measures are taken to ensure that the electrode potential does not move into a region where undesirable side reactions occur.

The general considerations and models employed in electroanalytical bulk electrolysis methods are also often applicable to large-scale and flow electrosynthesis, to galvanic cells (and batteries), to fuel cells, and to electroplating applications.

Bulk electrolysis methods for analysis frequently involve determination of the weight of a deposit on the electrode (*electrogravimetric methods*). In this case 100% current efficiency is not required but only the substance of interest (in a pure, known form) can be deposited. In *coulometric methods* the total quantity of electricity required to carry out an exhaustive electrolysis is determined. The quantity of material or number of electrons involved in the electrode reaction can then be determined by Faraday's laws, if the reaction occurred with 100% current efficiency. For *electroseparations*, electrolysis is used to remove selectively constituents from the solution.

Several related bulk electrolysis techniques should be mentioned. In *thin-layer electrochemical methods* (see Section 10.7) large A/V ratios are attained by trapping only a very small volume of solution in a thin (20–100 μm) layer against the working electrode. The current levels and time scale in these techniques are similar to those in voltammetric methods. The apparatus is more difficult to construct, and it is frequently difficult to avoid large resistive effects and nonuniform current distributions. *Flow electrolysis* techniques (see Section 10.6), in which a solution is exhaustively electrolyzed as it flows through a cell, can also be classified as bulk electrolysis methods. Finally there is *stripping analysis* (see Section 10.8), where bulk electrolysis is used to preconcentrate a material in a small volume or on the surface of an electrode, before a voltammetric analysis.

General treatments of bulk electrolysis techniques, as well as numerous examples of their application to analysis and separations, are contained in References 1 to 3.

10.2 GENERAL CONSIDERATIONS IN BULK ELECTROLYSIS

10.2.1 Extent or Completeness of an Electrode Process

The extent or degree of completion of a bulk electrolytic process can often be predicted (for nernstian reactions) from the applied electrode potential and a suitable form of the Nernst equation.

(a) Both Forms Soluble in Solution. Consider the overall reduction reaction

$$O + ne \rightleftharpoons R \tag{10.2.1}$$

$$E = E^{0'} + \left(\frac{RT}{nF}\right) \ln\left(\frac{C_O}{C_R}\right) \tag{10.2.2}$$

where both O and R are soluble. Suppose R is initially absent. Let C_i be the initial concentration of O, $E^{0'}$ be the formal potential of (10.2.1), V_s be the volume of the solution, and x be the fraction of O reduced to R at the electrode potential, E. Then

$$\text{moles O at equilibrium} = V_s C_i(1 - x) \tag{10.2.3}$$

$$\text{moles R at equilibrium} = V_s C_i x \tag{10.2.4}$$

and (10.2.2) to (10.2.4) yield

$$\boxed{E = E^{0'} + \left(\frac{RT}{nF}\right) \ln\left[\frac{(1 - x)}{x}\right]} \tag{10.2.5}$$

or

$$\boxed{\text{fraction of O reduced} = x = \{1 + 10^{[(E - E^{0'})n/0.059]}\}^{-1} \text{ (at 25°)}} \tag{10.2.6}$$

For example, for 99% completeness of reduction of O to R (i.e., $x = 0.99$) the potential of the working electrode should be

$$E = E^{0'} + \frac{0.059}{n} \log\left(\frac{0.01}{0.99}\right) \simeq E^{0'} - \frac{(0.059)(2)}{n} \tag{10.2.7}$$

or $118/n$ mV more negative than $E^{0'}$ at 25 °C.

(b) Deposition as an Amalgam. For the reaction

$$O + Hg + ne \rightleftharpoons R(Hg) \tag{10.2.8}$$

where R(Hg) represents an amalgam of R, that is, R dissolved in the mercury electrode (of volume, V_{Hg}), the situation is similar to that where both forms are soluble in solution, except that $E_a^{0'}$, the formal potential for the reaction in (10.2.8), replaces $E^{0'}$, and C_R represents the concentration of R in the mercury electrode (assumed to be below the saturation value). This results in the equation:

$$E = E_a^{0'} + \frac{RT}{nF} \ln \left[\frac{(V_s C_i (1 - x)/V_s)}{(V_s C_i x / V_{Hg})} \right] \tag{10.2.9}$$

$$\boxed{E = E_a^{0'} + \frac{RT}{nF} \ln \left(\frac{V_{Hg}}{V_s} \right) + \frac{RT}{nF} \ln \frac{1 - x}{x}} \tag{10.2.10}$$

(c) Deposition of a Solid. For the reaction

$$O + ne \rightleftharpoons R \text{ (solid)} \tag{10.2.11}$$

when more than a monolayer of R is deposited on an inert electrode (e.g., copper on a platinum electrode) or the deposition is carried out on an electrode made of R (e.g., copper on a copper electrode), the activity of R, a_R, is constant and equal to unity at the completion of the electrolysis. Thus the Nernst equation yields

$$E = E^0 + \frac{RT}{nF} \ln[\gamma_O C_i (1 - x)] \tag{10.2.12}$$

where γ_O is the activity coefficient of species O. When less than a monolayer of R is deposited on an inert substrate, $a_R \neq 1$, and an expression for a_R as a function of coverage by R must be used in the Nernst equation. It is sometimes assumed (4) that a_R is proportional to the fraction, θ, of the electrode surface covered with R. Thus, for example,

$$a_R \simeq \gamma_R \theta = \gamma_R \frac{A_R}{A} = \frac{\gamma_R N_R A_a}{A} \tag{10.2.13}$$

where A_R is the area occupied by R, A is the electrode area, A_a is the cross-sectional area of a molecule of R(cm^2), and N_R is the number of molecules of R deposited on the electrode. At equilibrium, N_R is given by

$$N_R = V_s C_i x N \tag{10.2.14}$$

where N is Avogadro's number. This then yields, when combined with (10.2.13) and the Nernst equation,

$$E = E^0 + \frac{RT}{nF} \ln \left(\frac{\gamma_O A}{\gamma_R V_s N A_a} \right) + \frac{RT}{nF} \ln \left[\frac{(1 - x)}{x} \right] \tag{10.2.15}$$

Thus, within this very simplified model, the shape of the deposition curve follows that of the case of soluble components or an amalgam [(10.2.5) or (10.2.10)] even though a solid is deposited (Figure 10.2.1). The deposition begins at potentials more positive than values where deposition of R occurs on bulk R. Consider, for example, the deposition of Ag on a 1-cm^2 Pt electrode from a 0.01-L solution containing 10^{-7} M Ag^+. Let $A_a = 1.6 \times 10^{-16}$ cm^2 and $\gamma_O = \gamma_R$. The potential for deposition of one-half of

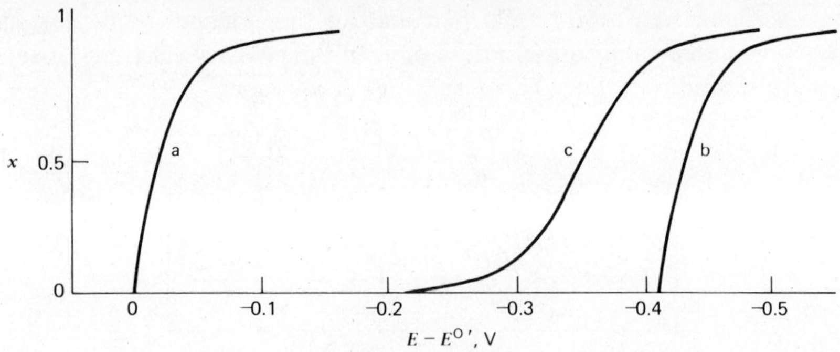

Figure 10.2.1

Fraction of a metal M^+ (e.g., Ag^+) deposited (x) as a function of potential. Curve a: 1 M Ag^+ on Ag. Curve b: 10^{-7} M Ag^+ on Ag. Curve c: 10^{-7} M Ag^+ on Pt, according to (10.2.15).

the silver (which forms about 0.05 monolayer) is $E = 0.35$ V, compared to $E = 0.43$ V required for the same amount of deposition on a silver electrode. Deposition at potentials before that predicted by the Nernst equation with $a_R = 1$ is called *underpotential deposition*. The situation is much more complicated than the above treatment suggests, since the deposition potential depends on the nature of the substrate (material and pretreatment) and on adsorption of O. Also, the treatment assumes that formation of a second layer does not start until the first is complete. However, this is frequently not the case; atoms of metal will often aggregate, rather than deposit on a foreign surface, and dendrites will form. Reviews on the nature of underpotential deposition and the deposition of solids in general are available (5–9).

For slow electron transfer (irreversible) processes, the eventual extent of the electrode process will be governed by equilibrium considerations and the Nernst equation, but the rate of electrolysis will be small at the potentials predicted in the previous sections and long-duration electrolyses would result. For these processes reduction must be carried out at somewhat more negative potentials; the actual potential is usually selected on the basis of experimental current-potential curves taken under conditions near those for the intended bulk electrolysis. Processes that are controlled by the rate of a homogeneous reaction [e.g., one preceding the electron transfer and forming the electroactive species as in (10.2.16)]

$$A \xrightarrow{k} O \qquad O + ne \rightleftharpoons R \tag{10.2.16}$$

may be slow and independent of the potential chosen. In this case the potential is selected for complete conversion of O to R, and other steps are taken, for example, an increase in temperature or addition of a catalyst, to increase the rate of reaction. Catalysts (or *mediators*) are sometimes added to carry out electrolyses of substances that themselves undergo very slow electron transfer rates at the electrode but react rapidly with the mediator. For example, most enzymes, (e.g., cytochromes) are reduced very slowly directly at an electrode. The addition of a mediator, such as methyl viologen, which is reduced reversibly at an electrode and whose reduced form reacts rapidly with the enzyme, can thus be used to carry out the reduction (10, 11). This strategy, which is related to the coulometric titration technique (see Section 10.4.2), can also be used with other irreversible electrode reactions.

10.2.2 Current Efficiency

It is generally desirable for bulk electrolytic processes to be carried out with high current efficiency. This requires that the working electrode potential and other conditions be chosen so that no side reactions (e.g., reduction or oxidation of solvent, supporting electrolyte, electrode material, or impurities) occur. In electrogravimetric methods, 100% current efficiency is usually not necessary, as long as the side reactions do not produce insoluble products. In coulometric titrations at constant current, 100% titration efficiency (rather than current efficiency) is required; the distinction is discussed in the appropriate section below.

10.2.3 Electrolysis Cells

Bulk electrolysis techniques, because of the longer duration of the experiments and the larger currents involved, usually present more problems in cell design than transient experiments. Typical bulk electrolysis cells are shown in Figure 10.2.2.

(a) Electrodes and Geometry. Solid electrodes usually are wire gauzes or foil cylinders, although packed beds of powders, slurries, or fluidized beds are sometimes used. The aim is to have as large a working electrode area as possible. Mercury electrodes generally take the form of pools. Proper orientation of the auxiliary electrode is needed to provide a uniform current density across the working electrode surface. Nonuniform current densities lead to different potential drops between electrode and solution at different locations across the electrode surface, and they cause undesired side reactions or ineffective use of the total electrode area (Figure 10.2.3). The proper location of the tip of the salt bridge from the reference electrode is also important, as is the long-term stability of the potential of the reference electrode (see also Section 13.6).

(b) Separators. Because the products of the electrode reaction at the auxiliary electrode, if they are soluble, will usually be reactive at the working electrode, the two electrodes are usually placed in separate compartments. These compartments are usually separated by sintered (or fritted) glass disks or ion-exchange membranes (or, less frequently, filter paper, asbestos mats, or porous ceramics). The proper choice of a separator that does not allow intermixing of *anolyte* and *catholyte* and yet that does not contribute appreciably to the cell resistance is often of major importance. Sometimes the judicious choice of the auxiliary electrode reaction, for example, when solid products or innocuous gaseous products are formed, allows cells without separators to be used. Examples of these are the use of silver anodes in halide media (e.g., $Ag + Cl^- \rightleftharpoons AgCl + e$) or hydrazine as an "anodic depolarizer" at platinum anodes ($N_2H_4 \rightarrow N_2 + 2H^+ + 2e$). Separator design is also important in the fabrication of galvanic cells and batteries.

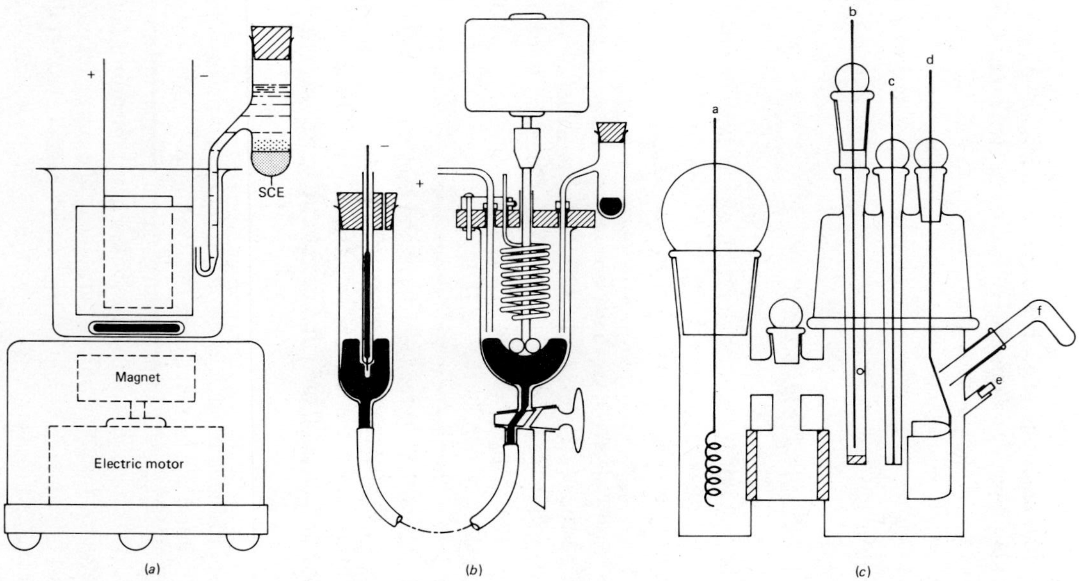

Figure 10.2.2
Typical cells for bulk electrolysis (*a*) Undivided cell for controlled potential separations and electrogravimetric analysis at a solid cathode. [From J. J. Lingane, *Anal. Chim. Acta*, **2**, 584 (1948), with permission.] (*b*) Undivided cell for coulometric analysis at mercury cathode with silver anode. [From J. J. Lingane, *J. Am. Chem. Soc.*, **67**, 1916 (1945), with permission.] (*c*) Three-compartment cell for coulometric and voltammetric studies on a vacuum line and provided with ground glass joints. (a) Platinum wire auxiliary electrode; (b) Silver wire reference electrode in separate compartment; (c) gold voltammetric working microelectrode; (d) Platinum foil coulometric working electrode; (e) silicone gum rubber septum for sample injection; (f) rotatable side arm for solid sample addition. Arm and joint for attachment of cell to the vacuum line are not shown. [Reprinted with permission from W. H. Smith and A. J. Bard, *J. Am. Chem. Soc.*, **97**, 5203 (1975). Copyright 1975, American Chemical Society.]

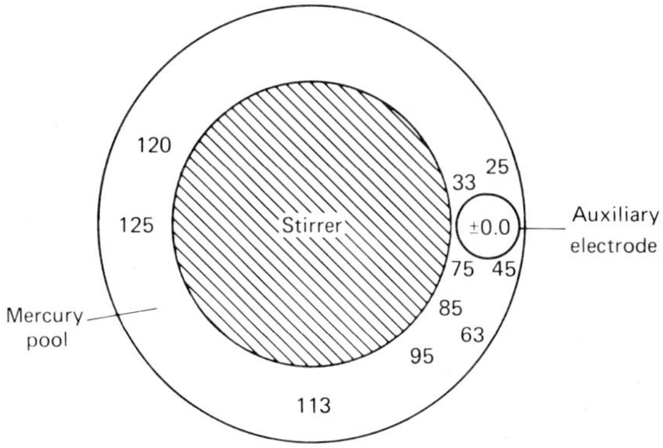

Figure 10.2.3

Potential distribution in millivolts at surface of ring-shaped mercury pool electrode using an unsymmetrical auxiliary electrode. Total cell current 40 mA, pool 1.5 in o.d., 1.0 in. i.d. Small circle shows position of the auxiliary electrode fritted-glass separator, situated 4 mm above the pool surface. [Reprinted with permission from G. L. Booman and W. B. Holbrook, *Anal. Chem.*, **35**, 1793 (1963). Copyright 1963, American Chemical Society.]

(c) Cell Resistance. High cell resistances are very deleterious in experiments involving high currents, because large values of i^2R mean wasted power, a need for high voltage output of potentiostat or power supply, and undesirable heat evolution. Moreover, in cells with high resistance, proper placement of the reference electrode tip near the working electrode, so that large amounts of uncompensated iR-drop are not obtained, may be difficult. Cell design for the minimization of cell resistance is especially important when nonaqueous solvents with lower dielectric constants, and hence inherently lower solution conductivities (e.g., acetonitrile, *N,N*-dimethylformamide, tetrahydrofuran, and ammonia) are employed.

10.3 CONTROLLED POTENTIAL METHODS

10.3.1 Current-Time Behavior

The current-potential characteristics described for stirred solutions (Section 1.4.2 and Chapter 8) generally apply to these electrolysis conditions as well, except that C_O^*, the bulk concentration, is a function of time, decreasing during the electrolysis. Thus i-E curves taken repeatedly during an electrolysis (assumed taken at such a rapid rate that no appreciable change of C_O^* occurs during the potential scan) will show a continuously decreasing i_l as C_O^* decreases (Figure 10.3.1). Consider the electrolysis of O, initially present in bulk solution at a concentration $C_O^*(0)$, by the reaction $O + ne \rightarrow R$, at an electrode of area A held at a potential E_c corresponding to the limiting current region. The current at any time is given by (1.4.9):

$$i_l(t) = nFAm_OC_O^*(t) \tag{10.3.1}$$

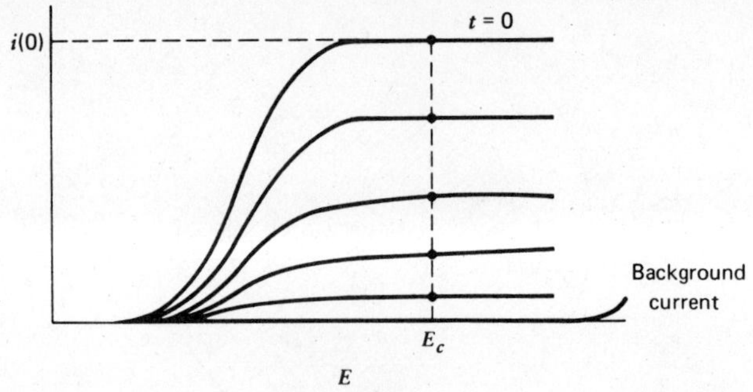

Figure 10.3.1
Current-potential curves at different times during a bulk controlled-potential electrolysis at $E = E_c$.

The current also indicates the total rate of consumption of O, dN_O/dt (mol/sec), due to electrolysis (assuming 100% current efficiency):

$$i_l(t) = -nF\left[\frac{dN_O(t)}{dt}\right] \tag{10.3.2}$$

where N_O is the total number of moles of O in the system. If one assumes that the solution can be considered completely homogeneous (i.e., one neglects the small volume of the diffusion layer $\delta_O A$ in the vicinity of the electrode surface, where C_O is smaller than C_O^*), then

$$C_O^*(t) = \frac{N_O(t)}{V} \tag{10.3.3}$$

where V is the total solution volume. Equations 10.3.2 and 10.3.3 yield

$$i_l(t) = -\,nFV\left[\frac{dC_O^*(t)}{dt}\right] \tag{10.3.4}$$

Equating the two relations for $i_l(t)$, we obtain

$$\frac{dC_O^*(t)}{dt} = -\left(\frac{m_O A}{V}\right)C_O^*(t) = -pC_O^*(t) \tag{10.3.5}$$

with the initial condition, $C_O^*(t) = C_O^*(0)$ at $t = 0$. Equation 10.3.5 is an equation characteristic of the kinetics of a first-order, homogeneous chemical reaction, where $p = m_O A/V$ is analogous to the first-order rate constant. The solution to this ordinary differential equation is

$$\boxed{C_O^*(t) = C_O^*(0)\exp(-pt)} \tag{10.3.6}$$

and using (10.3.1), we obtain the *i-t* behavior:

$$\boxed{i(t) = i(0)\exp(-pt)} \tag{10.3.7}$$

where $i(0)$ is the initial current (12). Thus a controlled-potential bulk electrolysis is like a first-order reaction, with the concentration and the current decaying exponentially with time during the electrolysis (Figure 10.3.2) and eventually attaining the background (residual) current level. Equation 10.3.7 can be used to determine the duration of the electrolysis for a given conversion:

$$\frac{-p}{2.3}\,t = \log\left[\frac{C_O^*(t)}{C_O^*(0)}\right] = \log\left[\frac{i(t)}{i(0)}\right] \tag{10.3.8}$$

For 99% completion of electrolysis, $C_O^*(t)/C_O^*(0) = 10^{-2}$, and $t = 4.6/p$; for 99.9% completion, $t = 6.9/p$. With effective stirring, $m_O \simeq 10^{-2}$ cm/sec, so that for $A(\text{cm}^2) \simeq V(\text{cm}^3)$, $p \simeq 10^{-2}\ \text{sec}^{-1}$, and a 99.9% electrolysis would require ~690 sec or ~12 minutes. Typically bulk electrolyses are slower than this, requiring 30 to 60 minutes, although cell designs with very large A/V and very effective stirring (e.g., using ultrasonics) with a p of $\sim 10^{-1}\ \text{sec}^{-1}$ have been described (13). For effective rates of electrolysis, A should be as large as possible and, in many practical devices (fuel cells and other power sources, preparative cells), porous electrodes and flow cells are employed (see Section 10.6).

The total quantity of electricity $Q(t)$ (in coulombs) consumed in the electrolysis is given by the area under the i-t curve (Figure 10.3.2c):

$$\boxed{Q(t) = \int_0^t i(t)\,dt} \tag{10.3.9}$$

Electrolysis at controlled potential is the most efficient method of carrying out a bulk electrolysis, because the current is always maintained at the maximum value (for given cell conditions) consistent with 100% current efficiency. Note that the rate of electrolysis is independent of $C_O^*(0)$, so that electrolysis of a 0.1 M solution of O and a 10^{-6} M solution of O, at the same values of E, A, V, and m_O, should require the same amount of time.

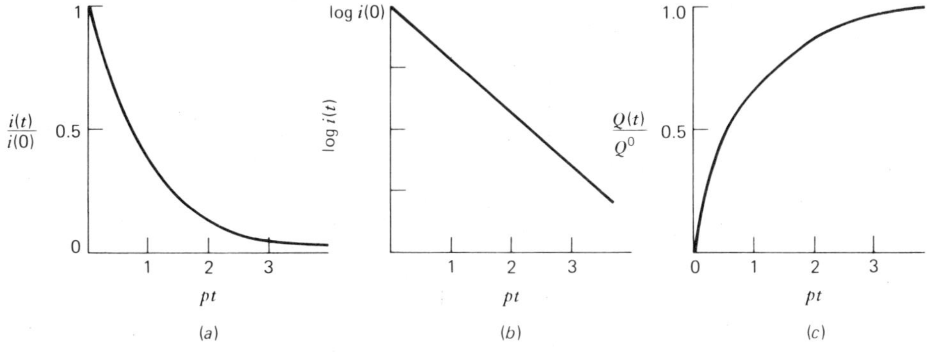

Figure 10.3.2
(a) Current-time curve (in dimensionless form) during controlled potential electrolysis. (b) log $i(t)$ $vs.$ t. (c) Q $vs.$ t.

10.3.2 Electrogravimetric Methods

The determination of a metal by selective deposition on an electrode, followed by weighing, is among the oldest of electroanalytical methods [Cruikshank (1801); W. Gibbs (1864)]. In controlled potential methods, the potential of the solid electrode is adjusted to a value where the desired plating reaction occurs and no interfering reaction leading to the deposition of another insoluble substance takes place.

The sensitivity of an electrogravimetric method is limited by the difficulty in determining the small difference in weight between the electrode itself and the electrode plus deposit. The technique also requires the washing and drying of the electrode and is, of course, limited to electrode reactions involving the formation of insoluble substances. For these reasons, many electrogravimetric determinations have been supplanted by coulometric ones (see Section 10.3.4) except for situations where 100% current efficiency cannot be attained.

Electrogravimetric determinations also require smooth and adherent metal platings and deposits. The physical characteristics of a deposit depend on the form of the metal ion in the solution, the presence of adsorbable surface-active agents in the solution, and other factors, some not completely understood. The reader is referred to References 14 to 16 and books on electroplating (17) for a detailed discussion of these factors. In general, depositions from solutions of complex ions frequently are smoother than those obtained from solutions containing only the aquo-form. For example, brighter deposits are obtained from solutions of Ag^+ in a CN^- medium [containing $Ag(CN)_2^-$] than from a nitrate medium. The addition of surface-active agents ("brighteners"), such as gelatin, often leads to improved deposits. It has been reported that organic additives are sometimes occluded with the metal deposit, leading to positive errors in electrogravimetric analysis. This error is apparently smaller when depositions are carried out under controlled potential conditions. Hydrogen evolution during deposition also leads to a rougher deposit. Deposits at very large current densities tend to be less adherent and rougher than those obtained at lower ones.

Some metals typically determined by electrogravimetric methods and their deposition potentials are given in Table 10.3.1. Detailed discussions of the methods and applications of electrogravimetric methods are available (1, 3, 18, 19).

10.3.3 Electroseparations

In an electrochemical separation the quantitative deposition of one metal (M_1) on a solid or mercury electrode is desired without appreciable deposition of a second metal (M_2). The considerations of Section 10.2.1 concerning the degree of completion of electrolysis as a function of potential apply. Thus for complete (i.e., $\geq 99.9\%$) deposition of M_1 as an amalgam, $E \leq E^{0'}_{a1} - 0.18/n_1$ V [at 25°, where $E^{0'}_{a1}$ is the formal potential for the n_1-electron reduction] with $V_{Hg} = V_s$. For essentially no (i.e., $\leq 0.1\%$) deposition of M_2, $E \geq E^{0'}_{a2} + 0.18/n_2$ V. Therefore, the separation between the formal potentials must be at least $0.18\,(n_1^{-1} + n_2^{-1})$ (Figure 10.3.3). If $|E^{0'}_{a2} - E^{0'}_{a1}|$ is smaller than this, a separation at the 99.9% level cannot be accomplished. In that case changing the supporting electrolyte to one that complexes one or both of the

Table 10.3.1

Deposition Potentials (volts vs. SCE) for Various Metals in Different Media at a Platinum Electrode[a]

	Supporting Electrolyte				
Metal	0.2 *M* H_2SO_4	0.4 *M* NaTart + 0.1 *M* NaHTart	1.2 *M* NH_3 + 0.2 *M* NH_4Cl	0.4 *M* KCN + 0.2 *M* KOH	EDTA + NH_4OAc[b]
Au	+0.70	(+0.50)[d]	—	−1.00	+0.40
Hg	+0.40	(+0.25)[d]	−0.05	−0.80	+0.30
Ag	+0.40	(+0.30)[b]	−0.05	−0.80	+0.30
Cu	−0.05	−0.30	−0.45	−1.55	−0.60
Bi	−0.08	−0.35	—	(−1.70)[d]	−0.60
Sb	−0.33	−0.75	—	−1.25	−0.70
Sn[c]	—	—	—	—	—
Pb	—	−0.50	—	—	−0.65
Cd	−0.80	−0.90	−0.90	−1.20	−0.65
Zn	—	−1.10	−1.40	−1.50	—
Ni	—	—	−0.90	—	—
Co	—	—	−0.85	—	—

[a] Adapted from table given in Reference 3.
[b] 5 g NH_4OAc + 200 ml H_2O(pH $\approx$ 5); [EDTA]:[metal] = 3:1.
[c] Tin can be deposited from solutions of Sn(II) in HCl or HBr media.
[d] Metal deposits obtained are not suitable for electrogravimetric analysis.

metals will often give an improved separation. The potential range for a successful separation can best be found by determining the *i-E* curve on a microelectrode under the same conditions (concentrations, supporting electrolyte, temperature) considered for the separation.

Although electrogravimetric analyses are rarely carried out with a mercury electrode, this electrode is often used for electroseparations. Metals that can be deposited at a mercury electrode are shown in Figure 10.3.4.

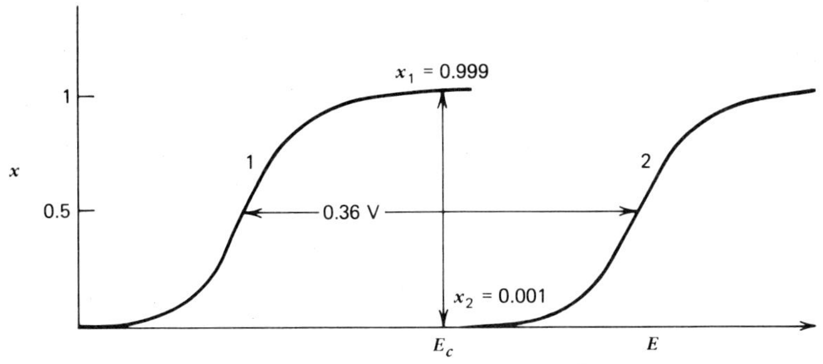

Figure 10.3.3
Conditions for complete separation of metals M_1 and M_2 at mercury electrode; $n_1 = n_2 = 1$.

Ia	IIa	IIIa	IVa	Va	VIa	VIIa	VIII			Ib	IIb	IIIb	IVb	Vb	VIb	VIIb	0
H																	He
Li	Be											B	C	N	O	F	Ne
Na	Mg											Al	Si	P	S	Cl	A
K	Ca	Sc	Ti	V	Cr	Mn	Fe	Co	Ni	Cu	Zn	Ga	Ge	As	Se	Br	Kr
Rb	Sr	Y	Zr	Nb	Mo	Tc	Ru	Rh	Pd	Ag	Cd	In	Sn	Sb	Te	I	Xe
Cs	Ba	La[a]	Hf	Ta	W	Re	Os	Ir	Pt	Au	Hg	Tl	Pb	Bi	Po	At	Rn
Fr	Ra	Ac[b]															

Note: Heavy solid lines enclose elements that can be quantitatively deposited in the mercury cathode. Broken lines enclose elements that are quantitatively separated from the electrolyte, but are not quantitatively deposited in the mercury. Light lines enclose elements that are incompletely separated.
[a] Also elements 58 to 71 (partial deposition of lanthanum and neodymium has been reported).
[b] Also elements 90 to 103.

Figure 10.3.4

Metals deposited at mercury electrode. [Reprinted with permission from J. A. Maxwell and R. P. Graham, *Chem. Revs.*, **46**, 471 (1950). Copyright 1950, Williams and Wilkins, Inc.]

10.3.4 Coulometric Measurements

In controlled potential coulometry the total number of coulombs consumed in an electrolysis is used to determine the amount of substance electrolyzed. To enable a coulometric method, the electrode reaction must satisfy the following requirements: (a) it must be of known stoichiometry; (b) it must be a single reaction or at least have no side reactions of different stoichiometry; (c) it must occur with close to 100% current efficiency. These requirements are similar to those for the titration reaction in usual titrimetric methods; the term coulometric titration is usually reserved, however, to constant current coulometric methods (see Section 10.4.2).

A block diagram of the apparatus used in controlled potential coulometry is shown in Figure 10.3.5. The potentiostats used generally have output powers of 100 W

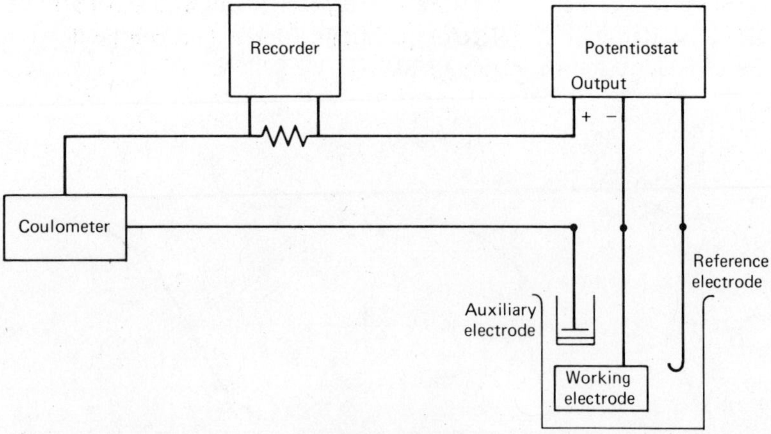

Figure 10.3.5

Block diagram of typical controlled potential coulometry apparatus. [From A. J. Bard and K. S. V. Santhanam, *Electroanal. Chem.*, **4**, 215 (1970), courtesy of Marcel Dekker, Inc.]

(e.g., 1 A at 100 V or 5 A at 20 V). Although electromechanical potentiostats have been used in the past, modern all-electronic potentiostats employing operational amplifier control circuits and solid-state output devices are available; these are more convenient to use and show much faster response time (see Chapter 13). The current is monitored during the electrolysis, usually with a strip-chart recorder, so that the background current can be determined and the completion of electrolysis observed. The shape of the *i-t* curve can be diagnostic of the mechanism of the electrode reaction and instrumental problems. For example, if the final current following electrolysis is constant, but appreciably higher than the preelectrolysis background current of the supporting electrolyte solution alone, a reaction of the electrolysis product may be regenerating starting material or another electroactive substance (see Section 11.6). This symptom can also indicate leakage of material from the auxiliary electrode compartment. If the current at the start of the electrolysis remains constant for some time before showing the usual exponential decay (Figure 10.3.2*a*), the output current or voltage of the potentiostat is probably insufficient for the electrolysis conditions (electrode area, C_O^*, cell resistance, stirring rate) to maintain the working electrode at the potential chosen.

A number of different types of coulometers are available. Formerly chemical types (gravimetric, titrimetric, and gas) were used. These can be related directly to a chemical primary standard (e.g., the silver coulometer) and are capable of high accuracy and precision. However, they are inconvenient and time-consuming to use and are now rarely employed. Electromechanical coulometers based on ball-and-disk integrators, sometimes operating as part of the current-measuring recorder, are also infrequently used and have been replaced by operational amplifier integrator circuits or digital circuits (e.g., based on voltage-to-frequency converters and counters). These give a direct readout in coulombs (or if desired, in equivalents) and can be employed to record *Q-t* curves during electrolysis; see, for example, Figure 10.3.2*c*. The shape of this curve for an uncomplicated electrolysis is immediately obtained from (10.3.7) and (10.3.9)

$$\boxed{Q(t) = \frac{i(0)}{p}(1 - e^{-pt}) = Q^0(1 - e^{-pt})} \qquad (10.3.10)$$

where Q^0 is the value of Q at the completion of the electrolysis ($t \to \infty$), and is given by

$$\boxed{Q^0 = nFN_O = nFVC_O^*(0)} \qquad (10.3.11)$$

where N_O represents the total number of moles of O initially present. Equation 10.3.11 is just a statement of Faraday's law and is the basis for any coulometric method of analysis.

There have been numerous applications of controlled potential coulometry to analysis. Many electrodeposition reactions that are the basis of electrogravimetric determinations can be employed in coulometric methods as well. However, some electrogravimetric determinations can be used when the electrode reactions occur with less than 100% current efficiency, for example, the plating of tin on a solid electrode.

Coulometric determinations can, of course, also be based on electrode reactions in which soluble products or gases are formed (e.g., reduction of Fe(III) to Fe(II), oxidation of I^- to I_2, oxidation of N_2H_4 to N_2, reduction of aromatic nitro-compounds). Many reviews concerned with controlled potential coulometric analysis have appeared (1, 19–21); some typical applications are given in Table 10.3.2.

Controlled potential coulometry is also a very useful method for studying the mechanisms of electrode reactions and for determining the n value for an electrode reaction without prior knowledge of electrode area or diffusion coefficient. (Note that in voltammetric methods if n is to be determined from the limiting current, D and A usually must be known. To determine n from potential measurements, knowledge about the reversibility of the reaction is required.) However, because the time scale of coulometric measurements (~10–60 min) is at least one or two orders of magnitude longer than that of voltammetric methods, perturbing homogeneous chemical reactions following the electron transfer, which might not affect the voltammetric measurement, may be important in coulometry (see Chapter 11). For example, consider the reaction sequence:

$$O + e \rightleftharpoons R \tag{10.3.12}$$

$$R \rightarrow A \quad (\text{slow, } t_{1/2} \sim 2\text{–}5 \text{ min}) \tag{10.3.13}$$

$$A + e \rightarrow B \quad (\text{A reduced at less negative potentials than O}) \tag{10.3.14}$$

Table 10.3.2

Typical Controlled Potential Coulometric Determinations

Substance	Working Electrode	Supporting Electrolyte[a]	Control Potential (Volts *vs.* SCE)	Overall Reaction
Li	Hg	0.1 *M* TBAP (CH_3CN)	−2.16	Li(I) → Li(Hg)
Cr	Pt	1 *M* H_2SO_4	+0.50	Cr(VI) → Cr(III)
Fe	Pt	1 *M* H_2SO_4	+0.20	Fe(III) → Fe(II)
Zn	Hg	2 *M* NH_3 1 *M* $(NH_4)_3$Citrate	−1.45	Zn(II) → Zn(Hg)
Te^{2-}	Hg	1 *M* NaOH	−0.60	$Te^{2-} \rightarrow Te$
Br^-	Ag on Pt	0.2 *M* KNO_3 (MeOH)	0.0	$Ag + Br^- \rightarrow AgBr$
I^-	Pt	1 *M* H_2SO_4	+0.70	$2I^- \rightarrow I_2$
U	Hg	0.5 *M* H_2SO_4	−0.325	U(VI) → U(IV)
Pu	Pt	1 *M* H_2SO_4	+0.70	Pu(III) → Pu(IV)
Ascorbic acid	Pt	0.2 *M* phthalate buffer, pH 6	+1.09	Oxdn. $n = 2$
DDT	Hg		−1.60	Redn. $n = 2$
Aromatic hydrocarbons (e.g., diphenylanthracene)	Hg or Pt	0.1 *M* TBAP (DMF)		Redn. $Ar \rightarrow Ar^{\cdot -}$
Aromatic nitro-compounds	Hg	0.5 *M* LiCl (DMSO)		Redn. $ArNO_2 \rightarrow ArNO_2^{\cdot -}$

[a] With water as solvent, unless indicated otherwise.

This sequence occurs, for example, in the reduction of *o*-iodonitrobenzene (O = $IPhNO_2$) in liquid ammonia (with 0.1 *M* KI as supporting electrolyte). A voltammetric experiment (e.g., cyclic voltammetry at scan rates of 50 to 500 mV/sec) shows a one-electron reaction with formation of the radical anion (R = $IPhNO_2^{\cdot -}$) which is stable on this time scale. However, controlled potential–coulometric reduction shows n values approaching 2 for reductions requiring 1-hour durations. In this time the radical anion loses an I^- to form the radical (A = $\cdot PhNO_2$) which is reduced at these potentials (to B = $^-{:}PhNO_2$); this then protonates to form nitrobenzene.

10.4 CONTROLLED CURRENT METHODS

10.4.1 Characteristics of Controlled Current Electrolysis

As before (Section 10.3.1), the course of a bulk electrolysis under controlled current conditions can be ascertained from consideration of the *i-E* curves (Figure 10.4.1). As long as the applied current i_{app} is less than the limiting current at a given bulk concentration $i_l(t)$, the electrode reaction proceeds with 100% current efficiency. As the electrolysis proceeds, the bulk concentration of O, $C_O^*(t)$, decreases and $i_l(t)$ decreases (linearly with time). When

$$C_O^*(t) = \frac{i_{app}}{nFAm_O} \qquad (10.4.1)$$

$i_{app} = i_l(t)$. At longer times, $i_{app} > i_l(t)$, and the potential shifts to a new, more negative value, where an additional electrode reaction can occur; this reaction contributes the additional current $i_{app} - i_l(t)$. The current efficiency thus drops below 100%. Since the potential is now sufficiently negative to be on the mass-transfer-controlled plateau of the O → R reduction, the electrolysis of O occurs as if it were being carried out under controlled potential conditions. Thus the current contribution for this reaction decays exponentially, as in (10.3.7) (Figure 10.4.2). If i_{app} is larger than the initial limiting current, the rate of electrolysis of O will be essentially the same as

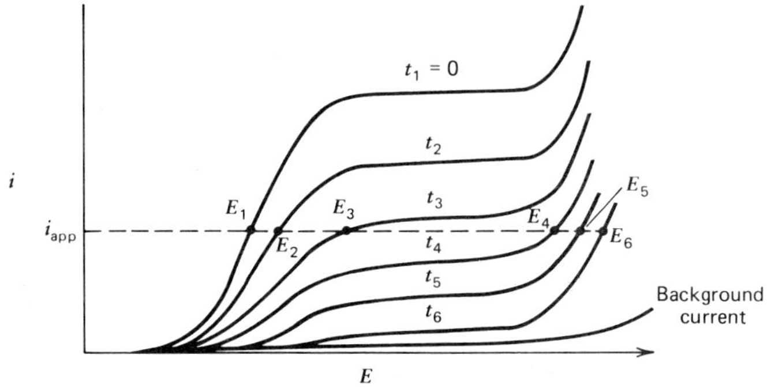

Figure 10.4.1
Current-potential curves at different times (increasing from t_1 to t_6) during a bulk electrolysis with an applied constant current, i_{app}. Note that the electrode potential shifts from E_1 to E_6 during the course of the electrolysis with the largest shift occurring (curve 3) when $i_{app} = i_l$.

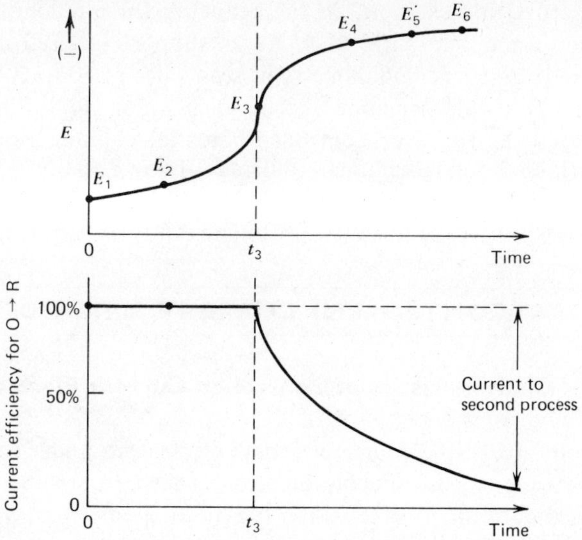

Figure 10.4.2

Potential and current efficiency for the electrolysis illustrated in Figure 10.4.1.

if the reduction were carried out under controlled potential conditions, but with much lower current efficiency. [Sometimes a constant current electrolysis will be somewhat faster than a controlled potential one under apparently identical conditions, because gas evolution (e.g., hydrogen or oxygen), occurring during the electrolysis, leads to effective stirring at the electrode surface and produces larger mass transfer rates (i.e., a larger m_O).] The selectivity of a constant current separation is obviously much poorer than the corresponding controlled potential method, since at some time during the electrolysis the potential must shift into a more negative region where a new electrode reaction occurs and, for example, a second metal could be deposited. One method of avoiding the interfering reaction would be to use an i_{app} less than 1% of the initial i_l (for a 99% complete electrolysis), so that 99% of the O would be reduced before the potential shift occurred. This method would lead to a very prolonged electrolysis duration, however. Another method that is sometimes employed involves the use of a "cathodic depolarizer," a substance that is introduced into the solution to be reduced more easily (i.e., at less negative potentials) than any interfering substance. For example, consider the reduction of Cu(II) in the presence of Pb(II) (Figure 10.4.3). If NO_3^- is added to the solution, it will be reduced preferentially before the Pb(II) reduction occurs, and will prevent the deposition of lead along with the copper. In this case NO_3^- is said to play the role of a cathodic depolarizer. (The term depolarizer implies that the substance fixes the potential, by its own reduction, at a certain desired value.) The hydrogen ion often also plays the role of a depolarizer.

Generally, except for the simpler apparatus involved, controlled current electrolysis offers no advantages over controlled potential methods. With the commercial availability of suitable potentiostats, controlled current methods are being used less

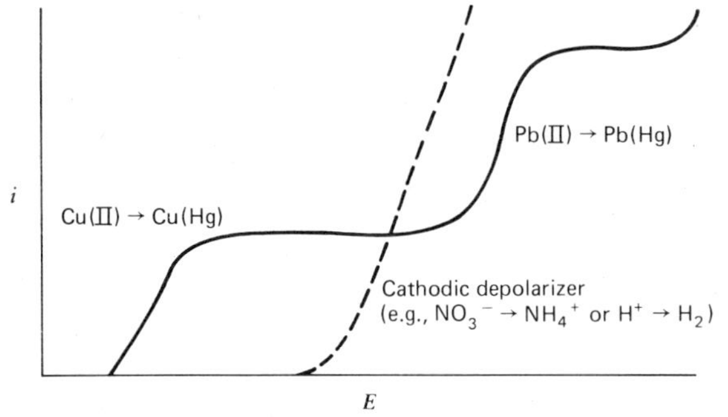

Figure 10.4.3
Schematic *i-E* curves illustrating action of a cathodic depolarizer in limiting the negative potential excursion of the working electrode and preventing codeposition of lead in separation of copper.

frequently in analysis and preparative electrolysis. For large-scale electrosynthesis or separations involving very high currents, especially in flow systems where the reactants are continually added to the cell and products removed, the simplicity of constant current methods is a great advantage. Here some degree of control of the working electrode potential can be obtained by regulation of the solution flow rate.

10.4.2 Coulometric Methods

A constant current coulometric method is attractive because a stable constant current source is easy to construct and the total number of coulombs consumed in an electrolysis can be calculated readily from the duration of the electrolysis, τ, by

$$Q = i_{\mathrm{app}}\tau \tag{10.4.2}$$

However, to use a coulometric method for a determination, the reaction of interest must proceed with close to 100% efficiency. To illustrate how this is accomplished in a constant current format, consider the coulometric determination of Fe^{2+} by oxidation at a platinum electrode to Fe^{3+} in an H_2SO_4 medium (Figure 10.4.4). If a constant current is applied to the Pt anode, then as described in Section 10.4.1, when the i_l for Fe^{2+} oxidation falls below i_{app}, the current efficiency would fall below 100% and part

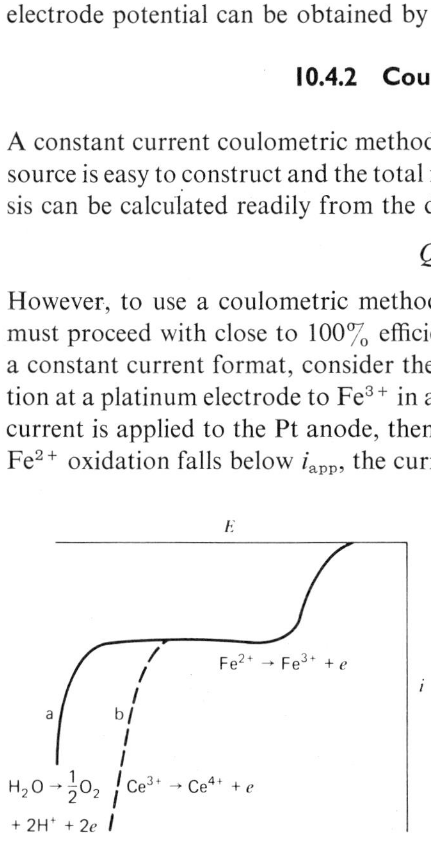

Figure 10.4.4
Schematic *i-E* curves for Fe^{2+} in 1 *M* H_2SO_4 in the absence (curve a) and in the presence (curve b) of excess Ce^{3+}.

of the applied current would go to a secondary process (e.g., oxygen evolution). However, if Ce^{3+} is added to the solution, when the current efficiency for the *direct* oxidation of Fe^{2+} falls below 100%, the next process to occur is $Ce^{3+} \rightarrow Ce^{4+} + e$. The Ce^{4+} so produced is capable of oxidizing any Fe^{2+} remaining in the bulk solution by the fast reaction

$$Ce^{4+} + Fe^{2+} \rightarrow Ce^{3+} + Fe^{3+} \tag{10.4.3}$$

Thus some Fe^{2+} is *indirectly* oxidized to Fe^{3+}, and the *titration efficiency* for the oxidation of Fe^{2+} is maintained. This then resembles an ordinary titration of Fe^{2+} with Ce^{4+}, in that a true equivalence point is reached. For this reason this technique is usually called a *coulometric titration* (of Fe^{2+} with the electrogenerated titrant, Ce^{4+}). Note that some end point detection technique (as is also required in an ordinary titration) must be used to indicate when the oxidation of Fe^{2+} is complete, since neither the current nor the potential of the working electrode is a good indicator of the course of the reaction.

The requirements for the coulometric *intermediate* or *titrant* (e.g., Ce^{4+}) are that it be generated with a high current efficiency and that it react rapidly and completely with the substance being determined (Fe^{2+}). In some cases, such as the Fe^{2+}–Ce^{4+} titration, the generation of the intermediate accounts for only part of the total electricity consumed. In others, such as the coulometric titration of olefins with electrogenerated Br_2 (generated by oxidation of Br^-), all of the current goes to the generation of the intermediate, which then reacts with the substance being titrated, that is,

$$2Br^- \rightarrow Br_2 + 2e \tag{10.4.4}$$

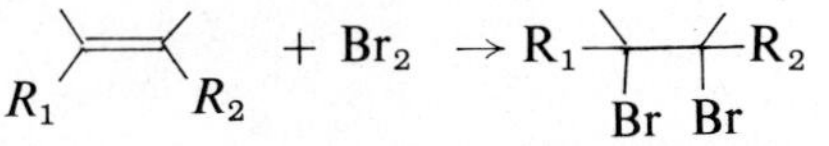

$$\underset{R_1 \quad R_2}{\searrow\!\!=\!\!\swarrow} + Br_2 \rightarrow R_1\underset{Br \quad Br}{-\!\!\!\!\top\!\!\!\!-\!\!\!\!\top\!\!\!\!-}R_2 \tag{10.4.5}$$

A block diagram of the apparatus used in coulometric titrations is shown in Figure 10.4.5. The cell is composed of a working electrode and an auxiliary electrode in

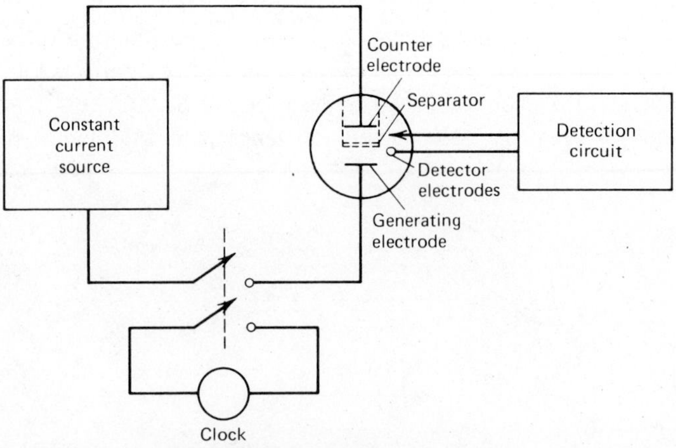

Figure 10.4.5
Apparatus for coulometric titrations.

separate compartments. End point detection is often made by an electrometric method (Section 10.5); hence indicator electrodes suited to the particular end point detection technique are also located within the cell. The constant current source can be simply a high-voltage (e.g., 400 V) power supply and a bank of resistors. This will produce essentially a constant current as long as the reversible cell potential and cell resistance are small compared to the applied voltage and circuit resistance. Electronic constant current sources (*amperostats* or *galvanostats*), for example, those based on operational amplifier circuitry, are also frequently used (see Chapter 13). Whenever current is switched to the cell, a timer is actuated, so that the total electrolysis time can be recorded. Typically the applied current is in the range of 10 μA to 200 mA and titration times are 10 to 100 sec.

The solution conditions and the end point detection system are usually chosen based on the same criteria for an ordinary titration (e.g., fast, definite, single, complete, titration reaction and sensitive endpoint detection). The current density range for generation of the titrant can be determined by taking *i-E* curves of the supporting electrolyte system with and without the titrant precursor species (A) present (22, 23) (Figure 10.4.6). The current efficiency for the generation of titrant (B) can be estimated by the equation

$$\text{Current efficiency} = \frac{100T}{T + S} \tag{10.4.6}$$

(assuming the process A $\pm$ $ne \rightarrow$ B does not affect the background electrolysis process). A plot of current efficiency as a function of current density can be prepared and, from this, the optimum region for titrant generation can be established. The current

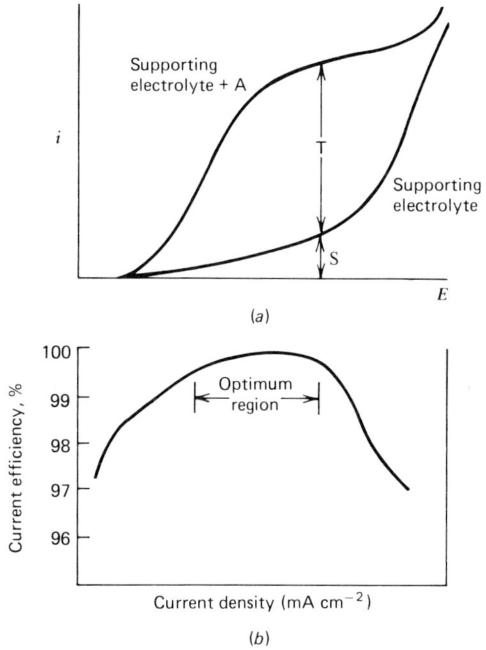

Figure 10.4.6
(*a*) Use of *i-E* curves for estimating current efficiency at given potential and current density. (*b*) Typical plot of current efficiency as function of current density for an electrogenerated titrant.

to be used is selected by consideration of the amount of substance to be determined and a convenient electrolysis time. Then an electrode area is calculated to give the needed current density. For example, the determination of 0.1 μequiv of a substance requires $\sim 10^4$ μC, or a current of 100 μA for 100 sec. Thus, if the optimum generation current density is 1 mA-cm^{-2}, a generating electrode area of 0.1 cm^2 would be used.

Coulometric titrations can be applied to a number of different types of determinations, including acid-base, precipitation, complexation, and redox titrations. Some typical examples are given in Table 10.4.1; detailed descriptions of the scope and nature of coulometric titrations are given in References 19 and 24 to 26. Coulometric titrations offer a number of advantages over conventional titrations with standard solutions: (a) Very small amounts of substances can be determined without the use of ultramicrovolumetric techniques. For example, a titration with $i_{app} = 10$ μA and $t = 100$ sec is quite easy and corresponds ($n = 1$) to about 10^{-8} mol or only a few micrograms of titratable material. (b) Standard solutions need not be prepared or stored. Standardizations using primary standards are not necessary. (c) Substances that are unstable or inconvenient to use because of volatility or reactivity can be employed as titrants, for example, Br_2, Cl_2, Ti^{3+}, Sn^{2+}, Cr^{2+}, Ag^{2+}, and Karl

Table 10.4.1

Typical Electrogenerated Titrants and Substances Determined by Coulometric Titration

Electrogenerated Titrant	Generating Electrode and Solution	Typical Substances Determined
		Oxidants
Bromine	Pt/NaBr	As(III), U(IV), NH_3, olefins, phenols, SO_2, H_2S, Fe(II)
Iodine	Pt/KI	H_2S, SO_2, As(III), water (Karl Fischer), Sb(III)
Chlorine	Pt/NaCl	As(III), Fe(II), various organics
Cerium(IV)	Pt/$Ce_2(SO_4)_3$	U(IV), Fe(II), Ti(III), I^-
Manganese(III)	Pt/$MnSO_4$	Fe(II), H_2O_2, Sb(III)
Silver(II)	Pt/$AgNO_3$	Ce(III), V(IV), $H_2C_2O_4$
		Reductants
Iron(II)	Pt/$Fe_2(SO_4)_3$	Mn(III), Cr(VI), V(V), Ce(IV), U(VI), Mo(VI)
Titanium(III)	Pt/$TiCl_4$	Fe(III), V(V,VI), U(VI), Re(VIII), Ru(IV), Mo(VI)
Tin(II)	Au/$SnBr_4$(NaBr)	I_2, Br_2, Pt(IV), Se(IV)
Copper(I)	Pt/Cu(II)(HCl)	Fe(III), Ir(IV), Au(III), Cr(VI), IO_3^-
Uranium(V),(IV)	Pt/UO_2SO_4	Cr(VI), Fe(III)
Chromium(II)	Hg/$CrCl_3$($CaCl_2$)	O_2, Cu(II)
		Precipitation and Complexation Agents
Silver(I)	Ag/$HClO_4$	Halide ions, S^{2-}, mercaptans
Mercury(I)	Hg/$NaClO_4$	Halide ions, xanthate
EDTA	Hg/$HgNH_3Y^{2-}$[a]	Metal ions
Cyanide	Pt/$Ag(CN)_2^-$	Ni(II), Au(III,I), Ag(I)
		Acids and Bases
Hydroxide ion	Pt(−)/Na_2SO_4	Acids, CO_2
Hydrogen ion	Pt(+)/Na_2SO_4	Bases, CO_3^{2-}, NH_3

[a] Y^{4-} is ethylenediamine-tetra-acetate anion.

Fischer reagent. (d) They are easily automated, since it is easier to control an electric current and monitor time than it is to control a buret valve and record volume. (e) They can be performed remotely (e.g., in the analysis of radioactive materials) and under an inert atmosphere more easily. (f) Dilution effects do not occur during the titration making end point location simpler. Another broad field of applications involves continuous coulometric titrators which are employed in process stream analyzers. In these the generating current is continuously adjusted to maintain a small excess of electrogenerated titrant to react with material in the incoming liquid or gaseous sample stream. The level of generating current is a measure of the instantaneous concentration of the titrated substance (27, 28). Coulometric titration methods have also been used in chromatographic detectors and for determination of homogeneous reaction rates (29).

10.5 ELECTROMETRIC END POINT DETECTION (30–35)

10.5.1 Classification

Electrometric methods are employed to detect the end points of conventional titrations, as well as the coulometric approaches described in Section 10.4. These methods usually involve measurement of the potential of the indicator electrode(s) (*potentiometric methods*) or the current passing in the indicator electrode circuit (*amperometric methods*). Further classification is based on the nature of the test electrodes. One may be nonpolarizable (i.e., a reference electrode), or both electrodes may be polarizable. Thus for potentiometric methods one has: (a) $i=0$, one nonpolarizable electrode (ordinary potentiometry); (b) constant applied current, one polarizable electrode ("one-electrode potentiometry"); (c) constant applied current, two polarizable electrodes ("two-electrode potentiometry"). Similarly for amperometric methods we can list (a) constant applied voltage, one polarizable electrode ("one-electrode amperometry"); (b) constant applied voltage, two polarizable electrodes ("two-electrode amperometry"). The polarizable, or indicator, electrode can be any steady-state voltammetric electrode, for example, a DME, a platinum microelectrode in stirred solution, an RDE, and so on. The reference electrode is usually a low-resistance one, such as an SCE with a low-resistance junction, so that with the usual small currents employed, iR drops are small and two-electrode systems can be employed. (Three-electrode potentiometric and amperometric systems are possible, but they are rarely used). The shapes and characteristics of the titration curves (E or i *vs.* fraction titrated, f) depend on the i-E curves for the indicator electrode(s) at different points during the titration. These depend on the composition of the solution at a given value of f and on the reversibility of the different couples in the solution.

10.5.2 Current-Potential Curves During Titration

Consider the titration of Fe^{2+} with Ce^{4+} by the reaction

$$Fe^{2+} + Ce^{4+} \rightarrow Fe^{3+} + Ce^{3+} \tag{10.5.1}$$

Assume that this is the usual manual titration of a solution initially containing Fe^{2+}

with addition of the titrant Ce^{4+}. For simplicity, however, we will neglect dilution effects. Both the Fe^{3+}/Fe^{2+} and Ce^{4+}/Ce^{3+} couples behave nearly reversibly at a platinum microelectrode. Schematic *i-E* curves obtained at different values of f (where f = moles Ce^{4+} added/moles Fe^{2+} initially present) are shown in Figure 10.5.1. Initially ($f = 0$) the cell contains only Fe^{2+}, and only the anodic wave for Fe^{2+} oxidation is observed. During the titration ($0 < f < 1$) the solution contains Fe^{2+}, Fe^{3+}, and Ce^{3+}, while after the equivalence point ($f > 1$) the solution contains Fe^{3+}, Ce^{3+}, and Ce^{4+}. The titration curves for the different potentiometric and amperometric methods can be derived from these *i-E* curves.

10.5.3 Potentiometric Methods

One-electrode potentiometry involves the measurement of the potential of an indicator electrode with respect to a reference (nonpolarizable) electrode at either open circuit or with a small anodic or cathodic current applied to the indicator electrode. These

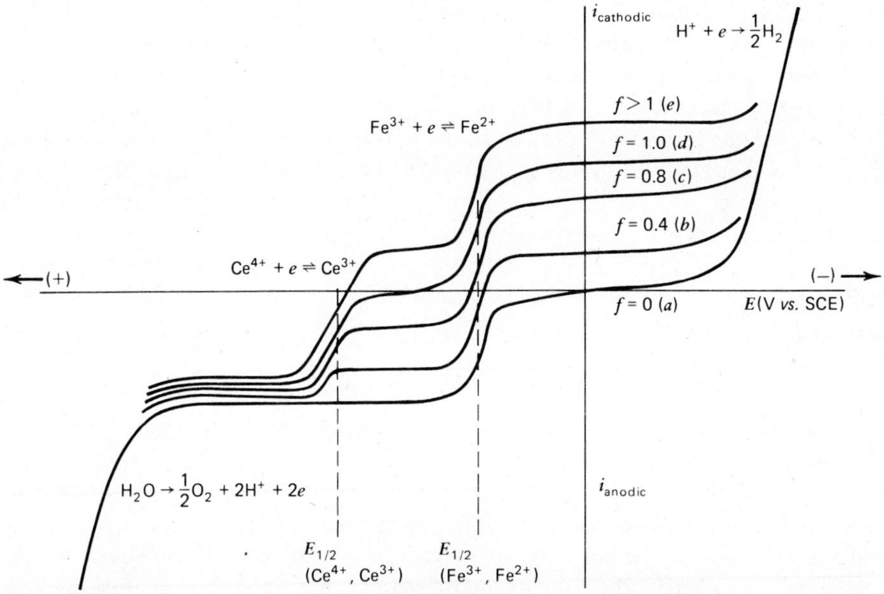

Figure 10.5.1

Current-potential curves obtained with a platinum electrode during titration of Fe^{2+} with Ce^{4+} at different fractions of Fe^{2+} titrated, f. (*a*) $f = 0$; only an anodic wave for $Fe^{2+} \rightarrow Fe^{3+}$ is observed. (*b*) (*c*) (*d*) $0 < f \leq 1$; the solution contains Fe^{2+}, Fe^{3+}, and Ce^{3+}. A composite wave for Fe^{3+}, Fe^{2+} couple, as well as an anodic wave for $Ce^{3+} \rightarrow Ce^{4+} + e$, are observed. (*e*) $f > 1$; the solution contains Fe^{3+}, Ce^{3+}, and Ce^{4+}. A composite wave for the Ce^{4+}, Ce^{3+} couple and a cathodic wave for $Fe^{3+} + e \rightarrow Fe^{2+}$ are observed. (The curves drawn are representative of those obtained for a steady-state voltammetric technique, for example, an RDE or a microelectrode in stirred solution. They are clearly idealized. Adapted from J. J. Lingane, "Electroanalytical Chemistry," 2nd ed., Wiley-Interscience, New York, 1958.)

three possibilities are shown in Figure 10.5.2 for the Fe^{2+}-Ce^{4+} titration, and the resulting titration curves are shown in Figure 10.5.3. The $i = 0$ curve, (*a*), is the usual potentiometric titration curve, showing the equilibrium potential of the solution (E_{eq}) as a function of f. When a small anodic current is impressed on the indicator electrode, the measured potential at a given f will be somewhat more positive than E_{eq} [curve (*c*)]. When a small cathodic current is applied, the potential will be more

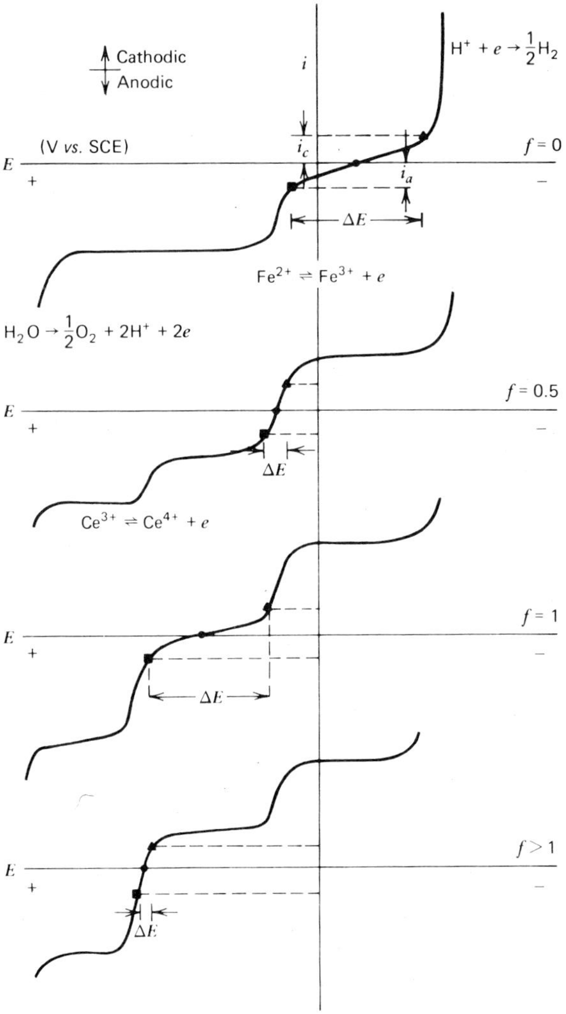

Figure 10.5.2

Current-potential curves at platinum electrode during titration of Fe^{2+} with Ce^{4+} at different fractions titrated, f, illustrating the potential attained by this indicator electrode (*vs.* SCE) at (*a*) zero current (●); (*b*) small applied cathodic current, i_c (▲); (*c*) small applied anodic current, i_a (■). (The magnitudes of the actual applied currents used in a titration would be much smaller than those shown here, which are exaggerated for clarity.)

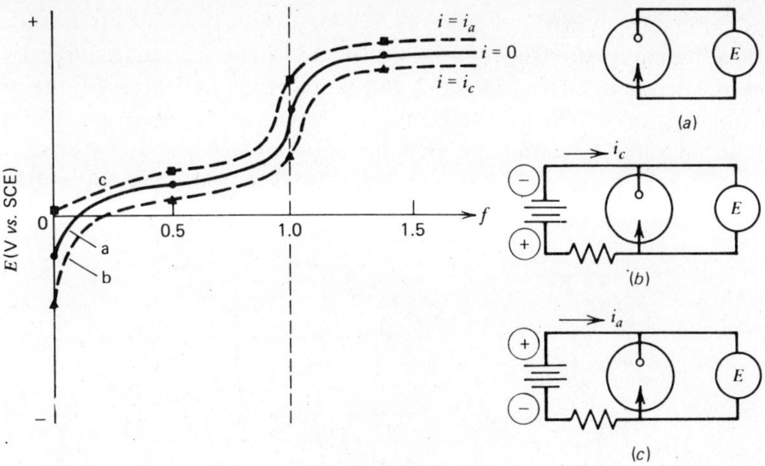

Figure 10.5.3

Potentiometric titration curves for platinum indicator electrode *vs.* SCE (reference) with impressed current of: curve a, —●—, 0; curve b, --▲--, i_c; curve c, --■--, i_a. Points on curves correspond to those in Figure 10.5.2. The corresponding circuits for these end point detection methods are also illustrated in (*a*) to (*c*). The meter, *E*, is assumed to have a high-input impedance and the battery-resistor combination is chosen so that the applied currents are essentially constant.

negative than E_{eq} [curve (*b*)]. If the applied currents are sufficiently small, the break in the *E vs. f* titration curve with applied current will only be slightly displaced from that of the E_{eq} *vs. f* curve. The advantage of potentiometric end point detection methods with applied current (or polarized electrodes) is that a steady potential is sometimes attained more rapidly under these conditions compared to measurements at open circuit. This is particularly true in titrations involving couples that show irreversible behavior. In these cases the shapes of the titration curves will be somewhat different from those depicted for reversible reactions. They are left as an exercise for the reader (see Problem 10.3).

Two-electrode potentiometry involves the measurement of the potential difference between two polarizable (i.e., indicator) electrodes when a small constant current is applied to them. The two electrodes are usually made of the sam material and are about the same size. The current through the cathode (i_c) must be equal to the current through the anode (i_a), and each adopts the potential governed by the appropriate *i-E* curve for that current density. If the electrodes are identical, the same *i-E* curve applies for each, and the difference in potential, ΔE, can be obtained from a single set of *i-E* curves such as those given in Figure 10.5.2. The two-electrode potentiometric curve for the Fe^{2+}-Ce^{4+} titration, obtained from either the difference between the one-electrode potentiometry curves (Figure 10.5.3) or the *i-E* curves themselves is given in Figure 10.5.4. The end point is signaled by a peak in ΔE. As in one-electrode potentiometry, the response will be different if one of the couples shows irreversible behavior at the indicator electrode.

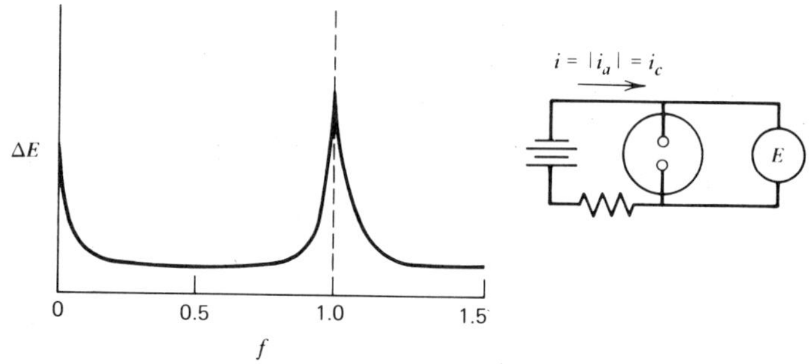

Figure 10.5.4
Titration curve for two-electrode potentiometric detection in the Fe^{2+}-Ce^{4+} titration. Two identical platinum electrodes with an impressed current $i = |i_a| = i_c$ is assumed. (See Figures 10.5.2 and 10.5.3.)

10.5.4 Amperometric Methods

One-electrode amperometry involves maintaining the potential of the indicator electrode at a constant value with respect to a reference electrode and determining the current as a function of f. Consider again the Ce^{4+}-Fe^{2+} titration, this time, however, with the potential of the indicator electrode maintained at a value on the plateau of the i-E curve for Fe^{2+} oxidation (E_1 in Figure 10.5.5). The current during the titration is shown in Figure 10.5.6. In this titration the current changes from anodic to cathodic at $f = 1$. Note that for this titration, holding the potential of the indicator electrode in the region of E_1 is the only way of obtaining an informative titration curve.† If the indicator electrode is held at other potentials, a useful titration curve usually would not result.

Another type of titration curve frequently observed by one-electron amperometry involves the formation of an insoluble substance. Consider the titration of Pb^{2+} with dichromate according to the reaction

$$Pb^{2+} + Cr_2O_7^{2-} \rightarrow \underline{PbCr_2O_7} \qquad (10.5.2)$$

using a DME as an indicator electrode. The sampled current (maximum)-potential curves are shown in Figure 10.5.7, and the resulting titration curves for the indicator electrode at two different potentials are shown in Figure 10.5.8.

Two-electrode amperometry involves the use of two indicator electrodes with a small constant potential impressed between them. Since they are in the same current loop, the anodic current in one will be equal in magnitude to the cathodic current in the other. The potential of each will thus shift during a titration to maintain the condition $i_c = |i_a|$. For example, for the Fe^{2+}-Ce^{4+} titration with two platinum electrodes held at a potential difference, ΔE (~ 50 mV) (Figure 10.5.5), the titration curve that results is shown in Figure 10.5.9.

† Unless the mass transfer coefficients for the various species (e.g., Fe^{2+} and Fe^{3+} *vs.* Ce^{3+} and Ce^{4+}) are quite different (see Problem 10.9).

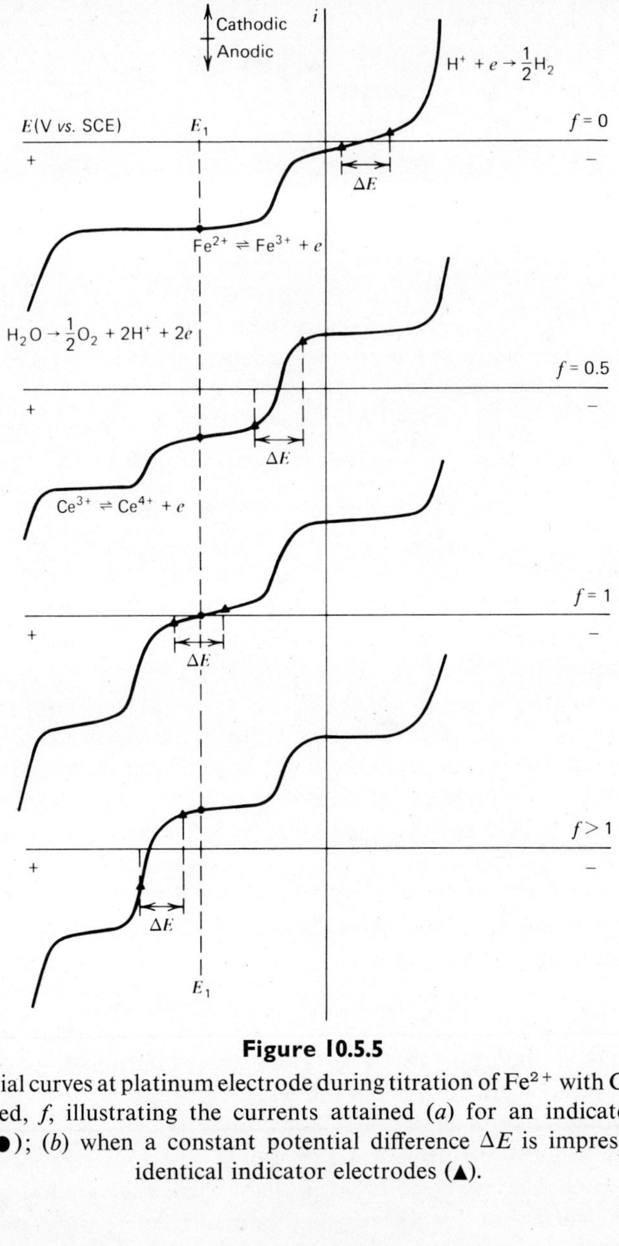

Figure 10.5.5

Current-potential curves at platinum electrode during titration of Fe^{2+} with Ce^{4+} at different fractions titrated, *f*, illustrating the currents attained (*a*) for an indicator electrode at potential E_1 (●); (*b*) when a constant potential difference ΔE is impressed across two identical indicator electrodes (▲).

The shape of the two-electrode, amperometric titration curve strongly depends on the reversibility of the electrode reactions of the titrant and titrate systems. For example, if the titrate involves reversible electrode reactions (e.g., I_3^-/I^-) and titrant electrode reactions are irreversible (e.g., $S_2O_3^{2-}/S_4O_6^{2-}$), the titration curve has the shape shown in Figure 10.5.10. The titration curves in this type of titration are sometimes said to involve "dead-stop" end points, since the current falls essentially to zero at the end point and remains there. This type of end point was applied in some of the

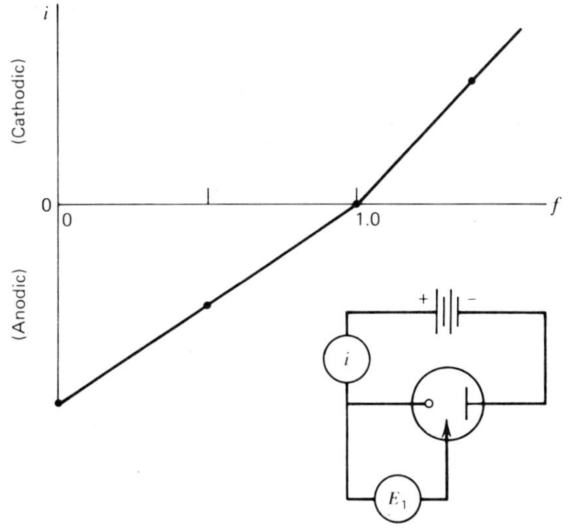

Figure 10.5.6
One-electrode amperometric titration curve for titration of Fe^{2+} with Ce^{4+} with the platinum indicator electrode held at E_1 (see Figure 10.5.5) (dilution neglected).

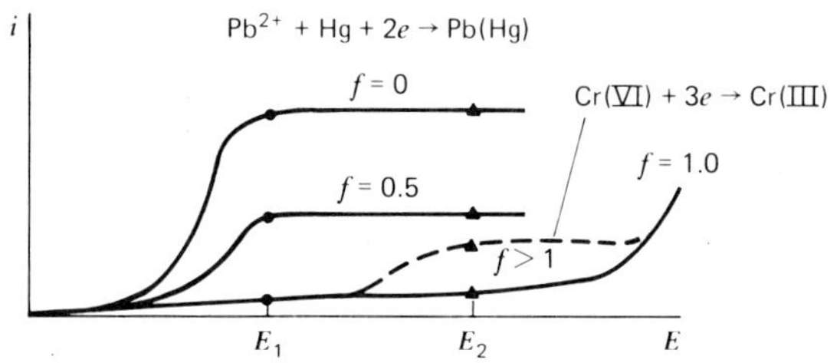

Figure 10.5.7
Current-potential curves at DME (only envelope of current maximum is shown) during titration of Pb^{2+} with $Cr_2O_7^{2-}$. Note that although thermodynamically $Cr_2O_7^{2-}$ should be reduced more easily than Pb^{2+}, the reaction rate is slow and the curve for Cr(VI) is displaced to more negative potentials.

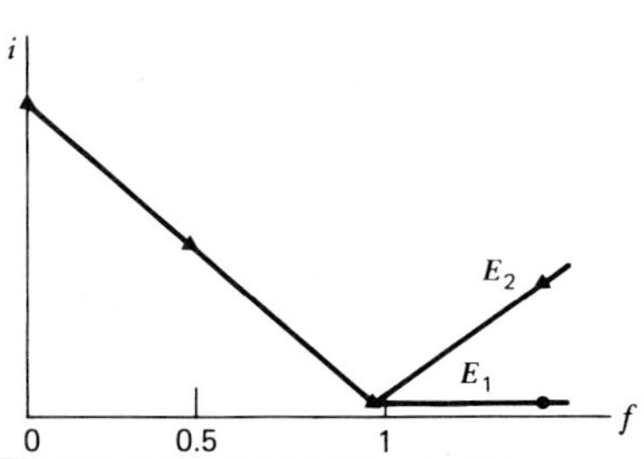

Figure 10.5.8
One-electrode amperometric titration curves for the titration of Pb^{2+} with $Cr_2O_7^{2-}$ with DME indicator electrode held at E_1 or E_2 (see Figure 10.5.7).

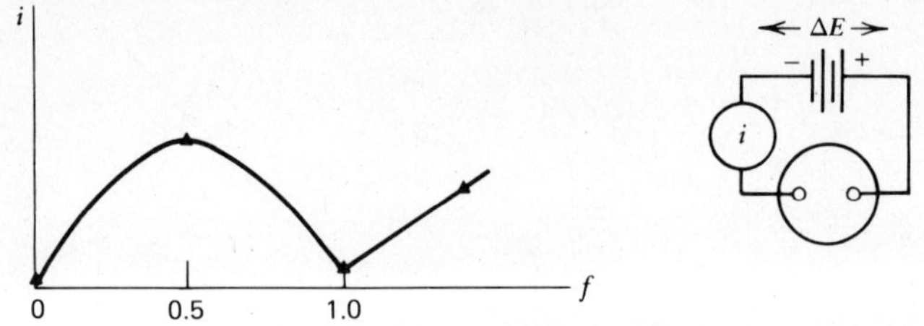

Figure 10.5.9

Two-electrode amperometric titration of Fe^{2+} with Ce^{4+} with a constant potential difference ΔE impressed between two platinum indicator electrodes (see Figure 10.5.5).

earliest amperometric methods; in fact, it was discovered and used before the theoretical basis for these titrations was established. We leave it to the reader to work out the titration curves (potentiometric and amperometric) that result for other different cases involving reversible and irreversible couples as titrants and titrates.

The amperometric titration curves for coulometric titrations will have somewhat different shapes than the ones for the manual titrations described, because usually one form of the couple exists in large excess. For example, in the titration of Fe^{2+} with electrogenerated Ce^{4+}, Ce^{3+} will be present from the start at a concentration that is large compared to that of the Fe^{2+}. Titrations for multicomponent systems can be treated in a similar manner. In all cases the curves can be derived by consideration of the *i-E* curves that arise during the titration.

10.6 FLOW ELECTROLYSIS

10.6.1 Introduction

An alternate method of bulk electrolysis involves flow electrolytic methods, where the solution to be electrolyzed flows continuously through a porous working electrode (36) of large surface area. These methods can result in high efficiencies and rapid electrolysis and are especially convenient where large amounts of solution are to be utilized. Flow methods are of use in industrial electrolytic processes (e.g., for removal of metals such as copper from waste streams) and have also been applied to electrosynthesis, separations, and analysis. The flow electrolysis cell (Figure 10.6.1) contains a working electrode of large surface area, composed, for example, of screens of fine mesh metal or beds of conductive material (e.g., graphite or glassy carbon grains, metal shot or

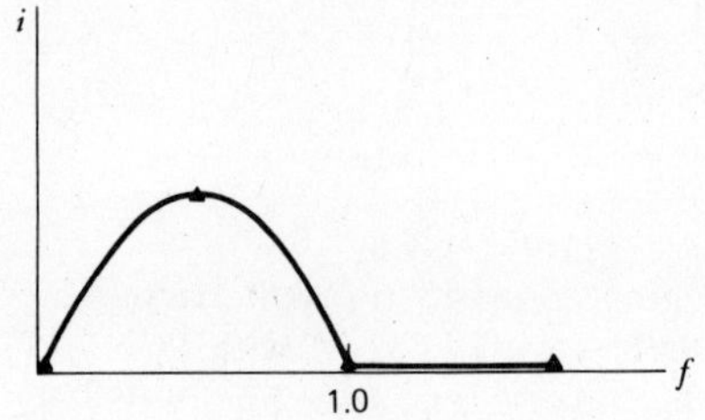

Figure 10.5.10

A "dead-stop" (two-electrode) amperometric titration resulting from the titration of a couple that shows reversible *i-E* behavior with a titrant that does not (e.g., $I_3^- + 2S_2O_3^{2-} \rightarrow 3I^- + S_4O_6^{2-}$).

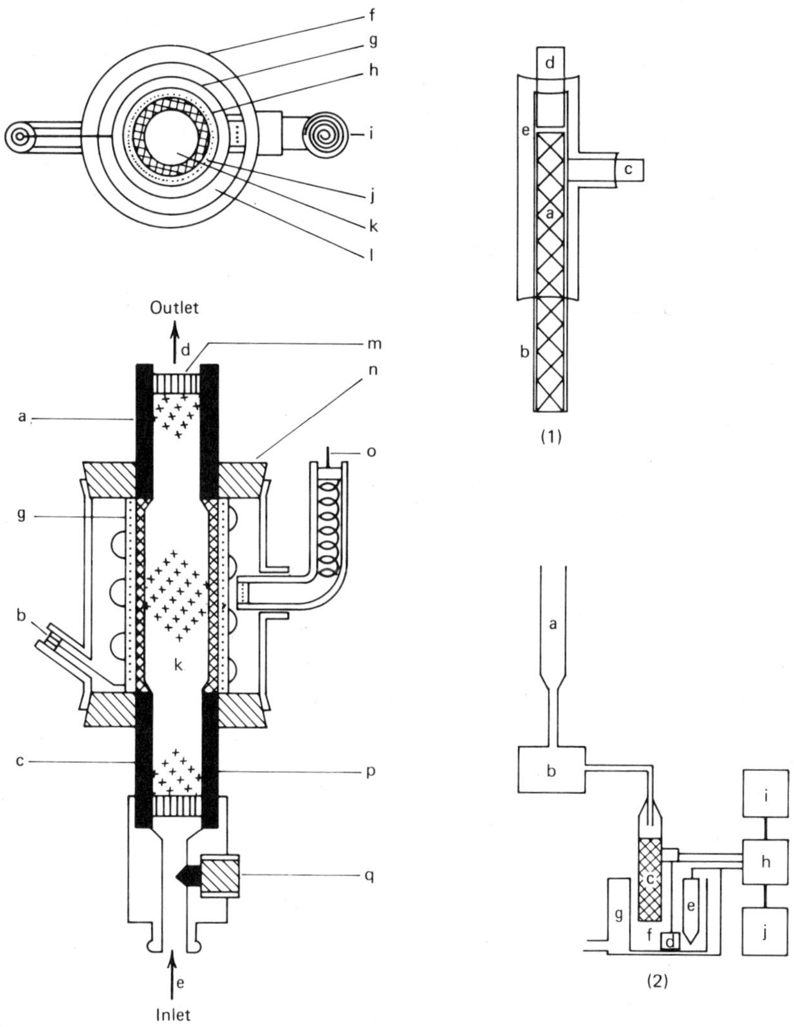

Figure 10.6.1

Flow electrolytic cells. *Left*: Cell utilizing glassy carbon granule working electrode (k), silver auxiliary electrode (g), Ag/AgCl reference electrode (o, i) with porous glass separator (h). Other components are (a, c) lead for working electrode; (b) lead to auxiliary electrode; (d) solution outlet; (e) solution inlet; (f) glass or plastic tube; (j, p) porous carbon tube; (l) saturated KCl solution; (m) silicone rubber. [Reprinted with permission from T. Fujinaga and S. Kihara, *CRC Crit. Rev. Anal. Chem.*, **6**, 223 (1977). Copyright The Chemical Rubber Co., CRC Press, Inc.] *Right*: Cell with Reticulated Vitreous Carbon (RVC) (trademark), a conductive foam-type material available in several porosities. (1) (A) RVC cylinder, (B) heat-shrink tubing, (C) graphite rod sidearm, (D) glass tube, (E) glass and epoxy support. (2) Schematic diagram of complete apparatus. (A) Solution reservoir, (B) pump, (C) RVC electrode, (D) platinum electrode, (E) SCE reference electrode, (F) downstream reservoir, (G) runover collector, (H) potentiostat, (I) recorder, (J) digital voltmeter. [Reprinted with permission from A. N. Strohl and D. J. Curran, *Anal. Chem.*, **51**, 353 (1979). Copyright 1979, American Chemical Society.]

powder). If a divided cell is not necessary, as in metal deposition, the counter electrode can be interleaved with the working electrode and insulated from it with simple separators. Divided cells require more complex dividers (porous glass or ceramics, ion-exchange membranes) and careful placement of the counter and reference electrodes to minimize iR drops. The cells are designed to show high conversions with a minimum length of electrode and maximum flow velocities.

10.6.2 Mathematical Treatment (37)

Consider a flow-through porous electrode of length L (cm) and cross-sectional area A (cm^2) immersed in a stream of flow velocity v (cm^3/sec) (Figure 10.6.2). The *linear flow velocity* of the stream, U (Cm/sec), is given by

$$U = \frac{v}{A} \tag{10.6.1}$$

The reaction being carried out at the electrode, O + $ne \rightarrow$ R, at a total current, i (A), is assumed to occur with 100% current efficiency. The inlet concentration of O is C_{O}(in) and C_{R}(in) is assumed to be zero. At the outlet, the concentrations are C_{O}(out) and C_{R}(out). The overall conversion of O to R in passage through the electrode is i/nF (mol/sec) or i/nFv (mol/cm^3). If R is the fraction of O converted ($R = 0$, no conversion; $R = 1$, 100% conversion), then

$$C_{\mathrm{O}}(\mathrm{out}) = C_{\mathrm{O}}(\mathrm{in})(1 - R) \tag{10.6.2}$$

$$C_{\mathrm{R}}(\mathrm{out}) = C_{\mathrm{O}}(\mathrm{in})\, R \tag{10.6.3}$$

$$C_{\mathrm{O}}(\mathrm{out}) = C_{\mathrm{O}}(\mathrm{in}) - \frac{i}{nFv} \tag{10.6.4}$$

$$R = \frac{i}{nFv\, C_{\mathrm{O}}(\mathrm{in})} = 1 - \frac{C_{\mathrm{O}}(\mathrm{out})}{C_{\mathrm{O}}(\mathrm{in})} \tag{10.6.5}$$

We desire an expression for the dependence of the current on flow velocity and electrode parameters. The total *internal area* of the electrode, which encompasses the sum of the areas of all of the pores, is a (cm^2), and the total electrode volume is LA

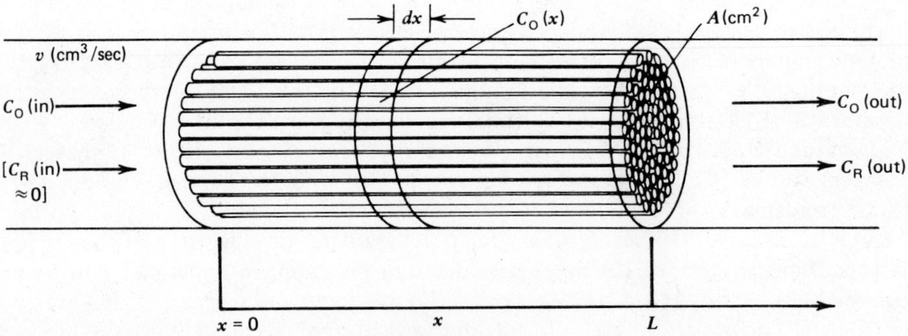

Figure 10.6.2
Schematic diagram of working electrode of flow electrolysis cell.

(cm³). Porous electrodes are frequently characterized by their *specific area*, s, given by

$$s\ (\text{cm}^{-1}) = a\ (\text{cm}^2)/LA\ (\text{cm}^3) \tag{10.6.6}$$

(see example, Figure 10.6.3). The concentration of O decreases continuously with distance from the front face of the electrode ($x = 0$), and the local current density at a given location, $j(x)$ (A/cm^2), varies with x. The net conversion in a slab of thickness dx is $j(x)sA\ dx/nF$ (mol/sec), where under mass-transfer-controlled, limiting-current conditions (see Section 1.4.2),

$$j(x) = nFm_OC_O(x) \tag{10.6.7}$$

The variation in concentration at x is then

$$-dC_O(x)\ (\text{mol/cm}^3) = \frac{j(x)sA\ dx}{nFv} \tag{10.6.8}$$

Combining (10.6.7) and (10.6.8) then yields

$$-\frac{dC_O(x)}{dx} = \frac{m_OC_O(x)sA}{v} \tag{10.6.9}$$

$$\int_{C_O(\text{in})}^{C_O(x)} \frac{dC_O(x)}{C_O(x)} = \frac{-m_OsA}{v}\int_0^x dx \tag{10.6.10}$$

$$C_O(x) = C_O(\text{in}) \exp\left(\frac{-m_OsA}{v}x\right) \tag{10.6.11}$$

$$j(x) = nFm_OC_O(\text{in}) \exp\left(\frac{-m_OsA}{v}x\right) \tag{10.6.12}$$

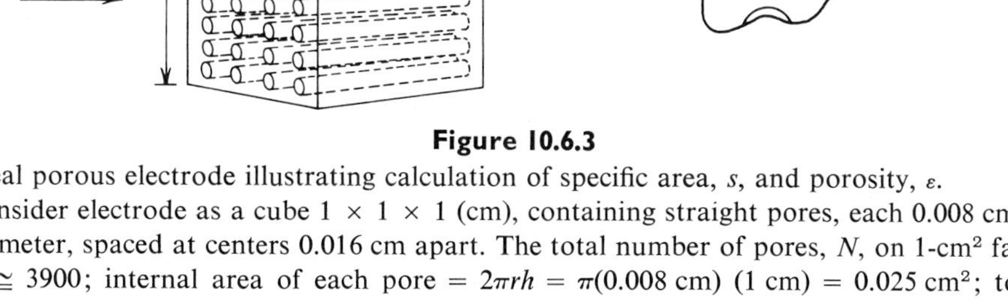

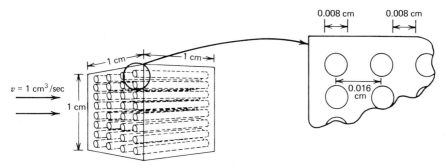

Figure 10.6.3

Ideal porous electrode illustrating calculation of specific area, s, and porosity, ε. Consider electrode as a cube 1 × 1 × 1 (cm), containing straight pores, each 0.008 cm in diameter, spaced at centers 0.016 cm apart. The total number of pores, N, on 1-cm² face: $N \cong 3900$; internal area of each pore $= 2\pi rh = \pi(0.008\ \text{cm})\ (1\ \text{cm}) = 0.025\ \text{cm}^2$; total internal electrode area $= (3900)\ (0.025\ \text{cm}^2) = 98\ \text{cm}^2$; total electrode volume $= 1\ \text{cm}^3$; specific area $= s = 98\ \text{cm}^2/1\ \text{cm}^3 = 98\ \text{cm}^{-1}$; facial area of each pore $= \pi r^2 = 5.0 \times 10^{-5}\ \text{cm}^2$; total open area on face $= (3900)(5.0 \times 10^{-5}\ \text{cm}^2) = 0.2\ \text{cm}^2$; porosity $= \varepsilon = 0.2\ \text{cm}^2/1\ \text{cm}^2 = 0.2$. If the solution velocity, $v = 1\ \text{cm}^3/\text{sec}$, linear flow velocity $= U = 1\ \text{cm}^3/\text{sec}/1\ \text{cm}^2 = 1\ \text{cm/sec}$. Interstitial velocity $= W = 1\ \text{cm/sec}/0.2 = 5\ \text{cm/sec}$.

The total current in the electrode is then

$$i = \int_0^L j(x)sA\,dx = nFm_OC_O(\text{in})sA\int_0^L \exp\left(\frac{-m_OsAx}{v}\right)dx \qquad (10.6.13)$$

$$i = nFC_O(\text{in})v\left[1 - \exp\left(\frac{-m_OsAL}{v}\right)\right] \qquad (10.6.14)$$

This can be combined with (10.6.5) to yield (37a)

$$\boxed{R = 1 - \exp\left(\frac{-m_OsAL}{v}\right)} \qquad (10.6.15)$$

The mass transfer coefficient m_O is a function of flow velocity U and is sometimes given as

$$m_O = bU^{\alpha} \qquad (10.6.16)$$

where b is a proportionality factor and α is a constant (frequently having values between 0.33 and 0.5 for laminar flow and increasing up to nearly 1 for turbulent flow). With this equation and (10.6.1), these equations take the form

$$i = nFAUC_O(\text{in})[1 - \exp(-bU^{\alpha-1}sL)] \qquad (10.6.17)$$

$$R = 1 - \exp(-bU^{\alpha-1}sL) \qquad (10.6.18)$$

Thus the conversion efficiency R increases with decreasing flow velocity and increasing specific area and length of electrode. From (10.6.11) it can be seen that the concentration of O varies exponentially with distance along the electrode with

$$\boxed{C_O(\text{out}) = C_O(\text{in})\exp\left(\frac{-m_OsAL}{v}\right)} \qquad (10.6.19)$$

The local current density $j(x)$ is highest at the front face of the electrode and decreases exponentially with x.

These equations can also be cast in a form comparable with those for batch bulk electrolysis. If the total front-surface, open area of the pores is a_p, then the *porosity*, ε, is defined as (see Figure 10.6.3)

$$\varepsilon = \frac{a_p}{A} \qquad (10.6.20)$$

The linear flow velocity, which is U in the liquid stream, increases upon entering the electrode to an *interstitial velocity*, W, given by

$$W = \frac{U}{\varepsilon} = \frac{v}{A\varepsilon} = \frac{v}{a_p} \qquad (10.6.21)$$

A volume element of solution moves down a pore at this velocity and, if it entered the electrode at time $t = 0$, then at time t it will be a distance x, given by

$$x = Wt = \frac{Ut}{\varepsilon} \qquad (10.6.22)$$

This allows the equations to be formulated in terms of time, so that substitution of (10.6.22) into (10.6.11) yields

$$C_O(t) = C_O(\text{in}) \exp\left(\frac{-m_O s}{\varepsilon} t\right) \tag{10.6.23}$$

This equation is of the same form as that for a batch electrolysis (10.3.6), with $m_O s/\varepsilon = p$ (compared to $p = m_O A/V$). Thus the cell factor p increases with increasing mass transfer rate, increasing specific area, and decreasing porosity. The length of porous electrode required for a given conversion R can be obtained from (10.6.15):

$$\boxed{L = -\frac{v}{m_O s A} \ln(1 - R)} \tag{10.6.24}$$

The time for an element of solution to transit the electrode, τ, from (10.6.22) and (10.6.24) (sometimes called the *residence time*) is given by

$$\tau = \frac{L\varepsilon}{U} = p^{-1} \ln(1 - R) \tag{10.6.25}$$

An alternate simplified approach to the efficiency of electrolysis in a porous electrode (38) considers the time t' required for O in the center of a pore of radius r to diffuse to the wall:

$$t' \simeq \frac{r^2}{2D_O} \tag{10.6.26}$$

The time required to move through the electrode down a pore of length L is given by [see (10.6.21) and (10.6.22)]

$$t = \frac{La_p}{v} \tag{10.6.27}$$

If this time is greater than or equal to t', a high conversion ($R \approx 1$) will be attained. By equating (10.6.26) and (10.6.27) we find that the flow velocity required for high conversion must satisfy the expression:

$$v \leq \frac{2a_p L D_O}{r^2} \tag{10.6.28}$$

For example, for a porous silver electrode with $A = 0.2\text{ cm}^2$, $\varepsilon = 0.5$, $L = 50\ \mu\text{m}$, $r = 2.5\ \mu\text{m}$, $D_O = 5 \times 10^{-6}\text{ cm}^2/\text{sec}$, $a_p = \varepsilon A = 0.1\text{ cm}^2$, the maximum flow velocity for $R \approx 1$ is $0.1\text{ cm}^3/\text{sec}$, with a residence time in the electrode of ~ 5 msec.

The simple treatment given here, which involves limiting current conditions and neglects (a) resistive drops in the electrode and in the solution in the pores, (b) kinetic limitations to the electron transfer reaction, and (c) the possibility of a current efficiency less than unity, can be employed to find general conditions of an efficient flow electrolysis. Treatments taking these other effects into account are available (39–41); these usually result in equations requiring numerical solution. Flow cells operating at $R = 1$ are convenient for the continuous analysis of liquid streams, since the measured current is directly proportional to the concentration of the substance undergoing electrolysis, that is, from (10.6.5), $C_O(\text{in}) = i/nFv$. Since this is actually a continuous coulometric analysis, such an analytical method is absolute and does not require calibration or knowledge of mass transfer parameters, electrode area, etc. (42). A chromatographic method based on flow electrolysis has also been described (43). Here, as in elution chromatography methods, a sample containing metal ions is introduced into a stream of flowing electrolyte solution with the potential of the porous working electrode maintained constant. Deposition of a metal ion on the column results in a current-time trace [see (10.6.23) for deposition under limiting current conditions], which allows determination of the amount of metal ion. Electrolytic chromatographic methods, where a gradient in potential is maintained along the length of the porous working electrode, are also available.

10.6.3 Dual-Electrode Flow Cells

Flow cells that incorporate two working electrodes in the flow channel have also been described (Figure 10.6.4*a*). These can be considered as the flow coulometric equivalent of the rotating ring-disk electrode where convective flow carries material from the first working electrode to the second. This has been used in the coulometric analysis of plutonium, where the two working electrodes were large beds of glassy carbon particles with the first electrode used to adjust the oxidation state of the plutonium to a single known level [Pu(IV)] and the second used for the coulometric analysis itself [Pu(IV) + $e \rightarrow$ Pu(III)] (43). This type of system can also be used to analyze for the products produced at the first electrode (the generator electrode), (e.g., in the reaction O + $ne \rightarrow$ R) by electrolysis at the second (the detector electrode), (e.g., R $\rightarrow$ O + ne). In this application, thin, but efficient, working electrodes separated by a small gap, g, are desired. A system involving porous silver disk working electrodes (50 μm average pore diameter) separated by a gap of 200 μm with porous Teflon material has been described (38) (Figure 10.6.4*b*). In this mode of operation, each working electrode is provided with its own auxiliary and reference electrode (so that this constitutes a six-electrode cell), and two separate potential control circuits must be used. The characteristics of these working electrodes are those given immediately after (10.6.28), and the estimated maximum flow velocity for good conversion ($\sim$0.1 cm^3/sec) was approximately confirmed experimentally by carrying out the reduction of Fe(III) to Fe(II) in an oxalate medium on one working electrode and noting where the current began to deviate from that predicted by (10.6.5) with $R = 1$. At this flow rate the transit time across the gap is about 40 msec and high collection efficiencies (i.e., $i_{\text{detector}}/i_{\text{generator}}$) were found even for flow rates where $R < 1$. The authors suggested that such a system might be useful for studying homogeneous reactions coupled to the electron transfer reaction. For example, if the product of the reaction at the generator electrode (R) decomposes (e.g., $R \rightarrow A$), then not only will the detector current for the oxidation of R be smaller but also product A will appear in the effluent and can be determined there by any of a number of analytical methods. Application of this cell to a study of the isomerization of the radical anion of diethyl maleate in N,N-dimethylformamide solution was reported.

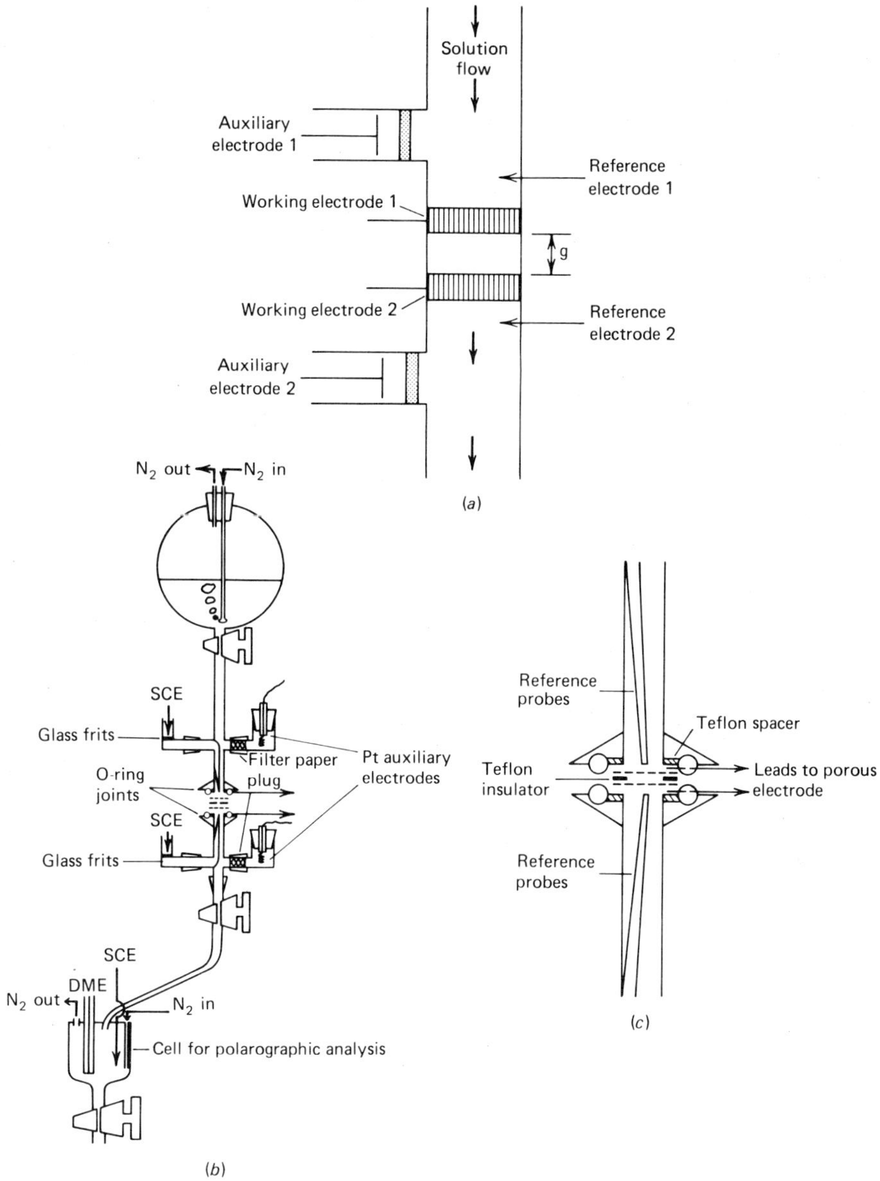

Figure 10.6.4

(*a*) Schematic representation of dual-electrode flow cell. (*b*) Actual complete dual-electrode flow cell assembly. Solution flows by gravity from upper reservoir. For greater clarity, the "O-ring joint" portion of the cell with the dual working electrodes is shown in exploded form. A close-up view of this portion with the porous silver electrodes is shown in (*c*). [From J. V. Kenkel and A. J. Bard, *J. Electroanal. Chem.*, **54**, 47 (1974), with permission.]

10.7 THIN-LAYER ELECTROCHEMISTRY

10.7.1 Introduction

An alternate approach to obtaining bulk electrolysis conditions and a large A/V ratio, even with no convective mass transfer, involves decreasing V, so that a very small solution volume (a few μL) is confined to a thin layer (2–100 μm) at the electrode surface. A schematic diagram of a thin-layer cell and some typical actual cell configurations are shown in Figure 10.7.1. As long as the cell thickness l is smaller than

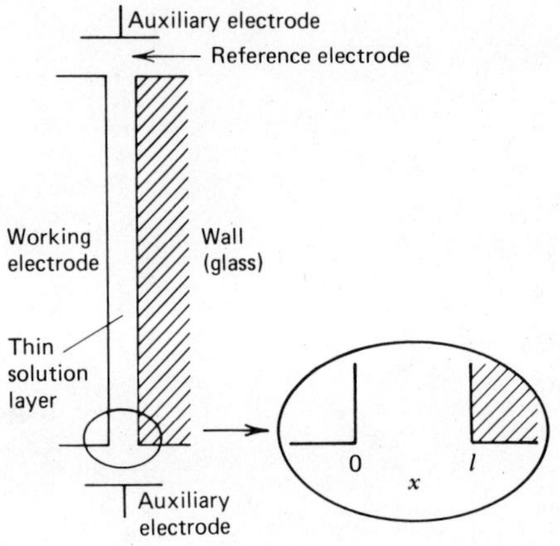

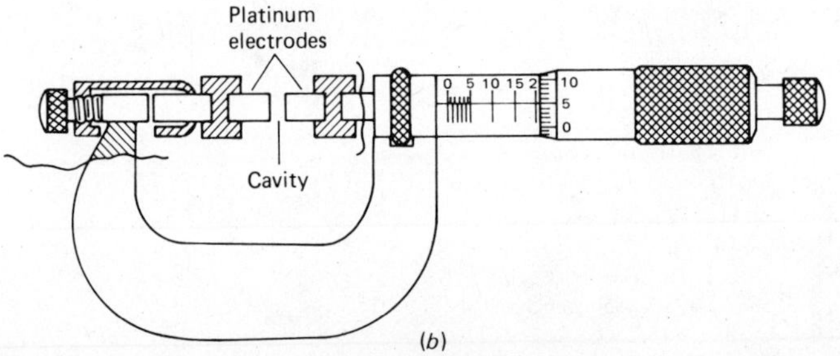

Figure 10.7.1

(*a*) Schematic diagram of single-electrode, thin-layer cell. (*b*) Micrometer, twin-electrode thin-layer cell with adjustable solution layer thickness. (*c*) Close-up of electrode portion for single-electrode configuration of (*b*). (*d*) Capillary-wire single-electrode thin-layer electrode. The solution layer is contained in the small space between the metal rod and the inner surface of the precision-bore capillary. The layer thickness is typically 2.5×10^{-3} cm. The metal rod may be positioned within the capillary to a high degree of concentricity, when necessary, by machining three small flanges onto the surface of the rod near each end. Highly reproducible rinsing and filling are accomplished by alternately applying and releasing nitrogen pressure. [From A. T. Hubbard and F. C. Anson, *Electroanal. Chem.*, **4**, 129 (1970), courtesy of Marcel Dekker, Inc.]

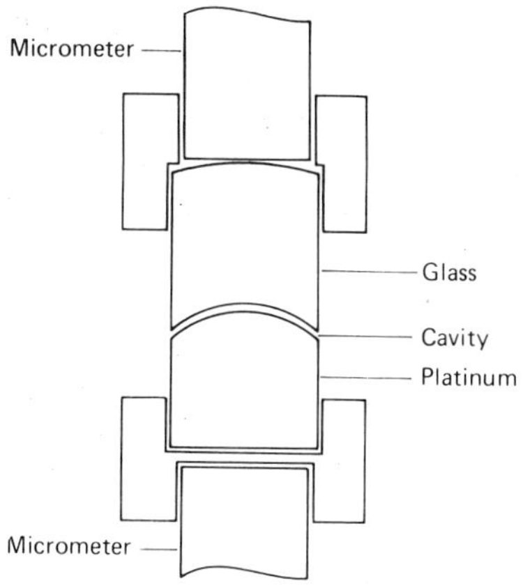

(c)

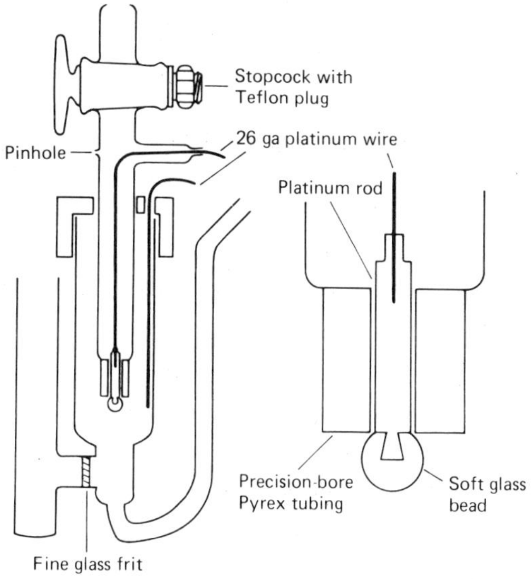

(d)

Figure 10.7.1—*Continued*

the diffusion layer thickness for a given experimental time, that is, $l \ll (2Dt)^{1/2}$, mass transfer within the cell can be neglected, and special bulk electrolysis equations result. At shorter times, diffusion in the cell must be considered. Thin-layer electrochemical cells were first utilized in the early 1960s, and the theory and applications of these have been reviewed in depth (44–49).

10.7.2 Potential Step (Coulometric) Methods

Consider the twin-working-electrode, thin-layer cell (Figure 10.7.1*b*), with the potential stepped from a value E_1, where no current flows, to E_2, where the reaction $O + ne \rightarrow R$ is virtually complete and the concentration of O at the electrode surface is essentially zero. To obtain the current-time behavior and the concentration profile one must solve the diffusion equation

$$\frac{\partial C_O(x, t)}{\partial t} = D_O\left(\frac{\partial^2 C_O(x, t)}{\partial x^2}\right) \tag{10.7.1}$$

with the boundary conditions

$$C_O(x, 0) = C_O^* \qquad t = 0; 0 \leq x \leq l \tag{10.7.2}$$

$$C_O(0, t) = C_O(l, t) = 0 \qquad t > 0 \tag{10.7.3}$$

Note that in this case the semi-infinite boundary condition used for the analogous experiment in Section 5.2 has been replaced by the condition for C_O at l.† Solution of these equations by the Laplace transform method yields (48)

$$C_O(x, t) = \frac{4C_O^*}{\pi}\sum_{m=1}^{\infty}\left(\frac{1}{2m - 1}\right)\exp\left[\frac{-(2m - 1)^2\pi^2 D_O t}{l^2}\right]\sin\frac{(2m - 1)\pi x}{l} \tag{10.7.4}$$

At later times, the concentration profile can be obtained from consideration of only the $m = 1$ term, since the $(2m - 1)^2$ factor in the exponential causes the terms for $m = 2, 3, \ldots$ to be small for $\pi^2 D_O t/l^2 \gg 1$. Then

$$C_O(x, t) \approx \frac{4C_O^*}{\pi}\exp\left(\frac{-\pi^2 D_O t}{l^2}\right)\sin\frac{\pi x}{l} \tag{10.7.5}$$

Typical concentration profiles are as given in Figure 10.7.2*a*. The current is always obtained from

$$i(t) = nFAD_O\left[\frac{\partial C_O(x, t)}{\partial x}\right]_{x=0} \tag{10.7.6}$$

$$\boxed{i(t) = \frac{4nFAD_O C_O^*}{l}\sum_{m=1}^{\infty}\exp\left[\frac{-(2m - 1)^2\pi^2 D_O t}{l^2}\right]} \tag{10.7.7}$$

or at later times,

$$i(t) \approx i(0)\exp(-pt) \tag{10.7.8}$$

with

$$p = \frac{\pi^2 D_O}{l^2} = \frac{\pi^2 D_O A}{Vl} = \frac{m_O A}{V}$$

$$m_O = \frac{\pi^2 D_O}{l} \quad \text{and} \quad i(0) = \frac{4nFAC_O^* m_O}{\pi^2}$$

Note that the form of (10.7.8) is the same as (10.3.7), which would be followed for the thin-layer cell if the concentration within the cell could be considered completely

---

† For a thin-layer cell with the solution constrained between a single working electrode and inert boundary, as in Figure 10.7.1*a*, the condition at this boundary is $[\partial C_O(x, t)/\partial x]_{x=l} = 0$.

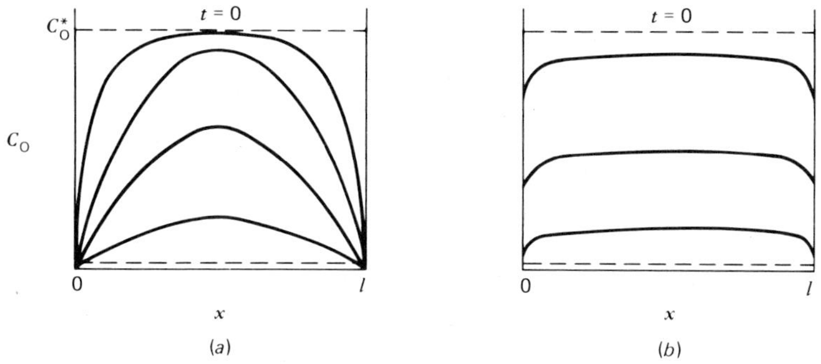

Figure 10.7.2
Concentration of O during reduction of O in twin-electrode, thin-layer cell. (*a*) Actual profiles. (*b*) Neglecting mass transfer in cell.

uniform throughout the electrolysis (as in Figure 10.7.2*b*). Finally, the total charge passed by the electrolytic reaction is

$$Q(t) = nFVC_O^*\left\{1 - \frac{8}{\pi^2}\sum_{m=1}^{\infty}\left(\frac{1}{2m-1}\right)^2 \exp\left[\frac{-(2m-1)^2\pi^2 D_O t}{l^2}\right]\right\} \quad (10.7.9)$$

$$Q(t) \approx nFVC_O^*\left(1 - \frac{8}{\pi^2}e^{-pt}\right) \quad \text{(later times)} \quad (10.7.10)$$

$$Q(t \to \infty) = nFVC_O^* = nFN_O \quad (10.7.11)$$

Equation 10.7.11 is the same as the coulometry equation (10.3.11), so that determinations of N_O or n are possible without the necessity of knowing D_O. The electrolysis rate constant in a thin-layer cell, p, can be quite large. For example, for $D = 5 \times 10^{-6}$ cm²/sec and $l = 10^{-3}$ cm, $p = 49$ sec^{-1} and the electrolysis will be 99% complete in less than 0.1 sec. In actual experiments the measured charge will be larger than that given by (10.7.9) to (10.7.11), because contributions from the double-layer charging and background reactions will be included.

10.7.3 Potential Sweep Methods

Consider again the reaction and initial conditions of Section 10.7.2 for the situation where the working electrode potential is swept from an initial value E_i, where no reaction occurs, toward negative values. Under conditions where the concentrations of O and R can be considered uniform [$C_O(x, t) = C_O(t)$ and $C_R(x, t) = C_R(t)$ for $0 \leq x \leq l$], the current is given, as in (10.3.4), by

$$i = -nFV\left[\frac{dC_O(t)}{dt}\right] \quad (10.7.12)$$

For a nernstian reaction,

$$E = E^{0'} + \frac{RT}{nF}\ln\frac{C_O(t)}{C_R(t)} \quad (10.7.13)$$

$$C_O^* = C_O(t) + C_R(t) \quad (10.7.14)$$

Combination of these two equations yields

$$C_O(t) = C_O^*\left\{1 - \left[1 + \exp\left(\frac{nF}{RT}(E - E^{0'})\right)\right]^{-1}\right\} \tag{10.7.15}$$

Differentiation of (10.7.15), and substitution into (10.7.12) with the sweep rate, $v = -(dE/dt)$, yields the expression for the current:

$$\boxed{i = \frac{n^2F^2vVC_O^*}{RT}\frac{\exp\left[\left(\frac{nF}{RT}\right)(E - E^{0'})\right]}{\left\{1 + \exp\left[\left(\frac{nF}{RT}\right)(E - E^{0'})\right]\right\}^2}} \tag{10.7.16}$$

The peak current occurs at $E = E^{0'}$ and is given by

$$\boxed{i_p = \frac{n^2F^2vVC_O^*}{4RT}} \tag{10.7.17}$$

A typical scan voltammogram in a thin-layer cell is shown in Figure 10.7.3. Note that the peak current is directly proportional to v, but the total charge under the i-E curve, given by (10.7.11), is independent of v.

The rigorous solution for this problem, accounting for nonuniform concentrations within the cell, can be derived (48). It has been shown that the approximate form (10.7.16) will hold at sufficiently small values of v, that is, when

$$|v| \le \frac{RT}{nF}\frac{\pi^2 D}{3l^2}\log\left(\frac{1-\varepsilon}{1+\varepsilon}\right) \tag{10.7.18}$$

where ε is the relative error tolerated in calculation of i_p.

For a totally irreversible reaction, the current is given by [see (3.2.7)]

$$\frac{i}{nFA} = k_f C_O(t) \tag{10.7.19}$$

where $k_f = k^0 \exp[-(\alpha n_a F/RT)(E - E^{0'})]$.

By combining (10.7.19) with (10.7.12) we obtain

$$\frac{dC_O(t)}{dt} = -\left[\frac{Ak_f(t)}{V}\right]C_O(t) \tag{10.7.20}$$

In a potential sweep experiment, $E(t) = E_i - vt$ (see equation 6.2.1); therefore, with $f = F/RT$,

$$k_f(t) = k^0 \exp[-\alpha n_a f(E_i - E^{0'})] \exp[\alpha n_a f v t] \tag{10.7.21}$$

By substitution of (10.7.21) into (10.7.20) and integration between $t = 0$ $(C_O = C_O^*)$ and $t\,[C_O = C_O(t)]$, under the conditions that $k^0 \exp[-\alpha n_a f(E_i - E^{0'})] \rightarrow 0$ (i.e., an

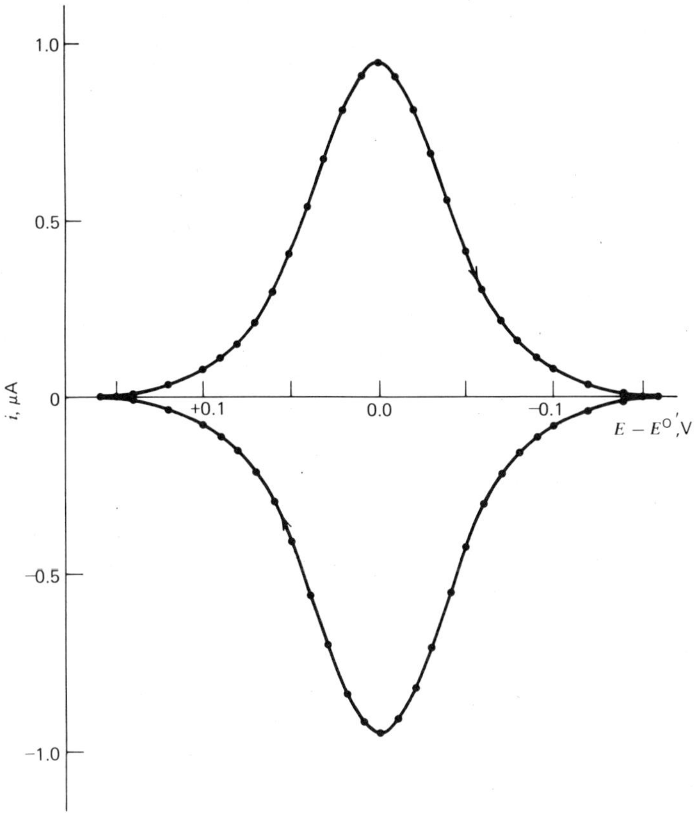

Figure 10.7.3
Cyclic current-potential curve for a reaction of the type, O + *ne* = R. The values of the experimental parameters assumed in making the graph are: $n = 1$; $V = 1.0\ \mu$L; $|v| = 1$ mV/sec; $C_O^* = 1.0$ m*M*; $T = 298$ K. [From A. T. Hubbard and F. C. Anson, *Electroanal. Chem.*, **4**, 129 (1970), courtesy of Marcel Dekker, Inc.]

initial potential well positive of $E^{0'}$), the following expressions for $C_O(E)$ and $i(E)$ are obtained (48, 49):

$$C_O(E) = C_O^* \exp\left(\frac{-RT}{\alpha n_a F}\frac{Ak_f}{Vv}\right) \tag{10.7.22}$$

$$i(E) = nFAk_f C_O^* \exp\left(\frac{-RT}{\alpha n_a F}\frac{Ak_f}{Vv}\right) \tag{10.7.23}$$

or, substituting for k_f,

$$i(E) = nFAC_O^* k^0 \exp\left\{-\alpha n_a f(E - E^{0'}) - \frac{Ak^0}{\alpha n_a f V v}\exp[-\alpha n_a f(E - E^{0'})]\right\} \tag{10.7.24}$$

Typical *i vs. E* curves for a totally irreversible reduction of O to R in a thin-layer cell are shown in Figures 10.7.4 and 10.7.5. The peak potential [obtained by differentiation

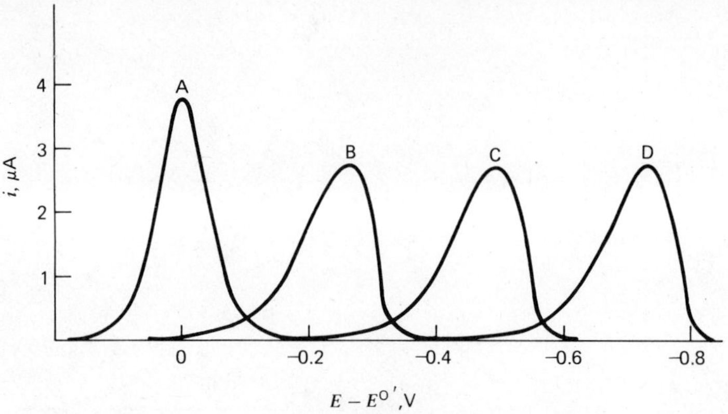

Figure 10.7.4
Theoretical cathodic current-potential curves for irreversible reactions (10.7.24) for several values of k^0. Curve A: reversible reaction (shown for comparison). Curve B: $k^0 = 10^{-6}$. Curve C: $k^0 = 10^{-8}$; Curve D: $k^0 = 10^{-10}$ cm/sec. The values assumed in making the plots were $|v| = 2$ mV/sec; $n = n_a = 1$; $A = 0.5$ cm^2; $C_O^* = 1.0$ mM; $\alpha = 0.5$; $V = 2.0$ μL. [From A. T. Hubbard, *J. Electroanal. Chem.*, **22**, 165 (1969), with permission.]

of (10.7.24) and setting the result equal to zero] occurs at

$$E_{pc} = E^{0'} + \frac{RT}{\alpha n_a F} \ln\left(\frac{ARTk^0}{\alpha n_a F v V}\right) \tag{10.7.25}$$

The peak current is still proportional to v and C_O^* and is

$$i_{pc} = \frac{n\alpha n_a F^2 V v C_O^*}{(2.718)RT} \tag{10.7.26}$$

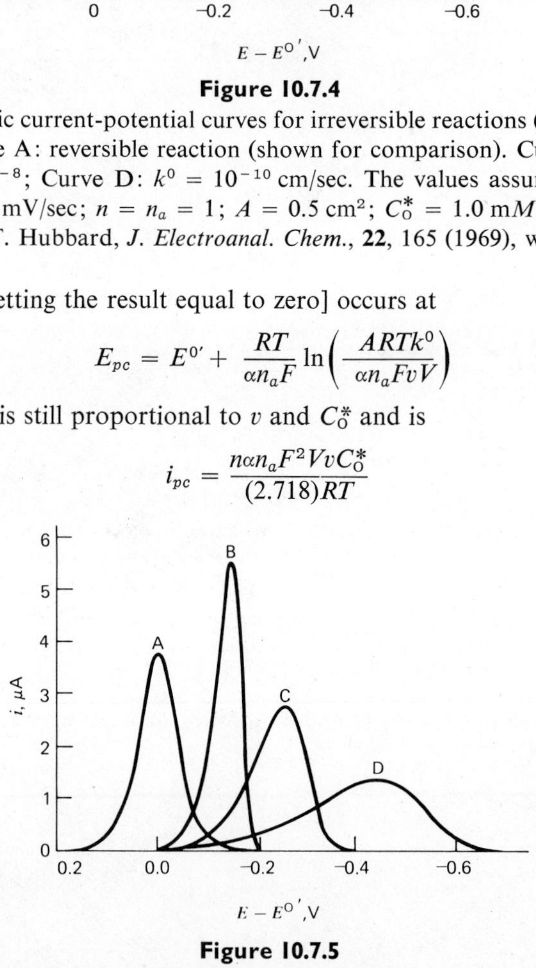

Figure 10.7.5
Theoretical cathodic current-potential curve for irreversible reactions (10.7.24) for several values of αn_a. Curve A: reversible reaction. Curve B: $\alpha n_a = 0.75$; $k^0 = 10^{-6}$ cm sec^{-1}. Curve C: $\alpha n_a = 0.5$; $k^0 = 10^{-6}$ cm sec^{-1}. Curve D: $\alpha n_a = 0.25$; $k^0 = 10^{-6}$ cm sec^{-1}. The values assumed in making the graphs were: $|v| = 2$ mV/sec; $n = 1$; $A = 0.5$ cm^2; $C_O^* = 1.0$ mM, $V = 2.0$ μL. [From A. T. Hubbard, *J. Electroanal. Chem.* **22**, 165 (1969), with permission.]

Thin-layer methods have been suggested for determination of kinetic parameters of electrode reactions (48–50), although they have not been widely used for this purpose. A difficulty in these methods, especially when nonaqueous solutions or very low supporting electrolyte concentrations are employed, is the high resistance of the thin layer of solution. Since the reference and auxiliary electrodes are placed outside the thin-layer chamber, one can have serious nonuniform current distributions and high uncompensated *iR* drops (producing for example, nonlinear potential sweeps) (51). Although cell designs that minimize this problem have been devised (48, 49), careful control of the experimental conditions is required in kinetic measurements. Thin-layer cells have been applied in a number of electrochemical studies, including investigations of adsorption, electrodeposition, complex reaction mechanisms, and *n*-value determinations. They have also become very popular in spectroelectrochemical studies (Chapter 14).

The theory and mathematical treatments used for thin-layer cells find application in other electrochemical problems; for example, the deposition of metals (as amalgams) into thin films of mercury and their subsequent stripping is fundamentally a thin-layer problem (see Section 10.8). Similarly the electrochemical oxidation or reduction of thin films (e.g., oxides, adsorbed layers, and precipitates) follows an analogous treatment (see Section 12.5). Thin-layer concepts are also directly applicable to synthetically modified electrodes featuring electroactive species bound to the surface. In many problems involving surface films, mass transfer truly is negligible over wide time domains and problems with uncompensated resistance are minimal; thus relatively fast experiments can be performed.

10.8 STRIPPING ANALYSIS

10.8.1 Introduction

Stripping analysis is a method for determination that utilizes a bulk electrolysis step (*preelectrolysis*) to preconcentrate a substance from solution into the small volume of a mercury electrode (a hanging mercury drop or a thin film) or onto the surface of an electrode. After this electrodeposition step, the material is redissolved ("stripped") from the electrode using some voltammetric technique (most frequently linear potential sweep voltammetry). If the conditions during the preelectrolysis step are maintained constant, exhaustive electrolysis of the solution is not necessary and, by proper calibration and with fixed electrolysis times, the measured voltammetric response (e.g., peak current) can be employed to find the solution concentration. This process is represented schematically in Figure 10.8.1. The major advantage of the method, as compared to direct voltammetric analysis of the original solution, is the preconcentration of the material to be analyzed on or within the electrode (by factors of 100 to >1000), so that the voltammetric (stripping) current is less perturbed by charging or residual impurity currents. The technique is especially useful for the analysis of very dilute solutions (down to 10^{-10} to 10^{-11} M). Stripping analysis is most frequently used for the determination of metal ions by cathodic deposition, followed by anodic stripping with a linear potential scan and, therefore, is sometimes called

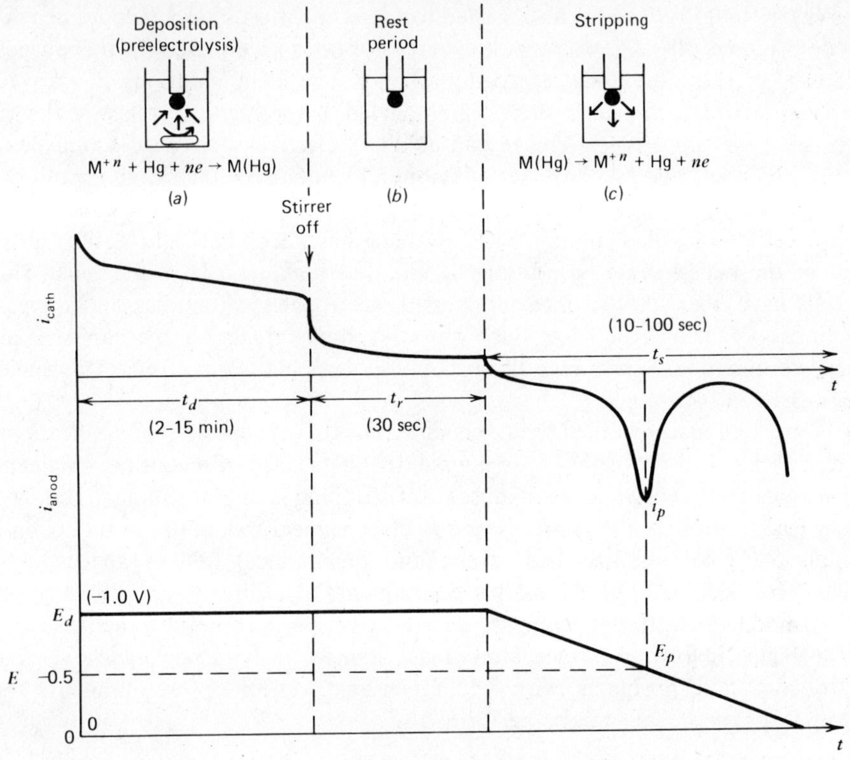

Figure 10.8.1

Principle of anodic stripping experiment. Values shown are typical ones used; potentials and E_p are typical of Cu^{2+} analysis. (*a*) Preelectrolysis at E_d; stirred solution. (*b*) Rest period, stirrer off. (*c*) Anodic scan (v = 10–100 mV/sec). [Adapted from E. Barendrecht, *Electroanal. Chem.*, **2**, 53 (1967), courtesy of Marcel Dekker, Inc.]

anodic stripping voltammetry (ASV) or, less frequently, *inverse voltammetry*. The basic theoretical principles and some typical applications will be described here. Several complete reviews describing the history, theory, and experimental methodology of this technique have appeared (53–58).

10.8.2 Principles and Theory

Typical mercury electrodes used in stripping analysis are shown in Figure 10.8.2. The hanging drops are essentially the same as those we have encountered in earlier discussions. The mercury film electrode (MFE) is more restricted to stripping work. In current practice, it is usually deposited onto a rotating glassy carbon or wax-impregnated graphite disk (Figure 10.8.2*b*). One usually adds mercuric ion (10^{-5}–10^{-4} *M*) directly to the analyte solution, so that during the preelectrolysis, the mercury codeposits with the species to be determined. The resulting mercury films are usually less than 100 Å thick. Since the MFE has a much smaller volume than the HMDE, the MFE shows a higher sensitivity. There is evidence that mercury electrodes with

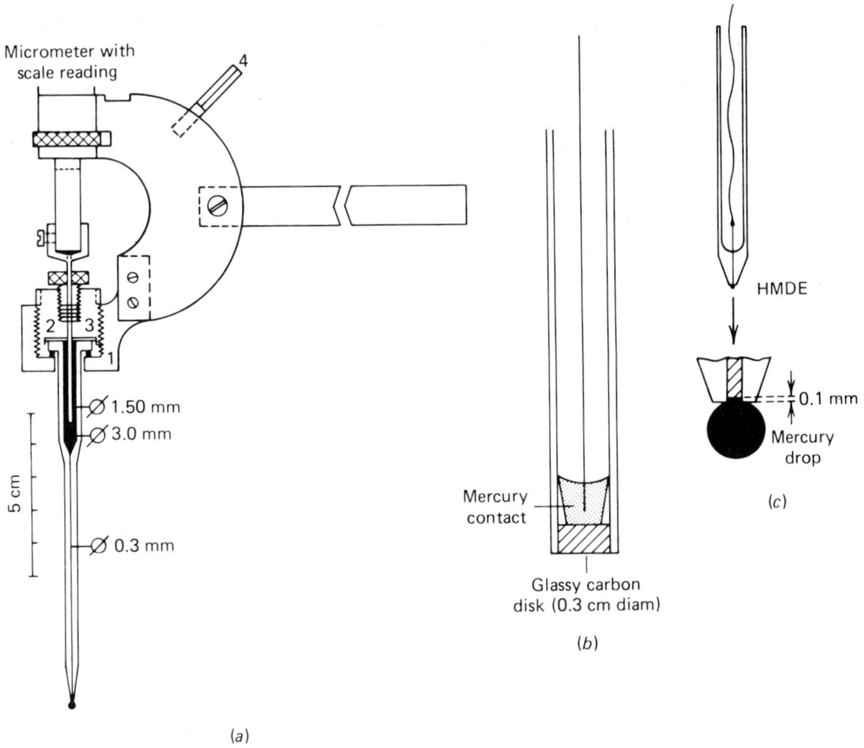

Figure 10.8.2

Typical mercury electrodes for stripping voltammetry. (*a*) Micrometer-type hanging mercury drop electrode (HMDE) [from E. Barendrecht, *Electroanal. Chem.*, **2**, 53 (1967), courtesy of Marcel Dekker, Inc.]; commercial versions are also available. (*b*) Mercury film electrode (MFE) on glassy carbon. A polished glassy carbon disk is sealed with epoxy cement into a glass tube with mercury film prepared in situ by electrodeposition [see T. M. Florence, *J. Electroanal. Chem.*, **27**, 273 (1970)]. (*c*) Platinum wire-type HMDE (0.04 cm diam platinum wire sealed in glass). Mercury drop from DME is caught on a small glass spoon and transferred to electrode [see W. L. Underkofler and I. Shain, *Anal. Chem.*, **33**, 1966 (1961)].

platinum contacts (e.g., Figure 10.8.2*c*) dissolve some platinum on prolonged contact, with possible deleterious effects; hence platinum is usually avoided. Solid electrodes (e.g., Pt, Ag, C) are used without mercury (less frequently) for ions that cannot be determined at mercury (e.g., Ag, Au, Hg).

The *electrodeposition step* is carried out in a stirred solution at a potential E_d, which is several tenths of a volt more negative than $E^{0'}$ for the least easily reduced metal ion to be determined. The relevant equations generally follow those for a bulk electrolysis (see Section 10.3.1). However, since the electrode area is so small, and t_d is much smaller than the time needed for exhaustive electrolysis, the current remains essentially constant (at i_d) during this step, and the number of moles of metal deposited is then $i_d t_d / nF$. Because the electrolysis is not exhaustive, the deposition conditions (stirring rate, t_d, temperature) must be the same for the sample and standards for high accuracy.

With an HMDE, one observes a *rest period*, when the stirrer is turned off, the solution is allowed to become quiescent, and the concentration of metal in the amalgam becomes more uniform. The *stripping step* is then carried out by scanning the potential linearly toward more positive values.

When an MFE is used, the stirring during deposition is controlled by rotation of the substrate disk. A rest period usually is not observed, and rotation continues during the stripping step.

The behavior governing the *i-E* curve during the anodic scan depends on the type of electrode employed. For an HMDE of radius r_0, the concentration of reduced form, *M*, at the start of the scan is uniform throughout the drop and is given by

$$C_M^* = \frac{i_d t_d}{nF(4/3)\pi r_0^3} \tag{10.8.1}$$

When the sweep rate v is sufficiently high that the concentration in the middle of the drop ($r = 0$) remains at C_M^* at the completion of the scan, then the behavior is essentially that of semi-infinite diffusion and the basic treatment of Section 6.2 applies (59). Correction must be made for the spherical nature of the drop [see (6.2.23) and (6.2.24)]. In this case the spherical correction term must be subtracted from the planar term, since the concentration gradient builds up inside the drop and the area of the extended diffusion field decreases with time. Thus, the equations that apply for a reversible stripping reaction at the HMDE are (59)

$$\boxed{i = nFAC_M^*\left[(\pi D_M\sigma)^{1/2}\chi(\sigma t) - \frac{D_M\phi(\sigma t)}{r_0}\right]} \tag{10.8.2}$$

$$\boxed{i_p = AD_M^{1/2}C_M^*\left[(2.69 \times 10^5)n^{3/2}v^{1/2} - \frac{(0.725 \times 10^5)nD_M^{1/2}}{r_0}\right]} \tag{10.8.3}$$

where i_p is in A, A in cm^2, D_M in cm^2/sec, C_M^* in mol/cm^3, v in V/sec, and r_0 in cm; the functions $\chi(\sigma t)$ and $\phi(\sigma t)$, where $\sigma = nFv/RT$, are tabulated in Table 6.2.1. These equations hold for the usual HMDE for $v > 20$ mV/sec, and clearly, under these conditions, a large fraction of the deposited metal remains in the drop. A comparison of the *i-E* curve predicted by (10.8.2) and a typical experimental stripping voltammogram at an HMDE is shown in Figure 10.8.3 (60). At very large scan rates the spherical term becomes negligible and linear diffusion scan behavior, with i_p proportional to $v^{1/2}$, results. Practical stripping measurements are usually carried out in this regime. At smaller rates, when the diffusion layer thickness exceeds r_0, the finite electrode volume and depletion of M at $r = 0$ must be considered. At the limit of very small v, when the drop is completely depleted of M during the scan, the behavior would approach that of a thin-layer cell or MFE (see below) with i_p proportional to v.

Because the volume and thickness of the mercury film on an MFE are small, the stripping behavior with this electrode follows more closely thin-layer behavior (see Section 10.7) and depletion effects predominate. The theoretical treatment for the

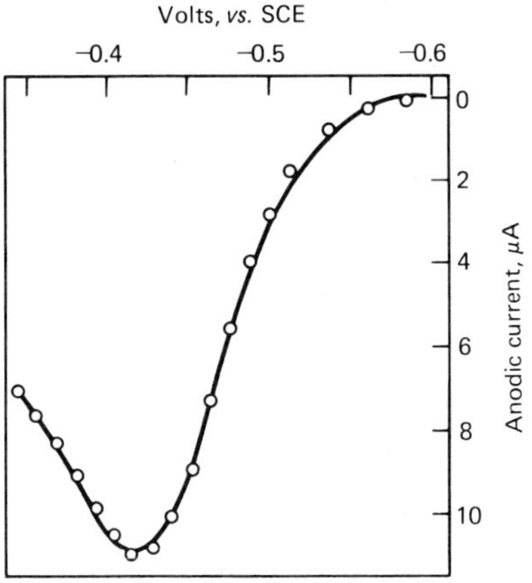

Figure 10.8.3
Experimental anodic stripping i-E curve for thallium. Experimental conditions: 1.0×10^{-5} M Tl^{+}, 0.1 M KCl solution, $E_d = -0.7$ V *vs.* SCE, $t_d = 5$ min, $v = 33.3$ mV/sec. ○—Theoretical points calculated from (10.8.2). [Reprinted with permission from I. Shain and J. Lewinson, *Anal. Chem.*, **33**, 187 (1961). Copyright 1961, American Chemical Society.]

MFE has appeared (61, 62); a diagram of the model employed is shown in Figure 10.8.4. If the stripping reaction is assumed to be reversible, the Nernst equation holds at the surface:

$$C_{M^{+n}}(0, t) = C_M(0, t) \exp\left[\frac{nF}{RT}(E_i - E^{0'} + vt)\right] \tag{10.8.4}$$

The solution of the diffusion equations with this condition and the initial and boundary conditions shown in Figure 10.8.4 leads to an integral equation that must be solved numerically. Typical results for i_p for films of different thicknesses (l) as a function of v are shown in Figure 10.8.5*a*. At small v and l, depletion or thin-layer behavior predominates and $i_p \propto v$. For high v and large l, semi-infinite linear diffusion behavior predominates and $i_p \propto v^{1/2}$. The limits of these zones are shown in Figure 10.8.5*b* Note that MFEs used in current practice fall within the region where thin-layer behavior can be expected for virtually all usual sweep rates (≤ 500 mV/sec). An

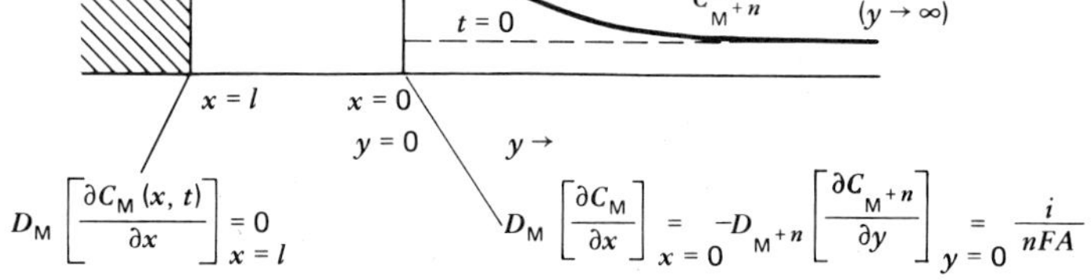

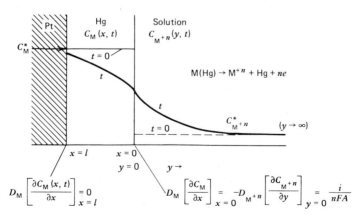

Figure 10.8.4
Notation, initial and boundary conditions for theoretical treatment of MFE.

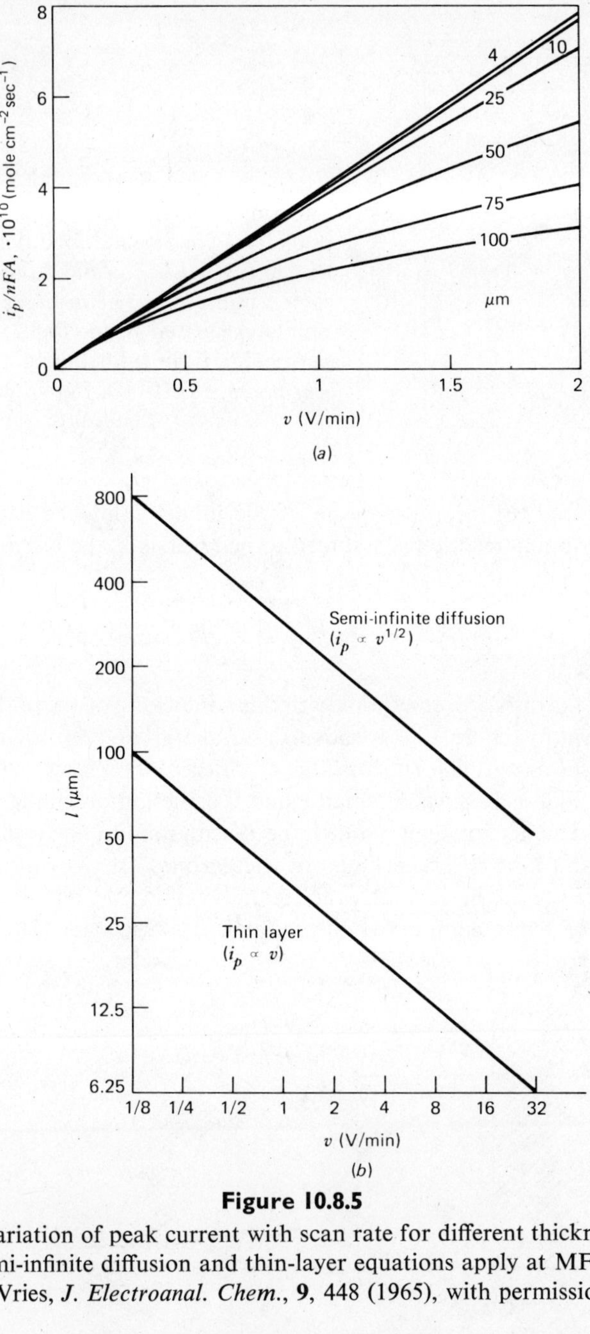

Figure 10.8.5

(*a*) Calculated variation of peak current with scan rate for different thickness of MFE. (*b*) Zones where semi-infinite diffusion and thin-layer equations apply at MFE. [From W. T. de Vries, *J. Electroanal. Chem.*, **9**, 448 (1965), with permission.]

approximate equation for the peak current in the thin-layer region based on a diffusion layer approximation in solution has also been proposed (63):

$$|i_p| = \frac{n^2F^2|v|lAC_M^*}{2.7RT} \tag{10.8.5}$$

Notice the similarity between this expression and the corresponding limiting thin-layer equation (10.7.17) (where $Al = V$).

10.8.3 Applications and Variations

The technique of controlled potential cathodic deposition followed by anodic stripping with a linear potential sweep has been applied to the determinations of a number of metals (e.g., Bi, Cd, Cu, In, Pb, and Zn) either alone or in mixtures (Figure 10.8.6). An increase in sensitivity can be obtained by using pulse polarographic, square wave, or coulostatic stripping techniques. Other variants, such as stripping by a potential step, current step, or more elaborate programs (e.g., an anodic potential step for a short time followed by a cathodic sweep) have also been proposed (53–58).

Important interferences that sometimes occur with mercury electrodes involve (a) reactions of the metals with the substrate material (e.g., Pt or Au) or with the mercury (e.g., Ni-Hg), or (b) intermetallic compound formation between two metals deposited into the mercury at the same time (e.g., Cu-Cd or Cu-Ni). These effects are much more serious with mercury films than with hanging drops, because MFEs feature fairly concentrated amalgams and a high ratio of substrate area to film volume. They can be an overriding concern in the choice between an MFE and an HMDE.

On the other hand, MFEs offer much better sensitivity in linear sweep stripping and better control of mass transfer during the deposition step. If an HMDE is chosen (e.g., to reduce interferences), one can use differential pulse stripping to obtain sensitivities comparable to those attained by LSV at an MFE.

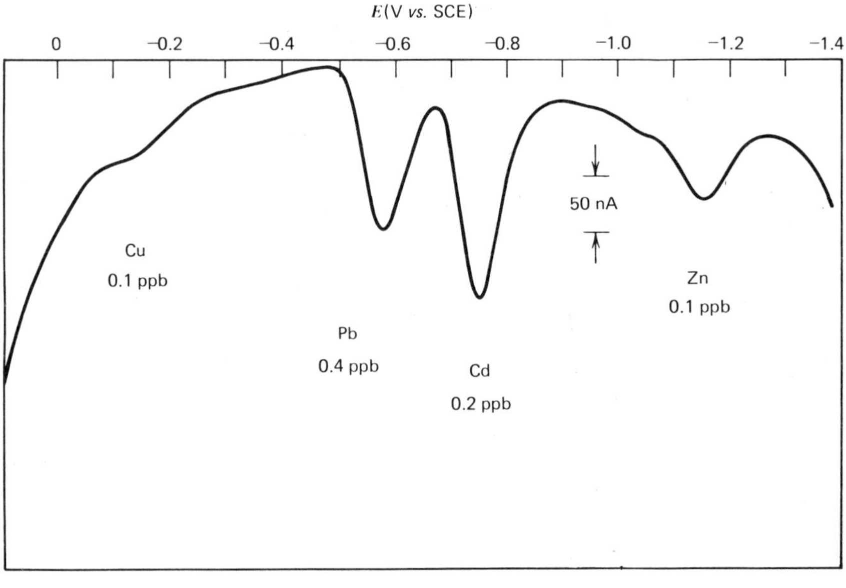

Figure 10.8.6
Anodic stripping analysis of solution containing 2×10^{-9} *M* Zn, Cd, Pb, and Cu at MFE (mercury-plated, wax-impregnated graphite electrode). Stripping carried out by differential pulse voltammetry.

Since stripping at an MFE gives total exhaustion of the thin film, the voltammetric peaks are narrow and can allow baseline resolution of multicomponent systems. Thin-layer properties and the sharpness of the peaks permit relatively fast stripping sweeps, which in turn shorten analysis times. In contrast, the falloff in current past the peak in a stripping voltammogram obtained at an HMDE comes from diffusive depletion, rather than exhaustion, and it continues for quite some time. Thus the peaks are broader, and overlap of adjacent peaks is more serious. (Compare, for example, Figures 10.8.7*a* and 10.8.7*d*.) This problem is usually minimized for the HMDE by using slow sweep rates, at a cost of lengthened analysis time.

Cathodic stripping analysis can also be carried out for the analysis of species (usually anions) that deposit in an anodic preelectrolysis. For example, the halides (X^-) can be determined at mercury by deposition as Hg_2X_2. Deposition on solid electrodes is also possible. In this case surface problems (e.g., oxide films) and underpotential deposition effects often appear. On the other hand, the sensitivity for stripping from a solid electrode is very high, since the deposit can be removed completely, even at high scan rates. Stripping of metal films has often been used to determine the thickness of coatings (e.g., Sn on Cu) and oxide layers (e.g., CuO on Cu).†

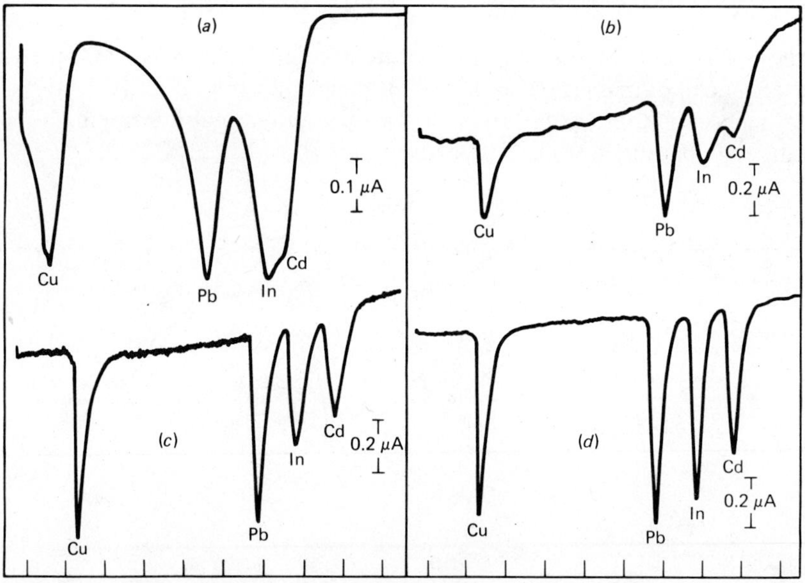

Figure 10.8.7

Stripping curves for 2×10^{-7} *M* Cd^{2+}, In^{3+}, Pb^{2+}, and Cu^{2+} in 0.1 *M* KNO_3. $|v| = 5$ mV/sec. (*a*) HMDE, $t_d = 30$ min. (*b*) Pyrolytic graphite, $t_d = 5$ min. (*c*) Unpolished glassy carbon, $t_d = 5$ min. (*d*) Polished glassy carbon, $t_d = 5$ min. For (*b*) to (*d*), $\omega/2\pi = 2000$ rpm and Hg^{2+} was added at 2×10^{-5} *M*. [From T. M. Florence, *J. Electroanal. Chem.*, **27**, 273 (1970), with permission.]

† In fact, one of the earliest electroanalytical (coulometric) methods was the determination of the thickness of tin coatings on copper wires (64).

10.9 REFERENCES

1. J. J. Lingane, "Electroanalytical Chemistry," 2nd. ed., Wiley-Interscience, New York, 1958, Chaps. 13–21.
2. P. Delahay, "New Instrumental Methods in Electrochemistry," Wiley-Interscience, New York, 1954, Chaps. 11–14.
3. N. Tanaka, in "Treatise on Analytical Chemistry," Part I, Vol. 4, I. M. Kolthoff and P. J. Elving, Eds., Wiley-Interscience, New York, 1963, Chap. 48.
4. L. B. Rogers and A. F. Stehney, *J. Electrochem. Soc.*, **95**, 25 (1949); J. T. Byrne and L. B. Rogers, *ibid.*, **98**, 452 (1951).
5. D. M. Kolb, *Adv. Electrochem. Electrochem. Engr.*, **11**, 125 (1978).
6. N. Tanaka, *op. cit.*, pp. 2241–2443.
7. H. Gerischer, D. M. Kolb, and M. Prazanyski, *Surf. Sci.*, **43**, 662 (1974); **51**, 323 (1975); *J. Electroanal. Chem.*, **54**, 25 (1974); and references therein.
8. V. A. Vincente and S. Bruckenstein, *Anal. Chem.*, **45**, 2036 (1973); *J. Electroanal. Chem.*, **82**, 187 (1977); and references therein.
9. E. Schmidt, M. Christen, and P. Beyelar, *J. Electroanal. Chem.*, **42**, 275 (1973) and references therein.
10. For example, M. Ito and T. Kuwana, *J. Electroanal. Chem.*, **32**, 415 (1971).
11. W. R. Heineman and T. Kuwana, *Biochem. Biophys. Res. Communs.*, **50**, 892 (1973).
12. J. J. Lingane, *J. Am. Chem. Soc.*, **67**, 1916 (1945); *Anal. Chim. Acta.*, **2**, 584 (1948).
13. A. J. Bard, *Anal. Chem.*, **35**, 1125 (1963).
14. J. A. Harrison and H. R. Thirsk, *Electroanal. Chem.*, **5**, 67–148 (1971).
15. J. O'M. Bockris and A. Damjanovic, *Mod. Asp. Electrochem.*, **3**, 224 (1964).
16. M. Fleischmann and H. R. Thirsk, *Adv. Electrochem. Electrochem. Engr.*, **3**, 123 (1963).
17. F. A. Lowenheim, "Modern Electroplating," 3rd ed., Wiley, New York, 1974; "Electroplating," McGraw-Hill, New York, 1978.
18. H. J. S. Sand, "Electrochemistry and Electrochemical Analysis," Blackie, London, 1940.
19. Biennial reviews in *Anal. Chem.* (up to 1974). S. E. Q. Ashley, **21**, 70 (1969); **24**, 91 (1952). D. D. Deford, **26**, 135 (1954); **28**, 660 (1956); **30**, 613 (1958); **32**, 31R (1960). A. J. Bard, **34**, 57R (1962); **36**, 70R (1964); **38**, 88R (1966); **40**, 64R (1968); **42**, 22R (1970). D. G. Davis, **44**, 79R (1972); **46**, 21R (1974).
20. G. A. Rechnitz, "Controlled Potential Analysis," Pergamon, New York, 1963.
21. J. E. Harrar, *Electroanal. Chem.*, **8**, 1 (1975).
22. Reference 1, *op. cit.*, pp. 488–495.
23. J. J. Lingane, C. H. Langford, and F. C. Anson, *Anal. Chim. Acta*, **16**, 165 (1959).
24. Ref. 1, *op. cit.*, Chap. 21.
25. H. L. Kies, *J. Electroanal. Chem.*, **4**, 257 (1962).
26. D. DeFord and J. W. Miller, in "Treatise on Analytical Chemistry," Part I,

Vol. 4, I. M. Kolthoff and P. J. Elving, Eds., Wiley-Interscience, New York, 1963, Chapter 49.
27. P. A. Shaffer, Jr., A. Briglio, Jr., and J. A. Brockman, Jr., *Anal. Chem.*, **20**, 1008 (1948).
28. R. S. Braman, D. D. DeFord, T. N. Johnston, and L. J. Kuhns, *Anal. Chem.*, **32**, 1258 (1960).
29. J. Janata and H. B. Mark, Jr., *Electroanal. Chem.*, **3**, 1 (1969).
30. Reference 1, *op. cit.*, Chaps. 5–8, 12 and references to the older literature contained therein.
31. C. N. Reilley and R. W. Murray (Chap. 43) and N. H. Furman (Chap. 45) in "Treatise on Analytical Chemistry," Part I, Vol. 4, I. M. Kolthoff and P. J. Elving, Eds., Wiley-Interscience, New York, 1963.
32. J. T. Stock, "Amperometric Titrations," Interscience, New York, 1965.
33. W. D. Cooke, C. N. Reilley, and N. H. Furman, *Anal. Chem.*, **23**, 1662 (1951).
34. I. M. Kolthoff and J. J. Lingane, "Polarography," 2nd ed., Vol. 2, Interscience, New York, 1952, Chap. 47.
35. L. Meites, "Polarographic Techniques," 2nd ed., Wiley-Interscience, New York, 1965, Chap. 9.
36. R. de Levie, *Adv. Electrochem. Electrochem. Engr.*, **6**, 329 (1967).
37. The treatment given here basically follows that outlined in the following papers and references contained therein:

    (a) R. E. Sioda, *Electrochim. Acta*, **13**, 375 (1968); **15**, 783 (1970); **17**, 1939 (1972); **22**, 439 (1977).
    (b) R. E. Sioda, *J. Electroanal. Chem.*, **34**, 399, 411 (1972); **56**, 149 (1974).
    (c) R. E. Sioda and T. Kambara, *J. Electroanal. Chem.*, **38**, 51 (1972).
    (d) I. G. Gurevich and V. S. Bagotsky, *Electrochim. Acta*, **9**, 1151 (1964).
    (e) I. G. Gurevich, V. S. Bagotsky, and Yu R. Budeka, *Electrokhimiya*, **4**, 321, 874, 1251 (1968).

38. J. V. Kenkel and A. J. Bard, *J. Electroanal. Chem.*, **54**, 47 (1974).
39. J. A. Trainham and J. Newman, *J. Electrochem. Soc.*, **124**, 1528 (1977).
40. R. Alkire and R. Gould, *ibid.*, **123**, 1842 (1976).
41. B. A. Ateya and L. G. Austin, *ibid.*, **124**, 83 (1977).
42. E. L. Eckfeldt, *Anal. Chem.*, **31**, 1453 (1959).
43. T. Fujinaga and S. Kihara, *CRC Crit. Rev. Anal. Chem.*, **6**, 223 (1977).
44. E. Schmidt and H. R. Gygax, *Chimia*, **16**, 156 (1962).
45. J. H. Sluyters, *Rec. Trav. Chim.*, **82**, 100 (1963).
46. C. R. Christensen and F. C. Anson, *Anal. Chem.*, **35**, 205 (1963).
47. C. N. Reilley, *Pure Appl. Chem.*, **18**, 137 (1968).
48. A. T. Hubbard and F. C. Anson, *Electroanal. Chem.*, **4**, 129 (1970).
49. A. T. Hubbard, *CRC Crit. Rev. Anal. Chem.*, **2**, 201 (1973).
50. A. T. Hubbard, *J. Electroanal. Chem.*, **22**, 165 (1969).
51. I. B. Goldberg and A. J. Bard, *ibid.*, **38**, 313 (1972).
52. G. M. Tom and A. T. Hubbard, *Anal. Chem.*, **43**, 671 (1971).
53. I. Shain, in "Treatise on Analytical Chemistry," Part I, Vol. 4, I. M. Kolthoff and P. J. Elving, Eds., Wiley-Interscience, New York, 1963, Chap. 50.

54. E. Barendrecht, *Electroanal. Chem.*, **2**, 53–109 (1967).
55. R. Neeb, "Inverse Polarographie and Voltammetrie," Akademie-Verlag, Berlin, 1969.
56. J. B. Flato, *Anal. Chem.*, **44** (11), 75A (1972).
57. T. R. Copeland and R. K. Skogerboe, *Anal. Chem.*, **46**, 1257A (1974).
58. F. Vydra, K. Stulik, and E. Julakova, "Electrochemical Stripping Analysis," Halsted Press, New York, 1977.
59. W. H. Reinmuth, *Anal. Chem.*, **33**, 185 (1961).
60. I. Shain and J. Lewinson, *Anal. Chem.*, **33**, 187 (1961).
61. W. T. de Vries and E. Van Dalen, *J. Electroanal. Chem.*, **8**, 366 (1964).
62. W. T. de Vries, *J. Electroanal. Chem.*, **9**, 448 (1965).
63. D. K. Roe and J. E. A. Toni, *Anal. Chem.*, **37**, 1503 (1965).
64. G. G. Grower, *Proc. Am. Soc. Testing Mater.*, **17**, 129 (1917).

10.10 PROBLEMS

10.1 Based on the curves in Figure 10.10.1, consider the titration of Sn^{2+} with I_2 using one-electrode amperometry. Sketch the resulting titration curves for a platinum indicator electrode maintained at (a) +0.2 V, (b) −0.1 V, and (c) −0.4 V *vs.* SCE.

10.2 Based on the curves in Figure 10.10.1, how could one determine a mixture of Br_2 and I_2 by titration with Sn^{2+} using one-electrode amperometry? Sketch the current-potential curves that would be obtained for various stages of the titration and the amperometric titration curves that would result from the method you propose. Sketch the titration curve of a mixture of Br_2 and I_2 by titration with Sn^{2+} using two-electrode amperometry with an impressed voltage of 100 mV.

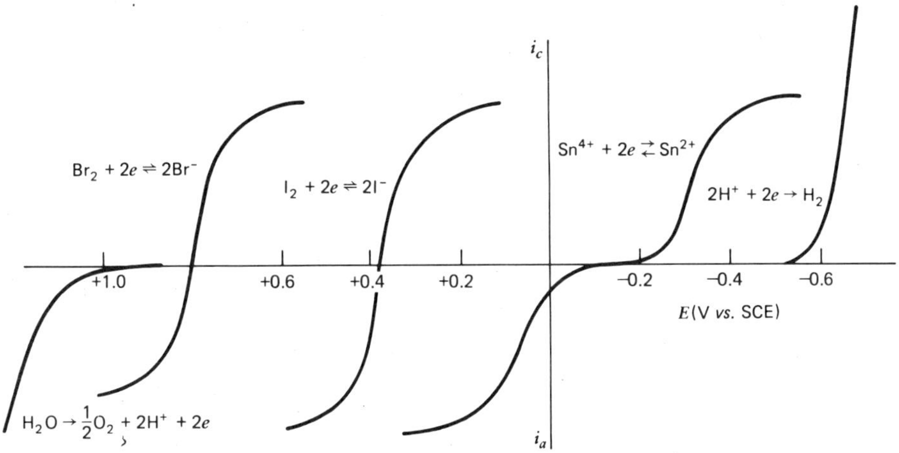

Figure 10.10.1
Idealized current-potential curves for several systems at a platinum electrode.

10.3 Based on the curves in Figure 10.10.1, sketch the titration curves for the titrations in Problems 10.1 and 10.2 for one- and two-electrode potentiometry with a small impressed current.

10.4 Fifty mL of a $ZnSO_4$ solution are transferred to an electrolytic cell with a mercury cathode and enough solid potassium nitrate is added to make the solution 0.1 *M* in KNO_3. The electrolysis of Zn^{2+} is carried to completion at a potential of -1.3 V *vs.* SCE with the passage of 241 C of electricity. Calculate the initial concentration of zinc ion.

10.5 Iodide is to be titrated coulometrically at constant current at a silver electrode. The sample is 1.0 m*M* NaI contained in a pH 4 acetic acid solution with 0.1 *M* sodium acetate with a total volume of 50 mL.

(a) Describe the course of the titration. What generating current do you recommend and what total titration time is expected?

(b) Consider the current-potential curves that would be recorded at a rotated platinum microelectrode upon scanning from the cathodic background limit (~ -0.5 V) to the anodic limit (at $\sim +1.5$ V *vs.* SCE). Draw curves for the 0%, 50%, 100% and 150% titration points. Label the waves with the electrode processes that cause them. All electrode reactions other than the background discharges are reversible. The following information is useful:

Reaction	E^0, volts *vs.* SCE
$Ag^+ + e^- \rightleftharpoons Ag$	+0.56
$I_3^- + 2e^- \rightleftharpoons 3I^-$	+0.30
$AgI + e \rightleftharpoons Ag + I^-$	−0.39

(c) Sketch amperometric titration curves for:

(1) One polarized electrode at -0.3 V *vs.* SCE

(2) One polarized electrode at $+0.4$ V *vs.* SCE

(3) Two polarized electrodes with 100-mV potential difference.

The indicator electrodes are rotated platinum microelectrodes.

10.6 Iodide in the solution in Problem 10.5 can also be determined by controlled potential oxidation to iodine at a platinum electrode. What potential should be used for this oxidation (see Figure 10.10.1)? How many coulombs will be passed?

10.7 The following is a standard procedure for the assay of uranium samples: (1) Dissolution of the sample in acid to produce UO_2^{2+} as the chloride. (2) Reduction of the UO_2^{2+} solution by passage through a Jones reductor (amalgamated zinc). This solution is perhaps 0.1 *M* in H_2SO_4. Reduction takes place to U^{3+}. (3) Stirring in air to give U^{4+}. (4) Addition of Fe^{3+} and Ce^{3+} in excess and coulometric titration to an end point.

Answer the following: (a) Suppose the solution after treatment (3) contains ~1 mM U^{4+}. Fe^{3+} and Ce^{3+} are added in quantities that yield 4 and 50 mM concentrations of iron and cerium species, respectively. Sulfuric acid is also added to bring its concentration to 1 M. Draw the current-potential curve that would be recorded at a rotating platinum microelectrode immersed in this solution. The anodic background limit is +1.7 V *vs.* NHE and the cathodic limit is at −0.2 V *vs.* NHE. (b) Explain the chemistry of step (4) and the setup for the coulometric titration. (c) Sketch current-potential curves for points at which the titration is 0%, 50%, 100%, and 150% complete. (d) Sketch amperometric titration curves for one polarized electrode operated at +0.3 V and at +0.9 V *vs.* NHE, and for two polarized electrodes separated by 100 mV. (e) Sketch the null current potentiometric responses on a quantitative potential scale.

The following information may be useful:

Reaction	Formal Potential, volts *vs.* NHE
1. $Ce^{4+} + e \rightleftharpoons Ce^{3+}$	1.44
2. $Fe^{3+} + e \rightleftharpoons Fe^{2+}$	0.77
3. $UO_2^{2+} + e \rightleftharpoons UO_2^{+}$	0.05
4. $UO_2^{+} + 4H^{+} + e \rightarrow U^{4+} + 2H_2O$	0.62
5. $U^{4+} + e \rightleftharpoons U^{3+}$	−0.61

All reactions except 4 are reversible. That process will not show a wave at platinum before the cathodic background limit is reached.

10.8 A molecule of interest in research is tetracyanoquinodimethane (TCNQ):

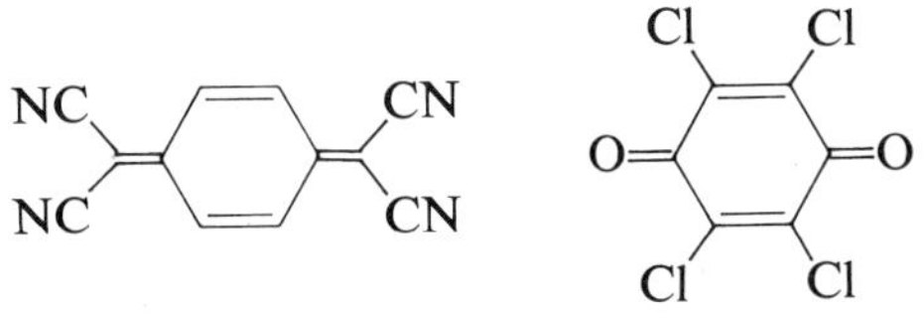

TCNQ $\quad\quad$ *p*-Chl

Samples of high purity are often required. Suppose you wish to develop a technique for assaying the purity of TCNQ samples. Describe a means for accomplishing this goal by coulometric titration with electrogenerated anion radicals of *p*-chloranil (*p*-Chl). Acetonitrile containing tetra-*n*-butyl ammonium perchlorate would be a suitable solvent. Its background limits at platinum are −2.5 V and +2 V *vs.* SCE. The following reduction potentials are relevant:

$$TCNQ + e \rightleftharpoons TCNQ^{\overline{\cdot}} \qquad E^0 = 0.20 \text{ V}$$
$$TCNQ^{\overline{\cdot}} + e \rightleftharpoons TCNQ^{2-} \qquad E^0 = -0.33 \text{ V}$$
$$p\text{-Chl} + e \rightleftharpoons p\text{-Chl}^- \qquad E^0 = 0.0 \text{ V}$$

All of these processes are reversible.

(a) Specify the details of the cell, the starting composition of the solution, and the chemical processes taking place at the electrodes and in homogeneous solution. (b) Draw the current-potential curves that would be recorded at a rotating platinum microelectrode if the titration were stopped at the 0%, 50%, 100%, and 150% points. (c) Sketch titration curves for: (1) amperometric detection with one polarized electrode at 1.0 V; (2) amperometric detection with one polarized electrode at 0.1 V; and (3) amperometric detection with two polarized electrodes separated by 100 mV.

10.9 Consider carrying out a one-electrode amperometric titration for the system Fe^{2+}–Ce^{4+} as shown in Figure 10.5.1 at several widely different potentials and sketch the amperometric titration curves that result. Consider, in each case, situations in which (a) the mass transfer coefficients for all species are equal and (b) those for iron ions are 25% larger than those for cerium species.

Which curves would be useful in a practical titration?

10.10 When a solution of volume 100 cm^3 containing metal ion, M^{2+}, at a concentration 0.010 *M* is electrolyzed with a rapid scan at a large-area (10 cm^2) rotating disk electrode, a limiting current of 193 mA is observed for reduction to metal M. Calculate the value of $m_{M^{2+}}$, the mass transport coefficient, in cm/sec. If an electrolysis of the solution is carried out at this electrode at controlled potential in the limiting current region, what time will be required for 99.9% of the M^{2+} to be plated out? How many coulombs will be required for this electrolysis?

10.11 If the solution in Problem 10.10 is electrolyzed at a constant current of 80 mA under the same conditions: (a) What is the concentration of M^{2+} remaining in solution when the current efficiency drops below 100%? (b) How long does it take to reach this point? (c) How many coulombs have been passed to this point? (d) How much longer will it take to decrease the M^{2+} concentration to 0.1% of its initial value? What is the overall current efficiency for removal of 99.9% of M^{2+} by this constant current electrolysis?

10.12 A solution of volume 200 cm^3 contains 1.0×10^{-3} *M* X^{2+} and 3.0×10^{-3} *M* Y^{2+}, where X and Y are metals. The solution is to be electrolyzed at a mercury pool electrode of area 50 cm^2 and volume 100 cm^3. Under the stirring conditions and cell geometry, both X^{2+} and Y^{2+} have mass transfer coefficients m of 10^{-2} cm/sec. The polarographic $E_{1/2}$ values for reduction of X^{2+} and Y^{2+} to the metal amalgams are -0.45 and -0.70 V *vs.* SCE, respectively. (a) A current-potential curve for the solution is taken under the above conditions. (Assume no changes in concentrations of X^{2+} and Y^{2+} during the scan.) Make a neat, labelled, quantitatively correct sketch of the *i-E* curve that would be obtained. (b) If the electrolysis is to be performed at a controlled potential, at what potentials can X^{2+} be quantitatively deposited (less than 0.1% left in solution) leaving Y^{2+} behind in solution (less than 0.1% Y^{2+} deposited in mercury)? (c) How long will it take to carry out this electrolysis at controlled potential?

10.13 Consider a chronopotentiometric experiment dealing with two components that are reversibly reduced in waves separated by 500 mV. Derive an expression for the second transition time in an experiment carried out in a thin-layer cell. Compare and contrast the properties of multicomponent systems in thin-layer chronopotentiometry with those of the semi-infinite method.

10.14 Suppose bromide ion is to be determined at very low concentrations. This is done by depositing bromide on a silver electrode, which is held at a potential where the following reaction occurs:

$$\text{Ag} + \text{Br}^- - e^- \rightarrow \text{AgBr}$$

(A typical deposition potential is +0.2 V *vs.* SCE.) Stripping is carried out by scanning in a negative direction to reverse the deposition. In general, it is observed that the response during stripping shows a complex dependence on deposition time, as shown in Figure 10.10.2. Explain this effect. What problems would be present in quantitative analysis? How could they be surmounted? [See H. A. Laitinen and N. H. Watkins, *Anal. Chem.*, **47**, 1352 (1975) for similar results in determinations of lead.]

10.15 A study of seawater by stripping analysis reveals an anodic copper peak having a height of 0.13 μA when deposition is carried out at −0.5 V. However, deposition at −1.0 V yields a larger peak of 0.31 μA. Account for these results.

Standard addition of 10^{-7} M Cu^{2+} elevates the peaks in both cases by 0.24 μA. Comment on the feasibility of obtaining polarograms of any type on this solution. What responses would you expect for dc, normal pulse, and differential pulse experiments? Would any of these supply useful analytical information?

10.16 An analysis for lead at the HMDE gives rise to a peak current of 1 μA under conditions in which the deposition time is held constant at 5 min and the sweep rate is 50 mV/sec. What currents would be observed for sweep rates of 25 and 100 mV/sec?

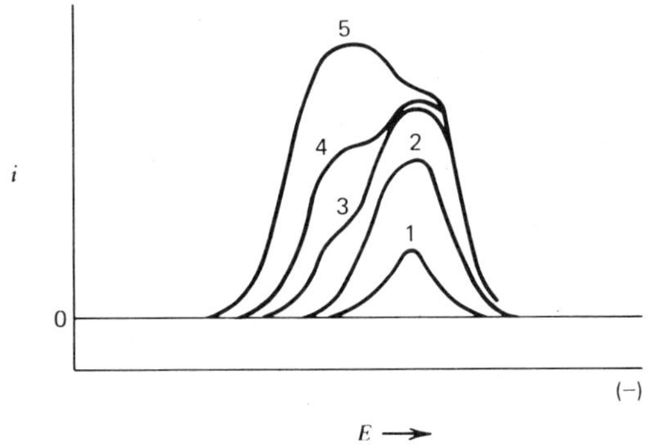

Figure 10.10.2

Cathodic stripping of AgBr from a silver electrode following anodic deposition. Curves 1 to 5 involve successively longer deposition times.

The same solution gives a peak current of 25 μA at a 100 Å thick mercury film electrode on glassy carbon when the deposition time is 1 min, the electrode rotation rate is 2000 rpm, and the sweep rate is 50 mV/sec. What currents would be observed for sweep rates of 25 and 100 mV/sec under otherwise unchanged conditions? Compare this situation to the one observed for a deposition time of 1 min, a sweep rate of 50 mV/sec, and a rotation rate of 4000 rpm? Suppose the film thickness were varied by the use of different concentrations of mercuric ion in the analyte. What effect would one see on the peak current under otherwise constant conditions?

chapter 11
Electrode Reactions with Coupled Homogeneous Chemical Reactions

11.1 INTRODUCTION

The previous chapters dealt with a number of electrochemical techniques and the responses obtained when the electroactive species (O) is converted in a heterogeneous electron transfer reaction at an electrode to the product (R). In many cases the electron transfer reaction is perturbed by a homogeneous chemical reaction which involves species O or R. For example, O may not be present initially but may be produced from another, nonelectroactive species, or R may react with solvent or supporting electrolyte species. Sometimes a substance that reacts with product R is intentionally added so that the rate of the reaction can be determined by an electrochemical technique or a new product will be produced. In this chapter we will survey the general classes of coupled homogeneous chemical reactions, describe how the electrochemical responses can be obtained for the different techniques, and discuss how electrochemical methods can be used to elucidate the mechanisms of these reactions.

The initial investigations of coupled chemical reactions were carried out by Brdička, Wiesner, and others of the Czechoslovakian polarographic school in the 1940s; since that time hundreds of papers dealing with the theory and application of different electroanalytical techniques to the study of coupled reactions have appeared. It is beyond the scope of this textbook to attempt to treat this area exhaustively. The reader is instead referred to monographs and review articles dealing with different aspects of this area (1–6).

11.1.1 Classification of Reactions

It is convenient to classify the different possible reaction schemes in which homogeneous reactions are associated with the heterogeneous electron transfer steps by using letters to signify the nature of the step. **E** represents an electron transfer at the electrode surface, and **C** represents a homogeneous **c**hemical reaction (7). Thus a reaction mechanism in which the sequence involves a chemical reaction of the product after the electron transfer would be designated an *EC reaction*. In the equations that follow, substances designated W, X, Y, and Z are assumed to be not electroactive in the potential range of interest.

(a) CE Reaction (Preceding Reaction)

$$Y \rightleftharpoons O \tag{11.1.1}$$

$$O + ne \rightleftharpoons R \tag{11.1.2}$$

Here the electroactive species O is generated by a reaction that precedes the electron transfer at the electrode. An example of the CE scheme is the reduction of formaldehyde at mercury in aqueous solutions. Formaldehyde exists as a nonreducible hydrated form, $H_2C(OH)_2$, in equilibrium with the reducible $H_2C\!=\!O$ form:

$$H_2C(OH)_2 \rightleftharpoons H_2C\!=\!O + H_2O \tag{11.1.3}$$

The equilibrium constant of (11.1.3) favors the hydrated form. Thus reaction (11.1.3) precedes the reduction of $H_2C\!=\!O$, and under some conditions the current will be governed by the kinetics of this reaction (yielding a so-called *kinetic current*). Other examples of this case involve reduction of some weak acids and the conjugate base anions, the reduction of aldoses, and the reduction of metal complexes.

(b) EC Reaction (Following Reaction)

$$O + ne \rightleftharpoons R \tag{11.1.4}$$

$$R \rightleftharpoons X \tag{11.1.5}$$

In this case the product of the electrode reaction, R, reacts (e.g., with solvent) to produce a species that is not electroactive at potentials where the reduction of O is occurring. An example of this scheme is the oxidation of *p*-aminophenol (PAP) at a platinum electrode in aqueous acidic solutions:

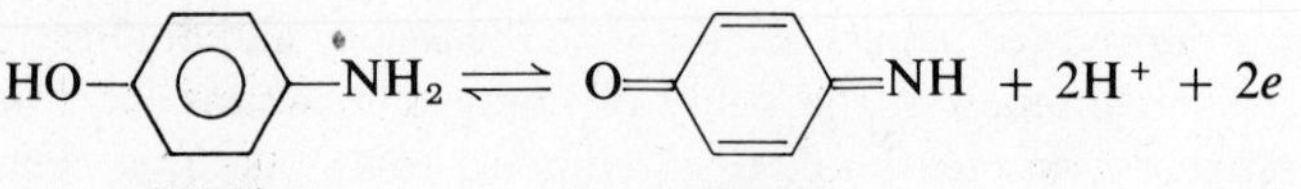

$$HO\text{-}C_6H_4\text{-}NH_2 \rightleftharpoons O\!=\!C_6H_4\!=\!NH + 2H^+ + 2e \tag{11.1.6}$$

(PAP)   (QI)

$$O\!=\!C_6H_4\!=\!NH + H_2O \longrightarrow O\!=\!C_6H_4\!=\!O + NH_3 \tag{11.1.7}$$

(BQ)

where the quinone imine (QI) formed in the initial electron transfer reaction undergoes a hydrolysis reaction to form benzoquinone (BQ), which is neither oxidized nor

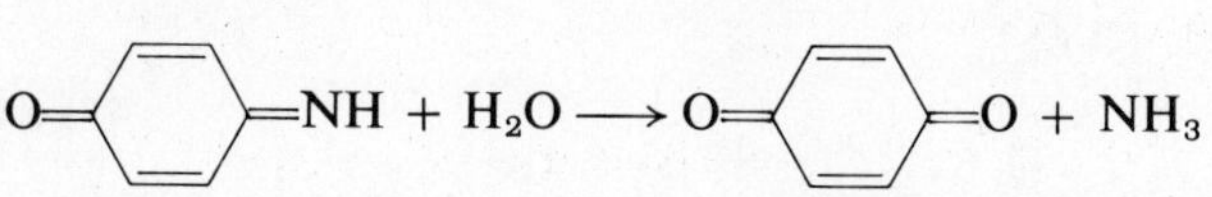

reduced at these potentials. This type of reaction sequence occurs quite frequently, since the electrochemical oxidation or reduction of a substance often produces a reactive species. For example, the one-electron reductions and oxidations that are characteristic of organic compounds in aprotic solvents [e.g., in acetonitrile (CH_3CN) or *N*,*N*-dimethylformamide ($Me_2NHC{=}O$)] produce radicals or radical ions that tend to dimerize:

$$R + e \rightleftharpoons R^{\overline{\cdot}} \tag{11.1.8}$$

$$2R^{\overline{\cdot}} \rightarrow R_2^{\,2-} \tag{11.1.9}$$

(e.g., where R is an activated olefin, such as diethyl fumarate). In this example, the reaction that follows the electron transfer is a second-order reaction, and this case is sometimes designated as an EC_2 reaction. Sometimes, yet another chemical reaction follows the first; thus in the dimerization of olefins, a protonation reaction follows:

$$R_2^{\,2-} + 2H^+ \rightarrow R_2H_2 \tag{11.1.10}$$

This sequence is an ECC (or EC_2C) reaction.

(c) Catalytic (EC′) Reaction

$$O + ne \rightleftharpoons R \tag{11.1.11}$$

$$R + Z \rightarrow O + Y \tag{11.1.12}$$

A special type of EC reaction involves reaction of the product with a nonelectroactive species, Z, in solution to regenerate O. If species Z is present in large excess compared to O, then (11.1.12) is a pseudo-first-order reaction. An example of this scheme is the reduction of Ti(IV) in the presence of a substance (e.g., NH_2OH or ClO_3^-) that can oxidize Ti(III):

$$\text{Ti(IV)} + e \rightarrow \text{Ti(III)} \tag{11.1.13}$$

$$ClO_3^-,\ NH_2OH$$

Since hydroxylamine or chlorate ion can be reduced by Ti(III), they should thermodynamically be reducible at the mercury electrode at these potentials. They are not, however, because the rates of their reductions at the electrode are very small. Other examples of EC′ reactions are the reduction of Fe(III) in the presence of H_2O_2 and the oxidation of I^- in the presence of oxalate. An important EC′ reaction involves reductions at mercury where the product can reduce protons or solvent (a so-called catalytic hydrogen reaction).

(d) ECE Reaction

$$O_1 + n_1e \rightleftharpoons R_1 \qquad E_1^0 \tag{11.1.14}$$

$$R_1 \xrightarrow{(+Z)} O_2 \tag{11.1.15}$$

$$O_2 + n_2e \rightleftharpoons R_2 \qquad E_2^0 \tag{11.1.16}$$

When the product of the following chemical reaction is electroactive at potentials of the O_1/R_1 electron transfer reaction, a second electron transfer reaction can take place. An example of this scheme is the reduction of a halonitroaromatic compound in an aprotic medium (e.g., in liquid ammonia or *N,N*-dimethylformamide), where the reaction proceeds as follows (X = Cl, Br, I):

$$XC_6H_4NO_2 + e \rightleftharpoons XC_6H_4NO_2^{\cdot -} \tag{11.1.17}$$

$$XC_6H_4NO_2^{\cdot -} \rightarrow X^- + \cdot C_6H_4NO_2 \tag{11.1.18}$$

$$\cdot C_6H_4NO_2 + e \rightleftharpoons \bar{:}C_6H_4NO_2 \tag{11.1.19}$$

$$\bar{:}C_6H_4NO_2 + H^+ \rightarrow C_6H_5NO_2 \tag{11.1.20}$$

Since protonation follows the second electron transfer step, this is actually an ECEC reaction sequence. The assignment of such an ECE sequence is not as straightforward as it might first appear, however. Because species O_2 is more easily reduced than O_1 (since $E_1^0 < E_2^0$), species R_1 diffusing away from the electrode is capable of reducing O_2. Thus, for the example mentioned above, the following reaction can occur:

$$XC_6H_4NO_2^{\cdot -} + \cdot C_6H_4NO_2 \rightleftharpoons XC_6H_4NO_2 + \bar{:}C_6H_4NO_2 \tag{11.1.21}$$

It is not simple to distinguish between this case, where the second electron transfer occurs in bulk solution [sometimes called the DISP mechanism], and the actual ECE case where the second electron transfer occurs at the electrode surface (8).

A second variety of this type of reaction scheme, which we will designate ECE′, occurs when the reduction of O_2 takes place at more negative potentials than O_1 (i.e., $E_1^0 > E_2^0$). In this case the reaction observed at the first reduction wave is an EC reaction; however, the second reduction wave will be characteristic of an ECE′ reaction.

A number of other reaction schemes are possible, for example, where the product couples with the reactant. Many of these can be treated as combinations or variants of the above cases. The observed behavior will depend on the reversibility or irreversibility of the electron transfer and homogeneous reactions (i.e., the importance of the back reactions). For example, subclasses of EC reactions can be distinguished depending on whether the reactions are reversible (r), quasi-reversible (q), or irreversible (i): for example, E_rC_r, E_rC_i, E_qC_i, etc. There has been much interest and success since the middle 1960s in the elucidation of complex reaction schemes by application of electrochemical methods, along with identification of intermediates by spectroscopic techniques (see Chapter 14) and judicious variation of solvent and reaction conditions. A complex example is the reduction of nitrobenzene ($PhNO_2$) to phenylhydroxylamine in liquid ammonia in the presence of proton donor (ROH), which has been analyzed as an EECCEEC process (9); that is,

$$PhNO_2 + e \rightleftharpoons PhNO_2^{\cdot -} \tag{11.1.22}$$

$$PhNO_2^{\cdot -} + e \rightleftharpoons PhNO_2^{2-} \tag{11.1.23}$$

$$PhNO_2^{2-} + ROH \rightleftharpoons Ph\overset{O}{N}OH^- + RO^- \tag{11.1.24}$$

$$Ph\overset{O}{N}OH^- \rightarrow PhNO \text{ (nitrosobenzene)} + OH^- \tag{11.1.25}$$

$$\mathrm{PhNO} + e \rightleftharpoons \mathrm{PhNO}^{\cdot-} \tag{11.1.26}$$

$$\mathrm{PhNO}^{\cdot-} + e \rightleftharpoons \mathrm{PhNO}^{2-} \tag{11.1.27}$$

$$\mathrm{PhNO}^{2-} + 2\mathrm{ROH} \rightarrow \mathrm{Ph\overset{H}{N}OH} + 2\mathrm{RO}^- \tag{11.1.28}$$

11.1.2 Effects of Coupled Reactions on Measurements

In general, a perturbing chemical reaction can affect the measured parameter of the forward reaction (e.g., the limiting or peak current in voltammetry), the forward reaction characteristic potentials (e.g., $E_{1/2}$ or E_p) and reversal parameters (e.g., i_{pa}/i_{pc}). A qualitative understanding of how different types of reactions affect the different parameters of a given technique is useful in choosing possible reaction schemes for more detailed and quantitative analysis. We assume here that the characteristics of the unperturbed electrode reaction ($\mathrm{O} + ne \rightleftharpoons \mathrm{R}$) have already been determined and consider how the perturbing coupled reaction affects these characteristics.

(a) Effect on Forward Parameters ($i, Q, \tau, \ldots$). The extent to which the limiting current for the forward reaction ($\mathrm{O} + ne \rightarrow \mathrm{R}$) is affected by the coupled reaction depends on the reaction scheme. For an EC reaction, the flux of O is not changed very much, so that the limiting current (or Q_f or τ_f) is only slightly perturbed. For a catalytic reaction (EC′), on the other hand, the limiting current will be increased, because O is being replenished by the reaction. The extent of this increase will depend on the duration (or characteristic time) of the experiment. For very short-duration experiments, i_l will be near that for the unperturbed reaction, since the regenerating reaction will not have sufficient time to regenerate O in appreciable amounts. For longer-duration experiments, i_l will increase continuously with time. Similar considerations apply to the ECE case, except that for longer-duration experiments an upper limit for i_l is reached.

(b) Effect on Characteristic Potentials ($E_{1/2}, E_p, \ldots$). The manner in which the potential of the forward reaction is affected depends not only on the type of coupled reaction and experimental duration but also on the reversibility of the electron transfer reactions. Consider the $\mathrm{E_rC_i}$ case; that is, a reversible (nernstian) electrode reaction followed by an irreversible chemical reaction:

$$\mathrm{O} + ne \rightleftharpoons \mathrm{R} \rightarrow \mathrm{X} \tag{11.1.29}$$

The potential of the electrode during the experiment is given by the Nernst equation:

$$E = E^{0\prime} + \frac{RT}{nF} \ln \frac{C_{\mathrm{O}}(x=0)}{C_{\mathrm{R}}(x=0)} \tag{11.1.30}$$

where $C_{\mathrm{O}}(x=0)/C_{\mathrm{R}}(x=0)$ is determined by the experimental conditions. The effect of the following reaction is to decrease $C_{\mathrm{R}}(x=0)$ and hence increase $C_{\mathrm{O}}(x=0)/C_{\mathrm{R}}(x=0)$. Thus the potential will be more positive at any current level than in the absence of the perturbation, and the wave will shift toward positive potentials (this case was considered with steady-state approximations in Section 1.5.2). For an EC reaction where the electron transfer is totally irreversible, the following reaction causes no change in E, because the i-E characteristic for this case contains no term involving $C_R(x = 0)$.

(c) Effect on Reversal Parameters (i_a/i_c, τ_r/τ_f, . . .). Reversal results are usually very sensitive to perturbing chemical reactions. For example, for the E_rC_i case, in the absence of the perturbation for cyclic voltammetry, i_{pa}/i_{pc} would be 1 (or for chronopotentiometry τ_r/τ_f would be 1/3). In the presence of the following reaction, $i_{pa}/i_{pc} < 1$ (or $\tau_r/\tau_f < 1/3$) because R is removed from near the electrode surface by reaction, as well as by diffusion. A similar effect will be found for a catalytic (EC′) reaction; here not only is the reverse contribution decreased, but the forward parameter is increased as well.

11.1.3 Time Windows and Accessible Rate Constants

As already implied in the previous discussion, the effect that a perturbing reaction will have on the measured parameters of the electrode reaction depends on the extent to which that reaction proceeds during the course of the electrochemical experiment. The characteristic lifetime of a chemical reaction with rate constant k can be taken as $t_1' = 1/k$ for a first-order reaction (where t_1' is the time required for the reactant concentration to drop to 37% of its initial value) or $t_2' \simeq 1/kC_i$ for a second-order (e.g., dimerization) reaction, with C_i the initial concentration of reactant. (t_2' is the time required for the concentration to drop to one-half of its initial value.) If the experimental duration of the electrochemical attack, τ, is small compared to t_1' or t_2', then the electron transfer reaction will be largely unperturbed. If $t' \ll \tau$, the perturbing reaction will have a large effect. For a given method with a particular apparatus, a certain range of available experimental times (a time window) exists. The shortest useful times are frequently limited by double-layer charging and instrumental response (which can be governed by the excitation apparatus, the measuring apparatus, or the cell design). The longest available times are often governed by the onset of natural convection or perturbation of the electrode surface. This time window is different for the different electrochemical techniques (Table 11.1.1). To study a coupled reaction, its characteristic lifetime must be within the time window of a given technique. Thus chronopotentiometry would be useful in studying first-order reactions with rate constants of ~0.02 to 10^3 sec^{-1}. The techniques suitable for study of the fastest reactions are rotating disk or ac methods. Coulometric methods are best for very slow reactions. The strategy adopted to study a reaction is thus to vary the experimental variable in the technique (e.g., sweep rate, rotation rate, or applied current) and determine how the forward parameters (e.g., $i_p/v^{1/2}C$, $i\tau^{1/2}/C$, and $i_l/\omega^{1/2}C$), the characteristic potentials (e.g., E_p, and $E_{1/2}$), and the reversal parameters (e.g., i_{pa}/i_{pc}, i_r/i_f, and Q_r/Q_f) vary. The directions and extents of variation of these provide *diagnostic criteria* for establishing the type of mechanism involved, and the measurements themselves provide data for evaluation of the magnitudes of the rate constants of the coupled reactions.

Table 11.1.1
Approximate Time Windows for Different Electrochemical Techniques

Technique	Time Parameter	Usual Range of Parameter[a]	Time Window (sec)[b]
ac polarography	$1/\omega = (2\pi f)^{-1}$ (sec) (f = freq., in Hz)	$\omega = 10\text{–}6000\ \text{sec}^{-1}$	2×10^{-4} to 0.1
Rotating disk electrode voltammetry	$1/\omega = (2\pi f)^{-1}$ (sec) (f = rotation rate, in r/sec)	$\omega = 30\text{–}1000\ \text{sec}^{-1}$	10^{-3} to 0.3
Chronopotentiometry Chronoamperometry Chronocoulometry	t (sec) τ (sec) (Forward phase duration)	10^{-3} to 50 sec	10^{-3} to 50
Linear scan voltammetry Cyclic voltammetry	RT/Fv (sec)	v = 0.02 to 100 V/sec	10^{-4} to 1
dc polarography	t_{max} (sec) (drop time)	1 to 5 (sec)	1 to 5
Coulometry Macroscale electrolysis	t(electrolysis duration) (sec)	100 to 3000 (sec)	100 to 3000

[a] This represents a readily available range; these limits can often be extended to shorter times under favorable conditions. For example, potential and current steps in the microsecond range and potential sweeps of 1000 V/sec have been reported.
[b] This time window should only be considered approximate. A better description of the conditions under which a chemical reaction will cause a perturbation of the electrochemical response can be given in terms of the dimensionless rate parameter, λ, as discussed in Section 11.3.

11.2 THEORETICAL TREATMENTS OF VOLTAMMETRIC METHODS

11.2.1 Basic Principles

The theoretical treatments for the different voltammetric methods (e.g., polarography, linear sweep voltammetry, and chronopotentiometry) and the various kinetic cases generally follow the procedures described previously. The appropriate partial differential equations (usually the diffusion equations modified to take account of the coupled reactions producing or consuming the species of interest) are solved with the requisite initial and boundary conditions. For example, consider the E_rC_i reaction scheme:

$$O + ne \rightleftharpoons R \text{ (at electrode)} \tag{11.2.1}$$

$$R \xrightarrow{k} Y \text{ (in solution)} \tag{11.2.2}$$

For species O, the unmodified diffusion equation still applies, since O is not involved directly in reaction (11.2.2); thus

$$\frac{\partial C_O(x, t)}{\partial t} = D_O\left[\frac{\partial^2 C_O(x, t)}{\partial x^2}\right] \tag{11.2.3}$$

For the species R, however, Fick's law must be modified because, at a given location in solution, R is removed not only by diffusion but also by the first-order chemical reaction. Since the rate of change of the concentration of R caused by the chemical reaction is

$$\left[\frac{\partial C_R(x, t)}{\partial t}\right]_{\text{chem.rxn.}} = -kC_R(x, t) \tag{11.2.4}$$

the appropriate equation for species R is

$$\frac{\partial C_R(x, t)}{\partial t} = D_R\left[\frac{\partial^2 C_R(x, t)}{\partial x^2}\right] - kC_R(x, t) \tag{11.2.5}$$

The initial conditions, assuming only O is initially present, are, as usual,

$$C_O(x, 0) = C_O^* \quad C_R(x, 0) = 0 \tag{11.2.6}$$

The usual boundary conditions for the electrode surface

$$D_O\left[\frac{\partial C_O(x, t)}{\partial x}\right]_{x=0} = -D_R\left[\frac{\partial C_R(x, t)}{\partial x}\right]_{x=0} \tag{11.2.7}$$

and as $x \to \infty$,

$$\lim_{x\to\infty} C_O^*(x, t) = C_O^* \qquad \lim_{x\to\infty} C_R(x, t) = 0 \tag{11.2.8}$$

also apply. The sixth needed boundary condition depends on the particular technique and the reversibility of the electron transfer reaction (11.2.1), just as described in Chapters 5 to 9. For example, for a potential step experiment to the limiting cathodic current region, $C_O(0, t) = 0$. For a step to an arbitrary potential, assuming (11.2.1) is reversible, the requisite condition is [see (5.4.6)]

$$\frac{C_O(0, t)}{C_R(0, t)} = \theta = \exp\left[\frac{nF}{RT}(E - E^{0'})\right] \tag{11.2.9}$$

and, for chronopotentiometry,

$$D_O\left[\frac{\partial C_O(x, t)}{\partial x}\right]_{x=0} = \frac{i}{nFA} \tag{11.2.10}$$

Note that equations need not be written for species Y, since its concentration does not affect the current or the potential. If reaction (11.2.2) were reversible, however, the concentration of species Y would appear in the equation for $\partial C_R(x, t)/\partial t$, and an equation for $\partial C_Y(x, t)/\partial t$ and initial and boundary conditions for Y would have to be supplied (see entry 3 in Table 11.2.1). Generally, then, the equations for the theoretical treatment are deduced in a straightforward manner from the diffusion equation and the appropriate homogeneous reaction rate equations. In Table 11.2.1, equations for several different reaction schemes and the appropriate boundary conditions for potential-step, potential-sweep, and current-step techniques are given.

Solutions of the equations listed in Table 11.2.1 are accomplished by standard Laplace transform techniques (for first-order coupled reactions) often with judicious substitutions and combinations of the equations. Complicated reaction schemes and systems involving higher-order reactions generally require numerical or digital simulation methods. Response function principles can be employed to obtain results for reversal experiments. For rotating disk electrode studies, the appropriate kinetic terms are added to the convective-diffusion equations. For ac techniques the equations in Table 11.2.1 are solved for $C_O(0, t)$ and $C_R(0, t)$ in a form obtained by convolution [equivalent to (9.2.14) and (9.2.15) for the appropriate case]. Substitution of the current expression (9.2.3) then yields the final expressions.

11.2.2 Solution of the E_rC_i Scheme in Current Step (Chronopotentiometric) Methods

To illustrate the analytical approach to solving problems involving coupled chemical reactions and the treatment of the theoretical results we consider the E_rC_i scheme for a constant current excitation. The equations governing this case are given as entry 4 in Table 11.2.1 and were discussed in Section 11.2.1.

(a) Forward Reaction. The equation for $C_O(x, t)$ is the same as that in the absence of the following reaction, that is, (7.2.13):

$$C_O(0, t) = C_O^* - \frac{2it^{1/2}}{nFAD_O^{1/2}\pi^{1/2}} \tag{11.2.11}$$

Thus, in this case, the forward transition time τ_f [when $C_O(0, t) = 0$] is unperturbed, and $i\tau_f^{1/2}/C_O^*$ is a constant given by (7.2.14). However, $C_R(x, t)$ is different, and this causes the E-t curve to be different. The Laplace transform of (11.2.5) with initial condition (11.2.6) yields

$$s\bar{C}_R(x, s) = D_R\left[\frac{d^2\bar{C}_R(x, s)}{dx^2}\right] - k\bar{C}_R(x, s) \tag{11.2.12}$$

$$\left[\frac{d^2\bar{C}_R(x, s)}{dx^2}\right] = \left[\frac{(s + k)}{D_R}\right]\bar{C}_R(x, s) \tag{11.2.13}$$

Solution of this equation with the boundary condition $\lim_{x \to \infty} \bar{C}_R(x, s) = 0$ yields

$$\bar{C}_R(x, s) = \bar{C}_R(0, s) \exp\left[-\left(\frac{s + k}{D_R}\right)^{1/2} x\right] \tag{11.2.14}$$

With the boundary condition

$$-D_R\left[\frac{\partial \bar{C}_R(x, s)}{\partial x}\right]_{x=0} = \frac{\bar{\imath}(s)}{nFA} \tag{11.2.15}$$

this finally yields

$$\bar{C}_R(0, s)(s + k)^{1/2}D_R^{1/2} = \frac{\bar{\imath}(s)}{nFA} \tag{11.2.16}$$

Table 11.2.1

Modified Diffusion Equations and Boundary Conditions for Several Different Coupled Homogeneous Chemical Reactions in Voltammetry

Case	Reactions	Diffusion Equations (all x and t)	General Initial and Semi-Infinite Boundary Conditions ($t = 0$ and $x \to \infty$)	Potential Step and Sweep Boundary Conditions (at $x = 0$)	Current Step Boundary Conditions (at $x = 0$)
1. C_rE_r	$Y \underset{k_b}{\overset{k_f}{\rightleftharpoons}} O$	$\frac{\partial C_Y}{\partial t} = D_Y \frac{\partial^2 C_Y}{\partial x^2} - k_f C_Y + k_b C_O$	$C_O/C_Y = K$ $C_O + C_Y = C^*$	$\frac{C_O}{C_R} = \theta S(t)$	$\frac{\partial C_O}{\partial x} = \frac{i}{nFAD_O}$
	$O + ne \rightleftharpoons R$	$\frac{\partial C_O}{\partial t} = D_O \frac{\partial^2 C_O}{\partial x^2} + k_f C_Y - k_b C_O$	$C_R = 0$ (Note 1)	(Note 2)	
		$\frac{\partial C_R}{\partial t} = D_R \frac{\partial^2 C_R}{\partial x^2}$			
2. C_rE_i	$Y \underset{k_b}{\overset{k_f}{\rightleftharpoons}} O$	(as above)	(as above)	$D_O\left(\frac{\partial C_O}{\partial x}\right) = k' C_O e^{bt}$ (Note 3)	(as above)
	$O + ne \rightarrow R$				
3. E_rC_r	$O + ne \rightleftharpoons R$	$\frac{\partial C_O}{\partial t} = D_O \frac{\partial^2 C_O}{\partial x^2}$	$C_O = C_O^*$	(as C_rE_r above)	(as above)
	$R \underset{k_b}{\overset{k_f}{\rightleftharpoons}} Y$	$\frac{\partial C_R}{\partial t} = D_R \frac{\partial^2 C_R}{\partial x^2} - k_f C_R + k_b C_Y$	$C_R = C_Y = 0$		
		$\frac{\partial C_Y}{\partial t} = D_Y \frac{\partial^2 C_Y}{\partial x^2} + k_f C_R - k_b C_Y$	(Note 1)		
4. E_rC_i	$O + ne \rightleftharpoons R$	(as above, with $k_b = 0$)	(as above)	(as C_rE_r above)	(as above)
	$R \xrightarrow{k_f} Y$	(equation for C_Y not required)			
5. E_rC_{2i}	$O + ne \rightleftharpoons R$	$\frac{\partial C_O}{\partial t} = D_O \frac{\partial^2 C_O}{\partial x^2}$	(as above)	(as C_rE_r above)	(as above)
	$2R \xrightarrow{k_f} X$	$\frac{\partial C_R}{\partial t} = D_R \frac{\partial^2 C_R}{\partial x^2} - k_f C_R^2$			

Case	Reactions	Diffusion Equations (all x and t)	General Initial and Semi-Infinite Boundary Conditions ($t = 0$ and $x \to \infty$)	Potential Step and Sweep Boundary Conditions (at $x = 0$)	Current Step Boundary Conditions (at $x = 0$)
6. E_rC_i'	$O + ne \rightleftharpoons R$	$\frac{\partial C_O}{\partial t} = D_O \frac{\partial^2 C_O}{\partial x^2} + k_f C_R$	$C_O = C_O^*$	(as C_rE_r above)	(as above)
	$R \xrightarrow{k_f} O$	$\frac{\partial C_R}{\partial t} = D_R \frac{\partial^2 C_R}{\partial x^2} - k_f C_R$	$C_R = 0$		
			(Note 1)		
7. $E_rC_iE_r$	$O_1 + n_1 e \rightleftharpoons R_1$	$\frac{\partial C_{O1}}{\partial t} = D_{O1} \frac{\partial^2 C_{O1}}{\partial x^2}$	$C_{O1} = C^*$	$\frac{C_{O1}}{C_{R1}} = \theta_1 S(t)$	$D_{O1} n_1 \left(\frac{\partial C_{O1}}{\partial x}\right) +$
	$R_1 \xrightarrow{k_f} O_2$	$\frac{\partial C_{R1}}{\partial t} = D_{R1} \frac{\partial^2 C_{R1}}{\partial x^2} - k_f C_{R1}$	$C_{R1} = C_{O2} = C_{R2} = 0$	$\frac{C_{O2}}{C_{R2}} = \theta_2 S(t)$	$D_{O2} n_2 \left(\frac{\partial C_{O2}}{\partial x}\right) = \frac{i}{FA}$
	$O_2 + n_2 e \rightleftharpoons R_2$	$\frac{\partial C_{O2}}{\partial t} = D_{O2} \frac{\partial^2 C_{O2}}{\partial x^2} + k_f C_{O2}$		(Note 4)	
		$\frac{\partial C_{R2}}{\partial t} = D_{R2} \frac{\partial^2 C_{R2}}{\partial x^2}$			

(Note 1) $D_O \left(\frac{\partial C_O}{\partial x}\right)_{x=0} = -D_R \left(\frac{\partial C_R}{\partial x}\right)_{x=0} \qquad D_Y \left(\frac{\partial C_Y}{\partial x}\right)_{x=0} = 0$

(Note 2) For potential sweep, $\theta = \exp\left[\frac{nF}{RT}(E_i - E^{0\prime})\right] \qquad S(t) = \exp\left(-\frac{nF}{RT} vt\right)$ $\qquad E_i$ = initial potential, v = scan rate

For potential step to potential E, $\theta = \exp\left[\frac{nF}{RT}(E - E^{0\prime})\right] \qquad S(t) = 1$

(Note 3) For sweep from E_i at scan rate v or for step to E_i with $v = 0$, $k' = k^0 \exp\left[\frac{-\alpha n_a F}{RT}(E_i - E^{0\prime})\right] \qquad b = \frac{\alpha n_a F}{RT} v$

(Note 4) For potential sweep, $\theta_j = \exp\left[\frac{n_j F}{RT}(E_i - E_j^{0\prime})\right] \qquad E_j^{0\prime}$ pertains to $O_j + n_j e \rightleftharpoons R_j$

For potential step, $\theta_j = \exp\left[\frac{n_j F}{RT}(E - E_j^{0\prime})\right]$

For the forward step at constant current,

$$\bar{i}(s) = \frac{i}{s} \tag{11.2.17}$$

$$\bar{C}_R(0, s) = \frac{i}{nFAD_R^{1/2}s(s + k)^{1/2}} \tag{11.2.18}$$

and, from the inverse transform,

$$C_R(0, t) = \frac{i}{nFAD_R^{1/2}k^{1/2}} \operatorname{erf}[(kt)^{1/2}] \tag{11.2.19}$$

(b) Potential-Time behavior. From the transition time equations 7.2.14 and 11.2.11,

$$C_O(0, t) = \frac{2i(\tau^{1/2} - t^{1/2})}{nFAD_O^{1/2}\pi^{1/2}} \tag{11.2.20}$$

For a reversible electron transfer reaction, the Nernst equation applies, that is,

$$E = E^{0'} + \left(\frac{RT}{nF}\right) \ln \frac{C_O(0, t)}{C_R(0, t)} \tag{11.2.21}$$

The E-t curve is obtained from (11.2.19) to (11.2.21):

$$E = E^{0'} + \frac{RT}{nF} \ln \frac{2}{\pi^{1/2}} \left(\frac{D_R}{D_O}\right)^{1/2} \frac{(kt)^{1/2}(\tau^{1/2} - t^{1/2})}{\operatorname{erf}[(kt)^{1/2}]t^{1/2}} \tag{11.2.22}$$

$$E = E^{0'} + \frac{RT}{2nF} \ln\left(\frac{D_R}{D_O}\right) + \frac{RT}{nF} \ln\left\{\frac{2}{\pi^{1/2}} \frac{(kt)^{1/2}}{\operatorname{erf}[(kt)^{1/2}]}\right\} + \frac{RT}{nF} \ln\left(\frac{\tau^{1/2} - t^{1/2}}{t^{1/2}}\right) \tag{11.2.23}$$

This can be written

$$\boxed{E = E_{1/2} - \frac{RT}{nF} \ln \Xi + \frac{RT}{nF} \ln\left(\frac{\tau^{1/2} - t^{1/2}}{t^{1/2}}\right)} \tag{11.2.24a}$$

$$\Xi = \frac{\pi^{1/2}}{2} \frac{\operatorname{erf}[(kt)^{1/2}]}{(kt)^{1/2}} \tag{11.2.24b}$$

The term $(RT/nF) \ln \Xi$ represents the perturbation caused by the chemical reaction. It is instructive to examine the limiting behavior of Ξ as a function of the dimensionless term (kt). For $(kt)^{1/2} < 0.1$, $\operatorname{erf}[(kt)^{1/2}] \approx 2(kt)^{1/2}/\pi^{1/2}$ (see Appendix A) or $\Xi = 1$. Thus the following reaction will have no effect for sufficiently small k or short times. This can be considered as the pure *diffusion-controlled zone.* As $(kt)^{1/2}$ increases, Ξ becomes smaller, so that the E-t curve is shifted toward more positive potentials. For example, when $(kt)^{1/2} = 1$, $\operatorname{erf}[(kt)^{1/2}] = 0.84$, $\Xi = 0.75$ and the wave is shifted at this point 7 mV on the potential axis in a positive direction. When $(kt)^{1/2} \geq 2$, $\operatorname{erf}[(kt)^{1/2}] \approx 1$, so that $\Xi = \frac{1}{2}(\pi/kt)^{1/2}$. This represents the limiting region for large k or t, and leads to the E-t equation for the pure *kinetic zone*:

$$E = E_{1/2} + \left(\frac{RT}{nF}\right) \ln\left(\frac{2k^{1/2}}{\pi^{1/2}}\right) + \left(\frac{RT}{nF}\right) \ln(\tau^{1/2} - t^{1/2}) \tag{11.2.25}$$

Note that this equation is very similar in form to that for a totally irreversible electron transfer reaction (with no coupled chemical reaction) (7.3.6) and will show a linear variation of E with $\ln(\tau^{1/2} - t^{1/2})$ in this zone. This equation can also be written as

$$E = E_{1/2} + \left(\frac{RT}{nF}\right) \ln\left(\frac{2}{\pi^{1/2}}\right) + \left(\frac{RT}{2nF}\right) \ln(kt) + \left(\frac{RT}{nF}\right) \ln\left(\frac{\tau^{1/2} - t^{1/2}}{t^{1/2}}\right) \quad (11.2.26)$$

or, at $t = \tau/4$, $E = E_{\tau/4}$, where

$$E_{\tau/4} = E_{1/2} + \left(\frac{RT}{nF}\right) \ln\left(\frac{2}{\pi^{1/2}}\right) + \left(\frac{RT}{2nF}\right) \ln(kt) \quad (11.2.27)$$

A plot of $E_{\tau/4}$ *vs.* log (kt) is shown in Figure 11.2.1. Note that the limiting diffusion and kinetic zones are given by the solid lines, and the dashed line is the exact equation (11.2.24). Of course the locations of the boundaries of these zones depend on the approximation employed, and the applicability of the limiting equations depends on the accuracy of the electrochemical measurements. For example, if potential measurements are made to the nearest 1 mV, the pure kinetic zone will be reached (for $n = 1$ and 25°) when $25.7 \ln[\text{erf}\,(kt)^{1/2}] \leq 1$ mV or when $(kt)^{1/2} \geq 1.5$.

(c) Current Reversal. The treatment involving current reversal generally employs the same equations and utilizes the zero-shift-theorem method, as in Section 7.4.2. Thus, for reversal of current at time t_1 $(t_1 \leq \tau_1)$,

$$i(t) = i - S_{t_1}(t)(2i) \quad (11.2.28)$$

where $S_{t_1}(t)$ is the step function, equal to 0 $(t \leq t_1)$ and 1 $(t > t_1)$. Then

$$\bar{i}(s) = \left(\frac{i}{s}\right)(1 - 2\,e^{-t_1 s}) \quad (11.2.29)$$

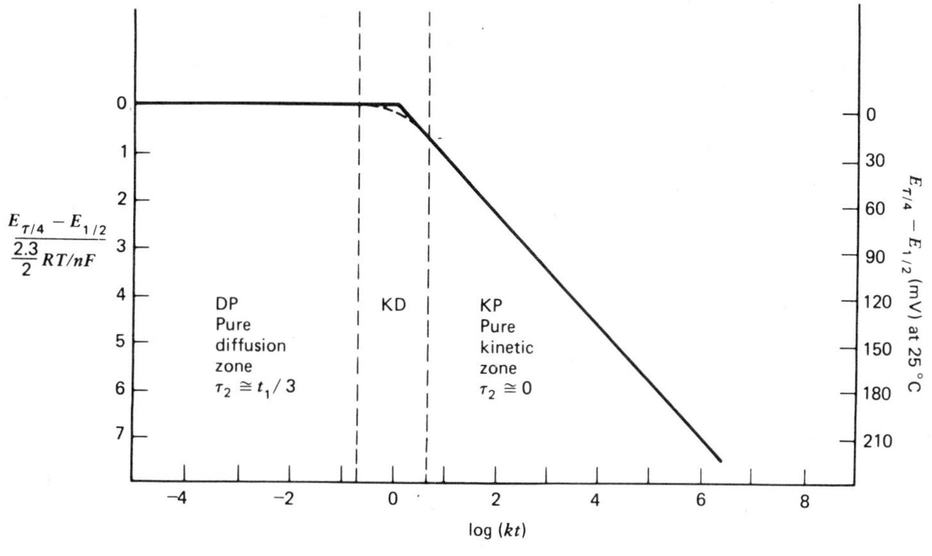

Figure 11.2.1
Variation of $E_{\tau/4}$ with log(kt) for chronopotentiometry with E_rC_i reaction scheme.

and from (11.2.16)

$$\bar{C}_R(0, s) = \frac{i}{nFAD_R^{1/2}}\left[\frac{1}{s(s+k)^{1/2}} - \frac{2\,e^{-t_1 s}}{s(s+k)^{1/2}}\right] \tag{11.2.30}$$

The inverse transform yields

$$C_R(0, t) = \frac{i}{nFAD_R^{1/2}}\left[\frac{\text{erf}(kt)^{1/2}}{k^{1/2}} - S_{t_1}(t)\,\frac{2}{k^{1/2}}\,\text{erf}\left\{[k(t - t_1)]^{1/2}\right\}\right] \tag{11.2.31}$$

At the reverse transition time, $t = t_1 + \tau_2$, $C_R(0, t) = 0$, so that

$$\boxed{\text{erf}\{[k(t_1 + \tau_2)]^{1/2}\} = 2\,\text{erf}[(k\tau_2)^{1/2}]} \tag{11.2.32}$$

Let us again examine the limiting behavior. When kt_1 is small (diffusion zone) $\text{erf}[(k\tau_2)^{1/2}]$ approaches $2(k\tau_2)^{1/2}/\pi^{1/2}$ and $\text{erf}\{[k(t_1 + \tau_2)]^{1/2}\}$ approaches $2[k(t_1 + \tau_2)]^{1/2}/\pi^{1/2}$. Under these conditions (11.2.32) becomes identical to the equation for unperturbed reversal chronopotentiometry and $\tau_2 = t_1/3$. When kt_1 is large (kinetic zone), τ_2 approaches 0. The variation of τ_2/t_1 with kt_1 is shown in Figure 11.2.2 (10–12). Note that kinetic information can be obtained from reversal measurements only in the intermediate zone ($0.1 \lesssim kt_1 \lesssim 5$). The actual value of k is obtained by determining τ_2/t_1 for different values of t_1 and fitting the data to the *working curve* shown in Figure 11.2.2 (13). Kinetic information can also be obtained from the shift of potential with τ_1; however, the reversible $E_{1/2}$ must be known before an actual value of k can be obtained.

The treatment given here is typical of that involved for other reaction schemes and voltammetric techniques. The results of these treatments are typically the establishment of diagnostic criteria for distinguishing one mechanistic scheme from another and working curves or tables that can be used in the evaluation of rate constants. A survey of results is given in Section 11.3.

11.3 RESULTS FOR TRANSIENT VOLTAMMETRY AND CHRONOPOTENTIOMETRY

We examine here the results of theoretical treatments for voltammetric and chronopotentiometric techniques for several of the important and frequently observed reaction schemes.

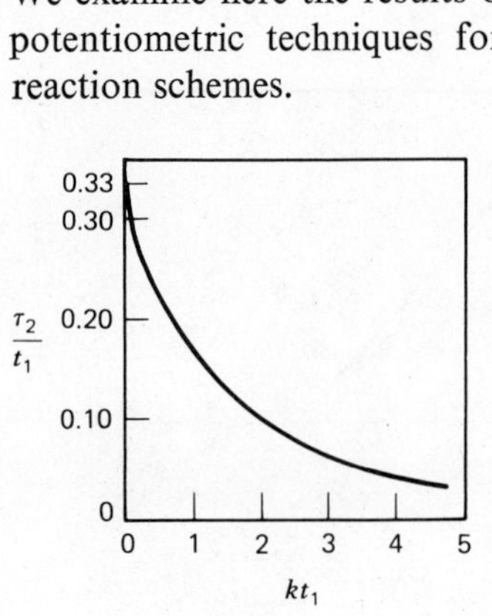

Figure 11.2.2
Variation of τ_2/t_1 with kt_1 for chronopotentiometry with E_rC_i reaction scheme.

11.3.1 Preceding Reaction—C_rE_r

$$Y \underset{k_b}{\overset{k_f}{\rightleftharpoons}} O \tag{11.3.1}$$

$$O + ne \rightleftharpoons R \tag{11.3.2}$$

$$K = k_f/k_b = C_O(x, 0)/C_Y(x, 0) \tag{11.3.3}$$

The behavior of this system depends on the magnitudes of both the first-order rate constants, k_f and k_b ($\sec^{-1}$), and the equilibrium constant, K. The relevant dimensionless parameters, λ, for the different techniques for this scheme are shown in Table 11.3.1. It is instructive to consider the "zone diagram" (14) (Figure 11.3.1), which defines where and how the electrochemical parameters are affected by λ and K, and when the limiting behavior will be observed within a given accuracy. When K is large (e.g., ≥ 20), the equilibrium in (11.3.1) lies so far to the right that most of the material is already in the form of electroactive substance O. The preceding reaction then has little effect on the electrochemical response, which appears as the unperturbed nernstian behavior. Similarly when k_f and k_b are small (e.g., $\lambda < 0.1$), the preceding reaction has little effect and a nernstian response results, but with the effective initial concentration of O, $C_O(x, 0)$, being given by

$$C_O(x, 0) = \frac{C^*K}{(K + 1)} \tag{11.3.4}$$

where

$$C_O(x, 0) + C_Y(x, 0) = C^* \tag{11.3.5}$$

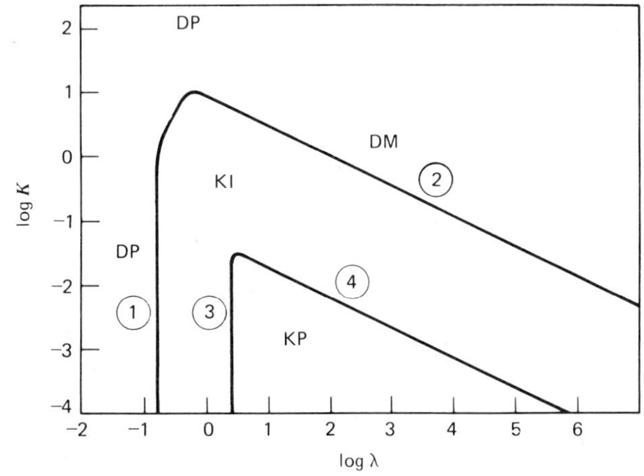

Figure 11.3.1

C_rE_r reaction diagram defining zones for different types of electrochemical behavior as a function of K and λ (defined in Table 11.3.1). The zones shown are DP—pure diffusion; DM—diffusion modified by equilibrium constant of preceding reaction; KP—pure kinetics; and KI—intermediate kinetics. [Adapted with permission from J. M. Saveant and E. Vianello, *Electrochim. Acta*, **8**, 905 (1963). Copyright by Pergamon Press, Inc.]

Table 11.3.1
Dimensionless Parameters for Voltammetric Methods

Technique	Time Parameter(s)	Dimensionless Kinetic Parameter, λ, for C_rE_r	E_rC_i	E_rC_i'
Chronoamperometry and polarography	t	$(k_f + k_b)t$	kt	$k'C_Z^*t$
Linear sweep and cyclic voltammetry	$1/v$	$\frac{(k_f + k_b)}{v}\left(\frac{RT}{nF}\right)$	$\frac{k}{v}\left(\frac{RT}{nF}\right)$	$\frac{k'C_Z^*}{v}\left(\frac{RT}{nF}\right)$
Chronopotentiometry	τ	$(k_f + k_b)\tau$	$k\tau$	$k'C_Z^*\tau$
Rotating disk electrode	$1/\omega$	$(k_f + k_b)/\omega$[a]	k/ω	$k'C_Z^*/\omega$

[a] Or $\delta/\mu = 1.61\, k^{1/2}\nu^{1/6}/\omega^{1/2}D^{1/6}$.

When λ is so large that (11.3.1) can always be considered in equilibrium, the reaction again shows nernstian behavior, but the wave is shifted along the potential axis from its unperturbed position by an extent that depends on the magnitude of K, as discussed in Sections 1.5.1 and 5.4.5. The realm of this zone, in the upper-right portion of Figure 11.3.1, depends on K and λ. When K is small and λ is large, the reaction is so fast that the reactants can be considered to be at steady-state values within the reaction layer near the electrode surface, and the differential equations governing the system can be solved by setting the derivatives with respect to time equal to zero (the "reaction-layer treatment"). This is the "pure kinetic zone." A more quantitative description of how the limits of these zones are chosen is described in Section 11.3.1(c).

(a) Polarographic and Chronoamperometric Methods. The current of interest is that for $C_O(0, t) = 0$, that is, at the limiting current plateau. For a planar electrode, assuming equal diffusion coefficients for all species ($D_O = D_Y = D$) and the chemical equilibrium favoring Y[$K \ll 1$, $C_Y(x, 0) \simeq C^*$], the current is given by (15)

$$i = nFAC^*D^{1/2}k_f^{1/2}K^{1/2}\exp(k_fKt)\,\mathrm{erfc}[(k_fKt)^{1/2}] \tag{11.3.6}$$

Letting $(k_fKt)^{1/2} = Z$, this can be written

$$i = nFAD^{1/2}C^*t^{-1/2}Z\exp(Z^2)\,\mathrm{erfc}(Z) \tag{11.3.7}$$

Note that this is of the same form as that for a totally irreversible wave [see (5.5.22)]. For large values of k_f, the function $Z\exp(Z^2)\,\mathrm{erfc}(Z)$ approaches $\pi^{-1/2}$ and the current becomes the diffusion-controlled value, i_d:

$$i_d = nFAD^{1/2}C^*/(\pi t)^{1/2} \tag{11.3.8}$$

Hence the behavior is in the diffusion zone on the right side in Figure 11.3.1. Now (11.3.6) can be written

$$\boxed{\frac{i}{i_d} = \pi^{1/2}Z\exp(Z^2)\,\mathrm{erfc}(Z)} \tag{11.3.9}$$

(Compare with (5.5.23) and results in Figure 5.5.2.) For small values of the argument, $Z \exp(Z^2) \operatorname{erfc}(Z) \simeq Z$, and (11.3.9) yields

$$i = i_d \pi^{1/2}(k_f K t)^{1/2} = nFAD^{1/2}C^*(k_f K)^{1/2} \tag{11.3.10}$$

This is the current in the pure kinetic region in Figure 11.3.1; it is independent of t and governed purely by the rate of conversion of Y to O. These equations hold for polarography as well (within the expanding plane approximation) with $t = t_{max}$ (the drop time) and the area A given by (5.3.3); and the approach of Section 5.5.5 applies. Treatments taking account of spherical diffusion and unequal diffusion coefficients have also been presented (16, 17). Note how the limiting current in polarography varies with t_{max} or the height of the mercury column, h_{corr}. For large k_f (in the diffusion region), i varies as $t_{max}^{1/6}$ or as $h_{corr}^{1/2}$. For small k_f, where (11.3.10) applies, i is independent of t_{max} and of h_{corr}.

(b) Linear Sweep and Cyclic Voltammetric Methods. The shape of the *i-E* curve depends on the values of K and λ; that is, on the region of interest in Figure 11.3.1 (14, 18). Some typical curves for different values of K and λ are shown in Figure 11.3.2. The behavior is diffusion-controlled in regions DP and DM. In the pure kinetic region (KP), the *i-E* curve no longer shows a peak, and i tends toward the limiting value defined in (11.3.10), which is independent of scan rate v. In this region, the shift of the half-peak potential $E_{p/2}$ (since a plateau is observed it is inconvenient to consider E_p) with v is given by

$$\frac{dE_{p/2}}{d \ln v} = \frac{RT}{2nF} \tag{11.3.11}$$

At 25° a tenfold increase in v causes the reduction peak to shift by $29/n$ mV in the positive direction as shown in (11.3.12).

$$\boxed{E_{p/2}(V) = E^{0'} - (0.007/n) - (0.029/n) \log k_b + (0.029/n) \log v} \tag{11.3.12}$$

As v increases (so that λ decreases) and the system enters the zone of intermediate kinetics (Figure 11.3.1, zone KI), the shift of $E_{p/2}$ with scan rate becomes smaller and finally is independent of v in the diffusion zone (DP). The shift of $E_{p/2}$ with the dimensionless parameter $K\lambda^{1/2}$ is shown in Figure 11.3.3 (18). A working curve showing the ratio of the kinetic peak current, i_K, to the diffusion-controlled current, i_d (attained at very slow scan rates), has been proposed (18) (Figure 11.3.4) and has been shown to fit the empirical equation

$$\frac{i_k}{i_d} = \frac{1}{1.02 + 0.471/K\sqrt{\lambda}} \tag{11.3.13}$$

In cyclic voltammetry, the anodic portion on the reverse scan is not affected as much as the forward scan by the coupled reaction (Figure 11.3.5). The ratio of i_{pa}/i_{pc} (with i_{pa} measured from the extension of the cathodic curve as described in Section 6.6) increases with increasing scan rate as shown in the working curve in Figure 11.3.6 (18). The actual *i-E* curves can be drawn using series solutions or a table given by Nicholson and Shain (18).

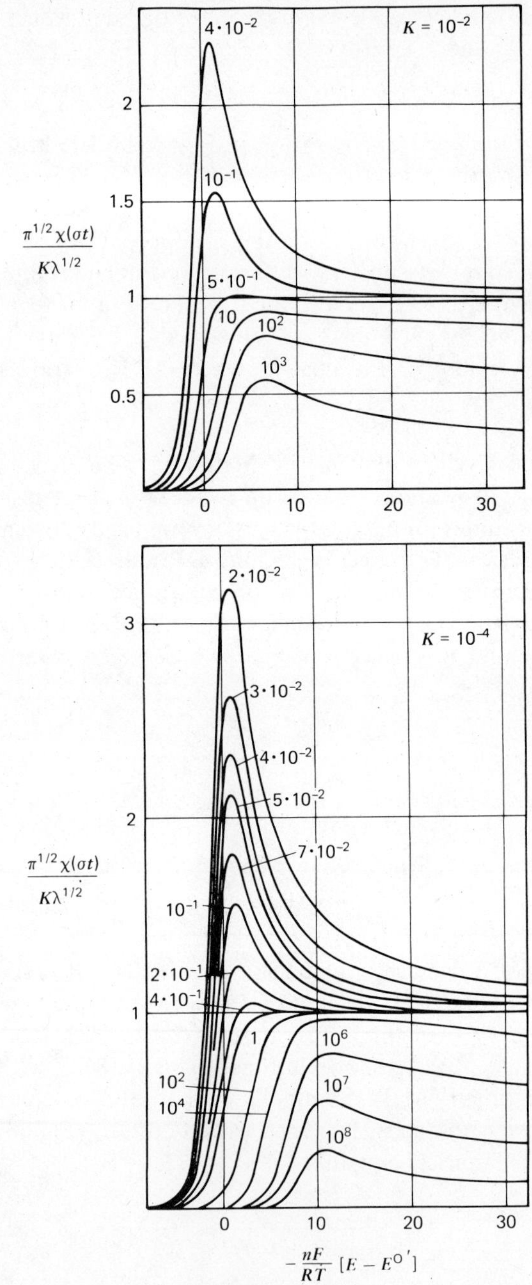

Figure 11.3.2

Curves of current [plotted as $\pi^{1/2}\chi(\sigma t)/K\lambda^{1/2}$, where $\chi(\sigma t)$ is defined as in (6.2.16)] *vs.* potential at $K = 10^{-2}$ (upper) and $K = 10^{-4}$ (lower), at different values of $\lambda = [(k_f + k_b)/v]/(RT/nF)$ shown on each curve for the C_rE_r reaction scheme. [Reprinted with permission from J. M. Saveant and E. Vianello, *Electrochim. Acta*, **8**, 905 (1963). Copyright by Pergamon Press, Inc.]

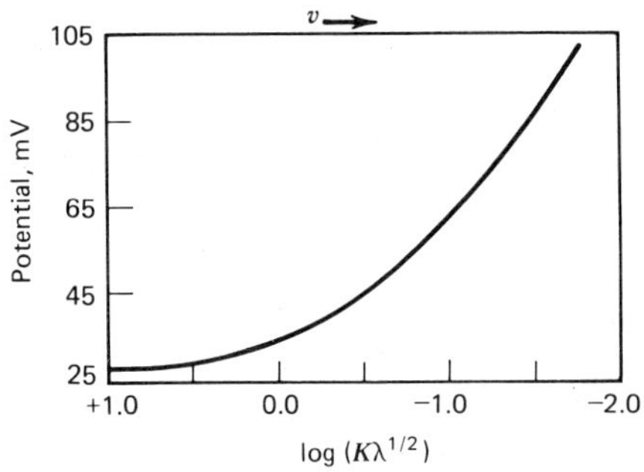

Figure 11.3.3
Variation of $E_{p/2}$ with $K\lambda^{1/2}$ for C_rE_r reaction scheme. Potential axis is $n(E_{p/2} - E_{1/2}) - (RT/F)\ln(K/1 + K)$. $v\rightarrow$ shows direction of increasing scan rate. [Reprinted with permission from R. S. Nicholson and I. Shain, *Anal. Chem.*, **36**, 706 (1964). Copyright 1964, American Chemical Society.]

(c) Chronopotentiometric Methods. The i-τ behavior is governed by the equation (19–21)

$$i\tau^{1/2} = \frac{nFAC^*(\pi D)^{1/2}}{2} - \frac{i}{2K}\left[\frac{\pi}{(k_f + k_b)}\right]^{1/2} \operatorname{erf}(\lambda^{1/2}) \tag{11.3.14}$$

The first term on the right is the value for the diffusion-controlled reaction, $i\tau_d^{1/2}$. Using the definition of λ (Table 11.3.1), this equation can then be written

$$i\tau^{1/2} = i\tau_d^{1/2} - \frac{i\tau^{1/2}\pi^{1/2}}{2K\lambda^{1/2}} \operatorname{erf}(\lambda^{1/2}) \tag{11.3.15}$$

or

$$i\tau^{1/2} = \frac{i\tau_d^{1/2}}{1 + \dfrac{0.886 \operatorname{erf}(\lambda^{1/2})}{K\lambda^{1/2}}} \tag{11.3.16}$$

The variation of $i\tau^{1/2}$ with $\lambda^{1/2}$ for several different values of K is shown in Figure 11.3.7. This equation is also useful in examining the limiting behavior of $i\tau^{1/2}$ and defining the different zones of interest as in Figure 11.3.1. Consider $\lambda^{1/2} \leq 0.4$. Then $\operatorname{erf}(\lambda^{1/2})/\lambda^{1/2}$ approaches the limiting value (within *ca.* 5%) of $2/\pi^{1/2}$ and (11.3.16) yields $(i\tau^{1/2}/i\tau_d^{1/2}) \simeq K/(1 + K)$ or the diffusion-controlled reaction corrected by

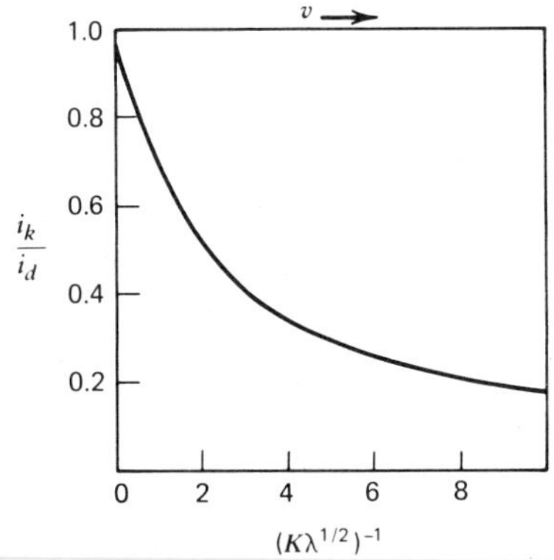

Figure 11.3.4
Working curve of i_k/i_d *vs.* $(k\lambda^{1/2})^{-1}$ for the C_rE_r reaction scheme. [Reprinted with permission from R. S. Nicholson and I. Shain, *Anal. Chem.*, **36**, 706 (1964). Copyright 1964, American Chemical Society.]

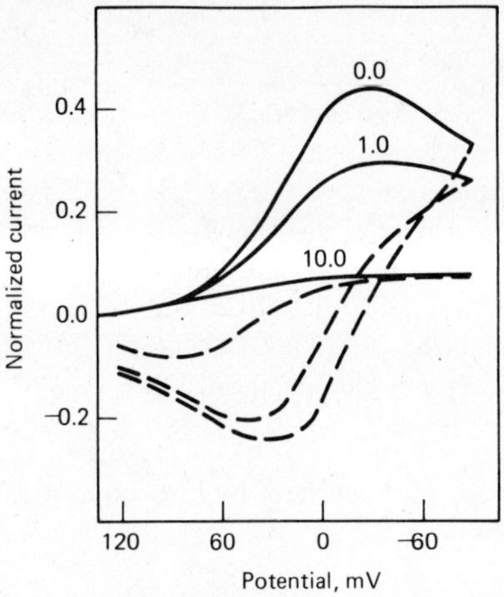

Figure 11.3.5
Cyclic voltammograms for the C_rE_r reaction scheme at several values of $(K\lambda^{1/2})^{-1}$ (shown on curves). Potential scale is that of Figure 11.3.3. [Reprinted with permission from R. S. Nicholson and I. Shain, *Anal. Chem.*, **36**, 706 (1964). Copyright 1964, American Chemical Society.]

calculating $C_O(x, 0)$ from C^*. Thus this condition, or $\log \lambda < -0.8$, defines the left boundary (line 1). For large λ (e.g., $\lambda^{1/2} \geq 1.4$), $\text{erf}(\lambda^{1/2}) \approx 1$ and (11.3.16) yields

$$\frac{i\tau^{1/2}}{i\tau_d^{1/2}} = \left(1 + \frac{0.886}{K\lambda^{1/2}}\right)^{-1} \tag{11.3.17}$$

This will yield diffusion-controlled behavior when $0.886/K\lambda^{1/2} \leq 0.05$, or $\log K = 1.25 - \frac{1}{2}\log \lambda$; this represents the right boundary (line 2). The pure kinetic region is also defined by large λ values, this time as $K \rightarrow 0$. One can set the boundary by using $\lambda^{1/2} \geq 1.4$ (line 3) and the condition in (11.3.17) that the second term on the right predominates. Thus $0.886/K\lambda^{1/2} \geq 10$ or $\log K = -\frac{1}{2}\log \lambda - 1.05$ (line 4). Note that the exact locations of these boundaries depend on the levels of approximations used. Moreover, in this pure kinetic region, (11.3.14) becomes

$$\boxed{i\tau^{1/2} = i\tau_d^{1/2} - \frac{i\pi^{1/2}}{2K(k_f + k_b)^{1/2}}} \tag{11.3.18}$$

so that a plot of $i\tau^{1/2}$ *vs.* i in this region is a straight line of slope $-\pi^{1/2}/2K(k_f + k_b)^{1/2}$. This behavior is evident in the plots shown in Figure 11.3.8.

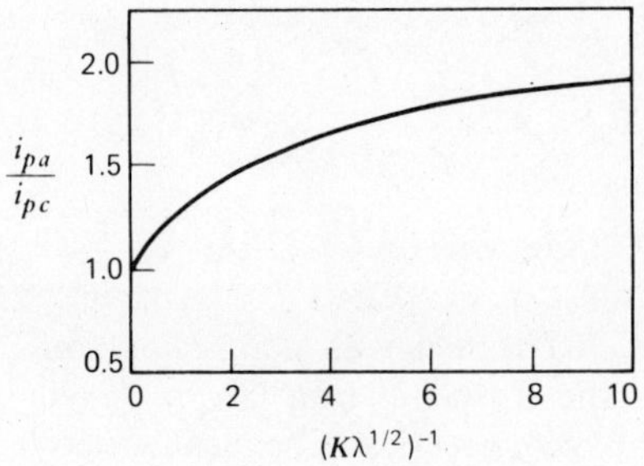

Figure 11.3.6
Ratio of anodic to cathodic peak currents as a function of the kinetic parameters for the C_rE_r reaction scheme. [Reprinted with permission from R. S. Nicholson and I. Shain, *Anal. Chem.*, **36**, 706 (1964). Copyright 1964, American Chemical Society.]

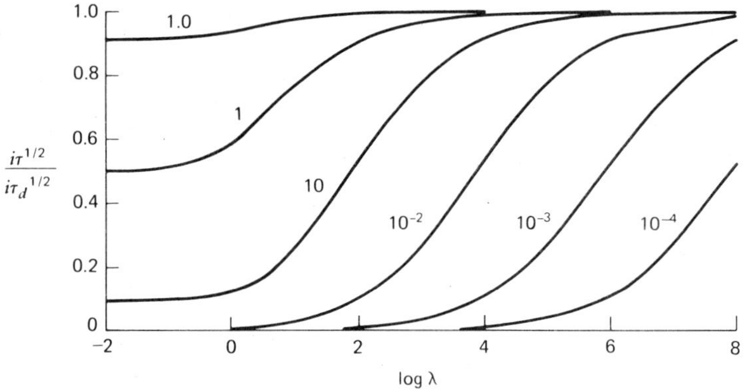

Figure 11.3.7
Variation of $i\,\tau^{1/2}/i\,\tau_d^{1/2}$ with λ for various values of K (indicated on curve) for chronopotentiometric study of the C_rE_r reaction scheme.

For simple reversal chronopotentiometry, the ratio of reversal transition time, τ_2, to the forward time, τ_1, is 1/3, just as in the diffusion-controlled case, independent of the rate constants. However, for cyclic chronopotentiometry the transition times for the third (τ_3) and subsequent reversals differ from those of the diffusion-controlled case (21). For example, for the C_rE_r reaction with $k\tau_1 = 1.00$, the relative transition times are: $a_1 = 1.000$, $a_2 = 0.333$, $a_3 = 0.665$, $a_4 = 0.385$, $a_5 = 0.631$, and $a_6 = 0.410$, compared to the diffusion-controlled values listed in Section 7.4.3.

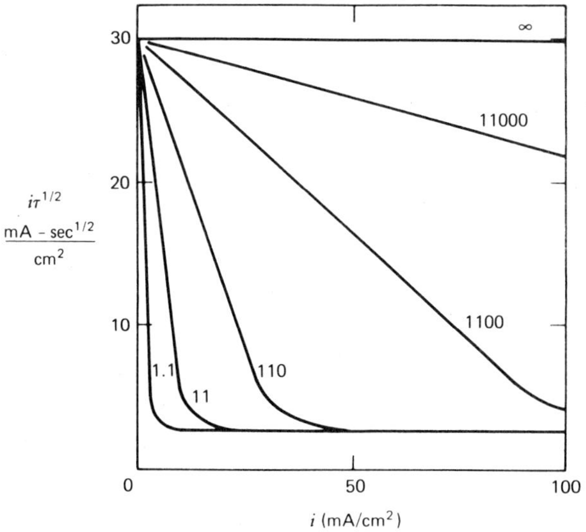

Figure 11.3.8
Variation of $i\,\tau^{1/2}$ with i, for various values of $(k_f + k_b)$ (in sec^{-1}). Calculated for $K = 0.1$, $C^* = 0.11$ mM, and $D = 10^{-5}$ cm^2/sec. [Reprinted with permission from P. Delahay and T. Berzins, *J. Am. Chem. Soc.*, **75**, 2486 (1953). Copyright 1953, American Chemical Society.]

11.3.2 Preceding Reaction—C_rE_i

This scheme is the same as that in Section 11.3.1, except that the electron transfer reaction (11.3.2) is totally irreversible and governed by the charge transfer parameters α and k^0. The limiting current behavior in chronoamperometric and polarographic methods will not be perturbed by irreversibility in the electron transfer reaction; since the potential is stepped to a value sufficiently beyond the equilibrium value that reaction (11.3.2) proceeds rapidly. Thus the results will be the same as in Section 11.3.1(a). This represents an important advantage of chronoamperometric methods—that the potential can be chosen to eliminate complexities in the analysis of the behavior caused by the heterogeneous electron transfer step. On the other hand, once the rate constants of the homogeneous reactions have been deduced, potential steps to less extreme potentials can provide information about α and k^0. This requires solution of the more complex problem where the boundary condition for $C_O(0, t)$ is governed by the heterogeneous reaction rate. This problem will not be examined here.

The C_rE_i case in linear sweep voltammetry has been treated (18); because of the irreversibility of the electron transfer, no anodic current is observed on the reverse scan and cyclic voltammetric behavior need not be considered. Typical *i-E* curves are shown in Figure 11.3.9. The limiting behavior again depends on the magnitude of the kinetic parameter $K\sqrt{\lambda_i}$, where λ_i is the λ factor of Table 11.3.1, with n replaced by αn_a:

$$\lambda_i = \frac{(k_f + k_b)}{v}\left(\frac{RT}{\alpha n_a F}\right) \tag{11.3.19}$$

When λ_i is small, the behavior is the same as that of the unperturbed irreversible reaction, as described in Section 6.3, except that the concentration of O is given by $C_O^*[K/(1 + K)]$. This represents the limiting behavior at high scan rates. For large λ_i and large values of $K\lambda_i^{1/2}$, the preceding reaction can be considered to be essentially at equilibrium at all times, and again the *i-E* behavior becomes that of the unperturbed irreversible case in Section 6.3, with the wave shifted (from the position it would have

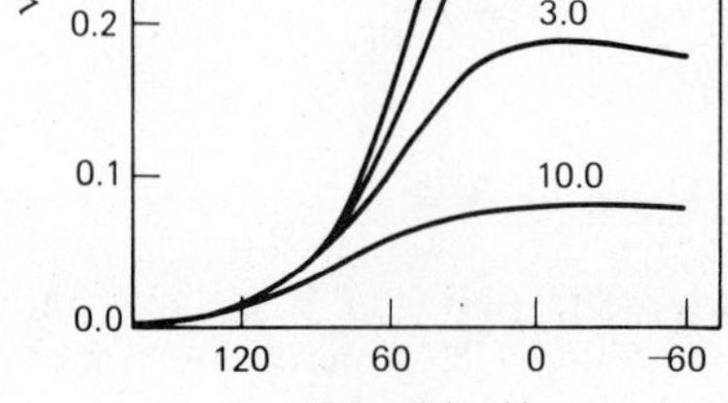

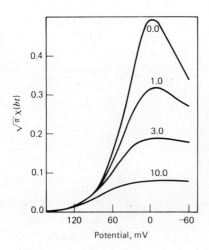

Figure 11.3.9
Curves of current [plotted as $\pi^{1/2}\chi(bt)$, where $\chi(bt)$ is defined as in (6.3.8)] versus potential for different values of $(K\lambda_i^{1/2})^{-1}$ (shown on curves). The potential scale is $(E - E^{0'})\alpha n_a + (RT/F)\ln(\sqrt{\pi D b}/k^0) - (RT/F)\ln[K/(1 + K)]$. $b = \alpha n_a F v/RT$; $\lambda_i = (k_f + k_b)/b$. [Reprinted with permission from R. S. Nicholson and I. Shain, *Anal. Chem.*, **36**, 706 (1964). Copyright 1964, American Chemical Society.]

had without the preceding reaction) in a negative direction by an amount $(RT/\alpha n_a F)$ ln $[K/(1 + K)]$. For small values of $K\lambda_i^{1/2}$ (with large λ_i), the behavior depends on k^0, as well as K and λ_i, and the i-E curve no longer shows a peak, but instead has an S-shape with a current plateau. This is the pure kinetic region where the limiting current becomes independent of v, as in the case of the C_rE_r scheme. Under these conditions the current is given by (18)

$$i = \frac{nFAC^* D^{1/2} K(k_f + k_b)^{1/2}}{1 + \left(\dfrac{\pi D^{1/2}(K+1)v(\alpha n_a F/RT)}{k^0(k_f + k_b)^{1/2}}\right) \exp\left[\dfrac{\alpha n_a F}{RT}(E - E^{0\prime})\right]} \qquad (11.3.20)$$

Nicholson and Shain (18) suggest that for all ranges of $K\sqrt{\lambda_i}$ the kinetic parameters can be obtained by fitting the kinetic peak (or plateau) current for the C_rE_i case, i_k, to that for the diffusion-controlled peak current for an irreversible charge transfer (6.3.13), i_d, by the empirical equation

$$\frac{i_k}{i_d} = \frac{1}{1.02 + 0.531/K\sqrt{\lambda_i}} \qquad (11.3.21)$$

[compare to (11.3.13)].

For chronopotentiometry, the behavior of τ is again like the C_rE_r case, since the wave will shift to sufficiently negative values to maintain the electron transfer reaction rate at the value required by the applied constant current, and the treatment in Section 11.3.1 applies.

11.3.3 Following Reaction—E_rC_i

This case for the chronopotentiometric method was treated in Section 11.2.2, and the zones for pure diffusion behavior (DP) and pure kinetic behavior (KP) in terms of the dimensionless kinetic parameter, λ (Table 11.3.1), were derived: DP, $\lambda < 0.1$; KP, $\lambda > 5$ (Figure 11.2.1). These zones generally apply with the other techniques as well.

(a) Chronoamperometric Methods. Since the forward reaction for a potential step to the limiting current region is unperturbed by the irreversible following reaction, no kinetic information can be obtained from the polarographic diffusion current or the limiting chronoamperometric i-t curve. Some kinetic information is contained in the rising portion of the i-E wave and the shift of $E_{1/2}$ with t_{max}. Since this behavior is similar to that found in linear potential sweep methods, these results will not be described separately. The reaction rate constant, k, can be obtained by reversal techniques (see Section 5.7) (22, 23). A convenient approach is the potential step method, where at $t = 0$ the potential is stepped to a potential where $C_O(x = 0) = 0$, and at $t = \tau$ it is stepped to a potential where $C_R(x = 0) = 0$. The equation for the ratio of i_a (measured at time t_r) to i_c (measured at time $t_f = t_r - \tau$) (see Figure 5.7.3) is

$$-\frac{i_a}{i_c} = \phi[k\tau, (t_r - \tau)/\tau] - \left[\frac{(t_r - \tau)/\tau}{1 + (t_r - \tau)/\tau}\right]^{1/2} \qquad (11.3.22)$$

where ϕ represents a rather complicated function involving a confluent hypergeometric series. Working curves can be derived from (11.3.22) showing i_a/i_c as a function of $k\tau$ (i.e., λ) and $(t_r - \tau)/\tau$ (Figure 11.3.10). Similar working curves have been obtained by digital simulation of this case for both chronoamperometry and chronocoulometry (23). These curves can be employed to obtain the rate constant k if a value of τ can be employed that yields a λ in the useful range.

(b) Linear Sweep and Cyclic Voltammetric Methods. Typical curves for this case are given in Figure 11.3.11 (18). Note that at small values of λ, essentially reversible behavior is found. For large values of λ (in the KP region), no current is observed on scan reversal and the shape of the curve is similar to that of a totally irreversible charge transfer (6.3.8). In this region the current function changes only slightly with scan rate (i.e., $i_p/v^{1/2}$ increases by about 5% for λ changing from 1 to 10). The peak, which has been shifted positive of the reversible E_p value by the following reaction, shifts in a negative direction (toward the reversible curve) with increasing v (Figure 11.3.12).

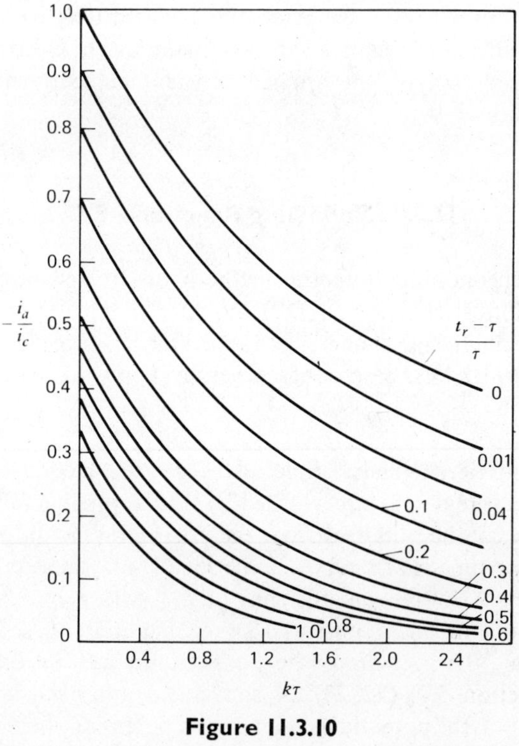

Figure 11.3.10

Working curves for double potential step chronoamperometry for E_rC_i case. i_a = anodic current, measured at time t_r; i_c = cathodic current, measured at time $t_f = t_r - \tau$; time of potential reversal = τ; k = rate constant of reaction. [Reprinted with permission from W. M. Schwarz and I. Shain, *J. Phys. Chem.*, **69**, 30 (1965). Copyright 1965, American Chemical Society.]

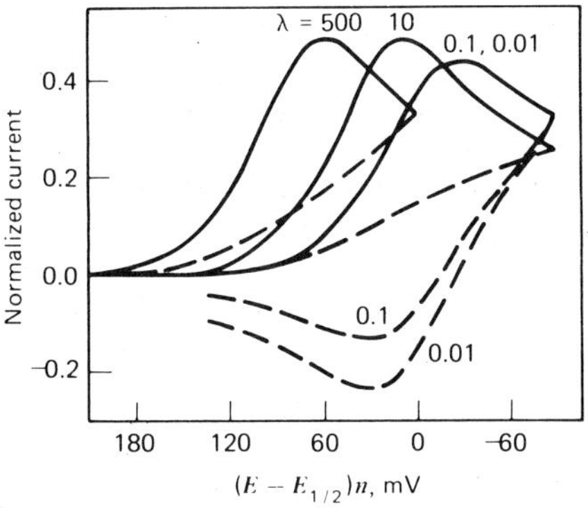

Figure 11.3.11
Cyclic voltammograms for E_rC_i case for several values of $\lambda = k(RT/nF)/v$. [Reprinted with permission from R. S. Nicholson and I. Shain, *Anal. Chem.*, **36**, 706 (1964). Copyright 1964, American Chemical Society.]

In the KP region, E_p is given by

$$\boxed{E_p = E_{1/2} - \frac{RT}{nF}0.780 + \frac{RT}{2nF}\ln \lambda} \tag{11.3.23}$$

so that the wave shifts toward negative potentials by about $30/n$ mV (at 25°) for a tenfold increase in v. In the intermediate region of λ (KO), that is, $5 > \lambda > 0.1$, information can be obtained from the ratios of anodic and cathodic peak currents (i_{pa} and i_{pc}, respectively, determined as described in Section 6.5.1). Nicholson and Shain (18) also plotted the ratio i_{pa}/i_{pc} as a function of $k\tau$, where τ is the time between $E_{1/2}$ and the switching potential, E_λ. By fitting the observed values to a working curve (Figure 11.3.13), a value of k_f can be estimated (assuming $E_{1/2}$ can be determined by experiments at sufficiently high scan rates). Note that reversal data yields kinetic information only over a small range of λ, however.

(c) Chronopotentiometric Methods. The equations governing τ, the E-t curve, and single reversal experiments are given in Section 11.2.2. Cyclic chronopotentiometry shows a continual decrease in the relative transition times on repeated reversals because of the irreversible loss of R during the course of the experiment (12). The relative transition times for the E_rC_i case depend on $k\tau_1$; for $k\tau_1 = 1.00$ they are

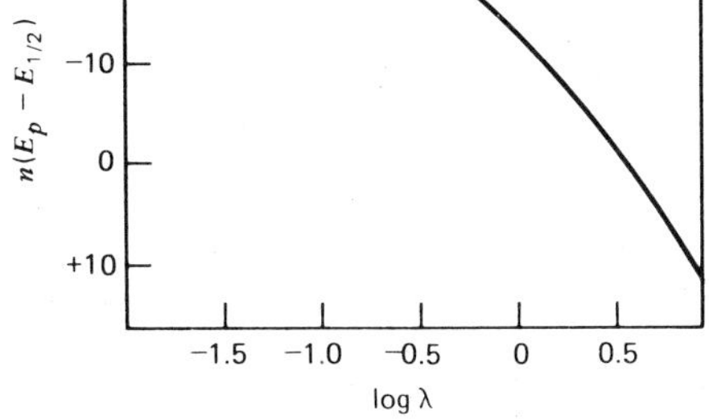

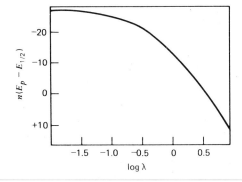

Figure 11.3.12
Variation of peak potential as a function of λ for E_rC_i case. [Reprinted with permission from R. S. Nicholson and I. Shain, *Anal. Chem.*, **36**, 706 (1964). Copyright 1964, American Chemical Society.]

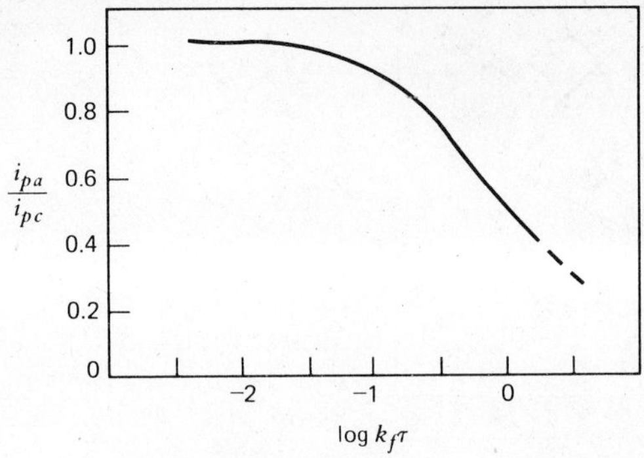

Figure 11.3.13
Ratio of anodic to cathodic peak current as a function of $k_f\tau$, where τ is the time between $E_{1/2}$ and the switching potential, E_λ. [Reprinted with permission from R. S. Nicholson and I. Shain, *Anal. Chem.*, **36**, 706 (1964). Copyright 1964, American Chemical Society.]

$a_1 = 1.000$, $a_2 = 0.167$, $a_3 = 0.384$, $a_4 = 0.138$, $a_5 = 0.292$, and $a_6 = 0.120$ (compare to diffusion-controlled values in Section 7.4.3 or those for the C_rE_r case in Section 11.3.1).

The E_rC case has been treated for a number of variations in addition to the irreversible first-order following reaction discussed here. For example, the case where the product R dimerizes:

$$2R \xrightarrow{k_2} R_2 \tag{11.3.24}$$

has been treated for several techniques (23–27). This case can be distinguished from the first-order one by the dependence of the electrochemical response on C_O^*. Also, the variation of $i_p/v^{1/2}$, E_p, . . . with the dimensionless kinetic parameter [which for this second-order reaction is $\lambda_2 = k_2C_O^*t_1$ or $\lambda_2 = k_2C_O^*\,(RT/nFv)$] is different. For example, for linear sweep voltammetry in the KP region, the peak potential equation is (25, 27)

$$E_p = E_{1/2} - \frac{RT}{nF}0.902 + \frac{RT}{3nF}\ln\left(\tfrac{2}{3}\lambda_2\right) \tag{11.3.25}$$

so that E_p shifts 20 mV (at 25°) for a tenfold change in scan rate. Other EC schemes, for example, for a reversible following reaction (E_rC_r) (18, 25), or where the product R can react with starting material O (23, 28), have also been discussed.

11.3.4 Following Reaction—E_qC_i

When the rate of the charge transfer reaction is sufficiently slow, the observed behavior depends on k^0 and α [for reaction (11.2.1] as well as the kinetic parameter λ for the following reaction. This case can be important even with the fast charge transfer reactions because, as shown in the discussion of E_rC_i reactions, the irreversible following reactions cause the voltammetric wave to shift toward positive values, and this shift away from the reversible $E_{1/2}$ value causes a decrease in the rate of the charge transfer reaction. We consider here only the cyclic voltammetric method (29, 30). It is convenient to define a dimensionless parameter, Λ, related to k^0 (30):

$$\Lambda = \frac{k^0}{D^{1/2}v^{1/2}}\left(\frac{RT}{nF}\right)^{1/2} \tag{11.3.26}$$

and to illustrate the general behavior by a zone diagram (Figure 11.3.14), showing the effect of Λ and λ. The different zones can be explained as follows. For $\lambda < 0.1$, the following reaction has no effect and the behavior is characteristic of reversible (DO), quasi-reversible (QR), or totally irreversible (IR) electron transfers, as described in Chapter 6. Similarly, for large Λ values (i.e., the upper portion of Figure 11.3.14), the reaction can be considered essentially reversible and the behavior corresponds to that in Section 11.3.3 (regions DP, KO, and KP). The effects of joint electron transfer and chemical irreversibility are mainly manifest in zone KG ($-0.7 < \log \Lambda < 1.3$, $-1.2 < \log \lambda < 0.8$). Tables with values of the electrochemical parameters as functions of Λ and λ are given in Reference 30. For example, E_p as a function of λ and Λ in this region is shown in Figure 11.3.15. The change in the value of $(\partial E_p/\partial \log v)$ at 25° from 0 at low Λ and λ values to $30/n$ mV with increasing λ and to $59/n$ mV with increasing Λ are clearly shown.

11.3.5 Catalytic Reaction—E_rC_i'

The catalytic reaction scheme almost always involves a nonelectroactive species (Z) in the following chemical reaction which regenerates starting material. Thus the

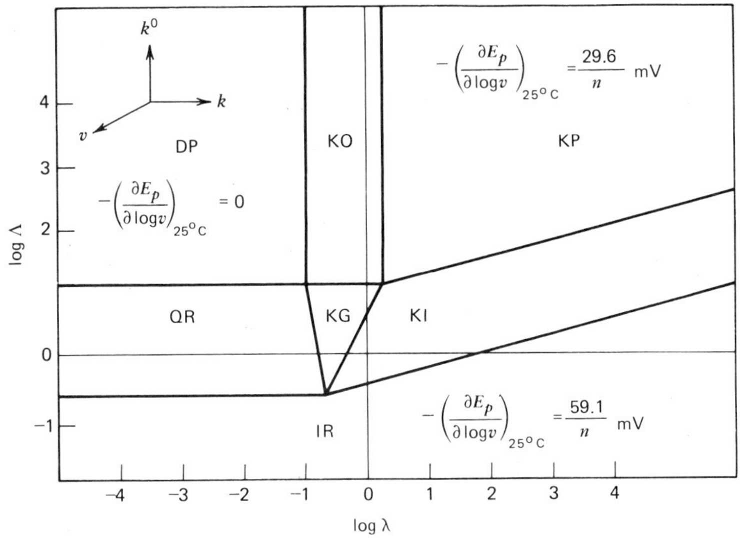

Figure 11.3.14
Zone diagram for the E_qC_i case: $\lambda = k(RT/nF)/v$; $\Lambda = k^0(RT/nF)^{1/2}/(Dv)^{1/2}$. [From L. Nadjo and J. M. Saveant, *J. Electroanal. Chem.*, **48**, 113 (1973), with permission.]

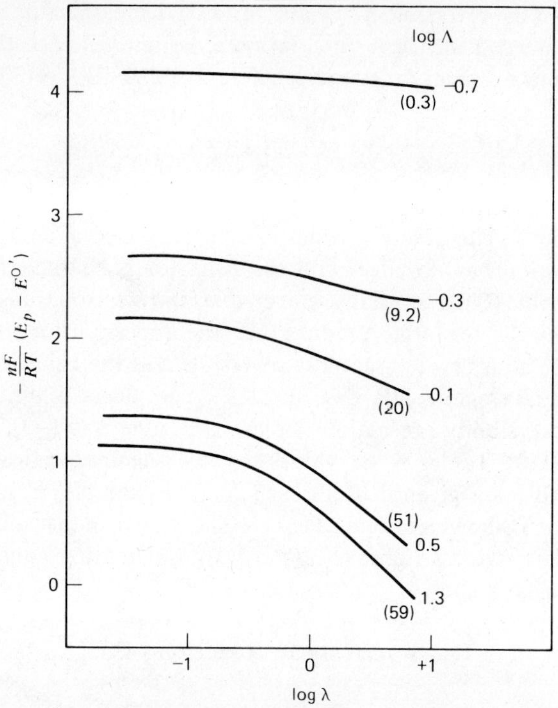

Figure 11.3.15

Variation of E_p for E_qC_i case in the intermediate kinetics region (KG) as a function of log λ, at several values of log Λ. Parenthesized numbers indicate limiting slope of each curve. [Data from L. Nadjo and J. M. Saveant, *J. Electroanal. Chem.*, **48**, 113 (1973).]

problem would involve consideration of a second-order reaction and the diffusion of species Z.

$$\text{O} + ne \rightleftharpoons \text{R} \tag{11.3.27}$$

$$\text{R} + \text{Z} \xrightarrow{k'} \text{O} + \text{Y} \tag{11.3.28}$$

In most treatments, however, it is assumed that Z is present in large excess, that is, $C_Z^* \gg C_O^*$, so that its concentration is essentially unchanged during the voltammetric experiment and (11.3.28) can be considered a pseudo-first-order reaction. Under these conditions the kinetic parameter of interest is

$$\lambda = k'C_Z^* t \quad \text{or} \quad \lambda = \frac{k'C_Z^*}{v}\left(\frac{RT}{nF}\right).$$

(a) Chronoamperometric Methods. The limiting current for a chronoamperometric experiment $[C_O(x = 0) = 0]$ is given by (31–33)

$$\boxed{\frac{i}{i_d} = \lambda^{1/2}\left[\pi^{1/2}\,\text{erf}(\lambda^{1/2}) + \frac{\exp(-\lambda)}{\lambda^{1/2}}\right]} \tag{11.3.29}$$

where i_d is the diffusion-controlled current in the absence of the following reaction [(11.3.8)]. This equation can be used to define the limiting regions of behavior. For small values of λ (e.g., $\lambda < 0.05$), $\text{erf}(\lambda^{1/2}) \simeq 2\lambda^{1/2}/\pi^{1/2}$ and $i/i_d \approx 1$ (DP region); here the catalytic reaction has no effect. For $\lambda > 1.5$, $\text{erf}(\lambda^{1/2}) \to 1$, $\exp(-\lambda)/\lambda^{1/2} \to 0$, and equation (11.3.29) becomes

$$\frac{i}{i_d} = \pi^{1/2}\lambda^{1/2} \tag{11.3.30}$$

This defines the pure kinetic region (KP). The chronoamperometric response can be employed to determine λ (or $k'C_Z^*$) from a suitable working curve based on (11.3.29) (Figure 11.3.16).

(b) Linear Sweep and Cyclic Voltammetric Methods. Typical linear sweep voltammograms for this case, treated in several papers (14, 18), are shown in Figure 11.3.17. Note that at sufficiently negative potentials all of the curves tend to a limiting value of current i_∞, independent of v, given by

$$i_\infty = nFAC_O^*(Dk'C_Z^*)^{1/2} \tag{11.3.31}$$

This limiting current arises when the rate of removal of O by the electrolysis is exactly compensated by the rate of production of O by (11.3.28), so that $C_O(x = 0)$ attains a value independent of time (or v). In region KP, when λ becomes large, the i-E curve loses its peak-shaped appearance and looks like a polarographic wave. The equation

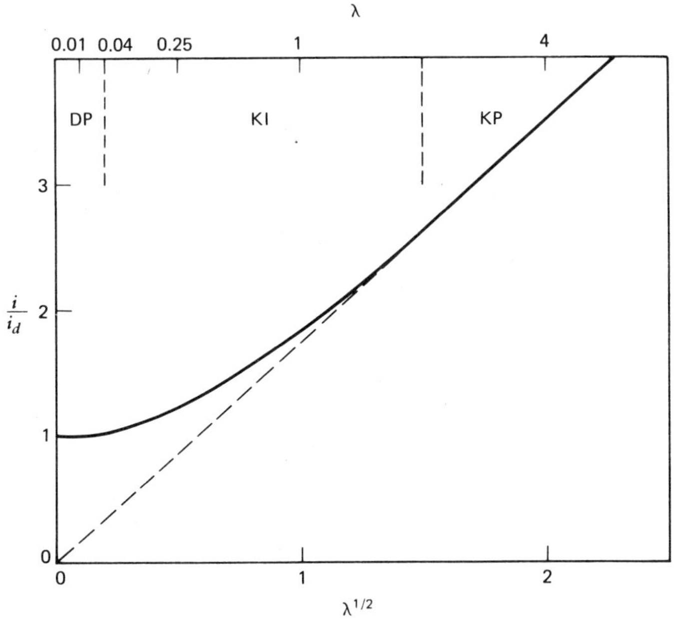

Figure 11.3.16
Chronoamperometric curve (i/i_d) for E_rC_i' case for various values of $\lambda^{1/2}$; $\lambda = k'C_Z^*t$ [(11.3.29)]. Dashed line is KP-region limiting line [(11.3.30)].

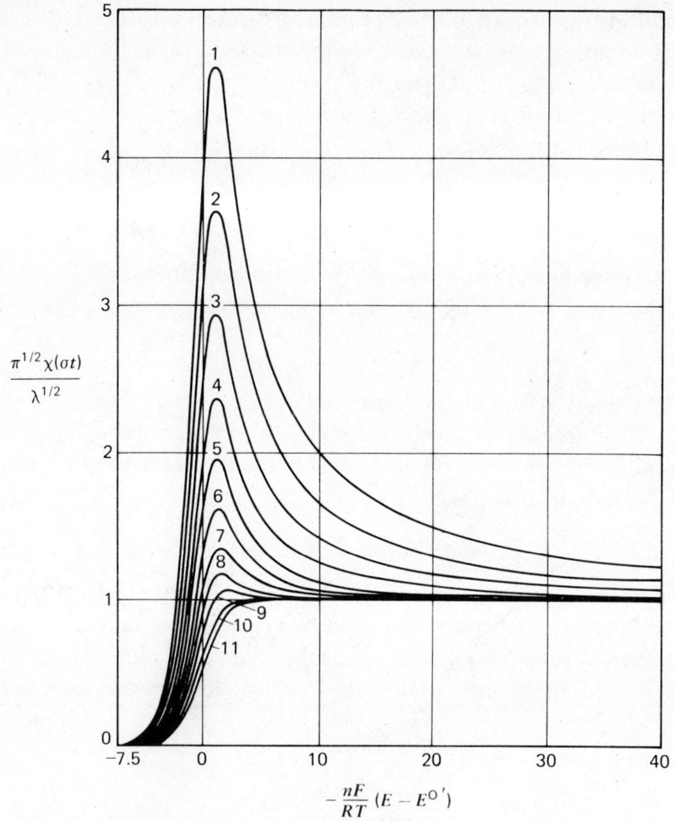

Figure 11.3.17

Linear sweep voltammograms for E_rC_i' case for various values of $\lambda = (RT/nFv)k'C_Z^*$:

1, 1.00×10^{-2};	7, 1.59×10^{-1};
2, 1.59×10^{-2};	8, 2.51×10^{-1};
3, 2.51×10^{-2};	9, 3.98×10^{-1};
4, 3.98×10^{-2};	10, 1.00;
5, 6.30×10^{-2};	11, ∞.
6, 1.00×10^{-1};	

[Reprinted with permission from J. M. Saveant and E. Vianello, *Electrochim. Acta*, **10**, 905 (1965). Copyright by Pergamon Press, Inc.]

for the wave in this region is

$$i = \frac{nFAC_O^*(Dk'C_Z^*)^{1/2}}{1 + \exp\left[\frac{nF}{RT}(E - E_{1/2})\right]} \tag{11.3.32}$$

or, from (11.3.31) and (11.3.32),

$$E = E_{1/2} + \frac{RT}{nF}\ln\left(\frac{i_\infty - i}{i}\right) \tag{11.3.33}$$

Thus in the KP region the analysis of the wave is quite easy and leads immediately to $E_{1/2}$ and k'. The dependence of the peak (or plateau) current on scan rate varies from a $v^{1/2}$ dependence in the DP zone to independence in the KP zone as shown by a plot of i/i_d *vs.* λ (Figure 11.3.18). The half-peak potential, $E_{p/2}$, is independent of λ at both high and low values of λ and shows a maximum value of $\Delta E_{p/2}/\Delta \log v$ in the KI region of about $24/n$ mV (25 °C) (Figure 11.3.19). For cyclic scans the ratio of i_{pa}/i_{pc} (i_{pa} measured from the extension of the cathodic curve) is always unity independent of λ, even in the KP region, where on the reverse scan the current tends to retrace the forward scan current (Figure 11.3.20).

(c) Chronopotentiometric Methods. The solution to this case (34) yields the following expressions for the concentrations ($t < \tau$) with $k = k'C_Z^*$:

$$C_O(0, t) = C_O^* - \frac{i}{nFAD^{1/2}k^{1/2}} \operatorname{erf}[(kt)^{1/2}] \qquad (11.3.34)$$

$$C_R(0, t) = C_O^* - C_O(0, t) \qquad (11.3.35)$$

At the transition time, $C_O(0, t) = 0$ and, from the expression for the transition time (τ_d) in the absence of the perturbing reaction (11.3.28),

$$\boxed{\left(\frac{\tau}{\tau_d}\right)^{1/2} = \frac{2\lambda^{1/2}}{\pi^{1/2}\operatorname{erf}(\lambda^{1/2})}} \qquad (11.3.36)$$

Note that the limiting values of τ/τ_d in the DP and KP regions can be obtained by consideration of the behavior at large and small λ values (see Problem 11.6). A plot showing this behavior is given in Figure 11.3.21. Note the similarity of the limiting behavior ($\lambda > 1.5$) (11.3.36) to the corresponding equation for chronoamperometry (11.3.30), as well as the similarity of the working curves. The *E-t* curves can be derived by substitution of the expressions for $C_O(0, t)$ and $C_R(0, t)$ into the Nernst equation (see Problem 11.7).

The equation for the reverse transition time in terms of τ_f in this case is the same as for the E_rC_i case (equation 11.2.32) (24, 35, 36), so that simple reversal experiments

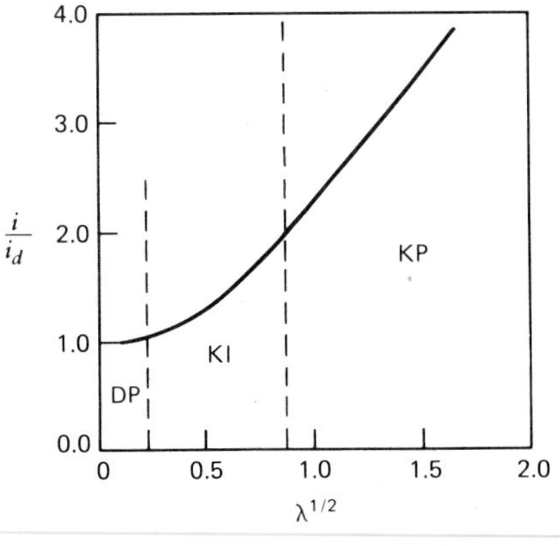

Figure 11.3.18
Ratio of kinetic peak current for E_rC_i' reaction scheme to diffusion-controlled peak current as a function of $\lambda^{1/2}$. [Reprinted with permission from R. S. Nicholson and I. Shain, *Anal. Chem.*, **36**, 706 (1964). Copyright 1964, American Chemical Society.]

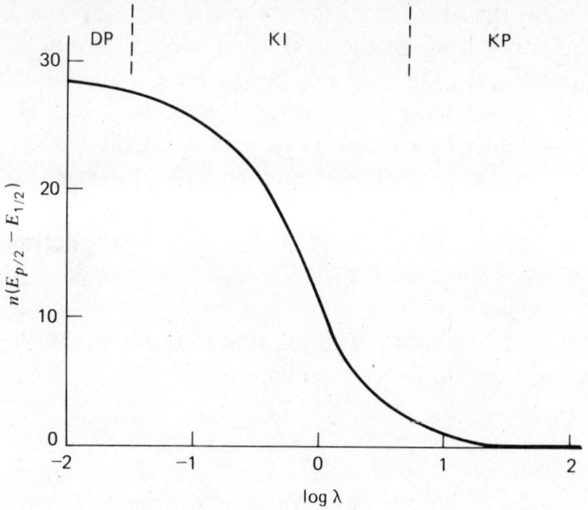

Figure 11.3.19
Variation of half-peak potential in E_rC_i' case with λ.

cannot distinguish between these cases. However, the variation of τ with i immediately differentiates between the E_rC_i and E_rC_i' cases. Moreover, the third and subsequent reversals in cyclic chronopotentiometry are very different (21), since rather than losing O continuously because of the irreversible reaction, O is regenerated. For $\lambda = 1$, the relative transition times are $a_1 = 1.00$, $a_2 = 0.167$, $a_3 = 0.673$, $a_4 = 0.169$, $a_5 = 0.658$, and $a_6 = 0.169$.

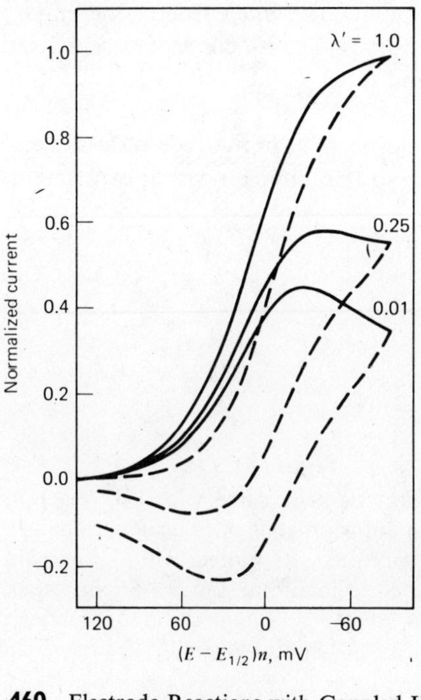

Figure 11.3.20
Cyclic voltammograms for E_rC_i' case for several values of λ. [Reprinted with permission from R. S. Nicholson and I. Shain, *Anal. Chem.*, **36**, 706 (1964). Copyright 1964, American Chemical Society.

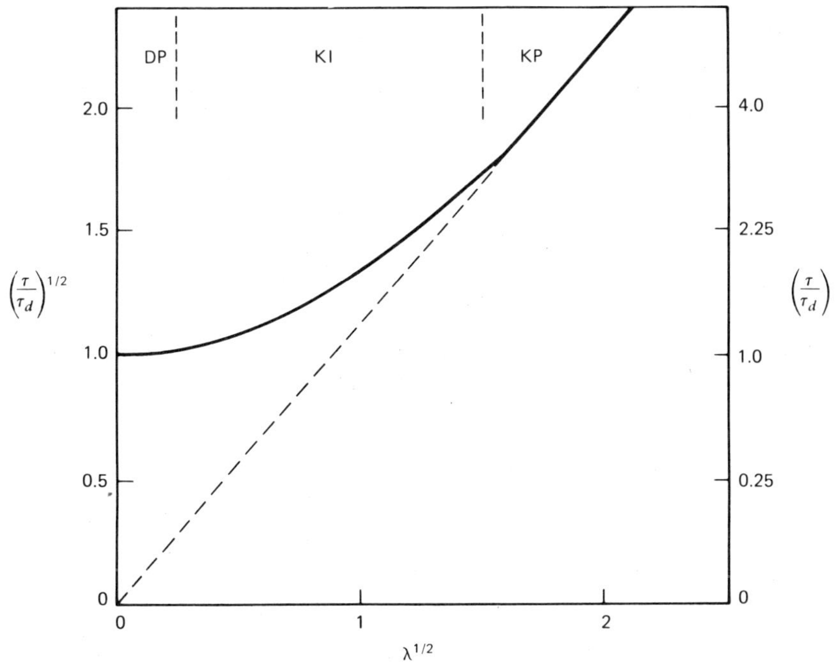

Figure 11.3.21

Variation of $(\tau/\tau_d)^{1/2}$ with $\lambda^{1/2}$ for $\mathrm{E_rC_i'}$ case in chronopotentiometry ($\lambda = k'C_Z^*\tau$). Dashed line is limiting behavior in KP region.

11.3.6 $\mathrm{E_rC_iE_r}$ Reactions

We consider here only the case

$$\mathrm{O_1} + n_1e \rightleftharpoons \mathrm{R_1} \qquad E_1^0 \tag{11.3.37}$$

$$\mathrm{R_1} \xrightarrow{k} \mathrm{O_2} \tag{11.3.38}$$

$$\mathrm{O_2} + n_2e \rightleftharpoons \mathrm{R_2} \qquad E_2^0 \tag{11.3.39}$$

where $E_2^0 \gg E_1^0$ (i.e., where $E_2^0 - E_1^0 \geq 180$ mV) and only a single voltammetric wave is observed on a forward cathodic scan. When the second reaction occurs at significantly more negative potentials, two waves result in voltammetry; the first wave involves the $\mathrm{E_rC_i}$ case and can be analyzed as described in Section 11.3.3. It is also assumed in the following sections that various nuances or variations to the basic ECE scheme that could lead to the same overall behavior, such as the path in (11.1.21) or by disproportionation of $\mathrm{R_1}$, are absent.

(a) Chronoamperometric Methods. There are clearly two limiting cases for this scheme assuming a step to potentials where $C_{\mathrm{O_1}}(x = 0) = C_{\mathrm{O_2}}(x = 0) = 0$. For $\lambda \to 0$, the behavior approaches the simple unperturbed reaction involving only (11.3.37), that is, (11.3.8) with $n = n_1$ (zone DP1). For $\lambda \to \infty$, the behavior again approaches diffusion control with both electron transfers occurring, that is, (11.3.8) with $n = n_1 + n_2$ (zone DP2):

$$(it^{1/2})_\infty = (n_1 + n_2)FAD^{1/2}C_O^*\pi^{-1/2} \tag{11.3.40}$$

For intermediate λ values the following equation applies (37):

$$\frac{(it^{1/2})}{(it^{1/2})_\infty} = 1 - \frac{n_2}{n_1 + n_2} e^{-\lambda} \tag{11.3.41}$$

Thus zone DP1 is attained (for $n_1 = n_2$) when $\lambda < 0.05$, while DP2 results when $\lambda > 3$. An alternate expression for this behavior, obtained by combining (11.3.40) and (11.3.41) and defining n_{app} as

$$n_{app} = \frac{it^{1/2}}{FAD^{1/2}C_O^*\pi^{-1/2}} \tag{11.3.42}$$

is

$$n_{app} = n_1 + n_2(1 - e^{-\lambda}) \tag{11.3.43}$$

or for $n_1 = n_2 = 1$, $n_{app} = 2 - \exp(-\lambda)$ (Figure 11.3.22). Discussions of the ECE reaction in polarography have also appeared (38–40), although the dc polarographic technique, with its limited time window, is not particularly useful in studying these kinds of reactions.

(b) Linear Scan and Cyclic Voltammetric Methods. For the ECE case under consideration here, with the second step occurring at less negative potentials than the first (i.e., $\Delta E^0 = E_1^0 - E_2^0 > 0$), only a single wave is observed on the first (cathodic) scan (Figure 11.3.23). On the reverse scan, a reversal wave (II) is observed at potentials of this wave if λ is not too large (i.e., in the KI region). As the reverse scan continues a second reversal wave (III) is observed representing the oxidation of R_2 to O_2. A second reversal reveals a corresponding cathodic wave (IV) for the reduction of O_2 to R_2. The appearance of waves II, III, and IV depends on the magnitude of λ. In the DP1 region, only II appears and the wave corresponds to a reversible n_1 electron transfer. In the DP2 region, II is absent and III and IV are prominent. Note, however, that

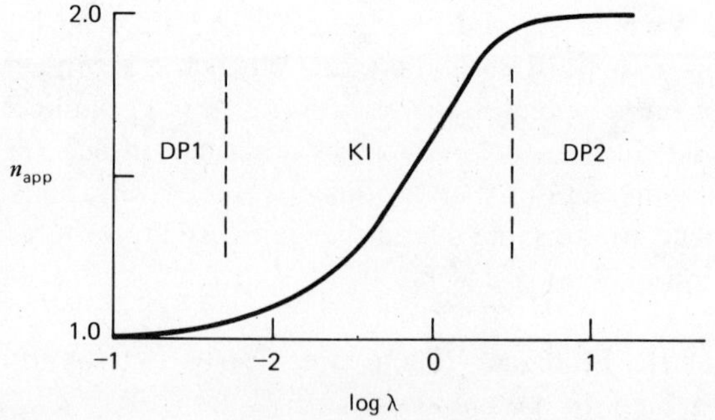

Figure 11.3.22
Variation of n_{app} with λ for $E_rC_iE_r$ reaction scheme, for $n_1 = n_2 = 1$; $\lambda = kt$.

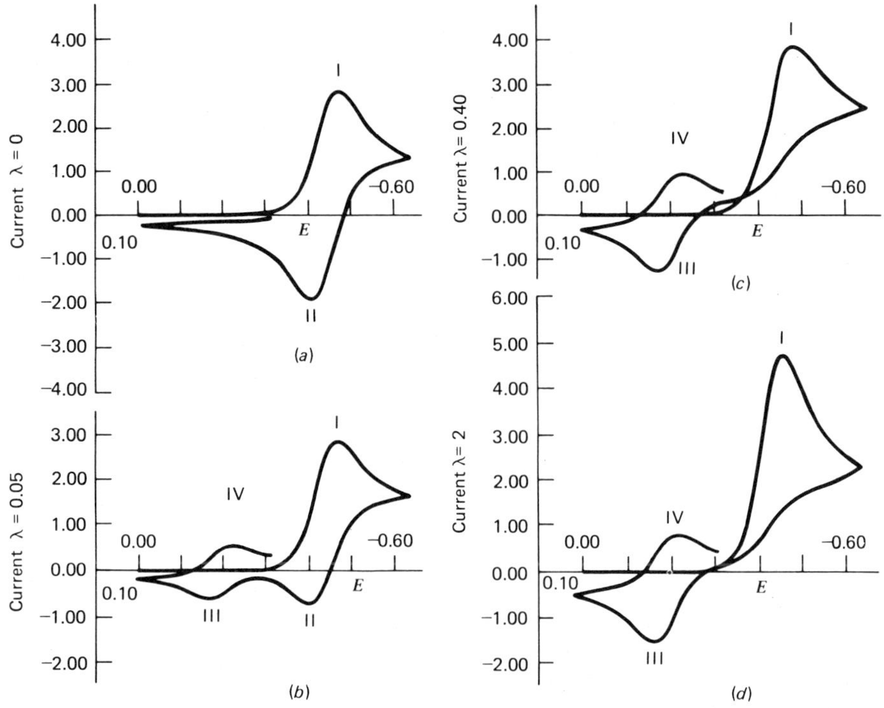

Figure 11.3.23
Cyclic voltammograms for the $E_rC_iE_r$ case obtained by digital simulation for $E_1^0 = -0.44$ V, $E_2^0 = -0.20$ V for different values of $\lambda = (k/v)(RT/nF)$; $n_1 = n_2 = 1$. (*a*) $\lambda = 0$ (unperturbed nernstian reaction). (*b*) $\lambda = 0.05$. (*c*) $\lambda = 0.40$. (*d*) $\lambda = 2$.

in many actual reactions, product R_2 may also be unstable and disappear by a fast following reaction (i.e., an $E_rC_iE_rC_i$ reaction); in this case waves III and IV will be absent but the I/II pair will show behavior near that of the $E_rC_iE_r$ scheme.

The quantitative treatment of this scheme has been described (41, 42). As in chronoamperometry the height of wave I changes from that of an unperturbed n_1 electron reaction (in the DP1 region) to one where the current limits at a value controlled by $n_1 + n_2$ (independent of λ) (in the DP2 region):

$$i = (n_1 + n_2)FAC^*(\pi D\sigma)^{1/2}\chi(\sigma t) \tag{11.3.44}$$

[compare to equation 6.2.17] (Figure 11.3.24). The relative sizes of waves I, II, III, and IV are clearly functions of λ in the KI region.

(c) Chronopotentiometric Methods. For constant current experiments on the $E_rC_iE_r$ case with $\Delta E^0 > 0$ the results are analogous to those found in chronoamperometry and linear scan voltammetry (7, 43). For small λ (zone DP1) (as $i \to \infty$), (7.2.14) (with $n = n_1$) applies. For very large λ (zone DP2) (as $i \to 0$), (7.2.14) again holds, with $n = n_1 + n_2$. The overall dependence of $i\tau^{1/2}$ on λ involves a rather complicated expression; the general trend is similar to that in Figure 11.3.22. On

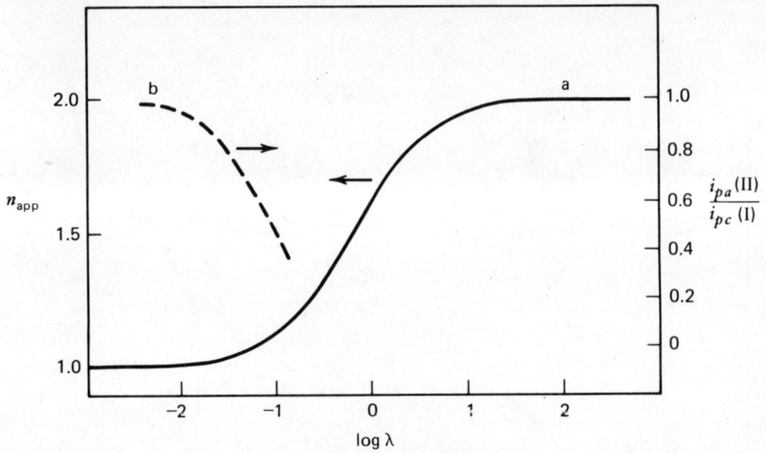

Figure 11.3.24
Variation of n_{app} and $i_{pa}(II)/i_{pc}(I)$ with λ for $E_rC_iE_r$ reaction in cyclic voltammetry for system shown in Figure 11.3.23. $n_{app} = i_p(I)(\lambda)/i_p(I)(\lambda = 0)$.

current reversal, only R_1 is oxidized near the forward wave and the ratio of transition times τ_r/τ_f as a function of λ is shown in Figure 11.3.25. This case has also been treated for cyclic chronopotentiometry (21, 24); the relative transition times for $\lambda = 1$ are $a_1 = 1.000$, $a_2 = 0.093$, $a_3 = 0.362$, $a_4 = 0.076$, $a_5 = 0.269$, and $a_6 = 0.066$.

(d) Other ECE Reaction Schemes. In general the overall ECE-reaction sequence involves many possible variations. The intermediate reaction can be reversible. In this case, both the equilibrium constant and the rate constants of the reaction must be

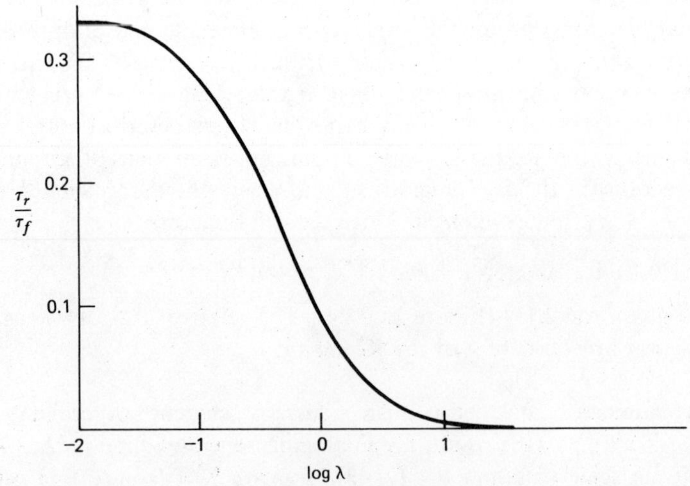

Figure 11.3.25
Variation of τ_r/τ_f with λ for $n_1 = n_2$ for chronopotentiometry in the $E_rC_iE_r$ reaction. [Data from H. B. Herman and A. J. Bard, *J. Phys. Chem.*, **70**, 396 (1966).]

considered, just as in the C_rE_i case in Section 11.3.2 (41, 44, 45). Variations of reversibility of the electron transfer steps yields four limiting cases (RR, RI, IR, II), (with quasi-reversibility of these steps also being possible). The behavior also depends on ΔE^0, with the three limiting regions ($\Delta E^0 > 0$, $\Delta E^0 = 0$, $\Delta E^0 < 0$) showing very different characteristics. Finally the general ECE mechanism, especially when the intervening chemical reaction is reversible, must consider the possible reaction of R_1 with O_2 (or R_2 with O_1) and its rate. Indeed, as discussed in Section 11.1.1(d), it is difficult to distinguish the ECE case from an EC case where electron transfer occurs in solution (the DISP mechanism) (8). It may be that many experimental systems that have been assigned to the ECE scheme actually proceed predominantly by this path. Clearly electrode reactions with coupled chemical steps can become quite complicated, and many precise measurements over a wide time range are required to establish the actual reaction path and determine the rate constants of all steps. Digital simulation techniques have been quite useful in treating this reaction sequence and a number of papers on ECE reactions and their nuances have appeared (8, 46).

11.4 ROTATING DISK AND RING-DISK METHODS

The rotating disk electrode (RDE) discussed in Chapter 8 has proven to be very useful in studies of coupled homogeneous reactions. The information content of the experimental results is high, since the coupled reactions cause perturbations in the limiting current, the half-wave potential, and the ring current at the RRDE; and an intuitive appreciation of the interpretation of the *i-E* curves is easy to attain. Also, by working at potentials in the limiting current region, complications caused by slow heterogeneous electron transfer steps can be avoided. Although the rigorous theoretical treatments at the RDE, and especially the RRDE, are rather difficult, a number of reaction schemes have been investigated, and the successful application of digital simulation methods has allowed even complex reaction schemes to be treated.

11.4.1 Theoretical Treatments

As with voltammetric methods the mass transfer equations for the various species must be changed to take account of loss or production of material because of the coupled reactions. Thus for the CE reactions [(11.3.1) and (11.3.2)] at the RDE (8.3.13) becomes

$$v_y\left(\frac{\partial C_{\rm O}}{\partial y}\right) = D_{\rm O}\left(\frac{\partial^2 C_{\rm O}}{\partial y^2}\right) + k_f C_{\rm Y} - k_b C_{\rm O} \tag{11.4.1}$$

This must be solved with the appropriate boundary conditions. For investigations at the RRDE, the radial convective terms must be included as well. Thus in the treatment of the E_rC_i scheme at the RRDE [(11.2.1) and (11.2.2)], where product R is oxidized at the ring electrode, the appropriate equation for R is

$$v_y\left(\frac{\partial C_{\rm R}}{\partial y}\right) + v_r\left(\frac{\partial C_{\rm R}}{\partial r}\right) = D_{\rm R}\left(\frac{\partial^2 C_{\rm R}}{\partial y^2}\right) - kC_{\rm R} \tag{11.4.2}$$

The solution of this equation would then follow as described in Section 8.4. The modifications of the mass transfer equations for the different cases generally follow those used in voltammetric methods as shown in Table 11.2.1. Appropriate dimensionless parameters are listed in Table 11.3.1. It is usually not possible to solve these equations analytically, and various approximations (e.g., the reaction layer approach, as described in Section 1.5.2), numerical methods, or digital simulations must be employed.

11.4.2 Preceding Reaction—C_rE_r

The solution for the limiting current at the RDE for the reaction scheme in (11.3.1) to (11.3.3) has been solved (48). To obtain a solution of the convective-diffusion equations it was assumed that the reactions were rapid so that $(D/\nu)^{1/3} \ll (k_f + k_b)/\omega$. With this condition, the limiting current is given by

$$i = \frac{nFADC^*}{1.61\ D^{1/3}\omega^{-1/2}\nu^{1/6} + D^{1/2}/Kk^{1/2}} \tag{11.4.3}$$

where $k = k_f + k_b$ and $C^* = C_O^* + C_Y^*$. The first term in the denominator is δ, the Nernst diffusion layer thickness:

$$\delta = 1.61\ D^{1/3}\omega^{-1/2}\nu^{1/6} \tag{11.4.4}$$

and the second term, which also has dimensions of length, contains the so-called reaction layer thickness μ.

$$\mu = (D/k)^{1/2} \tag{11.4.5}$$

Parameters for the reaction can be determined from the variation of i with ω. Note that at small ω, the first term of the denominator predominates ($\delta \gg \mu/K$), and the observed current is the mass-transfer-controlled limiting current i_l (8.3.21). At high $\omega(\delta \ll \mu/K)$, the pure kinetic current results:

$$i = nFAD^{1/2}C^*Kk^{1/2} \tag{11.4.6}$$

The current under these conditions is independent of ω and is identical to that found for chronoamperometry (11.3.10). Equation 11.4.3 can also be written in the form (49)

$$\boxed{\frac{i}{\omega^{1/2}} = \frac{i_l}{\omega^{1/2}} - i\left(\frac{D^{1/6}}{1.61\ \nu^{1/6}Kk^{1/2}}\right)} \tag{11.4.7}$$

so that a plot of $i/\omega^{1/2}$ *vs.* i can be used to obtain $K(k_f + k_b)^{1/2}$. This case for the rotating ring electrode has also been treated (50).

11.4.3 Following Reaction E_rC_i

A steady-state treatment of this case using the reaction layer concept was given in Section 1.5.2. The limiting cathodic current is not affected by the following reaction, and the curve is shifted toward positive potentials because of the following reaction, with $E_{1/2}$ a function of ω. The theoretical equation based on a more exact treatment of this case is (51)

$$E_{1/2} = E^0 + \left(\frac{RT}{nF}\right) \ln\left(\frac{\delta}{\mu}\right) \coth\left(\frac{\delta}{\mu}\right) \tag{11.4.8}$$

where $\delta/\mu = 1.61\, k^{1/2}\nu^{1/6}D^{-1/6}\omega^{-1/2}$. When δ/μ becomes larger (i.e., for $k/\omega > 100$) $\coth(\delta/\mu) \to 1$, and (11.4.8) becomes

$$\boxed{E_{1/2} = E^0 + \left(\frac{RT}{nF}\right) \ln(1.61\, D^{-1/6}\nu^{1/6}) + \frac{RT}{2nF} \ln\left(\frac{k}{\omega}\right)} \tag{11.4.9}$$

Note that the equation from the approximate treatment (1.5.25) becomes the same as (11.4.9), by choosing $\mu = (D/k)^{1/2}$ (the same reaction layer thickness defined in Section 11.4.2). The limits of applicability of these equations have been examined using analog computer simulations (51, 52). Note also the similarity between (11.4.9) and the voltammetric equation (11.3.23). (Recall for the RDE, $\lambda = k/\omega$.)

The RRDE is especially useful in the study of EC reactions. The ring is set at a potential where the mass-transfer-controlled oxidation of R back to O [the reverse of (11.2.1)] occurs, and the deviation of the measured (kinetic) collection efficiency, $N_K(= -i_r/i_d)$, from the value N found in the absence of the perturbing reaction (see Section 8.4.2), and the variation of N_K with ω, allows determination of k. An approximate treatment of steady-state kinetic collection efficiencies for thin ring–thin gap electrodes, later extended to a wider range of electrode geometries, has been presented (53–55). For those kind of electrodes two approximate equations were suggested (53): For $4.5 < \kappa < 3.5$

$$N_K \approx 1.75\kappa^{-3} \exp[-4\kappa^3(r_2 - r_1)/r_1] \tag{11.4.10}$$

For $\kappa < 0.5$

$$\frac{N}{N_K} = 1 + 1.28\left(\frac{\nu}{D}\right)^{1/3}\left(\frac{k}{\omega}\right) \tag{11.4.11}$$

where

$$\kappa = k^{1/2}\omega^{-1/2}D^{-1/6}\nu^{1/6}(0.51)^{-1/3} \tag{11.4.12}$$

and r_1 and r_2 are the radii of the disk and inner edge of the ring, respectively (see Figure 8.4.1). These expressions hold, however, only for electrodes where $r_2/r_1 \approx r_3/r_1$, which are very difficult to construct in practice. An alternate, more complicated expression for N_K was therefore suggested (54, 55):

$$N_K = N - (\beta')^{2/3}(1 - U_*A_1^{-1}) + \tfrac{1}{2}A_1^{-1}A_2^2\kappa^2U_*(\beta')^{4/3} - 2A_2\kappa^2T_2 \tag{11.4.13}$$

where $A_1 = 1.288$, $A_2 = 0.643\,\nu^{1/6}D^{1/3}$, $\beta' = 3\ln(r_3/r_2)$, $U_* = \kappa^{-1}\tanh(A_1\kappa)$, and T_2 is a rather complicated small correction factor approximately equal to $0.718\ln(r_2/r_1)$.

The problem of the E_rC_i reaction has also been attacked using digital simulation methods (56) (Appendix B). In this approach, no approximations need be made about electrode geometry. However, a separate simulation is required for each electrode geometry and each value of κ. Typical simulations of N_K as a function of the simulation parameter **XKT** ($= \kappa^2$) and results obtained from the approximate equations in the appropriate regions are shown in Figures 11.4.1 and 11.4.2. The range of rate constants that can be measured by the RRDE technique generally is, with rather generous limits, $0.3 < \kappa < 5$ (55). As can been seen in Figure 11.4.2 when $\kappa < 0.3$, $N_K \approx N$; while for $\kappa > 5$, $N_K \to 0$ and the ring current becomes too small to measure. For the usual range of ω, D, and ν, this yields a range of measurable rate constants of $0.03 < k < 10^3\ \text{sec}^{-1}$.

Theoretical treatments are also available for E_rC_i schemes involving second-order reactions (56, 57), as well as for the disk and ring transients in this reaction scheme (56).

11.4.4 Catalytic Reaction—E_rC_i'

A solution based on the reaction layer approach for this mechanism [see (11.3.27) and (11.3.28)] follows closely that given for the C_rE_r reaction discussed in Section 11.4.2 (48). Under the conditions that $\delta \gg \mu$ (i.e., that $\omega(D/\nu)^{1/3} \ll 3k'C_Z^*$), a limiting kinetic current, independent of ω, is found:

$$i = nFAD^{1/2}C_O^*(k'C_Z^*)^{1/2} \tag{11.4.14}$$

Note the similarity of this equation to (11.4.4). This limiting current only holds in the region of small ω. When $\lambda(= k'C_Z^*/\omega)$ becomes small, the behavior approaches the mass-transfer-controlled limiting current. Results of a digital simulation of the catalytic case (58) are shown in Figure 11.4.3. Other treatments of the E_rC_i' case at

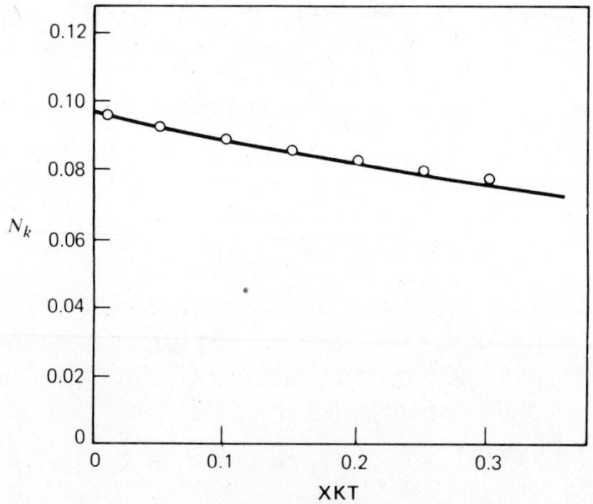

Figure 11.4.1

Collection efficiency *vs.* **XKT** [$= \kappa^2 = k\nu^{1/3}/\omega D^{1/3}(0.51)^{2/3}$] obtained from digital simulation of E_rC_i reaction at thin ring–thin gap electrode ($r_2/r_1 = 1.02$, $r_3/r_2 = 1.02$). ○ Comparison with N_K calculated from (11.4.11). [From K. B. Prater and A. J. Bard, *J. Electrochem. Soc.*, **117**, 335 (1970), reprinted by permission of the publisher, The Electrochemical Society, Inc.]

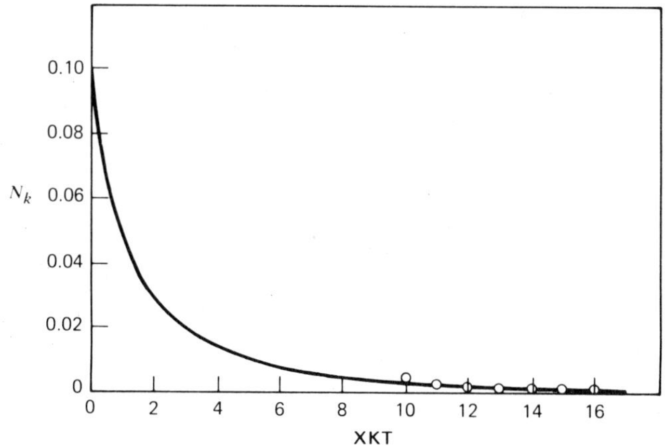

Figure 11.4.2

N_K *vs.* **XKT** as in Figure 11.4.1: ○ comparison with calculations from (11.4.10). [From K. B. Prater and A. J. Bard, *J. Electrochem. Soc.*, **117**, 335 (1970), reprinted by permission of the publisher, The Electrochemical Society, Inc.]

the RDE, as well as variations of this mechanism, have also appeared (59–61). The treatment of the E_rC_i' case for the RRDE by digital simulation techniques showed that the results (i.e., plots of N_K *vs.* **XKT**) are indistinguishable from those of the E_rC_i case for first- or pseudo-first-order reactions (58).

11.4.5 $E_rC_iE_r$ Reactions

This mechanism [see (11.3.5)] has also been treated at several different levels of approximation for the RDE. Karp (62), using the treatment for ECE reactions in bulk electrolysis, proposed the following equation for the limiting current:

$$i = i_l\left[2 - \frac{\tanh(\delta/\mu)}{\delta/\mu}\right] \tag{11.4.15}$$

$$\frac{n_{\text{app}}}{n} = 2 - \frac{\tanh(\delta/\mu)}{\delta/\mu} \tag{11.4.16}$$

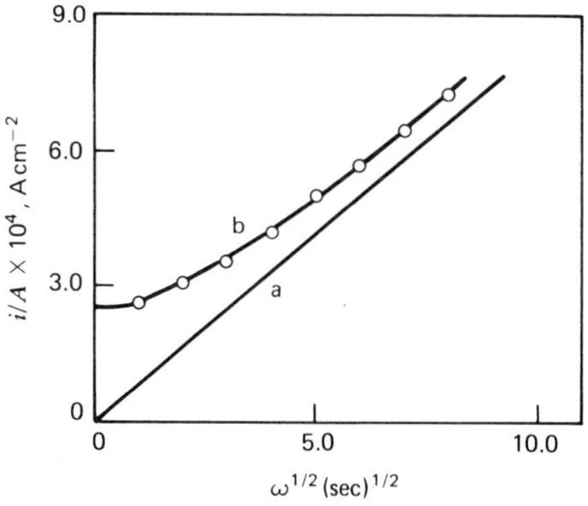

Figure 11.4.3

Simulated disk current for catalytic reaction (E_rC_i'). Curve a: In absence of following reaction. Curve b: In presence of following reaction with $k_2 = 145\ M^{-1}\ \text{sec}^{-1}$ and $C_Z^* = 10\ C_O^*$. ○ Experimental points for study of reduction of Fe^{3+} in presence of H_2O_2 (13). [From K. B. Prater and A. J. Bard, *J., Electrochem. Soc.*, **117**, 1517 (1970), reprinted by permission of the publishers, The Electrochemical Society, Inc.]

where $n_1 = n_2 = n$, i_l is the mass-transfer-limiting current given in (8.3.21), and δ and μ are as defined in (11.4.4) and (11.4.5). Filinovskii (63) proposed the equation

$$i = 0.94\, i_l \left[2 - \frac{(1 + \delta^2/1.9\, \mu^2)^{1/2}}{1 + \delta^2/\mu^2} \right] \tag{11.4.17}$$

Digital simulations of this reaction scheme at the RDE were also carried out (58, 64); these included cases with a second-order intervening reaction and the possibility of the homogeneous reaction between R_1 and O_2. A comparison of the results of the different treatments is given in Figure 11.4.4. A digital simulation of the RRDE behavior for this case for first- and second-order intervening reactions has been reported (58). As in other RRDE treatments of reactions with kinetic complications, the results depend on the electrode geometry. The variation of N_K with the kinetic parameters for a typical RRDE electrode (with an unperturbed collection efficiency, N, of 0.55) is shown in Figure 11.4.5.

11.5 SINE WAVE METHODS

Although there have been numerous theoretical treatments for the effects of coupled homogeneous chemical reactions on the measured faradaic impedance or ac polarographic response, these methods generally have not found very wide application to studies of chemical systems. This is probably mainly the result of the rather complicated and often cumbersome general equations that describe the measured ac response in terms of frequency (ω) and the kinetic parameters. In addition, the ac responses

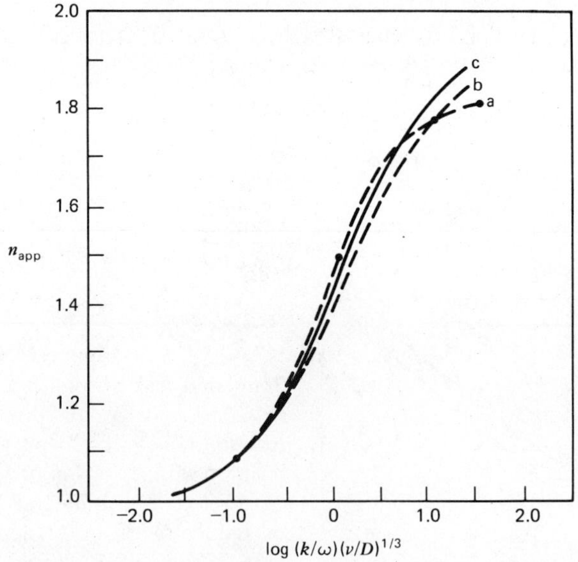

Figure 11.4.4
Variation of n_{app} with log $(k\omega^{-1}\nu^{1/3}D^{-1/3})$ [= log$(\delta^2/2.6\mu)$] for ECE mechanism. Curve a: Equation 11.4.17. Curve b: Equation 11.4.16. Curve c: Digital simulation (58).

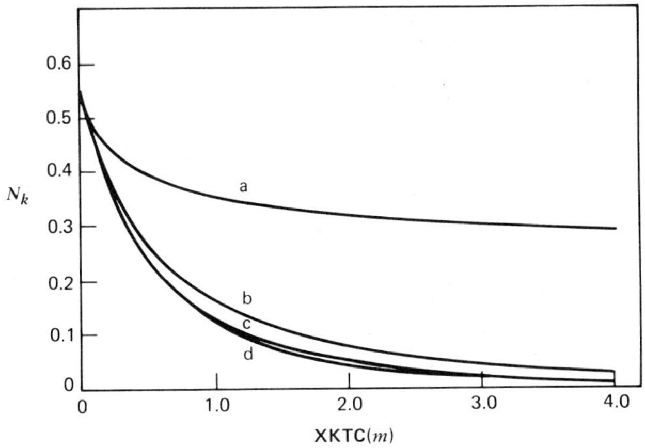

Figure 11.4.5

Variation of N_K with kinetic parameter for RRDE ($r_2/r_1 = 1.13$; $r_3/r_2 = 1.69$) for ECE mechanism $\mathbf{XKTC} = k_2C^*_{O_1}\omega^{-1}\nu^{1/3}D^{-1/3}(0.51)^{-2/3}$; $m = C^*_Z/C^*_{O_1}$, where intervening reaction is $R_1 + Z \xrightarrow{k_2} O_2 + Y$. Curve a: $m = 0.1$. Curve b: $m = 1.0$. Curve c: $m = 10$. Curve d: First order. [From K. B. Prater and A. J. Bard, *J. Electrochem. Soc.*, **117**, 1517 (1970), reprinted by permission of the publishers, The Electrochemical Society, Inc.]

such as ac current and phase angle, do not provide as clear a qualitative or semiquantitative feeling for the nature of the coupled reaction as do voltammetric responses. Finally the higher frequencies employed in ac methods frequently bring the electron transfer reactions from nernstian behavior into the quasi-reversible regime. While this can be helpful in determining these heterogeneous rate constants, it complicates the interpretation of the results and elucidation of the pathway of the overall reaction. However, ac methods have a high inherent accuracy and precision and, with the advent of computer-controlled data acquisition and Fourier transform techniques, they are capable of defining an electrode reaction over a wide frequency range in a rather short time. In this section, we will only discuss the results for one typical case, the C_rE_r reaction, studied by ac methods. They will illustrate the general approach and method of data treatment. Several extensive reviews on this subject are available (65–67).

11.5.1 The Faradaic Impedance

We consider the C_rE_r case (see Section 11.3.1), where the electron transfer reaction is nernstian with respect to the dc response but not necessarily with respect to the ac response. An equivalent circuit for the electrode reaction for the system uncomplicated by associated homogeneous reactions has been given (Figure 9.1.10), and it was shown that the faradaic impedance could be given by its real (Z_f') and imaginary (Z_f'') components [see (9.2.25) and (9.2.26).:

$$Z_f' = R_s = R_{ct} + \sigma/\omega^{1/2} \tag{11.5.1}$$

$$Z_f'' = 1/\omega C_s = \sigma/\omega^{1/2} \tag{11.5.2}$$

where R_{ct} represents the charge transfer resistance [(9.3.2)] and σ is a mass transfer parameter that depends on the diffusion coefficients and concentrations [e.g., see (9.3.6)], with the dc potential at its equilibrium value:

$$\sigma = \sigma_{\mathrm{O}} + \sigma_{\mathrm{R}} \tag{11.5.3}$$

$$\sigma_{\mathrm{O}} = \frac{RT}{n^2F^2A\sqrt{2}}\left(\frac{1}{D_{\mathrm{O}}^{1/2}C_{\mathrm{O}}^*}\right) \qquad \sigma_{\mathrm{R}} = \frac{RT}{n^2F^2A\sqrt{2}}\left(\frac{1}{D_{\mathrm{R}}^{1/2}C_{\mathrm{R}}^*}\right) \tag{11.5.4}$$

The presence of coupled chemical reactions adds additional complex impedances, which are functions of the rate constants, to the equivalent circuit. For the $\mathrm{C_rE_r}$ reaction the components of the faradaic impedance are (65, 67–69)

$$\boxed{Z_{\mathrm{f}}' = R_{ct} + \sigma_{\mathrm{R}}\omega^{-1/2} + \frac{K}{1+K}\sigma_{\mathrm{O}}\omega^{-1/2} + \frac{1}{1+K}\sigma_{\mathrm{O}}\left[\frac{(\omega^2+k^2)^{1/2}+k}{\omega^2+k^2}\right]^{1/2}} \tag{11.5.5}$$

$$\boxed{Z_{\mathrm{f}}'' = \sigma_{\mathrm{R}}\omega^{-1/2} + \frac{K}{1+K}\sigma_{\mathrm{O}}\omega^{-1/2} + \frac{1}{1+K}\sigma_{\mathrm{O}}\left[\frac{(\omega^2+k^2)^{1/2}-k}{\omega^2+k^2}\right]^{1/2}} \tag{11.5.6}$$

where $k = k_f + k_b$. Thus an analysis of Z_f' and Z_f'' as a function of ω and concentration can yield the rate constants for the preceding reaction as well as that for the charge transfer reaction. Let us examine some limiting cases (67): (a) At high frequencies, $\omega \gg k$ (i.e., $\omega > 10\,k$), the final term in these equations becomes $\sigma_{\mathrm{O}}/(1 + K)\omega^{1/2}$ and Z_f' and Z_f'' take the values in (11.5.1) and (11.5.2). This allows determination of R_{ct} and σ as discussed in Section 9.3. (b) At low frequencies, $\omega \ll k$ (i.e., $\omega < 0.3k$), the equations become

$$Z_f' = R_{ct} + \sigma_{\mathrm{R}}\omega^{-1/2} + \frac{K}{1+K}\sigma_{\mathrm{O}}\omega^{-1/2} + \frac{1}{1+K}\sigma_{\mathrm{O}}\left(\frac{2}{k}\right)^{1/2} \tag{11.5.7}$$

$$Z_f'' = \sigma_{\mathrm{R}}\omega^{-1/2} + \frac{K}{1+K}\sigma_{\mathrm{O}}\omega^{-1/2} \tag{11.5.8}$$

These resemble the impedances of the original equivalent circuit in their frequency dependence, (11.5.1) and (11.5.2), with R_{ct} replaced by $R_{ct} + (1 + K)^{-1}\sigma_{\mathrm{O}}(2/k)^{1/2}$ and σ by $\sigma_{\mathrm{R}} + (K/1 + K)\sigma_{\mathrm{O}}$.

11.5.2 ac Polarography

The general case for the $\mathrm{C_rE_r}$ reaction in ac polarography has been treated (65, 66, 70), taking account of factors such as growth and curvature of the mercury drop, which influence to some extent the magnitude of the ac current. The phase angle is less sensitive to these factors, and the stationary plane electrode model apparently works quite well, even for the DME. The general equations for the ac polarographic responses [$I(\omega t)$ and $\cot \phi$] with coupled chemical reactions are (65, 70)

$$I(\omega t) = I_{\text{rev}} F_{\text{dc}}(t) \left(\frac{2}{V^2 + U^2} \right)^{1/2} \sin(\omega t + \phi) \tag{11.5.9}$$

$$\cot \phi = V/U \tag{11.5.10}$$

where I_{rev} is the ac current for a reaction that is nernstian on both the dc and ac time scales, (9.4.11); $F_{\text{dc}}(t)$ is a measure of the effect of the heterogeneous charge transfer on the dc process; and V and U are functions of k, K, k_s, ω, and E:

$$V = \frac{(2\omega D)^{1/2}}{k^0(e^{-\alpha a} + e^{\beta a})} + \frac{1}{1 + e^a} \left\{ e^a + \frac{K}{1 + K} + \frac{\omega^{1/2}}{1 + K} \left[\frac{(\omega^2 + k^2)^{1/2} + k}{\omega^2 + k^2} \right]^{1/2} \right\} \tag{11.5.11}$$

$$U = \frac{1}{1 + e^a} \left\{ e^a + \frac{K}{1 + K} + \frac{\omega^{1/2}}{1 + K} \left[\frac{(\omega^2 + k^2)^{1/2} - k}{\omega^2 + k^2} \right]^{1/2} \right\} \tag{11.5.12}$$

$$a = \left(\frac{nF}{RT} \right) (E_{\text{dc}} - E_{1/2}) \tag{11.5.13}$$

Calculations based on these general equations have been given (70); typical results for the ac amplitude and $\cot \phi$ for totally nernstian conditions ($k^0 \to \infty$) (Figure 11.5.1) and nonnernstian conditions (Figure 11.5.2) are shown. It is again instructive to examine the limiting cases considered in Section 11.5.1 In the first case, where $\omega \gg k$, $U = 1$ and

$$V = 1 + \frac{(2\omega D)^{1/2}}{k^0(e^{-\alpha a} + e^{\beta a})} \tag{11.5.14}$$

Under these conditions, the $\cot \phi$ expression will be the same as that for a quasi-reversible reaction (9.4.23), allowing the determination of k^0. In the second case, where $k \gg \omega$,

$$U = \frac{1}{1 + e^a} \left(e^a + \frac{K}{1 + K} \right) \tag{11.5.15}$$

and

$$\cot \phi = 1 + \frac{(2\omega)^{1/2}}{U} \left[\frac{D^{1/2}}{k^0(e^{-\alpha a} + e^{\beta a})} + \frac{1}{(1 + K)(1 + e^a)k^{1/2}} \right] \tag{11.5.16}$$

Under these conditions a plot of $\cot \phi$ *vs.* ω will be linear, just as in the quasi-reversible case (Section 9.4), but the slope also depends on k and K.

The results here, when compared with those in Section 11.3.1, demonstrate the greater difficulty in applying ac methods to studies of coupled chemical reactions. The situation becomes even more complex when ECE-type reactions and the effects of adsorption are considered. At the present time ac techniques appear to be most useful where the reaction mechanism is already known, and curve-fitting procedures can then be applied (utilizing digital computer calculations from the theoretical equations) to find the rate constants.

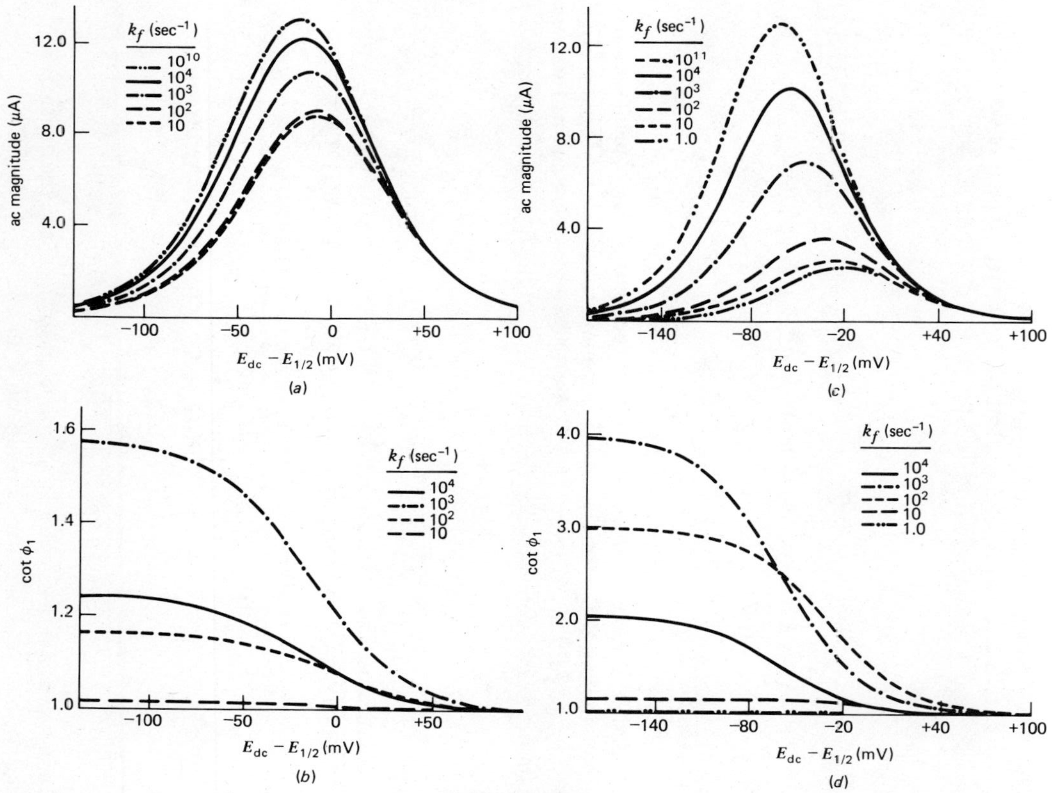

Figure 11.5.1

AC polarographic response for C_rE_r reaction with Nernstian electron transfer ($k^0 = \infty$). (*a*, *c*) AC Polarograms. (*b*, *d*) Phase angles calculated for $n = 1$, $T = 25\ °C$, $A = 3.50 \times 10^{-2}\ cm^2$; $\Delta E = 5.00$ mV, $C^* = 1.00 \times 10^{-3}\ M$, $D = 1.0 \times 10^{-5}\ cm^2/sec$, $\omega = 628\ sec^{-1}$. (*a*, *b*) $K = 1.0$, t_{max}(drop time) = 12 sec. (*c*, *d*) $K = 0.1$, $t = 6.00$ sec. Values of k_f shown on figure. [Reprinted with permission from T. G. McCord and D. E. Smith, *Anal. Chem.*, **41**, 116 (1969). Copyright 1969, American Chemical Society.]

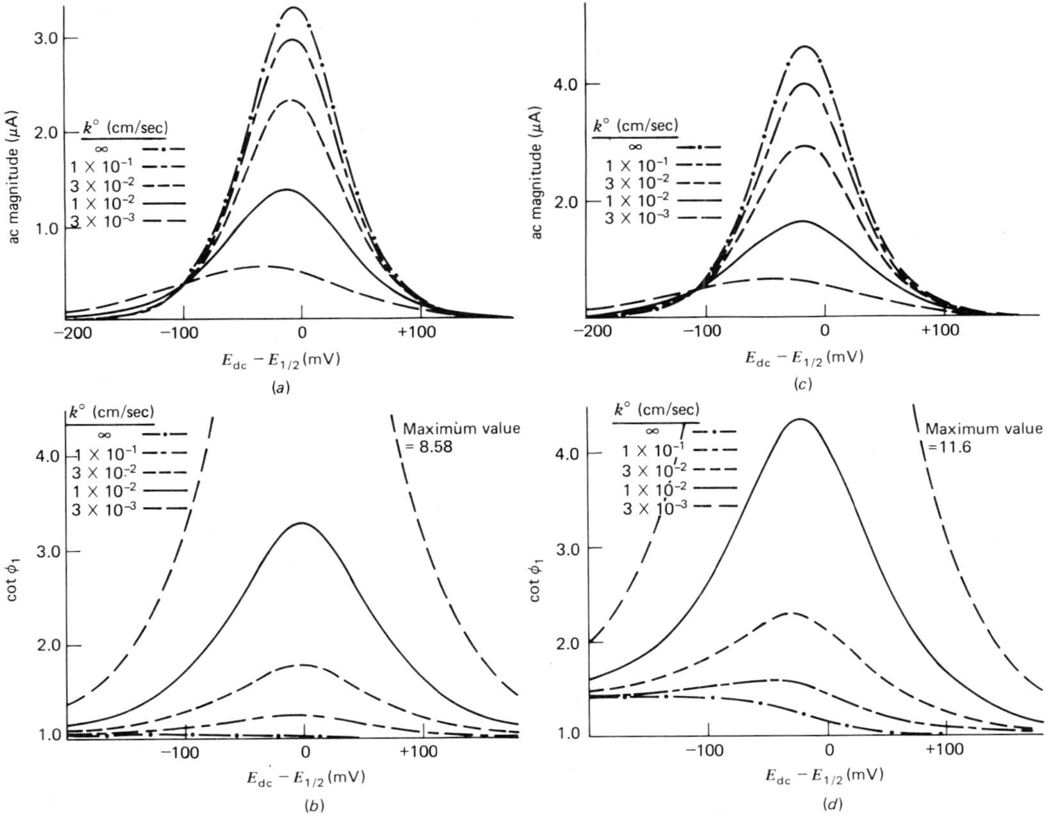

Figure 11.5.2

AC polarographic response for C_rE_r reaction with nonnernstian (ac time scale) electron transfer conditions as in Figure 11.5.1, except $K = 1.0$, $t = 3.0$ sec, $\alpha = 0.5$. (*a, b*) $k_f = 5.0 \text{ sec}^{-1}$. (*c, d*) $k_f = 500 \text{ sec}^{-1}$. k^0 values are given on figure. [Reprinted with permission from T. G. McCord and D. E. Smith, *Anal. Chem.*, **41**, 116 (1969). Copyright 1969, American Chemical Society.]

11.6 CONTROLLED POTENTIAL COULOMETRIC METHODS

The bulk electrolysis, controlled potential coulometric methods described in Section 10.3.4 are especially useful for examining the effects of slower reactions coupled to the electron transfer reaction. Since the time window of such methods is about 100 to 3000 sec, reactions with first-order rate constants of the order of 10^{-2} to 10^{-4} sec^{-1} can be studied. Moreover, by analysis of the solution following electrolysis (e.g., by spectroscopic, chromatographic, or electrochemical methods), the products of the reactions, and hence the overall reaction scheme, can be determined. The experiments are usually carried out at potentials corresponding to the limiting current plateau, so that the kinetics of the electron transfer reactions do not enter the analysis of results. Finally, the theoretical treatments for this technique and the analysis of the experimental results are frequently much simpler than those for the voltammetric methods.

11.6.1 Theoretical Treatments

The theoretical treatments for the different reaction schemes involve ordinary (rather than partial) differential equations, because the electrolysis solution is assumed to be essentially homogeneous (see Section 10.3.1). The concentrations are functions of time during the bulk electrolysis, but not of x. The measured responses in coulometry are the i-t curves and the quantity of electricity passed during the electrolysis, $Q(t)$, per mole of electroactive compound consumed, n_{app}, is

$$n_{app} = \frac{Q(t)}{FN_O} = \frac{Q(t)}{FV[C_O^*(0) - C_O^*(t)]} \tag{11.6.1}$$

where N_O is the moles of O consumed, $C_O^*(0)$, the initial concentration of O, $C_O^*(t)$, the concentration at time, t, and V, the total volume of solution. In coulometric experiments under limiting current conditions, electrode reactions are treated as first-order chemical reactions [see (10.3.5)] so that for the electron transfer step $O + ne \rightarrow R$,

$$\frac{dC_O^*(t)}{dt} = -pC_O^*(t) \tag{11.6.2}$$

where $p = m_O A/V$; m_O is the mass transfer constant for O (cm/sec), which is a function of D and the convective conditions during the electrolysis, and A is the electrode area (cm^2). The current at any time is calculated using (10.3.1):

$$i(t) = nFAm_O C_O^*(t) \tag{11.6.3}$$

As an example of a theoretical treatment involving a coupled reaction, let us consider the catalytic reaction (E_rC_i'):

$$O + ne \underset{}{\overset{p}{\rightleftharpoons}} R \tag{11.6.4}$$

$$R + Z \xrightarrow{k'} O + Y \tag{11.6.5}$$

where we assume $C_Z^*(0) \gg C_O^*(0)$, so that (11.6.5) represents a pseudo-first-order reaction with $k = k'C_Z^*(0)$. The equations governing the system are

$$\frac{dC_O^*(t)}{dt} = -pC_O^*(t) + kC_R^*(t) \tag{11.6.6}$$

$$\frac{dC_R^*(t)}{dt} = pC_O^*(t) - kC_R^*(t) \tag{11.6.7}$$

$$C_O^*(t) = C_O^*(0) = C_i \qquad \text{and} \qquad C_R^*(t) = 0 \text{ at } t = 0$$

From (11.6.6) and (11.6.7),

$$C_R^*(t) = C_i - C_O^*(t) \tag{11.6.8}$$

Substitution of this value of $C_R^*(t)$ into (11.6.6) and integration of the resulting equation yields

$$C_O^*(t) = C_i\left\{\frac{\gamma + \exp[-p(1+\gamma)t]}{1+\gamma}\right\} \tag{11.6.9}$$

where $\gamma = k/p$. The *i-t* behavior is obtained by combining (11.6.3) and (11.6.9) to yield

$$\boxed{\frac{i(t)}{i(0)} = \frac{\gamma + \exp[-p(1+\gamma)t]}{1+\gamma}} \tag{11.6.10}$$

Note that in this case the current does not decay to zero (or the background level), as in the case of the unperturbed reaction ($k \to 0$), but instead decays to a steady-state value, i_{ss}, where

$$\frac{i_{ss}}{i(0)} = \frac{\gamma}{(1+\gamma)} \tag{11.6.11}$$

(Figure 11.6.1). Thus γ can be determined from $i_{ss}/i(0)$. Since the current does not decay to background, no time-independent n_{app} value is attained for the EC′ case, and n_{app} keeps increasing during the electrolysis. From (11.6.1) and (11.6.10) and the fact that

$$Q(t) = \int_0^t i(t)\,dt \tag{11.6.12}$$

the n_{app} equation is derived:

$$n_{app} = n\left[\frac{1}{1+\gamma} + \frac{p\gamma t}{1 - e^{-p(1+\gamma)t}}\right] \tag{11.6.13}$$

The diagnostic criteria for the EC′ case in coulometry are thus a current that does not decay to background and a continually increasing n_{app} (with n_{app} values that are larger than expected). Further details about EC′ reactions in coulometry are available (71–73). We will now briefly consider the results of theoretical treatments for several cases of coupled chemical reactions in coulometry. Rather detailed reviews on this area have appeared (74, 75).

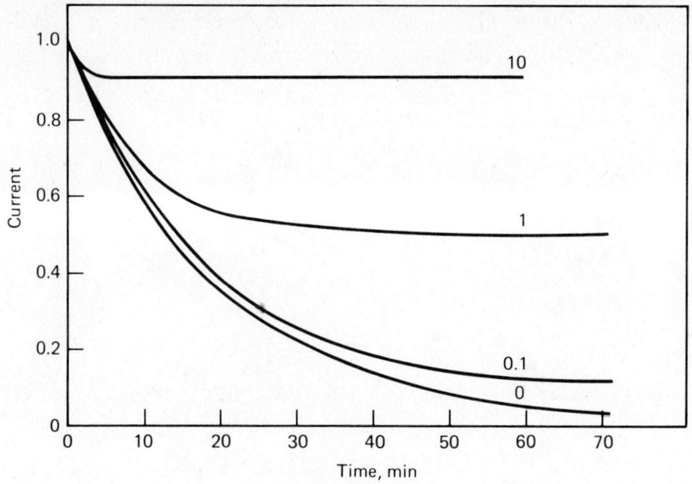

Figure 11.6.1

Current-time behavior for catalytic reaction case; $p = 0.05\ \text{min}^{-1}$ and k/p values indicated on the curves. [From A. J. Bard and K. S. V. Santhanam, *Electroanal. Chem.*, **4**, 215 (1970), courtesy of Marcel Dekker, Inc.]

11.6.2 Preceding Reaction—C_rE_r

When the electroactive species is generated by a preceding reaction [see (11.3.1) to (11.3.3)], the n_{app} value for complete electrolysis is unperturbed, that is, $n_{app} = n$. However, the i-t behavior is changed from the simple exponential decay behavior, because the current is partially controlled by the rate of conversion of Y to O. The current is given by (76, 77)

$$\boxed{\frac{i}{nFVp(C_O)_i} = C_1\, e^{-(L-G)t} + (1 - C_1)e^{-(L+G)t}} \qquad (11.6.14)$$

where

$$L = 0.5(k_f + k_b + p)$$
$$G = (L^2 - k_f p)^{1/2}$$
$$C_1 = G + L - p/2G$$
$$(C_O)_i = \frac{K}{1+K} C^*$$

Typical log i-t curves are shown in Figure 11.6.2. Their shapes can be explained as follows. The initial current is governed by the equilibrium concentration of O, $(C_O)_i$. The initial decay is governed by the rate of electrolysis of existing O, hence p. The limiting rate is determined by the rate of conversion of Y to O, hence k_f.

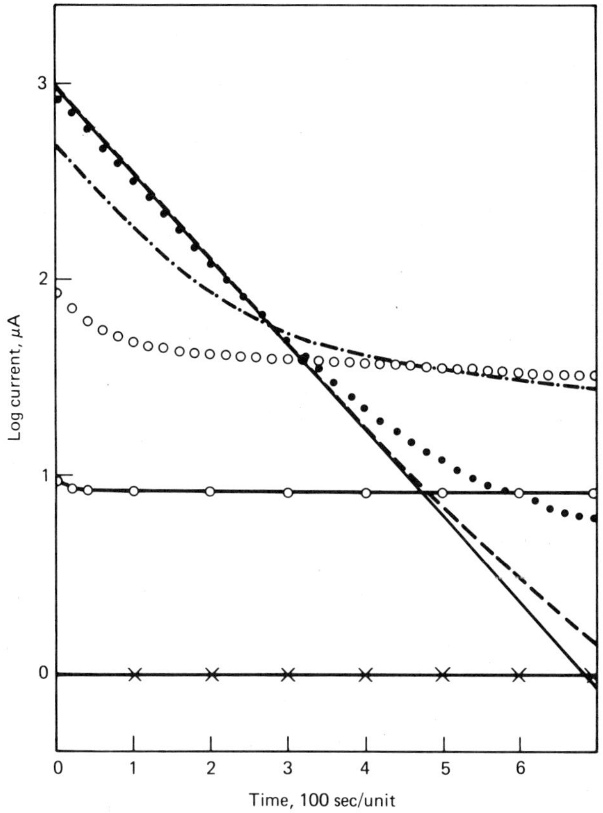

Figure 11.6.2
Current-time behavior for preceding reaction for different values of $K = k_f/k_b$, calculated for $(C_O)_i + (C_Y)_i = 10^{-4}$ M, $p = 0.01$ sec^{-1}, and $k_f = 10^{-3}$ sec^{-1}. (×—×) $K = 0.001$; (○—○) $K = 0.01$; (○) $K = 0.1$; (·—·—) $K = 1.0$; (····) $K = 10.0$; (----) $K = 100.0$; (——) $K = \infty$. [Reprinted with permission from A. J. Bard and E. Solon, *J. Phys. Chem.*, **67**, 2326 (1963). Copyright 1963, American Chemical Society.]

11.6.3 Following Reaction—E_rC_i; Reversal Coulometry

An irreversible following reaction does not perturb the forward *i-t* curve or the n_{app} value. However, by stepping the potential after the forward electrolysis ($O + ne \rightarrow R$) has proceeded for a time, t, to a value where the back oxidation occurs ($R \rightarrow O + ne$) and by determining the relative amounts of electricity passed during the forward step (Q_f) and the reversal step (Q_b), information about the rate of the following reaction ($R \xrightarrow{k} Y$) can be obtained. For the forward electrolysis (78),

$$Q_f = nFVC_O^*(0)[1 - \exp(-pt_1)] \tag{11.6.15}$$

This equation yields (10.3.11) as $t_1 \rightarrow \infty$. For the back electrolysis carried out for a time, t_2, in the absence of the perturbing reaction,

$$Q_b/Q_f = 1 - \exp[-p(t_2 - t_1)] \tag{11.6.16}$$

Table 11.6.1

Diagnostic Criteria in Controlled Potential Coulometry for Different Reaction Mechanisms

Reaction Mechanism	n_{app} [b]	Reversal Coulometry[c]	$\log i - t$ [d]
No kinetic effects $O + ne \rightarrow R$	$n_{app} = n$	$Q_b^0 = Q_f$	Linear, slope p
Catalytic reaction $O + ne \rightarrow R$ $R + Z \xrightarrow{k} O$	$n_{app} > n$	$0 \leqq Q_b^0 < Q_f$	Concave upward; current reaches steady-state value. If the electrolysis is interrupted and resumed, current will be higher than value at interruption.
$O + ne \rightarrow R$ $R + Z \xrightarrow{k_1} O$ $R + X \xrightarrow{k_2} Y$	$n_{app} > n$ $n_{app} = f(C_O)_t$	$0 \leqq Q_b^0 < Q_f$	Linear or concave downward
Preceding reaction $Y \underset{k_b}{\overset{k_f}{\rightleftharpoons}} O$	$n_{app} = n$	$Q_b^0 = Q_f$	Concave upward
Coupling reaction $O + ne \rightarrow R$ $R + O \xrightarrow{k} Y$	$n/2 < n_{app} < n$ $n_{app} = f(C_O)_t$	$Q_b^0 < Q_f$	Linear slope $> p$
$O + ne \rightarrow R$ $R + O \xrightarrow{k_1} Y$ $R + Z \xrightarrow{k_2} X$	$n/2 < n_{app} < n$ $n_{app} = f(C_O)_i$	$Q_b^0 < Q_f$	Linear slope $> p$
Competing reaction $O + ne \rightarrow R$ $O + Z \xrightarrow{k} X$	$n_{app} < n$ n_{app} is a function of time between mixing of solution and electrolysis, and $(C_O)_i$	$Q_b^0 = Q_f$	Linear slope $> p$

Following reaction			
$O + ne \rightarrow R$	$n_{app} = n$	$Q_b^0 < Q_f$	Linear slope = p
$R \xrightarrow{k} Y$			
or			
$R \underset{k_b}{\overset{k_f}{\rightleftharpoons}} Y$			
or			
$2R \xrightarrow{k} X$			
Consecutive reaction			
$O \xrightarrow{+n_1e} I \xrightarrow{+n_2e} R$	$n_{app} = n_1 + n_2$	$Q_b^0 = Q_f$	Linear or concave upward
$O \xrightarrow{+n_1e} I \xrightarrow{+n_2e} R$; $I \xrightarrow{k} X$	$n_1 < n_{app} < n_1 + n_2$	$Q_b^0 < Q_f$	Linear or concave upward
ECE reactions			
$O_1 \xrightarrow{n_1e} R_1 \xrightarrow{k_1} O_2 \xrightarrow{n_2e} R_2$	$n_{app} = n_1 + n_2$	$Q_b^0 = Q_f\left(\frac{n_2}{n_1 + n_2}\right)$ (irreversible intervening reaction)	Linear, concave upward or downward
$O_1 \xrightarrow{n_1e} R_1 \xrightarrow{k_1} O_2 \xrightarrow{n_2e} R_2$; $R_1 \xrightarrow{k_2} Z$	$n_1 < n_{app} < n_1 + n_2$	$Q_b^0 < Q_f\left(\frac{n_2}{n_1 + n_2}\right)$	Linear, concave upward or downward
$O_1 \xrightarrow{n_1e} R_1 \xrightarrow{k_1} O_2 \xrightarrow{n_2e} R_2$	$n_1 < n_{app} < n_1 + n_2$	$Q_b^0 < Q_f\left(\frac{n_2}{n_1 + n_2}\right)$	Linear, concave upward or downward
$2R_1 \xrightarrow{k_2} X$	$n_{app} = f(C_O)_i$	(irreversible intervening reactions)	

[a] From A. J. Bard and K. S. V. Santhanam, *Electroanal. Chem.*, **4**, 215–315 (1970) with permission.
[b] Where n_{app} is a function of initial concentration of O, $(C_O)_i$, it is so indicated.
[c] Assuming all electrochemical steps are reversible.
[d] Current decays to background, unless noted otherwise.

Clearly when $(t_2 - t_1) \rightarrow \infty$, $Q_b^0 = Q_f$. In the presence of the following reaction,

$$Q_b = \frac{nFVpC_R(t_1)}{(p + k)}\{1 - \exp[-(p + k)(t_2 - t_1)]\} \tag{11.6.17}$$

where $C_R(t_1)$ is the concentration of R at the start of the reverse electrolysis and is given by

$$C_R(t_1) = \frac{pC_O^*(0)}{(k - p)}[\exp(-pt_1) - \exp(-kt_1)] \qquad k \neq p \tag{11.6.18a}$$

$$C_R(t_1) = pC_O^*(0)t_1 \exp(-pt_1) \qquad k = p \tag{11.6.18b}$$

The rate constant k can be evaluated from (11.6.17) and (11.6.18) or, more simply, by allowing the back reaction to proceed to completion to yield Q_b^0 [when $(t_2 - t_1) \rightarrow \infty$]. Under these conditions,

$$\boxed{\frac{Q_b^0}{Q_f} = \frac{p^2}{k^2 - p^2}\left\{\frac{1 - \exp[(p - k)t_1]}{\exp(pt_1) - 1}\right\}} \qquad k \neq p \tag{11.6.19a}$$

$$\boxed{\frac{Q_b^0}{Q_f} = \frac{pt_1}{2[\exp(pt_1) - 1]} \qquad k = p} \tag{11.6.19b}$$

Similar treatments for reversal coulometry with reversible reactions and dimerizations following the electron transfer reaction have also been presented (78).

11.6.4 $E_rC_iE_r$ Reactions

As with voltammetric methods (Section 11.3.6) the ECE reaction where $\Delta E^0 > 0$ is characterized by an n_{app} value of n_1 at short times, when the intervening reaction cannot occur to an appreciable extent, and of $n_1 + n_2$ at longer times. The *i-t* curve for this case, assuming the same p value for each electron transfer step, is (77, 79, 80)

$$\frac{i}{FVC_O^*(0)p\exp(-pt)} = n_1 + \frac{n_2pk}{k - p}\left\{t - \frac{[1 - e^{-(k-p)t}]}{k - p}\right\} \tag{11.6.20}$$

Typical *i-t* curves for different values of k/p are shown in Figure 11.6.3. When $k/p \ll 1$ and $kt_1 \ll 1$, $n_{app} = n_1$ and the simple exponential decay of current is observed. For long electrolysis times, $n_{app} = n_1 + n_2$. When there are additional reactions that occur, the n_{app} values can be nonintegral even for long electrolysis times. For example, if the intermediate R_1 decomposes to a nonelectroactive species, Z, then

$$R_1 \xrightarrow{k_2} Z \tag{11.6.21}$$

in addition to its decomposition to R_2 (11.3.38); then (79)

$$n_{app}^0 = n_1 + n_2\frac{1}{1 + (k_2/k)} \tag{11.6.22}$$

This reaction path would lead to nonintegral n_{app}^0 values.

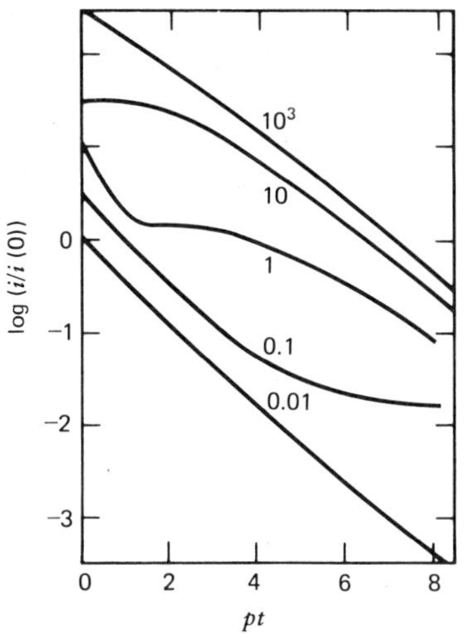

Figure 11.6.3
Variation of $i/i(0)$ with t for ECE mechanism with $p = 10^{-3}$ sec^{-1} for various values of k/p shown on each curve. The ordinate scale pertains to the lowermost curve, for which $k/p = 0.01$. Each successive curve above this is shifted upward 0.5 unit with respect to the ordinate axis. [From S. Karp and L. Meites, *J. Electroanal. Chem.*, **17**, 253 (1968), with permission.]

A number of other reaction schemes in coulometry have been treated (74, 75). The diagnostic criteria for a number of different reaction mechanisms are given in Table 11.6.1 (74).

11.7 REFERENCES

1. Z. Galus, "Fundamentals of Electrochemical Analysis," E. Harwood, Ltd., Chichester, 1976, pp. 255–395.
2. P. Delahay, "New Instrumental Methods in Electrochemistry," Wiley-Interscience, New York, 1954.
3. I. M. Kolthoff and J. J. Lingane, "Polarography," 2nd ed., Wiley-Interscience, New York, 1952.
4. D. D. Macdonald, "Transient Techniques in Electrochemistry," Plenum, New York, 1977.
5. D. Pletcher, *Chem. Soc. Rev.*, **4**, 471 (1975).
6. D. H. Evans, *Accts. Chem. Res.*, **10**, 313 (1977).
7. A. C. Testa and W. H. Reinmuth, *Anal. Chem.*, **33**, 1320 (1961).
8. C. Amatore and J. M. Saveant, *J. Electroanal. Chem.*, **85**, 27 (1977).
9. W. Smith and A. J. Bard, *J. Am. Chem. Soc.*, **97**, 5203 (1975).
10. A. C. Testa and W. H. Reinmuth, *Anal. Chem.*, **32**, 1512 (1960).
11. O. Dracka, *Collect. Czech. Chem. Communs.*, **25**, 338 (1960).
12. H. B. Herman and A. J. Bard, *Anal. Chem.*, **36**, 510 (1964).
13. D. A. Tryk and S. M. Park, *ibid.*, **51**, 585 (1979).
14. J. M. Saveant and E. Vianello, *Electrochim. Acta*, **8**, 905 (1963); **12**, 629 (1967).
15. J. Koutecky and R. Brdicka, *Collect. Czech. Chem. Communs.*, **12**, 337 (1947).
16. J. Koutecky and J. Cizek, *ibid.*, **21**, 836 (1956).
17. J. Koutecky, *ibid.*, **19**, 857 (1954).
18. R. S. Nicholson and I. Shain, *Anal. Chem.*, **36**, 706 (1964).

19. P. Delahay and T. Berzins, *J. Am. Chem. Soc.*, **75**, 2486 (1953).
20. W. H. Reinmuth, *Anal. Chem.*, **33**, 322 (1961).
21. A. J. Bard and H. B. Herman, "Polarography—1964," G. J. Hills, Ed., Wiley-Interscience, New York, 1966, p. 373.
22. W. M. Schwarz and I. Shain, *J. Phys. Chem.*, **69**, 30 (1965).
23. W. V. Childs, J. T. Maloy, C. P. Keszthelyi, and A. J. Bard, *J. Electrochem. Soc.*, **118**, 874 (1971).
24. M. L. Olmstead and R. S. Nicholson, *Anal. Chem.*, **41**, 851 (1969).
25. J. M. Saveant and E. Vianello, *Electrochim. Acta*, **12**, 1545 (1967).
26. M. L. Olmstead, R. T. Hamilton, and R. S. Nicholson, *Anal. Chem.*, **41**, 260 (1969).
27. R. S. Nicholson, *ibid.*, **37**, 667 (1965).
28. C. P. Andrieux, L. Nadjo, and J. M. Saveant, *J. Electroanal. Chem.*, **26**, 147 (1970).
29. D. H. Evans, *J. Phys. Chem.*, **76**, 1160 (1972).
30. L. Nadjo and J. M. Saveant, *J. Electroanal. Chem.*, **48**, 113 (1973).
31. P. Delahay and G. L. Steihl, *J. Am. Chem. Soc.*, **74**, 3500 (1952).
32. S. L. Miller, *ibid.*, **74**, 4130 (1952).
33. Z. Pospisil, *Collect. Czech. Chem. Communs.*, **12**, 39 (1947).
34. P. Delahay, C. C. Mattax, and T. Berzins, *J. Am. Chem. Soc.*, **76**, 5319 (1954).
35. O. Fischer, O. Dracka, and E. Fischerova, *Collect. Czech. Chem. Communs.*, **26**, 1505 (1961).
36. C. Furlani and G. Morpurgo, *J. Electroanal. Chem.*, **1**, 351 (1960).
37. G. S. Alberts and I. Shain, *Anal. Chem.*, **35**, 1859 (1963).
38. J. Koutecky, *Collect. Czech. Chem. Communs.*, **18**, 183 (1953).
39. R. S. Nicholson, J. M. Wilson, and M. L. Olmstead, *Anal. Chem.*, **38**, 542 (1966).
40. J. M. Saveant, *Bull. Chem. Soc. France*, **1967**, 91.
41. R. S. Nicholson and I. Shain, *Anal. Chem.*, **37**, 178 (1965).
42. J. M. Saveant, *Electrochim. Acta*, **12**, 753 (1967).
43. H. B. Herman and A. J. Bard, *J. Phys. Chem.*, **70**, 396 (1966).
44. M. Mastragostino, L. Nadjo, and J. M. Saveant, *Electrochim. Acta*, **13**, 721 (1968).
45. J. M. Saveant, C. P. Andrieux, and L. Nadjo, *J. Electroanal. Chem.*, **41**, 137 (1973).
46. M. K. Hanafey, R. L. Scott, T. H. Ridgway, and C. N. Reilley, *Anal. Chem.* **50**, 116 (1978).
47. S. W. Feldberg, *J. Phys. Chem.*, **75**, 2377 (1971).
48. J. Koutecky and V. G. Levich, *Zhur. Fiz. Khim.*, **32**, 1565 (1958); *Dokl. Akad. Nauk SSSR*, **117**, 441 (1957).
49. W. Vielstich and D. Jahn, *Z. Elektrochem.*, **32**, 2437 (1958).
50. H. Matsuda, *J. Electroanal. Chem.*, **35**, 77 (1972).
51. L. K. J. Tong, K. Liang, and W. R. Ruby, *ibid.*, **13**, 245 (1967).
52. S. A. Kabakchi and V. Yu. Filinovskii, *Electrokhim.*, **8**, 1428 (1972).
53. W. J. Albery and S. Bruckenstein, *Trans. Faraday Soc.*, **62**, 1946 (1966).
54. W. J. Albery, M. L. Hitchman, and J. Ulstrup, *ibid.*, **64**, 2831 (1968).

55. W. J. Albery and M. L. Hitchman, "Ring-Disc Electrodes," Clarendon Press, Oxford, 1971.
56. K. B. Prater and A. J. Bard, *J. Electrochem. Soc.*, **117**, 335 (1970).
57. W. J. Albery and S. Bruckenstein, *Trans. Faraday Soc.*, **62**, 2584 (1966).
58. K. B. Prater and A. J. Bard, *J. Electrochem. Soc.*, **117**, 1517 (1970).
59. P. Beran and S. Bruckenstein, *J. Phys. Chem.*, **72**, 3630 (1968).
60. D. Haberland and R. Landsberg, *Ber. Bunsenges. Phys. Chem.*, **70**, 724 (1966).
61. F. Kermiche-Aouanouk and M. Daguenet, *Electrochim. Acta*, **17**, 723 (1972).
62. S. Karp, *J. Phys. Chem.*, **72**, 1082 (1968).
63. V. Yu. Filinovskii, *Electrokhim.*, **5**, 635 (1969).
64. L. S. Marcoux, R. N. Adams, and S. W. Feldberg, *J. Phys. Chem.*, **73**, 2611 (1969).
65. D. E. Smith, *Electroanal. Chem.*, **1**, 1 (1966).
66. D. E. Smith, *CRC Crit. Rev. Anal. Chem.*, **2**, 247 (1971).
67. M. Sluyters-Rehbach and J. H. Sluyters, *Electroanal. Chem.*, **4**, 1 (1970).
68. H. Gerischer, *Z. Physik. Chem.*, **198**, 286 (1951).
69. H. Matsuda, P. Delahay, and M. Kleinerman, *J. Am. Chem. Soc.*, **81**, 6379 (1979).
70. T. G. McCord and D. E. Smith, *Anal. Chem.*, **41**, 116 (1969).
71. D. H. Geske and A. J. Bard, *J. Phys. Chem.*, **63**, 1957 (1959).
72. L. Meites and S. A. Moros, *Anal. Chem.*, **31**, 23 (1959).
73. G. A. Rechnitz and H. A. Laitinen, *Anal. Chem.*, **33**, 1473 (1961).
74. A. J. Bard and K. S. V. Santhanam, *Electroanal. Chem.*, **4**, 215 (1970).
75. L. Meites in "Techniques of Chemistry," A. Weissberger and B. W. Rossiter, Eds., Wiley-Interscience, New York, 1971, Part IIA, Vol. I, Chap. IX.
76. A. J. Bard and E. Solon, *J. Phys. Chem.*, **67**, 2326 (1963).
77. R. I. Gelb and L. Meites, *ibid.*, **68**, 630 (1964).
78. A. J. Bard and S. V. Tatwawadi, *J. Phys. Chem.*, **68**, 2676 (1964).
79. A. J. Bard and J. S. Mayell, *ibid.*, **66**, 2173 (1962).
80. S. Karp and L. Meites, *J. Electroanal. Chem.*, **17**, 253 (1968).

11.8 PROBLEMS

11.1 Consider the following system:

$$A + e \rightleftharpoons B \qquad E^{0\prime} = -0.5 \text{ V } vs. \text{ SCE}$$

$$B \rightarrow C$$

$$C + e \rightleftharpoons D \qquad E^{0\prime} = -1.0 \text{ V } vs. \text{ SCE}$$

The half-life of B is 100 msec. Both charge transfer reactions have large values of k^0.

Draw the expected cyclic voltammograms for scans beginning at 0.0 V *vs.* SCE and reversing at −1.2 V. Show curves for rates of 50 mV/sec, 1 V/sec, and 20 V/sec.

11.2 Draw a rough quantitative graph of collection efficiency versus rotation rate for an RRDE at which the electrode reaction for the previous problem is carried out. Assume electrolysis occurs in the mass-transfer-limited region. The collection efficiency for Fe(II) $\rightleftharpoons$ Fe(III) + e is 0.45.

11.3 The data below were recorded from a series of cyclic voltammetric experiments designed to elucidate the mechanism of the electrode reaction involving reduction of a certain compound. Formulate a mechanism to explain the behavior of the diagnostic functions, then briefly rationalize as many of the trends in the data as you can in terms of your mechanism. The switching potential was held constant at −1.400 V *vs.* SCE.

Scan Rate, v V/sec	$E_{p/2}$ (cathodic) volts *vs.* SCE	$E_{p/2}$ (anodic) volts *vs.* SCE	$\frac{i_p \text{ (anodic)}}{i_p \text{ (cathodic)}}$	i_p (cathodic)/$v^{1/2}$ $\mu A\text{-sec}^{1/2}\ V^{-1/2}$
0.1	−1.253	−1.17	0.1	35
2.0	−1.260	−1.185	0.51	34.4
10	−1.265	−1.197	0.84	33.0
20	−1.270	−1.208	0.91	32.8
100	−1.271	−1.212	1.01	32.6
200	−1.270	−1.212	1.01	32.7

11.4 Consider a material A, which can be reduced at a DME to substance B. A 1 m*M* solution of A in acetonitrile shows a wave at −1.90 V *vs.* SCE. The wave has a slope of 60.5 mV at 25° and it gives $(I_d)_{\max} = 1.95$. When phenol is added to the system, the polarographic behavior changes:

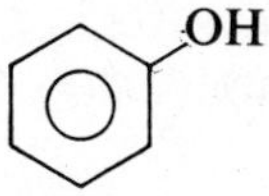

Phenol (C)

The following observations have been made:

Concentration of C, *M*	$E_{1/2}$	Wave Slope, mV	$(I_d)_{\max}$
10^{-3}	−1.88	61.4	2.81
10^{-2}	−1.85	61.9	3.85
10^{-1}	−1.82	61.8	3.87

Rationalize these facts. Can you suggest the chemical nature of A and B?

11.5 Cyclic voltammetry is often used to obtain information about standard potentials for correlative studies of molecular properties (e.g., ionization potentials, electron affinities, MO calculations, etc.). How is a standard potential extracted from data of this sort? Are assumptions involved? What kind of error would appear if the electrode process were EC? What if it involved slow electron transfer to a chemically stable product?

11.6 Derive the limiting values of $(\tau/\tau_d)^{1/2}$ for chronopotentiometry with a following catalytic reaction (E_rC_i') for large and small λ values; see (11.3.36). What is the slope of the line of $(\tau/\tau_d)^{1/2}$ *vs.* $\lambda^{1/2}$ in the KP region of Figure 11.3.21?

11.7 Derive the equation for the E-t curve for chronopotentiometry with a following catalytic reaction (E_rC_i'). See (11.3.34) and (11.3.35). Show that as $k \to 0$, the E-t curve for a nernstian electrode reaction is approached. What is the limiting (KP) behavior as $\lambda \to \infty$? Plot a curve showing how $E_{\tau/4}$ varies with λ.

11.8 The value of the limiting current, i_∞, in linear sweep voltammetry with an E_rC_i' reaction scheme (11.3.31) can be derived by noting that steady-state conditions apply, that is $[\partial C_O(x, t)/\partial t] = [\partial C_R(x, t)/\partial t] = 0$, and that $C_O(x, t) + C_R(x, t) = C_O^*$ (taking $D_O = D_R = D$). Do this derivation.

11.9 The controlled potential reduction of a 0.01 M M^{3+} solution in 1 M HCl produces M^{2+}. When the electrolysis is carried out in a volume of 100 cm^3 at a 50-cm^2 electrode with $m = 10^{-2}$ cm/sec, it is noticed that the current decays to a steady-state value, 24.5 mA, significantly higher than the pre-electrolysis residual current (500 μA) in 1 M HCl at this potential. This effect is attributed to the reaction

$$M^{2+} + H^+ \rightarrow M^{3+} + \tfrac{1}{2}H_2$$

which regenerates M^{3+}. What is the pseudo-first-order rate constant for this reaction? What is the steady-state concentration of M^{2+} in solution during electrolysis?

11.10 G. Costa, A. Puxeddu, and E. Reisenhofer (*J. Chem. Soc. Dalton*, **1973**, 2034) studied the cyclic voltammetric reduction of the complex Co^{II} (salen) in DMF at a mercury electrode [(salen) is the chelating ligand *N,N'*-ethylenebis(salicylideneiminate)]. In the absence of any additive for a 0.20 mM solution of complex, a typical reversible cyclic voltammogram results for the reaction

$$Co^{II}(salen) + e \rightleftharpoons Co^{I}(salen)^-$$

When ethyl bromide (EtBr) is added, the irreversible reaction below occurs:

$$Co^{I}(salen)^- + EtBr \xrightarrow{k} Et\text{-}Co^{III}(salen) + Br^-$$

where Et-Co^{III}(salen) is reduced at more negative potentials than Co^{II} (salen). To measure k, the ratio i_{pa}/i_{pc} was determined at concentrations of EtBr where the following reaction was pseudo-first-order, $k' = k$[EtBr]. When the time (τ) between $E_{1/2}$ and the switching potential, E_λ, was 32 msec, the i_{pa}/i_{pc} ratio was 0.7 at 0 °C with 13.3 mM EtBr. Calculate k and k'.

chapter 12

Double-Layer Structure and Adsorbed Intermediates in Electrode Processes

In Chapter 1, we introduced some elementary ideas about the double layer, including notions about its capacitance and its structure. In the remainder of the text to this point, we have made repeated references to its influence on electrode processes and electrochemical measurements. It is time now to delve into this aspect of electrochemical science in more detail. Here our goal is to examine the kinds of experimental measurements that can illuminate the structure of the double layer, as well as the important structural models, and their implications for electrode kinetics.

12.1 THERMODYNAMICS OF THE DOUBLE LAYER

A great deal of our knowledge about the double layer comes from measurements of macroscopic, equilibrium properties, such as interfacial capacitance and surface tension. In general, we are interested in the way in which these properties change with potential and with the activities of various species in the electrolyte. The next section will deal with experimental aspects in some detail. For the moment, we will concentrate on the theory that we use to suggest and interpret experiments. Since our concern now is with macroscopic, equilibrium properties, we can expect a thermodynamic treatment to describe the system rigorously without a postulated model. This is an important aspect, because it implies that we can obtain data that any successful structural model must rationalize.

We begin by developing the *Gibbs adsorption isotherm*, which describes interfaces in general, and from that we obtain the *electrocapillary equation* which describes the properties of electrochemical interfaces more particularly.

12.1.1 The Gibbs Adsorption Isotherm (1–4)

Suppose we have an interface of surface area A separating two phases α and β. A segment of the system is depicted in Figure 12.1.1. The region between the two solid lines is the interfacial zone, whose composition and properties concern us. To the right of BB′, there is pure phase β; and to the left of AA′, the system is purely α. Because intermolecular forces are exerted only over a short range, the interfacial zone is just a few hundred angstroms thick, and we can regard the interfacial perturbations on α and β as properties of a surface. The lines AA′ and BB′ can be defined anywhere, provided they contain all segments of the system that differ from the pure α and β phases as a result of the interfacial perturbation.

Let us now compare the real interfacial zone with an imaginary *reference* interfacial zone. In the reference zone, we will define a *dividing surface*, which is shown as the dotted line in Figure 12.1.1. The position we choose for the dividing surface is arbitrary and has no impact on the final results (see Problem 12.1), but it is convenient to think of it as coinciding with the actual interfacial surface. In this reference system, we imagine that the pure, unperturbed phase α extends to the dividing surface from the left, while the pure phase β extends to it from the right.

The reason for defining the reference system at all is that the properties of the interface are governed by *excesses* and *deficiencies* in the concentrations of components; that is, we are concerned with *differences* between the quantities of various species in the actual interfacial region, with respect to the quantities we would expect if the existence of the interface did not perturb the pure phases α and β. These differences are called *surface excess* quantities. For example, the surface excess in the number of moles of any species, such as potassium ions or electrons, would be

$$n_i^{\sigma} = n_i^{S} - n_i^{R} \tag{12.1.1}$$

where n_i^{σ} is the excess quantity and n_i^{S} and n_i^{R} are the numbers of moles of species i in the interfacial region for the actual system and the reference system, respectively. Surface excess quantities can be defined for any extensive variable.

One of those variables is the electrochemical free energy, which can be considered profitably in a general way. For the reference system, the free energy depends on the usual variables: temperature, pressure, and the molar quantities of all components. That is, $\overline{G}^{R} = \overline{G}^{R}(T, P, n_i^{R})$. The surface area has no impact on $\overline{G}^{R}$ because the interface does not perturb phases α and β. There is, therefore, no energy of interaction.

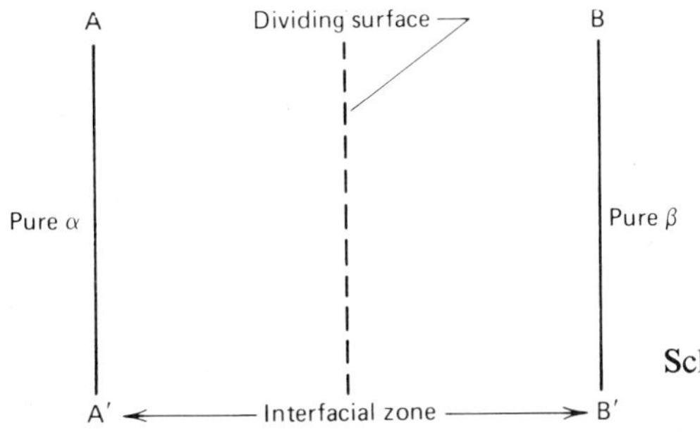

Figure 12.1.1
Schematic diagram of an interfacial region separating phases α and β.

On the other hand, we know from experience that real systems have a tendency to minimize or maximize the interfacial area; hence the free energy of the actual system $\bar{G}^S$ must depend on the area. Thus, $\bar{G}^S = \bar{G}^S(T, P, A, n_i^S)$.

The total differentials are

$$d\bar{G}^R = \left(\frac{\partial \bar{G}^R}{\partial T}\right) dT + \left(\frac{\partial \bar{G}^R}{\partial P}\right) dP + \sum_i \left(\frac{\partial \bar{G}^R}{\partial n_i^R}\right) dn_i^R \tag{12.1.2}$$

$$d\bar{G}^S = \left(\frac{\partial \bar{G}^S}{\partial T}\right) dT + \left(\frac{\partial \bar{G}^S}{\partial P}\right) dP + \left(\frac{\partial \bar{G}^S}{\partial A}\right) dA + \sum_i \left(\frac{\partial \bar{G}^S}{\partial n_i^S}\right) dn_i^S \tag{12.1.3}$$

We consider experiments only at constant temperature and pressure; hence the first two terms in each expression can be dropped. The partial derivatives $(\partial \bar{G}^R/\partial n_i^R)$ are the electrochemical potentials $\bar{\mu}_i$, which we encountered earlier in Section 2.2.4. Since equilibrium applies, this potential is constant throughout the system for any given species. Since it is the same in the interfacial zone as in the pure phases α and β, it must also be true that

$$\bar{\mu}_i = \left(\frac{\partial \bar{G}^R}{\partial n_i^R}\right) = \left(\frac{\partial \bar{G}^S}{\partial n_i^S}\right) \tag{12.1.4}$$

We can also give a name to the partial derivative $(\partial \bar{G}^S/\partial A)$. It is the *surface tension* γ.

Now we can write the differential *excess free energy* as

$$d\bar{G}^\sigma = d\bar{G}^S - d\bar{G}^R = \gamma dA + \sum_i \bar{\mu}_i\, d(n_i^S - n_i^R) \tag{12.1.5}$$

and from (12.1.1) we have

$$d\bar{G}^\sigma = \gamma\, dA + \sum_i \bar{\mu}_i\, dn_i^\sigma \tag{12.1.6}$$

This equation tells us that the interfacial free energy can be described (under our conditions of constant temperature and pressure) by the variables A and n_i, *all of which are extensive.* This feature allows us to invoke Euler's theorem,†, which yields

$$\bar{G}^\sigma = \left(\frac{\partial \bar{G}^\sigma}{\partial A}\right) A + \sum_i \left(\frac{\partial \bar{G}^\sigma}{\partial n_i^\sigma}\right) n_i^\sigma \tag{12.1.7}$$

$$\bar{G}^\sigma = \gamma A + \sum_i \bar{\mu}_i n_i^\sigma \tag{12.1.8}$$

Let us find the total differential, $d\bar{G}^\sigma$, from this expression.

$$d\bar{G}^\sigma = \gamma\, dA + \sum_i \bar{\mu}_i\, dn_i^\sigma + A\, d\gamma + \sum_i n_i^\sigma\, d\bar{\mu}_i \tag{12.1.9}$$

† We say that $\bar{G}$ is an *extensive function* of the *extensive variables* A and n_i. By this, we mean that the electrochemical free energy depends linearly on the physical extent of the system. If we double the size of the system by doubling A and all the n_i, then $\bar{G}$ doubles. Mathematically, such behavior implies that $\bar{G}(A, n_i)$ is a *linear homogeneous function* of A and n_i. The *Euler theorem* (5) applies generally to homogeneous functions and, for linear ones, it allows us to define the function itself in terms of derivatives and variables as in (12.1.7).

Clearly (12.1.6) and (12.1.9) must be equivalent, and hence the last two terms in (12.1.9) must sum to zero:

$$A\,d\gamma + \sum_i n_i^\sigma\, d\bar{\mu}_i = 0 \tag{12.1.10}$$

In general, it is more convenient to speak of excesses per unit area of surface; therefore, we now introduce the *surface excess concentration*, $\Gamma_i = n_i^\sigma/A$. Then (12.1.10) is reexpressed as

$$\boxed{-d\gamma = \sum_i \Gamma_i\, d\bar{\mu}_i} \tag{12.1.11}$$

which is the Gibbs adsorption isotherm. It already hints that measurements of surface tension will play an important role in elucidating interfacial structure, but to see the experimental ramifications we need to specialize it to an electrochemical situation. That is our next job.

12.1.2 The Electrocapillary Equation (1–4)

Let us now consider a specific chemical system in which a mercury surface contacts an aqueous KCl solution. The potential of the mercury is controlled with respect to a reference electrode having no liquid junction with the test solution. Suppose also that the aqueous phase contains a neutral species M, that might be interfacially active. For example, the cell could be

$$\text{Cu}'/\text{Ag}/\text{AgCl}/\text{K}^+, \text{Cl}^-, \text{M}/\text{Hg}/\text{Ni}/\text{Cu} \tag{12.1.12}$$

We will focus on the interface between the mercury electrode and the aqueous solution.

In writing the Gibbs adsorption isotherm for this case, it is useful to group terms relating separately to components of the mercury electrode, ionic components of the solution, and neutral components of the solution. Since we can obtain excesses of charge on the electrode surface, we need to consider a surface excess of electrons on the mercury. It could be either positive or negative. Thus

$$\begin{aligned} -d\gamma = {} & (\Gamma_{\text{Hg}}\, d\bar{\mu}_{\text{Hg}} + \Gamma_e\, d\bar{\mu}_e^{\text{Hg}}) \\ & + (\Gamma_{\text{K}^+}\, d\bar{\mu}_{\text{K}^+} + \Gamma_{\text{Cl}^-}\, d\bar{\mu}_{\text{Cl}^-}) \\ & + (\Gamma_{\text{M}}\, d\bar{\mu}_{\text{M}} + \Gamma_{\text{H}_2\text{O}}\, d\bar{\mu}_{\text{H}_2\text{O}}) \end{aligned} \tag{12.1.13}$$

where $\bar{\mu}_e^{\text{Hg}}$ refers to electrons in the mercury phase.

There are some important linkages between electrochemical potentials:

$$\bar{\mu}_e^{\text{Hg}} = \bar{\mu}_e^{\text{Cu}} \tag{12.1.14}$$

$$\bar{\mu}_{\text{KCl}} = \mu_{\text{KCl}} = \bar{\mu}_{\text{K}^+} + \bar{\mu}_{\text{Cl}^-} \tag{12.1.15}$$

Moreover,

$$\bar{\mu}_{\text{H}_2\text{O}} = \mu_{\text{H}_2\text{O}} \tag{12.1.16}$$

$$\bar{\mu}_{\text{M}} = \mu_{\text{M}} \tag{12.1.17}$$

By further recognizing that $d\bar{\mu}_{Hg} = d\mu_{Hg}^0 = 0$, we can reexpress (12.1.13) as

$$-d\gamma = \Gamma_e\, d\bar{\mu}_e^{Cu} + [\Gamma_{K^+}\, d\bar{\mu}_{KCl} - \Gamma_{K^+}\, d\bar{\mu}_{Cl^-} + \Gamma_{Cl^-}\, d\bar{\mu}_{Cl^-}] + [\Gamma_M\, d\mu_M + \Gamma_{H_2O}\, d\mu_{H_2O}] \tag{12.1.18}$$

Now we consider the important fact that our reference electrode responds to one of the components of the aqueous phase. From the equilibrium at the reference interface, we have

$$\bar{\mu}_{AgCl} + \bar{\mu}_e^{Cu'} = \bar{\mu}_{Ag} + \bar{\mu}_{Cl^-} \tag{12.1.19}$$

Since $d\bar{\mu}_{AgCl} = d\bar{\mu}_{Ag} = 0$,

$$d\bar{\mu}_e^{Cu'} = d\bar{\mu}_{Cl^-} \tag{12.1.20}$$

Substituting (12.1.20) into (12.1.18) and regrouping terms, we obtain

$$-d\gamma = \Gamma_e\, d\bar{\mu}_e^{Cu} - [\Gamma_{K^+} - \Gamma_{Cl^-}]\, d\bar{\mu}_e^{Cu'} + \Gamma_{K^+}\, d\mu_{KCl} + \Gamma_M\, d\mu_M + \Gamma_{H_2O}\, d\mu_{H_2O} \tag{12.1.21}$$

The excess charge density on the metallic side of the interface is

$$\sigma^M = -F\Gamma_e \tag{12.1.22}$$

An equal, but opposite charge density resides on the solution side:

$$\sigma^S = -\sigma^M = F(\Gamma_{K^+} - \Gamma_{Cl^-}) \tag{12.1.23}$$

In addition,

$$d\bar{\mu}_e^{Cu} - d\bar{\mu}_e^{Cu'} = -Fd(\phi^{Cu} - \phi^{Cu'}) = -FdE_- \tag{12.1.24}$$

where E_- is the potential of the mercury electrode with respect to the reference. We follow convention in attaching a negative subscript to signify that the reference electrode responds to an anionic component of our system. Invoking (12.1.22) to (12.1.24) converts (12.1.21) into

$$-d\gamma = \sigma^M\, dE_- + \Gamma_{K^+}\, d\mu_{KCl} + \Gamma_M\, d\mu_M + \Gamma_{H_2O}\, d\mu_{H_2O} \tag{12.1.25}$$

Now we must recognize that all of the parameters in this equation are not independent. We cannot change the chemical potentials of KCl, M, and H_2O separately (e.g., by changing the concentrations). If one of them changes, it affects the others. A consequence of this fact is that Γ_{K^+}, Γ_M, and Γ_{H_2O} are not independently measurable. Can we convert (12.1.25) into an expression of measurable and independently controllable quantities?

We can by recognizing the Gibbs-Duhem relation for the aqueous phase. It says that for any phase at constant T and P (6),

$$\sum_i X_i\, d\mu_i = 0 \tag{12.1.26}$$

where i ranges over all components and the X_i are mole fractions. For our aqueous phase,

$$X_{H_2O}\, d\mu_{H_2O} + X_{KCl}\, d\mu_{KCl} + X_M\, d\mu_M = 0 \tag{12.1.27}$$

Between (12.1.25) and (12.1.27), we can eliminate $d\mu_{H_2O}$ to give

$$-d\gamma = \sigma^M dE_- + \left[\Gamma_{K^+} - \frac{X_{KCl}}{X_{H_2O}}\Gamma_{H_2O}\right] d\mu_{KCl} + \left[\Gamma_M - \frac{X_M}{X_{H_2O}}\Gamma_{H_2O}\right] d\mu_M \tag{12.1.28}$$

The bracketed quantities are measurable parameters called *relative surface excesses.* They are symbolized separately as

$$\boxed{\Gamma_{K^+(H_2O)} = \Gamma_{K^+} - \frac{X_{KCl}}{X_{H_2O}}\Gamma_{H_2O}} \tag{12.1.29}$$

$$\boxed{\Gamma_{M(H_2O)} = \Gamma_M - \frac{X_M}{X_{H_2O}}\Gamma_{H_2O}} \tag{12.1.30}$$

Thus we now learn that we cannot measure the absolute surface excess of K^+, but only its excess relative to water. A zero relative excess does not imply a lack of adsorption of K^+, but only that K^+ and H_2O are adsorbed to the same degree. That is, K^+ and H_2O are adsorbed in the same mole ratio that they have in the bulk electrolyte. A positive relative excess means that K^+ is adsorbed to a greater degree than water, not in absolute molar quantities, but with respect to the amounts available in the bulk electrolyte.

Water is taken here as the *reference component.* It is advantageous to select the solvent S in any electrolyte as the reference component, because one then does not have to be concerned with its activity. Also, one can sometimes argue that the quantities $(X_i/X_S)\Gamma_S$ are negligibly small, so that measured relative surface excesses can be regarded as absolute surface excesses. This assumption is not rigorous, of course, but may be sound from a practical viewpoint in many experimental situations involving dilute solutions.

These considerations bring us to the final statement of the electrocapillary equation for our experimental system:

$$\boxed{-d\gamma = \sigma^M dE_- + \Gamma_{K^+(H_2O)} d\mu_{KCl} + \Gamma_{M(H_2O)} d\mu_M} \tag{12.1.31}$$

Other systems would have similar equations involving terms for other components. More general statements of the electrocapillary equation are available in the specialized literature (4).

Equation 12.1.31 is a relation involving experimentally significant quantities; that is, every quantity is either controllable or measurable. It is our key to an experimental attack on double-layer structure.

12.2 EXPERIMENTAL EVALUATION OF SURFACE EXCESSES AND ELECTRICAL PARAMETERS

12.2.1 Electrocapillarity and the Dropping Mercury Electrode (1–4, 7–9)

It is not obvious why (12.1.31) is called an electrocapillary equation. The name is a historic artifact derived from the early application of this equation to the interpretation of measurements of surface tension at mercury-electrolyte interfaces. The earliest measurements of this sort were carried out by Lippmann, who invented a device called a capillary electrometer for the purpose (10). Its principle involves null balance. The downward pressure on a mercury column is controlled so that the mercury/solution interface, which is confined to a capillary, does not move. In this balanced condition, the upward force exerted by the surface tension exactly equals the downward mechanical force. Because the method relies on null detection, it is capable of great precision. Elaborated approaches are still used. These instruments yield *electrocapillary curves*, which are simply plots of surface tension versus potential.

A more familiar device for achieving the same end is the *dropping mercury electrode* (DME), which actually was invented by Heyrovsky (11) for the measurement of surface tension. Of course, its utility far surpasses its original purpose (see Chapter 5). Figure 5.3.1 is a picture of a typical device.

The weight of the drop at the end of its life is $gmt_{\max}$ where m is the mass flow rate of mercury issuing from the capillary, g is the gravitational acceleration, and $t_{\max}$ is the lifetime of the drop. This force is counterbalanced by the surface tension γ acting around the circumference of the capillary, whose radius is r_c; thus

$$t_{\max} = \frac{2\pi r_c}{mg}\gamma \tag{12.2.1}$$

One can easily see that the drop time $t_{\max}$ is directly proportional to γ; hence a plot of $t_{\max}$ *vs.* potential has the same shape as the true electrocapillary curve. The ordinate is simply multiplied by a constant factor, which can be separately taken into account. Sometimes these plots of drop time are also called electrocapillary curves.

It is difficult to overestimate the importance of these devices to our current understanding of interfacial structure. We have already seen that the thermodynamic relations bearing on the issue emphasize surface tension. Since good measurements of surface tension are made far more conveniently at liquid-metal electrodes, work with mercury and amalgams has dominated research in this area. Mercury offers other advantages too. It has a large hydrogen overpotential, and hence there is a wide range for which only nonfaradaic processes are significant. It is a liquid; therefore, surface features such as grain boundaries do not enter the picture. At the DME a fresh surface is exposed every few seconds, so that problems with progressive contamination of the working surface are minimized. These advantages also extend to the use of mercury surfaces for faradaic electrochemistry, but they are overwhelmingly favorable for probing interfacial structure. Let us now see how electrocapillary curves can reveal part of that structure.

12.2.2 Excess Charge and Capacitance (1–4, 7–9)

Again, we take up the specific chemical system discussed in Section 12.1.2. From its electrocapillary equation, (12.1.31), it is clear that

$$\boxed{\sigma^M = -\left(\frac{\partial \gamma}{\partial E_-}\right)_{\mu_{KCl}, \mu_M}} \qquad (12.2.2)$$

hence the excess charge on the electrode is the slope of the electrocapillary curve at any potential. Figure 12.2.1 is a plot of the drop time of a DME in 0.1 *M* KCl *vs.* potential. It has the nearly parabolic shape that is usually characteristic of these curves, although there are significant variations in the curves as the electrolyte is changed. For example, see Figure 12.2.2

A common feature to all is the existence of a maximum in surface tension. The potential at which it occurs is the *electrocapillary maximum* (ECM) and is an extremely important point in the system. Since the slope of the curve is zero there, it is the *potential of zero charge* (PZC) for the system. That is, $\sigma^M = \sigma^S = 0$ at the PZC.

At more negative potentials, the electrode surface has a negative excess charge, and at more positive potentials there is a positive surface charge. The units of electronic charge comprising any excess repel each other; hence they counteract the usual tendency of the surface to contract, and they weaken the surface tension. Plots of surface charge can be made by differentiating electrocapillary curves. Some examples are shown in Figure 12.2.3.

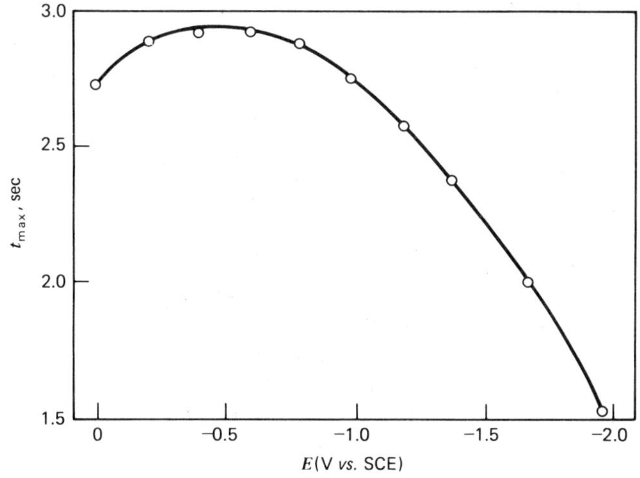

Figure 12.2.1
Electrocapillary curve of drop time versus potential at a DME in 0.1 *M* KCl. [Data of L. Meites, *J. Am. Chem. Soc.*, **73**, 2035 (1951).]

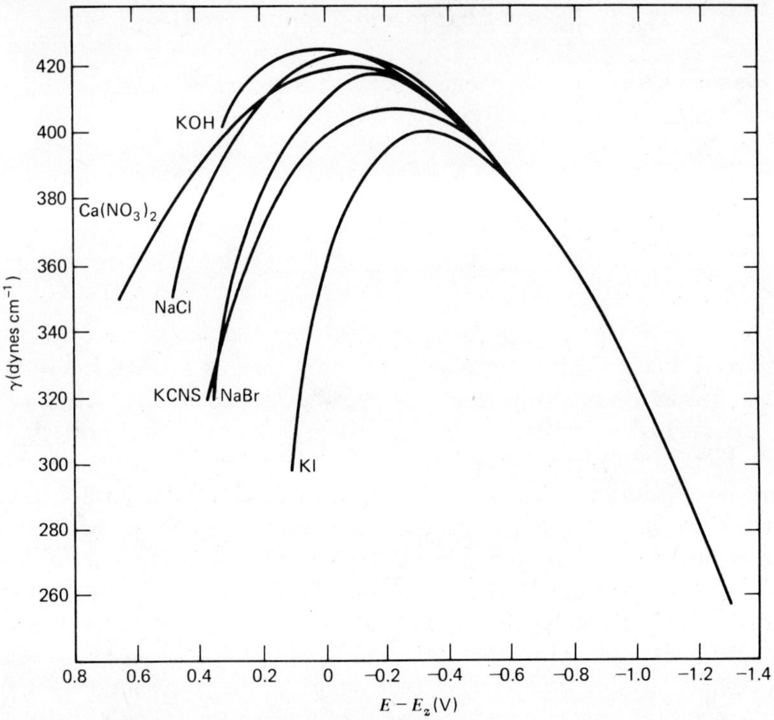

Figure 12.2.2

Electrocapillary curves of surface tension versus potential for mercury in contact with solutions of the indicated electrolytes at 18 °C. The potential is plotted with respect to the PZC for NaF. [Reprinted with permission from D. C. Grahame, *Chem. Rev.*, **41**, 441 (1947). Copyright 1947, Williams and Wilkins, Inc.]

The capacitance of the interface characterizes its ability to store charge in response to a perturbation in potential. One definition is based on the small change in charge density resulting from a small alteration in potential:

$$\boxed{C_d = \left(\frac{\partial \sigma^M}{\partial E}\right)} \tag{12.2.3}$$

This *differential capacitance* is obviously the slope of the plot of σ^M *vs.* E at any point. Figure 12.2.4 helps to clarify the definition. One can see there and in Figure 12.2.3 that C_d is not constant with potential, as it is for an ideal capacitor.

The variation in C_d with E causes us to define an *integral capacitance*, C_i (sometimes denoted K), which is the ratio of the total charge density σ^M at potential E to the total potential difference placing it there. That is,

$$\boxed{C_i = \frac{\sigma^M}{(E - E_z)}} \tag{12.2.4}$$

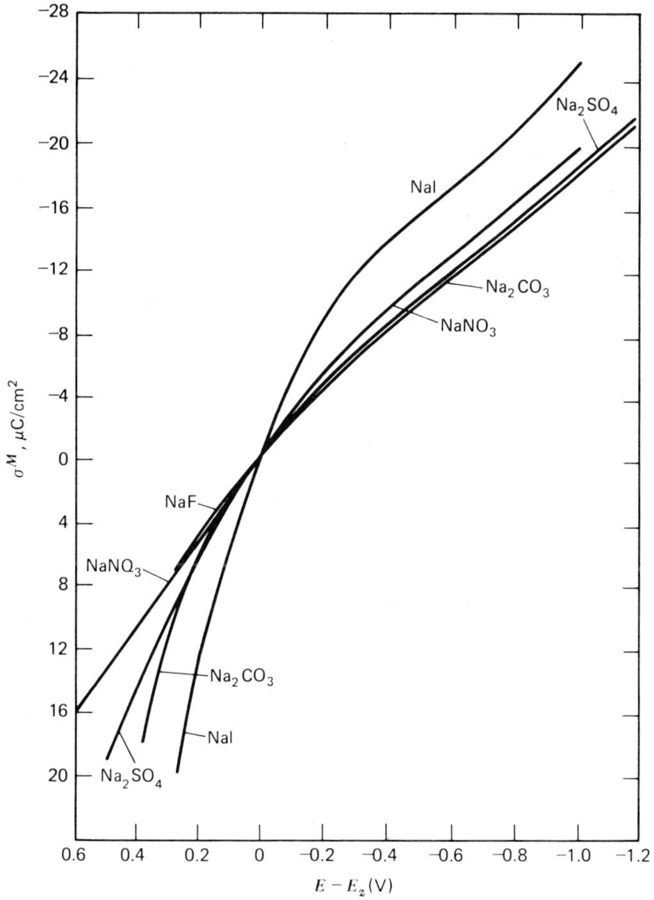

Figure 12.2.3

Charge density on the electrode versus potential for mercury immersed in 1 *M* solutions of the indicated electrolytes at 25°. The potentials are plotted with respect to the PZC for each electrolyte. [Reprinted with permission from D. C. Grahame, *Chem. Rev.*, **41**, 441 (1947). Copyright 1947, Williams and Wilkins, Inc.]

where E_z is the PZC. Figure 12.2.4 contains a graphical interpretation of C_i. It is related to C_d by the equation

$$C_i = \frac{\int_{E_z}^{E} C_d \, dE}{\int_{E_z}^{E} dE} \tag{12.2.5}$$

hence it is an average of C_d over the potential range from E_z to E.

The differential capacitance is the more useful quantity, in part because it is precisely measurable by impedance techniques (see Chapter 9). As we will see in Section 12.3, capacitance measurements have played crucial roles in the formulation of structural models for the double layer.

We can now understand that these measurements are largely equivalent to electrocapillary information. The capacitances can be obtained from the electrocapillary

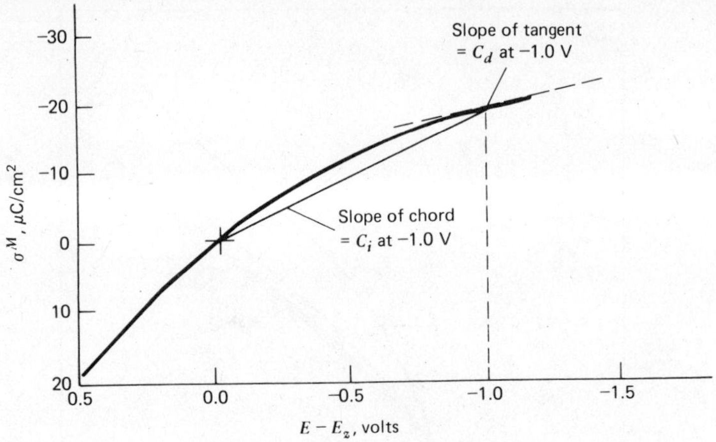

Figure 12.2.4
Schematic plot of charge density versus potential illustrating the definitions of the integral and differential capacitances.

curves by double differentiation, whereas the electrocapillary curves can be constructed from differential capacitances by double integration (12, 13):

$$\gamma = \iint_{E_z}^{E} C_d \, dE \tag{12.2.6}$$

The latter procedure requires separate knowledge of the PZC. Capacitances may be more generally useful primary data because the generation of σ^M *vs.* E and γ *vs.* E curves from them involves integration, which averages out experimental uncertainties. In contrast, differentiation of surface tension accentuates them. In addition, capacitance measurements can be made straightforwardly at solid electrodes, where γ is less accessible.

12.2.3 Relative Surface Excesses (1–4, 7–9)

Returning now to the electrocapillary equation (12.1.31), we find that the relative surface excess of potassium ion at the interface considered there is given by

$$\Gamma_{K^+(H_2O)} = -\left(\frac{\partial \gamma}{\partial \mu_{KCl}}\right)_{E_-, \mu_M} \tag{12.2.7}$$

Since

$$\mu_{KCl} = \mu^0_{KCl} + RT \ln a_{KCl} \tag{12.2.8}$$

we have

$$\boxed{\Gamma_{K^+(H_2O)} = \frac{-1}{RT}\left(\frac{\partial \gamma}{\partial \ln a_{KCl}}\right)_{E_-, \mu_M}} \tag{12.2.9}$$

This relation implies that we can evaluate $\Gamma_{K^+(H_2O)}$ at any potential E_- by measuring the surface tension for several KCl activities, while we hold the activity of M constant. The relative surface excess of chloride could then be evaluated from the charge balance in (12.1.23).

A relation analogous to (12.2.9) can be readily derived for the neutral species M; hence its relative surface excess can be evaluated by the effect of its activity on γ.

Figure 12.2.5 is a graph of relative surface excesses for the components of a 0.1 *M* KBr solution in contact with mercury. Note that bromide is strongly adsorbed in the positive region, whereas K^+ shows an excess in the negative range. One might expect such results from simple electrostatics. However other features of this plot have interesting chemical implications. We will discuss some of them in Section 12.3.4.

12.2.4 Studies at Solid Electrodes (8, 9, 13–21)

Examining interfacial structure at a solid surface is extremely difficult because there are great difficulties in reproducing a surface and in keeping it clean. Measurements of surface tension and surface stress are not easy, but a good deal of attention has been paid to them recently; and there is reason for optimism about the future in this

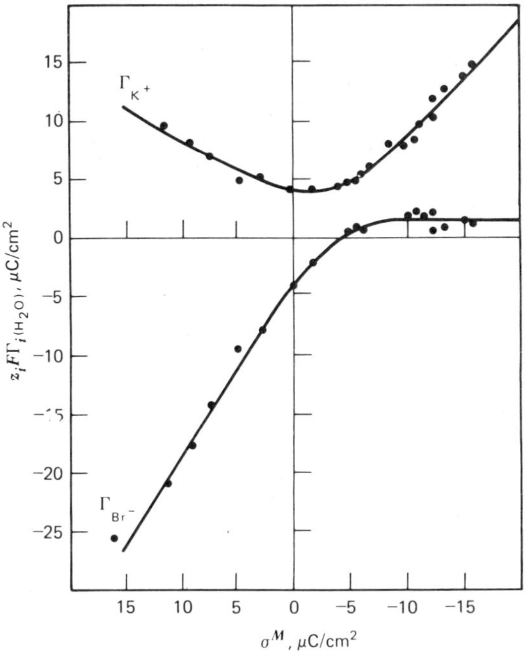

Figure 12.2.5
Surface excesses versus charge density on the electrode for mercury in 0.1 *M* KBr. [Reprinted with permission from M. A. V. Devanathan and S. G. Canagaratna, *Electrochim. Acta*, **8**, 77 (1963). Copyright by Pergamon Press, Inc.]

area. If these data are available, they can be interpreted in a manner similar to that discussed above.

Often there is a reliance on studies of differential capacitance. In principle, these measurements could provide all of the information needed to describe the surface charges and relative excesses; however, one must first know the PZC. Evaluating it for a solid electrode/electrolyte system is not straightforward. Many methods for attacking the problem have been proposed. Probably the most widely used approach is to evaluate the potential of minimum differential capacitance in a system involving dilute electrolyte. The identification of this potential as the PZC rests on the Gouy-Chapman-Stern theory, which is discussed in the next section.

Surface excesses are often examined by methods sensitive to the *faradaic* reactions of adsorbed species. Cyclic voltammetry, chronocoulometry, and polarography are all useful in this regard. Discussions of their application to this type of problem are provided in Section 12.5.

12.3 MODELS FOR DOUBLE-LAYER STRUCTURE

Now that we have seen how some of the basic facts about charge and molar excesses can be obtained for an interface, we would like to develop a picture of the way in which the excesses are arrayed. However, we cannot gain a structural view from purely thermodynamic quantities. Our recourse is to postulate a model, predict its properties, and compare them to the known facts of real systems. If significant differences are found, the model must be revised and tested again. Here we will consider several models that have been proposed for interfacial structure (2–4, 7–9, 14, 20, 22–26).

12.3.1 The Helmholtz Model

Since the metallic electrode is a good conductor it supports no electric fields within itself at equilibrium. In Chapter 2, we saw that this fact implies that any excess charge on a metallic phase resides strictly at the surface. Helmholtz, who was the first to think consequentially about charge separation at interfaces, proposed that the counter-charge in solution also resides at the surface. Thus there would be two sheets of charge, having opposite polarity, separated by a distance of molecular order. In fact, the name *double layer* arises from Helmholtz's early writings in this area (27–29).

Such a structure is equivalent to a parallel-plate capacitor, which has the following relation between the stored charge density σ and the voltage drop V between the plates (30):

$$\sigma = \frac{\varepsilon\varepsilon_0}{d} V \qquad (12.3.1)$$

where ε is the dielectric constant of the medium, ε_0 is the permittivity of free space, and

d is the interplate spacing. The differential capacitance is therefore†

$$\frac{\partial \sigma}{\partial V} = C_d = \frac{\varepsilon\varepsilon_0}{d} \tag{12.3.2}$$

The weakness of this model is immediately apparent in (12.3.2), which predicts that C_d is a constant. We know from our earlier discussion that it is not a constant in real systems. Figure 12.3.1 is a dramatic illustration for interfaces between mercury and sodium fluoride solutions of various concentrations. Variations in C_d with potential and concentration suggest that either ε or d depends on these variables; hence a more sophisticated model is clearly in order.

12.3.2 The Gouy-Chapman Theory (2–4, 7–9, 22, 23)

Even though the charge on the electrode is confined to the surface, the same is not necessarily true of the solution. Particularly at low concentrations of electrolyte, one has a phase with a relatively low density of charge carriers. It may take some significant thickness of solution to accumulate the excess charge needed to counterbalance σ^M. A finite thickness would arise essentially because there is an interplay between the tendency of the charge on the metallic phase to attract or repel the carriers according to polarity and the tendency of thermal processes to randomize them.

This model therefore involves a *diffuse layer* of charge in the solution like that described earlier in Section 1.2.3. The greatest concentration of excess charge would be adjacent to the electrode, where electrostatic forces are most able to overcome the thermal processes; and progressively lesser concentrations would be found at greater distances as those forces weakened. Thus, an average distance of charge separation replaces d in the capacitance expression (12.3.2). Also, we can expect that average distance to show dependences on potential and electrolyte concentration. As the electrode becomes more highly charged, the diffuse layer should become more compact

† Here and elsewhere in this book we use the electrical relations appropriate to the MKSA system of units, which involves the following definition of Coulomb's law (31):

$$F = \frac{qq'}{4\pi\varepsilon\varepsilon_0 r^2}$$

The force F (in newtons) between two charges q and q' (in coulombs) is therefore related to the distance of charge separation r (in meters), the dielectric constant of the medium ε (dimensionless) and the *permittivity of free space* ε_0. The last parameter is a measured constant equal to 8.85419×10^{-12} C^2N^{-1} m^{-2}. This system has the advantage that the electrical variables are measured in common units. An alternative is the electrostatic system, where Coulomb's law is

$$F = \frac{qq'}{\varepsilon r^2}$$

The force F (in dynes) is related here to the charges (in statcoulombs) by the dielectric constant ε and the separation distance (in cm). Equations for the electrostatic system can be converted to corresponding relations for the MKSA system by replacing ε with $4\pi\varepsilon\varepsilon_0$, and vice versa. Many treatments of interfacial structure involve electrostatic units. They are recognizable by the absence of ε_0 and the *appearance* of multiples of 4π in the results. In some treatments, $\varepsilon\varepsilon_0$ is denoted as a single quantity, usually ε, called the *permittivity* of the medium.

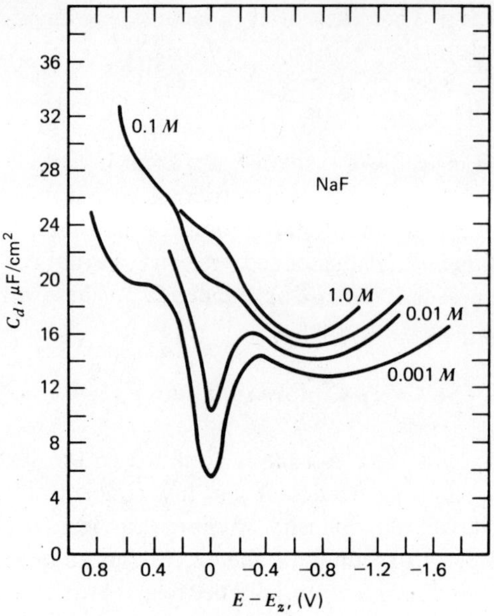

Figure 12.3.1
Differential capacitance versus potential for sodium fluoride solutions in contact with mercury at 25 °C. [Reprinted with permission from D. C. Grahame, *Chem. Rev.*, **41**, 441 (1947). Copyright 1947, Williams and Wilkins, Inc.]

and C_d should rise. As the electrolyte concentration rises, there should be a similar compression of the diffuse layer and a consequent rise in capacitance. Note that these qualitative trends are actually seen in the data of Figure 12.3.1.

Gouy and Chapman independently proposed the idea of a diffuse layer and offered a statistical mechanical approach to its description (32–34). We outline the attack here.

Let us start by thinking of the solution as being subdivided into laminae, parallel to the electrode and of thickness dx, as shown in Figure 12.3.2. All of these laminae are in thermal equilibrium with each other. However, the ions of any species i are not at the same energy in the various laminae, because the electrostatic potential ϕ varies. The laminae can be regarded as energy states with equivalent degeneracies; hence the number concentrations of species in two laminae have a ratio determined by a Boltzmann factor. If we take a reference lamina far from the electrode, where every ion is at its bulk concentration n_i^0, then the population in any other lamina is

$$n_i = n_i^0 \exp\left(\frac{-z_i e\phi}{kT}\right) \tag{12.3.3}$$

where ϕ is measured with respect to the bulk solution. The other quantities in (12.3.3) are the charge on the electron e, the Boltzmann constant k, the absolute temperature T, and the (signed) number of units of electronic charge z_i on ion i.

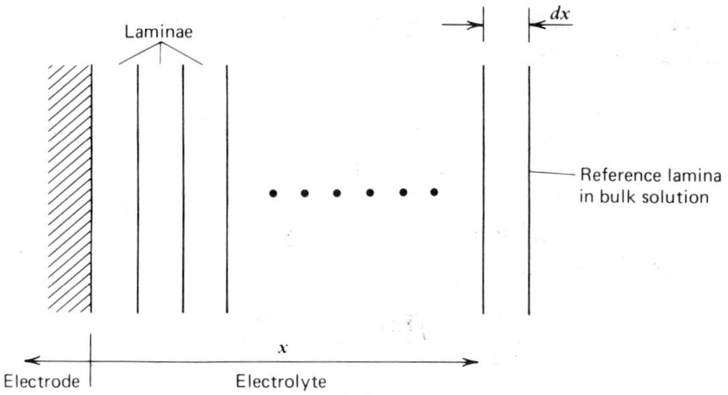

Figure 12.3.2
Schematic view of the solution near the electrode surface as a series of laminae.

The total charge per unit volume in any lamina is then

$$\begin{aligned}\rho(x) &= \sum_i n_i z_i e \\ &= \sum_i n_i^0 z_i e \exp\left(\frac{-z_i e\phi}{kT}\right)\end{aligned} \qquad (12.3.4)$$

where i runs over all ionic species. From electrostatics, we know that $\rho(x)$ is related to the potential at distance x by the Poisson equation (35):

$$\rho(x) = -\varepsilon\varepsilon_0 \frac{d^2\phi}{dx^2} \qquad (12.3.5)$$

hence (12.3.4) and (12.3.5) can be combined to yield the *Poisson-Boltzmann equation*, which describes our system:

$$\frac{d^2\phi}{dx^2} = -\frac{e}{\varepsilon\varepsilon_0}\sum_i n_i^0 z_i \exp\left(\frac{-z_i e\phi}{kT}\right) \qquad (12.3.6)$$

Equation 12.3.6 is treated by noting that

$$\frac{d^2\phi}{dx^2} = \frac{1}{2}\frac{d}{d\phi}\left(\frac{d\phi}{dx}\right)^2 \qquad (12.3.7)$$

hence,

$$d\left(\frac{d\phi}{dx}\right)^2 = -\frac{2e}{\varepsilon\varepsilon_0}\sum_i n_i^0 z_i \exp\left(\frac{-z_i e\phi}{kT}\right) d\phi \qquad (12.3.8)$$

Integration gives

$$\left(\frac{d\phi}{dx}\right)^2 = \frac{2kT}{\varepsilon\varepsilon_0}\sum_i n_i^0 \exp\left(\frac{-z_i e\phi}{kT}\right) + \text{constant} \qquad (12.3.9)$$

and the constant is evaluated by recognizing that at distances far from the electrode $\phi = 0$ and $(d\phi/dx) = 0$. Thus,

$$\left(\frac{d\phi}{dx}\right)^2 = \frac{2kT}{\varepsilon\varepsilon_0}\sum_i n_i^0\left[\exp\left(\frac{-z_i e\phi}{kT}\right) - 1\right] \tag{12.3.10}$$

Now it is useful to specialize the model to a system containing only a symmetrical electrolyte.† Applying this limitation yields

$$\frac{d\phi}{dx} = -\left(\frac{8kTn^0}{\varepsilon\varepsilon_0}\right)^{1/2}\sinh\left(\frac{ze\phi}{2kT}\right) \tag{12.3.11}$$

The details of the transformation from (12.3.10) to (12.3.11) are left to Problem 12.2. In (12.3.11), n^0 is the number concentration of each ion in the bulk and z is the *magnitude* of the charge on the ions.

(a) The Potential Profile in the Diffuse Layer. Equation (12.3.11) can be rearranged and integrated in the following manner:

$$\int_{\phi_0}^{\phi}\frac{d\phi}{\sinh(ze\phi/2kT)} = -\left(\frac{8kTn^0}{\varepsilon\varepsilon_0}\right)^{1/2}\int_0^x dx \tag{12.3.12}$$

where ϕ_0 is the potential at $x = 0$ relative to the bulk solution. In other words, ϕ_0 is the potential drop across the diffuse layer. The result is

$$\frac{2kT}{ze}\ln\left[\frac{\tanh(ze\phi/4kT)}{\tanh(ze\phi_0/4kT)}\right] = -\left(\frac{8kTn^0}{\varepsilon\varepsilon_0}\right)^{1/2}x \tag{12.3.13}$$

or,

$$\frac{\tanh(ze\phi/4kT)}{\tanh(ze\phi_0/4kT)} = e^{-\kappa x} \tag{12.3.14}$$

where

$$\kappa = \left(\frac{2n^0z^2e^2}{\varepsilon\varepsilon_0 kT}\right)^{1/2} \tag{12.3.15a}$$

For dilute aqueous solutions ($\varepsilon = 78.49$) at 25°, this equation can be expressed as

$$\kappa = (3.29 \times 10^7)zC^{*1/2} \tag{12.3.15b}$$

where C^* is the bulk $z{:}z$ electrolyte concentration in mol/L and κ is given in cm^{-1}.

† That is, an electrolyte having only one cationic species and one anionic species, both with charge magnitude z. Sometimes symmetrical electrolytes, example, NaCl, HCl, and $CaSO_4$, are called "$z{:}z$ electrolytes."

Equation 12.3.14 describes the potential profile in the diffuse layer in a general way, and in Figure 12.3.3 there are calculated profiles for several different values of ϕ_0. The potential always decays away from the surface. At large ϕ_0 (a highly charged electrode) the drop is precipitous, because the diffuse layer is relatively compact. As ϕ_0 becomes smaller, the decline is more gradual and approaches an exponential form.

In fact the exponential form holds strictly in the limit of small ϕ_0. If ϕ_0 is sufficiently low that $(ze\phi_0/4kT) \lesssim 0.5$, then $\tanh(ze\phi/4kT) \simeq ze\phi/4kT$ everywhere, and

$$\phi = \phi_0 e^{-\kappa x} \tag{12.3.16}$$

This relation is a good approximation for $\phi_0 \leq 50/z$ mV at 25°.

Note that the reciprocal of κ has dimensions of distance and characterizes the spatial decay of potential. It can be regarded as a kind of characteristic thickness of the diffuse layer. Table 12.3.1 provides values of $1/\kappa$ for several concentrations of a 1:1 electrolyte. The diffuse layer is clearly quite thin by comparison to the distance scale encountered in typical diffusion layers for faradaic experiments. It becomes thicker as the concentration of electrolyte falls, as we anticipated in the qualitative discussion above.

(b) The Relation Between σ^M and ϕ_0. Suppose we now imagine a Gaussian surface in the shape of a box placed in our system as shown in Figure 12.3.4. One end is placed at the interface. The sides are perpendicular to this end and extend far enough into the solution that the field strength $d\phi/dx$ is essentially zero. The box therefore contains all of the charge in the diffuse layer opposite the portion of the electrode surface adjacent to the end.

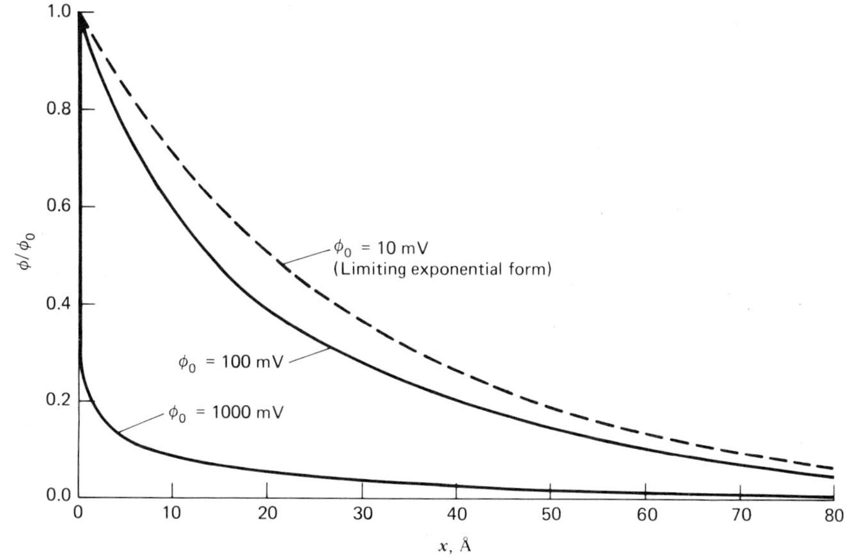

Figure 12.3.3
Potential profiles through the diffuse layer in the Gouy-Chapman model. Calculated for a 10^{-2} M aqueous solution of a 1:1 electrolyte at 25°. $1/\kappa = 30.4$ Å. See equations 12.3.14 to 12.3.16.

Table 12.3.1
Characteristic Thickness of the Diffuse Layer[a]

$C^*(M)$[b]	$1/\kappa$ (Å)
1	3.0
10^{-1}	9.6
10^{-2}	30.4
10^{-3}	96.2
10^{-4}	304

[a] For a 1:1 electrolyte at 25° in water.
[b] $C^* = n^0/N_A$ where N_A is Avogadro's number.

From the Gauss law (Section 2.2.1), this charge is

$$q = \varepsilon\varepsilon_0 \oint_{\text{surface}} \mathscr{E} \cdot d\mathbf{S} \tag{12.3.17}$$

Since the field strength, $\mathscr{E}$, is zero at all points on the surface except the end at the interface [where the magnitude of the field strength is $(d\phi/dx)_{x=0}$ at every point], we have

$$q = \varepsilon\varepsilon_0 \left(\frac{d\phi}{dx}\right)_{x=0} \int_{\substack{\text{end}\\\text{surface}}} dS \tag{12.3.18}$$

or

$$q = \varepsilon\varepsilon_0 A \left(\frac{d\phi}{dx}\right)_{x=0} \tag{12.3.19}$$

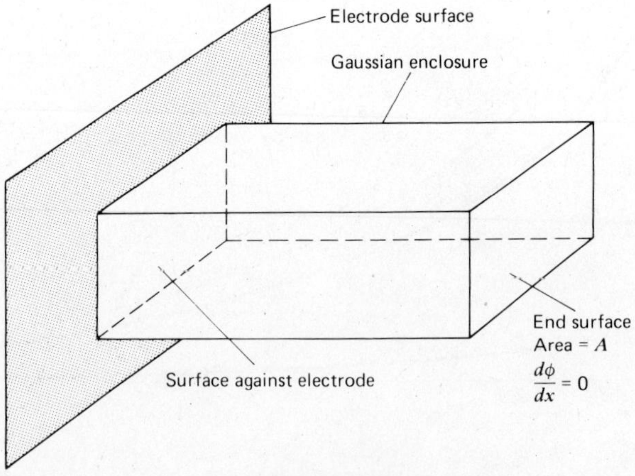

Figure 12.3.4
A Gaussian box enclosing the charge in the diffuse layer opposite an area A of the electrode surface.

Substituting from (12.3.11) and recognizing that q/A is the solution phase charge density σ^S, we obtain

$$\boxed{\sigma^M = -\sigma^S = (8kT\varepsilon\varepsilon_0 n^0)^{1/2} \sinh\left(\frac{ze\phi_0}{2kT}\right)} \tag{12.3.20a}$$

For dilute aqueous solutions at 25°, the constants can be evaluated to give

$$\boxed{\sigma^M = 11.7C^{*1/2} \sinh(19.5z\phi_0)} \tag{12.3.20b}$$

where C^* is in mol/L for σ^M in $\mu C/\text{cm}^2$. Note that ϕ_0 is related monotonically to the state of charge on the electrode.

(c) Differential Capacitance. Now we are in a position to predict the differential capacitance by differentiating (12.3.20):

$$\boxed{C_d = \frac{d\sigma^M}{d\phi_0} = \left(\frac{2z^2e^2\varepsilon\varepsilon_0 n^0}{kT}\right)^{1/2} \cosh\left(\frac{ze\phi_0}{2kT}\right)} \tag{12.3.21a}$$

For dilute aqueous solutions at 25°, this equation can be written,

$$\boxed{C_d = 228zC^{*1/2} \cosh(19.5z\phi_0)} \tag{12.3.21b}$$

where C_d is in μF/cm^2 and the bulk electrolyte concentration C^* is in mol/L. Figure 12.3.5 is a graph of the way in which C_d varies with potential according to the dictates of (12.3.21). There is a minimum at the PZC and a steep rise on either side.

The predicted V-shaped capacitance function does resemble the observed behavior in NaF at low concentrations and at potentials not too far from the PZC (see Figure 12.3.1). However, the actual system shows a flattening in capacitance at more extreme potentials, and the valley at the PZC disappears altogether at high electrolyte concentrations. Moreover, the actual capacitance is usually much lower than the predicted value. The partial success of the Gouy-Chapman theory suggests that it has elements of truth, but its failures are significant and indicate major defects. We will see in the next section that one of those defects is related to the finite size of the ions in the electrolyte.

12.3.3 Stern's Modification (2–4, 7–9, 22, 23)

The reason for the unlimited rise in differential capacitance with ϕ_0 in the Gouy-Chapman model is that the ions are not restricted with respect to location in the solution phase. They are considered as point charges that can approach the surface arbitrarily closely. Therefore, at high polarization, the effective separation distance between the metallic and solution-phase charge zones decreases continuously toward zero.

This view is not realistic. The ions have a finite size and cannot approach the

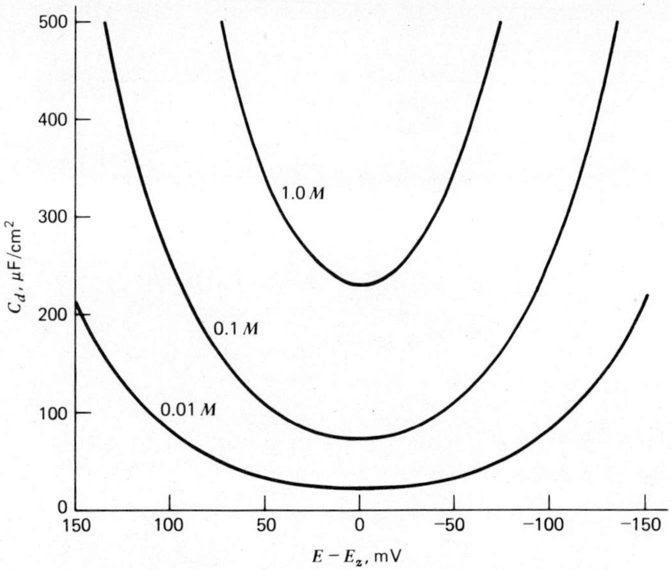

Figure 12.3.5
Predicted differential capacitances from the Gouy-Chapman theory. Calculated from (12.3.21) for the indicated concentrations of a 1:1 electrolyte in water at 25°. Note the very restricted potential scale. The predicted capacitance rises very rapidly at more extreme potentials relative to E_z.

surface any closer than the ionic radius. If they remain solvated, the thickness of the primary solution sheath would have to be added to that radius. Still another increment might be necessary to account for a layer of solvent on the electrode surface. For example, see Figure 1.2.3. In other words, we can envision a *plane of closest approach* for the centers of the ions at some distance x_2.

In systems with low electrolyte concentration, this restriction would have little impact on the predicted capacitance for potentials near the PZC, because the thickness of the diffuse layer is large compared to x_2. However, at larger polarizations or with more concentrated electrolytes, the charge in solution becomes more tightly compressed against the boundary at x_2, and the whole system begins to resemble the Helmholtz model. Then we can expect a corresponding levelling of the differential capacitance. The plane at x_2 is an important concept and is called the *outer Helmholtz plane* (OHP).

This interfacial model, first suggested by Stern (36), can be treated by extending the considerations of the last section. The Poisson-Boltzmann equation (12.3.6) and its solutions (12.3.10) and (12.3.11) still apply at distance $x \geq x_2$. Now the potential profile in the diffuse layer of a $z:z$ electrolyte is given by

$$\int_{\phi_2}^{\phi} \frac{d\phi}{\sinh(ze\phi/2kT)} = -\left(\frac{8kTn^0}{\varepsilon\varepsilon_0}\right)^{1/2} \int_{x_2}^{x} dx \tag{12.3.22}$$

or

$$\boxed{\frac{\tanh(ze\phi/4kT)}{\tanh(ze\phi_2/4kT)} = e^{-\kappa(x-x_2)}} \tag{12.3.23}$$

where ϕ_2 is the potential at x_2 with respect to the bulk solution and κ is defined by (12.3.15).

The field strength at x_2 is given from (12.3.11):

$$\left(\frac{d\phi}{dx}\right)_{x=x_2} = -\left(\frac{8kTn^0}{\varepsilon\varepsilon_0}\right)^{1/2} \sinh\left(\frac{ze\phi_2}{2kT}\right) \tag{12.3.24}$$

Since the charge density at any point from the electrode surface to the OHP is zero, we know from (12.3.5) that this same field strength applies throughout that interval. Thus the potential profile in the compact layer is linear. Figure 12.3.6*b* is a summary of the situation. Now we find the total potential drop across the double layer to be

$$\phi_0 = \phi_2 - \left(\frac{d\phi}{dx}\right)_{x=x_2} x_2 \tag{12.3.25}$$

Note also that all of the charge on the solution side resides in the diffuse layer, and its magnitude can be related to ϕ_2 by considering a Gaussian box exactly as we did above†:

$$\boxed{\sigma^M = -\sigma^S = -\varepsilon\varepsilon_0\left(\frac{d\phi}{dx}\right)_{x=x_2} = (8kT\varepsilon\varepsilon_0 n^0)^{1/2} \sinh\left(\frac{ze\phi_2}{2kT}\right)} \tag{12.3.26}$$

To find the differential capacitance, we substitute for ϕ_2 by (12.3.25)†:

$$\sigma^M = (8kT\varepsilon\varepsilon_0 n^0)^{1/2} \sinh\left[\frac{ze}{2kT}\left(\phi_0 - \frac{\sigma^M x_2}{\varepsilon\varepsilon_0}\right)\right] \tag{12.3.27}$$

Differentiation and rearrangement (Problem 12.4) gives

$$C_d = \frac{d\sigma^M}{d\phi_0} = \frac{(2\varepsilon\varepsilon_0 z^2 e^2 n^0/kT)^{1/2} \cosh(ze\phi_2/2kT)}{1 + (x_2/\varepsilon\varepsilon_0)(2\varepsilon\varepsilon_0 z^2 e^2 n^0/kT)^{1/2} \cosh(ze\phi_2/2kT)} \tag{12.3.28}$$

which is more simply stated as the inverse:

$$\boxed{\frac{1}{C_d} = \frac{x_2}{\varepsilon\varepsilon_0} + \frac{1}{(2\varepsilon\varepsilon_0 z^2 e^2 n^0/kT)^{1/2} \cosh(ze\phi_2/2kT)}} \tag{12.3.29}$$

This expression says that the capacitance is made up of two components that can be separated in the reciprocal, exactly as one would find for two capacitors in series.

---

† See (12.3.15b), (12.3.20b), and (12.3.21b) for evaluations of the constants for aqueous solutions at 25°.

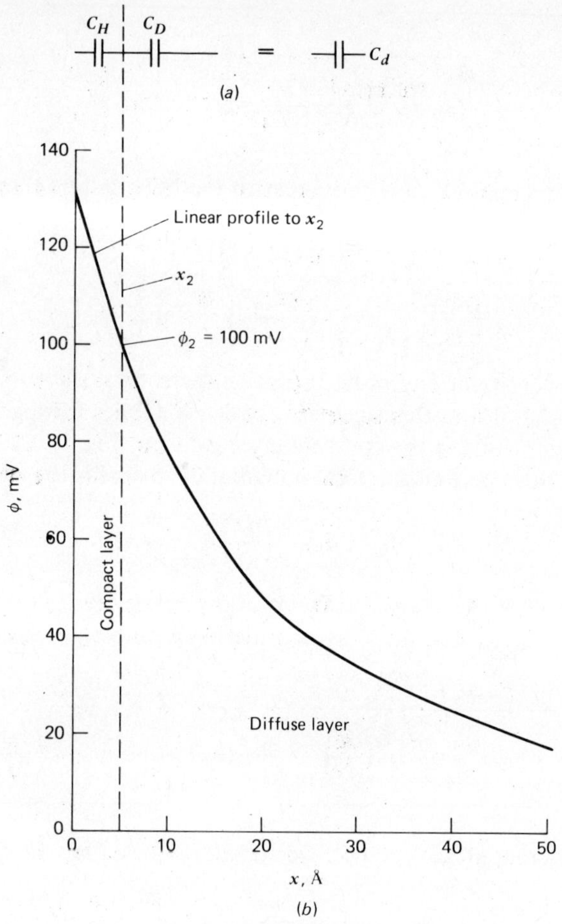

Figure 12.3.6
(*a*) A view of the differential capacitance in the GCS model as a series network of Helmholtz layer and diffuse layer capacitances. (*b*) Potential profile through the solution side of the double layer, according to GCS theory. Calculated from (12.3.23) for 10^{-2} *M* 1:1 electrolyte in water at 25°.

Thus we can identify the terms in (12.3.29) as the reciprocals of component capacitances C_H and C_D, which can be depicted as in Figure 12.3.6*a*:

$$\frac{1}{C_d} = \frac{1}{C_H} + \frac{1}{C_D} \tag{12.3.30}$$

By comparing the terms of (12.3.29) with (12.3.2) and (12.3.21), it is clear that C_H corresponds to the capacitance of the charges held at the OHP, whereas C_D is the capacitance of the truly diffuse charge.

The value of C_H is independent of potential, but C_D varies in the V-shaped fashion we found in the last section. The composite capacitance C_d shows a complex behavior and is governed by the *smaller* of the two components. Near the PZC in systems with low electrolyte concentration, we expect to see the V-shaped function characteristic of C_D. At larger electrolyte concentrations, or even at large polarizations in dilute media, C_D becomes so large that it no longer contributes to C_d and one sees only the constant capacitance of C_H. Figure 12.3.7 is a schematic picture of this behavior.

This model, known as the Gouy-Chapman-Stern (GCS) model, gives predictions that account for the gross features of behavior in real systems. There are still discrepancies, in that C_H is not truly independent of potential. Figure 12.3.1 is a plain illustration. This aspect must be handled by refinements to the GCS theory that take into account the structure of the dielectric in the compact layer, saturation (full polarization) of that dielectric in the strong interfacial field, differences in x_2 for anionic and cationic excesses, and other similar matters (2–4, 7–9, 14, 22–25). Often one must also consider the influence of charged or uncharged species that are adsorbed by chemical interactions with the electrode surface. That issue is our next concern.

12.3.4 Specific Adsorption (2–4, 7–9, 14, 23–26, 37)

In constructing models for interfacial structure, we have so far considered only long-range electrostatic effects as the basis for creating the excesses of charge found in the solution phase. Aside from the magnitude of the charges on the ions, and possibly their radii, we have been able to ignore their chemical identities. They are said to be *nonspecifically adsorbed.*†

However, there is more to the picture. Consider the data in Figure 12.2.2. Note that at potentials more negative than the PZC, the surface tension follows the kind of

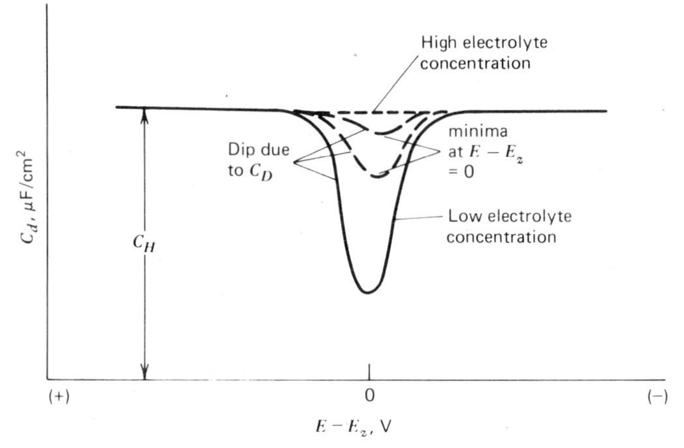

Figure 12.3.7
Schematic representation of the expected behavior of C_d (according to GCS theory) as the electrolyte concentration changes.

† Note that nonspecifically adsorbed species are not really adsorbed at all in the terms usually meant by the word adsorption. There is no close-range interaction in this case.

decline we have come to expect, and the decline is the same regardless of the composition of the system. This result is predictable from the GCS theory. On the other hand, the curves at potentials more positive than the PZC diverge markedly from each other. The behavior of the system in the positive region depends specifically on the composition. Since the deviations in behavior occur for potentials where anions must be in excess, we suspect that some sort of *specific adsorption* of anions takes place on mercury. Specific interactions would have to be very short-range in nature; hence we gather that specifically adsorbed species are tightly bound to the electrode surface in the manner depicted in Figure 1.2.3. The locus of their centers is the *inner Helmholtz plane* (IHP), at distance x_1 from the surface.

What experimental approaches are available for detecting and quantifying specific adsorption? Perhaps the most straightforward approach is through relative surface excesses. Turning now to Figure 12.2.5, we note several peculiarities. First, there are positive relative excesses of bromide at potentials negative of the PZC and potassium at values more positive than the PZC. At the PZC itself, positive excesses of both species are found. None of these features is accountable within an electrostatic model, such as the basic GCS theory.

The identity of the specifically adsorbed ion is revealed by considering the slopes of $z_iF\Gamma_{i(H_2O)}$ *vs.* σ^M in key regions. It is always true that

$$\sigma^M = -[F\Gamma_{K^+(H_2O)} - F\Gamma_{Br^-(H_2O)}] \tag{12.3.31}$$

Ordinarily, the charge on the electrode is counterbalanced by an excess of one ion and a deficiency of the other, as we see in the negative region of Figure 12.2.5. If the electrode becomes even more negative, the excess charge is accommodated by a growth in both the excess *and the deficiency*, so that $F\Gamma_{K^+(H_2O)}$ does not grow as fast as σ^M. In other words, the slope of $F\Gamma_{K^+(H_2O)}$ *vs.* σ^M in the negative region should have a magnitude no greater than unity. By similar reasoning, we conclude that the slope magnitude of $-F\Gamma_{Br^-(H_2O)}$ *vs.* σ^M should also be less than or equal to unity in the positive region.

The data of Figure 12.2.5 show that the system is well-behaved in this respect at potentials much more negative than the PZC. However, in the positive region, there is *superequivalent adsorption* of bromide. The slope $d(-F\Gamma_{Br^-(H_2O)})/d\sigma^M$ exceeds unity in magnitude, and hence a change in charge on the electrode is countered by more than an equivalent charge of Br^-. This evidence points strongly to specific adsorption of bromide at potentials more positive than the PZC. The existence of a positive excess of K^+ in the same region is explained by the necessity to compensate partially the superequivalence of adsorbed bromide. Apparently, the forces responsible for specific adsorption are strong enough to withstand the opposing coulombic field in at least part of the negative region, as one may infer from the positive excess of bromide in the zone of small negative σ^M.

Another indicator of specific adsorption of charged species is the *Esin-Markov effect*, which is manifested by a shift in the PZC with a change in electrolyte concentration (38). Table 12.3.2 provides data compiled by Grahame (2). The magnitude of the shift is usually linear with the logarithm of electrolyte activity, and the slope of the linear plot is the *Esin-Markov coefficient* for the condition of $\sigma^M = 0$. Similar results are obtained at nonzero, but constant, electrode charge densities; hence the Esin-Markov coefficient can be written generally as

$$\boxed{\frac{1}{RT}\left(\frac{\partial E_{\pm}}{\partial \ln a_{\text{salt}}}\right)_{\sigma^M} = \left(\frac{\partial E_{\pm}}{\partial \mu_{\text{salt}}}\right)_{\sigma^M}} \tag{12.3.32}$$

Nonspecific adsorption provides no mechanism for a dependence of the electrode potential on the concentration of the electrolyte, so the Esin-Markov coefficient should be zero in the absence of specific adsorption.

Now consider a system in which the anion is specifically adsorbed and the electrode is being held at the PZC. If we introduce more of the same electrolyte, more anions will be adsorbed, and hence σ^S becomes nonzero and must be balanced. Since the electrode is more polarizable than the solution, the countercharge is induced there. To regain the condition $\sigma^M = 0$, the potential must be shifted to a more negative value, so that the charge excess of specifically adsorbed anions is exactly counterbalanced by an opposing excess charge in the diffuse layer. Thus specific adsorption of anions is indicated by negative shifts in potential at constant charge density, whereas specific cationic adsorption is revealed by positive shifts.

From the data in Table 12.3.2, we see that chloride, bromide, and iodide all appear to be specifically adsorbed, but fluoride is not. It is clear now why sodium and potassium fluoride solutions in contact with mercury are the standard systems for testing the GCS theory of nonspecific adsorption.

It is obvious that specific adsorption will also introduce a capacitive component and should also be detectable by the study of C_d. In fact, we could anticipate that changes

Table 12.3.2
Potentials of Zero Charge in Various Electrolytes[a]

Electrolyte	Concentration, M	E_z, V *vs.* NCE[b]
NaF	1.0	−0.472
	0.1	−0.472
	0.01	−0.480
	0.001	−0.482
NaCl	1.0	−0.556
	0.3	−0.524
	0.1	−0.505
KBr	1.0	−0.65
	0.1	−0.58
	0.01	−0.54
KI	1.0	−0.82
	0.1	−0.72
	0.01	−0.66
	0.001	−0.59

[a] From D. C. Grahame, *Chem. Rev.*, **41**, 441 (1947).
[b] NCE = normal calomel electrode.

in the degree of specific adsorption with changes in potential would be highlighted by examining the derivative of C_d, that is, $\partial C_d/\partial E$. Some of the most general approaches to the analysis of interfacial structure are based on these ideas. Since their details are beyond our scope, the interested reader is urged to pursue them in the literature.

Specific ionic adsorption can alter the potential profile in the interfacial zone to an extreme degree. Figure 12.3.8 is a set of curves presented earlier by Grahame (2) for a mercury interface with 0.3 *M* NaCl. Note particularly the traces for the most positive potentials. These profiles can influence electrode kinetics by mechanisms considered in Section 12.7 below.

Neutral molecules are also interesting as adsorbates, because they influence or participate in faradaic processes (2–4, 7–9, 14, 23–26, 39). They can be detected and studied by the methods we have outlined above.

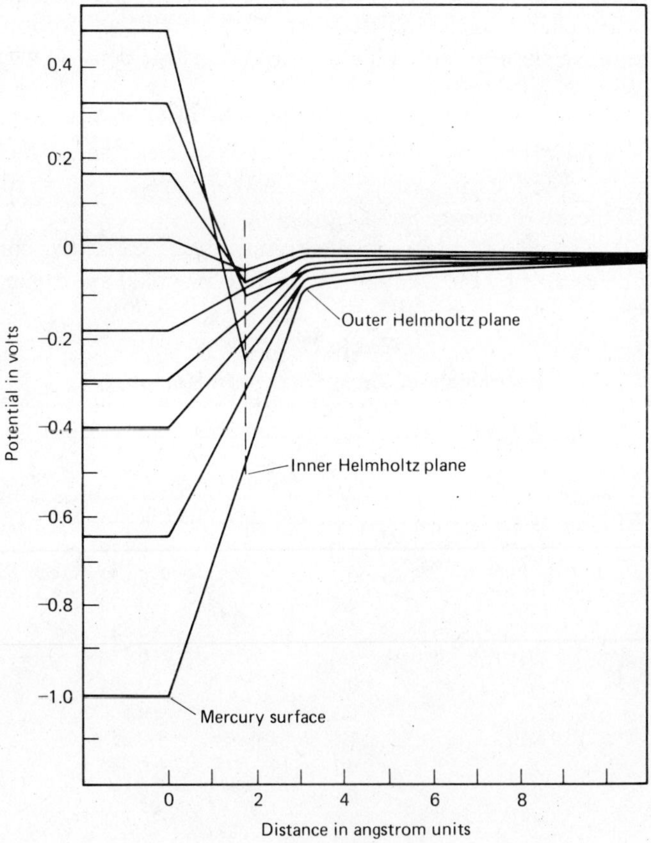

Figure 12.3.8
Potential profiles in the double layer for mercury in contact with 0.3 *M* NaCl in water at 25 °C. Potentials given with respect to the PZC in NaF. [Reprinted with permission from D. C. Grahame, *Chem. Rev.*, **41**, 441 (1947). Copyright 1947, Williams and Wilkins, Inc.]

An interesting aspect of their behavior is that adsorption from aqueous solutions is often effective only at potentials relatively near the PZC. The usual rationale for this phenomenon rests on a recognition that adsorption of a neutral molecule requires the displacement of water molecules from the surface. When the interface is strongly polarized, the water is tightly bound and its displacement by a less dipolar substance is energetically unfavorable. Adsorption can take place only near the PZC, where the water can be removed more easily. The applicability of this rationale in any given case obviously depends on the electrical properties of the specific neutral species at hand (26).

12.4 EXTENT AND RATE OF SPECIFIC ADSORPTION

We have discussed in the preceding section two types of adsorption: *nonspecific adsorption*, where long-range electrostatic forces perturb the distribution of ions near the electrode surface, and *specific adsorption*, where a strong interaction between the adsorbate and the electrode material causes the formation of a layer (partial or complete) on the electrode surface. The difference between nonspecific and specific adsorption is analogous to the difference between the presence of an ion in the ionic atmosphere of another oppositely charged ion in solution (e.g., as modeled by the Debye-Hückel theory) and the formation of a bond between the two solution species (as in a complexation reaction).

Nonspecific adsorption of an electroactive species can affect the electrochemical response, because it affects the concentration of the species, as well as the potential distribution, near the electrode. These effects are described in Section 12.7.

Specific adsorption can have several effects. If an *electroactive* species is adsorbed, the theoretical treatment of a given electrochemical method must be modified to account for the presence of the reactive species at the electrode surface in a relative amount higher than the bulk concentration at the start of the experiment. In addition, specific adsorption can affect the energetics of the reaction, for example, adsorbed O may be more difficult to reduce than dissolved O. The effects of specific adsorption in the different electrochemical methods are discussed in Section 12.5.

Specific adsorption of an *electroinactive* species can also alter the electrochemical response, for example, by forming a blocking layer on the electrode surface. However, adsorption may also increase the reactivity of a species, for example, by causing dissociation of a nonreactive material into reactive fragments, such as in the adsorption of aliphatic hydrocarbons on a platinum electrode. In this case, the electrode behaves as a catalyst for the redox reaction, and this phenomenon is usually called *electrocatalysis*. This subject is discussed briefly in Section 12.6.

12.4.1 Determination of Extent of Specific Absorption

Although the electrocapillary methods described in Sections 12.2 and 12.3 are very useful in the determination of relative surface excesses of specifically adsorbed species, they are mainly applicable to mercury electrodes and are not used often with solid electrodes. The measurement of the amount of electroactive substance adsorbed is frequently based on a faradaic response, as described in Section 12.5. The extent of adsorption of the product of an electrode reaction can often be determined by similar methods. Nonelectrochemical methods can also be applied. For example, the change in concentration of an adsorbable solution species, monitored by a spectrophotometric method, after immersion of a large-area electrode and application of different potentials, can give a direct measurement of the amount of substance that has left the bulk solution upon adsorption (8, 40). Radioactive tracers can also be employed to determine the change in adsorbate concentration in solution (41). Radioactivity measurements can also be applied to electrodes actually removed from the solution, with suitable corrections applied for bulk solution still wetting the electrode (41). A general problem with such direct methods is the sensitivity and precision required for accurate determinations. The amount of material in a monolayer of adsorbate depends on the size of the adsorbing molecule and its orientation on the electrode surface. However, for low-molecular-weight substances, this amount is usually of the order of 10^{-9} to 10^{-10} mol/cm^2. While this represents an easily measurable amount of charge ($>10\ \mu$C), bulk concentration changes caused by adsorption usually are rather small (see Problem 12.8). There has also been interest in studying the adsorbed layer directly on the electrode surface by spectrometric methods, either with the electrode immersed in the solution (e.g., by ellipsometry and Raman spectroscopy) or after removal of the electrode from solution. These methods are potentially very useful, since they can supply information about the structure of the adsorbed layer. They are discussed briefly in Chapter 14.

12.4.2 Adsorption Isotherms

The relationship between the amount of substance i adsorbed on the electrode per unit area, Γ_i, the activity in bulk solution, a_i^b, and the electrical state of the system, E or q^M, at a given temperature, is given by the *adsorption isotherm*. This is obtained from the condition of equality of electrochemical potentials for bulk and adsorbed species i at equilibrium

$$\bar{\mu}_i^A = \bar{\mu}_i^b \tag{12.4.1}$$

where the superscripts A and b refer to adsorbed i and bulk i, respectively. Thus

$$\bar{\mu}_i^{0,A} + RT \ln a_i^A = \bar{\mu}_i^{0,b} + RT \ln a_i^b \tag{12.4.2}$$

where the $\bar{\mu}_i^0$ terms are the standard electrochemical potentials. The standard free energy of adsorption, $\Delta\bar{G}^0$, which is a function of the electrode potential, is defined as

$$\Delta\bar{G}^0 = \bar{\mu}_i^{0,A} - \bar{\mu}_i^{0,b} \tag{12.4.3}$$

Thus

$$a_i^A = a_i^b\, e^{-\Delta\bar{G}_i^0/RT} = \beta_i a_i^b \tag{12.4.4}$$

where (4)

$$\beta_i = \exp\left(\frac{-\Delta\bar{G}_i^0}{RT}\right) \tag{12.4.5}$$

Equation 12.4.4 is a general form of an adsorption isotherm, with a_i^A a function of a_i^b and β_i. Different specific isotherms result from different assumptions or models for the relationship between a_i^A and Γ_i. A number have been proposed (4, 8, 39, 42); some frequently used ones are discussed below:

The *Langmuir isotherm* involves assumptions of (a) no interactions between the absorbed species on the electrode surface, (b) no heterogeneity of the surface, and (c) at high bulk activities, saturation coverage of the electrode by adsorbate (e.g., to form a monolayer) of amount Γ_s. Thus

$$\boxed{\frac{\Gamma_i}{\Gamma_s - \Gamma_i} = \beta_i a_i^b} \tag{12.4.6}$$

Isotherms are sometimes written in terms of the fractional coverage of the surface, $\theta = \Gamma_i/\Gamma_s$; the Langmuir isotherm in this form is

$$\frac{\theta}{1 - \theta} = \beta_i a_i^b \tag{12.4.7}$$

The Langmuir isotherm can be written in terms of the concentration of species i in solution by including activity coefficients in the β term. This yields

$$\Gamma_i = \frac{\Gamma_s \beta_i C_i}{1 + \beta_i C_i} \tag{12.4.8}$$

If two species, i and j, are adsorbed competitively, the appropriate Langmuir isotherms are

$$\Gamma_i = \frac{\Gamma_{i,s} \beta_i C_i}{1 + \beta_i C_i + \beta_j C_j} \tag{12.4.9}$$

$$\Gamma_j = \frac{\Gamma_{j,s} \beta_j C_j}{1 + \beta_i C_i + \beta_j C_j} \tag{12.4.10}$$

where $\Gamma_{i,s}$ and $\Gamma_{j,s}$ represent the saturation coverages of i and j, respectively. These equations can be derived from a kinetic model assuming independent coverages of θ_i and θ_j, with the rate of adsorption of each species proportional to the free area, $1 - \theta_i - \theta_j$ and the solution concentrations, C_i and C_j. The rate of desorption of each is assumed to be proportional to θ_i and θ_j.

Interactions between adsorbed species complicate the problem by making the energy of adsorption a function of surface coverage. Isotherms that include this possibility are the *logarithmic Temkin isotherm*:

$$\boxed{\Gamma_i = \frac{RT}{2g} \ln(\beta_i a_i^b) \qquad (0.2 < \theta < 0.8)} \tag{12.4.11}$$

and the *Frumkin isotherm*:

$$\boxed{\beta_i a_i^b = \frac{\Gamma_i}{\Gamma_s - \Gamma_i} \exp \frac{2g\Gamma_i}{RT}} \tag{12.4.12a}$$

The parameter g has dimensions of J/mol per mol/cm^2, and it expresses the way in which increased coverage changes the adsorption energy of species i. If g is positive, the interactions between two species i on the surface are attractive; and if g is negative, the interactions are repulsive. Note that as $g \rightarrow 0$ the Frumkin isotherm approaches the Langmuir isotherm. This isotherm can also be written in the form (including activity coefficients in the β term)

$$\beta_i C_i = \frac{\theta}{1-\theta} \exp(g'\theta) \tag{12.4.12b}$$

where $g' = 2g\Gamma_s/RT$. The range of g' is generally $2 \leq g' \leq -2$; g' may also be a function of potential (39).

12.4.3 Rate of Adsorption

The adsorption of a species i from solution upon creation of fresh electrode surface (e.g., at a fresh mercury drop at a DME) follows a general behavior analogous to that of an electrode reaction. If the rate of adsorption at the surface is rapid, equilibrium is established at the electrode surface, and the amount of substance adsorbed at a given time $\Gamma_i(t)$ is related to the concentration of the adsorbate at the electrode surface $C_i(0, t)$ by the appropriate isotherm. The rate of buildup of the adsorbed layer to its equilibrium value, Γ_i, is then governed by the rate of mass transfer to the electrode surface. This situation, for mass transfer by diffusion and convection (diffusion layer approximation), assuming a linearized isotherm, has been treated (43). When $1 \gg \beta C_i$, the isotherm in (12.4.8) can be linearized to yield (see Problem 12.9)

$$\Gamma_i = \Gamma_s \beta_i C_i = b_i C_i \tag{12.4.13}$$

where $b_i = \beta_i \Gamma_s$. This equation becomes the boundary condition for the problem, that is,

$$\Gamma_i(t) = b_i C_i(0, t) \tag{12.4.14}$$

The other equations are Fick's equation for i, and the conditions $C_i(x, 0) = C_i^*$, and $\lim_{x \to \infty} C_i(x, t) = C_i^*$. Moreover, the amount of material adsorbed at t is related to the flux of i at the electrode surface:

$$\Gamma_i(t) = \int_0^t D_i \left[\frac{\partial C_i(x, t)}{\partial x}\right]_{x=0} dt \tag{12.4.15}$$

The solution to this problem for a stationary plane electrode (semi-infinite linear diffusion) is (43)

$$\frac{C_i(x, t)}{C_i^*} = 1 - \exp\left(\frac{x}{b_i} + \frac{D_i t}{b_i^2}\right) \operatorname{erfc}\left[\frac{x}{2(D_i t)^{1/2}} + \frac{(D_i t)^{1/2}}{b_i}\right] \tag{12.4.16}$$

$$\frac{\Gamma_i(t)}{\Gamma_i} = 1 - \exp\left(\frac{D_i t}{b_i^2}\right) \operatorname{erfc}\left[\frac{(D_i t)^{1/2}}{b_i}\right] \tag{12.4.17}$$

This function is plotted in Figure 12.4.1.

Note that under the conditions of the linearized isotherm, $\Gamma_i(t)/\Gamma_i$ is independent of C_i^*. A consequence of this treatment is that for realistic values of D_i and b_i a rather

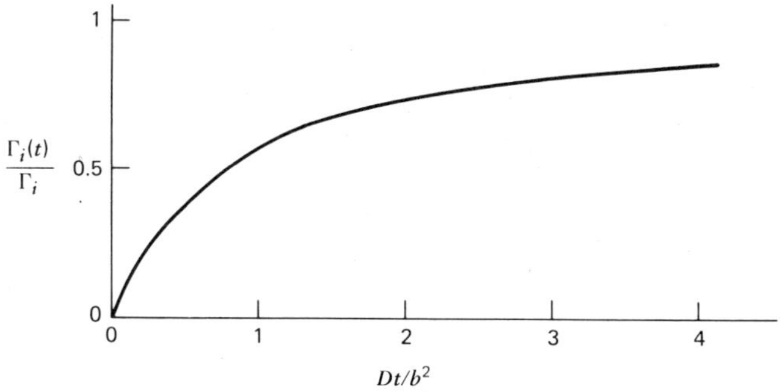

Figure 12.4.1
Rate of attainment of equilibrium coverage Γ_i for diffusion-controlled adsorption under conditions of a linearized isotherm; see (12.4.17). $b = \beta\Gamma_s$.

long time is required to attain equilibrium coverage [i.e., for $\Gamma_i(t)/\Gamma_i \cong 1$; see Problem 12.10]. Clearly, adsorption equilibrium may not be attained at the DME at the usual drop times or at a stationary electrode during a linear potential sweep at moderate rates from initial potentials where adsorption does not occur.

The assumption of a linear isotherm is, of course, valid only over a limited concentration range. The use of the full adsorption isotherm may require numerical solution of the problem; the results of such treatments are in qualitative agreement with that for the linearized isotherm (44, 45) (see Figure 12.4.2). The rate of attainment of equilibrium is clearly seen to depend on the bulk concentration C_i^*, however.

Of course, the rate of adsorption can be increased by stirring the solution. For the linearized isotherm in stirred solution (43),

$$\frac{\Gamma_i(t)}{\Gamma_i} = 1 - \exp\left(\frac{-m_i t}{b_i}\right) \tag{12.4.18}$$

where $m_i = D_i/\delta_i$ is the mass transfer coefficient. Other treatments of mass-transfer-controlled kinetics have been reviewed (23).

The case when the rate of adsorption at the electrode is governed by the adsorption process itself has also been treated, with the assumptions of a logarithmic Temkin isotherm and Temkin kinetics (4, 39, 46). The results of this approach have not been widely applied, although measurements of adsorption rates have been attempted (47). Delahay (39) concludes that the inherent rate of adsorption, at least on mercury from aqueous solution, usually is rapid, so that the overall rate is frequently governed by mass transfer.

12.5 EFFECT OF ADSORPTION OF ELECTROACTIVE SPECIES ON THE ELECTROCHEMICAL RESPONSE

12.5.1 Principles

The electrochemical response (e.g., the voltammetric *i-E* curves) for the electrode reaction $O + ne \rightarrow R$ can be affected quite significantly by the adsorption of O or R. The treatment of such problems is more complicated than those only involving

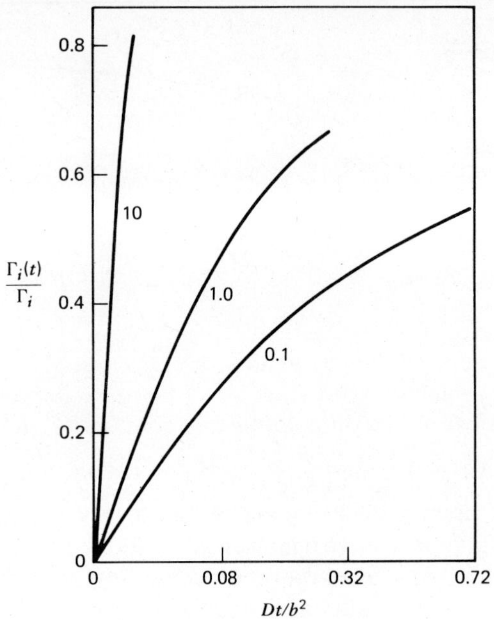

Figure 12.4.2
Rate of attainment of equilibrium coverage Γ_i for diffusion-controlled adsorption under conditions of Langmuir isotherm, at several values of bC^*/Γ_s indicated on curves. [Reprinted with permission from W. H. Reinmuth, *J. Phys. Chem.*, **65**, 473 (1961). Copyright 1961, American Chemical Society.]

dissolved species, because one must choose an adsorption isotherm, which involves the introduction of additional parameters and, in general, nonlinear equations. In addition, the treatment must include assumptions about (a) the degree to which adsorption equilibrium is attained before the start of the electrochemical experiment (i.e., how long after the formation of a fresh electrode surface the experiment is initiated) and (b) the relative rate of electron transfer to the adsorbed species compared to that for the dissolved species. These effects complicate the evaluation of the voltammetric data and make the extraction of desired mechanistic and other information more difficult. Thus adsorption is often considered a nuisance in an electrochemical experiment, to be avoided, when possible, by changing the solvent or changing concentrations. However, adsorption of a species is sometimes a prerequisite for rapid electron transfer (as in electrocatalysis), and can be of major importance in many processes of practical interest (e.g., the reduction of O_2, the oxidation of aliphatic hydrocarbons, or the reduction of proteins). Our discussion here will deal with the basic principles and several important cases.

The equations governing the voltammetric method (e.g., assuming only species O is present initially) include the same ones as used previously—the mass transfer equations [such as (5.4.2)] and the initial and boundary conditions (5.4.3) and (5.4.4). The flux condition at the electrode surface is different, however, because the net reaction involves the electrolysis of diffusing O as well as O adsorbed on the electrode, to produce R that diffuses away and R that remains adsorbed. The general flux equation is then

$$D_O\left[\frac{\partial C_O(x,t)}{\partial x}\right]_{x=0} - \frac{\partial \Gamma_O(t)}{\partial t} = -\left[D_R\left(\frac{\partial C_R(x,t)}{\partial x}\right)_{x=0} - \frac{\partial \Gamma_R(t)}{\partial t}\right] = \frac{i}{nFA} \quad (12.5.1)$$

where $\Gamma_O(t)$ and $\Gamma_R(t)$ are the amounts of O and R adsorbed at time t (mol/cm²). The introduction of these Γ terms requires additional equations relating Γ to C; they are most frequently the Langmuir (or linearized Langmuir) isotherms, for example, see (12.4.9) and (12.4.10):

$$\Gamma_O(t) = \frac{\beta_O \Gamma_{O,s} C_O(0,t)}{1 + \beta_O C_O(0,t) + \beta_R C_R(0,t)} \quad (12.5.2)$$

$$\Gamma_R(t) = \frac{\beta_R \Gamma_{R,s} C_R(0,t)}{1 + \beta_O C_O(0,t) + \beta_R C_R(0,t)} \quad (12.5.3)$$

Initial conditions must also be supplied, for example,

$$(t = 0) \qquad \Gamma_O = \Gamma_O^* \qquad \Gamma_R = 0 \quad (12.5.4)$$

The other equations appropriate to the given electrochemical method and the rates of electron transfer are then added and a solution of the problem is attempted.

12.5.2 Cyclic Voltammetry: Only Adsorbed O and R Electroactive–Nernstian Reaction.

Let us consider the case where only adsorbed O is electroactive (48–50). This could be the case where the sweep rate, v, is so large that no appreciable diffusion of O at the electrode surface occurs [i.e., $D_O(\partial C_O(0,t)/\partial x)_{x=0} \ll \partial\Gamma_O(t)/\partial t$]. Alternatively, the wave for adsorbed O could be shifted to potentials well before the reduction wave for dissolved O. The conditions for such behavior will be given below. There are also cases where adsorption is so strong that the adsorbed layer of O can form even when the solution concentration is so small that the contribution to the current from dissolved O is negligible. An assumption is also made that within the range of potentials of the wave, the Γ's are independent of E. Under these conditions, (12.5.1) becomes

$$-\frac{\partial \Gamma_O(t)}{\partial t} = \frac{\partial \Gamma_R(t)}{\partial t} = \frac{i}{nFA} \quad (12.5.5)$$

This equation along with (12.5.4) yields

$$\Gamma_O(t) + \Gamma_R(t) = \Gamma_O^* \quad (12.5.6)$$

From (12.5.2) and (12.5.3),

$$\frac{\Gamma_O(t)}{\Gamma_R(t)} = \frac{\beta_O \Gamma_{O,s} C_O(0,t)}{\beta_R \Gamma_{R,s} C_R(0,t)} = \frac{b_O C_O(0,t)}{b_R C_R(0,t)} \quad (12.5.7)$$

with $b_O = \beta_O \Gamma_{O,s}$, $b_R = \beta_R \Gamma_{R,s}$. If the reaction is nernstian, so that

$$\frac{C_O(0,t)}{C_R(0,t)} = \exp\left[\left(\frac{nF}{RT}\right)(E - E^{0\prime})\right] \quad (12.5.8)$$

then (12.5.7) yields

$$\frac{\Gamma_O(t)}{\Gamma_R(t)} = \left(\frac{b_O}{b_R}\right) \exp\left[\left(\frac{nF}{RT}\right)(E - E^{0\prime})\right] \tag{12.5.9}$$

From (12.5.5), (12.5.6), and (12.5.9), with

$$\frac{i}{nFA} = \frac{-\partial\Gamma_O(t)}{\partial t} = \left[\frac{\partial\Gamma_O(t)}{\partial E}\right]v \tag{12.5.10}$$

and $E = E_i - vt$, the equation for the *i-E* curve is obtained:

$$i = \frac{n^2F^2}{RT} \frac{vA\Gamma_O^*(b_O/b_R)\exp[(nF/RT)(E - E^{0\prime})]}{\{1 + (b_O/b_R)\exp[(nF/RT)(E - E^{0\prime})]\}^2} \tag{12.5.11}$$

Note the similarity between this equation and that derived for a thin-layer cell (10.7.16). This is readily understandable, since the sample is fully converted without mass transfer limitations in both cases. In the thin-layer cell, VC_O^* mol of O are electrolyzed during the potential sweep, compared to $A\Gamma_O^*$ mol of O on the electrode surface. Thus the *i-E* curve (Figure 12.5.1) has the same shape as that in Figure 10.7.3. The peak current is given by

$$i_p = \frac{n^2F^2}{4RT} vA\Gamma_O^* \tag{12.5.12}$$

and the peak potential by

$$E_p = E^{0\prime} - \left(\frac{RT}{nF}\right) \ln\left(\frac{b_O}{b_R}\right) \tag{12.5.13}$$

Note that the peak current, and indeed the current at all points on the wave, is proportional to v, in contrast to the $v^{1/2}$ dependence observed for nernstian waves of diffusing species. The proportionality between i and v is the same as that observed for a purely capacitive current [see (6.2.25)], and this fact has led to some treatments of adsorption in terms of pseudocapacitances (8, 49). The area under the reduction wave, corrected for any residual current, represents the charge associated with the reduction of the layer of adsorbed O, that is, $nFA\Gamma_O^*$. The anodic wave on scan reversal is the mirror image of the cathodic wave reflected across the potential axis. For an ideal nernstian reaction under Langmuir isotherm conditions, $E_{pa} = E_{pc}$, and the total width at half-height of either the cathodic or anodic wave is given by

$$\Delta E_{p,1/2} = 3.53\frac{RT}{nF} = \frac{90.6}{n} \text{ mV } (25\text{ °C}) \tag{12.5.14}$$

The location of E_p with respect to $E^{0\prime}$ depends on the relative strength of adsorption of O and R. If $b_O = b_R$, $E_p = E^{0\prime}$. If O is adsorbed more strongly ($b_O > b_R$) the wave is displaced toward negative potentials beyond the position where the reversible wave

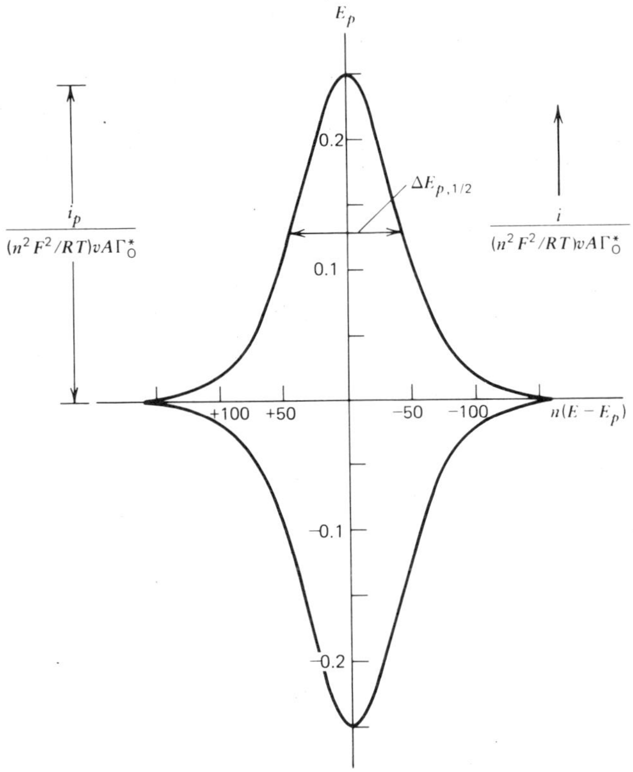

Figure 12.5.1
Current-potential curve for cyclic voltammetry for reduction of adsorbed O and sweep reversal; see (12.5.11). Current is given in normalized form and potential axis is shown for 25 °C.

of a diffusing species would occur. For this reason, it is termed a *postwave*. If R is adsorbed more strongly ($b_R > b_O$), the wave occurs at more positive potentials than $E^{0'}$ and is called a *prewave*. The wave shape observed for this case in experimental studies depends on the actual isotherm that applies to the reactants, and it can also be distorted by nonidealities (i.e., $a_i \neq C_i$), which necessitate the addition of activity coefficients to the expression for the *i-E* curve (50, 51) (Figure 12.5.2).

12.5.3 Cyclic Voltammetry: Only Adsorbed O Electroactive–Irreversible Reaction

For the case where adsorbed O is reduced in a totally irreversible reaction (49, 50), the langmuirian-nernstian boundary condition (12.5.9) is replaced by a kinetic one, similar to that used for dissolved reactants [e.g., (6.3.1)]:

$$\frac{i}{nFA} = k_f\Gamma_O(t) \qquad (12.5.15)$$

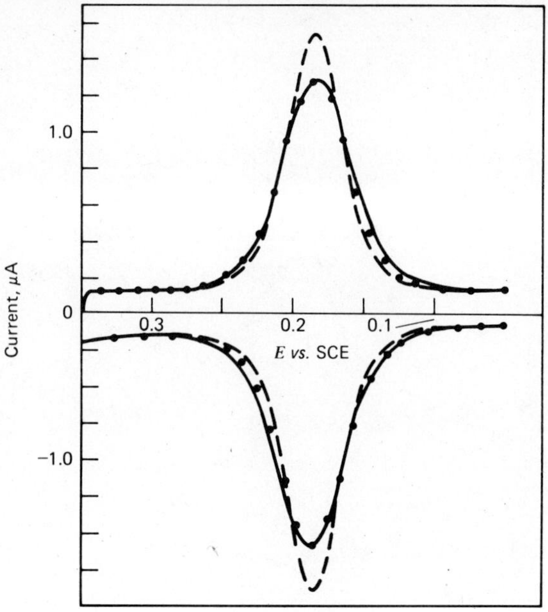

Figure 12.5.2
Experimental and theoretical cyclic voltammograms for reduction and reoxidation of 9,10-phenanthrenequinone irreversibly adsorbed on a pyrolytic graphite electrode. $\Gamma_O = 1.9 \times 10^{-10}$ mol/cm^2; $v = 50$ mV/sec in 1 M $HClO_4$. (——) experimental voltammogram; (----) theoretical voltammogram calculated from (12.5.11); (●) calculated including nonideality parameters. [Reprinted with permission from A. P. Brown and F. C. Anson, *Anal. Chem.*, **49**, 1589 (1977). Copyright 1977, American Chemical Society.]

Note that for adsorbed reactants, k_f has the units sec^{-1}, although it can still be given in the form of (3.2.9), or for the potential sweep experiment [see (6.3.4)]:

$$k_f = k_{fi}\, e^{at} \tag{12.5.16}$$

where $k_{fi} = k^0 \exp[-\alpha n_a f(E_i - E^{0\prime})]$ and $a = \alpha n_a f v$. By combining (12.5.10) with (12.5.15) and (12.5.16), we obtain

$$\frac{d\Gamma_O(t)}{dt} = -k_{fi}\, e^{at}\Gamma_O(t) \tag{12.5.17}$$

This is solved with the initial condition that at $t = 0$, $\Gamma_O(t) = \Gamma_O^*$, to yield the expressions for $\Gamma_O(t)$ and the i-E curve:

$$\Gamma_O(t) = \Gamma_O^* \exp\left(\frac{k_f}{a}\right) \tag{12.5.18}$$

and

$$\boxed{i = nFAk_f\Gamma_O^* \exp\left[\left(\frac{RT}{\alpha n_a F}\right)\left(\frac{k_f}{v}\right)\right]} \tag{12.5.19}$$

Note that these equations were obtained under the usual assumption that the sweep was started at sufficiently positive potentials that $k_{fi} \rightarrow 0$ (hence $e^{k_{fi}} \rightarrow 1$). The potential dependence of i is obtained by substitution for k_f. Again note the similarity between these equations and those in the thin-layer case, equations 10.7.22 and 10.7.23. The shapes of the i-E curves (Figure 12.5.3a) are independent of v and k^0 and follow closely those shown in Figures 10.7.4 and 10.7.5 with suitable minor modifications of parameters. The peak values are given by

$$\boxed{i_p = \frac{n\alpha n_a F^2 A v \Gamma_O^*}{2.718RT}} \tag{12.5.20}$$

$$\boxed{E_p = E^{0'} + \frac{RT}{\alpha n_a F} \ln\left(\frac{RT}{\alpha n_a F}\frac{k^0}{v}\right)} \tag{12.5.21}$$

$$\Delta E_{p,1/2} = 2.44\left(\frac{RT}{\alpha n_a F}\right) = \frac{62.5}{\alpha n_a} \text{ mV } (25\ ^\circ\text{C}) \tag{12.5.22}$$

Again i_p is proportional to v, but the wave is shifted from the reversible value and is distorted from its symmetrical parabolic shape. An experimental example of such a wave is shown in Figure 12.5.3b.

The treatment for the general case of a quasi-reversible reaction follows that given above, but involves consideration of the back reaction [i.e., use of (3.2.8)] as well as the adsorption isotherms for O and R. This case, as well as variants where coupled chemical reactions are associated with the electron transfer reactions, have been discussed rather extensively in the literature (49, 50, 52, 53 and references therein). Treatments of electroactive species that are totally constrained to the electrode surface are also of interest in connection with electrodes that have been modified by covalently binding, coating, or irreversibly adsorbing electroactive molecules at the surface (so-called surface-modified electrodes); representative papers in this area can be consulted (54–66).

12.5.4 Cyclic Voltammetry: Both Dissolved and Adsorbed Species Electroactive

When both the dissolved and adsorbed species are electroactive, the theoretical treatment involves the use of the full flux equation (12.5.1), along with adsorption isotherms and the other usual diffusion equations, initial and boundary conditions, as discussed in Chapter 6. Since the partial differential equations involving mass transfer must be employed, the mathematical treatment is more complicated and we consider here only the case for a nernstian electron transfer reaction where either O (reactant) or R (product) is adsorbed, but not both (67).

(a) Product (R) Strongly Adsorbed. For this case $\beta_O \rightarrow 0$ and β_R is reasonably large (i.e., $\beta_R C^* \geq 100$). Initially $C_O = C_O^*$, $C_R = 0$, $\Gamma_R^* = 0$. The equations to be solved are the diffusion equations for O and R, the total flux equation (12.5.1), the adsorption

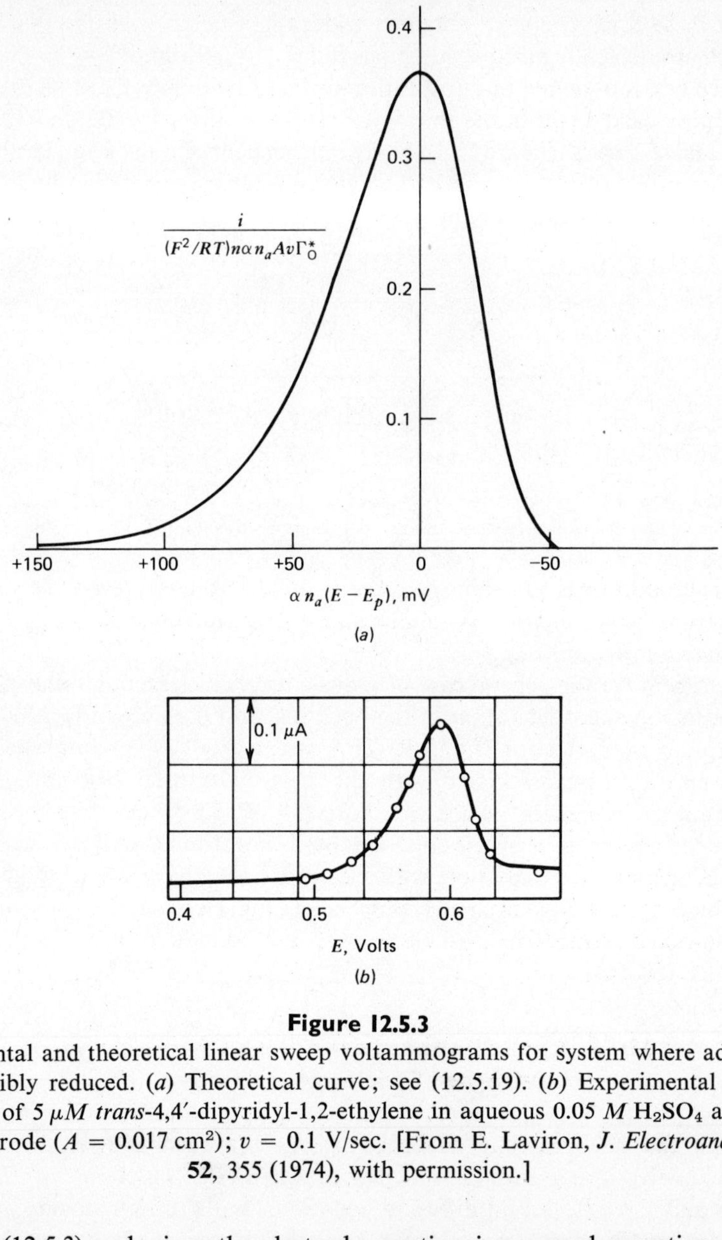

Figure 12.5.3
Experimental and theoretical linear sweep voltammograms for system where adsorbed O is irreversibly reduced. (*a*) Theoretical curve; see (12.5.19). (*b*) Experimental curve for reduction of 5 μM *trans*-4,4′-dipyridyl-1,2-ethylene in aqueous 0.05 M H_2SO_4 at mercury drop electrode (A = 0.017 cm^2); v = 0.1 V/sec. [From E. Laviron, *J. Electroanal. Chem.*, **52**, 355 (1974), with permission.]

isotherm (12.5.3) and, since the electrode reaction is assumed nernstian, equation 12.5.8. We assume that adsorption equilibrium is maintained at all times. The solution to the problem generally follows that described in Section 6.2 (67). In their treatment, Wopschall and Shain also consider the possibility of a variation of β_R with potential, that is,

$$\beta_R = \beta_R^0 \exp\left[\left(\frac{\sigma_R nF}{RT}\right)(E - E_{1/2})\right] \tag{12.5.23}$$

where σ_R represents a parameter that shows how $\Delta\bar{G}_i^0$ varies with potential; $\sigma_R = 0$ implies that β_R is independent of E.

The results of the numerical solution of the equations can be summarized as follows. A *prewave* (or *prepeak*) of the same shape and general properties as that described in Section 12.5.2 appears (Figure 12.5.4), representing the reduction of dissolved O to form a layer of adsorbed R. The wave occurs at potentials more positive than the diffusion-controlled wave, because the free energy of adsorption of R makes reduction of O to adsorbed R easier than to R in solution. This is followed by the wave for reduction of dissolved O to dissolved R. While this wave resembles that observed in the absence of adsorption, it is perturbed by the depletion of species O at the foot of the diffusion wave during reduction of O to adsorbed R. The larger the value of β_R, the more the prepeak precedes the diffusion peak (Figure 12.5.5). Since the peak current for the prepeak $(i_p)_{ads}$ increases with v, while that for the diffusion wave $(i_p)_{diff}$ varies with $v^{1/2}$, $(i_p)_{ads}/(i_p)_{diff}$ increases with increasing v (Figure 12.5.6). In a similar way, $(i_p)_{ads}/(i_p)_{diff}$ increases with $\Gamma_{R,s}$ at a given C_O^*. However, $(i_p)_{ads}/(i_p)_{diff}$ decreases with increasing C_O^* (Figure 12.5.7). At very low concentrations (assuming significant amounts of R are still adsorbed), only the prepeak is observed. As C_O^* increases, the prepeak increases, because Γ_R increases. However, $(i_p)_{ads}$ essentially attains a limiting value as Γ_R approaches $\Gamma_{R,s}$, and then the diffusion peak grows with respect to the adsorption peak. The width of the prepeak at half height, $\Delta E_{p,1/2}$, is a function of σ_R, and varies from $90.6/n$ to $7.5/n$ mV for $\sigma_R F/RT$ increasing from 0 to 0.4 mV^{-1}. Details concerning the derivation, results, and treatment of data are given in Reference 67. A general discussion of the effect of adsorption is given in Reference 68.

(b) Reactant (O) Strongly Adsorbed. ($\beta_R \rightarrow 0$, $\beta_O C_O^* \geq 100$). The adsorption of O results in a *postwave* (or *postpeak*), for the reduction of adsorbed O, following the peak for the diffusion-controlled reduction of O to R in solution (Figure 12.5.8). The postwave results from the greater stability with respect to reduction of adsorbed O compared to solution O. The general treatment and results are analogous to those discussed in (a). The diffusion wave on the forward scan is unperturbed by the adsorption of O, since it is assumed that adsorption equilibrium has been attained and $C_O(x, t) = C_O^*$ at all x before the scan is initiated. The reduction of dissolved O

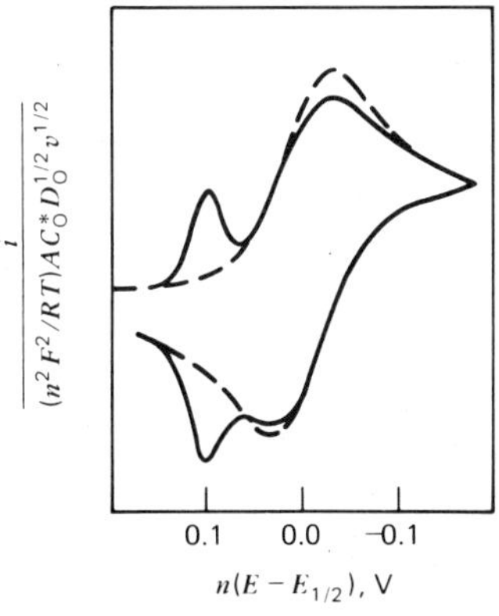

Figure 12.5.4
Cyclic voltammogram for reduction where product is strongly adsorbed showing prepeak; (----) behavior in absence of adsorption. [Reprinted with permission from R. H. Wopschall and I. Shain, *Anal. Chem.*, **39**, 1514 (1967). Copyright 1967, American Chemical Society.]

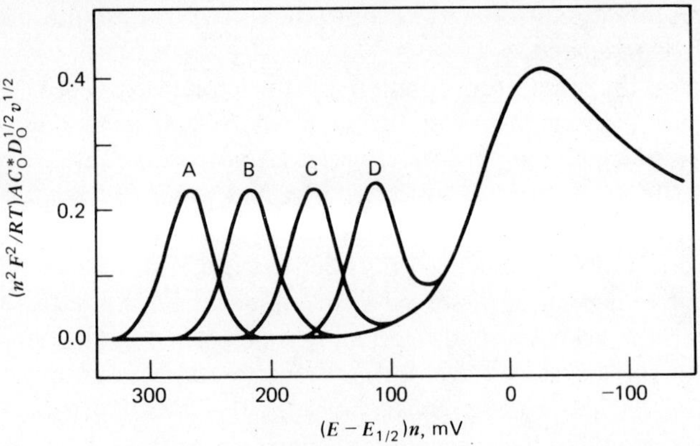

Figure 12.5.5

Variation of linear scan voltammograms for reduction where product is strongly adsorbed. Calculated for $C_O^*(\pi D_O)^{1/2}/4\Gamma_{R,s}$ $(nFv/RT)^{1/2} = 1$, $\sigma_R F/RT = 0.05$ mV^{-1}, and $4\Gamma_{R,s}\beta_R^0$ $(nFv/RT)^{1/2}/(\pi D_R)^{1/2}$ values of: curve A, 2.5×10^6; curve B, 2.5×10^5; curve C, 2.5×10^4; curve D, 2.5×10^3. [Reprinted with permission from R. H. Wopschall and I. Shain, *Anal. Chem.*, **39**, 1514 (1967). Copyright 1967, American Chemical Society.]

presumably occurs either through the adsorbed O film or at the free surface. The postwave has the typical parabolic shape and properties of the adsorption waves discussed in (a) and in Section 12.5.2, on the forward and reverse scans. The diffusion wave on reversal is only slightly perturbed.

(c) Reactant (O) Weakly Adsorbed. **($\beta_R \rightarrow 0$, $\beta_O C_O^* \leq 2$).** When adsorption is weak the difference in energies for reduction of adsorbed and dissolved O is small, and a

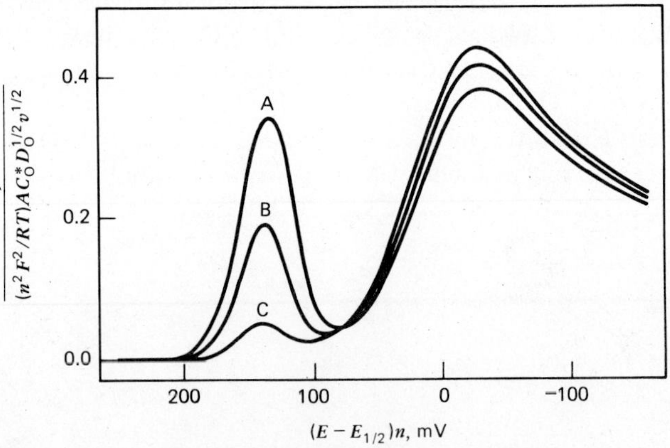

Figure 12.5.6

Effect of variation of scan rate and $\Gamma_{R,s}$ on linear scan voltammograms where the product is strongly adsorbed. Calculated for $\sigma_R F/RT = 0.05$ mV^{-1}, $\beta_R^0 C_O^*(D_O/D_R)^{1/2} = 2.5 \times 10^5$ and values of $\Gamma_{R,s}v^{1/2}$ $[4(nF/RT)^{1/2}/C_O^*(\pi D_O)^{1/2}]$ of curve A, 1.6; curve B, 0.8; curve C, 0.2. Note that with all parameters constant except v, relative scan rates are 64:16:1. [Reprinted with permission from R. H. Wopschall and I. Shain, *Anal. Chem.*, **39**, 1514 (1967). Copyright 1967, American Chemical Society.]

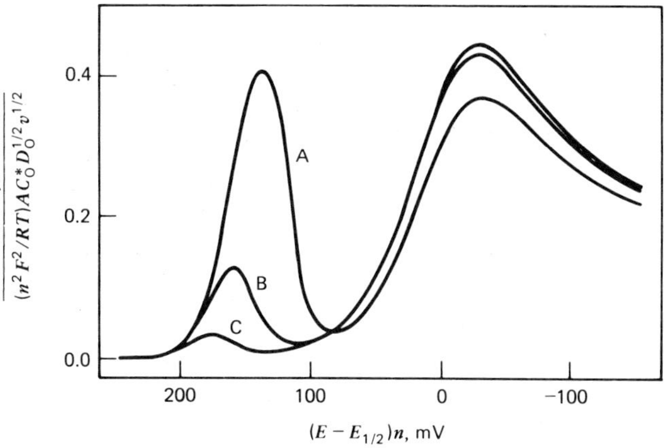

Figure 12.5.7

Effect of variation of C_O^* on linear scan voltammograms where the product is strongly adsorbed. Calculated for $\sigma_R F/RT = 0.05$ mV^{-1}, $4\Gamma_{R,s}\beta_R^0 v^{1/2}(nF/RT)^{1/2}/(\pi D_R)^{1/2} = 1.0 \times 10^6$, and $C_O^*(\pi D_O)^{1/2}/4\Gamma_{R,s}v^{1/2}(nF/RT)^{1/2}$ values of: curve A, 0.5; curve B, 2.0; curve C, 8.0. [Reprinted with permission from R. H. Wopschall and I. Shain, *Anal. Chem.*, **39**, 1514 (1967). Copyright 1967, American Chemical Society.]

separate postwave is not observed (Figure 12.5.9). The net effect is an increase in the height of the cathodic peak compared to that in the absence of adsorption, because both adsorbed and diffusing O contribute to the current. The anodic current on reversal is also increased, but to a smaller extent, because there is a larger amount of R near the electrode at the time of scan reversal. As in the case of strong adsorption of O, the relative contribution of adsorbed O increases at increasing scan rates (Figure 12.5.10). At the very high v limit, i_p approaches a proportionality with v, while at very low v, $i_p \sim v^{1/2}$ (see Problem 12.13). Similarly the ratio, i_{pa}/i_{pc}, is a function of v and is smaller than the value of unity, found in the absence of adsorption (Figure 12.5.11). As with strong adsorption, the relative contribution of the effect of adsorption decreases at high bulk concentrations of O.

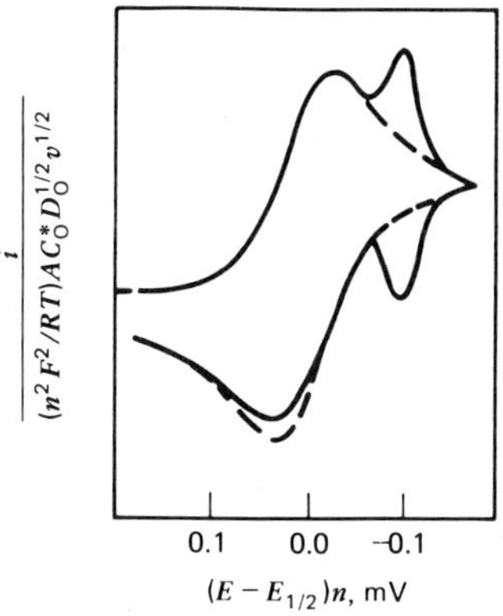

Figure 12.5.8

Cyclic voltammogram for reduction where reactant is strongly adsorbed, showing postpeak; (----) behavior in absence of adsorption. [Reprinted with permission from R. H. Wopschall and I. Shain, *Anal. Chem.*, **39**, 1514 (1967). Copyright 1967, American Chemical Society.

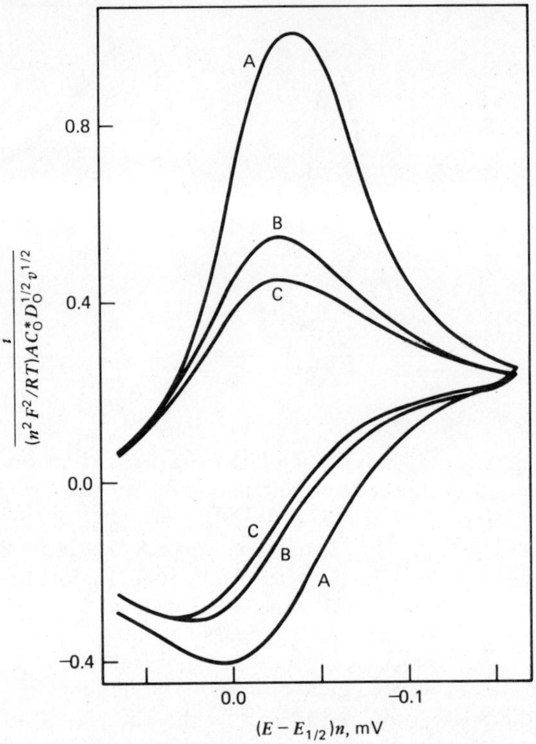

Figure 12.5.9
Effect of scan rate on cyclic voltammograms where reactant is weakly adsorbed. Calculated for $\beta_O C_O^* = 0.01$ and $v^{1/2}[4\Gamma_{O,s}\beta_O(nF/RT)^{1/2}/(\pi D_O)^{1/2}]$ values of: curve A, 5.0; curve B, 1.0; curve C, 0.1. (Curve C corresponds to an essentially unperturbed reaction). Note that relative scan rates are 2500:100:1. [Reprinted with permission from R. H. Wopschall and I. Shain, *Anal. Chem.*, **39**, 1514 (1967). Copyright 1967, American Chemical Society.]

(d) Product (R) Weakly Adsorbed. ($\beta_O = 0$, $\beta_R C_O^* \leq 2$). When R is weakly adsorbed, the cathodic current on the forward scan is only slightly perturbed, while the anodic current on reversal is enhanced (Figure 12.5.12). The cathodic peak current is only very slightly affected, but the wave shifts slightly toward more positive potentials with increasing v, representing a decrease in dissolved R near the electrode surface because of adsorption. The effect is similar to the positive shift observed in E_{pc} when R is involved in a following reaction (e.g., $E_r C_i$, see Section 11.3.3). In this case i_{pa}/i_{pc} is greater than unity and decreases with decreasing v.

(e) Digital Simulations—Irreversible Electron Transfer Reactions. A more general treatment of cyclic voltammetry involving adsorbed and dissolved reactants and products has been worked out with the aid of digital simulation techniques (69). This approach allowed application of the more general Frumkin isotherm, as well as consideration of rate limitations in the electron transfer reactions involving the dissolved or the adsorbed species. Several representative simulations showing the effects of

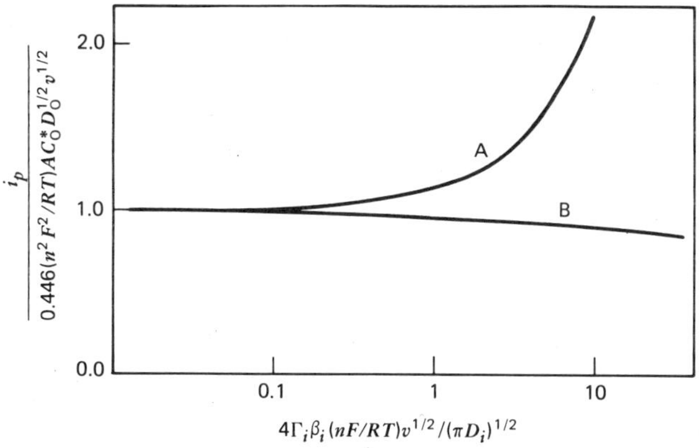

Figure 12.5.10
Variation of peak current with scan rate for linear scan voltammogram where reactant (A) or product (B) is weakly adsorbed. Curve A: $\Gamma_i = \Gamma_{O,s}$, $\beta_O C_O^* = 1$. Curve B; $\Gamma_i = \Gamma_{R,s}$, $\beta_R C_O^* = 1$. [Reprinted with permission from R. H. Wopschall and I. Shain, *Anal. Chem.*, **39**, 1514 (1967). Copyright 1967, American Chemical Society.]

interactions between adsorbed reactant or irreversibility are shown in Figure 12.5.13. Feldberg (69) has pointed out that irreversibility will begin to manifest itself when $k_{\text{diff}}^0/(\pi D_O vF/RT)^{1/2} + k_{\text{ads}}^0 \Gamma_{O,s} \beta_O^{1-\alpha} \beta_R^{\alpha}/(\pi D_O vF/RT)^{1/2} \lesssim \beta_O C_O^*$, where the k^0's refer to the diffusing and adsorbed species. Adsorption effects on cyclic voltammetric studies of the $E_r C_i$ reaction scheme (70) and the effects of rate-controlling adsorption have also been discussed (69, 71).

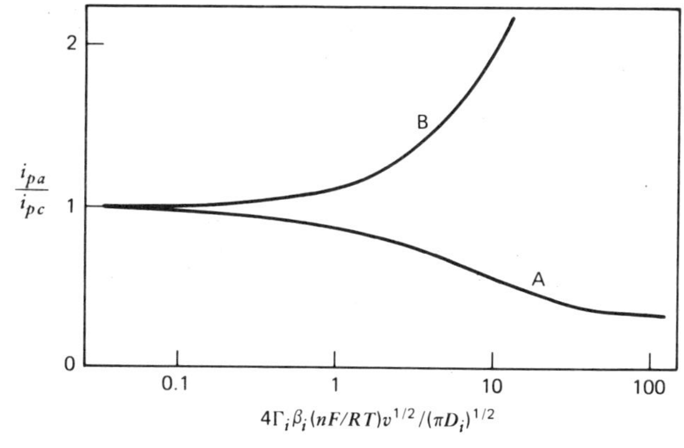

Figure 12.5.11
Variation of peak current ratio with scan rate for cyclic voltammetry where reactant (A) or product (B) is weakly adsorbed. Curve A: $\Gamma_i = \Gamma_{O,s}$, $\beta_O C_O^* = 1$. Curve B: $\Gamma_i = \Gamma_{R,s}$, $\beta_R C_O^* = 1$. Reversal potential $= E_{1/2} - (180/n)$ mV. [Reprinted with permission from R. H. Wopschall and I. Shain, *Anal. Chem.*, **39**, 1514 (1967). Copyright 1967, American Chemical Society.]

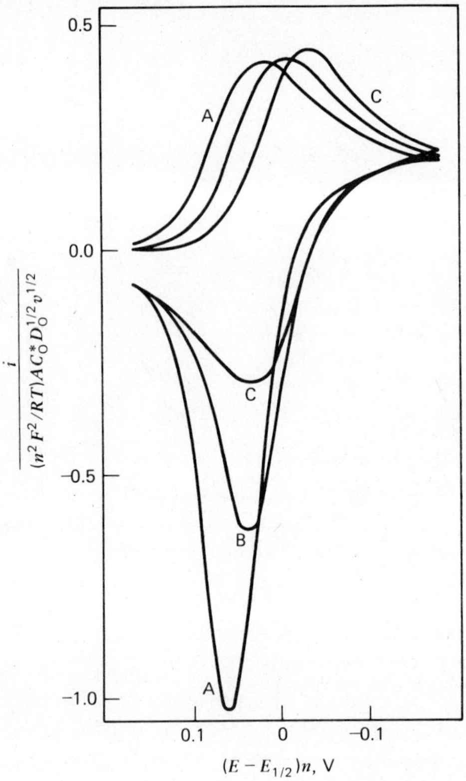

Figure 12.5.12
Effect of scan rate on cyclic voltammograms for initial reduction where product is weakly adsorbed calculated for $\beta_R C_O^* = 0.01$ and $v^{1/2}[4\Gamma_{R,s}\beta_R(nF/RT)^{1/2}/(\pi D_R)^{1/2}$ values of: curve A, 20; curve B, 5; curve C, 0.1. (Curve C corresponds to an essentially unperturbed reaction). Note that relative scan rates are 4×10^4:2500:1. [Reprinted with permission from R. H. Wopschall and I. Shain, *Anal. Chem.*, **39**, 1514 (1967). Copyright 1967, American Chemical Society.]

12.5.5 Adsorption in dc Polarography

While the treatment of adsorption at the DME generally follows that for linear sweep voltammetry at a stationary electrode, it is complicated by the growth of the drop with time and the continuous exposure of fresh surface. In this case, the rate of mass transfer of reactant and product (see Section 12.4.3) and the rate of adsorption could affect the height of the adsorption wave. Although the first explanation of adsorption in voltammetric methods and the explanation of prewaves and postwaves rest on the classic studies by Brdicka (72, 73), dc polarography may not be the method of choice in the study of adsorption. Only a brief discussion will be given here; more detailed treatments have appeared (50, 68, 74).

Let us consider the case where only product R is strongly adsorbed (i.e., the case of a prewave). After detachment of a drop, a new drop starts with a fresh surface. If the potential is in the region of the prewave, O is reduced to adsorbed R. The quantity of R adsorbed is given by [see (5.3.3)]

$$\text{moles R} = A(t)\Gamma_R(t) = (8.5 \times 10^{-3})m^{2/3}t^{2/3}\Gamma_R(t) \tag{12.5.24}$$

If the rate of reduction is diffusion limited, then

$$\text{moles R} = \frac{1}{nF}\int_0^t i_d\, dt \tag{12.5.25}$$

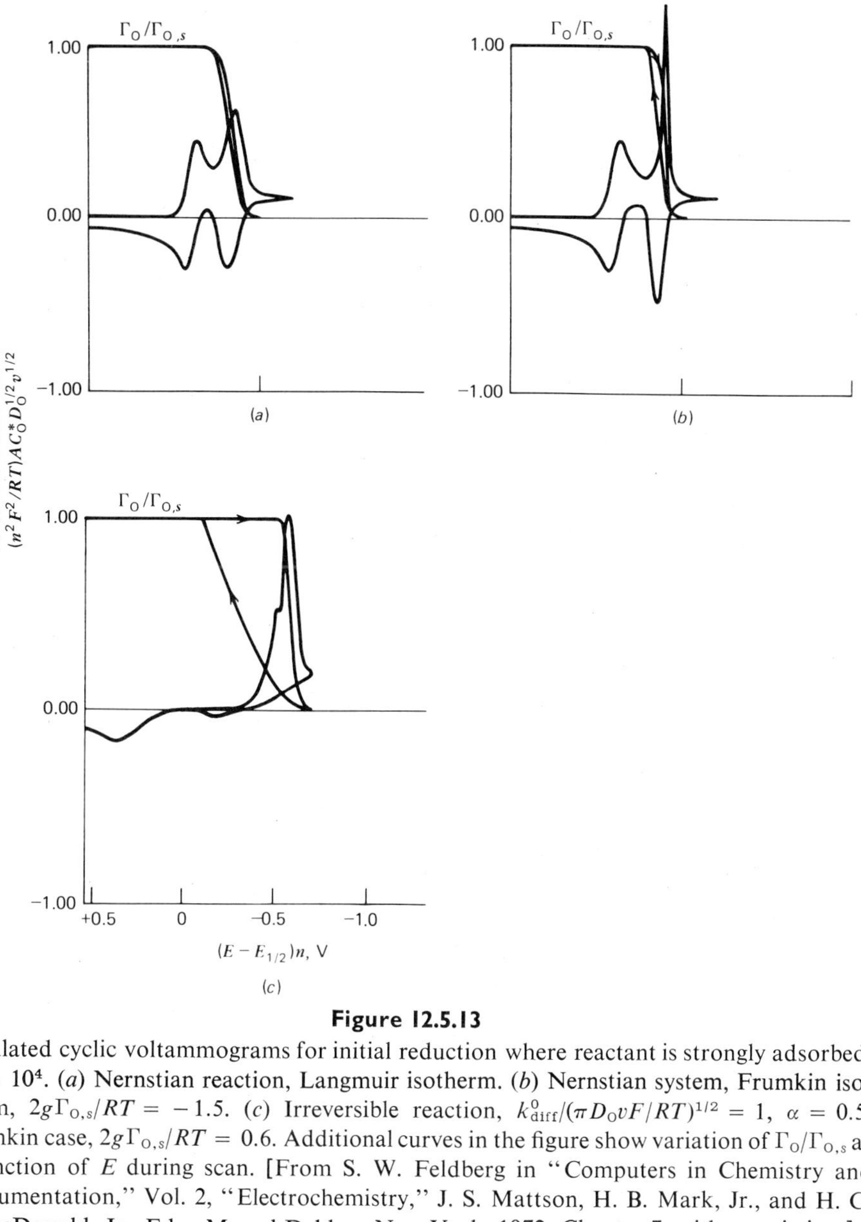

Figure 12.5.13

Simulated cyclic voltammograms for initial reduction where reactant is strongly adsorbed. $\beta_O = 10^4$. (*a*) Nernstian reaction, Langmuir isotherm. (*b*) Nernstian system, Frumkin isotherm, $2g\Gamma_{O,s}/RT = -1.5$. (*c*) Irreversible reaction, $k^0_{\text{diff}}/(\pi D_O vF/RT)^{1/2} = 1$, $\alpha = 0.5$, Frumkin case, $2g\Gamma_{O,s}/RT = 0.6$. Additional curves in the figure show variation of $\Gamma_O/\Gamma_{O,s}$ as a function of E during scan. [From S. W. Feldberg in "Computers in Chemistry and Instrumentation," Vol. 2, "Electrochemistry," J. S. Mattson, H. B. Mark, Jr., and H. C. MacDonald, Jr., Eds., Marcel Dekker, New York, 1972, Chapter 7, with permission.]

Substitution of the instantaneous polarographic current (5.3.6) for i_d, integration, and combination with (12.5.24) yields (75)

$$(8.5 \times 10^{-3})m^{2/3}t^{2/3}\Gamma_R(t) = (6.3 \times 10^{-3})D_O^{1/2}C_O^*m^{2/3}t^{7/6} \tag{12.5.26}$$

$$\Gamma_R(t) = 0.74D_O^{1/2}C_O^*t^{1/2} \tag{12.5.27}$$

Thus the time t_m needed to achieve saturation coverage $\Gamma_{R,s}$ at a given concentration (assuming adsorption itself is rapid) is

$$t_m = \frac{1.83\Gamma_{R,s}^2}{C_O^{*2} D_O} \tag{12.5.28}$$

When the drop time t_{max} is less than t_m, the height of the prewave is limited by diffusion, and is governed by the Ilkovic equation (5.3.6) (Figure 12.5.14*a*). When $t_{max} > t_m$, the surface becomes saturated, the current attains a limiting value i_a, independent of C_O^* and t_{max}, and excess O remains at the electrode surface. The value of i_a is obtained from the expression

$$i_a = \frac{nFd[A(t)\Gamma_{R,s}]}{dt} \tag{12.5.29}$$

$$i_a = (5.47 \times 10^2)nm^{2/3}\Gamma_{R,s}t^{-1/3} \tag{12.5.30}$$

(i_a in A; m in mg/sec; t in sec; $\Gamma_{R,s}$ in mol/cm²). Note that the form of (12.5.30) is the same as that for the charging current, equation 5.3.25. Again we see the analogy between adsorption and capacitance. When the potential is scanned into the region of the main wave, the excess O at the electrode surface is reduced and the diffusion wave appears (Figure 12.5.14*a*). The total current for this wave is also governed by the Ilkovic equation, since all O reaching the electrode surface is reduced, either to adsorbed or dissolved R. Note also that because of the different dependency of the current on time for $t < t_m$ and $t > t_m$ the *i-t* curve at potentials of the adsorption

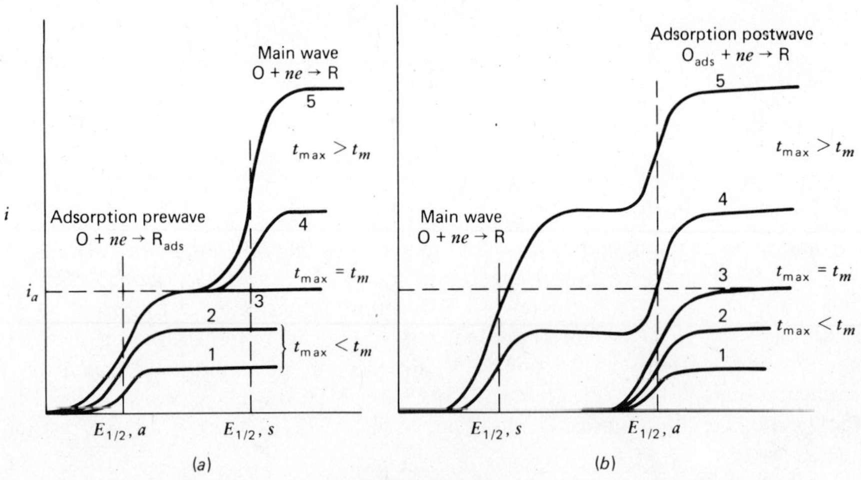

Figure 12.5.14
Polarographic current-potential curves showing (*a*) prewave and (*b*) postwave. Curves 1, 2: $t_{max} < t_m$, only adsorption wave observed. Curve 3: $t_{max} = t_m$, current attains i_a. Curves 4, 5: $t_{max} > t_m$, adsorption wave height remains at i_a, and main wave grows. This behavior is usually observed at constant t_{max} with increasing C_O^*. $t_m = 1.83\Gamma_{R,s}^2/C_O^{*2}D_O$; t_{max} = drop time.

prewave can have an unusual appearance (for $t_{\max} > t_m$). The current increases until $t \simeq t_m$, then falls off with a $t^{-1/3}$ dependence. For a nernstian reaction, both the prewave and the main wave will have the usual reversible shapes (50). Note also, from the dependence of m and t on the corrected mercury column height, h_{corr}, [i.e., $m \sim h_{\text{corr}}$ and $t \sim h_{\text{corr}}^{-1}$, (see Section 5.3.4)] that i_a is directly proportional to h_{corr} [see (12.5.30)], compared to the $h^{1/2}$ dependence for i_d.

A similar treatment holds for the polarographic behavior in the presence of adsorption of O, when a postwave appears (Figure 12.5.14*b*).

12.5.6 Chronocoulometry

The determination of the amount of adsorbed reactant, Γ_O, by integration of the area under the postwave on the linear sweep voltammogram is possible, in principle, when this wave is well-separated from the main wave. In practice this is difficult because it is difficult to subtract the main wave baseline and correct for double-layer charging. The result becomes more uncertain as the separation between the waves becomes smaller, although estimation of Γ_O at very high scan rates is still possible. Potential step chronocoulometry, discussed in Section 5.9, provides a method for the determination of Γ_O, independent of both the relative positions of the dissolved O and adsorbed O reductions and the kinetics of the reactions (76–78).

We consider the case where only O is adsorbed. The potential is stepped from a value E_i, where the amount of O adsorbed per unit area is Γ_O (which may be a function of E_i), to a value sufficiently negative that all O on the electrode surface is reduced and $C_O(0, t) \approx 0$. As shown in (5.9.2), the total charge at time t is

$$Q_f(t \leq \tau) = 2nFAC_O^*\left(\frac{D_O t}{\pi}\right)^{1/2} + nFA\Gamma_O + Q_{dl} \tag{12.5.31}$$

where the terms on the right side of the equation represent the contributions of dissolved O, adsorbed O, and double-layer charging, respectively. As shown in Figure 5.9.1, a plot of Q_f *vs.* t yields an intercept Q_f^0 given by

$$Q_f^0 = nFA\Gamma_O + Q_{dl} \tag{12.5.32}$$

The determination of Γ_O requires an independent estimate of Q_{dl}. While the amount of charge for the supporting electrolyte solution alone, Q'_{dl}, can be obtained in an experiment involving a potential step over the same region in the absence of O, often the adsorption of O will perturb C_d, so that $Q'_{dl} \neq Q_{dl}$. The proper correction can be obtained, however, by a double potential step experiment, in which the potential is returned to E_i at $t = \tau$. The charge during this reverse step, Q_r, measured as shown in Figure 5.9.2, is given by

$$Q_r(t > \tau) = 2nFAC_O^* D_O^{1/2}\pi^{-1/2}\theta + nFA\Gamma_O\left(1 - \frac{2}{\pi}\sin^{-1}\sqrt{\frac{\tau}{t}}\right) + Q_{dl} \tag{12.5.33}$$

where $\theta = \tau^{1/2} + (t - \tau)^{1/2} - t^{1/2}$. Christie et al. have shown (77) that, to a good approximation, the plot of Q_r *vs.* θ is linear and follows the equation

$$Q_r(t > \tau) = 2nFAC_O^* D_O^{1/2}\pi^{-1/2}\left(1 + \frac{a_1 nFA\Gamma_O}{Q_c}\right)\theta + a_0 nFA\Gamma_O + Q_{dl} \tag{12.5.34}$$

where Q_c is the total charge arising from the diffusing species during the forward step, that is,

$$Q_c = 2nFAC_O^*\left(\frac{D_O\tau}{\pi}\right)^{1/2} \tag{12.5.35}$$

and the values of a_0 and a_1 depend slightly on the range of $\theta/\tau^{1/2}$, but are usually taken as $a_0 = -0.069$ and $a_1 = 0.97$. Thus a plot of Q_r *vs.* θ has an intercept, Q_r^0, given by

$$Q_r^0 = a_0 nFA\Gamma_O + Q_{dl} \tag{12.5.36}$$

The value of Q_{dl} in the presence of adsorption is thus near Q_r^0, or more exactly

$$Q_{dl} = \frac{Q_r^0 - a_0 Q_f^0}{1 - a_0} \tag{12.5.37}$$

Once Q_{dl} is determined, $nFA\Gamma_O$ can be obtained from (12.5.32).

The results of a typical experiment involving the reduction of Cd(II) at an HMDE are shown in Figure 12.5.15 (78). In the absence of SCN^-, Cd^{2+} is not adsorbed on Hg, and the chronocoulometric responses (A) show equal intercepts of Q'_{dl}. In the presence of SCN^-, Cd^{2+} is adsorbed and the $Q_f - t^{1/2}$ and $Q_r - \theta^{1/2}$ curves (B) have significantly different intercepts, which allow calculation of Γ_O by using the treatment given above. The variation of Γ_O with potential can be studied by changing E_i. Similarly the variation of Γ_O at different concentrations of O (or supporting electrolyte) is often of interest. The adsorption of Cd(II) in the presence of SCN^- is an example of *anion-induced adsorption*, in which a specifically adsorbed substance (e.g., SCN^-, N_3^-, halide ion) bonds to a metal ion in solution and thus promotes specific adsorption of the metal [e.g., Cd(II), Pb(II), Zn(II)] as well (79, 80).

Chronocoulometry can also be applied to the cases discussed in Section 12.5.2 where only adsorbed species are electroactive (81). In this case the potential step causes only double-layer charging and the electrolysis of the adsorbed species. One can estimate Q_{dl} by steps between a potential at the foot of the adsorption wave, E_i, and potentials, E_j, beyond the adsorption wave. If C_d is not a function of E in the region of the wave, the following equation results (81):

$$Q = Q_{dl} + Q_{\text{ads}} = AC_d(E_i - E_j) + nFA\Gamma_O \tag{12.5.38}$$

so that a plot of Q *vs.* $(E_i - E_j)$ can be employed to determine C_d and Γ_O.

12.5.7 Other Methods

Adsorption of electroactive reactant or product also affects the response in the other methods considered in previous chapters. In constant current step (chronopotentiometric) methods (82–85), the treatment depends on the order in which the adsorbed and diffusing species are electrolyzed. If only adsorbed O is electrolyzed, then the transition time τ follows the relation

$$i\tau = nFA\Gamma_O \tag{12.5.39}$$

A similar equation holds for a prewave where dissolved species O is reduced to adsorbed R. If both dissolved and adsorbed O are reduced, but with adsorbed O

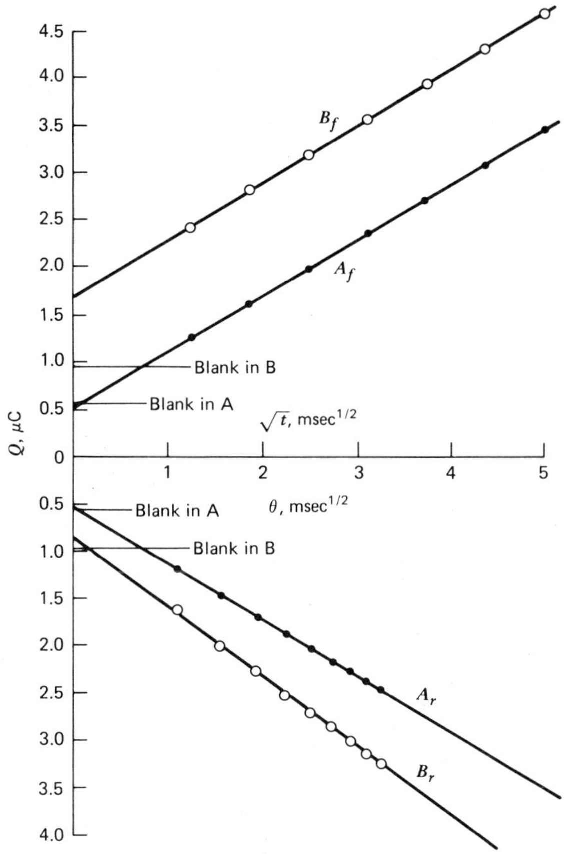

Figure 12.5.15

Double potential step chronocoulometric experiment in study of the induced adsorption of Cd(II) by SCN^- at an HMDE. Potential stepped from $E_i = -0.200$ V to -0.900 V *vs.* SCE and back to E_i. A (●) 1 mM Cd(II) in 1 M $NaNO_3$. The lines have the following slopes (S) and intercepts (Q^0): $S_f = S_r = 0.58\ \mu C/msec^{1/2}$; $Q_f^0 = 0.54\ \mu C$, $Q_r^0 = 0.55\ \mu C$. B (○) 1 mM Cd(II) in 0.2 M NaSCN + 0.8 M $NaNO_3$. $S_f = 0.60\ \mu C/msec^{1/2}$; $Q_f^0 = 1.67\ \mu C$; $Q_r^0 = 0.86\ \mu C$. The points labelled blank in A and B refer to the coulombs required to charge the double layer in the cadmium-free supporting electrolyte solutions alone. The area of HMDE was 0.032 cm^2. [From F. C. Anson, J. H. Christie, and R. A. Osteryoung, *J. Electroanal. Chem.*, **13**, 343 (1967), with permission.]

being reduced almost completely before reduction of the dissolved O, then the $i\tau$ equation is

$$i\tau = \frac{n^2F^2\pi D_O A^2 C_O^{*2}}{4i} + nFA\Gamma_O \tag{12.5.40}$$

If adsorbed O is reduced last (i.e., a postwave), the overall τ is given by $\tau = \tau_1 + \tau_2$, where τ_1 is the transition time due only to diffusing species:

$$\tau_1 = \frac{n^2F^2\pi DA^2C_O^{*2}}{4i^2} \tag{12.5.41}$$

The equation governing this case is

$$\frac{nF\pi A\Gamma}{i} = \tau \cos^{-1}\left(\frac{\tau_1 - \tau_2}{\tau}\right) - 2(\tau_1\tau_2)^{1/2} \tag{12.5.42}$$

For simultaneous reduction of adsorbed and dissolved species, the behavior is more complex and depends on the form of the adsorption isotherm as well the manner in which the current divides between reduction of adsorbed and diffusing O. The problem is, in many ways, similar to that concerning the effect of double-layer charging in chronopotentiometry (see Section 7.3.6). For example, with the assumption that the fraction of the current contributing to the reduction of adsorbed O remains constant for $0 \leq t \leq \tau$, (7.3.19) holds, or

$$i\tau = \frac{nFA(\pi D_O)^{1/2}C_O^*}{2} + nFA\Gamma_O \tag{12.5.43}$$

Clearly, constant current methods are not as useful as chronocoulometric ones for determination of adsorbed reactant. However, once Γ_O is determined, the chronopotentiometric response can yield information about the relative order of reduction of dissolved and adsorbed O.

The effect of adsorption of electroactive species in ac methods is taken into account by modification of the equivalent circuit representing the electrode reaction (68, 86–90). This is usually accomplished by adding an "adsorption impedance" in parallel to the Warburg impedance and double-layer capacitance. Expressions for this impedance for reversible (87, 88) and irreversible (89, 90) systems have been suggested, but the general complexity of the resulting analysis has not led to widespread application of these techniques. The situation in ac polarography is even more complicated, since the effect of adsorption on the dc process, as well as the rate of attainment of adsorption equilibrium at the DME, must be taken into account (68, 91). The general effect of adsorption of O or R on the ac polarogram is an enhancement of the peak height and an increase in the phase angle ϕ, sometimes to values greater than the 45° characteristic of a reversible process. (Recall that slow electron transfer kinetics or coupled chemical reactions lead to ϕ values below 45°.) The method has not been applied widely to studies of adsorption itself, but one should be aware of the complications adsorption can cause in interpretation of ac (and dc) polarographic results (68, 92).

12.6 EFFECT OF ADSORPTION OF ELECTROINACTIVE SPECIES

The adsorption of electroinactive species (sometimes referred to as the "formation of a film of foreign substance") on an electrode surface occurs frequently. Such adsorption can inhibit (or poison) an electrode reaction (e.g., by formation of an impervious layer that blocks a portion of the electrode surface), or it can accelerate the electrode reaction (e.g., by double-layer effects, as discussed in Section 12.7, or as in the anion-induced adsorption of metal ions discussed previously). Indeed, in many studies with solid electrodes, a slow change in the electrochemical response with time is observed, which can be ascribed by the buildup of adsorbed impurities on the electrode surface at a rate limited by their diffusion from the bulk solution. Moreover, in aqueous solutions, metals form layers of adsorbed oxygen (or equivalently, oxide film monolayers) or adsorbed hydrogen, which can affect the electrochemical behavior. A great advantage of mercury electrodes is the possibility of easily renewing the surface, and thus allowing repetitive measurements at surfaces essentially free of adsorbed films. At solids, reproducible surface behavior can sometimes be attained by preceding an experiment with a program of potential steps to values where desorption of impurities occurs or where oxide films are formed and then reduced; applying such a program is sometimes described as "activating" the electrode surface (93). Several reviews have appeared dealing with these topics (39, 94–98).

An adsorbed film may inhibit an electrode reaction by completely blocking the electrode surface so that reaction only occurs at the uncovered fraction, $(1 - \theta)$. Alternatively, the reaction could occur at the filmed portion of the electrode, for example, by penetration of the reactive species or transfer of electrons through the film, but at a reduced rate compared to the free surface. It is also possible that the sites of adsorbed material promote the electrode reaction. These effects are sometimes treated by assuming that the rate constant for the overall electrode process, k^0, is a linear function of coverage θ (94):

$$k^0 = k^0_{\theta=0}(1 - \theta) + k^0_c\theta \tag{12.6.1}$$

where $k^0_{\theta=0}$ is the standard rate constant at the bare surface and k^0_c that at the filmed portions. For complete blockage by the film, $k^0_c = 0$; while for catalysis by the filmed portion, $k^0_c > k^0_{\theta=0}$. Actual attempts at experimental verification of this equation under conditions of equilibrium coverage are few. A study of the reduction of Zn(II) on Zn(Hg) with blockage by adsorption of alcohols such as amyl alcohol, thymol, and cyclohexanol showed such a linear dependence at small coverages, but significant deviations were observed at higher coverages (99). In this case, correction for ϕ_2 effects (see Section 12.7) and the determination of the extent of coverage at higher alcohol concentrations caused some difficulty in obtaining corrected rate constants. Other studies at mercury electrodes that also failed to confirm (12.6.1) have been reported (100, 101).

The effect of adsorbed substances on solid electrodes has been the subject of numerous investigations because of the technological implications of such adsorption. This is especially true for noble metals that are used as electrodes or electrocatalysts in fuel cells and other applications (93, 102). For example, the current-potential curve for a platinum electrode in an aqueous solution shows peaks for the formation and oxidation of both adsorbed hydrogen and adsorbed oxygen (Figure 12.6.1). Measurement of the areas under the peaks for adsorbed hydrogen and oxygen, assuming they represent monolayer coverage, have been suggested as a means of determining the "true" (as opposed to "geometric" or "projected") area of the electrode. Many substances can adsorb onto a platinum electrode and inhibit the hydrogen electrode reactions. Evidence for this effect is the decrease in the area under the adsorbed hydrogen region of the *i-E* curve, when such substances (e.g., compounds of mercury and arsenic, carbon monoxide, and many organic substances) are added to the system. Alternatively, the formation of an adsorbed oxygen (or oxide) layer on platinum inhibits many oxidation processes (e.g., the oxidation of hydrogen, oxalic

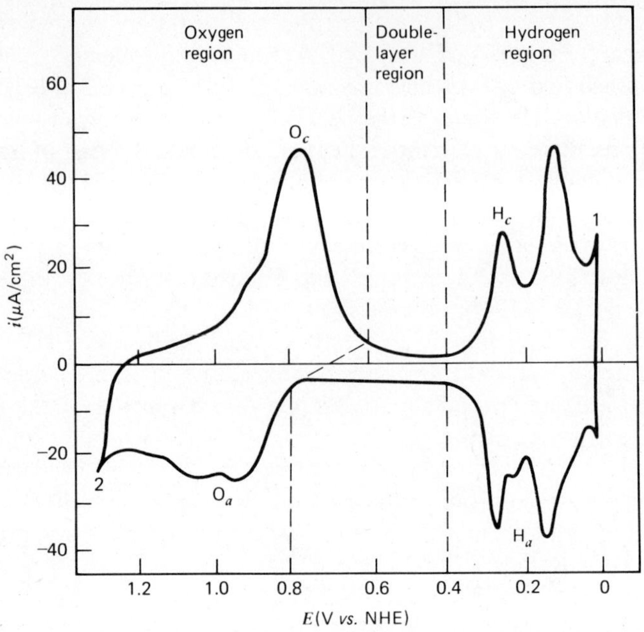

Figure 12.6.1
Cyclic voltammetric current-potential curve for a smooth platinum electrode in 0.5 M H_2SO_4. Peaks H_c show formation of adsorbed hydrogen; H_a, oxidation of adsorbed hydrogen; O_a, formation of adsorbed oxygen or platinum oxide layer; O_c, reduction of oxide layer. Point 1: start of bulk hydrogen evolution. Point 2: start of bulk oxygen evolution. The shape, number, and size of the peaks for adsorbed hydrogen depend on the crystal faces of platinum exposed (103), pretreatment of electrode, solution impuritics, and supporting electrolyte.

acid, hydrazine, and a number of organic substances). Adsorption of electroinactive substances plays an important role in electrodeposition processes where they can act as brighteners (see Section 10.3.2). Adsorbed organic molecules (such as acridine, or quinoline derivatives) can also act as corrosion inhibitors by decreasing the rates of any of the reactions that may occur at a metal surface (e.g., metal dissolution or oxygen reduction).

12.7 DOUBLE-LAYER EFFECTS ON ELECTRODE REACTION RATES

12.7.1 Introduction and Principles

The fact that the structure of the double layer and the specific adsorption of ions can affect the kinetics of electrode reactions was recognized as early as 1933 (104). Such effects can give rise to a number of apparent anomalies. For example, the rate constant k^0 for a given electrode reaction might be a function of the nature of the supporting electrolyte ions or the supporting electrolyte concentration, even when no apparent bulk reaction involving electrolyte ions (e.g., complexation or ion pairing) occurs. Nonlinear Tafel plots (see Section 3.5.3) can be observed. Sometimes rather spectacular effects can be observed in the *i-E* curves. For example, in the reduction of anionic species (e.g., $S_2O_8^{2-}$), the current on the diffusion plateau can drop at a certain potential, with a resulting minimum in the *i-E* curve.

These effects can be understood and interpreted in terms of the variation of potential in the double-layer region, as discussed in Section 12.3. The basic concepts were described by Frumkin (8), and this effect is sometimes called the *Frumkin effect*. If we assume that the species undergoing reduction, O^z, in the electrode reaction

$$O^z + ne \rightleftharpoons R^{z'} \tag{12.7.1}$$

is not specifically adsorbed, then its position of closest approach to the electrode is the OHP ($x = x_2$) (see Section 12.3.3). The potential at the OHP, ϕ_2, is not equal to the potential in solution, ϕ^s, because of the potential drop through the diffuse layer (and possibly because some ions are specifically adsorbed). These potential differences in the double layer, as shown for example in Figure 12.3.6, can affect the electrode reaction kinetics in two ways.

(a) If $z \neq 0$, the concentration of O^z at x_2 will be different from that immediately outside the diffusion layer, $C_O{}^b$, which for our calculations can be regarded as the concentration "at the electrode surface."† Thus from (12.3.3),

$$C_O(x_2, t) = C_O{}^b\, e^{-zF\phi_2/RT} \tag{12.7.2}$$

An alternate way of expressing this idea is to note that when the electrode has a positive charge (i.e., $q^M > 0$), $\phi_2 > 0$, and anions (e.g., $z = -1$) will be attracted to the electrode surface, while cations (e.g., $z = +1$) will be repelled. For $q^M < 0$, the opposite effect will hold, while at the PZC, $q^M = 0$, $\phi_2 = 0$, and $C_O(x_2, t) = C_O{}^b$.

(b) The potential difference driving the electrode reaction is not $\phi^M - \phi^S$ (as in Section 3.4), but instead $\phi^M - \phi^S - \phi_2$; thus the effective electrode potential is $E - \phi_2$.

Consider the rate equation for a totally irreversible reaction as written previously (e.g., in Section 3.3):

$$\frac{i}{nFA} = k^0 C_O(0, t)\, e^{-(\alpha nF/RT)(E - E^{0'})} \tag{12.7.3}$$

Let us now apply the correction in (12.7.2) and that for E. The equation written in terms of the true rate constant, k_t^0, is then

$$\frac{i}{nFA} = k_t^0 C_O{}^b e^{-zF\phi_2/RT}\, e^{-(\alpha nF/RT)(E - \phi_2 - E^{0'})} \tag{12.7.4}$$

or

$$\boxed{\frac{i}{nFA} = k_t^0\, e^{(\alpha n - z)F\phi_2/RT} C_O{}^b\, e^{-\alpha nF(E - E^{0'})/RT}} \tag{12.7.5}$$

---

† Since the diffuse layer thickness ($\sim 1/\kappa$, see Table 12.3.1) is much smaller than the diffusion layer thickness, even for dilute solutions and very short experimental durations, our previous specification "at $x = 0$" can be taken as outside the diffuse layer.

By comparison of (12.7.3) and (12.7.5), noting that $C_O{}^b \approx C_O(0, t)$, we find

$$\boxed{k^0 = k_t^0 \exp\left[\frac{(\alpha n - z)F\phi_2}{RT}\right]} \tag{12.7.6}$$

This important relationship, sometimes called the *Frumkin correction*, allows the calculation of the true (or corrected) standard rate constant k_t^0 from the apparent one k^0. In a similar way a true exchange current $i_{0,t}$ can be defined as in (3.5.6):

$$i_{0,t} = nFAk_t^0 C_O^{*(1-\alpha)} C_R^{*\alpha} \tag{12.7.7}$$

$$\boxed{i_0 = i_{0,t} \exp\left[\frac{(\alpha n - z)F\phi_2}{RT}\right]} \tag{12.7.8}$$

An alternate, and somewhat more rigorous derivation of the relations in (12.7.6) and (12.7.8) can be obtained using the approach based on electrochemical potentials, as outlined in Section 3.4 (39, 105). For the derivation including the double-layer effect, the process expressed by (12.7.1) is represented as proceeding through four states plus the transition state:

State I. O^z in bulk solution and n electrons in metal. Thus, as in (3.4.8),

$$\bar{G}_I^0 = \mu_O^{0S} + n\mu_e^{0\,M} - nF\phi^M \tag{12.7.9}$$

(with ϕ^S being taken as a zero reference value.)

State II. O^z at OHP (x_2) and n electrons in metal (the so-called *preelectrode state*):

$$\bar{G}_{II}^0 = \mu_O^{0S} + n\mu_e^{0M} + zF\phi_2 - nF\phi^M \tag{12.7.10}$$

The Transition State. Parameters denoted by ‡.

State III. $R^{z'}$ at OHP:

$$\bar{G}_{III}^0 = \mu_R^{0S} + z'F\phi_2 \tag{12.7.11}$$

State IV. $R^{z'}$ in bulk solution; see eq. (3.4.9):

$$\bar{G}_{IV}^0 = \mu_R^{0S} \tag{12.7.12}$$

The derivation then follows that shown in Section 3.4 and yields (12.7.6) and (12.7.8).

The overall effect of the double layer on kinetics (sometimes also referred to as "the ϕ_2-effect" or, in the Russian literature, as "the ψ-effect") is that the apparent quantities, k^0 and i^0, are functions of potential, through the variation of ϕ_2 with E. They are functions of supporting electrolyte concentration, C, as well, since ϕ_2 depends on C. Correction of apparent rate data to find the potential and concentration-independent k_t^0 or i_t^0 values thus involves obtaining a value of ϕ_2 for the given experimental conditions based on some model for the double-layer structure (see Section 12.3).

12.7.2 Double-Layer Effects in the Absence of Specific Adsorption of Electrolyte

Corrections for the ϕ_2 effect can be made most readily for the mercury electrode, since the variation of σ^M with E and C can be obtained from electrocapillary curves, as discussed in Section 12.2. In the absence of specific adsorption of electrolyte, ϕ_2 can then be calculated by assuming that the GCS model applies [12.3.26)]. Such corrections are rarely attempted at solid electrodes because of the general lack of data about the double-layer structure at them. Typical results showing such corrections for the reduction of Zn(II) at a Zn(Hg) electrode in aqueous solution (99) and for the reduction of several aromatic compounds in *N,N*-dimethylformamide solution (106) are shown in Table 12.7.1. Note that for the Zn(II) reduction, where $n = 2$, $z = 2$, and $\alpha = 0.30$, the i^0 value is larger than i_t^0 for negative ϕ_2 values, since the negative charge on the electrode attracts the positively charged zinc ion and this effect outweighs the effect of the potential drop in the diffuse double layer. On the other hand, for reduction of the uncharged aromatic compounds, $n = 1$, $z = 0$, and $\alpha \sim 0.5$, k_t^0 is larger than k^0. The size of the correction depends on ϕ_2, which is a function of $E - E_z$ and the concentration of supporting electrolyte, as well as on α, n, and z. The correction

Table 12.7.1

Typical Experimental Results Showing Corrections of Heterogeneous Electron Transfer Rate Data for Double-Layer Effects

A. Zn(II) Reduction at Zn(Hg)[a]

Supporting Electrolyte (M)	ϕ_2 (mV)	i_0 mA/cm^2	$i_{0,t}$ mA/cm^2
0.025 $Mg(ClO_4)_2$	−63.0	12	0.40
0.05	−56.8	9	0.43
0.125	−46.3	4.7	0.37
0.25	−41.1	2.7	0.38

B. Reduction of Aromatic Compounds at Mercury Electrode in 0.5 *M* tetra-*n*-butylammonium perchlorate in *N,N*-dimethylformamide[b]

Compound	$E_{1/2}$ (V *vs.* SCE)	α	ϕ_2 (mV)	k^0 (cm/sec)	k_t^0 (cm/sec)
Benzonitrile	−2.17	0.64	83	0.61	4.9
Phthalonitrile	−1.57	0.60	71	1.5	7.4
Anthracene	−1.82	0.55	76	5.0	27.0
p-Dinitrobenzene	−0.55	0.61	36	0.9	2.2

[a] From Reference 99, $n = 2$, $z = 2$, $T = 26 \pm 1$ °C. For $C_{Zn(II)} = 2$ mM, $C_{Zn(Hg)} = 0.048$ M. Exchange currents determined by galvanostatic method. $\alpha = 0.30$.
[b] From Reference 106, $n = 1$, $z = 0$, $T = 22 \pm 2$ °C. Concentrations of compounds $\sim$1 mM. Rate constants measured by ac impedance method.

factors in several different possible cases for the actual ϕ_2 values observed for a mercury electrode in NaF (107) are shown in Table 12.7.2. Clearly these factors can be quite large, especially at low concentrations of supporting electrolyte and at potentials distant from E_z. A number of other cases and details about the treatment of experimental data are discussed in more extensive reviews (23, 39).

While these results involving double-layer corrections are very useful in explaining supporting electrolyte effects on rate constants, we must be aware of several limitations in this treatment. The absence of specific adsorption of electrolyte, reactants, and products is a rather rare occurrence. The limitations of the GCS model, as well as the general lack of a single "plane-of-closest approach" when the electrolyte contains a number of different ions, leads to uncertainties in the correct values for ϕ_2 and x_2. Indeed these uncertainties often lead to sufficient differences in correction factors as to hinder a comparison of measured apparent rate constants to those predicted by different theories of electron transfer (106). In addition, the GCS model involves average potentials in the vicinity of the electrode and ignores the discrete nature of charges in solution. Such "discreteness-of-charge effects" have been treated and invoked to account for failures in the usual double-layer corrections (108).

Table 12.7.2

Double-Layer Data for Mercury Electrode in NaF solutions and Frumkin Correction Factors for Several Different Cases[a]

			Frumkin Correction Factors ($\alpha = 0.5$)[b]		
$E - E_z$ (V)	σ^M (μC/cm^2)	ϕ_2 (V)	$n = 1$, $z = 0$	$n = 1$, $z = 1$	$n = 1$, $z = -1$
0.010 M NaF ($E_z = -0.480$ V *vs.* NCE)					
−1.4	−23.2	−0.189	0.025	39.5	1.6×10^{-5}
−1.0	−16.0	−0.170	0.037	27.3	4.9×10^{-5}
−0.5	−8.0	−0.135	0.072	13.8	3.8×10^{-4}
0	0	0	1.0	1.0	1.0
+0.5	11.5	0.153	19.6	0.051	7.5×10^{3}
0.10 M NaF ($E_z = -0.472$ V *vs.* NCE)					
−1.4	−24.4	−0.133	0.075	13.3	4.3×10^{-4}
−1.0	−17.0	−0.114	0.11	9.2	1.3×10^{-3}
−0.5	−8.9	−0.083	0.20	5.0	7.9×10^{-3}
0	0	0	1.0	1.0	1.0
+0.5	13.2	0.102	7.3	0.14	3.8×10^{2}
1 M NaF ($E_z = -0.472$ V *vs.* NCE)					
−1.4	−25.7	−0.078	0.22	4.6	1.1×10^{-2}
−1.0	−18.0	−0.062	0.30	3.3	2.6×10^{-2}
−0.5	−9.8	−0.039	0.47	2.1	0.10
0	0	0	1.0	1.0	1.0
+0.5	14.9	0.054	2.9	0.35	23

[a] σ^M and ϕ_2 data taken from compilation (107) based on Grahame's data.
[b] Correction factor $= \exp[(\alpha n - z)\phi_2 F/RT]$.

12.7.3 Double-Layer Effects with Specific Adsorption of Electrolyte

When an ion from the supporting electrolyte (e.g., Cl^-, I^-) is specifically adsorbed, the value of ϕ_2 is perturbed from the value calculated strictly from diffuse double-layer corrections. Specific adsorption of an anion will cause ϕ_2 to be more negative, while specific adsorption of a cation will cause ϕ_2 to be more positive. In principle these effects could be taken into account using the Frumkin correction factor; however, the question of the location of the plane-of-closest approach for the reacting species and the actual potential at the preelectrode state often cannot be answered, and qualitative, rather than quantitative, explanations of these effects are usually given. Specific adsorption of an ion may also result in blocking of the electrode surface, as discussed in Section 12.6, and decrease the reaction rate, independent of the ϕ_2 effect. Consider the case of the polarographic reduction of CrO_4^{2-} at the DME. Because $z = -2$, the rate of reaction is very sensitive to ϕ_2 effects (109). The addition of quaternary ammonium (R_4N^+) hydroxides at low concentrations greatly accelerates the reduction rate, because R_4N^+ is specifically adsorbed from aqueous solutions, and this adsorption makes ϕ_2 more positive (Figure 12.7.1). At higher concentrations, however, the rate is decreased. This effect is attributed to blocking of the electrode surface and is clearly of more importance as the size of the R-group increases

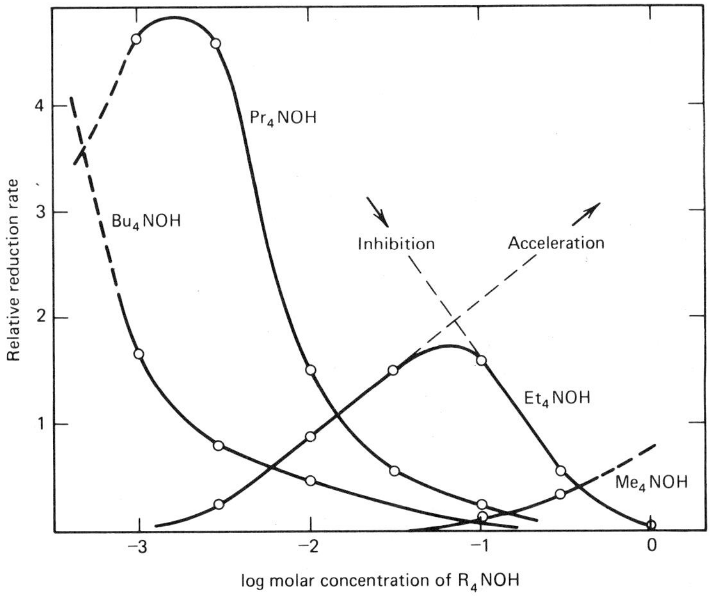

Figure 12.7.1

Variation of the rate of reduction of chromate (0.2 mM) in different tetra-alkylammonium hydroxides (R_4NOH) at −0.75 V *vs.* SCE and 25 °C. (Me, methyl; Et, ethyl; Pr, propyl; Bu, butyl). [From L. Gierst, J. Tondeur, R. Cornelissen, and F. Lamy, *J. Electroanal. Chem.*, **10**, 397 (1965), with permission.]

(Bu > Pr > Et > Me). Studies of the effects of double-layer structure on reaction rates, although frequently complicated, can provide information about details of the electrode reaction mechanism, the location of the reacting species, and the nature of the reacting site. See, for example, recent studies on the electroreduction of complex ions at a mercury electrode (110).

12.8 REFERENCES

1. P. Delahay, "Double Layer and Electrode Kinetics," Wiley-Interscience, 1965, Chap. 2.
2. D. C. Grahame, *Chem. Rev.*, **41**, 441 (1947).
3. R. Parsons, *Mod. Asp. Electrochem.*, **1**, 103 (1954).
4. D. M. Mohilner, *Electroanal. Chem.*, **1**, 241 (1966).
5. I. M. Klotz, "Chemical Thermodynamics," 2nd. ed., Benjamin, New York, 1964, pp. 14–16.
6. *Ibid.*, p. 253.
7. D. C. Grahame, *Ann. Rev. Phys. Chem.*, **6**, 337 (1955).
8. B. E. Conway, "Theory and Principles of Electrode Processes," Ronald, New York, 1965, Chaps. 4 and 5.
9. R. Payne in "Techniques of Electrochemistry," Vol. 1, E. Yeager and A. J. Salkind, Eds., Wiley-Interscience, New York, 1972, pp. 43ff.
10. G. Lippmann, *Compt. Rend.*, **76**, 1407 (1873).
11. J. Heyrovsky, *Chem. Listy*, **16**, 246 (1922).
12. J. Lawrence and D. M. Mohilner, *J. Electrochem. Soc.*, **118**, 259, 1596 (1971).
13. D. M. Mohilner, J. C. Kreuser, H. Nakadomari, and P. R. Mohilner, *ibid.*, **123**, 359 (1975).
14. D. J. Schiffrin in "Electrochemistry" (A Specialist Periodical Report), Vols. 1–3, G. J. Hills, Senior Reporter, Chemical Society, London, 1971–1973.
15. J. Clavilier and N. V. Huong, *J. Electroanal. Chem.*, **41**, 193 (1973).
16. R. A. Fredlein and J. O'M. Bockris, *Surf. Sci.*, **46**, 641 (1974).
17. R. Yu. Bek, N. V. Makhnyr, and A. G. Zelinski, *Sov. Electrochem.*, **11**, 1503 (1975).
18. A. V. Gokhshtein, *Russ. Chem., Rev.*, **44**, 921 (1975).
19. K. F. Lin and T. R. Beck, *J. Electrochem. Soc.*, **123**, 1145 (1976); **126**, 252 (1979).
20. I. Morcos in "Electrochemistry" (A Specialist Periodical Report), Vol. 6, H. R. Thirsk, Senior Reporter, Chemical Society, London, 1978, p. 65ff.
21. R. E. Malpas, R. A. Fredlein, and A. J. Bard, *J. Electroanal. Chem.*, **98**, 339 (1979).
22. P. Delahay, *op. cit.*, Chap. 3.
23. R. Parsons, *Adv. Electrochem. Electrochem. Engr.*, **1**, 1 (1961).
24. R. Payne, *J. Electroanal. Chem.*, **41**, 277 (1973).
25. R. M. Reeves, *Mod. Asp. Electrochem.*, **9**, 239 (1974).
26. F. C. Anson, *Accounts Chem. Res.*, **8**, 400 (1975).

27. H. L. F. von Helmholtz, *Ann. Physik*, **89**, 211 (1853).
28. G. Quincke, *Pogg. Ann.*, **113**, 513 (1861).
29. H. L. F. von Helmholtz, *Ann. Physik*, **7**, 337 (1879).
30. D. Halliday and R. Resnick, "Physics," 3rd ed., Wiley, New York, 1978, p. 664.
31. E. M. Pugh and E. W. Pugh, "Principles of Electricity and Magnetism," 2nd ed., Addison-Wesley, Reading, Mass., 1970, Chap. 1.
32. G. Gouy, *J. Phys. Radium*, **9**, 457 (1910).
33. G. Gouy, *Compt. Rend.*, **149**, 654 (1910).
34. D. L. Chapman, *Phil. Mag.*, **25**, 475 (1913).
35. E. M. Pugh and E. W. Pugh, *op. cit.*, pp. 69, 146.
36. O. Stern, *Z. Elektrochem.*, **30**, 508 (1924).
37. P. Delahay, *op. cit.*, Chap. 4.
38. O. A. Esin and B. F. Markov, *Acta Physicochem. USSR*, **10**, 353 (1939).
39. P. Delahay, *op. cit.*, Chap. 5.
40. B. E. Conway, T. Zawidzki, and R. G. Barradas, *J. Phys. Chem.*, **62**, 676 (1958).
41. N. A. Balashova and V. E. Kazarinov, *Electroanal. Chem.*, **3**, 135 (1969).
42. R. Parsons, *Trans. Faraday Soc.*, **55**, 999 (1959); *J. Electroanal. Chem.*, **7**, 136 (1964).
43. P. Delahay and I. Trachtenberg, *J. Am. Chem. Soc.*, **79**, 2355 (1957).
44. P. Delahay and C. T. Fike, *J. Am. Chem. Soc.*, **80**, 2628 (1958).
45. W. H. Reinmuth, *J. Phys. Chem.*, **65**, 473 (1961).
46. P. Delahay and D. M. Mohilner, *J. Am. Chem. Soc.*, **84**, 4247 (1962).
47. W. Lorenz, *Z. Elektrochem.*, **62**, 192 (1958).
48. E. Laviron, *Bull. Soc. Chim. France*, 3717 (1967).
49. S. Srinivasan and E. Gileadi, *Electrochim. Acta*, **11**, 321 (1966).
50. E. Laviron, *J. Electroanal. Chem.*, **52**, 355, 395 (1974).
51. A. P. Brown and F. C. Anson, *Anal. Chem.*, **49**, 1589 (1977).
52. H. Angerstein-Kozlowska and B. E. Conway, *J. Electroanal. Chem.*, **95**, 1 (1979).
53. V. Plichon and E. Laviron, *ibid.*, **71**, 143 (1976).
54. B. F. Watkins, J. R. Behling, E. Kariv, and L. L. Miller, *J. Am. Chem. Soc.*, **97**, 3549 (1975).
55. R. J. Lenhard and R. W. Murray, *J. Electroanal. Chem.*, **78**, 195 (1977).
56. M. Fujihira, A. Tamura, and T. Osa, *Chem. Lett.*, 361 (1977).
57. G. J. Leigh and C. J. Pickett, *J. Chem. Soc., Dalton Trans.*, 1797 (1977).
58. A. Diaz, *J. Am. Chem. Soc.*, **99**, 5838 (1977).
59. L. Horner and W. Birch, *Justus Liebigs Ann. Chem.*, 1354 (1977).
60. R. F. Lane and A. T. Hubbard, *J. Chem. Phys.*, **77**, 1401 (1973).
61. A. P. Brown, C. Koval, and F. C. Anson, *J. Electroanal. Chem.*, **72**, 379 (1976).
62. K. S. V. Santhanam, N. Jespersen, and A. J. Bard, *J. Am. Chem. Soc.*, **99**, 274 (1977).
63. M. R. Van De Mark and L. L. Miller, *J. Am. Chem. Soc.*, **100**, 639, 3223 (1978).

64. A. Merz and A. J. Bard, *ibid.*, **100**, 3222 (1978).
65. A. M. Yacynych and T. Kuwana, *Anal. Chem.*, **50**, 640 (1978).
66. M. S. Wrighton, R. G. Austin, A. B. Bocarsly, J. M. Bolts, O. Haas, K. D. Legg, L. Nadjo, and M. C. Palazzotto, *J. Electroanal. Chem.*, **87**, 429 (1978).
67. R. H. Wopschall and I. Shain, *Anal. Chem.*, **39**, 1514 (1967).
68. M. Sluyters-Rehbach and J. H. Sluyters, *J. Electroanal. Chem.*, **65**, 831 (1975).
69. S. W. Feldberg, in "Computers in Chemistry and Instrumentation," Vol. 2, "Electrochemistry", J. S. Mattson, H. B. Mark, Jr., and H. C. MacDonald, Jr., Eds., Marcel Dekker, New York, 1972, Chap. 7.
70. R. H. Wopschall and I. Shain, *Anal. Chem.*, **39**, 1535 (1967).
71. M. H. Hulbert and I. Shain, *ibid.*, **42**, 162 (1970).
72. R. Brdicka, *Z. Elektrochem.*, **48**, 278, 686 (1942).
73. R. Brdicka, *Collect. Czech. Chem. Commun.*, **12**, 522 (1947).
74. M. Sluyters-Rehbach, C. A. Wijnhorst, and J. H. Sluyters, *J. Electroanal. Chem.*, **74**, 3 (1976).
75. J. Koryta, *Collect. Czech. Chem. Commun.*, **18**, 206 (1953).
76. F. C. Anson, *Anal. Chem.*, **38**, 54 (1966).
77. J. H. Christie, R. A. Osteryoung, and F. C. Anson, *J. Electroanal. Chem.*, **13**, 236 (1967).
78. F. C. Anson, J. H. Christie, and R. A. Osteryoung, *ibid.*, **13**, 343 (1967).
79. F. C. Anson and D. J. Barclay, *Anal. Chem.*, **40**, 1791 (1968) and references therein.
80. H. B. Herman, R. L. McNeely, P. Surana, C. M. Elliot, and R. W. Murray, *Anal. Chem.*, **46**, 1268 (1974) and references therein.
81. M. T. Stankovich and A. J. Bard, *J. Electroanal. Chem.*, **86**, 189 (1978).
82. F. C. Anson, *Anal. Chem.*, **33**, 1123 (1961).
83. W. Lorenz, *Z. Elektrochem.*, **59**, 730 (1955).
84. W. H. Reinmuth, *Anal. Chem.*, **33**, 322 (1961).
85. S. V. Tatwawadi and A. J. Bard, *Anal. Chem.*, **36**, 2 (1964).
86. M. Sluyters-Rehbach and J. H. Sluyters, *Electroanal. Chem.*, **4**, 1 (1970).
87. B. Timmer, M. Sluyters-Rehbach, and J. H. Sluyters, *J. Electroanal. Chem.*, **18**, 93 (1968).
88. P. Delahay and K. Holub, *ibid.*, **16**, 131 (1968).
89. P. Delahay, *ibid.*, **19**, 61 (1968).
90. I. Epelboin, C. Gabrielli, M. Keddam, and H. Takenouti, *Electrochim. Acta*, **20**, 913 (1975) and references therein.
91. D. E. Smith, *Electroanal. Chem.*, **1**, 1 (1966).
92. A. M. Bond and G. Hefter, *J. Electroanal. Chem.*, **35**, 343 (1972); **42**, 1 (1973).
93. S. Gilman, *Electroanal. Chem.*, **2**, 111 (1967).
94. J. Heyrovsky and J. Kuta, "Principles of Polarography," Academic Press, New York, 1966.
95. C. N. Reilley and W. Stumm, in "Progress in Polarography", P. Zuman and I. M. Kolthoff, Eds., Wiley-Interscience, New York, 1962, Vol. 1, pp. 81–121.
96. H. W. Nurnberg and M. von Stackelberg, *J. Electroanal. Chem.*, **4**, 1 (1962).

97. A. N. Frumkin, *Dokl. Akad. Nauk. S.S.S.R.*, **85**, 373 (1952); *Electrochim. Acta*, **9**, 465 (1964).
98. R. Parsons, *J. Electroanal. Chem.*, **21**, 35 (1969).
99. A. Aramata and P. Delahay, *J. Phys. Chem.*, **68**, 880 (1964).
100. T. Biegler and H. A. Laitinen, *J. Electrochem. Soc.*, **113**, 852 (1966).
101. K. K. Niki and N. Hackerman, *J. Phys. Chem.*, **73**, 1023 (1969); *J. Electroanal. Chem.*, **32**, 257 (1971).
102. R. Woods, *Electroanal. Chem.*, **9**, 1 (1976).
103. P. N. Ross, Jr., *J. Electrochem. Soc.*, **126**, 67 (1979).
104. A. N. Frumkin, *Z. Physik. Chem.*, **164A**, 121 (1933).
105. D. M. Mohilner and P. Delahay, *J. Phys. Chem.*, **67**, 588 (1963).
106. H. Kojima and A. J. Bard, *J. Am. Chem. Soc.*, **97**, 6317 (1975).
107. C. D. Russell, *J. Electroanal. Chem.*, **6**, 486 (1963).
108. W. R. Fawcett and S. Levine, *ibid.*, **43**, 175 (1973).
109. L. Gierst, J. Tondeur, R. Cornelissen, and F. Lamy, *ibid.*, **10**, 397 (1965).
110. M. J. Weaver and T. L. Satterberg, *J. Phys. Chem.*, **81**, 1772 (1977).

12.9 PROBLEMS

12.1 Prove that relative surface excesses are independent of the position of the dividing surface used in the reference system chosen for the thermodynamic treatment of the interface.

12.2 Derive the special case (12.3.11) from (12.3.10).

12.3 Present an argument, based only on Gaussian boxes, for a linear potential profile inside the compact layer.

12.4 Obtain (12.3.28) from (12.3.27).

12.5 Why do we view adsorbed neutral species as being intimately bound to the electrode surface, rather than being collected in the diffuse layer?

12.6 Interpret the data in Figures 12.9.1 and 12.9.2. How do the traces in Figure 12.9.2 relate to those in Figure 12.9.1? What implications can be derived from the flat region in the electrocapillary curves in the presence of *n*-heptyl alcohol? Construct a chemical model to explain the very low differential capacitance from -0.4 to -1.4 V in the presence of *n*-heptyl alcohol. Can you provide a formal (i.e., mathematical) rationale for the sharp peaks in C_d? Can you rationalize them chemically?

12.7 Elliott and Murray have performed chronocoulometric experiments to measure the surface excess of Tl^+ at a mercury-electrolyte interface. Of interest was the influence of bromide on the adsorption. Explain how such measurements would be carried out. The results are summarized in Figure 12.9.3. Explain the results in terms of chemical processes.

12.8 A solution containing a certain organic compound, Z, at a concentration of 1.00×10^{-4} M shows a UV absorbance ($\mathscr{A}$) of 0.500 when measured at 330 nm in a spectrophotometric cell of path length 1.00 cm. Into 50 cm^3 of

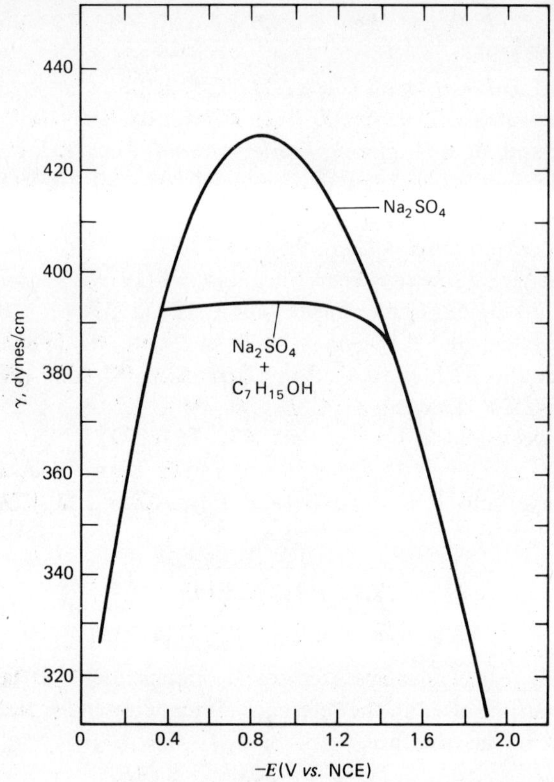

Figure 12.9.1
Electrocapillary curves for mercury in contact with 0.5 *M* Na_2SO_4 in the presence and absence of *n*-heptanol. Data from G. Gouy, *Ann. Chim. Phys.*, **8**, 291 (1906). [Reprinted with permission from D. C. Grahame, *Chem. Rev.*, **41**, 441 (1947). Copyright 1947, Williams and Wilkins, Inc.]

this solution, a platinum electrode with a surface area of 100 cm^2 is immersed. If the amount of Z adsorbed corresponds to 1.0×10^{-9} mol/cm^2, what will be the absorbance of this solution after adsorption equilibrium occurs?

12.9 The adsorption of a certain substance, X, follows a Langmuir isotherm. Saturation coverage of the material is 8×10^{-10} mol/cm^2 and $\beta = 5 \times 10^7$ cm^3/mol (assuming $a_i = C_i$). At what concentration of X will the electrode surface be half covered (i.e., $\theta = \frac{1}{2}$)? Sketch the adsorption isotherm of the substance. At what concentrations of X will the linearized isotherm be valid (to ~1%)?

12.10 For the substance X in Problem 12.9, under linearized conditions and taking $D = 10^{-5}$ cm^2/sec, how long after immersion will be required for the surface of a plane electrode to attain half of the equilibrium coverage (see Figure 12.4.1)? How long will be required to attain half of equilibrium coverage if the solution is stirred and $m = 10^{-2}$ cm/sec?

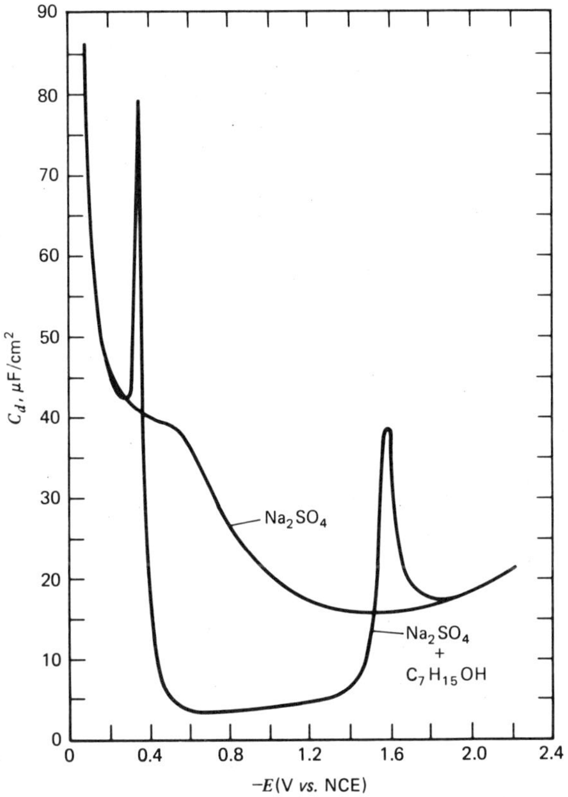

Figure 12.9.2
Differential capacitance curves corresponding to the systems of Figure 12.9.1. [Reprinted with permission from D. C. Grahame, *Chem. Rev.*, **41**, 441 (1947). Copyright 1947, Williams and Wilkins, Inc.]

12.11 Derive, using a kinetic model, the Langmuir isotherms for the simultaneous adsorption of two species, *i* and *j* [see (12.4.9) and (12.4.10)].

12.12 From the curves in Figure 12.5.3*b*, estimate the amount of *trans*-4,4′-dipyridyl-1,2-ethylene adsorbed per cm². Assume $n = 2$.

12.13 From Figure 12.5.10 for weak adsorption of a reactant, estimate the ranges of v where (a) i_p is proportional to v and (b) where i_p is proportional to $v^{1/2}$, in terms of β_O, $\Gamma_{O,s}$, and D_O, at 25 °C and $\beta_O C_O^* = 1$.

12.14 The amount of adsorbed O, Γ_O, can also be determined in a double potential step chronocoulometric experiment from the ratio of the slopes of the curves for the forward (S_f) and reverse (S_r) curves. Explain how.

12.15 Using the data in Figure 12.5.15, calculate D_O and Γ_O [O is Cd(II)]. Also calculate Q_{dl} and C_d in absence and presence of SCN^-.

12.16 Calculate the values of σ^M corresponding to various values of ϕ_2 (from -0.2 to $+0.2$ V) for a mercury electrode in 0.01 *M* NaF based on the GCS

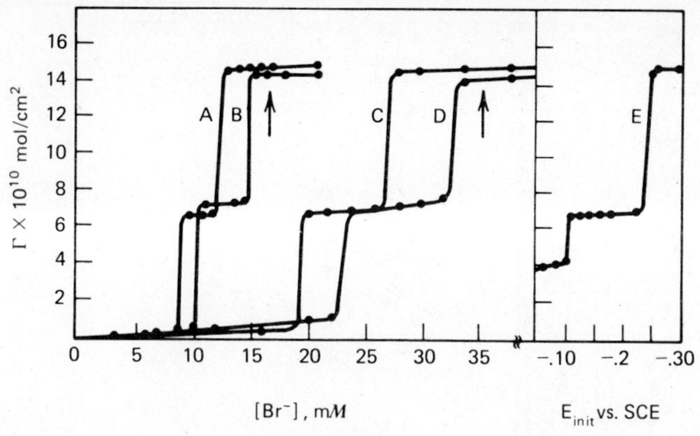

Figure 12.9.3

Surface excesses of Tl^+ at mercury in the presence of Br^-. The step potential was −0.70 V *vs.* SCE in every case. Curve A: 1 m*M* Tl^+, initial potential = −0.30 V. Curve B: 1 m*M* Tl^+, initial potential = −0.20 V. Curve C: 0.5 m*M* Tl^+, initial potential = −0.30 V. Curve D: 0.5 m*M* Tl^+, initial potential = −0.20 V; Curve E: 1 m*M* Tl^+, 14 m*M* Br^-. Arrows show saturation with respect to precipitation of TlBr from bulk solution. [Reprinted with permission from C. M. Elliott and R. W. Murray, *J. Am. Chem. Soc.*, **96**, 3321 (1974). Copyright 1974, American Chemical Society.]

model. (a) Plot ϕ_2 *vs.* σ^M. (b) From the variation of σ^M with $E - E_z$ shown in Table 12.7.2, prepare a plot of ϕ_2 *vs.* $E - E_z$.

12.17 Because of the ϕ_2 or Frumkin effect, Tafel plots are not linear and have the following varying slopes in the cathodic region:

$$\frac{F}{RT}\left[-\alpha n_a + (\alpha n_a - z)\left(\frac{\partial \phi_2}{\partial \eta}\right)\right]$$

(a) Derive this equation. (b) Asada, Delahay, and Sundaram [J. Am. Chem. Soc., **83**, 3396 (1961)] suggested that a plot of $\ln[i \exp(zF\phi_2/RT)]$ against $\phi_2 - \eta$ (a "corrected Tafel plot") is linear and has a slope of $\alpha nF/RT$. Show, by suitable manipulation of the equations, that this is so.

12.18 Aramata and Delahay (99) found that for a solution containing 2 m*M* Zn(II) and 0.025 *M* $Ba(ClO_4)_2$ at a Zn(Hg) electrode containing 0.048 *M* Zn, the apparent exchange current density was 9.1 mA/cm². At the equilibrium potential for this system, $\phi_2 = -60.8$ mV. Calculate $i_{0,t}$ and k_t^0, given $\alpha = 0.30$, $n = 2$, and $z = +2$.

chapter 13
Electrochemical Instrumentation

In this chapter, we will consider the electronic means for enforcing a controlled potential at an electrode or for controlling the current through a cell. Our goal is to explore the basis for the usual approaches, not to review exhaustively all types of electrochemical instruments.

The chief electrochemical variables are all *analog* (or continuous) quantities (at least in the ranges of interest), so our concern is with circuitry for *controlling* and *measuring* voltages, currents, and charges in the analog domain. The circuit elements best suited to these jobs are *operational amplifiers*. We will have to explore their properties before we can understand the way in which the amplifiers are assembled into instruments.

13.1 OPERATIONAL AMPLIFIERS

13.1.1 Ideal properties (1–7)

Operational amplifiers are devices with special properties, and are almost always used as purchased, packaged circuit components. Most are now integrated circuits, although many are hermetically sealed packages containing discrete transistors, resistors, capacitors, and so on. We have no interest in the contents of the amplifier; our concern is strictly with its behavior as a unit in a circuit.

In Figure 13.1.1*a* we note that several connections must be made to the amplifier. First there are the power lines. Usually these devices require two supplies, one at +15 V and the other at −15 V relative to a common circuit point, defined at the power supplies, called *ground*. Many measurements are made with respect to this point, which may or may not be related to *earth ground*. In addition to the power lines, there are input and output connections as shown. Usually one side of the output is ground. Most amplifiers are constructed so that neither input terminal must be at ground; hence both can be *floating* away from ground. The important parameter is the difference in voltage between the two input terminals. In circuit diagrams, the

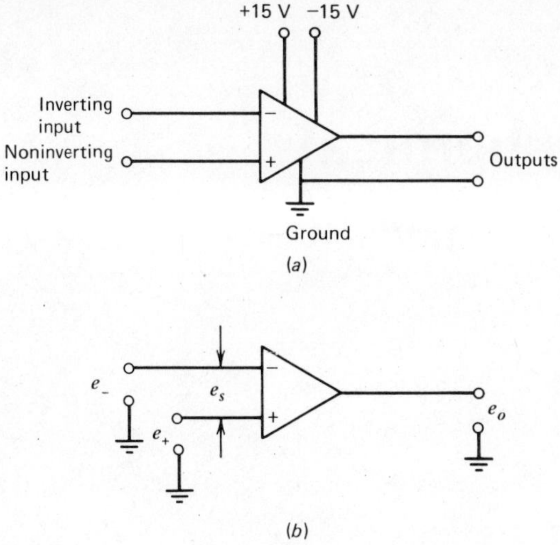

Figure 13.1.1
Schematic diagrams of operational amplifiers.

power connections are always understood, and the amplifier is depicted as shown in Figure 13.1.1*b*.

The two input terminals are labelled with signs in the manner depicted there. The top one is called the *inverting input* and the bottom one is the *noninverting input*. The fundamental property of the amplifier is that the output, e_o, is the inverted, amplified voltage difference, e_s, where e_s is the voltage of the inverting input *with respect to the noninverting input*. That is,

$$e_o = -Ae_s \tag{13.1.1}$$

where A is the *open-loop gain*.

The names of the inputs come from a different way of looking at e_s. We could picture the system as having two independent inputs, e_- and e_+, both measured with respect to ground. The output is then

$$e_o = -Ae_- + Ae_+ \tag{13.1.2}$$

which is the sum of the inverted, amplified signal, e_-, and the noninverted, amplified signal, e_+. Equation 13.1.2 is equivalent to (13.1.1), since $e_s = e_- - e_+$.

The ideal operational amplifier has several important properties. First, its open-loop gain A is effectively infinite so that the slightest input voltage e_s will drive the output to the limit deliverable by the power supply (usually ± 13–14 V). The reasons for desiring the highest possible amplification factor will become clear in Section 13.2. For now, let us note that if the ideal amplifier is operating in any circuit with its output anywhere in the range between the voltage limits, then the two input terminals *must be at the same voltage.*

Ideal amplifiers also have infinite input impedance, so that they can accept input voltages without drawing any current from the voltage sources. This feature enables us to measure voltages without perturbing them. On the other hand, our ideal device can supply any desired current to its load; hence it has effectively zero output impedance. Finally, we regard the ideal amplifier as having infinite bandwidth; that is, it responds faithfully to a signal of any frequency.

In most discussions of circuitry we assume ideal behavior because it simplifies the approach. For most electrochemical applications, available devices perform so well that nonidealities are negligible. However, in demanding circumstances, nonideal properties may have to be recognized.

13.1.2 Nonidealities (1–7)

(a) Open-Loop Gain. Actual devices have A values for dc signals ranging from 10^4 to 10^8. A typical figure for a general-purpose amplifier is 10^5. The open-loop gain is frequency dependent. It declines at high frequencies, and this feature is one aspect bearing on the useful operating range.

(b) Bandwidth. The degradation in performance of real devices at high frequencies can be measured in several ways. The frequency at which the open-loop gain becomes unity for an input signal of small amplitude is called the *unity-gain bandwidth*. Depending on the purpose for which the device is designed, this bandwidth may be as low as 100 Hz or as high as 100 MHz. Typical values for general-purpose amplifiers are near 1 MHz. Since most applications of operational amplifiers are based on a high open-loop gain, the useful bandwidth is usually one or two orders of magnitude lower than the unity-gain bandwidth.

Another parameter describing an amplifier's limitations at high frequency is the *slew rate*, which is the maximum rate of change in the output voltage in response to a large amplitude step at the input. Real values range from 100 V/sec to 1000 V/μsec. General-purpose devices have slew rates on the order of 1 V/μsec; hence the minimum time required for a transition over their full output range would be about 30 μsec.

Still a third characterization of high-frequency response is the *settling time*. This figure applies to an amplifier operating in a given feedback-stabilized circuit. Often a unity-gain inverter is used (Section 13.2.2). An essentially ideal step function is applied at the input. The settling time is then measured as the time required for the output to settle within some defined error range (usually 0.1 to 0.01%) around the new equilibrium output value. The settling time is dependent on the circuit within which the amplifier is used.

The characteristics of present amplifiers are such that one can easily obtain accurate, reliable performance on time scales of 100 μsec or greater (i.e., bandwidths less than 10 kHz). Time scales between 10 to 100 μsec (bandwidths between 100 to 10 kHz) can be reached with care in circuit design and choice of components. Building reliable operational amplifier circuits like those described below for time scales under 10 μsec is very difficult.

(c) Input Impedance. The range of input impedance in real devices is 10^5 to 10^{13} Ω. General-purpose amplifiers typically offer about 10^6 Ω. Higher impedances are specifically sought for more demanding purposes, such as monitoring resistive voltage sources (like glass electrodes) and service in integrators.

(d) Output Limits. The *voltage limits* of the amplifier are controlled by the power supplies. They usually are quite close to the supply values. For most devices, the limits are $\pm \sim 13$ V. Currents will be supplied freely to a load until the limits of current are reached, typically at ± 5 to 100 mA. Special devices with larger current or voltage output limits are available, but high output power in operational amplifier circuits is usually obtained by booster stages, as described below.

(e) Offset voltage. In general, a zero input voltage e_s will not produce zero output voltage in a practical device. Instead there is a nonzero *offset* at the output. Most amplifiers have a provision for nulling the offset by an external adjustable resistor.

(f) Other Properties. In some applications, noise and drift characteristics of the devices and their stability with temperature may be of concern. Usually these aspects are of secondary importance in electrochemical instrumentation.

13.2 CURRENT FEEDBACK

We have already noted that a negligibly small voltage differential at the inputs will drive a practical amplifier to its limit; thus we almost never use the amplifier to deal with an input signal without modification. Normally the amplifier is stabilized by feeding back part of its output to the *inverting* input. The manner in which the feedback is accomplished determines the operational properties of the whole circuit. Here our concern is with circuits involving the routing of a *current* from the output to the input (1–7).

13.2.1 Current Follower

Consider the circuit shown in Figure 13.2.1. The resistor, R_f, is the feedback element, through which there is a feedback current, i_f. The input is a current i_{in}, which might be from a working electrode or a photomultiplier tube. From the conservation of charge (Kirchhoff's law) the sum of all the currents into the *summing point S* must be zero and, since a negligibly small current passes between the inputs,

$$i_f = -i_{\text{in}} \tag{13.2.1}$$

From Ohm's law,

$$\frac{e_o - e_s}{R_f} = -i_{\text{in}} \tag{13.2.2}$$

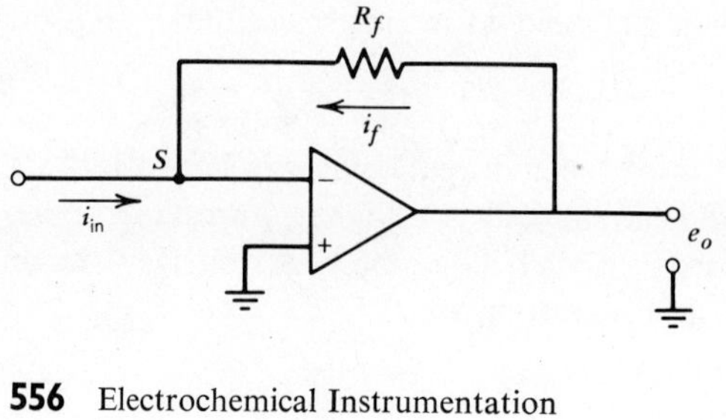

Figure 13.2.1
A current follower.

and, by substitution from (13.1.1),

$$e_o\left(1 + \frac{1}{A}\right) = -i_{in} R_f \tag{13.2.3}$$

Since the value of A is very high, the parenthesized quantity is virtually unity, and

$$\boxed{e_o \simeq -i_{in}R_f} \tag{13.2.4}$$

Thus the output voltage is proportional to the input current by a scale factor determined by R. The circuit is called a *current follower* or a *current-to-voltage* (i/E or i/V) *converter*.

The voltage of the summing point e_s is $-e_o/A$, which for a typical device lies between ± 15 V/10^5, or ± 150 μV. In other words, S is a *virtual ground.* It is not a true ground in that there is no direct connection, but it has virtually the same potential as ground. This feature is important because it allows currents to be converted to equivalent voltages while the current source is maintained at ground potential. We will utilize that virtue later in building a potentiostat.

There is an easier way to analyze this circuit than we have used here. Since we recognize that the two inputs are always at virtually the same potential, it is intuitive that e_s is a virtual ground. From (13.2.1), we can therefore write immediately

$$\frac{e_o}{R_f} = -i_{in} \tag{13.2.5}$$

which is the final result.

13.2.2 Scaler/Inverter

The circuit in Figure 13.2.2 differs from the current follower only in that the input current is driven through an input resistor by the voltage e_i. Our previous analysis holds exactly, but we can now reexpress i_{in} in (13.2.5) as e_i/R_i; hence

$$\boxed{e_o = -e_i\left(\frac{R_f}{R_i}\right)} \tag{13.2.6}$$

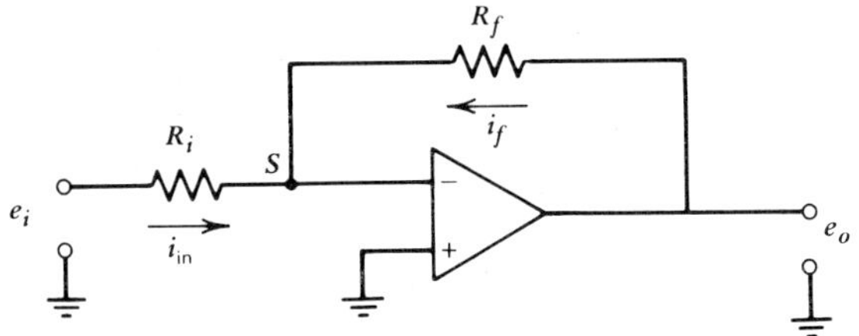

Figure 13.2.2
A scaler/inverter.

This circuit is therefore a scaler in which the output is simply the inverted input multiplied by the factor (R_f/R_i). By choosing precision resistors, (R_f/R_i) can be set at any desired value, although the practical ratios for a single stage lie between ~0.01 and ~200. When $R_f = R_i$, the circuit is an *inverter*.

Note that the voltage source e_i must be able to supply the input current i_{in}, so that the effective input impedance of the whole circuit is R_i. Typical values are 1 to 100 kΩ.

13.2.3 Adders

In Figure 13.2.3, we consider a circuit in which three different voltage sources e_1, e_2, and e_3 supply three input currents i_1, i_2, i_3 to the summing junction S through separate input resistors. The feedback arrangement is just as before. Now we write

$$i_f = -(i_1 + i_2 + i_3) \tag{13.2.7}$$

and, since the summing point is a virtual ground,

$$\frac{e_o}{R_f} = -\left(\frac{e_1}{R_1} + \frac{e_2}{R_2} + \frac{e_3}{R_3}\right) \tag{13.2.8}$$

or,

$$\boxed{e_o = -\left[e_1\left(\frac{R_f}{R_1}\right) + e_2\left(\frac{R_f}{R_2}\right) + e_3\left(\frac{R_f}{R_3}\right)\right]} \tag{13.2.9}$$

The output is therefore the sum of independently scaled input voltages. The scale factors are again set by selecting appropriate resistors. If all the resistors are equal, we have a simple inverting adder:

$$e_o = -(e_1 + e_2 + e_3) \tag{13.2.10}$$

Note that the fundamental basis of the addition is the summing of *currents* at point S. That, in turn, is possible because S is a virtual ground.

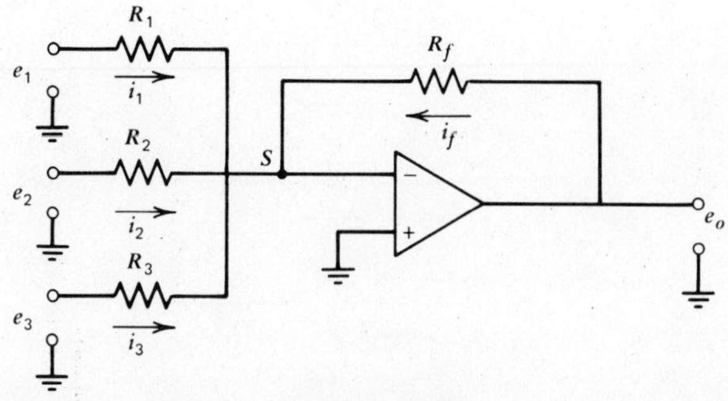

Figure 13.2.3
An adder circuit.

13.2.4 Integrators

In Figure 13.2.4, we consider a capacitor C as a feedback element. The input is a current i_{in}. Equation 13.2.1 still holds and S is still a virtual ground; therefore, we can write by substitution into (13.2.1):

$$C\frac{de_o}{dt} = -i_{\text{in}} \qquad (13.2.11)$$

or

$$\boxed{e_o = -\frac{1}{C}\int i_{\text{in}}\,dt} \qquad (13.2.12)$$

The output is a voltage proportional to the integrated input current, which is actually the charge stored on the capacitor C. Current integrators are useful in coulometric and chronocoulometric experiments.

Usually one desires to discharge the capacitor before starting a new measurement. The *reset* switch in Figure 13.2.4 allows that.

If the charges are to be stored on C for more than a few seconds, one must take care to minimize losses due to leakage. They mainly occur through the dielectric in the capacitor and through the input impedance in the amplifier. One can minimize them by choosing special capacitors and using amplifiers with very high input impedances.

An input voltage can be integrated by the circuit shown in Figure 13.2.5, where the input current is driven by e_i through the resistor R. Equation 13.2.12 still holds, and we can substitute there to obtain

$$\boxed{e_o = \frac{-1}{RC}\int e_i\,dt} \qquad (13.2.13)$$

A special type of voltage integrator is the *ramp generator*, which involves a constant e_i. If the experiment begins with a reset condition, then

$$e_o = \frac{-e_i}{RC}t \qquad (13.2.14)$$

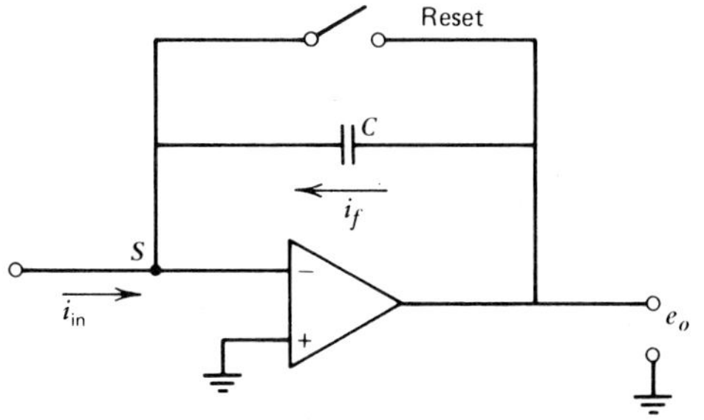

Figure 13.2.4
A current integrator.

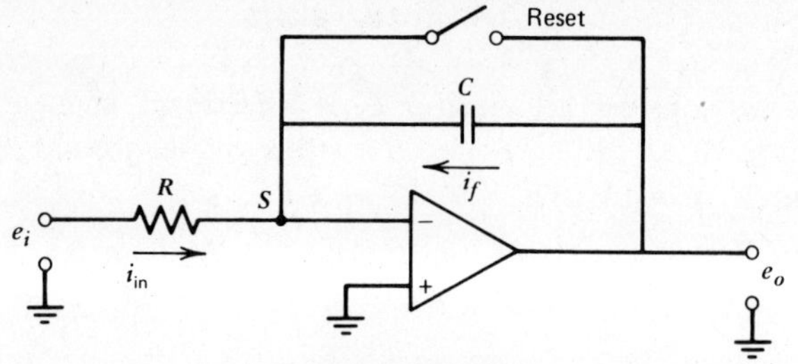

Figure 13.2.5
A voltage integrator.

Such an arrangement is often used to generate waveforms for linear sweep experiments. The sweep rate is controlled by the combination of e_i, R, and C; and the direction of sweep is governed by the polarity of e_i.

13.2.5 Differentiator

In Figure 13.2.6 one sees an input capacitor and a feedback resistor, which pass currents i_{in} and i_f, respectively. Starting as usual with equation 13.2.1 and substituting for the currents we obtain,

$$\frac{e_o}{R} = -C\frac{de_i}{dt} \tag{13.2.15}$$

or

$$\boxed{e_o = -RC\left(\frac{de_i}{dt}\right)} \tag{13.2.16}$$

The output is therefore a scaled derivative of e_i with respect to time.

This kind of circuit finds use in highlighting inflections in voltage-time functions. However, differentiation of experimental data tends to degrade the signal-to-noise ratio (Problem 13.5) and is usually avoided.

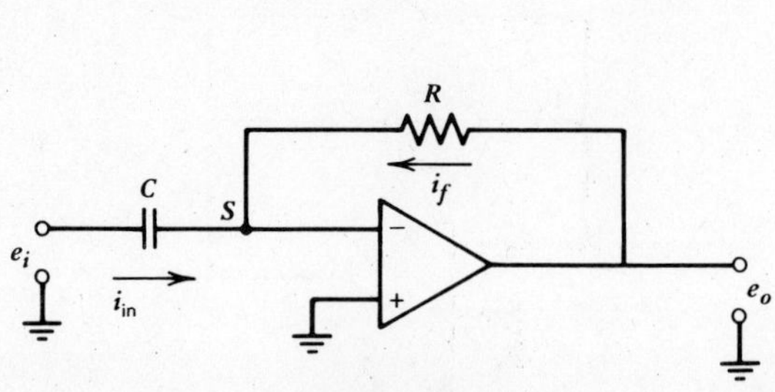

Figure 13.2.6
A differentiator.

13.3 VOLTAGE FEEDBACK

An alternative to feeding back a current from the output is stabilization by returning part of the output *voltage* to the inverting output (1–7). These circuits generally require negligible input currents and are especially well-suited to control functions and the measurement of voltages. Circuits based on current feedback are, by contrast, often better suited to signal processing in the manner discussed earlier.

13.3.1 Voltage Follower

Figure 13.3.1 contains an important circuit in which the whole output voltage is returned to the input. We treat it by invoking (13.1.1) and noting that $e_s = e_o - e_i$; thus

$$e_o = -A(e_o - e_i) \tag{13.3.1}$$

or

$$e_o = \frac{e_i}{(1 + 1/A)} \tag{13.3.2}$$

Since A is very large,

$$\boxed{e_o \simeq e_i} \tag{13.3.3}$$

This result could have been obtained intuitively by noting that the two inputs are at virtually the same potential.

The circuit is called a *voltage follower* because the output is the same as the input. Its function is to match impedances. It offers a very high input impedance and a very low output impedance; hence it can accept an input from a device that cannot supply much current (such as a glass electrode) and offer the same voltage to a significant load (e.g., a recorder). It is an intermediary that allows the *measurement* of a voltage without *perturbing* that voltage significantly.

13.3.2 Control Functions

Consider the arrangement shown in Figure 13.3.2. An input voltage is applied so that the top terminal is e_i with respect to the bottom one. However, the inverting input is a virtual ground; therefore, the lower terminal and point A are at $-e_i$ *vs*. ground.

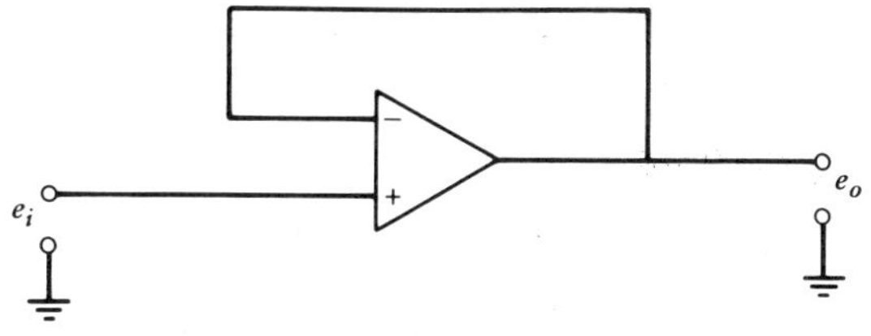

Figure 13.3.1
A voltage follower.

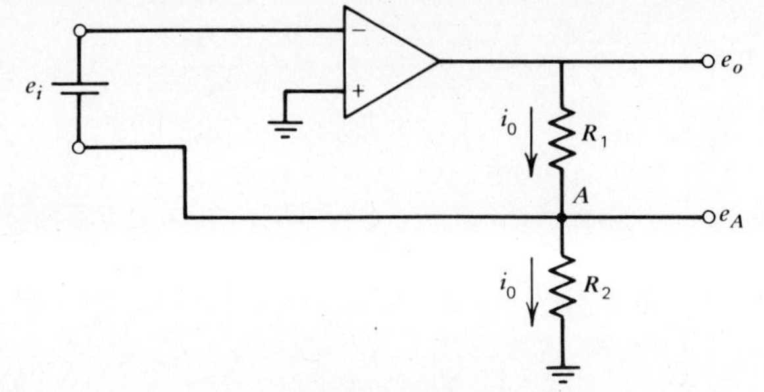

Figure 13.3.2
A circuit for controlling the potential at point A regardless of changes in R_1 and R_2. Note that the feedback circuit passes through the voltage source e_i, which is shown for simplicity as a battery.

The amplifier will adjust its output to control the currents through the resistors so that this condition is maintained. We therefore have a means for controlling the voltage at a fixed point in a network of resistances, even if the resistances (or, more generally, impedances) fluctuate during the experiment. Note that this job is precisely what we ask a potentiostat to do.

Since the current through R_1 must also pass through R_2, the total output e_o is $i_0(R_1 + R_2)$. Since $i_0 = -e_i/R_2$,

$$e_o = -e_i\left(\frac{R_1 + R_2}{R_2}\right) \tag{13.3.4}$$

This basic design can also be used to control the current through a load. Consider the circuit in Figure 13.3.3, which has an arbitrary load impedance Z_L. Since the voltage at point A is $-e_i$, the current through the resistor R is $i_0 = -e_i/R$. It passes through the load, too, and is independent of the value of Z_L or fluctuations in it. Such a circuit could be employed as a galvanostat. The cell would simply replace the load impedance (see Section 13.5).

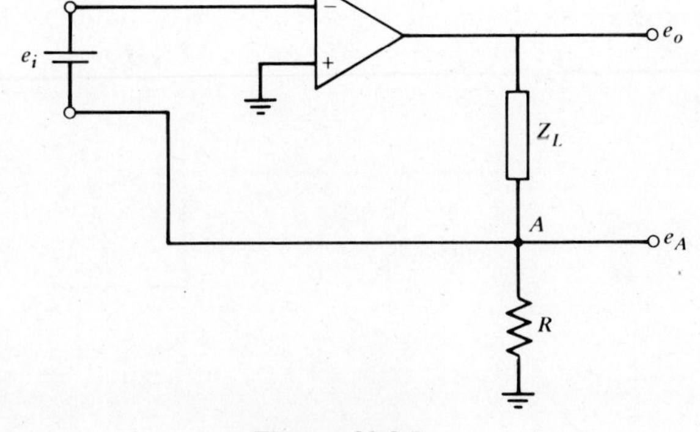

Figure 13.3.3
A circuit for controlling the current through the arbitrary load Z_L.

13.4 POTENTIOSTATS

13.4.1 Basic Considerations (5–8)

From an electronic standpoint, an electrochemical cell can be regarded as a network of impedances like those shown in the equivalent circuit, Figure 13.4.1*a*, where Z_c and Z_{wk} represent the interfacial impedances at the counter and working electrodes, and the solution resistance is divided into two fractions, R_Ω and R_u, depending on the position of the reference electrode's probe in the current path. This representation can be distilled further into that of Figure 13.4.1*b*.

Suppose we now incorporate the cell into the circuit of Figure 13.4.2. If the cell is equivalent to the network in Figure 13.4.1*b*, then we can immediately see that the overall circuit bears a strong analogy to the control system in Figure 13.3.2. The current through the cell is controlled by the amplifier so that the reference electrode is $-e_i$ *vs.* ground. Since the working electrode is grounded,

$$e_{wk}(vs.\ \mathrm{ref}) = e_i \tag{13.4.1}$$

regardless of fluctuations in Z_1 and Z_2.

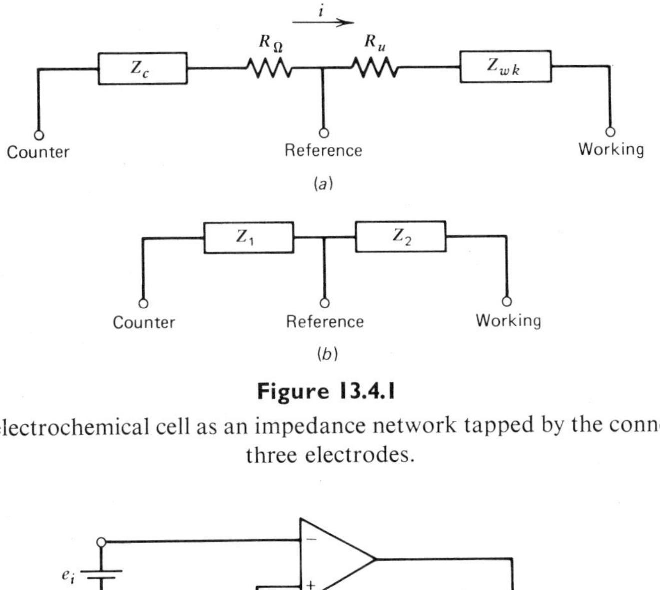

Figure 13.4.1
Views of an electrochemical cell as an impedance network tapped by the connections to the three electrodes.

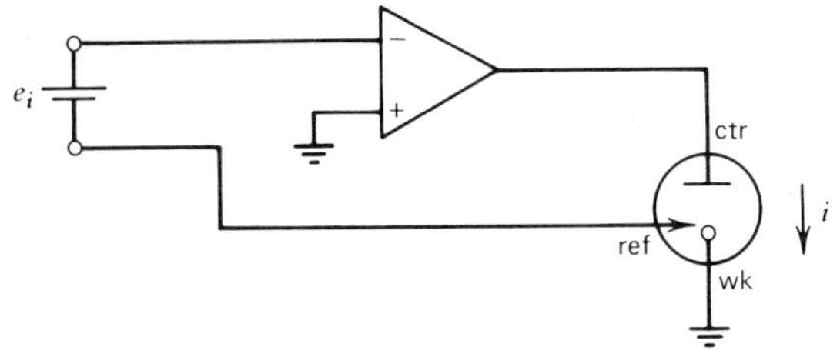

Figure 13.4.2
A simple potentiostat based on the control circuit of Figure 13.3.2.

Figure 13.4.1 shows that the controlled voltage, e_{ref} *vs.* ground, contains a portion iR_u of the total voltage drop in the solution. The presence of this *uncompensated resistance* loss keeps the circuit from giving accurate control over the true *potential* of the working electrode with respect to the reference, but in many cases iR_u can be made negligibly small by careful placement of the reference electrode (see Section 1.3.4). At other times, the uncompensated resistance is a major factor in understanding experimental results. We will have more to say about it later.

13.4.2 The Adder Potentiostat (5–8)

The potentiostat of Figure 13.4.2 illustrates the basic principles of potential control and will accomplish that task as well as any of several other designs. Its drawbacks concern its input requirements. First, note that neither input terminal is a true ground; hence the function generator supplying the waveform for potential control would have to possess a differential floating output. Most waveform sources would not meet that demand.

Consider also the *form* of the desired control function. Suppose, for example, we wish to carry out an ac polarographic experiment involving a scan from -0.5 V. The waveform needed at e_i is shown in Figure 13.4.3. It is a complicated function and could not be supplied simply. We must synthesize it by adding together a ramp function, a sinusoidal perturbation, and a constant offset. It is generally true that electrochemical waveforms are syntheses of several simpler signals, and therefore we need a general facility for accepting and adding basic inputs at the potentiostat itself.

The *adder potentiostat* shown in Figure 13.4.4 remedies both drawbacks of the control circuit considered above, and is by far the most widely used design. Since the currents into the summing point S must add to zero,

$$-i_{\text{ref}} = i_1 + i_2 + i_3 \tag{13.4.2}$$

Since S is a virtual ground,

$$-e_{\text{ref}} = e_1\left(\frac{R_{\text{ref}}}{R_1}\right) + e_2\left(\frac{R_{\text{ref}}}{R_2}\right) + e_3\left(\frac{R_{\text{ref}}}{R_3}\right) \tag{13.4.3}$$

Note that, as before, $-e_{\text{ref}}$ is the potential of the working electrode with respect to the reference. Thus the circuit maintains the working electrode at a potential equal

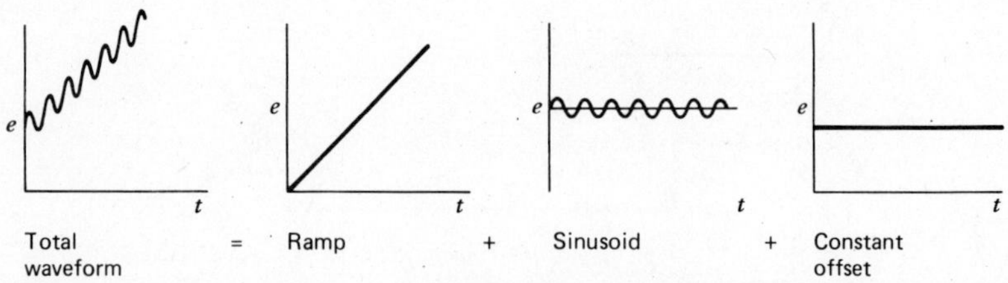

Figure 13.4.3
Synthesis of a complex waveform. For clarity, the magnitude of the sinusoid has been exaggerated and its frequency has been lowered, relative to the values usually employed.

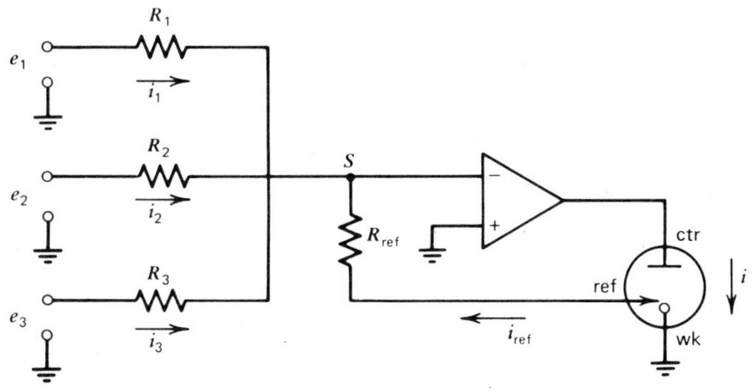

Figure 13.4.4
A basic adder potentiostat.

to the weighted sum of the inputs. Usually all the resistors have the same value, and one has

$$\boxed{e_{wk}(vs.\ \text{ref}) = e_1 + e_2 + e_3} \tag{13.4.4}$$

The facility for addition of input signals allows the straightforward synthesis of complex waveforms, and each input signal is individually referred to circuit ground. Any reasonable number of signals can be added at the input. One simply requires a resistor into the summing point for each of them.

13.4.3 Refinements to the Adder Potentiostat (5–8)

There are three important deficiencies in the design of Figure 13.4.4: (a) The reference electrode must supply a significant current i_{ref} to the summing point; (b) there is no facility for measuring the current through the cell; and (c) the power that is available at the cell is only that available from the output of the operational amplifier. Figure 13.4.5 is a schematic of a potentiostat that remedies these deficiencies. It is a design in very common use.

A voltage follower has been inserted into the feedback loop, so that the reference electrode is not loaded by the current fed into the summing point. The follower's output, e_F, is also available externally for use with a recording device. It is a convenient continuous monitor of $-e_{wk}$ ($vs.$ ref).

The working electrode now feeds a current follower, whose output is proportional to the current. Note that the current follower allows the working electrode to remain at virtual ground, which is an essential condition for the operation of the system.

Increased power has been achieved by inserting a booster amplifier in the output loop. A booster is simply a noninverting amplifier, usually having low gain, capable of delivering higher currents or higher voltages, or both, than the operational amplifier itself. Since it is noninverting, one can consider it as an extension of the operational amplifier such that the overall open-loop gain of the combination is $A = A_{OA}A_B$,

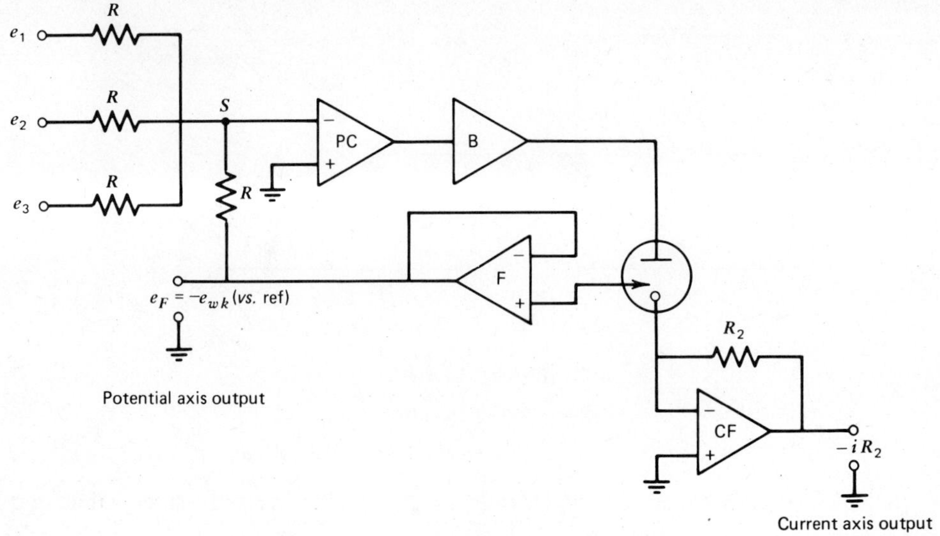

Figure 13.4.5
A full potentiostatic system based on an adder control amplifier (PC). The booster (B) might be included to improve available output voltage. If it is intended to boost the available current then a second booster would have to be added to the current follower (CF), which must handle the total cell current.

where the open-loop gain of the operational amplifier is A_{OA} and that of the booster is A_B. Then (13.1.1) applies directly and feedback principles apply as before.

13.4.4 Bipotentiostats (5, 9, 10)

Some electrochemical experiments, such as those involving rotating ring-disk electrodes, require simultaneous control of two working interfaces. A device that will meet this demand is called a *bipotentiostat*.

The usual approach is shown in Figure 13.4.6. One electrode is controlled in exactly the manner discussed in the previous section. The circuitry devoted to it is shown in the left half of the figure. The second electrode is controlled by the elements in the right half. There one finds a current follower (CF2) with a summing point held away from ground by some voltage difference Δe, because its noninverting input is away from ground by Δe. This circuit has the effect of using the first electrode as a reference point for the second. We can set the first at any desired potential e_1 with respect to the reference, then the second working electrode is offset with respect to the first by $\Delta e = e_2 - e_1$, where e_2 is the potential of the second electrode with respect to the reference. The counter electrode passes the sum of the currents i_1 and i_2.

The remaining amplifiers (I2 and Z2) serve as inverting and zero-shifting stages. They allow one to supply the desired potential e_2 at the input without concern for the value of e_1. (see Problem 13.7). Such a convenience is rather valuable when one wishes to vary e_1 and e_2 independently in time.

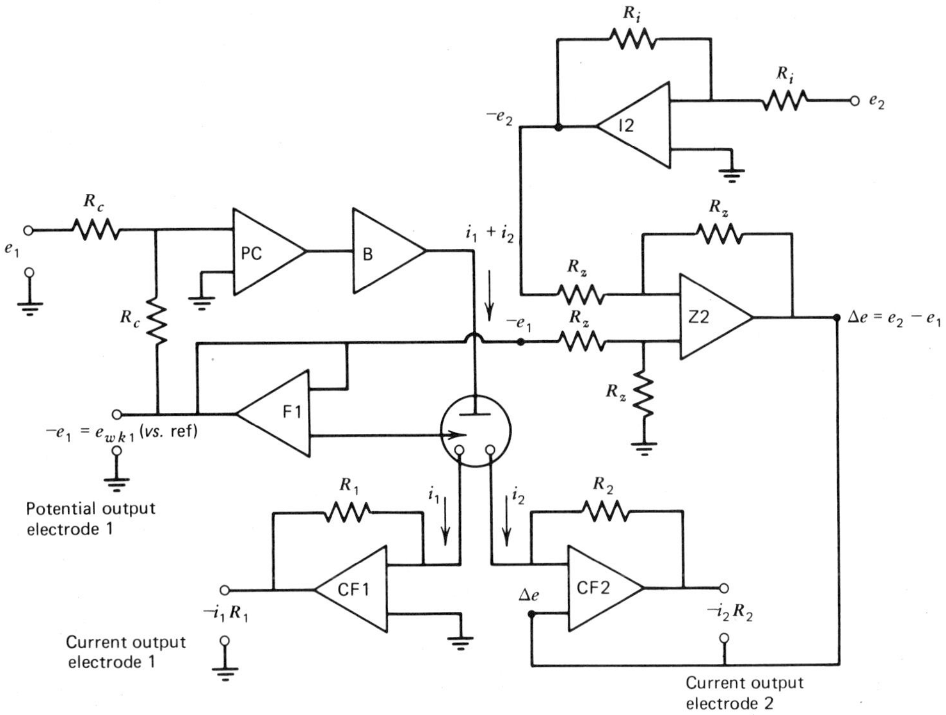

Figure 13.4.6

A bipotentiostat based on the adder concept. On the left is essentially the system of Figure 13.4.5, which is devoted to electrode 1. On the right is a network for controlling electrode 2. For large currents at both electrodes, boosters might have to be added to CF1 and CF2.

13.5 GALVANOSTATS

Controlling the current through a cell is simpler than controlling the potential at an electrode because only two elements of the cell, the working and counter electrodes, are involved in the control circuit. In galvanostatic experiments one is usually interested in the potential of the working electrode with respect to a reference electrode, and circuitry is usually added to permit that measurement, but it makes no contribution to the control function.

Two different galvanostats can be derived from operational amplifier circuits that we have considered above (5–7). The device shown in Figure 13.5.1 is strongly reminiscent of the scaler/inverter discussed in Section 13.2.2. The cell has replaced the feedback resistance R_f. Summing currents at S, we have

$$i_{\text{cell}} = -i_{\text{in}} = \frac{-e_i}{R} \tag{13.5.1}$$

hence the cell current is controlled by the input voltage. It can be constant or vary in any arbitrary fashion, and the cell current will follow suit.

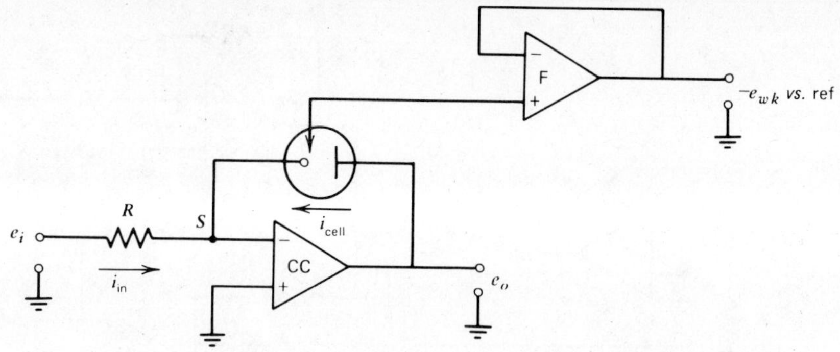

Figure 13.5.1
A simple galvanostat based on the scaler/inverter circuit.

This design holds the working electrode at virtual ground, and that feature is a convenience for the measurement of the potential difference between the reference and working electrodes. The voltage follower F gives the reference electrode's potential versus ground, which is $-e_{wk}$ (*vs.* ref). Note by comparing Figures 13.5.1 and 13.4.1 that the follower's output has a contribution from uncompensated resistance equal to $i_{cell}R_u$.

The input network can be expanded by adding resistors into the summing point to create a system that will provide a cell current equal to the sum of input currents in the fashion of an adder. Each input voltage source must be capable of supplying its contribution to the cell current, as one can see from Figure 13.5.1. This requirement can create problems in systems intended for applying high currents.

In that case, the galvanostat shown in Figure 13.5.2 may be more useful. It is based on the design of Figure 13.3.3. The arbitrary impedance Z_L has been replaced by the cell. The current through the cell is

$$i_{cell} = \frac{-e_i}{R} \qquad (13.5.2)$$

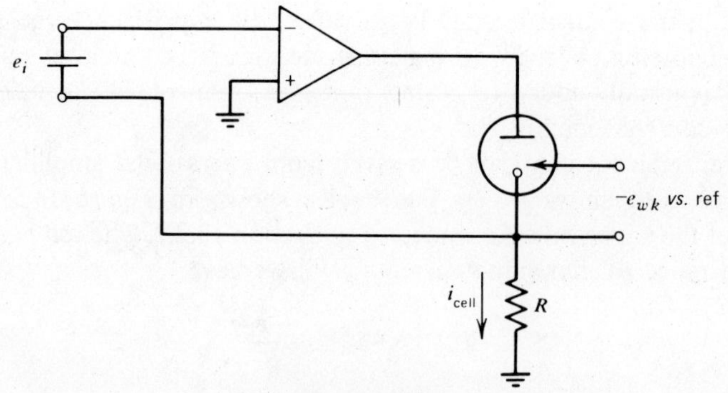

Figure 13.5.2
A galvanostat based on the circuit of Figure 13.3.3.

and this current does not have to be supplied by the voltage source e_i. One drawback to the circuit is that the working electrode is off ground by $-e_i$, and hence the potential of the working electrode with respect to the reference must be measured differentially. In addition, the input voltage e_i is subject to the inflexibilities discussed earlier for Figure 13.4.2.

13.6 DIFFICULTIES IN POTENTIAL CONTROL

The foregoing sections have outlined the principles of potential control, but they should not be construed as a guide to practice. Here we would like to examine some of the difficulties that can arise in measurements with real systems. Those complications mostly pertain to fast measurements or experiments, such as bulk electrolysis, involving the long-term passage of large currents.

13.6.1 The Effects of Solution Resistance (5–7, 11–16)

The impact of compensated and uncompensated solution resistance on those experiments is substantial and involves several considerations. Note first that both categories involve high currents in some manner. The long-term current flow takes place at a high level in bulk electrolysis because electrode areas are large and mass transfer is effective. In fast transient experiments, high current pulses are encountered because dE/dt is high at some time during the measurement. At the very least, there is a capacitive component to the current. Suppose, for example, we wish to impose a 1-V step in 1 μsec on an electrode having an interfacial capacitance of 2 μF. The average current in that period is 2 μC/μsec or 2 A. The peak current would be higher.

Thus in both types of experiments, the potentiostat must have an adequate *power* reserve. It has to be able to supply the necessary currents (even if they are demanded only momentarily), and it must be able to force those currents through the cell. In high-current situations, the output voltage of the potentiostat is mostly dropped across the solution resistance $R_\Omega + R_u$, and the voltage requirement can easily exceed 100 V.

Whenever currents are passed, there is always a potential control error due to the uncompensated resistance. It was seen above to be iR_u. If a cathodic current flows, the true working electrode potential will be *less negative* than the nominal value by that amount. The opposite holds for an anodic current. Even small values of R_u, such as 1 to 10 Ω, can cause a large control error when substantial currents flow. This is one reason why very large-scale electrosynthesis is not usually carried out potentiostatically. In that instance, controlling the current density may be more meaningful.

The control error in a fast experiment may be a transient problem existing only during brief periods of high current flow. Consider a step experiment on the equivalent circuit shown in Figure 13.6.1*a*, which has only a capacitance representing the double layer at the working interface. Even if an ideal control circuit exists so that e_{ref} is instantaneously stepped (from e.g., 0V), there will be a lag in the true potential, e_{true}, because iR_u is nonzero while the double layer is charging. The actual relation (see Problem 13.8) is

$$e_{true} = e_{ref}(1 - e^{-t/R_uC_d}) \tag{13.6.1}$$

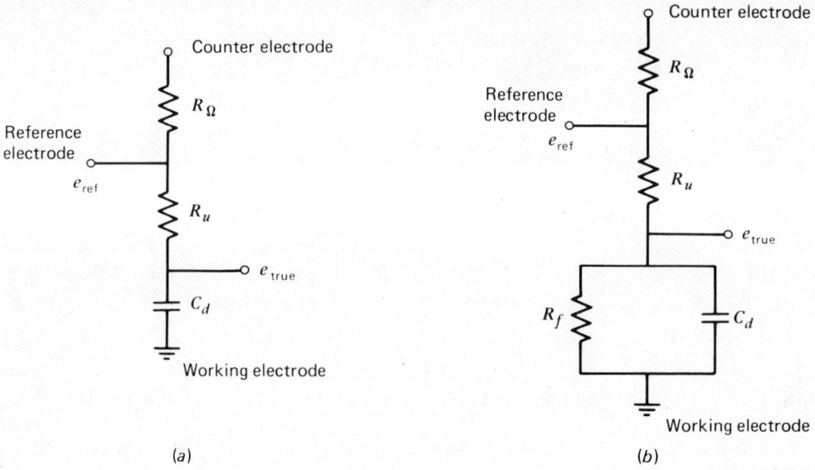

Figure 13.6.1
Simple dummy cells. (*a*) For a nonfaradaic system: C_d is the double-layer capacitance and $R_u + R_\Omega$ is the solution resistance, where R_u is uncompensated. (*b*) For a system passing faradaic currents through R_f, as well as nonfaradaic ones through C_d.

The relationship between e_{true} and e_{ref} is shown in Figure 13.6.2. Eventually e_{true} would reach e_{ref} as C_d became fully charged and the current dropped to zero. However, the rise in potential at the working interface is governed by the exponential, which is controlled by the *cell time constant*, R_uC_d. This time constant defines the shortest time domain over which the cell will accept a significant perturbation. The picture would not change drastically if a faradaic impedance were placed in parallel with C_d, but then e_{true} would never equal e_{ref} because a current would always leak through Z_f and cause a control error iR_u. This error may or may not be significant, depending on the sizes of i and R_u (see Problem 13.9 and Figure 13.6.1*b*).

These considerations show that transient experiments will not be meaningful unless the cell time constant is small compared to the time scales of the measurements, regardless of the high-frequency characteristics of the control circuitry.

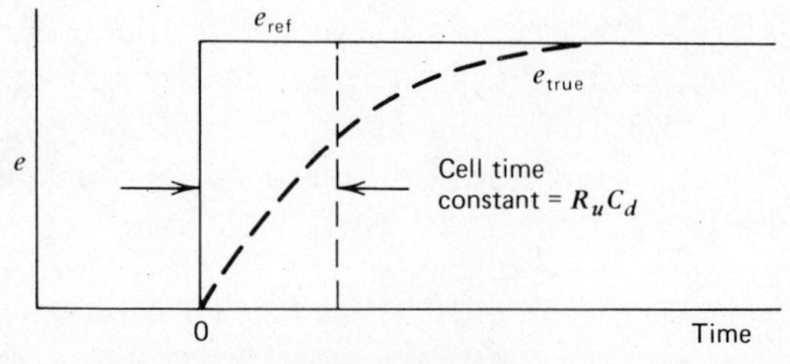

Figure 13.6.2
Schematic representation of the effect of the cell time constant on the rise of the true working electrode potential after an instantaneous step is applied.

13.6.2 Cell Design and Electrode Placement (11, 13, 15)

The time constant R_uC_d can be reduced in at least three ways: (a) One can reduce the total resistance $R_\Omega + R_u$ by increasing the conductivity of the medium through an increase in supporting electrolyte concentration or solvent polarity, or through a decrease in the viscosity. (b) One can shrink the size of the working electrode to reduce C_d proportionally. (c) One can move the reference electrode tip as close as possible to the working electrode so that R_u is a smaller fraction of the total resistance $R_\Omega + R_u$, which would remain the same. All of these steps should be considered in any application, although (a) and (b) may be restricted by other experimental concerns. For example, the chemistry of the system may dictate the nature of the medium, and aspects of electrode fabrication may place significant constraints on the size of the working electrode.

When high currents pass through the electrolyte, that phase is not an equipotential volume. Thus the interfacial potential difference between the working electrode and the solution varies across the surface of the working electrode. One can therefore expect a *nonuniform current density* over the interface. In general, current densities will be higher at points on the working electrode at closest proximity to the counter electrode. Nonuniform current densities imply that the effective working area is less than the actual area by an amount related to the absolute magnitude of the current. Such a condition is clearly unacceptable for most work depending on correlations of theory and experiment.

The remedy is to design the cell so that the current paths from all points on the working electrode are equivalent. Symmetry in the design and placement of the working and counter electrodes is important in this regard. (See Section 10.2.3.)

The resistance between the working and counter electrodes directly controls the power levels required from the potentiostat and the resistive heating in bulk electrolysis that might have to be dissipated by cooling. It should be minimized by shortening the gap between the electrodes and removing impediments to current flow (such as frits and other separators) to the extent that is feasible within the constraints imposed by a desire for possible chemical isolation of the counter electrode or by a need for a spatial relationship concordant with uniform current density on the working surface.

Designing a cell for a demanding experiment is a task requiring optimization of many factors. We have space here only to outline some of the important considerations. The interested reader is referred to more specialized literature on the subject.

13.6.3 Electronic Compensation of Resistance (5, 7, 12–16)

If we know that uncompensated resistance causes a potential control error equal to iR_u, then it might be reasonable to attempt some correction by adding into the input of the potentiostat a correction voltage proportional to the current flow. If we were lucky, we might use a proportionality factor equal to R_u, so that no potential control errors occurred. This idea is the basis for *positive feedback compensation* schemes, the most common version of which is implemented in the circuit of Figure 13.6.3.

The system is identical to the refined adder potentiostat of Figure 13.4.5 with the exception of the new feedback loop connecting the current follower to the potential

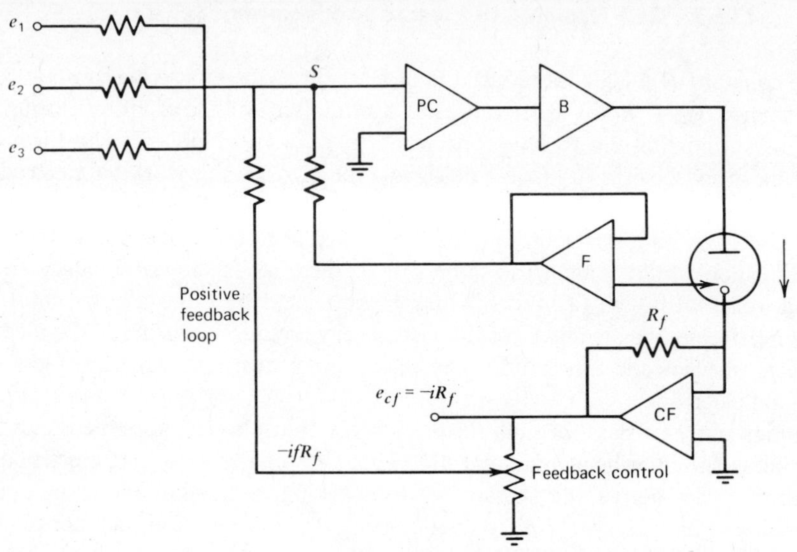

Figure 13.6.3
An adder potentiostat with positive feedback compensation.

control amplifier. The potentiometer selects some fraction f of the current follower's output for application to the input network; hence the feedback voltage is $-ifR_f$.

From the discussion in Section 13.4.2, we know that the working electrode's potential is then†

$$e_{wk}\ (vs.\ \text{ref}) = e_1 + e_2 + e_3 - ifR_f \tag{13.6.2}$$

The true working electrode potential versus the reference is

$$\boxed{e_{\text{true}} = e_1 + e_2 + e_3 - ifR_f + iR_u} \tag{13.6.3}$$

and this differs from the desired sum of the signal inputs $e_1 + e_2 + e_3$ by the control error $i(R_u - fR_f)$. The effect of the feedback loop has been to reduce the uncompensated resistance by the amount fR_f.

These considerations suggest that we might be able to set fR_f exactly equal to R_u and achieve total compensation. They also indicate that almost any degree of undercompensation or overcompensation is available.

In practice there are problems with this scheme because the elements of the cell and the amplifiers in the control circuit introduce phase shifts. Thus, there are significant time lags in the application of a correction signal, the establishment of a correction, and the sensing that the correction has been applied. These delays can cause the whole feedback system to overcorrect for changes in the input signal $e_1 + e_2 + e_3$. Effects such as overshoot and ringing are manifestations. In severe cases the potentiostat will break into a high-frequency oscillation and therefore lose control completely over the

† Note that i adheres to our usual definition for this discussion. Cathodic currents are positive.

cell. For reasons that are not obvious and that are too detailed for this treatment, most potentiostats require some uncompensated resistance to achieve stability. Total compensation is usually impractical.

In addition, there is a problem in knowing the value of R_u. It can be measured in ac experiments, but in most cases it is unknown. The usual approach to this problem is to set the electrode potential at a value where faradaic processes do not occur and increase the value of f with the potentiometer until the potentiostat oscillates. Then f is decreased to a value about 10 to 20% below the critical figure, so that stability is reestablished. One usually assumes that the critical f corresponds to total compensation; however, the critical point may lie either below or above full compensation, depending on the electronic properties of the whole system. The method must therefore be used with care.

The details of this subject and discussions of alternative approaches to compensation are covered in several good reviews (12, 13, 15, 16). A reader involved in experiments requiring compensation should consult them.

13.7 DIGITAL INSTRUMENTATION

Despite the essential analog nature of most electrochemical instrumentation, computers (including microprocessors) have come to play big roles in the acquisition and analysis of electrochemical data (5, 17–20). They offer enormous flexibility and sophistication in the execution and control of experiments, and their influence will doubtlessly be more and more widely felt. In fact, it seems likely that new types of essentially digital electrochemical instrumentation will be created for greater compatibility with the data-processing equipment; but to date only a small amount of progress has been made along that line.

One of the tasks that a computer does best is to synthesize complex waveforms. A very good example is described in Section 9.8. The waveform in Figure 9.8.2f would be extremely difficult to synthesize in the analog domain (18, 19). Another case is the ideal potential program for differential pulse polarography, shown in Figure 5.8.8. Analog generation of this function usually involves adding a slow ramp to the desired voltage pulses. The value of dE/dt is then never zero, and one always has a charging current from that source. Newer instruments for differential pulse measurements incorporate microprocessors capable of the more ideal waveform. These potential programs are created as numeric arrays within a digital memory, then the numbers of the array are presented sequentially to a *digital-to-analog* (D/A) *converter* (21, 22), which produces an analog voltage proportional to the input number. The analog voltage is then applied to an ordinary potentiostat in the manner we have discussed.

Computers are also used to control timing of different phases of experiments, simplify operation, and anticipate, prevent, or warn against operator errors. These functions are widely utilized in instruments intended for routine analysis.

Electrochemical responses, such as current or charge, are basically continuous, but they are often represented digitally as lists of numbers in a memory. For example, a voltage representing the current (at the output of an i/E converter) might be *digitized* at some fixed interval. This could be done by presenting the voltage to an *analog-to-digital* (A/D) *converter* (21, 22), which (upon receiving a start signal) will produce a number at its output proportional to the input voltage. The number is then stored by the computer. Small computers now in use can receive a data point in this manner at rates of ~10 kHz to 1 MHz. Faster rates can be obtained with a *transient recorder*, which basically is just a fast A/D converter and streamlined memory logic. The signal is digitized and stored in the memory, then it can be read out of memory by a computer at a rate suited to the computer. A *digital oscilloscope* is a type of transient recorder.

Digital data acquisition can offer high precision and, through automatic signal scaling, a very large measurement range for single transients. However, it usually is used in order to obtain the benefits of *analyzing* the data in the computer. Very complex schemes can be applied (17–20). A good example is the Fourier analysis discussed in Section 9.8. In newer instruments for electroanalysis, sophisticated routines for subtracting baselines, comparing responses with those from standards, calculating unknown concentrations, identifying peaks, and plotting rescaled results are incorporated as standard features.

13.8 REFERENCES

1. H. V. Malmstadt, C. G. Enke, and S. R. Crouch, "Instrumentation for Scientists Series," Module 2, "Control of Electrical Quantities in Instrumentation," W. A. Benjamin, Inc., Menlo Park, Calif., 1973, Chap. 5.
2. *Ibid.*, Module 3, "Digital and Analog Data Conversions," Chap. 2.
3. A. J. Diefenderfer, "Principles of Electronic Instrumentation," 2nd ed., W. B. Saunders, Philadelphia, 1979, Chap. 9.
4. J. G. Graeme, G. E. Tobey, and L. P. Huelsman, Eds., "Operational Amplifiers—Design and Applications," McGraw-Hill, New York, 1971.
5. D. T. Sawyer and J. L. Roberts, "Experimental Electrochemistry for Chemists," Wiley, New York, 1974, Chap. 5.
6. D. E. Smith, *Electroanal. Chem.*, **1**, 1 (1966).
7. R. R. Schroeder in "Computers in Chemistry and Instrumentation," Vol. 2, "Electrochemistry," J. S. Mattson, H. B. Mark, Jr., and H. C. MacDonald, Jr., Eds., Marcel Dekker, New York, 1972, Chap. 10.
8. W. M. Schwarz and I. Shain, *Anal. Chem.*, **35**, 1770 (1963).
9. D. T. Napp, D. C. Johnson, and S. Bruckenstein, *Anal. Chem.*, **39**, 481 (1967).
10. B. Miller, *J. Electrochem. Soc.*, **116**, 1117 (1969).
11. D. T. Sawyer and J. L. Roberts, *op. cit.*, Chap. 3.
12. D. E. Smith, *Crit. Rev. Anal. Chem.*, **2**, 247 (1971).
13. J. E. Harrar and C. L. Pomernacki, *Anal. Chem.*, **35**, 47 (1973).
14. D. Garreau and J. M. Saveant, *J. Electroanal. Chem.*, **86**, 63 (1978).
15. D. Britz, *ibid.*, **88**, 309 (1978).
16. D. K. Roe in "Laboratory Techniques in Electroanalytical Chemistry," P. T. Kissinger, Ed., Marcel Dekker, New York, in press.
17. See, for example, J. S. Mattson, H. B. Mark, and H. C. MacDonald, Eds.,

"Computers in Chemistry and Instrumentation," Vol. 2, "Electrochemistry," Marcel Dekker, New York, 1972, Chap. 11 (by R. A. Osteryoung), Chap. 12 (by D. E. Smith), Chap. 13 (by S. P. Perone), Chap. 1 (by P. R. Mohilner and D. M. Mohilner), Chap. 2 (by H. C. MacDonald, Jr.), Chap. 4 (by R. F. Martin and D. G. Davis), and Chap. 6 (by A. A. Pilla).

18. S. C. Creason, J. W. Hayes, and D. E. Smith, *J. Electroanal. Chem.*, **47**, 9 (1973).
19. D. E. Smith, *Anal. Chem.*, **48**, 221A, 517A (1976).
20. J. W. Hayes, D. E. Glover, D. E. Smith, and M. W. Overton, *ibid.*, **45**, 277 (1973).
21. H. V. Malmstadt, C. G. Enke, and S. R. Crouch, *op. cit.*, Module 3, Chap. 4.
22. A. J. Diefenderfer, *op. cit.*, Chap. 11.

13.9 PROBLEMS

13.1 Consider a voltage follower circuit with the input leads reversed so that the feedback loop involves the noninverting input. Derive a formula linking the output e_0 to the input e_i. Is e_0 defined for any condition (e.g., any frequency)? Suppose the amplifier is at an equilibrium condition and a sudden positive change is made in e_i. Would e_o converge on a new equilibrium value, given a finite delay in the response of e_o to e_i? Answer these same questions for the conventional voltage follower. Is it now clear why the feedback involves the inverting input?

13.2 Devise an operational amplifier circuit that will integrate the sum of two input signals. Only one amplifier is required.

13.3 Suppose you wanted a ramp generator with a facility for stopping the sweep at any point and holding a constant output until the sweep is to be resumed. How could you fabricate such a device?

13.4 Current followers often feature a capacitor in parallel with the feedback resistor. What is its effect? Is it useful?

13.5 Suppose you have an input signal with information at $\omega/2\pi = 10$ Hz and noise at $\omega/2\pi = 60$ Hz; for example,

$$e_i = 10 \sin 2\pi(10)t + 0.1 \sin 2\pi(60)t$$

What is the signal-to-noise ratio in e_i? To what extent is the ratio degraded by an analog differentiation? Calculate the improvement upon integration. Is there an optimal RC product for either differentiation or integration?

13.6 Consider the adder potentiostat of Figure 13.4.5. What would be the effect of adding a capacitor between the summing point and the booster output? Explain the mechanism for the effect by considering currents at the summing point. When might this arrangement be useful?

13.7 Show that amplifiers I2 and Z2 in Figure 13.4.6 place a voltage $e_2 - e_1$ at the noninverting input of CF2. What is the output of CF2?

13.8 Derive a formula describing current flow in the dummy cell shown in Figure 13.6.1*a* on application of a step in e_{ref} from 0V to an arbitrary value e_{ref}. Derive equation 13.6.1 from your result.

13.9 Derive a formula describing current flow in the dummy cell shown in Figure 13.6.1*b* upon application of a step in e_{ref} from 0V to an arbitrary value e_{ref}. Derive an equation for the true potential difference between the reference and working electrodes corrected for the drop through R_u. Is the cell time constant still a factor controlling the rise of e_{true}?

13.10 What would happen to the working electrode potential if the output of the current follower in Figure 13.4.5 reached its voltage limit under a heavy current load? Suppose this happened during a potential step. What would the effect be on the rise of the true potential difference between the working and reference electrodes?

13.11 An alternative potentiostatic circuit is shown in Figure 13.9.1. Explain its operation. What simple amplifier circuit is it based on? Evaluate its strong and weak points relative to the simple circuit of Figure 13.4.2 and the adder design of Figure 13.4.4. Design a potentiostatic system equivalent to that of Figure 13.4.5 on the basis of this circuit.

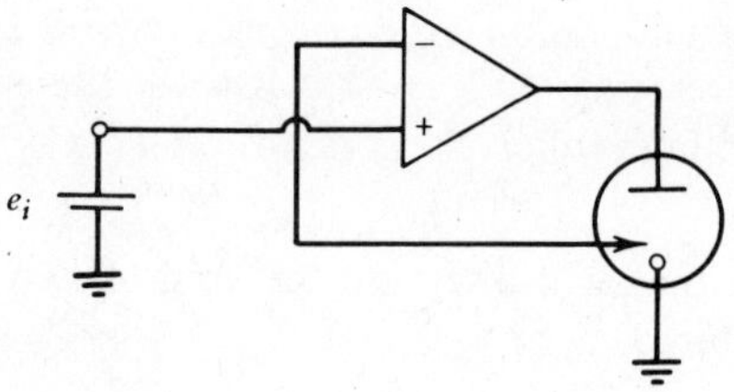

Figure 13.9.1
An alternative potentiostatic circuit.

chapter 14
Spectrometric and Photochemical Experiments

Recent years have brought much interest in studying electrode processes by experiments that involve more than the usual electrochemical variables of current, charge, and potential. Part of the motivation for this work has been to provide ways for obtaining information about electrochemical systems that could not be gathered in purely electrochemical experiments. Other efforts have been driven by interest in using stimuli such as photons to influence electrode reactions. Both lines have yielded a good deal of progress already, and they remain among the most active domains of electrochemical research.

In this chapter, we will examine some of the more important topics in spectroelectrochemistry and photoelectrochemistry. The first three sections concern methods for studying electrode systems by a variety of spectrometric techniques, and the last three deal with experiments in which photons actually participate in the electrode processes of interest. With such a great variety of material involved here, we cannot delve into any topic with much depth. Instead, we will simply review the basic principles, consider some typical experimental arrangements, and outline the types of chemical information that can be obtained in each case. More extensive reviews are available in the cited literature.

14.1 OPTICAL SPECTROELECTROCHEMISTRY

14.1.1 Transmission Experiments

Perhaps the simplest spectroelectrochemical experiment is to direct a light beam through the electrode surface, as shown in Figure 14.1.1, and to measure absorbance changes resulting from species produced or consumed in the electrode process. Figure 14.1.2 is an illustration of two types of cells in which such experiments could be carried out.

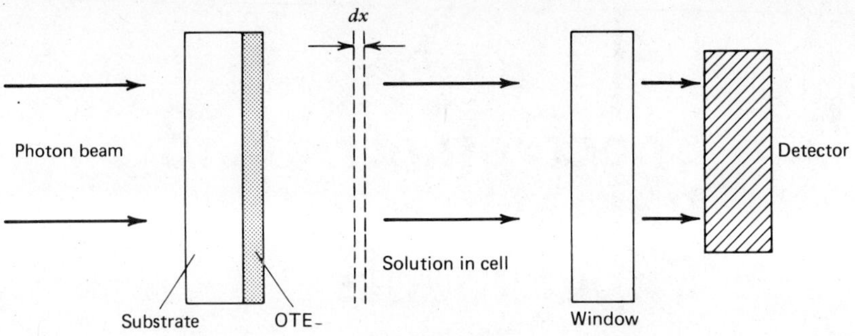

Figure 14.1.1
Schematic view of the experimental arrangement for transmission spectroelectrochemistry.

The obvious prerequisite is an *optically transparent electrode* (OTE). Several types of OTEs have been reported (1–7). They may be thin films of a semiconductor (e.g., SnO_2 or In_2O_3) or a metal (e.g., Au or Pt) deposited on a glass, quartz, or plastic substrate; or they may be fine wire mesh "minigrids" with perhaps several hundred wires per centimeter. The films are quite planar, uniform electrode surfaces, but the minigrids are not planar and have a structure involving alternating regions of opaque bulk metal and transparent openings. On the other hand, the minigrids will behave

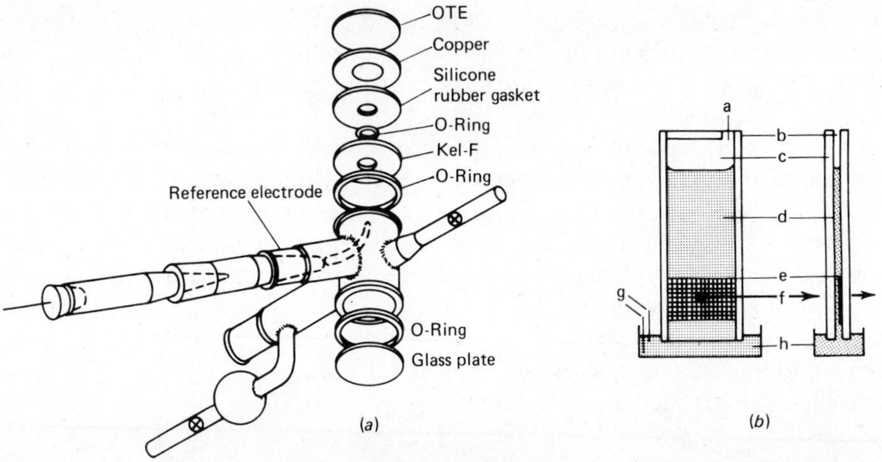

Figure 14.1.2
Cells for transmission spectroelectrochemistry. (*A*) Cell for experiments involving semi-infinite linear diffusion. Light beam passes along vertical axis. [From N. Winograd and T. Kuwana, *Electroanal. Chem.*, **7**, 1 (1974), Courtesy of Marcel Dekker, Inc.] (*B*) Optically transparent thin-layer system: front and side views. (a) Point of suction application in changing solutions; (b) Teflon tape spacers; (c) 1 × 3 in. microscope slides; (d) test solution; (e) gold minigrid, 1 cm high; (f) optical beam axis; (g) reference and auxiliary electrodes; (h) cup containing test solution. [Reprinted with permission from W. R. Heineman, B. J. Norris, and J. F. Goelz, *Anal. Chem.*, **47**, 79 (1975). Copyright 1975, American Chemical Society.]

like planar electrodes if the electrochemical experiment has a characteristic time long enough to allow the diffusion layer thickness to become much larger than the size of these openings. Then the diffusion field feeding the electrode is one-dimensional and has a cross-sectional area equal to the projected area of the whole electrode, including the spaces (8, 9).

Transmission experiments may involve the study of absorbance versus time as the electrode potential is stepped or scanned; or they may involve wavelength scans to provide spectra of electrogenerated species. Either of these experimental goals can be attained with cells fitting into conventional spectrophotometers; but if one wishes to follow spectral evolutions over comparatively short time scales, then a rapid scanning system is needed. Apparatus permitting the acquisition of as many as 1000 spectra per second has been utilized (1, 3). With such a capability for high repetition rates, signal averaging is a useful method for improving spectral quality.

The cell in Figure 14.1.2*a* is designed for experiments involving semi-infinite linear diffusion of the electroactive species to the electrode surface (1). It is normally used for experiments in which one applies large-amplitude steps in order to carry out electrolysis in the diffusion-limited region, and one then records the change in absorbance $\mathscr{A}$ *vs.* time. From an electrochemical standpoint, the result is the same as that of the Cottrell experiment described in Section 5.2.

The absorbance change can be described by considering a segment of solution of thickness dx and cross-sectional area A, as shown in Figure 14.1.1. The differential absorbance registered upon passage of the light through this segment is $d\mathscr{A} = \varepsilon_R C_R(x, t)\, dx$, if species R is the only species absorbing at the monitored wavelength. Its molar absorptivity is ε_R. The total absorbance is then

$$\mathscr{A} = \varepsilon_R \int_0^\infty C_R(x, t)\, dx \tag{14.1.1}$$

Note, however, that the integral in (14.1.1) is the total amount of R produced per unit area (if R is a stable species); hence the integral is equal to Q_d/nFA, where Q_d is the charge passed in electrolysis. Since Q_d is given by the integrated Cottrell equation, (5.9.1), we have

$$\boxed{\mathscr{A} = \frac{2\varepsilon_R C_O^* D_O^{1/2} t^{1/2}}{\pi^{1/2}}} \tag{14.1.2}$$

which shows that the absorbance should be linear with $t^{1/2}$, as shown in Figure 14.1.3. Note that the slope of the $\mathscr{A}$-$t^{1/2}$ plot affords a way of measuring diffusion coefficients without independent knowledge of the area A.

Since the usual transmission experiment directly monitors the electrolytic product, it offers many of the diagnostic features of reversal chronoamperometry or reversal chronocoulometry. In effect, $\mathscr{A}$ is a continuous index of the total amount of the monitored species still remaining in solution at the time of observation. Equation 14.1.2 describes the limiting case in which the product is completely stable. If homogeneous chemistry tends to deplete the concentration of R, different absorbance-time relations will be seen. They can be predicted (e.g., by digital simulations; see Appendix B), and curves for many mechanistic cases have been reported (10).

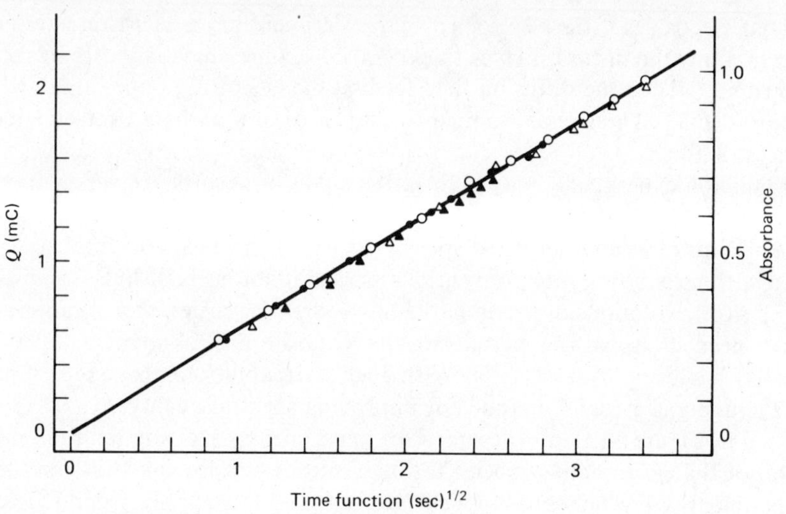

Figure 14.1.3

Responses at a 2000 wire-per-inch gold minigrid during a double potential step experiment. The solution contained 0.8 mM o-tolidine in 1 M $HClO_4$-0.5 M acetic acid. In the forward step o-tolidine was oxidized in a diffusion-controlled, two-electron process. The stable product was rereduced in the reversal. Open circles, forward step charge $vs.$ $t^{1/2}$; filled circles, forward step absorbance $vs.$ $t^{1/2}$; open triangles, $Q_r(t > \tau)$ $vs.$ θ (see Section 5.9.2); filled triangles, $\mathscr{A}(\tau) - \mathscr{A}$ $vs.$ θ. The forward step duration is denoted by τ. [Reprinted with permission from M. Petek, T. E. Neal, and R. W. Murray, *Anal. Chem.*, **43**, 1069 (1971). Copyright 1971, American Chemical Society.]

Another popular mode for transmission experiments involves a thin-layer system (3, 4, 7, 11) like that shown in Figure 14.1.2*b*. The working electrode is sealed into a chamber (e.g., between two microscope slides spaced perhaps 0.05–0.5 mm apart) containing the electroactive species in solution. The chamber is usually filled by capillarity, and it contacts solution in a larger container, which also holds the reference and counterelectrodes. The electrolytic characteristics of the cell are naturally similar to those of the conventional thin-layer systems discussed in Section 10.7. One can do cyclic voltammetry, bulk electrolysis, and coulometry in the ordinary way, but there is also a facility for obtaining absorption spectra of species in the cell.

The particular advantage of this *optically transparent thin-layer electrode* (OTTLE) is that bulk electrolysis is achieved in a few seconds, so that (for a chemically reversible system) the whole solution reaches an equilibrium with the electrode potential, and spectral data can be gathered on a static solution composition.

Figure 14.1.4 is a display of spectra obtained for the cobalt complex with the Schiff base ligand bis(salicylaldehyde)-ethylenediimine (12):

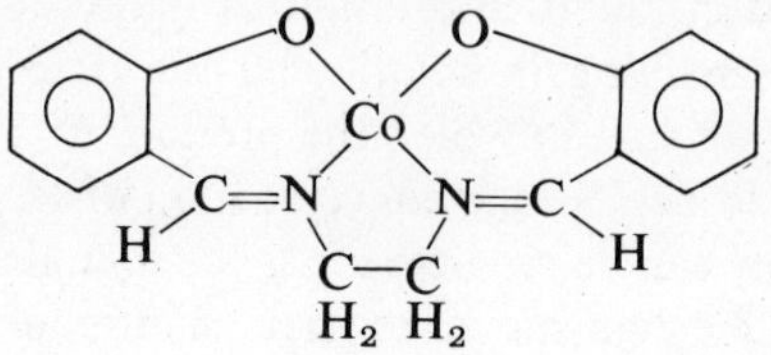

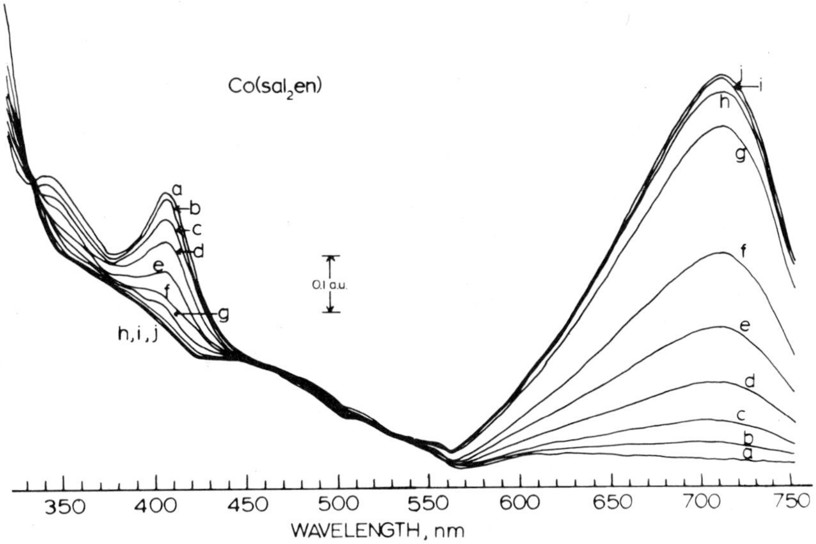

Figure 14.1.4

Spectra of the cobalt complex with the ligand bis(salicylaldehyde)ethylenediimine, obtained at an OTTLE. Applied potentials: (a) −0.900, (b) −1.120, (c) −1.140, (d) −1.160, (e) −1.180, (f) −1.200, (g) −1.250, (h) −1.300, (i) −1.400, and (j) −1.450 V *vs.* SCE. [From D. F. Rohrbach, E. Deutsch, and W. R. Heineman in "Characterization of Solutes in Nonaqueous Solvents," G. Mamantov, Ed., Plenum, New York, 1978, with permission.]

At −0.9 V *vs.* SCE the complex involves Co(II), and at −1.45 V it corresponds to Co(I). Spectra obtained in this way may be intrinsically interesting for characterizing the electronic properties of the species under scrutiny, or they may be used to obtain precise standard potentials in the manner elicited in Problem 14.1.

Spectroelectrochemical methods can be especially useful for unraveling a complex sequence of charge transfers. Figure 14.1.5 is a display from a classic example (13). The sample is a mixture of cytochrome *c* and cytochrome *c* oxidase which is initially fully oxidized. The experiment is a coulometric titration by the electrogenerated radical cation of methyl viologen:

$$MV^{2+} + e \rightleftharpoons MV^{\ddagger} \tag{14.1.3}$$

where methyl viologen corresponds to

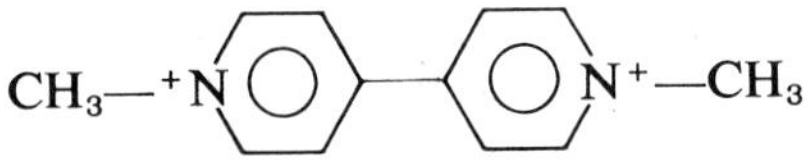

In solution, one $MV^{\ddagger}$ ion can reduce a single heme site in cytochrome *c* or one of two in the oxidase. The spectra were recorded after 5-nanoequivalent increments of charge, and they indicate that one of the heme groups in cytochrome *c* oxidase is reduced first. Then $MV^{\ddagger}$ reduces the heme in cytochrome *c* before it deals with the second heme of the oxidase.

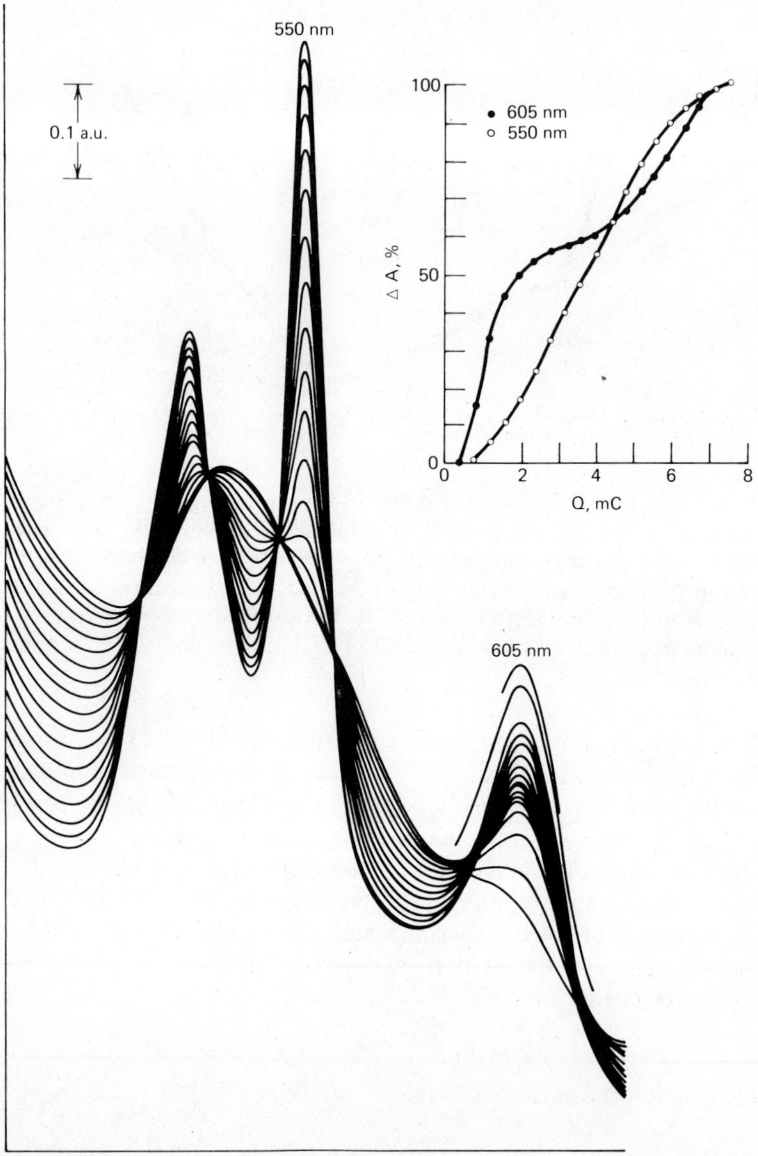

Figure 14.1.5

Coulometric titration of cytochrome *c* (17.5 μM) and cytochrome *c* oxidase (6.3 μM) by MV^{+} generated at an SnO_2 OTE. Each spectrum was recorded after 5×10^{-9} equivalents of charge were passed. [From W. R. Heineman, T. Kuwana, and C. R. Hartzell, *Biochem. Biophys. Res. Commun.*, **50**, 892 (1973), with permission.]

This example is a good illustration of the indirect electrochemical measurements that often have to be applied with biological macromolecules, which usually will not undergo direct charge exchange with an electrode (possibly for steric reasons). Instead, one uses smaller molecules to exchange charge heterogeneously with the electrode, and homogeneously with the macromolecules. These species are called *mediator titrants* (3, 4, 11, 13). They provide a mechanism for enforcing and maintaining electrochemical equilibrium with the macromolecules, and hence they are especially useful for characterizing standard potentials for redox centers on these species.

14.1.2 Specular Reflectance and Ellipsometry

One might guess that changes in the surface of an electrode, even minute changes, such as the adsorption of submonolayer amounts of iodide, would affect the reflecting properties of the surface, and therefore might provide information about the kinetics of surface chemistry and the nature of surface films. This approach is indeed useful and has given rise to several different experimental techniques. For the two areas covered in this section, we want to concentrate on the surface itself. We are interested in the properties of a light beam that is reflected from that surface, and we seek specifically to avoid any changes in the optical properties of the solution during the course of the experiment. Thus these methods are well-suited to the direct observation of surface chemistry in an operating cell, but not in the presence of simultaneous homogeneous reactions involving absorbing species.

(a) Optical Principles. Before we proceed further, we must review some basic concepts of optics. It is clearly beyond our scope to go very far along this line, but we will try to develop a basis for understanding the methods at hand. The reader who requires more complete information should consult more comprehensive sources (14–17).

Optical reflection is best understood in terms of the wavelike properties of light. The electric field vector associated with the wave oscillates in a plane as the wave propagates (Figure 14.1.6), and the intensity of the light is proportional to the square of the electric field amplitude. A magnetic field vector, oscillating in a perpendicular plane, as shown in Figure 14.1.6, accompanies the electric vector; but generally we do not have to consider its presence.

Most light sources comprise independent emitters yielding rays that are uncorrelated with each other; thus the planes of oscillation of the electric field are randomized with respect to the angle about the axis of propagation. This light is said to be *unpolarized.* However, reflection of the light from a surface, or even transmission through a window, is generally more efficient for rays having certain particular orientations with respect to the actual physical surfaces involved; hence the reflected and transmitted beams are *partially polarized,* because particular directions of oscillation predominate. By processing the beam carefully, one can achieve a *linearly* (or *plane*) *polarized* beam in which all rays have the same angle for the plane of electric field oscillation.

In reflectance studies, one usually wants to control the state of polarization with

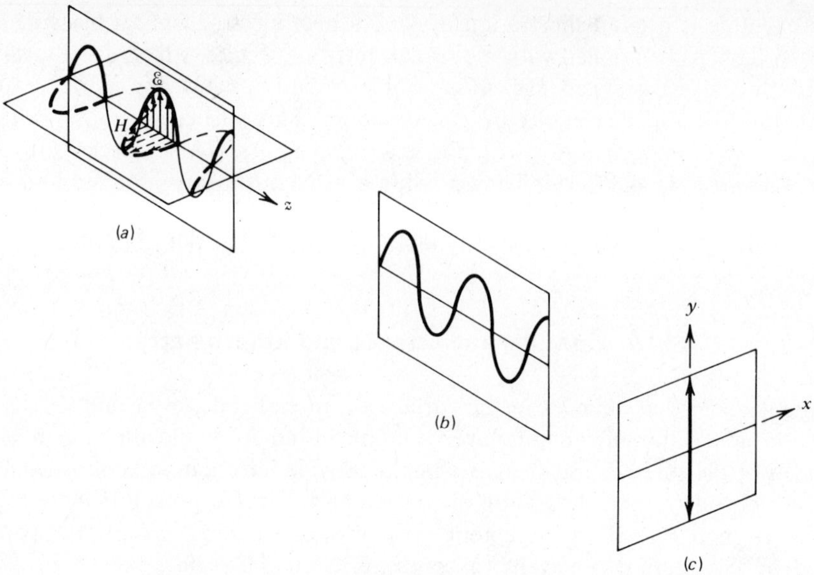

Figure 14.1.6

(*a*) Electric field and magnetic field vectors of a light wave propagating along the z direction. (*b*) The *plane of polarization* contains the electric field vectors. (*c*) To an observer at a fixed point, a train of these waves, all oriented in the same way, would have electric field vectors along the indicated line. Such a light beam is *linearly polarized.* [From R. H. Muller, *Adv. Electrochem. Electrochem. Engr.*, **9**, 167 (1973), reprinted by permission of John Wiley & Sons, Inc.]

respect to the experimental apparatus. Measurements are always referred to the physical *plane of incidence*, as defined in Figure 14.1.7. If the polarization is *parallel* to this plane, then it and parameters related to it are traditionally denoted by the subscript p. For polarization *perpendicular* to the plane of incidence a subscript s is employed.

If some other angle of polarization with respect to the plane of incidence is employed, such as 45°, then one usually resolves the electric field vector into the parallel and perpendicular components. Thus, any linearly polarized beam incident on a surface can be regarded as a combination of separate beams with parallel and perpendicular polarization. Each ray in the parallel-polarized beam has a partner in the beam with perpendicular polarization, and they are locked in phase so that the orientation of the resultant electric field vector is always at the constant angle of polarization of the beam as a whole.

If a linearly polarized beam is reflected from a surface, one usually finds that the parallel and perpendicular components undergo different changes in amplitude and phase. Thus individual pairs of rays in the two beams are in phase upon incidence, but are out of phase upon reflection. This effect has interesting consequences, as shown in Figure 14.1.8. Note that now the resultant electric field vector for a ray pair is not at a fixed angle, as it was upon incidence, but instead it traces out a spiral as the wave propagates. The projection of the spiral is an ellipse; hence the light is *elliptically*

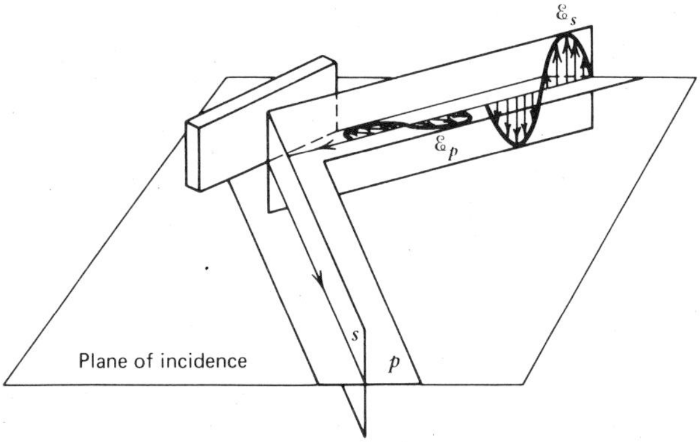

Figure 14.1.7

Reflection of polarized light from a surface. [From R. H. Muller, *Adv. Electrochem. Electrochem. Engr.*, **9**, 167 (1973), reprinted by permission of John Wiley & Sons, Inc.]

polarized. The shape of the ellipse is controlled by the relative amplitudes and the phase differences of the two beams. *Circular polarization* represents the special case in which the amplitudes are equal and the phase shift is 90°.

The optical properties of any material are controlled by its *optical constants.* One of these is the *index of refraction, n,* which is

$$n = \frac{c}{v} = (\varepsilon\mu)^{1/2} \tag{14.1.4}$$

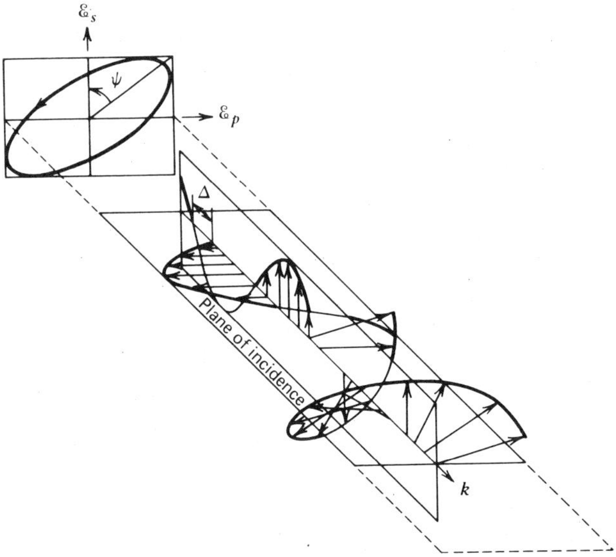

Figure 14.1.8

Elliptic polarization arising from a phase shift Δ between parallel and perpendicular components. [From R. H. Muller, *Adv. Electrochem. Electrochem. Engr.*, **9**, 167 (1973), reprinted by permission of John Wiley & Sons, Inc.]

where c is the speed of light in vacuo, v is the speed of propagation in the medium, ε is the *optical-frequency dielectric constant*, and μ is the *magnetic permeability*. If the medium is light-absorbing, we must add the *extinction coefficient*, k, which is proportional to the *absorption coefficient* α, which in turn characterizes the exponential falloff in light intensity upon passage through the medium. The absorbance of a thickness x of the medium is $\alpha x/2.303$, and k is related to α by

$$\alpha = \frac{4\pi k}{\lambda} \tag{14.1.5}$$

where λ is the wavelength of incident light.†

In the literature on reflectance, one finds that it is often convenient to work with a *complex refractive index*, $\hat{n}$, defined by

$$\hat{n} = n - jk \tag{14.1.6}$$

where $j = \sqrt{-1}$. Thus the real and imaginary components of $\hat{n}$ are n and $-k$. By analogy to (14.1.4), one then defines a *complex optical-frequency dielectric constant*, $\hat{\varepsilon}$, adhering to

$$\hat{n} = (\mu\hat{\varepsilon})^{1/2} \tag{14.1.7}$$

with $\hat{\varepsilon}$ expressed in terms of the real and imaginary components as

$$\hat{\varepsilon} = \varepsilon' - j\varepsilon'' \tag{14.1.8}$$

where

$$\varepsilon' = \frac{n^2 - k^2}{\mu} \quad \text{and} \quad \varepsilon'' = \frac{2nk}{\mu} \tag{14.1.9}$$

The magnetic permeability μ is essentially unity at optical frequencies for most materials. The basic optical characteristics of the phase are defined by μ, n, and k, or, alternatively, by μ, ε', and ε''. Both sets are used in the analysis of experimental results.

(b) Specular Reflectance Spectroscopy (16–22). Measurements of specular reflectance involve the intensities of light reflected from the surface of interest. Usually the incident light is polarized either parallel (p) or perpendicular (s) to the plane of incidence, as shown in Figure 14.1.7, and a detector such as a photomultiplier monitors the intensity of the reflected beam. The light is usually monochromatic, but is often tuned over large wavelength ranges. The surface under examination must be flat and, preferably, smooth. For our purpose, it is an electrode in an electrochemical cell.

The *reflectance*, R, is defined as the ratio of the reflected light intensity to the intensity of the incident beam. Absolute reflectances are difficult to measure and are not necessarily of interest. Instead one is usually interested in the change in reflectance ΔR induced by some change in the system, for example, in electrode potential. Experimentally, one measures only the intensity of the reflected beam, I_R. Then if the incident

† Note that neither k nor α is the same as the *molar absorptivity* ε, which is also frequently called the "extinction coefficient."

intensity remains constant, a change ΔI_R in the reflected beam gives $\Delta R/R = \Delta I_R/I_R$. The basic data of reflectance experiments are plots of $\Delta R/R$ *vs.* the variable of interest, which may be frequency of incident light, potential, concentration of an electroactive species, etc.

Values of $\Delta R/R$ typically range from 10^{-6} to unity, hence quite small effects are often involved. To make them experimentally accessible, various modulation schemes are used and lock-in detection is employed (16, 20, 21). In the simplest case, the light beam is chopped. In other circumstances, an experimental variable, such as the potential, may be simultaneously modulated.

Specular reflectance measurements are attractive for the evaluation of the optical constants of metals and other materials, particularly in the form of films, whose properties may differ markedly from those of bulk solids. The strength of the method is the relative ease in obtaining results as a function of wavelength. Optical constants of materials may be of interest in their own right (e.g., in defining band structure or locating surface states; see Section 14.5), or they may be needed for a later analysis (e.g., in determining the thickness of an anodic film growing in situ). Figure 14.1.9 is an illustration of typical data for gold (23).

The *electroreflectance* method features a modulation of the potential of the electrode, perhaps sinusoidally, and lock-in detection of resulting reflected light intensity changes, which are proportional to dR/dE. The reported response is usually (1/R) ×

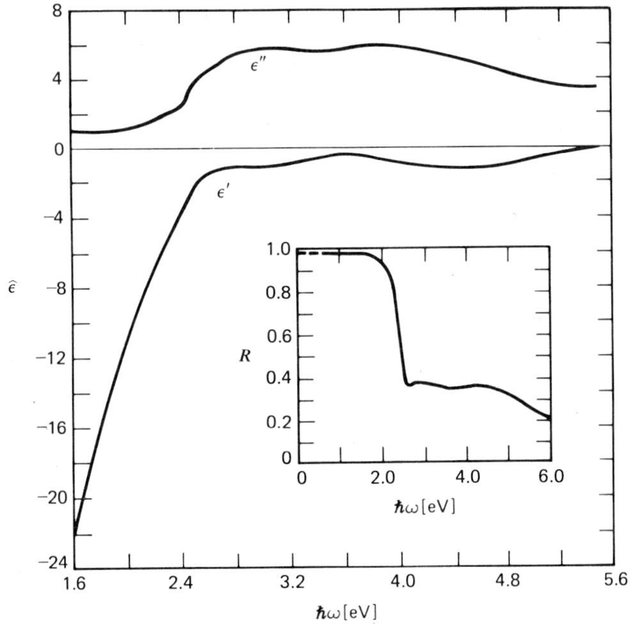

Figure 14.1.9

Dependence of the real and imaginary components of the optical frequency dielectric constant of gold on the energy of incident photons. The normal-incidence reflectivity of gold in air is shown in the inset. [From D. M. Kolb and J. D. E. McIntyre, *Surf. Sci.*, **28**, 321 (1971), with permission.]

(dR/dE). Figure 14.1.10 illustrates typical results for electroreflectance as a function of incident light energy. These kinds of spectra reveal the electronic structure of the system in the interfacial region. The sharp peak in Figure 14.1.10 is ascribed to the effect of the high-double-layer field on the optical properties of the metallic phase (16).

Probably the most important applications of specular reflectance spectroscopy in electrochemistry involve the monitoring of surface films and adsorption layers. In Figure 14.1.11 one can see the effect of specific adsorption of anions on the reflectance properties (24). Figure 14.1.12 contains data demonstrating the formation of anodic oxide films on platinum (20). Note that on platinum, the film forms at potentials more positive than 0.5 V, but the original state of the electrode is restored on the return sweep. The optical constants and thicknesses of films can be evaluated by these methods, and that information can be useful in identifying the chemical nature of the films.

(c) Ellipsometry (17, 25–27). We saw above that reflection of linearly polarized light from a surface generally produces elliptically polarized light, because the parallel and perpendicular components are reflected with different efficiencies and with different phase shifts. We can measure these changes in intensity and phase angle,

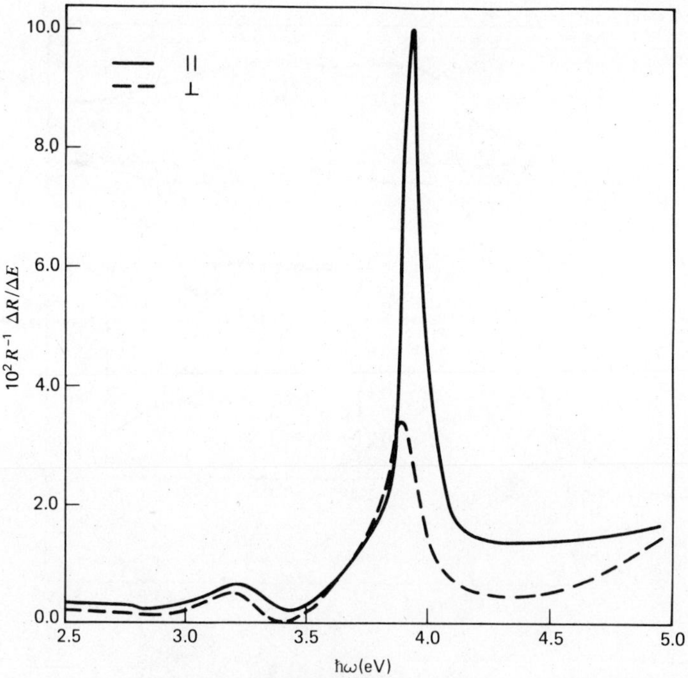

Figure 14.1.10

Electroreflectance spectra of Ag in 1 M $NaClO_4$. $E_{dc} = -0.5$ V *vs.* SCE. The potential was modulated with $\Delta E = 100$ mV rms at 27 Hz. [From J. D. E. McIntyre, *Adv. Electrochem. Electrochem. Engr.*, **9**, 61 (1973), reprinted by permission of John Wiley & Sons, Inc.]

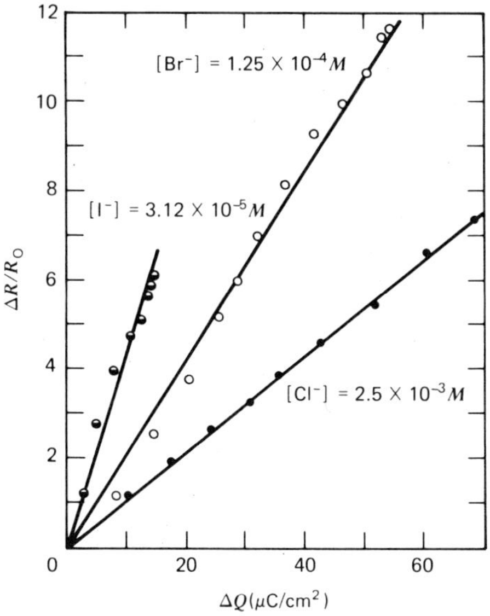

Figure 14.1.11

Reflectance changes caused by halide adsorption on gold in 0.2 *M* $HClO_4$. [From T. Takamura, K. Takamura, and E. Yeager, *Symp. Faraday Soc.*, **4**, 91 (1970), with permission.]

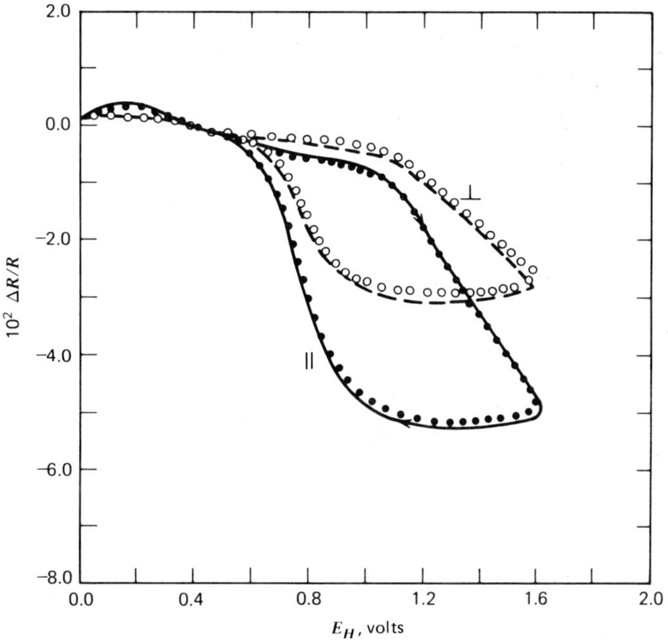

Figure 14.1.12

Reflectance change versus potential for a platinum electrode in 1.0 M $HClO_4$. v = 30 mV/sec. Separate data are shown for parallel and perpendicular polarization. The curves are for Ar saturated solutions, and the points are for O_2 saturated solutions. Note that the reflectance changes are independent of the faradaic reduction of O_2, which takes place in the negative part of this range. [From J. D. E. McIntyre and D. M. Kolb, *Symp. Faraday Soc.*, **4**, 99 (1970), with permission.]

and we can use them to characterize the reflecting system. That approach is called *ellipsometry*.

The basic ellipsometric parameters are defined in Figure 14.1.8. The difference in phase angle between the fast and slow components is given by Δ, and the ratio of electric field amplitudes:

$$\frac{|\mathscr{E}_p|}{|\mathscr{E}_s|} = \tan \psi \tag{14.1.10}$$

defines the second parameter ψ. The values of Δ and ψ may be recorded as functions of other experimental variables, such as potential or time.

Several methods for evaluating Δ and ψ exist, (16, 17) but the most precise approaches rely on a null balance like that depicted in Figure 14.1.13. Light that is polarized linearly at 45° with respect to the plane of incidence impinges on the sample. It has $|\mathscr{E}_p| = |\mathscr{E}_s|$ and $\Delta = 0$. After reflection the beam is passed through a *compensator*, which is adjusted to restore the original condition of $\Delta = 0$. The position of the compensator required for this restoration is a measure of the value of Δ induced by reflection. The resulting linearly polarized beam is then passed through a second polarizer (an *analyzer*), which is rotated until its axis of transmission is at right angles to the plane of polarization of the oncoming light. Then no light passes through the analyzer to the detector, and the condition of *extinction* is reached. The angular position of the analyzer then provides a measure of ψ. Note that extinction will not be achieved unless the compensator and the analyzer are both correctly adjusted. These

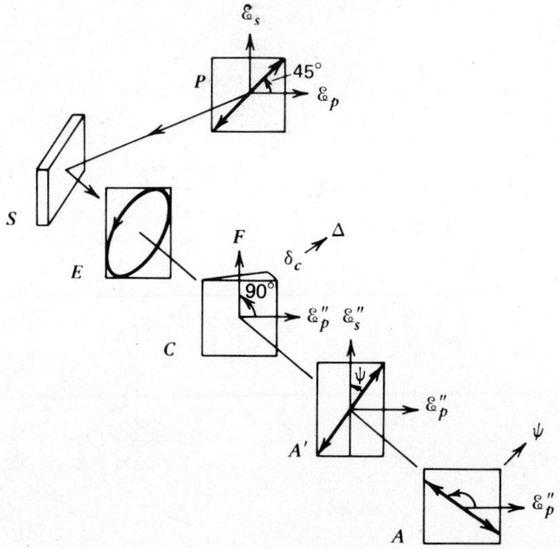

Figure 14.1.13

Schematic layout of one type of ellipsometer. Linearly polarized light (P) is incident on the sample (S). Reflection produces elliptic polarization (E), which is restored to linear polarization (A') by the compensator (C). The analyzer (A) is adjusted to achieve extinction. [From R. H. Muller, *Adv. Electrochem. Electrochem. Engr.*, **9**, 167 (1973), reprinted by permission of John Wiley & Sons, Inc.]

adjustments usually require tens of seconds with manual instruments, but shorter balance times can be achieved with automated equipment.

Ellipsometry is very widely used to study film growth on electrode surfaces. Results for a typical case, the formation of an anodic film on aluminum (28), are shown in Figure 14.1.14. Initial measurements on the substrate are found at the point marked 0.0 Å, and subsequent measurements made at various stages of film growth are shown as crosses. They trace out a closed figure, and then with still greater thicknesses (shown as circles), they begin to retrace the figure. With two parameters Δ and ψ and known optical constants for aluminum, one can, from an optical model, calculate two fundamental parameters for the film at any stage of growth. In this case, since the film is assumed to be nonabsorbing ($k = 0$), the refractive index n and the thickness d can be calculated. The curve in Figure 14.1.14 is the predicted response for $n = 1.62$ and the various indicated thicknesses.

The kinetics of film growth can be studied in this manner without removing the electrode from the cell or interrupting the electrolysis. Figure 14.1.15 contains data

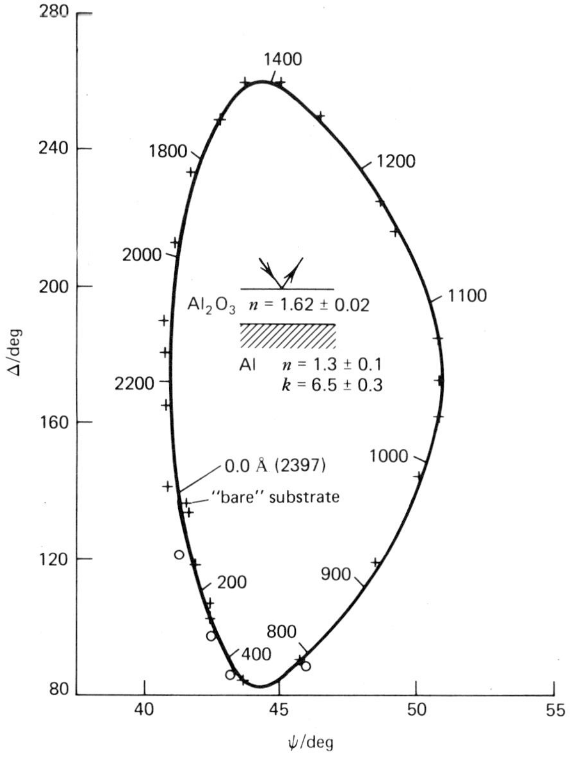

Figure 14.1.14

Ellipsometric results for anodization of aluminum in 3% tartaric acid (pH 5.5). Numbers along fitted curve indicate film thickness in angstroms. [From C. J. Dell'Oca and P. J. Fleming, *J. Electrochem. Soc.*, **123**, 1487 (1976), reprinted by permission of the publishers, The Electrochemical Society, Inc.]

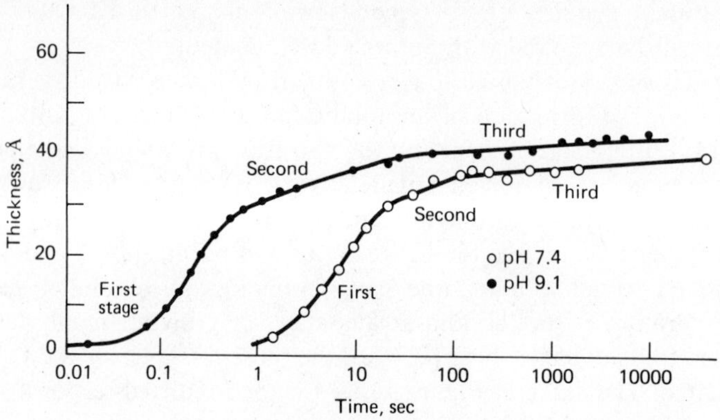

Figure 14.1.15

Growth of passive film on iron at 0.8 V *vs.* SCE. [From J. Kruger and J. P. Calvert, *J. Electrochem. Soc.*, **114**, 43 (1967), reprinted by permission of the publishers, The Electrochemical Society, Inc.]

showing three regimes of growth kinetics for the formation of a passive film on iron (29).

Faster measurements can be obtained in transient experiments (25, 27), but they may require a sacrifice in one's ability to interpret results quantitatively. Figure 14.1.16 contains data (30) relating to the growth of a passivating film on *p*-type GaAs (see also Section 14.5). The ellipsometric signals shown there represent the intensity of light reaching the detector under *constant* settings of the compensator and the analyzer. The resulting traces would be difficult to interpret quantitatively, but they have diagnostic value in showing that rapid film growth begins when the current reaches a maximum. This kind of information can be useful in understanding the mechanism of film formation.

There has also been interest in using ellipsometric methods to study double-layer structure. Clearly detectable signals are seen (26), but quantitative interpretation in terms of optical models is still in its infancy.

In general, specular reflectance methods and ellipsometric approaches offer essentially the same information about the sample at any given wavelength. Ellipsometry offers higher precision, whereas specular reflectance affords a readier approach to transient experiments and greater ease in obtaining data as a function of wavelength.

14.1.3 Internal Reflection Spectroelectrochemistry (1–4, 31–35)

Another approach to the optical sampling of the electrochemical interface is to utilize an optically transparent electrode (OTE, Section 14.1.1) as an internal reflection element in the manner depicted in Figure 14.1.17. The light beam is directed parallel to the plane of the electrode and enters a prism that refracts it so that it passes into the electrode substrate at an angle greater than the critical angle. Then the beam travels by internal reflection through the glass until it reaches the second prism, which allows it to exit, after refraction, along its original line of propagation. The

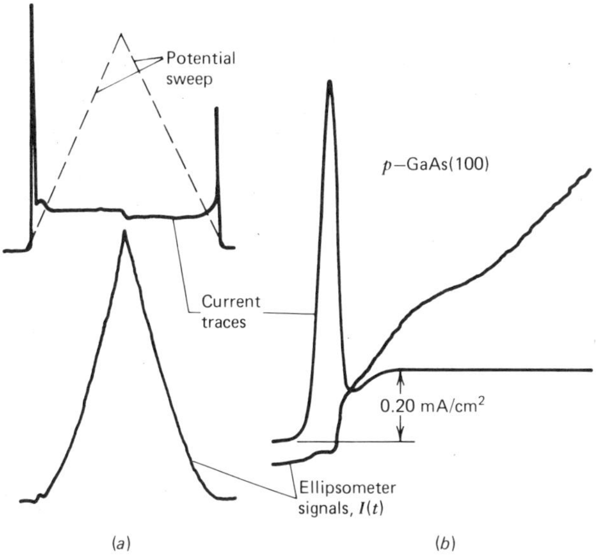

Figure 14.1.16

Anodization of single-crystal, *p*-type GaAs with (100) crystallographic orientation. The solution was a pH 7.4 borate buffer, and potential followed the indicated triangular waveform. (*a*) Current and ellipsometer signals. Ellipsometer nulled at rest potential. (*b*) Expanded time base; view of the early segments of the curves in (*a*). [Reprinted with permission from W. W. Harvey and J. Kruger, *Electrochim. Acta*, **16**, 2017 (1971). Copyright by Pergamon Press, Ltd.]

intensity of the beam is then measured photometrically. While the light is trapped within the electrode/substrate assembly, it is reflected several times from the electrode/electrolyte interface, and at each point of reflection it can interact optically with that region. Changes induced there, for example by electrolysis, cause detectable changes in the beam intensity measured by the detector, and they are the basis for *internal reflection spectroelectrochemistry* (IRS).

Mostly these experiments are concerned with light absorption by species at the interface, although other effects such as changes in refractive index can also be measured (1, 2, 31, 32). Absorption is possible because the electric field accompanying the light wave is not wholly contained within the electrode/substrate assembly. At the points of reflection, the field extends into the solution for a short distance. Its strength falls exponentially with distance from the interface according to the relation:

$$\langle \mathscr{E}^2 \rangle = \langle \mathscr{E}_0^2 \rangle \, e^{-x/\delta} \tag{14.1.11}$$

where $\langle \mathscr{E}_0^2 \rangle$ is the average square field amplitude at the interface, $\langle \mathscr{E}^2 \rangle$ is the average square amplitude at distance x, and δ is the *penetration depth*. This field is a manifestation of the *evanescent wave* existing on the solution side of the interface. It interacts with absorbing species, with a probability of absorption proportional to $\langle \mathscr{E}^2 \rangle$.

The penetration depth defines the distance into the solution over which optical

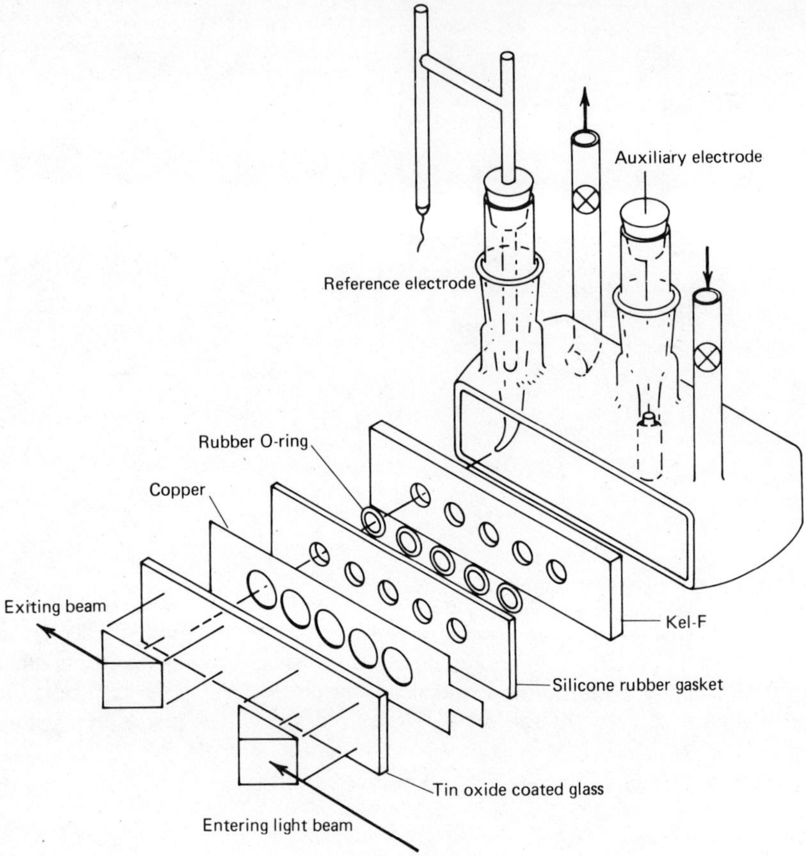

Figure 14.1.17

Cell assembly for internal reflection spectroelectrochemistry. [Adapted from N. Winograd and T. Kuwana, *J. Electroanal. Chem.*, **23**, 333 (1969), with permission.]

sampling occurs. It is calculable from the optical parameters of the system (2, 31). For the usual three-phase case (e.g., SnO_2 on glass in contact with solution),

$$\delta = \frac{\lambda}{4\pi \operatorname{Im} \xi} \tag{14.1.12}$$

where λ is the wavelength of incident light and $\operatorname{Im} \xi$ is the imaginary part of $\sqrt{\hat{n}_3^2 - n_1^2 \sin^2 \theta_1}$, in which $\hat{n}_3$ is the complex refractive index of the solution, n_1 is the index of refraction of the substrate (usually glass), and θ_1 is the angle of incidence within the substrate. In general, the absorbing properties of the solution do not affect δ very much (2, 31, 32); hence $\hat{n}_3 \cong n_3$ for this calculation. Since typical values of δ (Problem 14.7) are 500 to 2000 Å, the IRS method is strictly sensitive to the part of the solution very near the interface.

Suppose now that only the reduced form of the O/R couple absorbs. Its absorbance is usually expressed as

$$\mathscr{A}_R(t) = N_{eff}\varepsilon_R \int_0^\infty C_R(x, t) \exp\left(\frac{-x}{\delta}\right) dx \tag{14.1.13}$$

where the integral arises because the probability of light absorption is proportional to $\langle \mathscr{E}^2 \rangle$ and $C_R(x, t)$ at any value of x. At all points the absorbance is proportional to the molar absorptivity ε_R. The parameter N_{eff} is a sensitivity factor incorporating the number of reflections and the relationship between the incident intensity and $\langle \mathscr{E}_0^2 \rangle$. It depends on the materials used in the electrode/substrate assembly, the beam geometry, and its state of polarization. In general, N_{eff} must be evaluated empirically. Values on the order of 50 to 100 are common.

If the electrolysis time is longer than about 1 msec, the diffusion layer is much thicker than δ; hence $C_R(x, t) = C_R(0, t)$ at all values of x where the exponential factor has a significant value, and

$$\boxed{\mathscr{A}_R(t) = N_{eff}\varepsilon_R \delta C_R(0, t)} \tag{14.1.14}$$

Thus the absorbance is a measure of the *surface concentration* of R. The spectrum of absorbance will not be the same as the conventional absorption spectrum (ε_R *vs.* λ), because N_{eff} and δ vary with λ and because various optical effects associated with the electrode/substrate system complicate the picture (1, 31, 32).

For electrolyses involving time scales shorter than about 500 μsec, the diffusion layer is of the same order as δ, and the absorbance is then sensitive to the evolving concentration profile of R (1, 32, 33). The resulting optical transients can be useful means for characterizing rather fast electrochemical processes, which are otherwise complicated severely by nonfaradaic contributions to current and charge functions. Theoretical absorbance transients can be computed from (14.1.13), once the diffusion-kinetic equations defining the concentration profile of R have been solved, either analytically or by numeric methods, such as digital simulation.

Figure 14.1.18 contains data obtained in experiments of this sort (33). The electrode reaction for curve a is the oxidation of tri-*p*-anisylamine (TAA) to its cation radical in acetonitrile:

$$\text{TAA} - e \rightarrow \text{TAA}^{+} \tag{14.1.15}$$

and the transient absorbance is due to TAA^{+} generated in a step experiment involving an 800-μsec width. To improve the signal-to-noise ratio, the pulses were repeated at 30 Hz for 2 min, and the 3600 transients were averaged. The results are reported in terms of *normalized absorbance*, which is the value of $\mathscr{A}(t)$ divided by the limiting $\mathscr{A}$ at long times [essentially provided by (14.1.14) with $C_R(0, t) = C_R^*$, assuming $D_R = D_O$]. This procedure allows cancellation of N_{eff} and ε.

Experiments b and c were performed to evaluate the rate constant for the electron transfer from TAA to acetylferricenium cation (AF^+):

$$\text{TAA} + \text{AF}^+ \underset{k_2}{\overset{k_1}{\rightleftharpoons}} \text{TAA}^{+} + \text{AF} \tag{14.1.16}$$

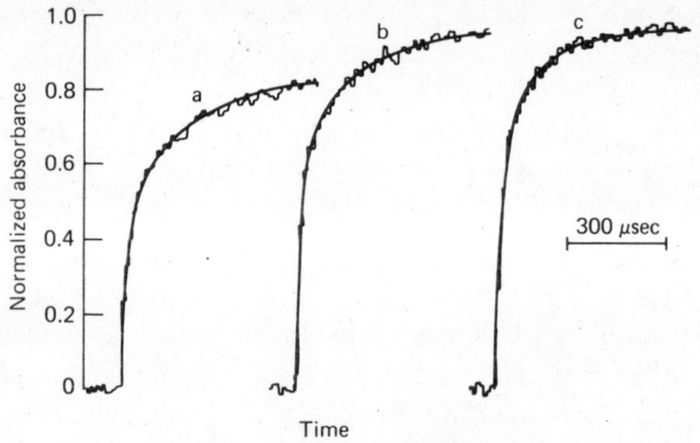

Figure 14.1.18

Transient absorbances caused by potential steps of 800-μsec width from 0.40 to 0.80 V *vs.* SCE. Pulses were repeated at 30 Hz for 2 min and transients were averaged. Curve a: 0.182 m*M* TAA in acetonitrile. Curve b: 0.182 m*M* TAA and 0.182 m*M* AF. Curve c: 0.167 m*M* TAA and 0.333 m*M* AF. Solid curves are fitted results yielding the rate constant given in the text. [Reprinted with permission from N. Winograd and T. Kuwana, *J. Am. Chem. Soc.*, **93**, 4343 (1971). Copyright 1971, American Chemical Society.]

The ratio k_1/k_2 is the equilibrium constant, which is available from standard potentials. The step in potential causes oxidation of *both* TAA and AF, then the AF^+ diffuses out into solution and reacts with TAA, so that $TAA^{\ddagger}$ is produced faster than in the absence of AF^+. From the shape of the absorbance rise, one can evaluate the rate constant k_1 as 3.8×10^8 M^{-1} sec^{-1}. Note that this figure is quite large, and implies a reaction time scale that will severely tax any purely electrochemical method.

The IRS method has obvious possibilities for examining adsorbed species and surface films (2, 34, 35), although these aspects have not been extensively exploited in electrochemical situations.

14.1.4 Photoacoustic and Photothermal Spectroscopy

The transmission and reflection techniques, while very powerful in characterizing changes occurring at or near the electrode surface, make rather severe demands on the type of electrode that can be used. For example, electrodes with very rough surfaces cannot be examined easily by reflection techniques, because the impinging light is largely scattered, and of course, only transparent electrodes can be employed in transmission experiments. There has thus been interest in developing techniques for the optical investigation of electrodes (and solids in general), either in situ or after removal from the electrochemical cell, by detecting directly the amount of absorbed radiation through temperature changes in the electrode, rather than by detection of properties of the transmitted or reflected beam. The most direct way of accomplishing this is by attaching a thermistor directly to the electrode surface, near, but out of the path of the irradiating beam, in a technique called *photothermal spectroscopy* (PTS)

(36). Typically the thermistors, whose resistances are inversely proportional to temperature, are used in a differential bridge arrangement, with the reference thermistor near the detector thermistor but away from the electrode surface; a typical experimental arrangement is shown in Figure 14.1.19. The irradiation source in this technique must have a rather high intensity, since the measured temperature change, ΔT, will be proportional to the light intensity; thus xenon lamps or lasers are used. The light beam can be chopped rather slowly (e.g., with a period of ~ 20 sec), with ΔT measured directly on a recorder (Figure 14.1.20*a*), or more rapidly (e.g., at ~ 8 Hz, employing a thermistor with a fast response time) with lock-in amplifier detection. A plot of ΔT *vs.* λ then yields a type of absorption spectrum (Fig. 14.1.20*b*). So far, such measurements with metal electrodes have only been used to demonstrate the applicability of the technique in electrochemical studies (37, 38). With semiconductor electrodes (see Section 14.5), however, PTS has been of value in determinations of the band gaps of materials, as well as in measurements of the efficiency of conversion of light to electrical and chemical energy in photoelectrochemical cells (39, 40).

In *photoacoustic spectroscopy* (PAS), the thermal changes in a solid sample produced by an intense chopped light beam induce periodic pressure fluctuations in the medium in contact with the sample, and they are detected with a microphone or

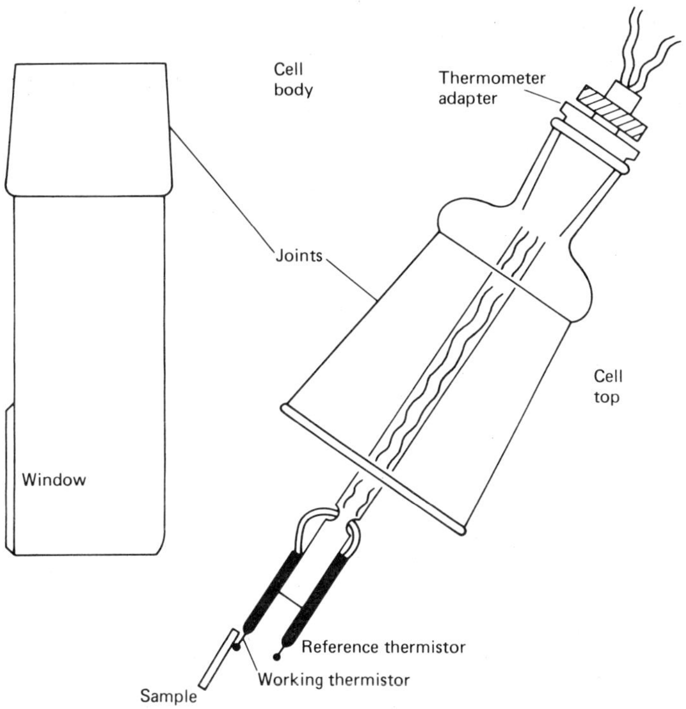

Figure 14.1.19

Cell and working electrode for photothermal spectroscopic electrochemical studies (counter and reference electrodes not shown). [Reprinted with permission from G. H. Brilmyer and A. J. Bard, *Anal. Chem.*, **52**, 685 (1980). Copyright 1980, American Chemical Society.]

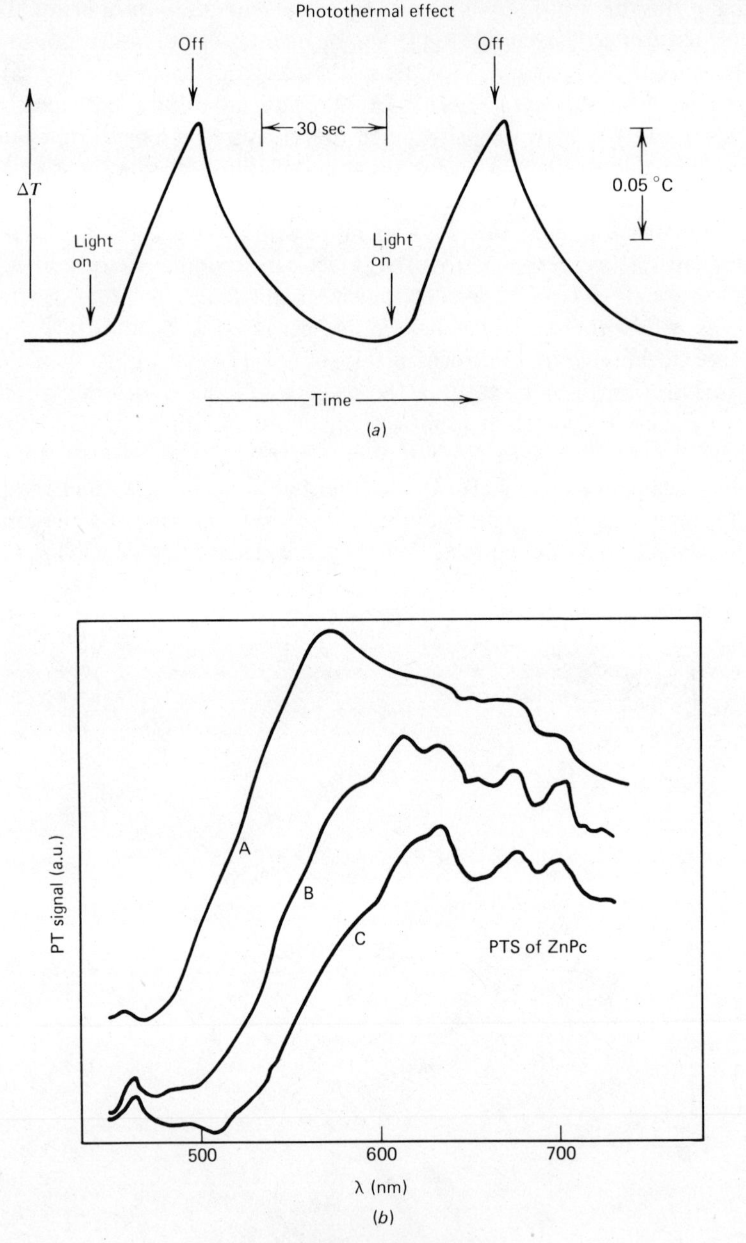

Figure 14.1.20

(*a*) Typical photothermal (thermistor) response with light-absorbing substance on electrode and irradiation with monochromatic light of about 7 mW/cm^2. (*b*) Photothermal spectra (corrected for power spectrum of irradiating source) for zinc phthalocyanine films on a platinum electrode: Curve A, ~3000 Å; Curve B, ~1000 Å; Curve C, ~200 Å. [Reprinted with permission from G. H. Brilmyer and A. J. Bard, *Anal. Chem.*, **52**, 685 (1980). Copyright 1980, American Chemical Society.]

piezoelectric detector (41–44). Although such techniques show promise for the examination of the electrode/solution interface, few applications to electrochemistry have been reported so far. Preliminary studies (45) have been carried out on the direct detection of light absorption by a metal electrode directly bonded to a piezoelectric crystal (Figure 14.1.21). Modulated light absorbed by the electrode as a result of electrochemically induced surface changes leads to small modulated changes in the dimensions of the electrodes. The induced modulated stress on the piezoelectric crystal results in a varying voltage that is detected with a lock-in amplifier. Thus when a colored solid, such as the heptyl viologen radical cation bromide ($HV^{\ddagger}Br^{-}$), is electrodeposited on an electrode and then stripped (14.1.17) in a cyclic voltammetric scan:

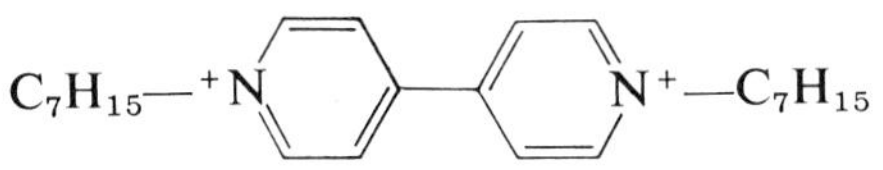

C_7H_{15}—$^{+}$N
N^{+}—C_7H_{15}

$$+ \; Br^- + e \rightleftharpoons \underline{HV^{\ddagger}Br^-} \qquad (14.1.17)$$

(HV^{2+})

the piezoelectric detector signal under irradiation with 550 nm light chopped at 96 Hz shows the coloring and bleaching of the electrode (Figure 14.1.22).

It seems appropriate to add in this section brief mention of techniques in which thermal or stress measurements are made at electrodes without illumination. Direct measurements of temperature changes at electrodes either with the whole cell in a calorimeter or with attached thermistors can yield information about the thermodynamics of the electrode reactions and indicate processes that occur with the evolution or consumption of heat (46–50). Similarly, stress changes that occur at a

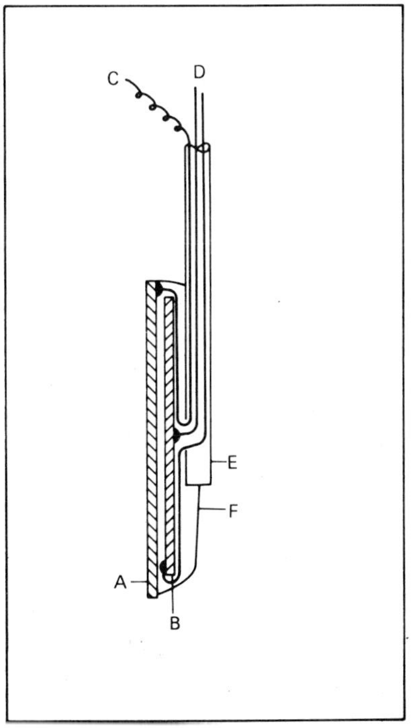

Figure 14.1.21

Working electrode with attached piezoelectric crystal for detection of absorption of radiation. (A) Platinum foil electrode. (B) Piezoelectric ceramic. (C) Lead to potentiostat. (D) Leads to lock-in amplifier. (E) Pyrex glass tube. (F) Epoxy cement. [Reprinted with permission from R. E. Malpas and A. J. Bard, *Anal. Chem.*, **52**, 109 (1980). Copyright 1980, American Chemical Society.]

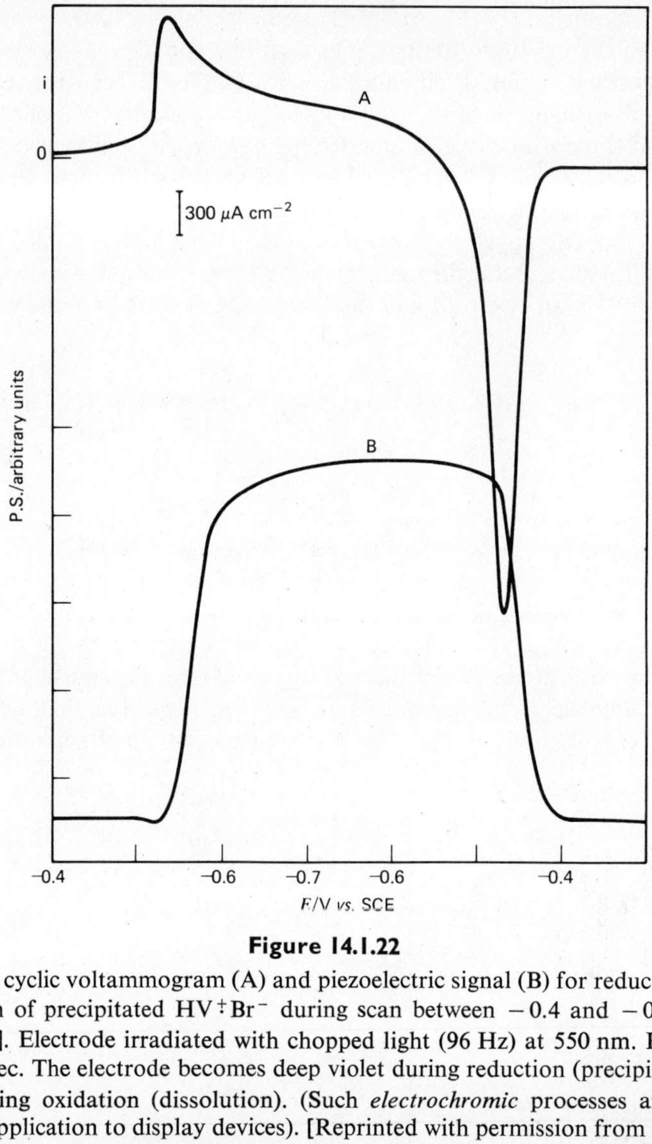

Figure 14.1.22

Simultaneous cyclic voltammogram (A) and piezoelectric signal (B) for reduction of HV^{2+} and oxidation of precipitated $HV^{\ddagger}Br^{-}$ during scan between -0.4 and -0.7 V *vs.* SCE [see (14.1.17)]. Electrode irradiated with chopped light (96 Hz) at 550 nm. Potential scan rate, 20 mV/sec. The electrode becomes deep violet during reduction (precipitation) and is bleached during oxidation (dissolution). (Such *electrochromic* processes are of interest for possible application to display devices). [Reprinted with permission from R. E. Malpas and A. J. Bard, *Anal. Chem.*, **52**, 109 (1980). Copyright 1980, American Chemical Society.]

solid electrode because of surface processes or double-layer capacitance changes can be detected with an attached piezoelectric crystal (51, 52).

14.1.5 Raman Spectroscopy

Raman scattering experiments usually involve excitation of a sample with light that is not absorbed by the sample. Most of this light passes directly through the system or is *elastically* scattered (the *Rayleigh effect*); that is, it is scattered without a change in photon energy. However, some photons exchange energy with the sample and are *inelastically* scattered, with a change in wavelength reflecting the loss or gain in energy. This process is the *Raman effect* (53, 54), and it provides much qualitative information about the sample from the characteristic changes in energy observed in the scattered photons.

The scattering process can be viewed in the manner depicted in Figure 14.1.23*a*. The incident photon can be imagined as raising a molecule to a "virtual state," which is a nonstationary state of the system. Immediate reemission without loss of energy effects Rayleigh scattering, and reemission to a final state other than the original state gives Raman scattering. Note that the Raman effect will produce light with *discrete energy differences* relative to the energy of incident light. These differences correspond to quanta of the vibrational normal modes of the molecule. Usually one studies the *Stokes lines*, which are Raman emissions at lower energy than the excitation energy. However, the scattered photon can also have more energy than the incident light by being scattered from a system with some initial vibrational activation. This *anti-Stokes* branch is generally less useful because it is usually of lower intensity.

The probability of Raman scattering depends on certain selection rules, but under most circumstances is quite small; hence experiments must involve intense light sources and high sample concentrations.

Recently, there has been much interest in the *resonance Raman effect* (54, 55) which yields very large enhancements in the scattering efficiency. A schematic view is provided in Figure 14.1.23*b*. Excitation is made within an absorption band to a virtual state nearly of the same energy as one of the stationary states of the system. The near-resonance electronic interaction enables the molecule to interact much more effectively with the light and provides enhancement factors of 10^4 to 10^6 in scattering probability.

Since Raman experiments always involve the measurement of small energy shifts on the order of 100 to 3000 cm^{-1} from the excitation energy, a monochromatic source

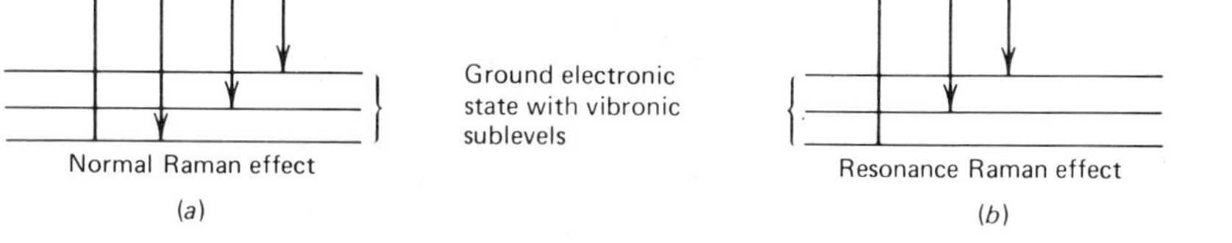

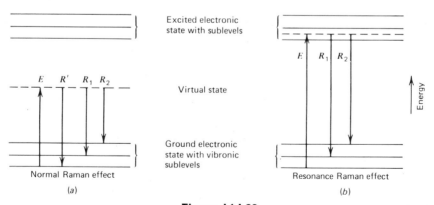

Figure 14.1.23

Schematic views of Raman scattering. Excitation (E) to a nonstationary virtual state is followed by Rayleigh scattering (R'), with no change in energy, or Raman scattering (R_1 and R_2) with energy changes equal to vibrational quanta. (*a*) Normal Raman effect involves excitation in a strictly nonabsorbing region. (*b*) Resonance Raman effect involves excitation very near an allowed absorption transition.

is essential. Since high intensity is also required, lasers are universally used. A high-resolution double monochromator is employed to separate the Raman lines from the intense Rayleigh line. In electrochemical situations, measurements are usually made on species within the operating cell. Considerable geometric freedom can be exercised in cell design (56), since a transparent electrode is not required.

Van Duyne and co-workers have pioneered the development of *resonance Raman spectroelectrochemistry* (56–60), which usually is applied to the examination of electrolytically produced substances. A good example is represented by the spectra in Figure 14.1.24, which were recorded for the system (57),

$$\text{TCNQ} + e \rightleftharpoons \text{TCNQ}^{\overline{\cdot}} \qquad (14.1.18)$$

where TCNQ is tetracyanoquinodimethane:

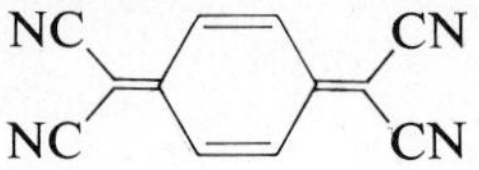

The anion radical was generated by coulometric bulk reduction of a TCNQ solution. The extremely high information content of these spectra is readily apparent. They

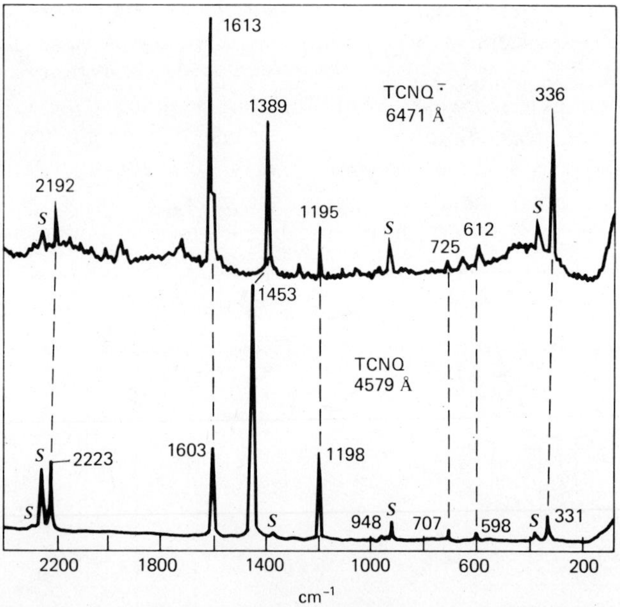

Figure 14.1.24

Resonance Raman spectra of TCNQ and electrogenerated TCNQ$^{\overline{\cdot}}$, which was coulometrically produced by reduction at -0.10 V *vs.* SCE. Initially TCNQ was present at 10.9 m*M* in acetonitrile containing 0.1 *M* tetra-*n*-butylammonium perchlorate. Excitation wavelengths are indicated. Abscissa shows frequency shift with respect to excitation line. *S* denotes a normal Raman band of the solvent. [Reprinted with permission from D. L. Jeanmaire and R. P. Van Duyne, *J. Am. Chem. Soc.*, **98**, 4029 (1976). Copyright 1976, American Chemical Society.]

can be used as diagnostics, and interpreted in much the same manner as infrared spectra, to characterize unknown electrolysis products (58). Alternatively, they can be interpreted in fundamental terms for the information they contain about the electronic and vibrational properties of the species under examination (56, 57, 60).

Raman data can also be obtained on species in the diffusion layer at a faradaically active electrode (56, 59). Whole spectra are gathered by scanning the monochromator slowly as the electrode is cycled through a repeated double-step waveform involving, for example, a short period of forward electrolysis and a long reversal step.

Alternatively, transients in Raman intensity are available by holding the monochromator on a selected line for the duration of the experiment. An example (59) is shown in Figure 14.1.25*a*. Since this result is the average signal resulting from 1000 cycles of a 50-msec period of forward electrolysis and a 950-msec reversal, the whole experiment required 1000 sec. The Raman intensity quantifies the total amount of the product generated, thus it is analogous to the absorbance-time transient observed in a transmission experiment and the charge-time curve in chronocoulometry. The forward phase should yield a signal proportional to $t^{1/2}$, and reversal should produce an intensity proportional to $t^{1/2} - (t - \tau)^{1/2}$, where τ is the duration of the

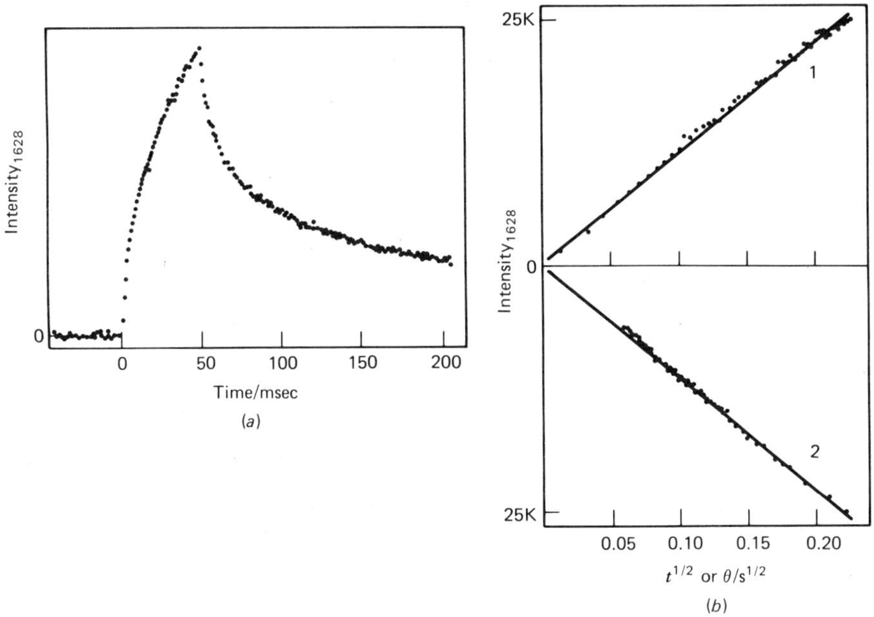

Figure 14.1.25

(*a*) Resonance Raman intensity transient for the cation radical of *N,N,N′,N′*-tetramethyl-*p*-phenylenediamine (TMPD) produced in a 50-msec step and rereduced in a 950-msec step. Average of 1000 experiments. Intensity is for the 1628 cm^{-1} line of TMPD‡, under excitation at 6120 Å. [TMPD] = 3.0 mM in CH_3CN. (*b*) Plots of the forward-phase intensity (1) *vs.* $t^{1/2}$ and reverse-phase intensity (2) *vs.* $\theta = t^{1/2} - (t - \tau)^{1/2}$. Data from transient in (*a*). [From D. L. Jeanmaire and R. P. Van Duyne, *J. Electroanal. Chem.*, **66**, 235 (1975), with permission.]

forward phase. The two plots should have the same slope (Problem 14.8). The graph in Figure 14.1.25*b* verifies the expectations. Note that these experiments are extremely selective, because the monitored Raman line is so narrow that interference from another species in solution is quite improbable. This technique is still in the development stage, but it offers exciting possibilities for mechanistic diagnosis of electrode processes, since it combines extremely good chemical selectivity with a facility for quantitative comparison with diffusion-kinetic models.

Another area of interest is the use of Raman spectroscopy to characterize adsorbed substances (61–69). If this approach can be developed into a general one, there will be substantial benefits resulting from the detailed information that one could obtain about bonding and relative orientations of chemisorbed species. These possibilities are being investigated intensively in view of the success reported first by Fleischmann, Hendra, and McQuillan (61), who studied pyridine adsorbed on a silver electrode. That system still is the most widely studied case. Albrecht and Creighton (65) reported the results shown in Figure 14.1.26, which show the great enhancements in the Raman lines of pyridine that are observed as the silver electrode is first anodized, then reduced to produce a fresh silver surface. Jeanmaire and Van Duyne first recognized that the signal of the adsorbed species is observable because the scattering efficiency is increased enormously (perhaps by a factor of 10^6) in the adsorbed state. The origin of the enhancement is not fully understood at present and is the subject of considerable research (66, 70, 71). If it turns out to be a fairly general phenomenon, then one can expect rapidly increasing studies of adsorbates by Raman methods.

The principal drawbacks to all Raman experiments relate to the intensity of the signal. Low intensities and high-resolution requirements require expensive optics and

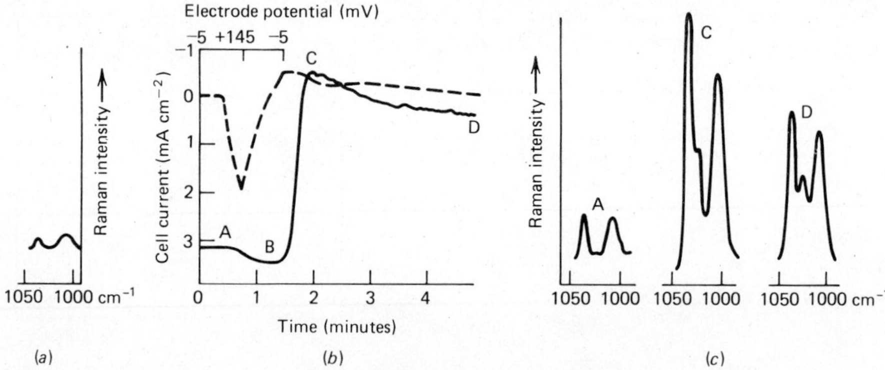

Figure 14.1.26

(*a*) Raman spectrum of a freshly cleaned silver electrode at open circuit in an aqueous solution of 0.01 *M* pyridine and 0.1 *M* KCl. Laser power at 5145 Å = 100 mW. (*b*) Variation of the signal at 1025 cm^{-1} (——) and the cell current (----) during the application of triangular potential waveform. Laser power = 9 mW. Same intensity scale as for (*a*). (*c*) Raman spectra of adsorbed species recorded near points A, C, and D of (*b*). The spectrum in A has been amplified by a factor of 10. [Reprinted with permission from M. G. Albrecht and J. A. Creighton, *J. Am. Chem. Soc.*, **99**, 5215 (1977). Copyright 1977, American Chemical Society.]

slow scanning times or, in transient experiments, a facility for averaging signals. Thus there is now a limited ability to study chemically unstable systems. In addition, fluorescence from the sample can provide serious background interferences. Improvements in instrumentation, for example, introduction of multielement optical detectors, and new experimental techniques can be expected to reduce some of these problems in the near future.

14.2 ELECTRON AND ION SPECTROMETRY

There has been very rapid growth in the development of techniques for surface and thin-film analysis based on the detection of charged particles derived from or interacting with a sample (72–76). Much of the progress along this line has been driven by demands for characterization of thin-film systems used in the manufacture of electronic devices, such as integrated circuits. Research in other areas of science and technology that is concerned with surface problems has benefited from these tools, and electrochemistry has shared. These methods are very powerful and are sure to be more extensively applied to electrochemical problems in the future.

A common feature of all of them is that the measurement is carried out in *high vacuum* (10^{-5}–10^{-7} torr) or *ultrahigh vacuum* ($< 10^{-8}$ torr); thus any electrode surface to be examined must be removed from the cell, possibly rinsed, dried of solvent, and then placed in vacuo. It is not usually possible to examine a system in situ. The necessity for transferring the sample into a system where there is no electrolyte always raises the possibility that the analyzed interface differs significantly from the one in the cell, which is actually the point of interest. Special apparatus has been designed to minimize these problems (77–80), but one must always be alert for artifacts engendered by the transfer.

14.2.1 X-Ray Photoelectron Spectrometry (73–75, 81–83)

If one irradiates a sample with monochromatic X rays (e.g., the Al K_α line at 1486.6 eV or the Mg K_α line at 1253.6 eV), electrons will be ejected from the sample into the surrounding vacuum. Some of these electrons will have been removed from deep core levels of atoms making up the lattice, and they are of particular interest to us now. If the atoms are sufficiently close to the surface ($\lesssim 20$ Å), then there is a high probability that the electrons will escape without being inelastically scattered, and suffering the consequent loss in kinetic energy. We are interested in the distribution of unscattered electrons versus their kinetic energy in vacuo, that is, the *photoelectron spectrum*. This approach is called *X-ray photoelectron spectroscopy* (XPS) or, sometimes, ESCA (for "electron spectroscopy for chemical analysis").

The energy of the photon that ejects an electron must be conserved and is devoted to four terms (82); that is,

$$h\nu = \mathbf{E}_b + T + \mathbf{E}_r + \phi_{sp} \tag{14.2.1}$$

The two most important of these are the kinetic energy T of the electron in the spectrometer and the energy required to remove the electron from the initial state, that is, the *binding energy*, $\mathbf{E}_b$. Since the value of $\mathbf{E}_b$ is discrete and is well-defined for different atomic levels, one can expect *discrete kinetic energies* T corresponding to these levels; hence the photoelectron spectrum shows a peak corresponding to each level. The binding energy associated with a given peak is approximately $h\nu - T$. Minor corrections for the *recoil energy*, $\mathbf{E}_r$, at the site of ejection (usually very small) and for the *spectrometer work function*, ϕ_{sp} (3–4 eV), must be applied for accurate binding energy assignment.

From an analytical standpoint, the utility of XPS is that it provides atomic information about the surface region without seriously damaging that region. Some information about oxidation states is also available, because the binding energy of an electron in a given orbital is affected slightly by its electronic environment. Thus, one can see, for example, separate peaks for 1s electrons derived from nitrogen in its amide and nitro forms (see below). In general, the surface and thin-film analytical tools discussed here are not very informative about the chemical forms in which atoms are present, and the ability of XPS to supply some such information has made it useful for electrochemical applications.

XPS signals can be detected for atoms throughout the periodic table, except helium and hydrogen. The sensitivity limits are on the order of 0.1 atomic percent, except for lighter elements, which are often detectable only above 1 to 10%.

The characterization of anodic oxide films is one area where XPS has been extremely useful. Figure 14.2.1 contains spectra for platinum samples that have been oxidized in three different ways (84). Curve a represents a sample reduced in H_2, then exposed to O_2 at ambient temperature. The two peaks arise from Pt 4f (7/2) and 4f (5/2) orbitals, and each is resolved into two components. The larger is assigned to platinum, and the smaller to platinum associated with adsorbed oxygen atoms. The electrochemically oxidized samples (curves b and c) show structure at higher binding energies, manifesting a more positive platinum center. This feature is assigned to the oxides PtO and PtO_2. Resolution of the curves permits an estimate of the relative contributions of the various forms as shown in Table 14.2.1.

XPS peaks are often broad and show severe overlap like that present in Figure 14.2.1, and curve resolution is widely practiced. Obviously it must be done with care,

Table 14.2.1

Estimated Compositions of Oxidized Platinum Surfaces[a]

Species	Binding Energy, eV 4f (7/2)	4f (5/2)	Relative Peak Areas[b] +0.7 V	+1.2 V	+2.2 V
Pt	70.7	74.0	56	39	34
PtO_{ads}	71.6	74.9	39	37	24
PtO	73.3	76.6	<5	24	22
PtO_2	74.1	77.4	0	0	20

[a] From K. S. Kim, N. Winograd, and R. E. Davis, *J. Am. Chem. Soc.*, **93**, 6296 (1971).

[b] Oxidation carried out at indicated potential (*vs.* SCE) for 3 min.

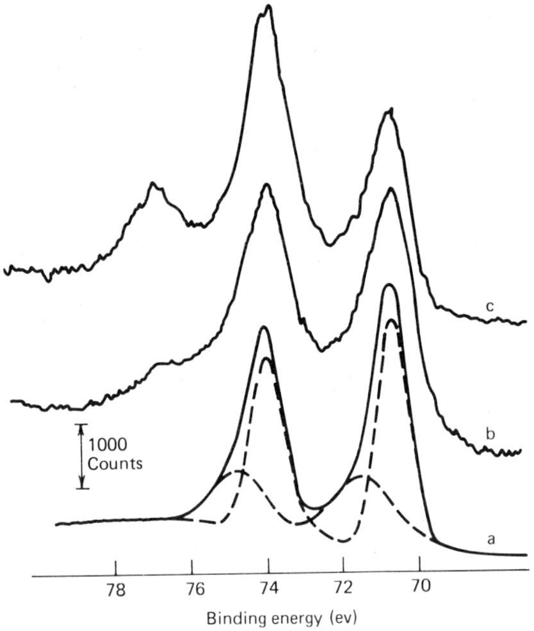

Figure 14.2.1

XPS responses for Pt 4*f* levels. Platinum foil treated by (a) H_2 reduction at 400° for 10 hr, H_2 desorption at 400° (10^{-5} torr) for 5 hr, then exposure to pure O_2 (1 atm) at ambient temperature, (b) electrochemical oxidation at +1.2 V, and (c) at +2.2 V *vs.* SCE. For (b) and (c) electrolyte was 1 *M* $HClO_4$. Curves have been displaced vertically for clarity. [Reprinted with permission from K. S. Kim, N. Winograd, and R. E. Davis, *J. Am. Chem. Soc.*, **93**, 6296 (1971). Copyright 1971, American Chemical Society.]

preferably with foreknowledge of the actual single-component spectra of the substances to which components in mixtures are assigned.

Another electrochemical domain to which XPS has made major contributions is surface modification. Figure 14.2.2*a* contains data showing the effect of treating a glassy carbon surface with γ-aminopropyltriethoxysilane to produce an "amine functionalized" carbon surface (85). The rise of the silicon and nitrogen peaks and the drop in carbon response show the presence of the reagent on the surface. This kind of information is extremely useful in following a surface synthesis.

A similar case (85) is involved in Figure 14.2.2*b*. Dinitrophenylhydrazine (DNPH) was reacted with the surface to produce what is thought to be a hydrazone derivative of a quinoidal surface site:

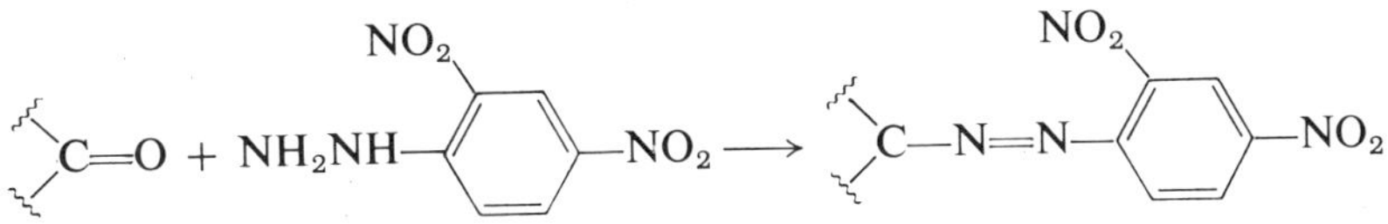

The XPS spectrum (Curve B) shows separate peaks due to nitro nitrogen at high binding energy and the less oxidized nitrogen at lower energy. Holding the electrode

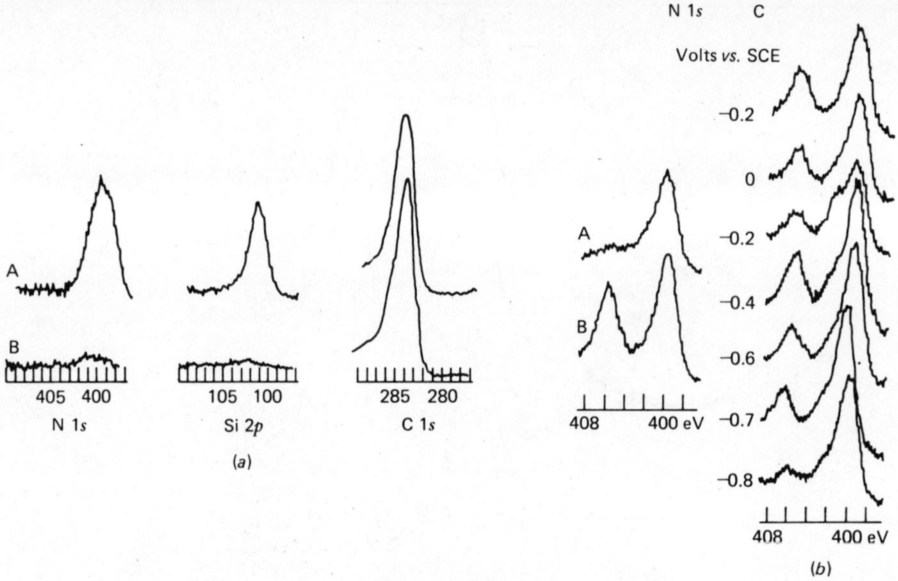

Figure 14.2.2

XPS responses for derivatized glassy carbon electrodes. (*a*) Curves A, following treatment with γ-aminopropyltriethoxysilane. Curves B, unreacted surfaces. (*b*) Nitrogen 1*s* spectra for surfaces treated with DNPH: A, derivatized electrode cycled between 0 and -1.2 V *vs*. SCE; B, fresh modified electrode; C, a series of samples held at indicated potentials for 3 min. [Reprinted with permission from C. M. Elliott and R. W. Murray, *Anal. Chem.*, **48**, 1247 (1976). Copyright 1976, American Chemical Society.]

at potentials more negative than about -0.8 V *vs*. SCE eliminates the peak due to the nitro form and elevates the remaining peak; thus it appears that the nitro functions are reduced in a faradaic process.

Considerable attention has been devoted recently to the phenomenon of *underpotential deposition* of metal atoms (Section 10.2.1) and to the nature of the interaction between the adatom and its substrate. In Figure 14.2.3, one can see that the binding energy of Cu 2*p* (3/2) electrons for Cu adatoms on Pt differs markedly from the value for bulk copper (86, 87). The *negative* shift in binding energy suggests that the deposited copper is not in an oxidized form, but instead is metallic atom in a distinctive electronic environment.

14.2.2 Auger Electron Spectrometry (73–76, 83, 88)

If a vacancy is created in an atomic core level, for example, by irradiation with X rays, as above, or with electrons, then an electron from an upper level can be expected to fill the hole. Figure 14.2.4 is a schematic view of the process in which a K-shell vacancy in silicon is filled by an L_1 electron. The energy difference liberated by this relaxation is 1690 eV, and it can be released wholly in the form of a photon (*X-ray fluorescence*) or by ejecting an *Auger electron* from the atom. In the example of Figure 14.2.4, the

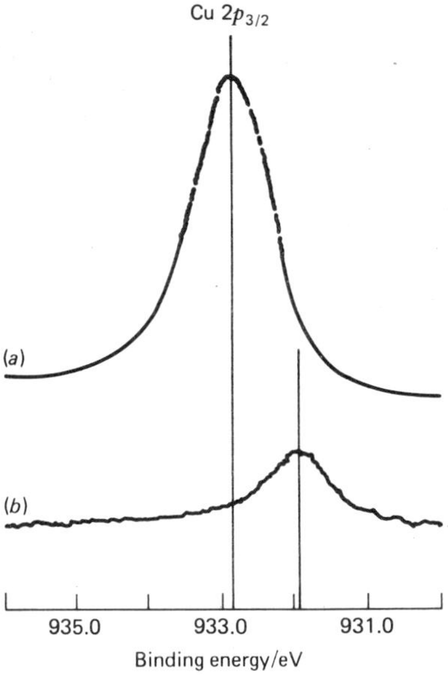

Figure 14.2.3

XPS responses for copper. (*a*) Bulk metal cleaned by etching with beam of Ar^+ in situ. (*b*) Copper deposited at underpotential on platinum. [From J. S. Hammond and N. Winograd, *J. Electroanal. Chem.*, **80**, 123 (1977), with permission.]

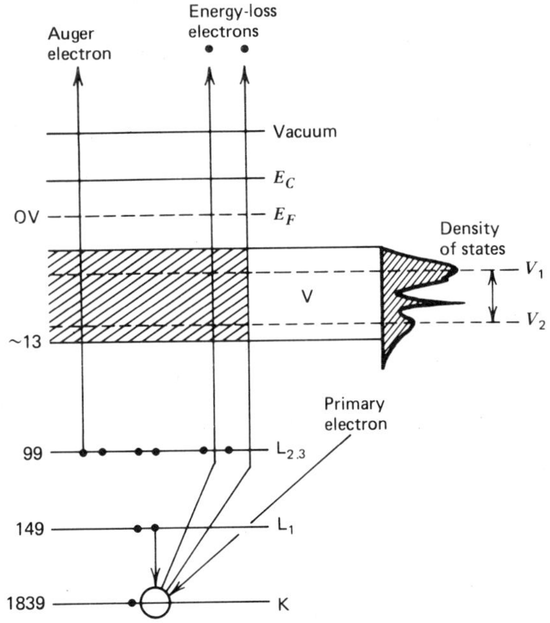

Figure 14.2.4

Schematic view of the Auger emission process from silicon. The atom is ionized initially by an incident electron. That electron and one from the K shell leave the sample as "energy-loss electrons." The energies of the levels relative to the Fermi level E_F (Section 14.5) are given on the left. [From C. C. Chang in "Characterization of Solid Surfaces," P. F. Kane and G. B. Larrabee, Eds., Plenum, New York, 1974, Chap. 20, with permission.]

Auger electron comes from the $L_{2,3}$ shell. The required energy loss of 1690 eV will be divided into the energy required to remove the electron from the sample (mostly its binding energy) and the kinetic energy it retains upon entering the vacuum. Since the total energy of the transition and the binding energy are well-defined values, the kinetic energy of the Auger electron in vacuo is also well-defined. By measuring the distribution of electrons versus kinetic energy, we can obtain a spectrum that should show sharp peaks at the discrete Auger energies. Each Auger line is characteristic of the originating atom and can be used analytically to indicate the presence of that species.

Auger transitions are conventionally labelled with a three-letter notation indicating, respectively, the shell of the primary vacancy, the shell of the filling electron, and the shell from which the Auger electron was emitted. Thus the transition in Figure 14.2.4 would be called the KL_1L_2 or KL_1L_3 process. Any given atom may show several Auger transitions, hence several lines in the spectrum.

If the electron is scattered inelastically during its passage through the sample, its kinetic energy in vacuo will differ from the characteristic Auger energy, and it will contribute only to the broad continuum on which the Auger lines are superimposed. Thus, Auger electron spectrometry (AES) is strictly a surface technique in that atoms only within about 20 Å from the surface can contribute unscattered electrons.

In most instruments, an electron beam is used to excite the sample. The spectrum of emitted and scattered electrons, including Auger electrons, is analyzed according to kinetic energy in a manner that produces a derivative readout, so that the sharp Auger structure is more easily seen on the broad continuum (Figure 14.2.5).

AES signals can be seen for all elements except hydrogen and helium, but the line positions are insufficiently resolved to indicate any information about oxidation state except in very few cases. Detection limits are usually about 0.1 to 1 atomic percent. The electron beam can damage the sample in some cases.

Some instruments, called *scanning Auger microprobes* (SAM) offer two-dimensional scan control of the electron beam, so that analysis can be carried out as a function of

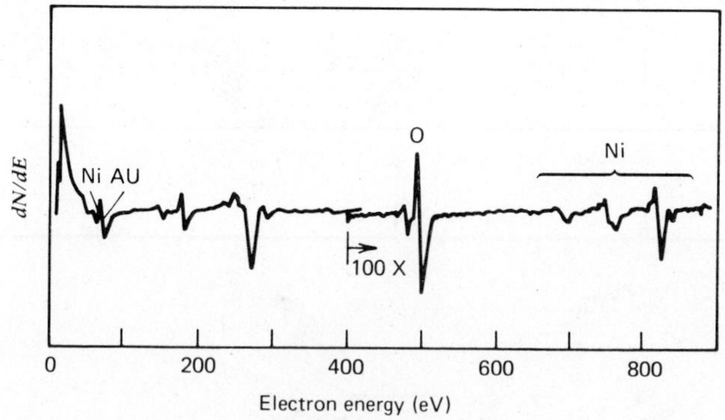

Figure 14.2.5

Derivative Auger spectrum of nickel oxide on the surface of gold-plated nickel. Oxide is sufficiently thin that the 69-eV peak of Au is visible. Additional peaks from 150 to 300 eV are from S, Cl, and C contamination of the surface. [From S. H. Kulpa and R. P. Frankenthal, *J. Electrochem. Soc.*, **124**, 1588 (1977), reprinted by permission of the publisher, The Electrochemical Society, Inc.]

surface position. The spatial resolution is controlled by the beam diameter, which typically is 1 to 10 μm.

A very useful feature on most equipment is a facility for obtaining Auger response as a function of depth into a sample. This technique, called *depth profiling*, is carried out by etching the sample with a beam of high-energy ions (e.g., Ar^+) through a sputtering process. After etching for some period, an Auger spectrum can be recorded; or one can record an Auger line intensity versus etching time to follow the distribution of a particular element with depth. The chief artifacts that can arise with this procedure are homogenization of the sample by the high-energy ion beam and differential sputtering, which involves removal of one component at a faster rate than another.

AES is quite widely employed for the characterization of anodic films, particularly among investigators interested in corrosion. The spectrum in Figure 14.2.5 relates to a study of the tarnishing of nickel in air, and the Auger depth profiling method was used to measure film composition and thickness as a function of ambient conditions during exposure of the sample (89). Figure 14.2.6 contains depth profiles for an anodic film formed on GaAs (90). The results in (*a*) show that the electrochemically formed oxide region is actually quite complex, in that it comprises four distinct zones with varying arsenic-to-gallium ratios. Heat treatment (Figure 14.2.6*b*) changes the profiles considerably and particularly enhances the gallium-rich surface zone. Results of this sort are useful in advancing technology that depends on the properties of films such as passivating layers or insulating barriers.

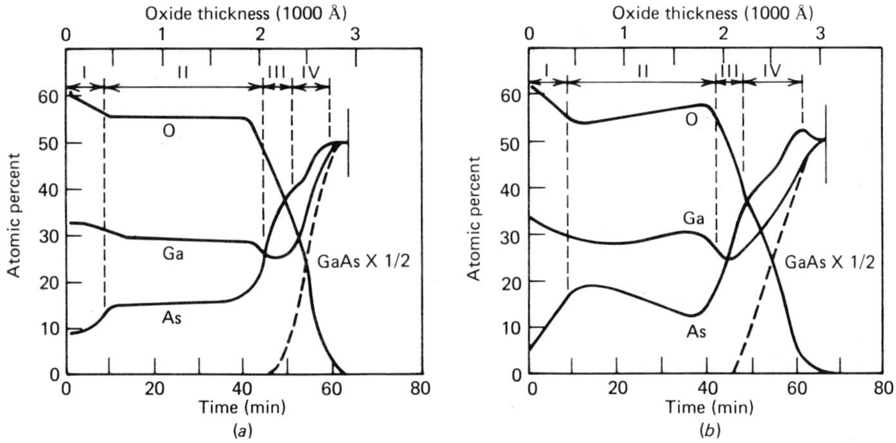

Figure 14.2.6

AES depth profiles for GaAs anodized in H_3PO_4 solutions. Ordinate has been corrected for relative Auger intensities and differential sputtering rates. Abscissa is sputtering time. Thickness scales are approximate. Roman numerals indicate different compositional regions in the oxide layer. Bulk GaAs is at rightmost limit. (*a*) From electrochemical treatment only. (*b*) With added annealing step at 250°. [From C. C. Chang, B. Schwartz, and S. P. Murarka, *J. Electrochem. Soc.*, **124**, 922 (1977), reprinted by permission of the publisher, The Electrochemical Society, Inc.]

Auger techniques have also proven useful in the characterization of electrochemically induced changes in thin-film electrodes. An example concerns films (500–2000 Å thick) of magnesium phthalocyanine (MgPc):

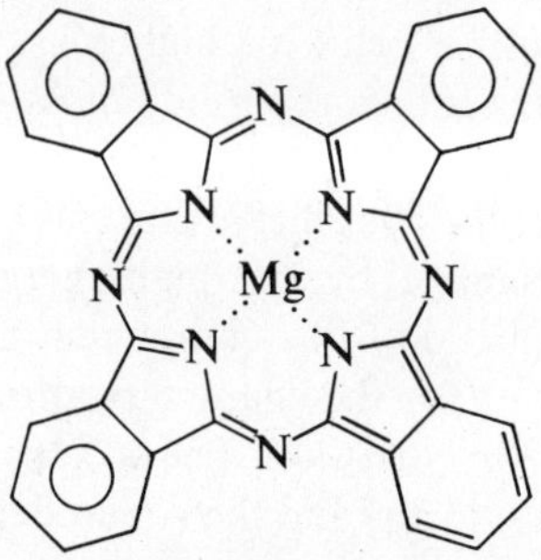

deposited over a gold contact layer on a glass substrate (91). Since the phthalocyanines may be useful in electrocatalytic systems, their properties as electrode materials are of interest and have been studied in this manner (92, 93). At potentials more positive than about 0.6 V *vs.* SCE, one generally finds that MgPc films undergo large-scale oxidation and change color. Figure 14.2.7 is a set of Auger depth profiles showing that the charges created within the film upon oxidation are counterbalanced by anions extracted from the electrolyte. Accommodating the ions probably requires rather substantial changes in lattice properties.

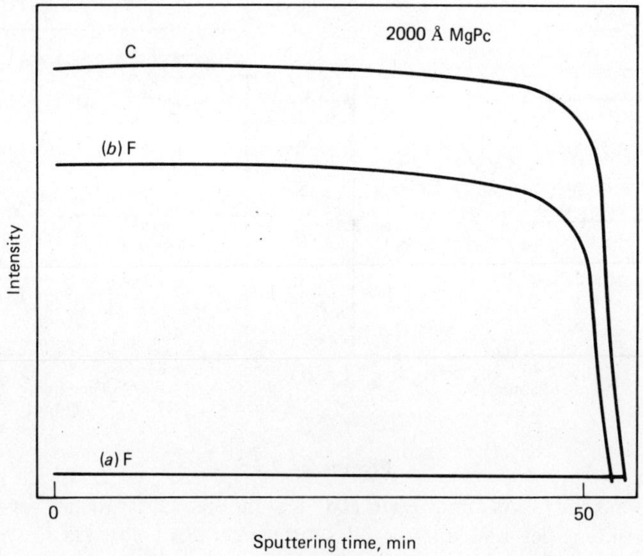

Figure 14.2.7

AES depth profiles for carbon and fluorine in 2000-Å-thick MgPc films. Carbon profiles of separate samples were normalized to a common value. (*a*) MgPc film immersed in aqueous 0.1 *M* KPF_6. (*b*) MgPc film oxidized in 0.1 *M* KPF_6 solution.

14.2.3 Low-Energy Electron Diffraction (94, 95)

Electrons travelling in vacuo at kinetic energies in the range from 10 to 500 eV have de Broglie wavelengths on the order of angstroms; hence one could expect a monochromatic beam of these electrons to be reflected from an ordered solid in a diffraction pattern that provides information about the structure of the solid. This effect is the basis of *low-energy electron diffraction* (LEED).

A LEED experiment differs significantly from other types of diffraction experiments, in that the probing beam cannot penetrate the sample to a distance greater than a few angstroms without being scattered inelastically and losing energy. Thus it is incapable of sampling the three-dimensional order of the solid, and any observed diffraction is due to the *two-dimensional order of the surface.* Thus LEED is a very specific tool for examining the geometric pattern of atoms on a surface, and it has been widely used for studies of adsorption from the gas phase and catalysis of gas-phase/solid-surface reactions. Recently, electrochemists have begun to use it for the characterization of surfaces of interest to them (77–79, 96, 97).

Figure 14.2.8 is a schematic diagram of a typical apparatus. The chamber is always at ultrahigh vacuum ($<10^{-8}$ torr), so that the surface remains clean during the experiment. Electrons are directed toward the sample in a beam and are reflected diffractively along certain well-defined lines. The grids filter out inelastically scattered electrons (at lower energy) and then allow the diffracted ones to accelerate toward a luminescent screen. Bright spots on the screen can be observed and photographed from the viewing port. The arrangement of this system also allows AES experiments to be done by changing the signals on the grids and the energy of the electron beam. One very frequently finds combined LEED/Auger systems in use, because it is convenient to be able to monitor surface contamination or adsorption by AES during LEED studies.

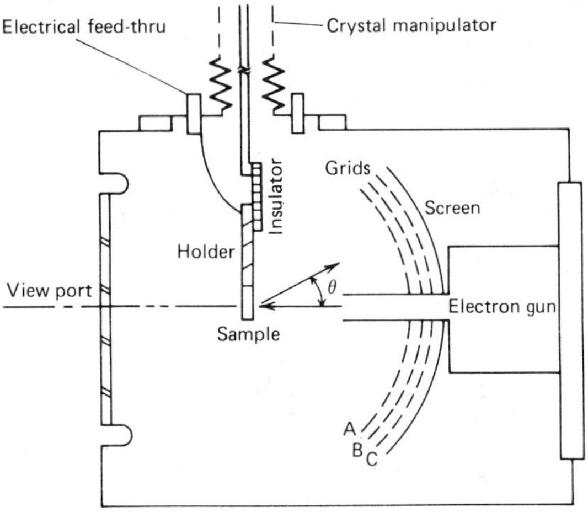

Figure 14.2.8

Schematic diagram of an LEED apparatus. [From G. A. Somorjai and H. H. Farrell, *Adv. Chem. Phys.*, **20**, 215 (1971), reprinted by permission of John Wiley & Sons, Inc.]

Different spot patterns can be interpreted in terms of different surface structures in a fairly straightforward manner. There is a standard notation for describing the structures and their corresponding patterns (94, 95), but it is beyond our scope to delve into it here.

In electrochemical experiments, LEED is used to define the structure of a single-crystal electrode surface [e.g., the (100) face of platinum] before its use in a cell, and to monitor changes that may have taken place upon immersion or electrochemical treatment. One often finds, for example, that a single-crystal surface will *reconstruct* itself, to yield a new surface arrangement, upon contact with an electrochemical medium (77, 96).

14.2.4 Mass Spectrometry

Electrochemists often use mass spectrometry as a tool for the identification of electrolysis products, but the approach is conventional and requires no amplification here.

More extraordinary is the arrangement reported by Bruckenstein and Rao Gadde for monitoring gaseous electrolysis products by a mass spectrometer interfaced directly to an operating cell (98). A porous platinum electrode was mounted onto a glass frit closing the end of a piece of tubing connecting to the inlet port of the spectrometer. Gaseous products, such as O_2, NO, or NO_2, pass through the porous separator into the spectrometer where they can be detected dynamically.

Also of interest to electrochemistry is *secondary-ion mass spectrometry* (SIMS), which is another method for surface and thin-film characterization (75, 76, 99). This approach involves the bombardment of a surface with a high-energy primary ion beam (e.g., 15 keV Cs^+), which etches the surface by sputtering and produces *secondary ions* derived from the surface constituents. These ions are detected mass spectrometrically. Two-dimensional characterization can be carried out by scanning the primary beam, and depth profiles are obtained by monitoring a single ion intensity versus sputtering time. SIMS offers much better detection limits (10^{-4}–10^{-8} atomic percent) than XPS or AES. However, it is not a true surface technique, because the efficiency of secondary-ion production is determined by the three-dimensional properties of the thin ion-implanted layer created by the primary beam (100). Artifacts in depth profiles arise at interfaces from this aspect. SIMS has not been widely used by electrochemists in the past, but a considerable growth in interest is likely because the secondary ions include molecular fragments or clusters of atoms that can offer useful clues to the chemical aspects of surface structure.

14.3 ELECTRON SPIN RESONANCE

14.3.1 Introduction and Principles

Electron spin resonance (ESR) (also known as electron paramagnetic resonance, EPR) spectroscopic methods are used for the detection and identification of electrogenerated products or intermediates that contain an odd number of electrons; that is, radicals, radical ions, and certain transition metal species. Because ESR spectroscopy is a very sensitive technique, allowing detection of radical ions at about the 10^{-8} *M* level under favorable circumstances, and because it produces information-rich, distinctive, and easily interpretable spectra, it has found extensive application to electrochemistry, especially in studies of aromatic compounds in nonaqueous solutions. Also, electrochemical methods are particularly convenient for the generation of radical ions; thus they have been used frequently by ESR spectroscopists for the preparation of materials for study. Several reviews dealing with the principles of ESR and the application to electrochemical investigations have appeared (101–104).

ESR measurements are based on detection of the absorption of radiation of frequency, ν, by a paramagnetic species contained in a magnetic field, H. The magnetic field serves to cause a splitting of the unpaired electron energy levels by an amount $g\beta H$, where g is the spectroscopic splitting or *g factor* (which depends on the orbital and electronic environment of the electron; equals ~ 2 for a free electron and most organic radical species), and β is a constant called the *Bohr magneton* (Figure 14.3.1*a*). When the field is such that the relation

$$\Delta E = h\nu = g\beta H \qquad (14.3.1)$$

is satisfied, transitions between these levels are observed. The structure found in ESR spectra (*hyperfine structure*) arises from additional splittings of these energy levels by protons and other nuclei (e.g., N^{14}, P^{31}) contained in the molecule, with magnetic moments that interact with the unpaired electron. Detailed descriptions of the principles of ESR and the interpretation of ESR spectra are given in many reviews and monographs (105–107).

14.3.2 Apparatus

Commercial ESR spectrometers are available from Varian, JEOL Analytical, and Bruker Instrument Companies. A block diagram of a typical spectrometer is shown in Figure 14.3.2. An outline of the operation of such a spectrometer is as follows. The radiation source is a klystron, which for most analytical spectrometers operates in the X-band region of the microwave spectrum at about 9.5 GHz with an intensity of 150 to 700 mW. The radiation detector is a diode that converts the microwave radiation to direct current. The sample is contained in a cell in a microwave cavity that is held between the poles of an electromagnet. This cavity forms one arm of a magic-T or circulator bridge (which is the microwave equivalent of a resistance or impedance bridge), so that absorption of microwave radiation by the sample causes the bridge to become unbalanced and allows radiation to flow to the detector diode. Sensitivity is improved by modulating the overall dc magnetic field (about 3.2 kG) at 100 kHz with a modulation amplitude of 0.05 to 40 G and using lock-in-amplifier detection methods. The spectrum is recorded by maintaining the microwave radiation frequency constant and slowly scanning the dc magnetic field. The spectrum that results is a first-derivative presentation of the absorption spectrum (because of the detection method) as a function of field (Figure 14.3.1).

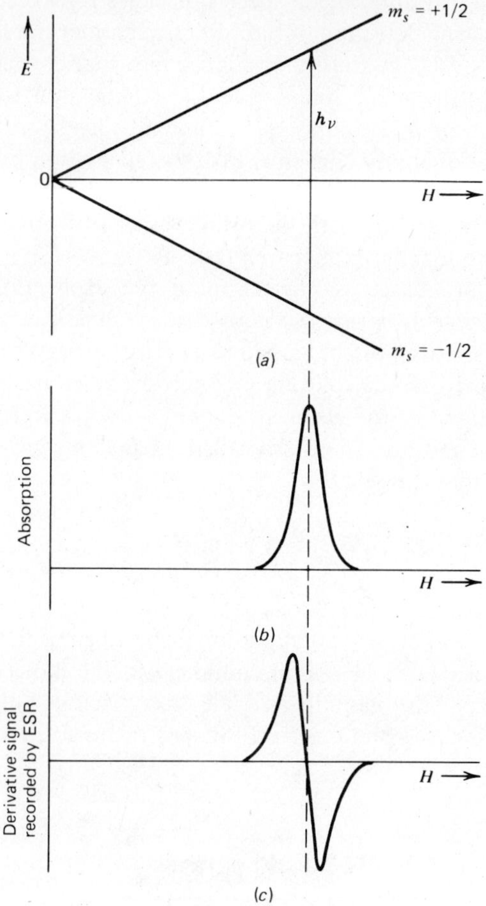

Figure 14.3.1

Principles of the ESR experiment. (*a*) Energy-level diagram of a free electron in a magnetic field. (*b*) ESR absorption versus magnetic field. (*c*) Derivative ESR signal obtained after phase-sensitive detection.

Electrochemical-ESR experiments utilize a number of different sample cell arrangements (101–103). For very stable radical ion or radical species generated in a controlled potential coulometric or bulk electrolysis experiment, the sample can be withdrawn under an inert atmosphere into an ordinary flat or cylindrical ESR sample tube (Figure 14.3.3*a*; "external" generation). Cells in which the working electrode (e.g., platinum gauze or mercury pool) is positioned inside the ESR cavity, with the counter and reference electrodes placed in a portion of the assembly outside of the cavity ("internal" or "intra muros" generation; Figure 14.3.3*b*) are also frequently used. However, because the working electrode is positioned in what is effectively a thin-layer cell arrangement (see Section 10.7) with distant counter and reference electrodes, solution resistance effects are significant. They produce a current density distribution

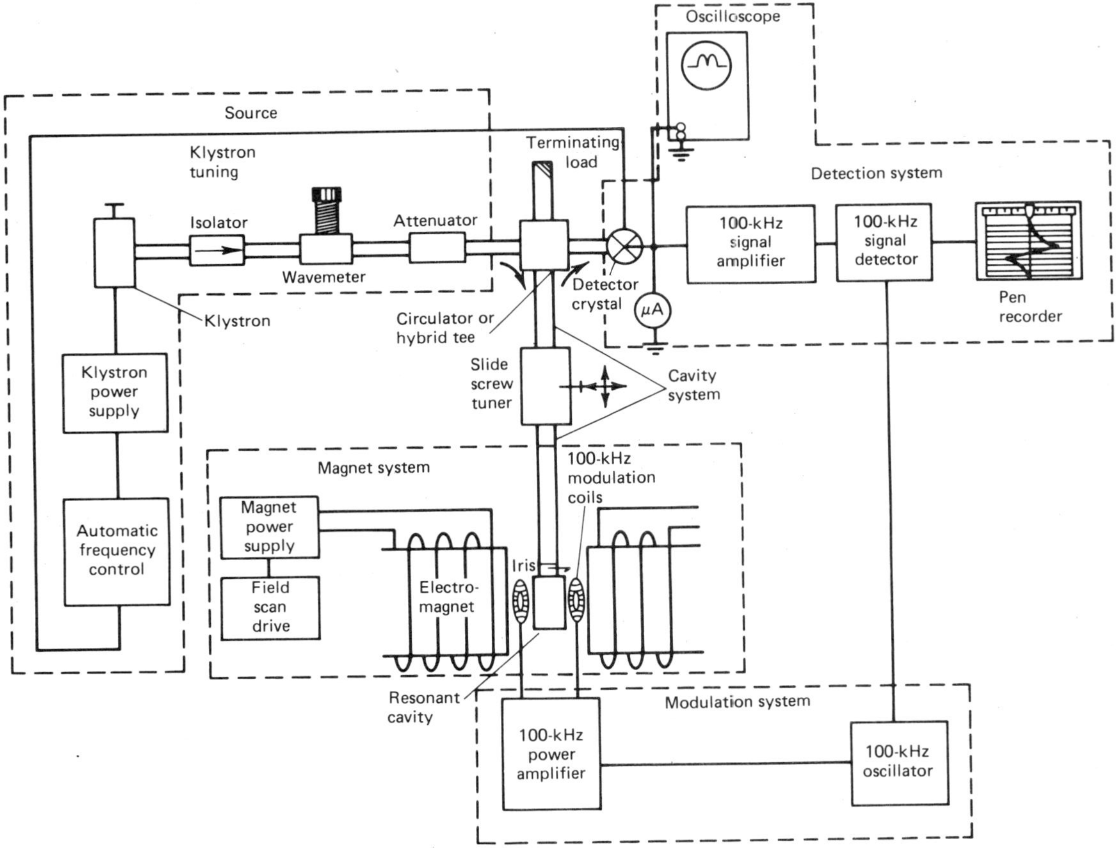

Figure 14.3.2

Block diagram of typical ESR spectrometer employing 100-kHz field modulation. [Adapted from J. E. Wertz and J. R. Bolton, "Electron Spin Resonance: Elementary Theory and Practical Applications," McGraw-Hill, New York, 1972, with permission.]

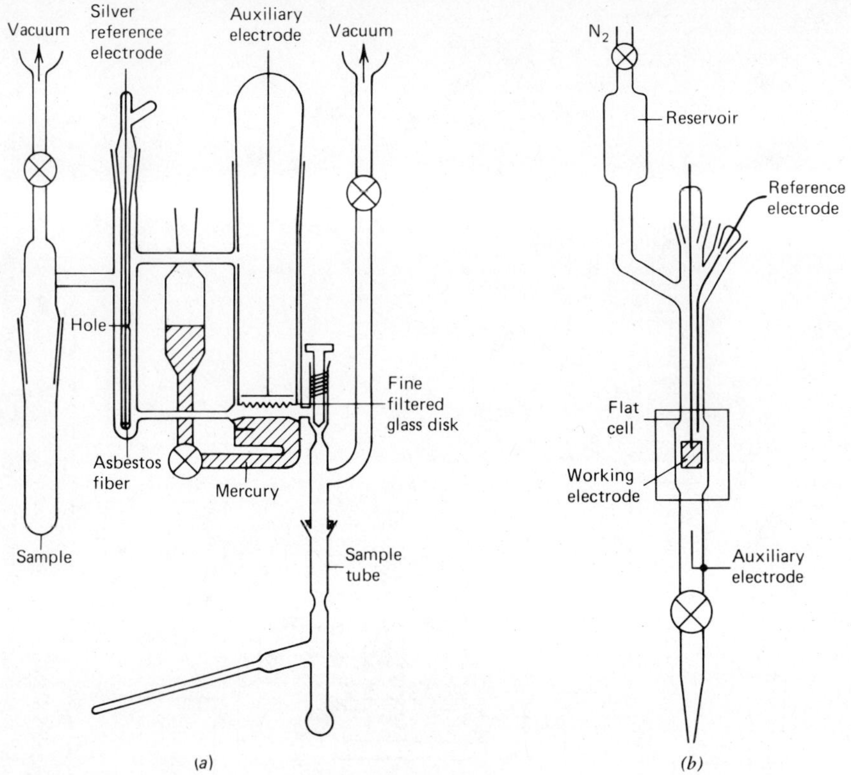

Figure 14.3.3

ESR-Electrochemical cells. (*a*) Cell for external generation of radical ions at mercury pool under vacuum. Sample can be transferred to ESR tube and sealed under vacuum. [From J. R. Bolton and G. K. Fraenkel, *J. Chem. Phys.*, **40**, 3307 (1964), with permission.] (*b*) Cell for "internal" or "intra muros" generation. [From I. B. Goldberg and A. J. Bard, in "Magnetic Resonance in Chemistry and Biology," J. N. Herak and K. J. Adamic, Eds., M. Dekker, Inc., New York, 1975, Chap. 10, Courtesy of Marcel Dekker, Inc.]

over the surface of the working electrode that is nonuniform and varies with electrolysis time, so that good control of the working electrode potential is not possible (102, 108). To circumvent this problem, cells in which all three electrodes are contained in the small portion inside the cavity have been proposed (109, 110) (Figure 14.3.4). While these cells are more difficult to construct, they allow close control of the working electrode potential and allow experiments in which the ESR signal and the electrolysis current can be monitored simultaneously as functions of potential or time [in so-called *simultaneous electrochemical-ESR* (SEESR) experiments].

14.3.3 Types of Electrochemical-ESR Studies

In many electrochemical experiments just the appearance of an ESR signal is sufficient to provide the desired information, that is, that the product of an electrode reaction is a radical or radical ion. The persistence of the ESR signal after cessation of elec-

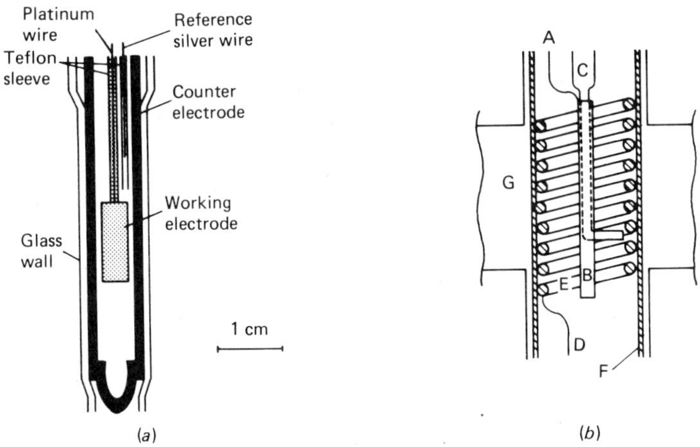

Figure 14.3.4

Cells for simultaneous electrochemical-ESR experiments. (*a*) Flat cell with platinum working and tungsten auxiliary electrodes for use in rectangular cavity. [Reprinted with permission from I. B. Goldberg and A. J. Bard, *J. Phys. Chem.*, **75**, 3281 (1971). Copyright 1971, American Chemical Society.] (*b*) Cell with gold helical working electrode (*E*) which forms the center conductor of a coaxial cylindrical microwave cavity (*G*). A, auxiliary electrode lead; B, central platinum auxiliary electrode; C, Luggin capillary for reference electrode; D, working electrode lead; F, quartz tube. [Reprinted with permission from R. D. Allendoerfer, G. A. Martinchek, and S. Bruckenstein, *Anal. Chem.*, **47**, 890 (1975). Copyright 1975, American Chemical Society.]

trolysis then provides qualitative or semiquantitative information about the stability and lifetime of a species. The earliest studies (111) of the formation of radical anions of aromatic nitro-compounds, for example,

$$PhNO_2 + e \rightarrow PhNO_2^{\overline{\cdot}} \qquad (14.3.2)$$

were of this type. Stable radical anions of quinones, aromatic hydrocarbons, carbonyl compounds, and activated olefins are often observed by these methods. Similarly aromatic hydrocarbons, heterocyclic compounds (e.g., phenothiazines and thianthrenes), and some amines form stable radical cations. In such experiments, one must verify that the signal arises from the major product, rather than from a radical ion resulting from a small amount of side reaction, by having some estimate of the signal level expected for the concentrations employed.

Analysis of the ESR spectrum of the resulting radical ion can be used for identification of the species and also can provide detailed information about effects such as ion pairing (by noting the effects of the electrolyte counter ion on the spectrum), solvation, restricted internal rotation in the molecule (leading to "alternating linewidth" effects), and molecular conformations. Moreover, a detailed comparison of *hyperfine coupling constants* (representing the extent of the interaction of the unpaired electron with nuclei having magnetic moments) with calculated values from molecular orbital (MO) calculations can provide information about the electron density distribution in the molecule and yield the proper MO parameters for calculation of molecular energy levels. These in turn should correlate with the potentials at which the molecule is oxidized or reduced, as well as with the optical absorption and emission spectra (112, 113). Finally, studies of the changes in linewidths in the ESR spectra of stable radical ions (e.g., $A^{\overline{\cdot}}$) as a function of parent (A) concentration, can provide the rate constants k_{11} for homogeneous electron transfer reactions (114–117):

$$A^{\overline{\cdot}} + A \rightleftharpoons A + A^{\overline{\cdot}} \qquad (14.3.3)$$

According to modern theories of electron transfer reactions, for example, by Marcus (118, 119), this constant should correlate with the rate constant for the heterogeneous outer-sphere electron transfer reaction (at an electrode):

$$A + e \rightleftharpoons A^{\overline{\cdot}} \qquad (14.3.4)$$

by the expression

$$k^0 = Z_{el} \left(\frac{k_{11}}{Z_{soln}} \right)^{1/2} \approx \left(\frac{k_{11}}{10^3} \right)^{1/2} \qquad (14.3.5)$$

where Z_{el} and Z_{soln} are the collision frequencies for the electrode and solution, respectively. The constants k^0 and k_{11} are expressed in cm/sec and $M^{-1}\,\text{sec}^{-1}$, respectively.

As an example of an application to studies of reaction rates and mechanism, consider the SEESR studies of hydrodimerization [see (11.1.8) and (11.1.9)]; typical results are given in Figure 14.3.5 (120). The radical anion of an activated olefin A was

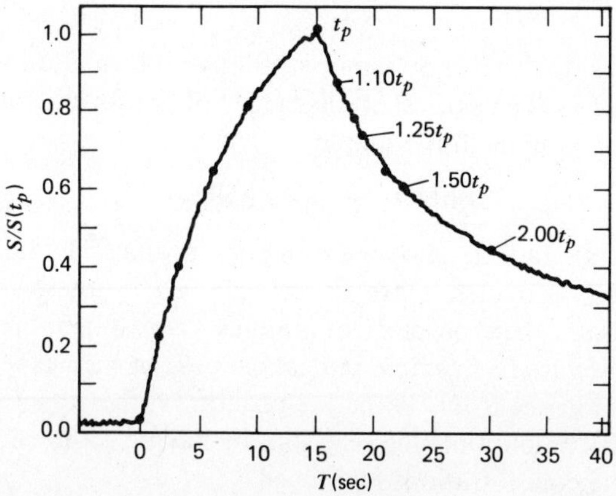

Figure 14.3.5

Example of kinetic study with SEESR cell. ESR signal, S, vs. time for the reduction of 7.5 mM diethyl fumarate (in N,N-dimethylformamide–0.1 M tetrabutylammonium iodide) with a constant current pulse of 0.21 mA for a pulse width t_p of 15 sec. Points shown are calculated values based on a dimerization rate constant of 38 $M^{-1}\,\text{sec}^{-1}$. With signal-averaging techniques, pulse times as short as 200 msec have been used. [Reprinted with permission from I. B. Goldberg, D. Boyd, R. Hirasawa, and A. J. Bard, *J. Phys. Chem.*, **78**, 295 (1974). Copyright 1974, American Chemical Society.]

generated by a constant current pulse in a cell such as that shown in Figure 14.3.4*a*. The rate of buildup and decay of the ESR signal at a fixed magnetic field, which is proportional to the concentration of $A^{\overline{\cdot}}$ near the electrode, was monitored. Analysis of this signal-time transient demonstrated that the reaction was second-order in $A^{\overline{\cdot}}$, so that dimerization of $A^{\overline{\cdot}}$ was suggested, and yielded a rate constant for the reaction that was in general agreement with values obtained by chronoamperometric and RRDE studies. Simultaneous monitoring of the ESR signal and the electrolysis current during a potential sweep can be employed to identify electrochemical waves that are associated with the production of odd-electron species. Such SEESR studies have not yet been used very widely, perhaps because of the difficulty of cell construction. Other kinetic studies involving, for example, the protonation of radical anions by in situ or external flow methods have been reported (121–123).

Very unstable electrogenerated radicals cannot be detected directly by ESR, because their short lives preclude the buildup of sufficient concentrations near the electrode for direct observation. The detection of such species has been accomplished by using the technique of *spin-trapping* (124). In this method, the unstable radical R· reacts rapidly with the spin trap X to form a stable radical, which is then detected by ESR:

$$R\cdot + X \rightarrow RX\cdot \tag{14.3.6}$$

A typical spin trap is phenyltertbutylnitrone (PBN); other nitrones or nitroso compounds are also employed, since these form stable, paramagnetic, nitroxides upon reaction with radicals. For example, phenyl radical was detected in the electroreduction of phenyldiazonium by such a technique (125).

$$PhN_2^+ + e \rightarrow Ph\cdot + N_2 \tag{14.3.7}$$

$$Ph\cdot + \underset{(PBN)}{Ph\overset{H}{C}{=}\overset{O}{N}C(CH_3)_3} \rightarrow Ph_2\overset{H}{C}{-}\overset{O\cdot}{N}C(CH_3)_3 \tag{14.3.8}$$

14.4 ELECTROGENERATED CHEMILUMINESCENCE

Electrochemistry is very well-suited to studies of the solution chemistry of radical ions, because, as we have just seen, one can readily generate these reactive species by oxidizing or reducing stable precursors, such as aromatic amines, nitriles, nitro compounds, or polycyclic hydrocarbons. A particularly striking facet of their chemistry is the chemiluminescence that arises from some of their homogeneous electron transfer reactions. Even though this light almost always comes from reactions in solution, it is usually studied by experiments involving electrolytic production of the participants; hence it is called *electrogenerated chemiluminescence* (or *electrochemiluminescence*, ECL). The topic has been extensively studied and thoroughly reviewed (126–140). Here we will simply outline the basic chemical and experimental aspects. The interested reader can pursue details in the more recent reviews.

14.4.1 Chemical Fundamentals

Typical reactions producing ECL are the following ones involving radical ions of rubrene (R), *N,N,N',N'*-tetramethyl-*p*-phenylenediamine (TMPD), and *p*-benzoquinone (BQ):

$$R^{\bar{\cdot}} + R^{\dagger} \rightarrow {}^1R^* + R \tag{14.4.1}$$

$$R^{\bar{\cdot}} + TMPD^{\dagger} \rightarrow {}^1R^* + TMPD \tag{14.4.2}$$

$$R^{\dagger} + BQ^{\bar{\cdot}} \rightarrow {}^1R^* + BQ \tag{14.4.3}$$

The emission in all cases is the yellow fluorescence of rubrene, which arises from the first excited singlet, ${}^1R^*$:

$${}^1R^* \rightarrow R + h\nu \tag{14.4.4}$$

These reactions are typically carried out in acetonitrile or DMF.

The basic rationale for forming an excited state as a result of electron transfer involves a kinetic manifestation of the Franck-Condon principle (136, 138). The reactions are very energetic (typically 2–4 eV) and very fast (perhaps on the timescale of molecular vibration for the actual transfer). Since it is difficult for the molecular frames to accept such a large amount of released energy in a mechanical form (e.g., vibration) on so short a timescale, there is a significant probability that an excited product will be produced, with consequently smaller mechanical excitation.

Research in this area mainly addresses the fundamental aspects of energy disposition in fast, very exergonic reactions (136), and it serves as a test for theories of electron transfer (Section 3.7). More recently, it has become interesting to a wider audience, in view of demonstrations by Schuster and coworkers (141) that light production in many peroxide-based chemiluminescent systems probably is due to electron transfer.

The free energy available in a redox process producing ground-state products, e.g.,

$$R^{\dagger} + R^{\bar{\cdot}} \rightarrow 2R \tag{14.4.5}$$

is essentially the energy available for exciting a product (132, 136). This number is readily computed from reversible standard potentials for ion/precursor couples, and it can be compared with excited state energies obtained via spectroscopy. States which are lower than the available energy (including perhaps a few kT from thermal activation) are accessible and may be populated in the reaction. Higher states are energetically inaccessible.

Figure 14.4.1 is a diagram of the energetics of reactions (14.4.1) to (14.4.3). Note first that since all states of BQ and TMPD are inaccessible, only the rubrene singlet (${}^1R^*$) and triplet (${}^3R^*$) species can be produced. In addition, we see that (14.4.2) and (14.4.3) are *energy-deficient* in that the ultimate emitter ${}^1R^*$ is not accessible to the electron transfer process. Thus, the reactions written in (14.4.2) and (14.4.3) are overall processes involving more complex mechanisms. In contrast, (14.4.1) is *energy-sufficient*, in that the emitting state is marginally accessible, hence direct production of ${}^1R^*$ could be considered. Such a path is usually called the *S* (for singlet) *route*.

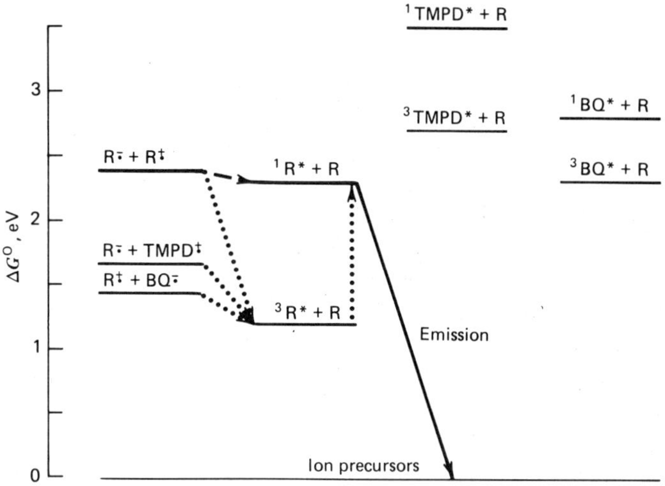

Figure 14.4.1

Energetics for chemiluminescent reactions of rubrene radical ions. All energies measured with respect to ground-state neutral species. Dashed arrow shows *S* route. Dotted arrows show *T* route. Promotion from $^3R^* + R$ to $^1R^* + R$ requires another rubrene triplet. [From L. R. Faulkner, *Meth. Enzymol.*, **57**, 494 (1978), with permission from Academic Press, Inc.]

To rationalize the production of emitters in energy-deficient cases, one usually invokes a mechanism involving triplet intermediates, for example,

$$R^{\cdot -} + TMPD^{\cdot +} \rightarrow {}^3R^* + TMPD \tag{14.4.6}$$

$$^3R^* + {}^3R^* \rightarrow {}^1R^* + R \tag{14.4.7}$$

The second step here is called *triplet-triplet annihilation*, and it allows the energy from two electron transfers to be pooled into the production of a singlet. There is much evidence favoring the operation of this mechanism, which is usually called the *T* (for triplet) *route*. Note that it may operate in energy-sufficient systems, as well as energy-deficient ones.

Hundreds of ECL reactions have been reported, and many are spectroscopically simple enough to be understood in these terms. Others offer emission bands due to *excimers* (excited dimers such as $(DMA)_2^*$, where DMA is 9,10-dimethylanthracene) (135, 139), exciplexes [excited-state complexes, such as $(TPTA^{\cdot +}BP^{\cdot -})$, where TPTA is tri-*p*-tolylamine and BP is benzophenone] (135, 139), or simply decay products of the radical ions. More complicated mechanisms are obviously needed to deal with such situations. Most studies involve radical ions, but others have dealt with metal complexes such as $Ru(bipy)_3{}^{2+}$ [bipy = 2,2′-bipyridine], superoxide, solvated electrons, and classical chemiluminescent reagents, such as lucigenin (126).

The primary experimental goals of ECL experiments are to define the nature of the emitting state, the mechanism by which it is produced, and the efficiency of excited-state production.

14.4.2 Apparatus and Reactant Generation (137)

ECL experiments are carried out in fairly conventional electrochemical apparatus, but procedures must be modified to allow the electrogeneration of two reactants, rather than one, as is more commonly true. In addition, one must pay scrupulous attention to the purity of the solvent/supporting electrolyte system. Water and oxygen are particularly harmful to these experiments. Thus, most apparatus is constructed to allow transfer of solvent and degassing on a high-vacuum line. Other constraints may be imposed by optical equipment used to monitor the light.

Most experiments have been carried out by generating the reactants sequentially at a single electrode. For example, one might start with a solution of rubrene and TMPD in DMF. A platinum working electrode stepped to -1.6 V *vs.* SCE produces R^- in the diffusion layer. After a forward generation time t_f, which could be 10 μsec to 10 sec, the potential would be changed to perhaps $+0.35$ V to produce $TMPD^{\dagger}$, which diffuses outward. Since $R^{\bar{\cdot}}$ is oxidized at this potential, its surface concentration drops effectively to zero, and $R^{\bar{\cdot}}$ in the bulk begins to diffuse back toward the electrode. Thus $TMPD^{\dagger}$ and $R^{\bar{\cdot}}$ move together and react. If the reaction rate constant is very large, their concentration profiles do not overlap, and the reaction occurs in the plane where they meet, as shown in Figure 14.4.2. As the experiment proceeds, the $R^{\bar{\cdot}}$ gradually is used up, and the reaction plane moves farther from the electrode. The light appears as a pulse that decays with time because of the depletion of $R^{\bar{\cdot}}$. Experiments like these may be carried out in a double- or triple-step format to produce one pulse of light; or a train of alternating steps can be used to produce a sequence of light pulses.

Another fruitful approach to reactant generation (137, 142, 143) involves the rotating ring-disk electrode (RRDE). For example, the disk might be set at -1.6 V to generate $R^{\bar{\cdot}}$ at the mass-transfer-limited rate, and the ring could be set to produce $TMPD^{\dagger}$ at $+0.35$ V. The convective flow pattern (Chapter 8) would sweep $R^{\bar{\cdot}}$ out to the ring,

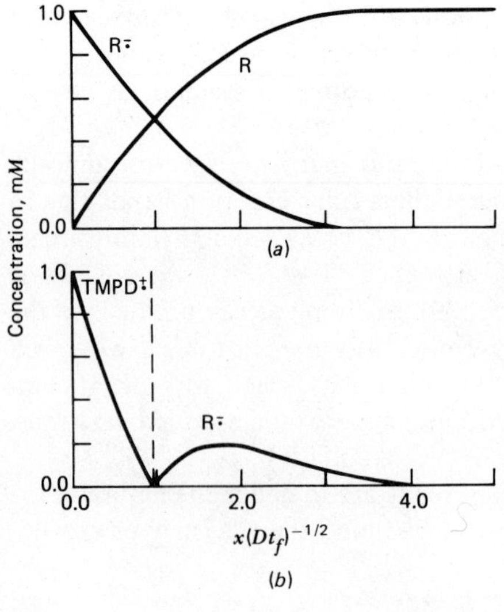

Figure 14.4.2

Concentration profiles near an electrode during an ECL step experiment. Data apply to 1 mM R and 1 mM TMPD. (*a*) Profiles of R and $R^{\bar{\cdot}}$ at the end of the forward step. (*b*) Concentrations of $TMPD^{\dagger}$ and $R^{\bar{\cdot}}$ during the second step. Reaction boundary is shown by the dotted line. Curves apply for a time 0.4 t_f into the second step, where t_f is the forward step width. For this illustration all diffusion coefficients are taken as equal.

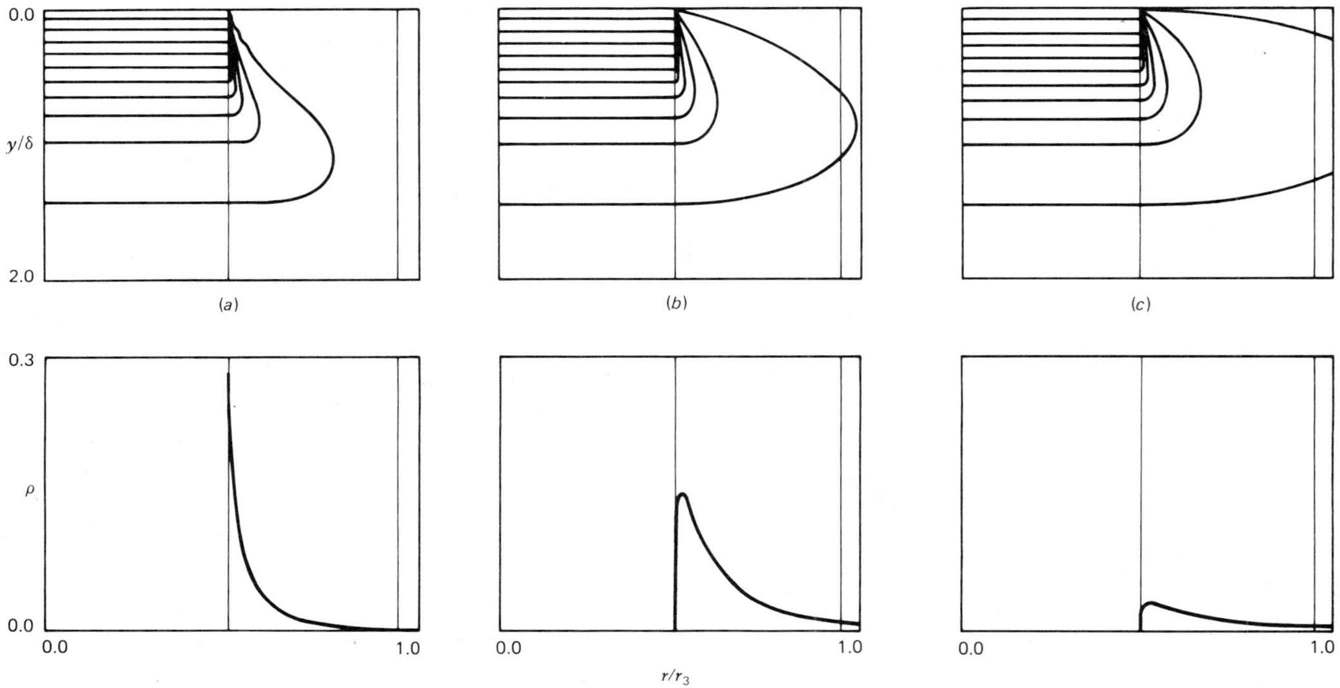

Figure 14.4.3

Concentration profiles (top row) and distribution of homogeneous redox events (bottom row) at the RRDE. Abscissa is distance in the radial direction normalized by the outer radius of the ring. Vertical lines near $r/r_3 = 0.5$ mark the disk radius and the inner ring radius. In the concentration profiles, y/δ describes distance below the RRDE in units of $\delta = 1.8D^{1/3}\nu^{1/6}\omega^{-1/2}$. The lines are contours of equal concentration of disk-generated reactant. Columns (*a*), (*b*), and (*c*) correspond to $kC^*/\omega = 10, 0.625$, and 0.016, respectively, where C^* is the bulk precursor concentration, ω is the angular rotation rate, and k is the bimolecular homogeneous electron transfer rate constant. [Reprinted with permission from J. T. Maloy, K. B. Prater, and A. J. Bard, *J. Am. Chem. Soc.*, **93**, 5959 (1971). Copyright 1971, American Chemical Society.]

where it would react with $TMPD^{+}$ over the face of the ring. Concentration profiles and light distributions are shown in Figure 14.4.3 for several different rate constants of electron transfer (143). For a fast reaction, the $R^{\overline{\cdot}}$ is intercepted without penetrating very far into the region below the ring; hence the light is seen along the inner radius. Slower reactions would allow $R^{\overline{\cdot}}$ to penetrate further and produce a broader distribution of light. Real systems correspond more closely to the case in Figure 14.4.3*a*.

Most studies of ECL have been carried out with variants of these two approaches, although occasional investigations have involved sequential current steps at a single electrode (144) or dual-working-electrode systems with thin-layer geometry or with flow streams to move the reactants together (145).

Many investigations involve special optical procedures, such as calibrating the spectral response of a detection system or measuring the absolute total emission rate of an ECL process. These techniques are beyond our purview; discussions of them are available in the review literature (137).

14.4.3 Kinds of Experiments

The most obvious ECL experiments involve recording spectra of the emitted light, which are essential for identification of emitting species. In some cases ECL produces emission from states that may play only a small role in the fluorescence of the electrolyzed solution. Figure 14.4.4 offers an example (143). The fluorescence of a solution

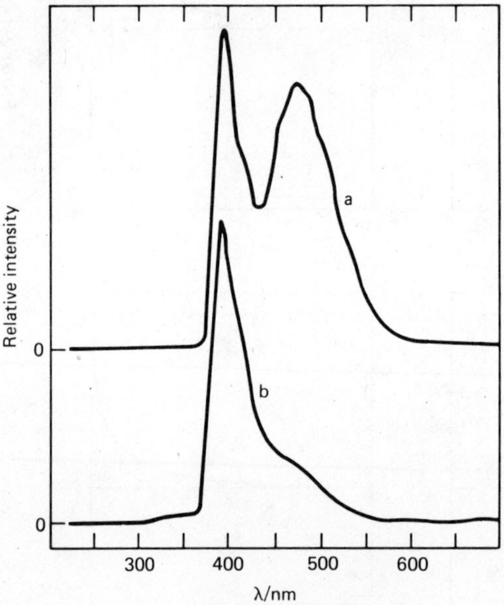

Figure 14.4.4

Curve a: Chemiluminescence from the reaction between $Py^{\overline{\cdot}}$ and $TMPD^{+}$ in DMF. Electrogeneration of ions was carried out at an RRDE in a solution of 1 m*M* TMPD and 5 m*M* Py. Curve b: Fluorescence spectrum of the same solution under excitation at 350 nm. [Original data from Reference 143. Figure from L. R. Faulkner, *Int. Rev. Sci.*: *Phys. Chem. Ser. Two*, **9**, 213 (1975). Reproduced with permission of the American Chemical Society.]

of pyrene (Py) and TMPD excited at 350 nm shows a sharp band at 400 nm, ascribed to $^1Py^*$, and a minor shoulder at 450 nm, due in part to the excimer $^1Py_2^*$, which emits in the dissociative process:

$$^1Py_2^* \rightarrow 2Py + h\nu \tag{14.4.8}$$

In contrast, the ECL from the reaction between $Py^{\cdot-}$ and $TMPD^{\cdot+}$, produced by electrolysis at an RRDE, shows predominant emission from the excimer. Thus the chemiluminescent system has a specific path for relatively efficient production of the excimer. It is believed to be the triplet-triplet annihilation involving $^3Py^*$.

Another useful experiment for deciphering the basic chemistry of ECL can be illustrated by the data in Figure 14.4.5, which shows light intensity from the TMPD/Py system versus disk potential (143). In the upper frame $Py^{\cdot-}$ was generated at the ring, and one sees that light results from the oxidation (at the disk) of TMPD to either $TMPD^+$ (first wave) or $TMPD^{2+}$ (second wave). On the other hand, oxidation products generated at very positive potentials (perhaps Py^+ or its decay products) quench the ECL. Interpretation of the lower frame is left as Problem 14.14.

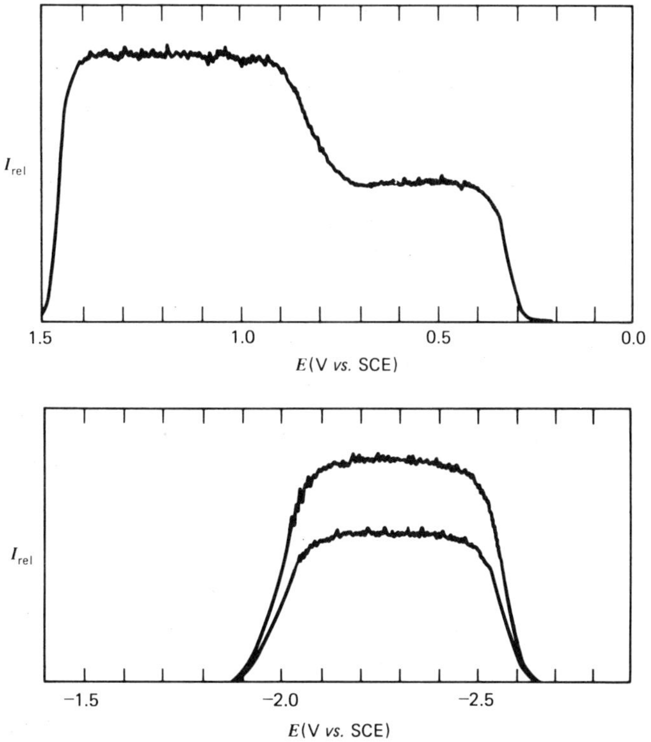

Figure 14.4.5

Steady-state ECL behavior versus disk potential in the pyrene-TMPD system. Upper frame: $Py^{\cdot-}$ generated at ring (emission detected at 393 nm); lower frame: $TMPD^{\cdot+}$ generated at ring [emission detected at 393 nm (curve a) and 470 nm (curve b)]. [Reprinted with permission from J. T. Maloy and A. J. Bard, *J. Am. Chem. Soc.*, **93**, 5968 (1971). Copyright 1971, American Chemical Society.]

The mechanism of light production is always of interest, and many experiments have been devised to probe it. One approach is based on shapes of single pulses of light produced in a step sequence like that described above (136–138, 140). The basic idea is to find the dependence of light intensity on the rate of redox reaction between the oxidant and reductant. For example, the *S* route calls for a linear dependence, whereas the *T* route generally would yield a relationship of higher order. The diffusion-kinetic problem for step generation of ECL has been solved (146), and the time decay of the redox reaction rate can be calculated for a given system. It can then be compared with the observed intensity transient. Data (147) for the reaction between the energy-sufficient reaction between the cation radical of thianthrene (TH) and the anion radical of 2,5-diphenyl-1,3,4-oxadiazole (PPD) are shown in Figure 14.4.6. These results show that the observed intensity exactly follows the *square* of the reaction rate through a rather complex passage, and this fact alone has some useful things to say about the mechanism of light production. The reason for the dip is part of Problem 14.15.

Still other experiments are designed to intercept intermediates such as triplets or singlet oxygen. Results for such a case are given in Figure 14.4.7, which arose from a study of the energy-deficient reaction between the cation radical of 10-methylphenothiazine and the anion radical of fluoranthene (148). The system is believed to produce light by the annihilation of $^3FA^*$. Addition of anthracene (An) could be accomplished without disturbing the electrochemistry needed to produce the reactants, but it transformed the emission spectrum from that of $^1FA^*$ to that of $^1An^*$. Apparently, this result was caused by the energy transfer:

$$An + {}^3FA^* \rightarrow FA + {}^3An^* \qquad (14.4.9)$$

followed by annihilation of $^3An^*$.

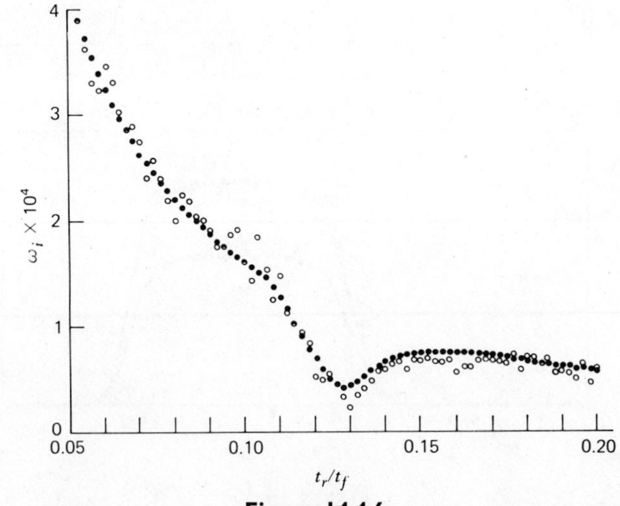

Figure 14.4.6

Light intensity (open circles) and square of redox reaction rate (filled circles) versus time t_r measured into the second step of a transient experiment. Forward step width t_f was 500 msec. The ECL process involved $TH^{\dot{+}}$ and $PPD^{\dot{-}}$. See Problem 14.15 for a discussion of the dip. [Reprinted with permission from P. R. Michael and L. R. Faulkner, *J. Am. Chem. Soc.*, **99**, 7754 (1977). Copyright 1977, American Chemical Society.]

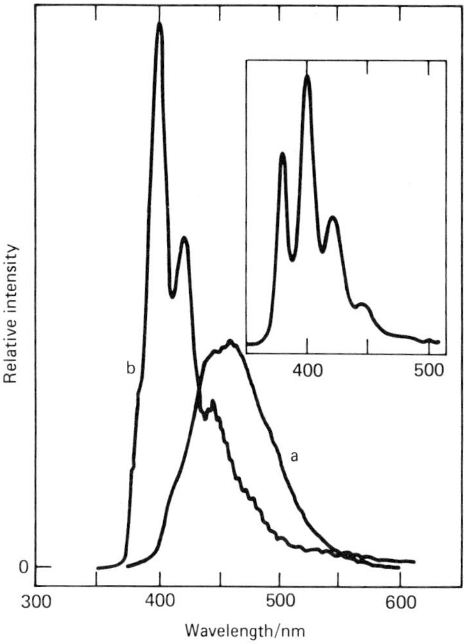

Figure 14.4.7

Curve a: Chemiluminescence from $FA^{\cdot-}$ and $10\text{-}MP^{\cdot+}$ in DMF. Reactants were generated from solutions containing 1 m*M* FA and 10-MP. Curve b: Spectrum upon addition of anthracene. Inset shows fluorescence from anthracene at 10 μM in DMF. Shortest wavelength peak is not seen in ECL because of self-absorption. [Original data from Reference 148. Figure from L. R. Faulkner, *Int. Rev. Sci.*: *Phys. Chem. Ser. Two.*, **9**, 213 (1975). Reproduced with permission of the American Chemical Society.]

Sometimes magnetic fields enhance ECL intensities, and studies along this line have been used for mechanistic diagnosis (136–139). The effects seem to arise from field-dependent rate constants for certain reactions involving triplets, and hence they are associated with the *T* route.

Generally, ECL is regarded as a solution-phase process, on the basis of both direct evidence (Problem 14.15) and a belief that metal electrodes quench excited states (149, 150). The band structure of semiconductor electrodes sometimes removes the latter difficulty (see Section 14.5), and there have been recent reports of emission from excited states produced directly in *heterogeneous* charge transfer at semiconductors (151–153).

14.5 PHOTOELECTROCHEMISTRY AT SEMICONDUCTORS

14.5.1 Introduction

In photoelectrochemical experiments, irradiation of an electrode with light that is absorbed by the electrode material causes the production of a current (the *photocurrent*). The dependence of the photocurrent on wavelength, electrode potential, and solution composition provides information about the nature of the photoprocess, its energetics, and its kinetics. Photocurrents at electrodes can also arise because of photolytic processes occurring in the solution near the electrode surface; these are discussed in Section 14.6. Photoelectrochemical studies are frequently carried out to obtain a better understanding of the nature of the electrode/solution interface.

However, because the production of a photocurrent can represent the conversion of light energy to electrical and chemical energy, such processes are also investigated for their potential practical applications. Since most of the studied photoelectrochemical reactions occur at semiconductor electrodes, we will review briefly the nature of semiconductors and their interfaces with solutions. Several detailed reviews of this area are available (154–159).

14.5.2 Semiconductor Electrodes

The electronic properties of solids are usually described in terms of the *band model*, in which the behavior of an electron moving in the field of the atomic nuclei and all of the other electrons is treated (156, 160–163). Consider the formation of a lattice of a solid (e.g., Cu, Si, TiO_2). When the isolated atoms, which are characterized by filled and vacant orbitals, are assembled into a lattice containing $\sim 5 \times 10^{22}$ atoms/cm^3, new molecular orbitals form. These orbitals are so closely spaced that they form essentially continuous bands; the filled bonding orbitals form the *valence band* and the vacant antibonding orbitals form the *conduction band* (Figure 14.5.1). These bands are separated by a *forbidden region* or *band gap* of energy $\mathbf{E}_g$, which is usually given in units of electron volts. When $\mathbf{E}_g \ll kT$, or when the conduction and valence bands overlap, the material is a good conductor of electricity (e.g., Cu, Ag). Under these circumstances, there exist in the solid filled and vacant electronic energy levels at virtually the same energy, so that an electron can move from one level to another with only a small energy of activation. For larger values of $\mathbf{E}_g$ (e.g., for Si, where $\mathbf{E}_g$ = 1.1 eV), the valence band (VB) is almost filled and the conduction band (CB) almost vacant. The only carriers arise because of thermal excitation of electrons from the

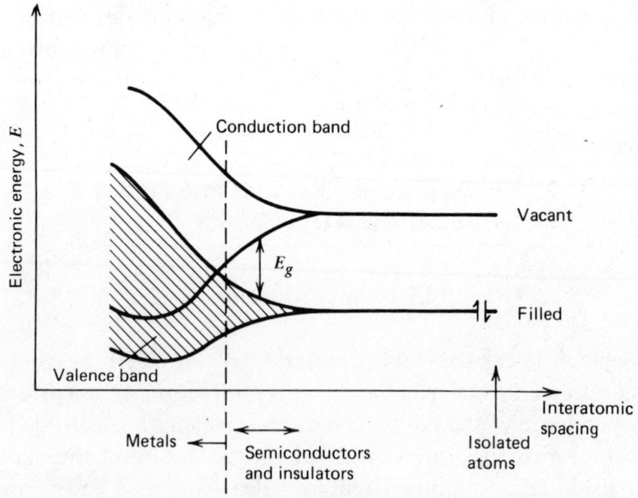

Figure 14.5.1
Formation of bands in solids by assembly of isolated atoms (characterized by orbitals at far right of X-axis) into a lattice.

VB into the CB (Figure 14.5.2). This process produces an electron in the CB, which is capable of moving freely to vacant levels in the CB, and leaves behind a hole in the VB that can be filled by VB electrons. Such a material is called an *intrinsic semiconductor* in which the density of CB electrons, n_i, and of VB holes, p_i, is approximately given by the expression (156)

$$n_i = p_i \approx 2.5 \times 10^{19} \exp\left(\frac{-\mathbf{E}_g}{2kT}\right) \text{cm}^{-3} \text{ (near } 25^\circ) \tag{14.5.1}$$

For example, for silicon, $n_i = p_i \approx 1.4 \times 10^{10}$ cm^{-3}. The mobile carriers (electrons and holes) move in the semiconductor in a manner analogous to the movement of ions in solution. The mobilities of these species, u_n and u_p, are orders of magnitude larger than for ions in solution, however. For example, for silicon, $u_n = 1350$ and $u_p =$ 480 cm^2 V^{-1} sec^{-1}. For materials with $\mathbf{E}_g > 1.5$ eV, so few carriers are produced by thermal excitation at room temperature that in the pure state such solids are electrical insulators (e.g., GaP and TiO_2, with $\mathbf{E}_g$ equal to 2.2 and 3.0 eV, respectively).

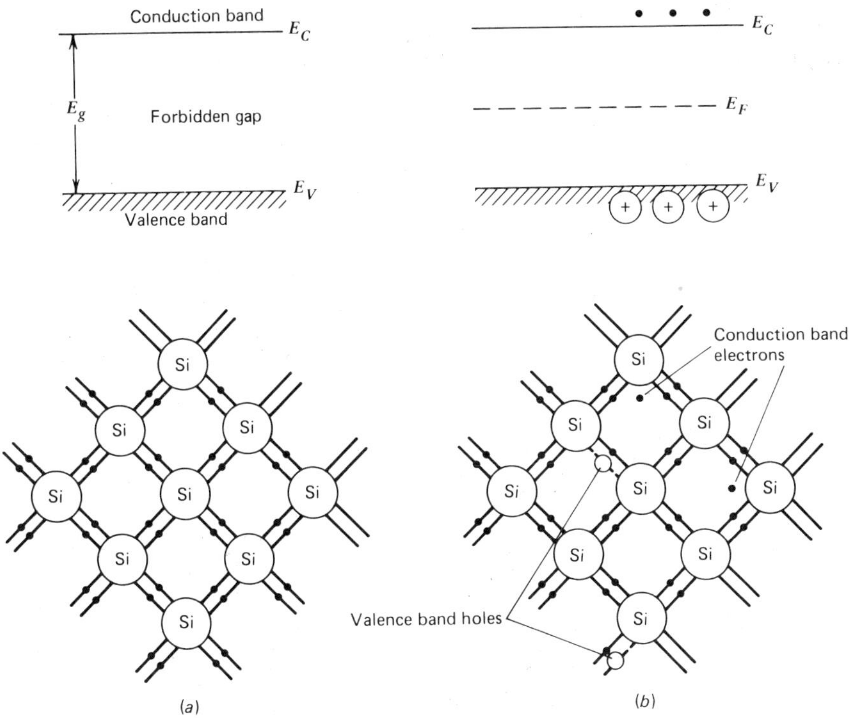

Figure 14.5.2
Energy bands and two-dimensional representation of an intrinsic semiconductor lattice. (*a*) At absolute zero (or $\mathbf{E}_g \gg kT$), assuming a perfect lattice; no holes or electrons exist. (*b*) At a temperature where some lattice bonds are broken, yielding electrons in the conduction band and holes in the valence band. $\mathbf{E}_F$ represents the Fermi level in this intrinsic semiconductor.

Electrons in the CB and holes in the VB can also be introduced by the addition of acceptor and donor species (called *dopants*) into the semiconductor to produce *extrinsic* materials. Thus arsenic (a group V element) behaves as an electron donor material when added to silicon (a group IV element) and introduces an energy level at $\mathbf{E}_D$ near (within ~ 0.05 eV) the bottom of the CB. At room temperature most of the donor atoms will be ionized, each one yielding a CB electron and leaving behind isolated positive sites at the donor atom nuclei (Figure 14.5.3*a*). For example, if the amount of dopant is about 1 ppm, the donor density N_D will be $\sim 5 \times 10^{16}$ cm^{-3}; this will essentially be the CB electron density n. The hole density p is much smaller and is given by

$$p = \frac{n_i^2}{N_D} \tag{14.5.2}$$

Thus for this example of As-doped Si, $p \approx 4000$ cm^{-3} at 25°. Clearly in such a material, most of the electrical conductivity can be attributed to the CB electrons, which are thus the *majority carriers*. The holes, which only make a small contribution to the conductivity, are called the *minority carriers*. A material doped with donor atoms is called an *n-type* semiconductor.

If an acceptor element (e.g., gallium, a group III element) is added to the silicon, an energy level at $\mathbf{E}_A$ near the top of the VB is introduced (Figure 14.5.3*b*). In this case

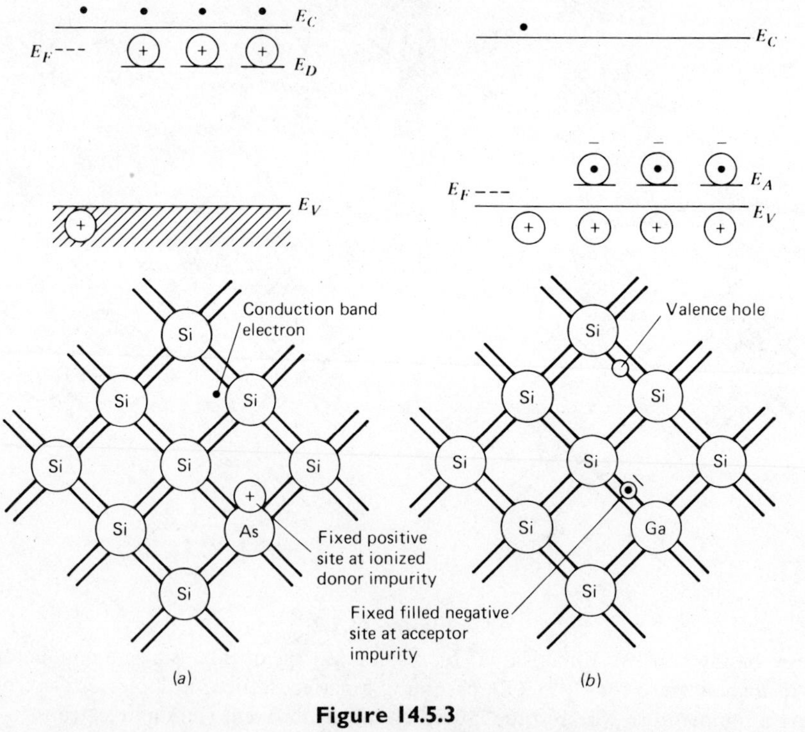

Figure 14.5.3
Energy bands and two-dimensional representation of extrinsic semiconductor lattices. (*a*) *n*-type semiconductor. (*b*) p-type semiconductor.

electrons are thermally excited from the VB into these acceptor sites, leaving behind mobile holes in the VB with the formation of isolated negatively charged acceptor sites. Thus the acceptor density N_A for the addition of 5×10^{16} atoms of acceptor/cm^3, is essentially the same as the hole density p. In this case the CB electron density n is given by

$$n = \frac{n_i^2}{N_A} \tag{14.5.3}$$

or, in the above example, $n \approx 4000$ cm^{-3}. Thus, the holes are the majority carriers, the electrons are the minority carriers, and the material is called a *p-type* semiconductor.

An important concept in the description of semiconductor electrodes is that of the *Fermi level*, $\mathbf{E}_F$, which is defined as that energy where the probability of a level being occupied by an electron is 1/2 (i.e., where it is equally probable that the level is occupied or vacant). For an intrinsic semiconductor at room temperature, $\mathbf{E}_F$ lies essentially midway between the CB and VB within the forbidden gap region. Note that in contrast to metals, where both occupied and vacant states are present at energies near $\mathbf{E}_F$, for an intrinsic semiconductor neither electrons nor unfilled levels exist near $\mathbf{E}_F$. For a doped material, the location of $\mathbf{E}_F$ depends on the doping level, N_A or N_D; for moderately or heavily doped n-type solids ($N_D > 10^{17}$ cm^{-3}), $\mathbf{E}_F$ lies slightly below the CB edge (Figure 14.5.3*a*). Similarly, for moderately or heavily doped p-type materials, $\mathbf{E}_F$ lies just above the VB edge (Figure 14.5.3*b*).

It is convenient to identify $\mathbf{E}_F$ in a more thermodynamic way, so that electronic properties of the semiconductor can be correlated with those of solutions. This is easily accomplished, since the Fermi level of a phase α, $\mathbf{E}_F^{\alpha}$, can be identified as the electrochemical potential (see Section 2.2.4) of an electron in α, $\bar{\mu}_e^{\alpha}$ (154, 155):

$$\mathbf{E}_F^{\alpha} = \bar{\mu}_e^{\alpha} = \mu_e^{\alpha} - e\phi^{\alpha} \text{ (in eV)} \tag{14.5.4}$$

The absolute value of $\mathbf{E}_F$ depends on the choice made for the reference state. Frequently this is taken as zero for a free electron in a vacuum, and $\mathbf{E}_F$ levels in metals and semiconductors can be determined from measurements of work functions or electron affinities (Figure 14.5.4). Since an electron is at a lower energy in almost all materials than in vacuum, $\mathbf{E}_F$ values are usually negative (e.g., about -5.1 eV for Au or -4.8 eV for intrinsic Si).

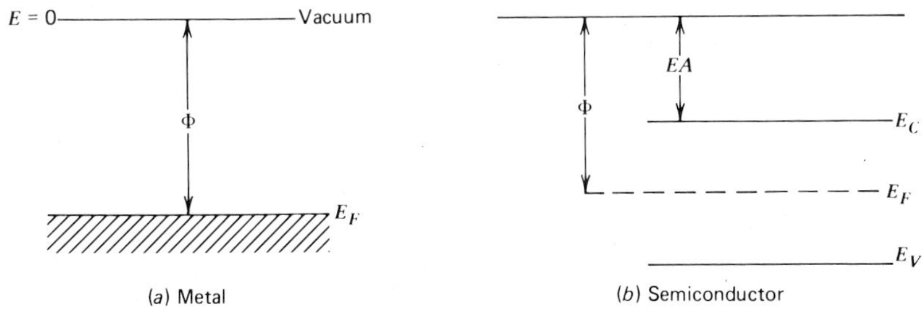

Figure 14.5.4

Relationships between energy levels and Φ (work function) and EA (electron affinity) for (*a*) metal; (*b*) semiconductor.

Let us now consider the formation of the semiconductor/solution interface. The Fermi level in the solution phase, s, can be identified as $\bar{\mu}_e^s$ by (14.5.4) and is calculated in terms of E^0 values by the procedures described in Section 2.2. While for most electrochemical purposes it is convenient to refer E^0 values to the NHE (or other reference electrodes), in this case it is more instructive to estimate them with respect to the vacuum level. This can be accomplished by theoretical and experimental means with relaxation of thermodynamic rigor, and one obtains an E^0 value for the NHE at about -4.6 ± 0.1 eV on the zero vacuum-level scale (164, 165) (Figure 14.5.5*a*). Consider the formation of the junction between an *n*-type semiconductor and a solution containing a redox couple O/R, as shown in Figure 14.5.5*b*. When the semiconductor and the solution are brought into contact, if electrostatic equilibrium is attained, $\bar{\mu}_e$ in both phases must become equal (or equivalently the Fermi levels must become equal), and this can occur by charge transfer between the phases. In the case illustrated in Figure 14.5.5*b*, where $\mathbf{E}_F$ of the semiconductor lies above that in solution, electrons will flow from the semiconductor (which becomes positively charged) to the solution phase (which becomes negatively charged).† The excess charge in the semiconductor does not reside at the surface, as it would in a metal, but instead is distributed in a *space charge region*. This charge distribution is analogous to that found in the diffuse double layer that forms in solution (see Section 12.3). The resulting electric field that forms in the space charge region is represented by a bending of the bands. The bands are bent upward (with respect to the level in the bulk semiconductor) when the semiconductor charge is positive with respect to the solution. An excess electron in the space charge region would thus move toward the bulk semiconductor in the direction consistent with the existing electric field. An excess hole in the space charge region would move toward the interface. The potential at which no excess charge exists in the semiconductor is clearly the potential of zero charge, E_z. Since there is no electric field and no space charge region under these conditions, the bands are not bent. For this reason this electrode potential is called the *flat-band potential*, E_{fb}.‡

The derivations for the relationships between the excess charge in the semiconductor and the potential, the potential distribution in the space charge region, and the differential capacitance all follow closely those for the diffuse double layer in solution (156, 160) (see Section 12.3). Thus for an intrinsic semiconductor the relationship between the space charge per unit area, σ^{SC}, and the potential of the surface with respect to the bulk of the semiconductor, $\Delta\phi$, is given by (12.3.20), with the following replacements: σ^M by σ^{SC}, ϕ_0 by $\Delta\phi$, and n^0 by n_i, with $z = 1$, and ε now referring to the dielectric constant of the semiconductor. Similarly the space charge capacitance, C_{SC},

† Although this description is frequently given of the semiconductor/solution junction, in fact, such reversible behavior of a semiconductor electrode is rarely found, especially for aqueous solutions. This lack of equilibration can be ascribed to corrosion of the semiconductor, surface film (e.g., oxide) formation, or inherently slow electron transfer across the interface. Under such conditions, the behavior of the semiconductor electrode approaches ideal polarizability (see Section 1.2).

‡ For the ideally polarized semiconductor electrode, a space charge region in the semiconductor forms when a potential is applied across the semiconductor/solution interface so that the electrode potential is displaced from E_{fb}.

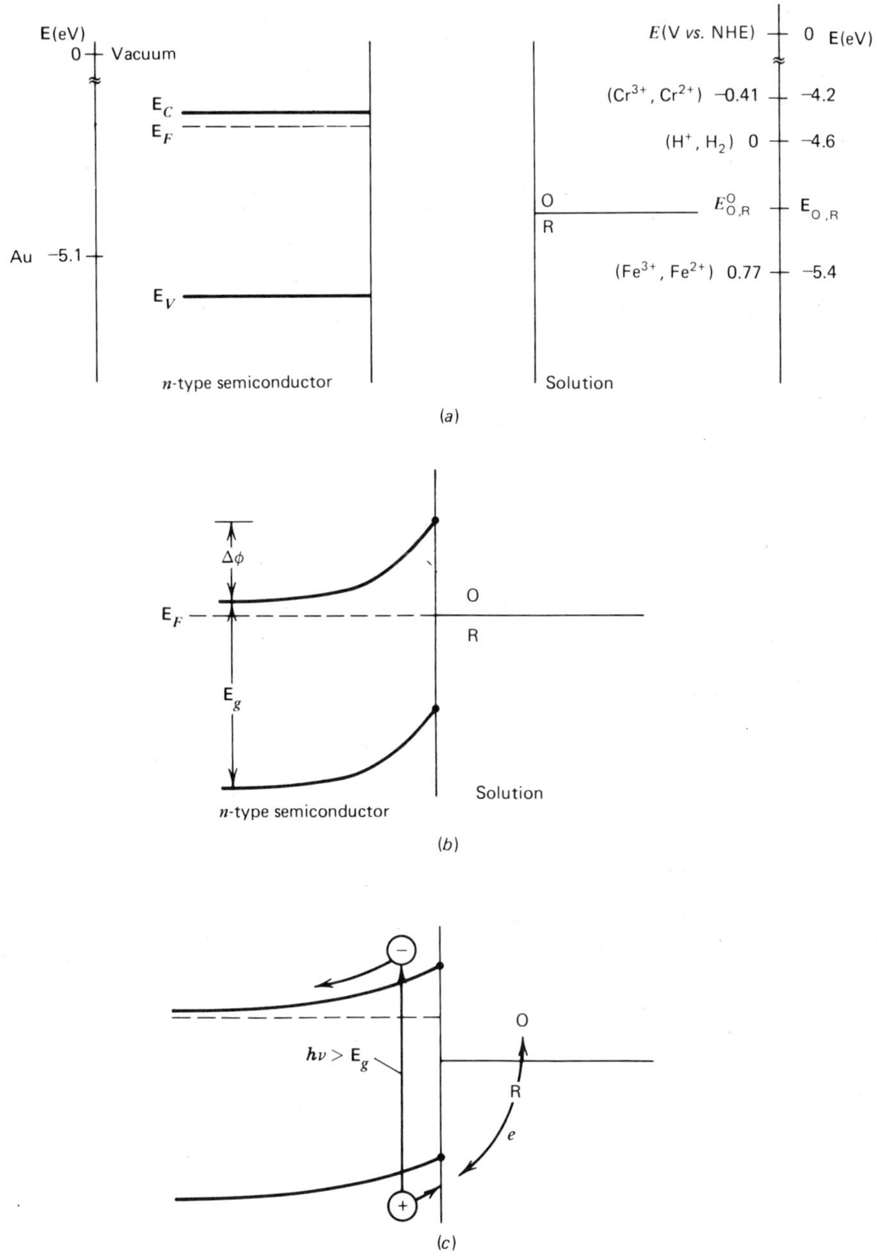

Figure 14.5.5

Representation of the formation of the junction between an *n*-type semiconductor and a solution containing a redox couple O/R. (*a*) Before contact in the dark. Typical values for energy levels shown referenced to NHE (E^0) and to vacuum (**E**). (b) After contact in dark and electrostatic equilibration. (*c*) Junction under irradiation. [Adapted from A. J. Bard, *J. Photochem.*, **10**, 59 (1979), with permission].

is given by (12.3.21) with the same replacements. For an extrinsic (e.g., n-type) semiconductor, C_{SC} is given by

$$C_{SC} = \frac{(2kTn_i\varepsilon\varepsilon_0)^{1/2}e}{2kT}\left\{\frac{|-\lambda e^{-Y} + \lambda^{-1}e^{Y} + (\lambda - \lambda^{-1})|}{[\lambda(e^{-Y} - 1) + \lambda^{-1}(e^{Y} - 1) + (\lambda - \lambda^{-1})Y]^{1/2}}\right\} \quad (14.5.5)$$

where $Y = e\Delta\phi/kT$ and $\lambda = n_i/N_D$.

A particular case of interest for the semiconductor-liquid interface is that shown in Figure 14.5.5*b*, where the surface layer of the n-type semiconductor becomes depleted of majority carriers (electrons), so that a *depletion layer* forms and the bands are bent upward. Under these conditions several simplifications to (14.5.5) can be made. Since this is an n-type material, $\lambda^{-1} \gg \lambda$. Under depletion-layer conditions, with $\Delta\phi$ negative so that electrons are repelled from the surface, $\lambda e^{-Y} \ll \lambda^{-1}$, then (14.5.5) becomes

$$C_{SC} = \left(\frac{e^2\varepsilon\varepsilon_0 N_D}{2kT}\right)^{1/2}\left(-\frac{e\Delta\phi}{kT} - 1\right)^{-1/2} \quad (14.5.6)$$

Rearrangement of this equation yields a very useful relationship (which was first derived for the metal/semiconductor junction) called the *Mott-Schottky equation* (166, 167):

$$\frac{1}{C_{SC}^2} = \left(\frac{2}{e\varepsilon\varepsilon_0 N_D}\right)\left(-\Delta\phi - \frac{kT}{e}\right) \quad (14.5.7)$$

which at 298 K (N_D in cm^{-3}, $\Delta\phi$ in V, C_{SC} in $\mu F\ cm^{-2}$) becomes

$$\boxed{\frac{1}{C_{SC}^2} = \left[\frac{1.41 \times 10^{20}}{\varepsilon N_D}\right][-\Delta\phi - 0.0257]} \quad (14.5.8)$$

Since $-\Delta\phi = E - E_{fb}$, a plot of $1/C_{SC}^2$ *vs.* potential, E, should be linear. The potential where the line intersects the potential axis yields the value of E_{fb}, and the slope can be used to obtain the doping level N_D. While such "Mott-Schottky plots" have been useful in characterizing the semiconductor/solution interface, they must be used with caution, because perturbing effects, such as those attributable to surface states,† can cause deviations from the predicted behavior (168). One should verify that the parameters obtained from such plots are independent of the frequency employed in the capacitance measurements.

14.5.3 Photoeffects at Semiconductor Electrodes

Let us return to the situation of the n-type semiconductor in contact with the solution containing couple O/R, as illustrated in Figure 14.5.5. As described in the preceding section, a space charge region of the order of 50 to 2000 Å wide (depending on the

† *Surface states* are energy levels arising from orbitals localized on atoms of the lattice near a surface. It is easy to see, for example, that silicon atoms in a surface plane cannot be surrounded with the tetrahedral symmetry found in the bulk solid. Thus the electronic properties of these atoms differ. Often the surface states are found in the bandgap, and they have a big effect on the electronic properties of any junction made with the surface.

doping level and $\Delta\phi$) has formed in the semiconductor at the interface. The direction of the electric field in this region is such that any excess holes created here would move toward the surface and any excess electrons would move toward the bulk semiconductor. When the interface is irradiated with light of energy greater than the band gap $\mathbf{E}_g$, photons are absorbed and electron-hole pairs are created (Figure 14.5.5*c*). Some of these, especially those formed beyond the space charge region, recombine with the evolution of heat. However, the space charge field promotes the separation of electrons and holes. The holes, delivered to the surface at an effective potential equivalent to the valence band edge, cause the oxidation of R to O while the electrons move into the external circuit through the semiconductor electrode lead. Thus irradiation of an *n*-type semiconductor electrode promotes photo-oxidations (or causes a photoanodic current). This is illustrated in the *i-E* curves in Figure 14.5.6*a*. In the dark (curve 1), when the potential of the semiconductor electrode is made more and more positive, essentially no current flows, because there are few holes in the semiconductor to accept electrons from the reduced form of the redox couple located at potentials within the gap. (At very positive potentials, however, dark anodic current can flow from breakdown phenomena.) Under irradiation (curve 2), a photoanodic current (i_{ph}) flows as long as the potential of the electrode is more positive than E_{fb}, so that electron/hole pair separation can occur. Thus the onset of the photocurrent is near E_{fb} (although surface recombination processes† can move this onset potential toward more positive values). The photo-oxidation of R to O occurs at less positive applied potentials than those required to carry out this process at an inert metal electrode (curve 3). This apparently easier oxidation represents the contribution of the light energy to promotion of the oxidation process and, for this reason, such processes are frequently called "photoassisted" electrode reactions.

The behavior of a *p*-type semiconductor with a couple with a redox potential located in the gap region is analogous to that of the *n*-type material (Figure 14.5.7). Here the field in the space charge region moves electrons toward the surface and holes toward the bulk material. Thus, upon irradiation of a *p*-type material, a photocathodic current flows and photoreductions are photoassisted. Typical *i-E* curves under these conditions are shown in Figure 14.5.6*b*.

Note that photoeffects at *n*-type materials are generally not observed for redox couples located at potentials negative of E_{fb}. In this case the bands are bent downwards, the majority carrier tends to accumulate near the surface (i.e., an *accumulation layer* forms), and the semiconductor behavior approaches that of an inert metal electrode. Similarly, a (hole) accumulation layer forms in a *p*-type material for couples located positive of E_{fb}.

Photoelectrochemical cells usually consist of a semiconductor electrode and a suitable counter electrode. Such cells, currently being investigated because of their possible use in the conversion of radiant energy to electrical or chemical energy, will be discussed briefly here. Three types of photoelectrochemical cells can be devised (Figure 14.5.8) (169). In *photovoltaic* cells, (*a*) and (*b*), the reaction that occurs at the counter electrode is simply the reverse of the photoassisted process at the semiconductor. Ideally the cell operates, with conversion of light to electricity, with no change in

† *Recombination* is the annihilation of an electron and a hole. It is often promoted by surface states at interfaces.

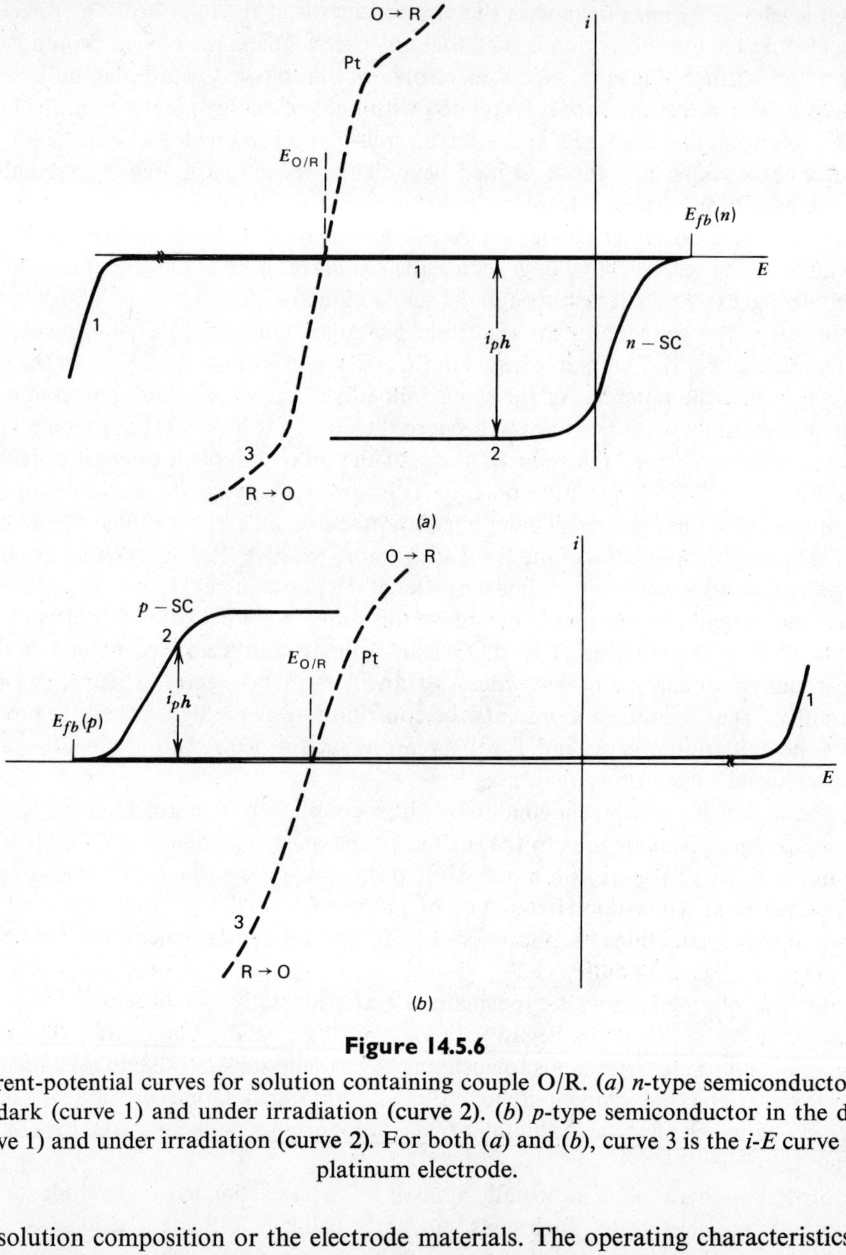

Figure 14.5.6

Current-potential curves for solution containing couple O/R. (*a*) *n*-type semiconductor in the dark (curve 1) and under irradiation (curve 2). (*b*) *p*-type semiconductor in the dark (curve 1) and under irradiation (curve 2). For both (*a*) and (*b*), curve 3 is the *i-E* curve at a platinum electrode.

the solution composition or the electrode materials. The operating characteristics of such a cell can be deduced from *i-E* curves such as those in Figure 14.5.6. In *photoelectrosynthetic* cells, the reaction at the counter electrode is different than that at the semiconductor (so that a separator might be necessary in the cell to keep the products apart). In this case, the net cell reaction is driven by light in the nonspontaneous direction ($\Delta G > 0$), so that radiant energy is stored as chemical energy. The necessary condition for driving the reaction in such a cell (e.g., with an *n*-type semiconductor, see Figure 14.5.8*c*) is that the potential of couple O/R lies above the valence band edge,

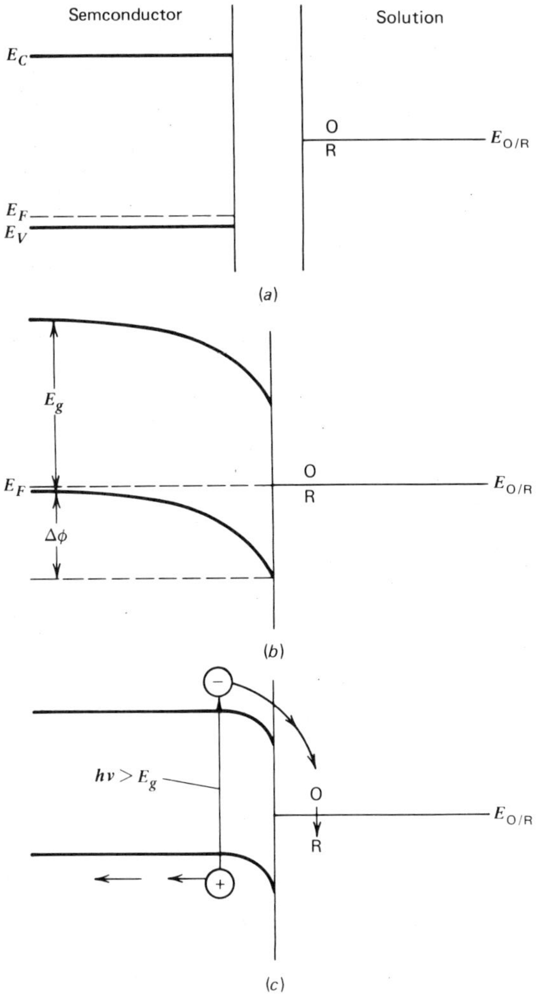

Figure 14.5.7

Representation of the formation of the junction between a *p*-type semiconductor and a solution containing a redox couple O/R. (*a*) Before contact in the dark. (*b*) After contact (in the dark) and electrostatic equilibration. (*c*) Junction under irradiation.

while that of O′/R′ lies below E_{fb}. If this condition does not exist, it is still possible to drive the reaction in the desired direction by applying an external bias to the cell. *Photocatalytic* cells (Figure 14.5.8*e*,*f*) are similar to these, except that the relative locations of the potentials of the O/R and O′/R′ couples are changed. In this case the reaction is driven in the spontaneous direction (which is presumably very slow in the dark) ($\Delta G < 0$), with the light energy used to overcome the energy of activation of the process. Several examples of such processes are listed in the caption of Figure 14.5.8. While some of these are interesting, the efficiencies of many of the processes carried out in the photoelectrosynthetic and photocatalytic cells are often rather low.

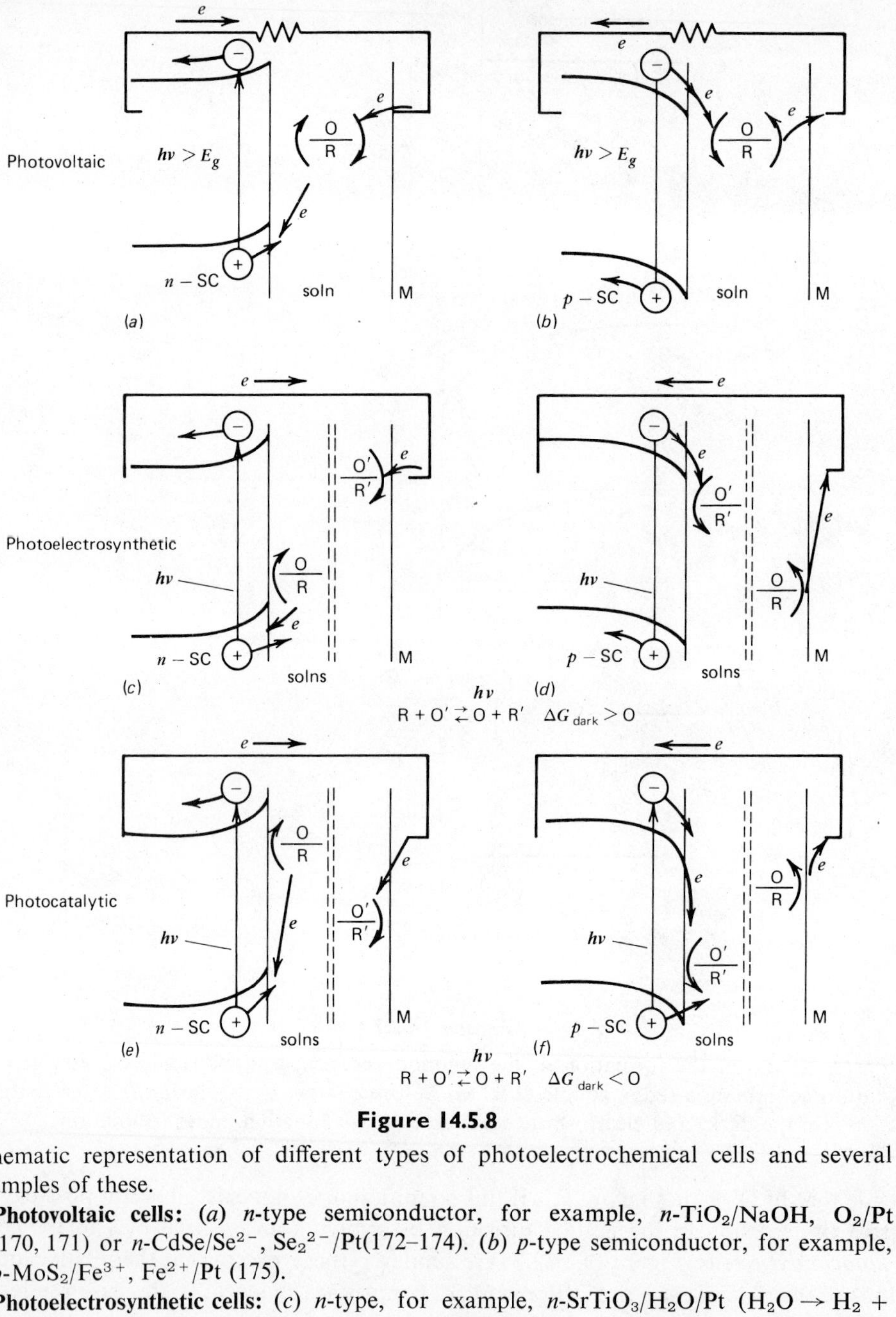

Figure 14.5.8

Schematic representation of different types of photoelectrochemical cells and several examples of these.

Photovoltaic cells: (*a*) *n*-type semiconductor, for example, *n*-TiO_2/NaOH, O_2/Pt (170, 171) or *n*-CdSe/Se^{2-}, Se_2^{2-}/Pt(172–174). (*b*) *p*-type semiconductor, for example, *p*-MoS_2/Fe^{3+}, Fe^{2+}/Pt (175).

Photoelectrosynthetic cells: (*c*) *n*-type, for example, *n*-$SrTiO_3$/H_2O/Pt ($H_2O \rightarrow H_2 + \frac{1}{2}O_2$) (176–178). (*d*) *p*-type, for example, *p*-GaP/CO_2(pH = 6.8)/C (reduction of CO_2) (179).

Photocatalytic cells: (*e*) *n*-type, for example, *n*-TiO_2/CH_3COOH/Pt ($CH_3COOH \rightarrow \frac{1}{2}C_2H_6 + CO_2 + H_2$) (180). (*f*) *p*-type, for example, *p*-GaP/DME, $AlCl_3$, N_2/Al (reduction of N_2 by Al) (181).

The light employed to carry out a photoassisted reaction at a semiconductor electrode must have a photon energy greater than that of the band gap E_g, since light of lower energy is not absorbed by the semiconductor. Thus a plot of photocurrent versus the wavelength of irradiating light can be employed to determine E_g. For example, for n-TiO_2(rutile), only light of photon energy above 3.0 eV is useful. Since more than 95% of the total energy in the solar spectrum at the earth's surface lies below this value, sunlight is not utilized very effectively by TiO_2. An approach to better utilization of longer wavelength light (and also to interesting experiments in their own right) involves *dye sensitization* of a semiconductor (158, 182, 183). The principles of this experiment are illustrated in Figure 14.5.9. Assume a thin layer of dye, D, is coated on the semiconductor. When the dye is excited (step 1), it injects an electron into the semiconductor band (step 2), in the process becoming oxidized to D^+. In the absence of a suitable couple in solution, this photoprocess would cease when all of D is consumed. If a species R exists in solution that is capable of reducing D^+, then electron transfer from R to D^+ can occur, regenerating D (step 3). Thus dye sensitization allows longer-wavelength light to be employed, but the hole generated by the light is at a less positive potential than the one produced in the valence band of the semiconductor. An analogous situation arises for a dye whose energy levels bracket the valence band; in this case hole injection into the semiconductor takes place. The depiction of the energy-level diagram for this case and a description of the processes involved is left as an exercise for the reader (see Problem 14.18).

Photoeffects are also observed at metal electrodes, although in these cases the resulting photocurrents are much smaller. For example, the irradiation of a metal electrode can cause the photoejection of an electron into the solvent. If this electron is scavenged by some reactant in solution, a net cathodic photocurrent results (186, 187) (see Section 14.6). These electron photoinjection studies are of interest, because they can provide information about the nature of an electron at the instant of injection into a medium as well as the energetics and kinetics of its relaxation to equilibrium solvation. Excitation of dyes adsorbed on metals can also lead to photocurrents, but these

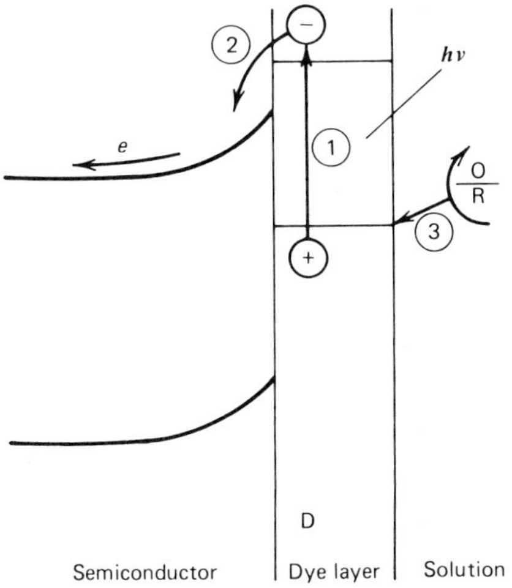

Figure 14.5.9
Dye sensitization of photoprocess at semiconductor electrode. For example, n-ZnO/rose bengal/I^-/Pt (184).

are usually in the nanoampere range even with intense irradiation (183). This low efficiency of net conversion of photons to external photocurrent is attributed to the ability of a metal to act as a quencher of excited states at or very near the surface by either electron or energy transfer (149, 150).

14.6 ELECTROCHEMICAL DETECTION OF PHOTOLYTIC AND RADIOLYTIC PRODUCTS

14.6.1 Photoemission of Electrons (187–192)

A metal surface exposed to light generally will eject electrons, which travel 20 to 100 Å into the electrolyte and then become solvated. These electrons are reactive and produce some interesting chemistry if *scavengers* are available to interact with them. In the absence of such a species, the electrons return to the electrode by diffusion, and no net loss of charge is detected. If a scavenger exists, for example, N_2O in water, they react and fail to return, and therefore the faradaic charge transfer can be detected:

$$e_{aq} + N_2O + H_2O \rightarrow N_2 + OH^- + OH\cdot \qquad (14.6.1)$$

In fact the $OH\cdot$ diffuses to the electrode and, depending on the potential, may withdraw additional charge.

Since the time for recollection of the electrons is only on the order of 100 nsec, this method is well-suited to the study of fast reactions that may not be readily examined by purely electrochemical methods. Usually the stimulus is a pulsed (10–20 nsec) laser. Since the quantum yield for photoemission is low, an intense source is needed.

The detection method resembles that of the coulostatic impulse technique (Section 7.7), in that one observes the potential shift caused by charge ejection and the following relaxation back to the original state. Figure 7.9.2 offers some typical data for the N_2O system, and Problem 7.10 deals with its interpretation.

Photoemission is useful for studying the electrochemistry of radicals that may be of interest as intermediates in an electrochemical process, but cannot be isolated from that process by other methods. A good example is $H\cdot$, whose chemistry bears on electrolytic hydrogen evolution. This species can be generated cleanly in acidic solutions by photoemission, and its chemistry can be studied without complication from other steps in the hydrogen evolution reaction. Problem 7.11 concerns this case.

14.6.2 Monitoring Products of Pulse Radiolysis (193)

The system shown in Figure 14.6.1 provides another approach to the electrochemical study of species produced by reactive electrons. The excitation of the system is carried out by a pulsed (~20 nsec) beam of high-energy electrons. Their passage through the cell creates solvated electrons, radicals, and ions, whose distribution can often be controlled by known chemistry. Significant concentrations of these species are produced, and they can be detected faradaically.

If the radiolytic species of interest has a fairly long lifetime, then a potentiostatic experiment will yield a Cottrell-like response after the pulse, because the faradaic

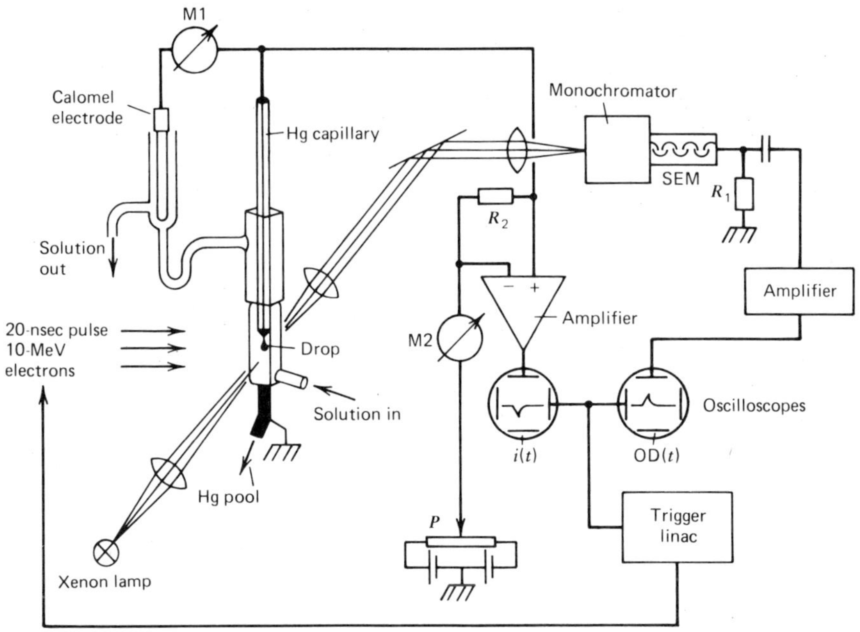

Figure 14.6.1

Apparatus for electrochemical and optical monitoring of pulse radiolysis. The electrode is essentially an HMDE. [From A. Henglein, *Electroanal. Chem.*, **9**, 163 (1976), courtesy of Marcel Dekker, Inc.]

current is controlled by diffusion, or possibly in part by the electrode kinetics. Data for this case are shown in Figure 14.6.2. The radiolytically generated radical of ascorbic acid decays on a millisecond time scale. Current transients following separate radiolytic pulses were recorded for different potentials, and samplings were made at a fixed delay time to produce the sampled-current voltammogram displayed there. Obtaining this kind of thermodynamic data for such a reactive species is not an easy matter by other methods.

If the radicals are short-lived, then the current transients are controlled by their homogeneous decay, as well as by diffusion and heterogeneous kinetics. This complication must be taken into account if the current-time curves are to be interpreted.

14.6.3 Electrochemistry of Photolytic Products (126, 192, 194)

Similar in concept to the approach described above is the use of electrochemistry to monitor the products of flash photolysis. The apparatus resembles that of Figure 14.6.1, except that the excitation beam comprises photons from a pulsed discharge. Most work has involved flashlamps dissipating perhaps 100 to 500 J of energy and producing a light pulse 10 to 30 μsec wide (194–196). The current-time curves are controlled by reactant decay and by diffusion and heterogeneous charge transfer

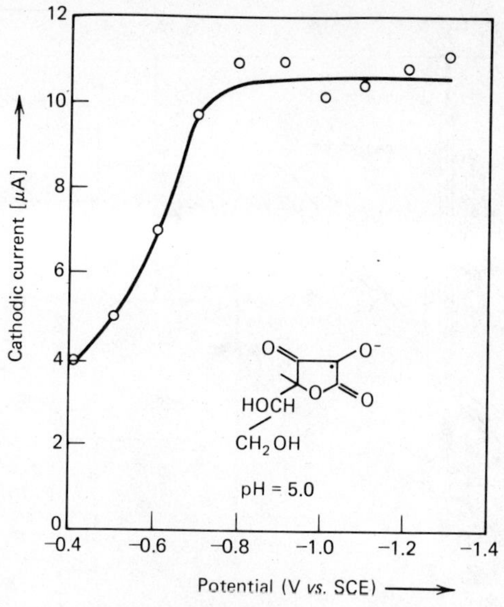

Figure 14.6.2

Sampled-current voltammogram for the ascorbic acid radical at 15 μsec after the radiolytic pulse. [From A. Henglein, *Electroanal. Chem.*, **9**, 163 (1976), courtesy of Marcel Dekker, Inc.]

exactly as described above. At very short times, the current transient can have substantial contributions from photoemission of electrons from the electrode. An interesting application of this approach concerns the intermediates formed in the photolysis of ferrioxalate (197, 198).

Photolytic intermediates can also be monitored in steady-state experiments (199–201) involving apparatus like that shown in Figure 14.6.3. There a photon beam is directed downward through a window replacing the disk of an RRDE. Photolysis products are then swept past the ring, where they can be detected. The advantage of this approach is that products are generated specifically near the electrode and measurements are made at steady state. On the other hand, the intermediates to be detected must have longer lifetimes than for transient approaches.

14.6.4 Photogalvanic Cells (202)

It is possible to drive many homogeneous redox systems in a nonspontaneous direction by using light energy. For example, the complex $Ru(bipy)_3^{2+}$, where bipy is 2,2′-bipyridine, can absorb light to produce an excited state that is a fairly good reductant. Thus one observes the reaction:

$$[Ru(bipy)_3^{2+}]^* + Fe^{3+} \rightarrow Fe^{2+} + Ru(bipy)_3^{3+} \qquad (14.6.2)$$

The products Fe^{2+} and $Ru(bipy)_3^{3+}$ will react to give back the starting materials:

$$Ru(bipy)_3^{3+} + Fe^{2+} \rightarrow Fe^{3+} + Ru(bipy)_3^{2+} \qquad (14.6.3)$$

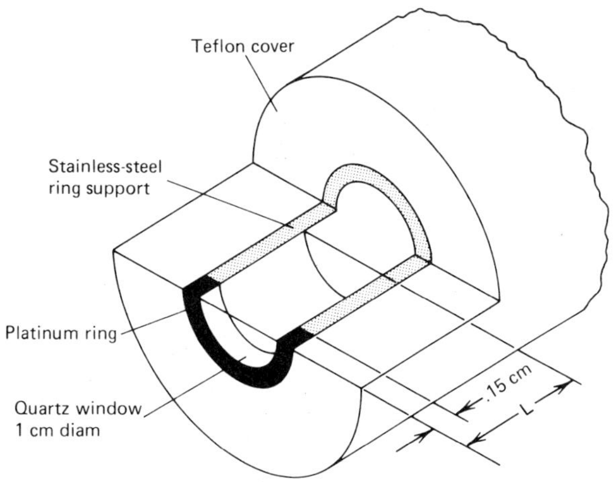

Figure 14.6.3
Cut-away view of a rotating photoelectrode [Reprinted with permission from D. C. Johnson and E. W. Resnick, *Anal. Chem.*, **44**, 637 (1972). Copyright 1972, American Chemical Society.]

hence the net effect is a quenching of the excited complex by a reversible electron transfer mechanism. The light energy is thermalized.

On the other hand, (14.6.3) is not very fast (203). Thus, one can build up appreciable concentrations of $Ru(bipy)_3{}^{3+}$ and Fe^{2+} in the system. By inserting electrodes into the solution, one can effect reaction (14.6.3) faradaically by reducing the complex at one electrode and oxidizing ferrous iron at the other. Usually one electrode is chosen to be reversible only toward one of the half reactions, in order to enforce specific behavior at each electrode. Since the energy of (14.6.3) is available in the external circuit, these devices, called *photogalvanic cells*, have been of interest for solar energy conversion. A number of specific chemical systems have been studied. Obviously the efficiency of conversion depends strongly on the degree to which the kinetics can be optimized (202, 204–206).

14.7 REFERENCES

1. N. Winograd and T. Kuwana, *Electroanal. Chem.*, **7**, 1 (1974).
2. W. N. Hansen, *Adv. Electrochem. Electrochem. Engr.*, **9**, 1 (1973).
3. T. Kuwana and W. R. Heineman, *Accounts Chem. Res.*, **9**, 241 (1976).
4. W. R. Heineman, *Anal. Chem.*, **50**, 390A (1978).
5. T. P. DeAngelis, R. W. Hurst, A. M. Yacynych, H. B. Mark, Jr., W. R. Heineman, and J. S. Mattson, *ibid.*, **49**, 1395 (1977).
6. R. Cieslinski and N. R. Armstrong, *ibid.*, **51**, 565 (1979).
7. R. W. Murray, W. R. Heineman, and G. W. O'Dom, *ibid.*, **39**, 1666 (1967).
8. T. Gueshi, K. Tokuda, and H. Matsuda, *J. Electroanal. Chem.*, **89**, 247 (1978).
9. M. Petek, T. E. Neal, and R. W. Murray, *Anal. Chem.*, **43**, 1069 (1971).

10. M. K. Hanafey, R. L. Scott, T. H. Ridgway, and C. N. Reilley, *ibid.*, **50**, 116 (1978).
11. W. R. Heineman, B. J. Norris, and J. F. Goelz, *ibid.*, **47**, 79 (1975).
12. D. F. Rohrbach, E. Deutsch, and W. R. Heineman in "Characterization of Solutes in Nonaqueous Solvents," G. Mamantov, Ed., Plenum, New York, 1978.
13. W. R. Heineman, T. Kuwana, and C. R. Hartzell, *Biochem. Biophys. Res. Commun.*, **50**, 892 (1973).
14. D. Halliday and R. Resnick, "Physics," 3rd ed., Wiley, New York, 1978.
15. F. A. Jenkins and H. E. White, "Fundamentals of Optics," 4th ed., McGraw-Hill, New York, 1976.
16. J. D. E. McIntyre, *Adv. Electrochem. Electrochem. Engr.*, **9**, 61 (1973).
17. R. H. Muller, *ibid.*, p. 167.
18. B. D. Cahan, J. Horkans, and E. Yeager, *Symp. Faraday Soc.*, **4**, 36 (1970).
19. M. A. Barrett and R. Parsons, *ibid.*, p. 72.
20. J. D. E. McIntyre and D. M. Kolb, *ibid.*, p. 99.
21. A. Bewick and A. M. Tuxford, *ibid.*, p. 114.
22. W. J. Plieth, *ibid.*, p. 137.
23. D. M. Kolb and J. D. E. McIntyre, *Surf. Sci.*, **28**, 321 (1971).
24. T. Takamura, K. Takamura, and E. Yeager, *Symp. Faraday Soc.*, **4**, 91 (1970).
25. J. Kruger, *Adv. Electrochem. Electrochem. Engr.*, **9**, 227 (1973).
26. M. Stedman, *Symp. Faraday Soc.*, **4**, 64 (1970).
27. J. O'M. Bockris, M. Genshaw, and V. Brusic, *ibid.*, p. 177.
28. C. J. Dell'Oca and P. J. Fleming, *J. Electrochem. Soc.*, **123**, 1487 (1976).
29. J. Kruger and J. P. Calvert, *ibid.*, **114**, 43 (1967).
30. W. W. Harvey and J. Kruger, *Electrochim. Acta*, **16**, 2017 (1971).
31. W. N. Hansen, T. Kuwana, and R. A. Osteryoung, *Anal. Chem.*, **38**, 1810 (1966).
32. N. Winograd and T. Kuwana, *J. Electroanal. Chem.*, **23**, 333 (1969).
33. N. Winograd and T. Kuwana, *J. Am. Chem. Soc.*, **93**, 4343 (1971).
34. W. N. Hansen, *Symp. Faraday Soc.*, **4**, 27 (1970).
35. H. B. Mark and E. N. Randall, *ibid.*, p. 157.
36. G. H. Brilmyer, A. Fujishima, K. S. V. Santhanam, and A. J. Bard, *Anal. Chem.*, **49**, 2057 (1977).
37. G. H. Brilmyer and A. J. Bard, *Anal. Chem.*, **52**, 685 (1980).
38. A Fujishima, H. Masuda, K. Honda, and A. J. Bard, *Anal. Chem.*, **52**, 682 (1980).
39. A. Fujishima, G. H. Brilmyer, and A. J. Bard, in "Semiconductor Liquid-Junction Solar Cells," A. Heller, Ed., Electrochemical Society, Princeton, N.J. (Proceedings Vol. 77-3), 1977.
40. A. Fujishima, Y. Maeda, K. Honda, G. H. Brilmyer, and A. J. Bard, *J. Electrochem. Soc.*, **127**, 840 (1980).
41. A. Rosencwaig, *Anal. Chem.*, **47**, 592A (1975).
42. W. R. Harshbarger and M. B. Robin, *Accounts. Chem. Res.*, **6**, 329 (1973).
43. Y. H. Pao, Ed., "Optoacoustic Spectroscopy and Detection," Academic Press, New York, 1977.

44. R. B. Somoano, *Angew. Chem., Int. Ed.*, **17**, 238 (1978).
45. R. E. Malpas and A. J. Bard, *Anal. Chem.*, submitted.
46. J. M. Sherfey and A. Brenner, *J. Electrochem. Soc.*, **105**, 665 (1958).
47. M. J. Joncich and H. F. Holmes, in "Proceedings of First Australian Conference on Electrochemistry," J. A. Friend and F. Gutmann, Eds., Pergamon, New York, 1965, pp. 138–146.
48. B. B. Graves, *Anal. Chem.*, **44**, 993 (1972).
49. S. L. Cooke and B. B. Graves, *Chem. Instr.*, **1**, 119 (1968).
50. R. Tamamushi, *J. Electroanal. Chem.*, **65**, 263 (1975).
51. R. E. Malpas, R. A. Fredlein, and A. J. Bard, *ibid.*, **98**, 171, 339 (1979).
52. A. V. Gokhshtein, *Russ. Chem. Rev.*, **44**, 921 (1975).
53. W. H. Flygare, "Molecular Structure and Dynamics," Prentice-Hall, Englewood Cliffs, N.J., 1978, Chap. 8.
54. H. A. Szmanski, Ed., "Raman Spectroscopy," Plenum, New York, 1970.
55. W. Kiefer and H. J. Bernstein, *Mol. Phys.*, **23**, 835 (1972).
56. D. L. Jeanmaire, M. R. Suchanski, and R. P. Van Duyne, *J. Am. Chem. Soc.*, **97**, 1699 (1975).
57. D. L. Jeanmaire and R. P. Van Duyne, *ibid.*, **98**, 4029 (1976).
58. M. R. Suchanski and R. P. Van Duyne, *ibid.*, **98**, 250 (1976).
59. D. L. Jeanmaire and R. P. Van Duyne, *J. Electroanal. Chem.*, **66**, 235 (1975).
60. R. P. Van Duyne, M. R. Suchanski, J. M. Lakovits, A. R. Siedle, K. D. Parks, and T. M. Cotton, *J. Am. Chem. Soc.*, **101**, 2832 (1979).
61. M. Fleischmann, P. J. Hendra, and A. J. McQuillan, *J. Chem. Soc., Chem. Commun.*, **1973**, 80.
62. M. Fleischmann, P. J. Hendra, and A. J. McQuillan, *Chem. Phys. Letters*, **26**, 163 (1974).
63. R. L. Paul, A. J. McQuillan, P. J. Hendra, and M. Fleischmann, *J. Electroanal. Chem.*, **66**, 248 (1975).
64. R. P. Cooney, E. S. Reid, P. J. Hendra, and M. Fleischmann, *J. Am. Chem. Soc.*, **99**, 2002 (1977).
65. M. G. Albrecht and J. A. Creighton, *ibid.*, **99**, 5215 (1977).
66. D. L. Jeanmaire and R. P. Van Duyne, *J. Electroanal. Chem.*, **84**, 1 (1977).
67. G. Hagen, B. Simic-Glavaski, and E. Yeager, *ibid.*, **88**, 269 (1978).
68. J. A. Creighton, M. G. Albrecht, R. F. Hester, and J. A. D. Matthew, *Chem. Phys. Letters*, **55**, 55 (1978).
69. B. Pettinger and U. Wenning, *ibid.*, **56**, 253 (1978).
70. M. R. Philpott, *J. Chem. Phys.*, **62**, 1812 (1975).
71. F. K. King, R. P. Van Duyne, and G. C. Schatz, *J. Chem. Phys.*, **69**, 4472 (1978).
72. P. F. Kane and G. B. Larrabee, Eds., "Characterization of Solid Surfaces," Plenum, New York, 1974.
73. T. A. Carlson, "Photoelectron and Auger Spectroscopy," Plenum, New York, 1975.
74. A. W. Czanderna, Ed., "Methods of Surface Analysis," Elsevier, Amsterdam, 1975.
75. C. A. Evans, Jr., *Anal. Chem.*, **47**, 818A, 855A (1975).

76. J. W. Mayer and J. M. Poate, in "Thin Films—Interdiffusion and Reactions," J. M. Poate, K. N. Tu, and J. W. Mayer, Eds., Wiley-Interscience, New York, 1978, Chap. 6.
77. R. M. Ishikawa and A. T. Hubbard, *J. Electroanal. Chem.*, **69**, 317 (1976).
78. W. E. O'Grady, M. Y. C. Woo, P. L. Hagans, and E. Yeager, *J. Vac. Sci. Technol.*, **14**, 365 (1977).
79. P. N. Ross, Jr., *J. Electroanal. Chem.*, **76**, 139 (1977).
80. J. S. Hammond and N. Winograd, *ibid.*, **78**, 55 (1977).
81. D. A. Shirley, Ed., "Electron Spectroscopy," North Holland, Amsterdam, 1972.
82. S. H. Hercules and D. M. Hercules in P. F. Kane and G. B. Larrabee, Eds., *op. cit.*, Chap. 13.
83. B. G. Baker, *Mod. Asp. Electrochem.*, **10**, 93 (1975).
84. K. S. Kim, N. Winograd, and R. E. Davis, *J. Am. Chem. Soc.*, **93**, 6296 (1971).
85. C. M. Elliott and R. W. Murray, *Anal. Chem.*, **48**, 1247 (1976).
86. J. S. Hammond and N. Winograd, *J. Electroanal. Chem.*, **80**, 123 (1977).
87. J. S. Hammond and N. Winograd, *J. Electrochem. Soc.*, **124**, 826 (1977).
88. C. C. Chang in P. F. Kane and G. B. Larrabee, Eds., *op. cit.*, Chap. 20.
89. S. H. Kulpa and R. P. Frankenthal, *J. Electrochem. Soc.*, **124**, 1588 (1977).
90. C. C. Chang, B. Schwartz, and S. P. Murarka, *ibid.*, **124**, 922 (1977).
91. J. L. Kahl and L. R. Faulkner, unpublished results.
92. H. Tachikawa and L. R. Faulkner, *J. Am. Chem. Soc.*, **100**, 4379 (1978).
93. F.-R. Fan and L. R. Faulkner, *ibid.*, **101**, 4779 (1979).
94. G. A. Somorjai and H. H. Farrell, *Adv. Chem. Phys.*, **20**, 215 (1971).
95. J. B. Pendry, "Low Energy Electron Diffraction," Academic Press, New York, 1974.
96. E. Yeager, W. E. O'Grady, M. Y. C. Woo, and P. Hagans, *J. Electrochem. Soc.*, **125**, 348 (1978).
97. P. N. Ross, Jr., *ibid.*, **126**, 67, 78 (1979).
98. S. Bruckenstein and R. Rao Gadde, *J. Am. Chem. Soc.*, **93**, 793 (1971).
99. J. A. McHugh in A. W. Czanderna, Ed., *op. cit.*
100. V. R. Deline, W. Katz, C. A. Evans, Jr., and P. Williams, *Appl. Phys. Letters*, **38**, 832 (1978).
101. T. M. McKinney, *Electroanal. Chem.*, **10**, 97 (1977).
102. I. B. Goldberg and A. J. Bard, in "Magnetic Resonance in Chemistry and Biology," J. N. Herak and K. J. Adamic, Eds., M. Dekker, Inc., New York, 1975, Chap. 10.
103. B. Kastening, *Progr. Polarogr.*, **3**, 195 (1972); *Chem. Ing. Tech.*, **42**, 190 (1970).
104. R. N. Adams, *J. Electroanal. Chem.*, **8**, 151 (1964).
105. J. E. Wertz and J. R. Bolton, "Electron Spin Resonance: Elementary Theory and Practical Applications," McGraw-Hill, New York, 1972.
106. R. S. Alger, "Electron Paramagnetic Resonance: Techniques and Applications," Interscience, New York, 1968.
107. N. M. Atherton, "Electron Spin Resonance: Theory and Applications," Halsted Press, New York, 1973.

108. I. B. Goldberg, A. J. Bard, and S. W. Feldberg, *J. Phys. Chem.*, **76**, 2550 (1972).
109. I. B. Goldberg and A. J. Bard, *J. Phys. Chem.*, **75**, 3281 (1971); **78**, 290 (1974).
110. R. D. Allendoerfer, G. A. Martinchek, and S. Bruckenstein, *Anal. Chem.*, **47**, 890 (1975).
111. D. H. Geske and A. H. Maki, *J. Am. Chem. Soc.*, **82**, 2671 (1960).
112. M. E. Peover, *Electroanal. Chem.*, **2**, 1 (1967).
113. A. J. Bard, K. S. V. Santhanam, J. T. Maloy, J. Phelps, and L. O. Wheeler, *Disc. Faraday Soc.*, **45**, 167 (1968).
114. R. L. Ward and S. I. Weissman, *J. Am. Chem. Soc.*, **76**, 3612 (1954); **79**, 2086 (1957).
115. R. Chang and C. S. Johnson, Jr., *J. Am. Chem. Soc.*, **88**, 2338 (1966).
116. A. E. J. Forno, M. E. Peover, and R. Wilson, *Trans. Faraday Soc.*, **66**, 1322 (1970).
117. B. A. Kowert, L. Marcoux, and A. J. Bard, *J. Am. Chem. Soc.*, **94**, 5538 (1972).
118. R. A. Marcus, *Electrochim. Acta*, **13**, 995 (1968) and references cited therein.
119. J. M. Hale and M. E. Peover, in "Reactions of Molecules at Elcctrodes," N. S. Hush, Ed., Wiley-Interscience, New York, 1971, Chaps. 4 and 5.
120. I. B. Gøldberg, D. Boyd, R. Hirasawa, and A. J. Bard, *J. Phys. Chem.*, **78**, 295 (1974).
121. R. Koopmann and H. Gerischer, *Ber. Bunsenges Phys. Chem.*, **70**, 118, 127 (1966).
122. B. Kastening, *ibid.*, **72**, 20 (1968).
123. K. Umemoto, *Bull. Chem. Soc. Japan*, **40**, 1058 (1967).
124. E. Janzen, *Accounts Chem. Res.*, **4**, 31 (1971).
125. A. J. Bard, J. C. Gilbert, and R. D. Goodin, *J. Am. Chem. Soc.*, **96**, 620 (1974).
126. T. Kuwana, *Electroanal. Chem.*, **1**, 197 (1966).
127. A. J. Bard, K. S. V. Santhanam, S. A. Cruser, and L. R. Faulkner in "Fluorescence," G. G. Guilbault, Ed., Marcel Dekker, New York, 1967, Chap. 14.
128. A. Zweig, *Adv. Photochem.*, **6**, 425 (1968).
129. E. A. Chandross, *Trans. N.Y. Acad. Sci., Ser. 2*, **31**, 571 (1969).
130. D. M. Hercules, *Accounts Chem. Res.*, **2**, 301 (1969).
131. D. M. Hercules in "Physical Methods of Organic Chemistry," Part II, Chap. 13, A. Weissberger and B. Rossiter, Eds., Academic, New York, 1971.
132. A. Weller and K. Zachariasse in "Chemiluminescence and Bioluminescence," M. Cormier, D. M. Hercules, and J. Lee, Eds., Plenum, New York, 1973, pp. 169–181.
133. A. J. Bard, C. P. Keszthelyi, H. Tachikawa, and N. E. Tokel, *ibid.*, p. 193.
134. K. A. Zachariasse in "The Exciplex," M. Gordon, W. R. Ware, P. DeMayo, and W. R. Arnold, Eds., Academic Press, New York, 1975, p. 275.
135. A. J. Bard and S. M. Park, *ibid.*, p. 305.
136. L. R. Faulkner, *Int. Rev. Sci.: Phys. Chem. Ser. Two*, **9**, 213 (1975).
137. L. R. Faulkner and A. J. Bard, *Electroanal. Chem.*, **10**, 1 (1977).

138. L. R. Faulkner, *Meth. Enzymol.* **57**, 494 (1978).
139. S. M. Park, *Rev. Chem. Intermed.*, in press.
140. R. S. Glass and L. R. Faulkner in "Chemi- and Bio-Energized Processes," W. Adam and G. Cilento, Eds., Academic, New York, in press.
141. G. B. Schuster, *Accounts Chem. Res.*, **12**, 366 (1979).
142. J. T. Maloy in "Computers in Chemistry and Instrumentation," Vol. 2, "Electrochemistry," J. S. Mattson, H. B. Mark, Jr., and H. C. MacDonald, Jr., Marcel Dekker, New York, 1972, Chap. 9.
143. J. T. Maloy and A. J. Bard, *J. Am. Chem. Soc.*, **93**, 5968 (1971).
144. J. A. Seckel and J. T. Maloy, *J. Phys. Chem.*, **83**, 1293 (1979).
145. G. H. Brilmyer and A. J. Bard, *J. Electrochem. Soc.*, **127**, 104 (1980).
146. L. R. Faulkner, *J. Electrochem. Soc.*, **124**, 1725 (1977).
147. P. R. Michael and L. R. Faulkner, *J. Am. Chem. Soc.*, **99**, 7754 (1977).
148. D. J. Freed and L. R. Faulkner, *J. Am. Chem. Soc.*, **93**, 2097 (1971).
149. E. A. Chandross and R. E. Visco, *J. Phys. Chem.*, **72**, 378 (1968).
150. H. Kuhn, *J. Chem. Phys.*, **53**, 101 (1970).
151. M. Gleria and R. Mcmming, *Z. Phys. Chem.*, **101**, 171 (1976).
152. L. S. R. Yeh and A. J. Bard, *Chem. Phys. Letters*, **44**, 339 (1976).
153. J. D. Luttmer and A. J. Bard, *J. Electrochem. Soc.*, **125**, 1423 (1978).
154. H. Gerischer, *Adv. Electrochem. Electrochem. Engr.*, **1**, 139 (1961).
155. H. Gerischer, in "Physical Chemistry—An Advanced Treatise," Vol. IXA, H. Eyring, D. Henderson, and W. Jost, Eds., Academic Press, New York, 1970, p. 463.
156. V. A. Myamlin and Yu. V. Pleskov, "Electrochemistry of Semiconductors," Plenum Press, New York, 1967.
157. P. J. Holmes, "The Electrochemistry of Semiconductors," Academic Press, New York, 1961.
158. R. Memming, *Electroanal. Chem.*, **11**, 1 (1979).
159. A. J. Nozik, *Annu. Rev. Phys. Chem.*, **29**, 189 (1978).
160. A. Many, Y. Goldstein, and N. B. Grover, "Semiconductor Surfaces," No. Holland Publ. Co., Amsterdam, 1965.
161. A. K. Jonscher, "Principles of Semiconductor Device Operation," Wiley, New York, 1960.
162. D. Madelung, in "Physical Chemistry—An Advanced Treatise," Vol. X, W. Jost, Ed., Academic Press, New York, 1970, Chap. 6.
163. G. Ertl and H. Gerischer, *ibid.*, Chap. 7.
164. F. Lohmann, *Z. Naturforsch., Teil A*, **22**, 843 (1967).
165. R. Gomer and G. Trypson, *J. Chem. Phys.*, **66**, 4413 (1977).
166. W. Schottky, *Z. Phys.*, **113**, 367 (1939); **118**, 539 (1942).
167. N. F. Mott, *Proc. Roy. Soc.* (*London*), **A171**, 27 (1939).
168. E. C. Dutoit, F. Cardon, and W. P. Gomes, *Ber. Bunsenges. Phys. Chem.*, **80**, 1285 (1976).
169. A. J. Bard, *J. Photochem.*, **10**, 59 (1979).
170. A. Fujishima and K. Honda, *Bull. Chem. Soc. Jpn.*, **44**, 1148 (1971); *Nature* (London), **238**, 37 (1972).
171. D. Laser and A. J. Bard, *J. Electrochem. Soc.*, **123**, 1027 (1976).

172. B. Miller and A. Heller, *Nature* (London), **262**, 680 (1976).
173. G. Hodes, D. Cahen, and J. Manassen, *ibid.*, **260**, 312 (1976).
174. M. S. Wrighton, A. B. Bocarsly, J. M. Bolts, A. B. Ellis, and K. D. Legg in "Semiconductor Liquid-Junction Solar Cells", A. Heller, Ed., Electrochem. Soc., Princeton, N.J., Proc. Vol. 77-3, 1977, p. 138.
175. H. Tributsch, *Ber. Bunsenges. Phys. Chem.*, **81**, 361 (1977).
176. M. S. Wrighton, A. B. Ellis, P. T. Wolczanski, D. L. Morse, H. B. Abrahamson, and D. S. Ginley, *J. Am. Chem. Soc.*, **98**, 2774 (1976).
177. J. G. Mavroides, J. A. Kafalas, and D. F. Kolesar, *Appl. Phys. Lett.*, **28**, 241 (1976).
178. T. Watanabe, A. Fujishima, and K. Honda, *Bull. Chem. Soc. Jpn.*, **49**, 355 (1976).
179. M. Halman, *Nature*, **275**, 115 (1978).
180. B. Kraeutler and A. J. Bard, *J. Am. Chem. Soc.*, **99**, 7729 (1977).
181. C. R. Dickson and A. J. Nozik, *ibid.*, **100**, 8007 (1978).
182. H. Gerischer and F. Willig, "Topics in Current Chemistry," Springer Verlag, Berlin, Vol. 61, 1976, p. 31.
183. H. Gerischer, *J. Electrochem. Soc.*, **125**, 218C (1978).
184. H. Tsubomura, M. Matsumura, Y. Nomura, and T. Amamiya, *Nature*, **261**, 402 (1976).
185. Yu. Y. Gurevich, Yu. V. Pleskov, and Z. A. Rotenberg, "Photoelectrochemistry," Plenum, New York, 1978.
186. G. A. Kenney and D. C. Walker, *Electroanal. Chem.*, **5**, 1 (1971).
187. G. C. Barker, *Ber. Bunsenges. Phys. Chem.*, **75**, 728 (1971).
188. G. C. Barker, D. McKeown, M. J. Williams, G. Bottura, and V. Concialini, *Faraday Discuss. Chem. Soc.*, **56**, 41 (1974).
189. Yu. V. Pleskov, Z. A. Rotenberg, V. V. Eletsky, and V. I. Lakomov, *ibid.*, p. 52.
190. A. Brodsky and Yu. V. Pleskov, in "Progress in Surface Sciences," Vol. 2, Part 1, S. G. Davidson, Ed., Pergamon Press, Oxford, 1972.
191. Yu. V. Pleskov and Z. A. Rotenberg, *Adv. Electrochem. Electrochem. Engr.*, **11**, 1 (1978).
192. H. Tachikawa and L. R. Faulkner in "Laboratory Techniques in Electroanalytical Chemistry," P. T. Kissinger, Ed., Marcel Dekker, New York, in press.
193. A. Henglein, *Electroanal. Chem.*, **9**, 163 (1976).
194. S. P. Perone and H. D. Drew in "Analytical Photochemistry and Photochemical Analysis: Solids, Solutions, and Polymers," J. Fitzgerald, Ed., Marcel Dekker, New York, 1971, Chap. 7.
195. S. P. Perone and J. R. Birk, *Anal. Chem.*, **38**, 1589 (1966).
196. G. L. Kirschner and S. P. Perone, *ibid.*, **44**, 443 (1972).
197. R. A. Jamieson and S. P. Perone, *J. Phys. Chem.*, **76**, 830 (1972).
198. J. I. H. Patterson and S. P. Perone, *J. Phys. Chem.*, **77**, 2437 (1973).
199. D. C. Johnson and E. W. Resnick, *Anal. Chem.*, **44**, 637 (1972); J. R. Lubbers, E. W. Resnick, P. R. Gaines, and D. C. Johnson, *ibid.*, **46**, 865 (1974).
200. H. Debrodt and K. E. Heusler, *Ber. Bunsenges. Phys. Chem.*, **81**, 1172 (1977).

201. W. J. Albery, M. D. Archer, N. J. Field, and A. D. Turner, *Faraday Discuss. Chem. Soc.*, **56**, 28 (1974).
202. M. D. Archer, *J. Appl. Electrochem.*, **5**, 17 (1975).
203. C. T. Lin and N. Sutin, *J. Phys. Chem.*, **80**, 97 (1976).
204. W. J. Albery and M. D. Archer, *Electrochim. Acta*, **21**, 1155 (1976).
205. W. J. Albery and M. D. Archer, *J. Electrochem. Soc.*, **124**, 688 (1977).
206. W. J. Albery and M. D. Archer, *J. Electroanal. Chem.*, **86**, 1, 19 (1978).

14.8 PROBLEMS

14.1 The absorbance values at 710 nm in Figure 14.1.4 are 0.040, 0.072, 0.111, 0.179, 0.279, 0.411, 0.633, 0.695, 0.719, and 0.725 for curves a to j. Calculate the ratio of concentrations of the Co(II) complex and Co(I) complex for the potentials corresponding to curves a to j. Plot E *vs.* the logarithm of the ratio, and from the plot verify n and find $E^{0\prime}$.

14.2 Given $D = 6.2 \times 10^{-6}$ cm^2/sec for *o*-tolidine and its oxidation product, calculate the molar absorptivity ε for the product from the slope of the absorbance plot in Figure 14.1.3. Calculate the effective area of the minigrid.

14.3 Calculate absorbance-time curves for a gold film OTE at which a reduction product is produced with $\varepsilon_R = 10^2$, 10^3, and 10^4 M^{-1} cm^{-1}. Let $D_O = 1 \times 10^{-5}$ cm^2/sec, and $C_O^* = 1$ mM. Draw graphs for times ranging from 1 to 100 msec. Comment on the magnitudes of the absorbances and their experimental implications.

14.4 Calculate the extinction coefficient k corresponding to a 10^{-3} M solution of a compound with $\varepsilon = 10^4$ M^{-1} cm^{-1}.

14.5 From data in Figure 14.1.9, calculate the values of n and k for gold at 2.0, 2.4, 2.8, 3.2, 3.6, and 4.0 eV. Plot them versus the wavelengths corresponding to these photon energies. Can you explain gold's color on the basis of your plots?

14.6 An absorbing species R is monitored in an IRS experiment featuring a forward step, in which species O is reduced to R at the diffusion-controlled rate, and a reversal step in which R is converted back to O, also under diffusion control. Show that the normalized absorbance is given by

$$\mathscr{A}(t \leq \tau)/\mathscr{A}(\tau) = 1 - \exp(a^2 t)\,\mathrm{erfc}(a t^{1/2})$$
$$\mathscr{A}(t > \tau)/\mathscr{A}(\tau) = \exp[a^2(t - \tau)]\,\mathrm{erfc}[a(t - \tau)^{1/2}]$$

where τ is the forward step width and $a = D_R^{1/2}/\delta$. Assume that τ is large enough that (14.1.14) applies. For simplicity, you may assume $D_O = D_R = D$. How many parameters are needed to fit the transient? Plot points for about 10 values of t/τ in the range $0 \leq t/\tau \leq 2$.

14.7 Calculate the value of δ for an IRS system involving an aqueous solution ($n_3 = 1.34$) in contact with a platinum film on glass ($n_1 = 1.55$). The angle of incidence θ_1 is 75°, and the wavelength of incident light is 400 nm. What values of δ would apply at 600 and 800 nm? Suppose the angle of incidence were increased to 80°. What value of δ would apply at 600 nm?

14.8 Derive equations describing the Raman intensity as a function of time for the forward and reversal phases of the experiment of Figure 14.1.25. Prove that the linear relations observed in the plots of Figure 14.1.25*b* are expected. Could any information be obtained from the actual magnitude of the slopes? How could such transients be used for mechanistic diagnosis? Note that this problem can be approached in a manner similar to that employed in the derivation of (14.1.2).

14.9 Suppose the XPS bands in Figure 14.2.3 were excited by the Al K_α line at 1486.6 eV. What were the kinetic energies of the photoelectrons? What kinetic energies would be measured for excitation by the Mg K_α line at 1253.6 eV?

14.10 Why is it necessary to normalize the carbon responses in drawing Figure 14.3.7?

14.11 (a) Suppose the experiment illustrated in Figure 14.3.5 is carried out with an electrogenerated radical anion that is stable, but otherwise under the same conditions as described (t_p, concentration, etc.). Draw the expected signal-time curve under these conditions.

(b) The rate constant k_2 for the dimerization of the radical anions in the system of Figure 14.3.5 can be estimated from the half-life $t_{1/2}$ and the approximate concentration of radical anion at t_p. Estimate k_2 based on the results in Figure 14.3.5 and compare with the value listed in the caption, obtained by a rigorous solution of the problem.

14.12 It is possible to measure the efficiency of triplet production in some redox processes by using *trans*-stilbene as an interceptor. For example, the fluoranthene triplet undergoes the reaction:

$$^3\text{FA}^* + trans\text{-S} \rightarrow {}^3(\text{Stilbene})^* + \text{FA}$$

then the stilbene triplet decays to *cis* and *trans* ground-state forms with a known partition ratio. Devise a bulk electrolysis experiment to measure the efficiency of triplet formation (triplets per redox event) in the reaction between 10-$\text{MP}^{\dot{+}}$ and $\text{FA}^{\dot{-}}$. Derive an equation relating the measured quantities to the triplet yield. Why is bulk electrolysis needed?

14.13 From the half-wave potentials in Figure 14.4.5, the values of $E^{0\prime}$ for the electrode reactions:

$$\text{TMPD}^{\dot{+}} + e \rightleftharpoons \text{TMPD}$$
$$\text{Py} + e \rightleftharpoons \text{Py}^{\dot{-}}$$

can be estimated. What are they? What is the free energy released in the reaction:

$$\text{TMPD}^{\dot{+}} + \text{Py}^{\dot{-}} \rightarrow \text{TMPD} + \text{Py}$$

From the fluorescence spectrum in Figure 14.4.4, estimate the energy of $^1\text{Py}^*$

relative to Py. Comment on the probability of forming $^1Py^*$ in the reaction between $TMPD^{+\cdot}$ and $Py^{-\cdot}$, and account for the light.

14.14 Interpret the lower frame of Figure 14.4.5.

14.15 The transient in Figure 14.4.6 was generated by a more complex waveform than usually applies. The first step generated $PPD^{-\cdot}$ for 500 msec, then the second step produced $TH^{+\cdot}$. However, for the period $0.10 \leq t_r/t_f \leq 0.12$, the potential was stepped to 0V, where $TH^{+\cdot}$ is reduced heterogeneously, then electrogeneration of $TH^{+\cdot}$ resumed. The fact that the dip in light output lagged the application of the zero pulse was cited as evidence that the light-producing reaction took place in solution at some distance from the electrode. Argue this point. Can you estimate the distance of the reaction zone from the surface?

14.16 The properties of a semiconductor photovoltaic cell can be deduced (neglecting the effect of internal resistance) from the *i-E* curves, such as those in Figure 14.5.6*a*. Assume that couple O/R is $Fe(CN)_6^{3-}$ (0.1 *M*)/ $Fe(CN)_6^{4-}$ (0.1 *M*), that $E_{fb} = -0.20$ V *vs.* SCE, and that the limiting photocurrent is governed by the light flux, 6.2×10^{15} photons/sec, which is completely absorbed and converted to separated electron/hole pairs. For a cell comprising the *n*-type semiconductor and a platinum electrode in the electrolyte, what are the maximum open-circuit voltage and the short-circuit current under illumination? Sketch the expected output current versus output voltage plot for this cell. What is the maximum output power for the cell?

14.17 An expression frequently used for the "thickness of the space charge region," L_1, is (154–156)

$$L_1 = \left(\frac{2\varepsilon\varepsilon_0}{eN_D}\,\Delta\phi\right)^{1/2} \approx \left(1.1 \times 10^6\,\frac{\Delta\phi}{N_D}\,\varepsilon\right)^{1/2}\ \text{cm}$$

(units: N_D, cm^{-3}; $\Delta\phi$, V). Sketch the variation of L_1 with $\Delta\phi$ for several values of N_D for a semiconductor with $\varepsilon = 10$. For efficient utilization of light, most of the radiation should be absorbed within the space charge region. The absorption of light follows a Beer's law relationship with an absorption coefficient, α (cm^{-1}), so that a "penetration depth" of $\sim 1/\alpha$ can be estimated. If $\alpha = 10^5\ cm^{-1}$ and a band-bending of 0.5 V is to be used, what is the recommended doping level, N_D? Why would much lower doping in a semiconductor photoelectrochemical cell be undesirable?

14.18 Consider the dye sensitization of a semiconductor electrode with the energy level situation depicted in Figure 14.8.1. Explain how this system operates under illumination.

14.19 The Mott-Schottky plots in Figure 14.8.2 for *n*- and *p*-type InP in 1 *M* KCl, 0.01 *M* HCl were reported by A. M. Van Wezemael, et al. From these results estimate the flat-band potentials and doping levels of the two semiconductors. How does the difference in E_{fb} for *n*- and *p*-type InP compare to E_g for this material (1.3 eV)?

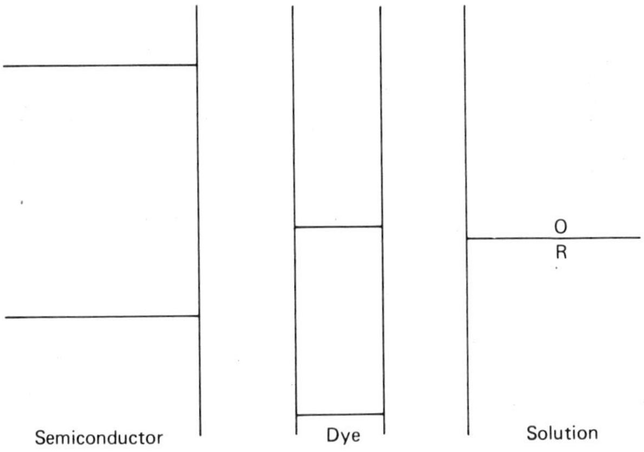

Figure 14.8.1

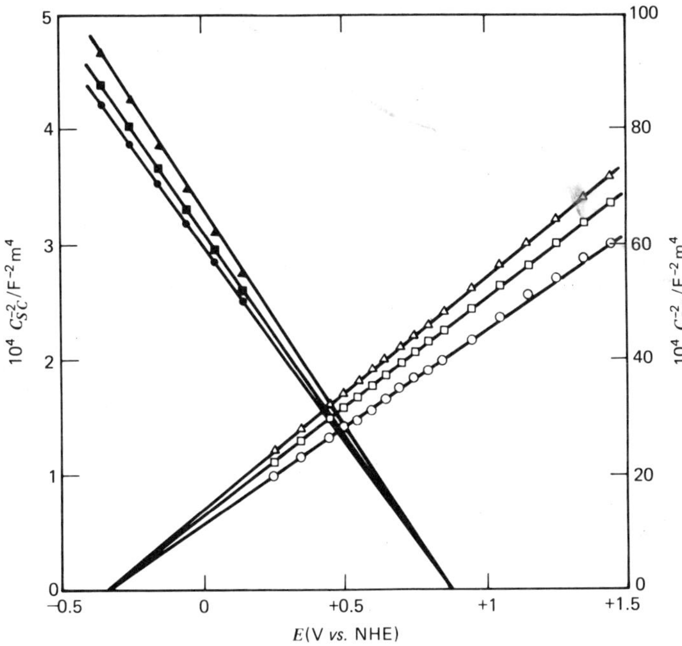

Figure 14.8.2

Mott-Schottky plots for *n*- and *p*-type InP, $(\overline{111})$ face. Electrolyte composition: 1 *M* KCl + 0.01 *M* HCl, *n*-type: (○) 200 Hz, (□) 2500 Hz, (△) 20000 Hz; *p*-type: (●) 200 Hz, (■) 2500 Hz, (▲) 20000 Hz. The vertical axis to the left applies to *p*-type, the one to the right to *n*-type. [From A. M. Van Wezemael, W. H. Laflere, F. Cardon, and W. P. Gomes, *J. Electroanal. Chem.*, **87**, 105 (1978), with permission.]

14.20 The following cell is proposed as a photoelectrochemical storage battery:

$$n\text{-TiO}_2/0.2\ M\ \text{Br}^-(\text{pH} = 1)//0.1\ M\ \text{I}_3^-(\text{pH} = 1)/\text{Pt}$$

The E_{fb} for n-TiO_2 under these conditions is -0.30 V *vs.* SCE. Under irradiation, a photo-oxidation producing Br_2 occurs at TiO_2. (a) Write the half-cell reactions that occur at both electrodes during irradiation. What is the maximum open-circuit voltage under illumination (assume no liquid junction potential)? (b) During the "photo-charge cycle" Br_2 and I^- accumulate in the cell. If half of the I_3^- is converted during charging and a platinum electrode is used in the Br_2/Br^- cell for dark discharge, what is the cell voltage of the charged cell? Write the half reactions that take place during dark discharge.

Appendix A
Mathematical Methods

An understanding of many electrochemical phenomena depends strongly on an ability to solve certain differential equations, so it is worthwhile to summarize some of the important mathematical tools. This appendix is meant only as a review or an introduction. It is not a mathematically rigorous exposition. If more complete information is desired, the references should be consulted. Nevertheless, the level of the presentation will enable the reader to solve the accompanying problems and to follow the derivations in the text.

A.1 SOLVING DIFFERENTIAL EQUATIONS BY THE LAPLACE TRANSFORM TECHNIQUE

A.1.1 Partial Differential Equations

Our main encounter with partial differential equations (PDEs) arises in the treatment of diffusion near the surface of an electrode at which a heterogeneous reaction is under way. Solute concentrations are therefore functions both of time t and of distance from the electrode x. Thus, the concentration $C(x, t)$ ordinarily will obey some form of Fick's diffusional laws (Section 4.3), such as

$$\frac{\partial C(x, t)}{\partial t} = D \frac{\partial^2 C(x, t)}{\partial x^2} \tag{A.1.1}$$

where D is the diffusion coefficient of the substance at hand. This equation is a *linear* PDE, because it contains only first or zeroth powers of $C(x, t)$ and its derivatives. The *order* of the equation is that of the highest derivative; hence the example above is of the second order.

Much of the difficulty in solving partial differential equations derives from the fact that the PDE does not fix even the functional form of its *general* solution. Indeed, there usually are many such solutions to a given PDE; for example, the equation:

$$\frac{\partial z}{\partial x} - \frac{\partial z}{\partial y} = 0 \tag{A.1.2}$$

is satisfied by any of the following relations:

$$z = A\,e^{(x+y)} \quad \text{(A.1.3)}$$

$$z = A\sin(x + y) \quad \text{(A.1.4)}$$

$$z = A(x + y) \quad \text{(A.1.5)}$$

This feature contrasts with the properties of ordinary differential equations (ODEs), which contain only derivatives of functions of a single variable. Usually the *form* of the solution to an ODE is dictated by the ODE itself. Thus, the linear, first-order ODE describing unimolecular decay:

$$\frac{dC(t)}{dt} = -kC(t) \quad \text{(A.1.6)}$$

has the single general solution:

$$C(t) = (\text{constant})\,e^{-kt} \quad \text{(A.1.7)}$$

and the boundary condition applied to a specific problem serves only to supply the integrating constant.

The solution to a PDE usually depends on the boundary conditions for its form as well as the evaluation of constants; and the same PDE solved under different boundary conditions will often yield different functional relations.

A.1.2 Introduction to the Laplace Transformation (1–3)

The Laplace transformation is of great value in solving certain kinds of differential equations—especially those encountered in electrochemistry—because it enables a conversion of the problem into a domain where simpler mathematical manipulation is possible. Probably the most familiar transformation of this sort is the use of logarithms for solving complex multiplication problems. One begins the procedure with a transformation of the operands into their images, the logarithms. In the transform domain, the problem is solved by addition. Thus, one attains the transform of the desired result, and inverse transformation yields the result itself.

The application of the Laplace transformation is analogous. Transformation of an ODE yields an expression from which *algebraic* manipulation will provide the transform of the ODE's solution. Inverse transformation then completes the solution. By similar means, PDEs can be transformed into ODEs, which are then solved conventionally or by further application of transform techniques. This method is extremely convenient, but it is restricted almost entirely to linear differential equations.

The Laplace transform in t of the function $F(t)$ is symbolized by $L\{F(t)\}$, $f(s)$, or $\bar{F}(s)$, and is defined by

$$\boxed{L\{F(t)\} \equiv \int_0^\infty e^{-st}F(t)\,dt} \quad \text{(A.1.8)}$$

The existence of the transform is conditional. It requires (a) that $F(t)$ be bounded at all *interior* points on the interval $0 \le t < \infty$; (b) that it have a finite number of discontinuities; and (c) that it be of exponential order. This last condition means that $e^{-\alpha t}|F(t)|$ must be bounded for some constant α as $t \to \infty$; thus, it requires the function's magnitude to rise more slowly than some exponential $e^{\alpha t}$ as t becomes very large. It is plain that e^{-2t} is of exponential order, whereas e^{t^2} is not. The first condition clearly rules out $(t-1)^{-1}$ as having a transform, but it says nothing about $t^{-1/2}$ or t^{-1}. It turns out that $F(t)$ may possess an infinite discontinuity at $t = 0$ if $|t^n F(t)|$ is bounded there for some positive value of n less than unity. Thus, $t^{-1/2}$ does have a Laplace transform, but t^{-1} does not. In real practical applications, (a) and (c) do occasionally offer obstacles, but (b) rarely does.

Many transforms are obtained directly from the integral of definition, but others are more conveniently extracted from indirect approaches. Table A.1.1 gives a short list of some commonly encountered functions and their transforms. Much more extensive tables can be found in the references (1, 2, 4–6).

A.1.3 Fundamental Properties of the Transform (1–3)

The Laplace transformation is linear in that

$$L\{aF(t) + bG(t)\} = af(s) + bg(s) \tag{A.1.9}$$

This property follows directly from the definition and the basic properties of integral calculus.

Table A.1.1
Laplace Transforms of Common Functions[a]

$F(t)$	$f(s)$
A(constant)	A/s
e^{-at}	$1/(s+a)$
$\sin at$	$a/(s^2+a^2)$
$\cos at$	$s/(s^2+a^2)$
$\sinh at$	$a/(s^2-a^2)$
$\cosh at$	$s/(s^2-a^2)$
t	$1/s^2$
$t^{(n-1)}/(n-1)!$	$1/s^n$
$(\pi t)^{-1/2}$	$1/s^{1/2}$
$2(t/\pi)^{1/2}$	$1/s^{3/2}$
$\dfrac{x}{2(\pi k t^3)^{1/2}}[\exp(-x^2/4kt)]$	$e^{-\beta x}$, where $\beta = (s/k)^{1/2}$
$\left(\dfrac{k}{\pi t}\right)^{1/2}[\exp(-x^2/4kt)]$	$e^{-\beta x}/\beta$
$\mathrm{erfc}[x/2(kt)^{1/2}]$	$e^{-\beta x}/s$
$2\left(\dfrac{kt}{\pi}\right)^{1/2}\exp(-x^2/4kt) - x\,\mathrm{erfc}[x/2(kt)^{1/2}]$	$e^{-\beta x}/s\beta$
$\exp(a^2 t)\,\mathrm{erfc}(a t^{1/2})$	$\dfrac{1}{s^{1/2}(s^{1/2}+a)}$

[a] Adapted from R. V. Churchill, "Operational Mathematics," 2nd ed., McGraw-Hill, New York, 1958, pp. 323ff.

The value of the transformation for solving differential equations issues from its conversion of derivatives with respect to the transformation variable into algebraic expressions in s. For example,

$$L\left\{\frac{dF(t)}{dt}\right\} = sf(s) - F(0) \tag{A.1.10}$$

A proof rests upon integration by parts:

$$L\left\{\frac{dF(t)}{dt}\right\} = \int_0^\infty e^{-st}\frac{dF(t)}{dt}\,dt \tag{A.1.11}$$

$$= \left[e^{-st}F(t)\right]_0^\infty + s\int_0^\infty e^{-st}F(t)\,dt \tag{A.1.12}$$

$$= -F(0) + sf(s) \tag{A.1.13}$$

One can show similarly that

$$L\{F''\} = s^2f(s) - sF(0) - F'(0) \tag{A.1.14}$$

and generally that

$$\boxed{L\{F^{(n)}\} = s^nf(s) - s^{n-1}F(0) - s^{n-2}F'(0) - \cdots - F^{(n-1)}(0)} \tag{A.1.15}$$

The transformation is oblivious to differential operators other than those involving t:

$$L\left\{\frac{\partial F(x,t)}{\partial x}\right\} = \frac{\partial f(x,s)}{\partial x} \tag{A.1.16}$$

because variables other than t are regarded as constants for purposes of conversion.

Other useful properties involve the transforms of integrals and the effect of multiplication by an exponential:

$$\boxed{L\left\{\int_0^t F(x)\,dx\right\} = \frac{1}{s}f(s)} \tag{A.1.17}$$

$$\boxed{L\{e^{at}F(t)\} = f(s-a)} \tag{A.1.18}$$

For example,

$$L\{\sin bt\} = \frac{b}{s^2 + b^2} \tag{A.1.19}$$

$$L\{e^{at}\sin bt\} = \frac{b}{(s-a)^2 + b^2} \tag{A.1.20}$$

When inversion of the transform cannot be carried out from tabulated functions, one can sometimes obtain it from the *convolution integral*:

$$L^{-1}\{f(s)g(s)\} = F(t)*G(t)$$
$$= \int_0^t F(t-\tau)G(\tau)\,d\tau \qquad \text{(A.1.21)}$$

Note that $F(t)*G(t)$ merely symbolizes the convolution integral. It does not imply a multiplication.

A.1.4 Solving Ordinary Differential Equations by Laplace Transformation

As an example, let us determine the time-dependent position of a mass on a spring, relative to its equilibrium position, after release from an initial displacement A. This is, of course, the linear harmonic oscillator problem. Let y be the displacement and k be the spring's force constant. Thus, with $y'(0) = 0$,

$$m\frac{d^2y}{dt^2} = -ky \qquad \text{(A.1.22)}$$

Under transformation, we have

$$s^2\bar{y} - sy(0) - y'(0) = -\frac{k}{m}\bar{y} \qquad \text{(A.1.23)}$$

$$s^2\bar{y} - As = -\frac{k}{m}\bar{y} \qquad \text{(A.1.24)}$$

$$\bar{y} = \frac{As}{s^2 + k/m} \qquad \text{(A.1.25)}$$

The inverse transform then gives the solution:

$$L^{-1}\{\bar{y}\} = y(t) = A\cos\left(\frac{k}{m}\right)^{1/2} t \qquad \text{(A.1.26)}$$

As a second case, let us find $i(t)$ resulting from closure of the switch in the circuit of Figure A.1.1. If we assume the initial charge on C to be zero, basic considerations yield the following description of the system:

$$E = iR + \frac{1}{C}\int_0^t i(\tau)\,d\tau + L\frac{di}{dt} \qquad \text{(A.1.27)}$$

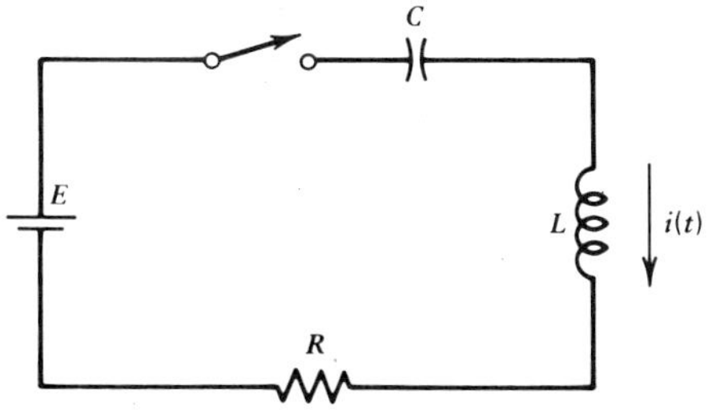

Figure A.1.1

Proceeding to the solution as before, we find

$$\frac{E}{s} = \bar{\imath}R + \frac{1}{sC}\bar{\imath} + Ls\bar{\imath} \tag{A.1.28}$$

$$\bar{\imath} = \frac{E/L}{s^2 + Rs/L + 1/LC} \tag{A.1.29}$$

This has the form of

$$L\{A\, e^{-at} \sin bt\} = \frac{Ab}{(s + a)^2 + b^2} \tag{A.1.30}$$

and algebraic manipulation identifies a as $R/2L$ and b^2 as $(1/LC - R^2/4L^2)$. The constant A is therefore E/Lb, and the solution is

$$i = \frac{E}{Lb} e^{-at} \sin bt \tag{A.1.31}$$

Several times in our study we will need to solve equations of the following form:

$$\frac{d^2C(x)}{dx^2} - a^2C(x) = -b \tag{A.1.32}$$

Transformation and further manipulation yields

$$s^2\bar{C}(s) - sC(0) - C'(0) - a^2\bar{C}(s) = -b/s \tag{A.1.33}$$

$$\bar{C}(s) = \frac{-b + s^2C(0) + sC'(0)}{s(s - a)(s + a)} \tag{A.1.34}$$

It is inconvenient to invert this expression without first segmenting it into partial fractions; hence we will suspend our consideration of this solution while we summarize the technique.

The expression to be expanded must first be factored as far as possible into real linear and real quadratic factors; for example,

$$\frac{s + 3}{(s - 1)^2(s - 2)(s - 3)(s^2 + 2s + 2)} \tag{A.1.35}$$

then a series-sum representation of the expression is written according to the following rules (7):

1. If the linear factor $as + b$ occurs n times as a factor in the denominator, there corresponds to this factor a sum of n partial fractions:

$$\frac{A_1}{as + b} + \frac{A_2}{(as + b)^2} + \cdots + \frac{A_n}{(as + b)^n} \tag{A.1.36}$$

where the A_i are constants and $A_n \neq 0$.

2. If a quadratic factor $as^2 + bs + c$ occurs n times as a factor in the denominator, there corresponds to this factor a sum of n partial fractions:

$$\frac{A_1s + B_1}{as^2 + bs + c} + \frac{A_2s + B_2}{(as^2 + bs + c)^2} + \cdots + \frac{A_ns + B_n}{(as^2 + bs + c)^n} \tag{A.1.37}$$

where the A's and B's are constants and $A_ns + B_n$ is not identically zero.

Thus, the example above expands to

$$\frac{s + 3}{(s - 1)^2(s - 2)(s - 3)(s^2 + 2s + 2)} = \frac{A}{s - 1} + \frac{B}{(s - 1)^2} + \frac{C}{s - 2} + \frac{D}{s - 3} + \frac{Es + F}{s^2 + 2s + 2} \tag{A.1.38}$$

Ordinarily there are two ways for evaluating the constants. One can either multiply by the denominator on the left-hand side, expand both sides into polynomials, and equate coefficients of like powers of s; or one can substitute values for s and solve simultaneous equations.

Returning now to our suspended problem, we find that we can expand the solution transform as follows.

$$\bar{C}(s) = \frac{-b + s^2C(0) + sC'(0)}{s(s - a)(s + a)} \tag{A.1.39}$$

$$= \frac{A'}{s + a} + \frac{B'}{s - a} + \frac{D'}{s} \tag{A.1.40}$$

By multiplying the equation with s and then setting $s = 0$, one easily determines the constant D' to be b/a^2. Evaluation of A' and B' must await definition of the boundary conditions $C(0)$ and $C'(0)$, but we nevertheless can invert the transform to a general form:

$$C(x) = \frac{b}{a^2} + A'\, e^{-ax} + B'\, e^{ax} \tag{A.1.41}$$

A.1.5 Solutions of Simultaneous Linear Ordinary Differential Equations

To illustrate the usefulness of transform methods for solving simultaneous linear ODEs, we consider the following kinetic scheme.

$$\begin{aligned} \mathrm{A} &\xrightarrow{k_1} \mathrm{B} + \mathrm{C} \\ \mathrm{B} &\xrightarrow{k_2} \mathrm{D} \\ \mathrm{C} + (\mathrm{Z}) &\xrightarrow{k_3'} \mathrm{D} \end{aligned} \tag{A.1.42}$$

Our goal is to determine the time profiles of the concentrations of A, B, C, and D; and in so doing, we will regard k_3' as a pseudo-first-order rate constant. Suppose at $t = 0$, $[\mathrm{A}] = A^*$, and $[\mathrm{B}] = [\mathrm{C}] = [\mathrm{D}] = 0$. The ODEs describing the system are written straightforwardly:

$$\frac{d[\mathrm{A}]}{dt} = -k_1[\mathrm{A}] \tag{A.1.43}$$

$$\frac{d[\mathrm{B}]}{dt} = k_1[\mathrm{A}] - k_2[\mathrm{B}] \tag{A.1.44}$$

$$\frac{d[\mathrm{C}]}{dt} = k_1[\mathrm{A}] - k_3'[\mathrm{C}] \tag{A.1.45}$$

$$\frac{d[\mathrm{D}]}{dt} = k_2[\mathrm{B}] + k_3'[\mathrm{C}] \tag{A.1.46}$$

Denoting $L\{[\mathrm{A}]\} = a$, *etc.*, we can write the following simultaneous *algebraic* equations:

$$sa - A^* = -k_1 a \tag{A.1.47}$$

$$sb = k_1 a - k_2 b \tag{A.1.48}$$

$$sc = k_1 a - k_3' c \tag{A.1.49}$$

$$sd = k_2 b + k_3' c \tag{A.1.50}$$

These relations obviously can be solved easily for a, b, c, and d, which can in turn be transformed to the desired concentrations.

A.1.6 Solutions of Partial Differential Equations (1, 2)

We have already anticipated above our continuing need for solving the diffusion equation:

$$\frac{\partial C(x, t)}{\partial t} = D\frac{\partial^2 C(x, t)}{\partial x^2} \tag{A.1.51}$$

in a variety of circumstances with differing boundary conditions. The solution requires an initial condition ($t = 0$), and two boundary conditions in x. Typically one takes $C(x, 0) = C^*$ for the initial state, and one uses the semi-infinite limit:

$$\lim_{x \to \infty} C(x, t) = C^* \tag{A.1.52}$$

as one x condition. Thus, we usually will require only one additional boundary condition to define a problem completely. Even with the information at hand, a partial solution can be obtained, and it is instructive for us to carry it through.

Transforming the PDE on the variable t, we obtain

$$s\overline{C}(x, s) - C^* = D\frac{d^2\overline{C}(x, s)}{dx^2} \tag{A.1.53}$$

$$\frac{d^2\overline{C}(x, s)}{dx^2} - \frac{s}{D}\overline{C}(x, s) = -\frac{C^*}{D} \tag{A.1.54}$$

In considering ODEs above, we were able to solve this equation, and we can immediately write from (A.1.41)

$$\overline{C}(x, s) = \frac{C^*}{s} + A'(s)\exp[-(s/D)^{1/2}x] + B'(s)\exp[(s/D)^{1/2}x] \tag{A.1.55}$$

The semi-infinite limit can itself be transformed to

$$\lim_{x \to \infty} \overline{C}(x, s) = \frac{C^*}{s} \tag{A.1.56}$$

hence, $B'(s)$ must be zero for the conditions at hand. Therefore,

$$\boxed{\bar{C}(x, s) = \frac{C^*}{s} + A'(s) \exp[-(s/D)^{1/2}x]} \tag{A.1.57}$$

and

$$C(x, t) = C^* + L^{-1}\{A'(s) \exp[-(s/D)^{1/2}x]\} \tag{A.1.58}$$

Final evaluation depends on the third boundary condition.

A.1.7 The Zero-Shift Theorem (1)

In electrochemical experiments, one often encounters abrupt changes in boundary conditions. Simple step techniques are the most obvious examples. Theoretical treatments of these experiments are often simplified by application of the *unit step function*, $S_\kappa(t)$, which rises from zero to unity at $t = \kappa$. More precisely,

$$S_\kappa(t) = 0 \qquad t \leq \kappa \tag{A.1.59}$$

$$S_\kappa(t) = 1 \qquad t > \kappa \tag{A.1.60}$$

It can be regarded as a mathematical "switch" that is "closed" at time $t = \kappa$, and it allows compact expression of complex boundary conditions. For example, consider a potential that is held at E_1 until $t = \kappa$, then is changed abruptly to E_2. The whole sequence can be written

$$E(t) = E_1 + S_\kappa(t)(E_2 - E_1) \tag{A.1.61}$$

Similarly, the potential program representing a linear scan following a step could be expressed as

$$E(t) = E_1 + S_\kappa(t)v(t - \kappa) \tag{A.1.62}$$

Once the boundary conditions are written out, they must usually be transformed. The *zero-shift theorem* provides the necessary basis. It is summarized by

$$\boxed{L\{S_\kappa(t)F(t - \kappa)\} = e^{-\kappa s}f(s)} \tag{A.1.63}$$

The proof rests on the definition:

$$L\{S_\kappa(t)F(t - \kappa)\} \equiv \int_0^\infty e^{-ts}S_\kappa(t)F(t - \kappa)\, dt = \int_\kappa^\infty e^{-ts}F(t - \kappa)\, dt \tag{A.1.64}$$

Defining $\theta = t - \kappa$ and rearranging, we obtain the desired result:

$$\int_\kappa^\infty e^{-ts}F(t - \kappa)\, dt = e^{-\kappa s}\int_0^\infty e^{-\theta s}F(\theta)\, d\theta = e^{-\kappa s}f(s) \tag{A.1.65}$$

Equation A.1.63 is called the zero-shift theorem because it shows that multiplication by $e^{-\kappa s}$ in transform space corresponds to a shift in the real time axis by an amount κ. This effect is shown in Figure A.1.2 for the simple function $F(t) = 2t$.

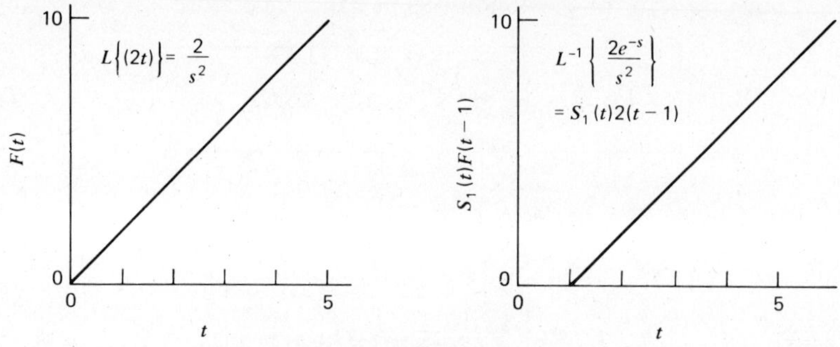

Figure A.1.2
Effect of the zero-shift theorem on $F(t) = 2t$.

A.2 TAYLOR EXPANSIONS

It is often useful to expand a function into a series (2, 8) when direct usage of the function is impractical because of complexity or when a linear approximation is sought. In general, some point is chosen as a central location, and the expansion represents the function at points in its neighborhood. Truncated series can be expected to give accurate descriptions near a central value and less accurate representations at more remote points.

A.2.1 Expansion for a Function of Several Variables

A function of three variables, $f(x, y, z)$, can be expanded about the point (x_0, y_0, z_0) by the *Taylor formula*:

$$f(x, y, z) = f(x_0, y_0, z_0) + \sum_{j=1}^{\infty} \frac{1}{j!}\left[\left(\delta x \frac{\partial}{\partial x} + \delta y \frac{\partial}{\partial y} + \delta z \frac{\partial}{\partial z}\right)^j f(x, y, z)\right]_{(x_0, y_0, z_0)} \tag{A.2.1}$$

where $\delta x = x - x_0$, $\delta y = y - y_0$, and $\delta z = z - z_0$. The expression in parentheses is a differential operator that is raised to the jth power. The various powers of $\partial/\partial x$, $\partial/\partial y$, and $\partial/\partial z$ are symbols for repeated differentiation. After the operator acts on $f(x, y, z)$, the limits are taken for (x_0, y_0, z_0).

As an example, consider the expansion of the current-overpotential equation (3.5.10):

$$g[C_O(0, t), C_R(0, t), \eta] = i/i_0 = \frac{C_O(0, t)}{C_O^*} e^{-\alpha n f \eta} - \frac{C_R(0, t)}{C_R^*} e^{(1-\alpha) n f \eta} \tag{A.2.2}$$

The central point is chosen at $(C_O^*, C_R^*, 0)$, where $g = 0$. If only the terms for $j = 1$ are kept, then we have

$$\frac{i}{i_0} = \left[\delta C_O(0,t)\frac{\partial}{\partial C_O(0,t)} + \delta C_R(0,t)\frac{\partial}{\partial C_R(0,t)} + \delta\eta\frac{\partial}{\partial\eta}\right] g\left[C_O(0,t), C_R(0,t), \eta\right]$$

$$\text{(derivatives evaluated at } C_O^*, C_R^*, \eta = 0) \tag{A.2.3}$$

Substituting for the derivatives and evaluating at the central point, one obtains

$$\frac{i}{i_0} = \frac{\delta C_O(0,t)}{C_O^*} - \frac{\delta C_R(0,t)}{C_R^*} - nf\delta\eta \tag{A.2.4}$$

or

$$i = i_0\left[\frac{C_O(0,t)}{C_O^*} - \frac{C_R(0,t)}{C_R^*} - \frac{nF}{RT}\eta\right] \tag{A.2.5}$$

which is equivalent to (3.5.33). By truncating the series at $j = 1$, we have derived a simple linear approximation to the more complex relation (3.5.10). It is valid for small excursions from the central point. For larger values of $\delta C_O(0, t)$, $\delta C_R(0, t)$, and $\delta\eta$, additional terms of the expansion would have to be included. The complete series is readily derived, but it is left as an exercise for the reader.

A.2.2 Expansion for a Function of a Single Variable

If the function of interest has only a single independent variable, the Taylor formula is a simplified version of (A.2.1):

$$f(x) = f(x_0) + \sum_{j=1}^{\infty}\frac{1}{j!}(x - x_0)^j\left[\frac{\partial^j}{\partial x^j}f(x)\right]_{x=x_0} \tag{A.2.6}$$

In this case the expansion is made about the point $x = x_0$.

A.2.3 Maclaurin Series

When a Taylor expansion of $f(x)$ is carried out for $x = 0$, it is called a *Maclaurin series*. The general formula is

$$f(x) = f(0) + \sum_{j=1}^{\infty}\frac{1}{j!}x^j\left[\frac{\partial^j}{\partial x^j}f(x)\right]_{x=0} \tag{A.2.7}$$

A.3 THE ERROR FUNCTION

In treating diffusion problems, one frequently encounters the integrated normal error curve, that is, the *error function* (2, 8):

$$\text{erf}(x) \equiv \frac{2}{\pi^{1/2}}\int_0^x e^{-y^2}\,dy \tag{A.3.1}$$

This relation approaches a unit limit as x becomes very large; hence, its *complement*, which also arises often, is straightforwardly defined as

$$\operatorname{erfc}(x) \equiv 1 - \operatorname{erf}(x) \tag{A.3.2}$$

Both functions are illustrated graphically in Figure A.3.1. Note that erf(x) rises steeply and essentially reaches its limit for any x greater than 2.

Evaluations are made by series representations (9, 10). The Maclaurin expansion is

$$\operatorname{erf}(x) = \frac{2}{\pi^{1/2}}\left(x - \frac{x^3}{3} + \frac{x^5}{5\cdot 2!} - \frac{x^7}{7\cdot 3!} + \frac{x^9}{9\cdot 4!} - \cdots\right) \tag{A.3.3}$$

It is convenient for $0 \leq x \leq 2$. For values of x less than 0.1, one can take the linear approximation from its first term:

$$\operatorname{erf}(x) \cong \frac{2x}{\pi^{1/2}} \qquad (x < 0.1) \tag{A.3.4}$$

Large arguments ($x > 2$) are better evaluated from

$$\operatorname{erf}(x) = 1 - \frac{e^{-x^2}}{\pi^{1/2}x}\left[1 - \frac{1}{2x^2} + \frac{1\cdot 3}{(2x^2)^2} - \frac{1\cdot 3\cdot 5}{(2x^2)^3} + \cdots\right] \tag{A.3.5}$$

The derivative of erf(x) can be evaluated by the Leibnitz rule as described in the next section.

A.4 LEIBNITZ RULE

The Leibnitz rule (8) furnishes a basis for differentiating a definite integral with respect to a parameter:

$$\frac{d}{d\alpha}\int_{u_1(\alpha)}^{u_2(\alpha)} f(x, \alpha)\, dx = f[u_2(\alpha), \alpha]\frac{du_2}{d\alpha} - f[u_1(\alpha), \alpha]\frac{du_1}{d\alpha} + \int_{u_1(\alpha)}^{u_2(\alpha)} \frac{\partial f(x, \alpha)}{\partial \alpha}\, dx \tag{A.4.1}$$

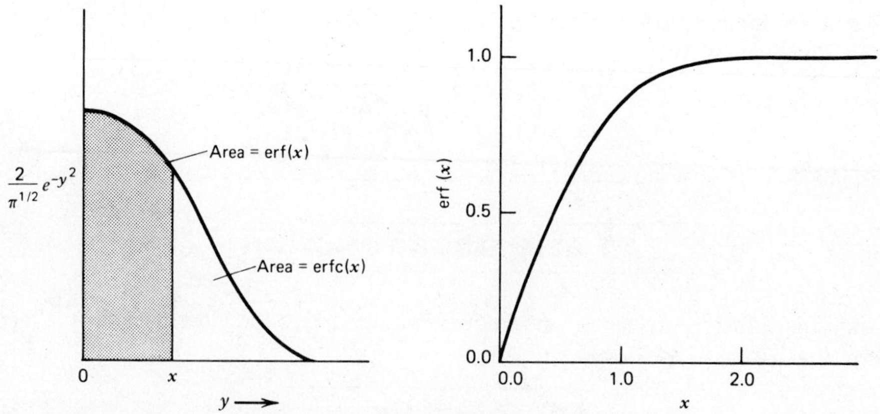

Figure A.3.1
Definition and behavior of erf(x) and erfc(x).

For example,

$$\frac{d}{dx}\,\mathrm{erf}(x) = \frac{2}{\pi^{1/2}}\frac{d}{dx}\int_0^x e^{-y^2}\,dy = \frac{2}{\pi^{1/2}}\,e^{-x^2} \tag{A.4.2}$$

A.5 COMPLEX NOTATION

In many problems involving vector-like variables, such as ac circuit analysis (Chapter 9), it is useful to represent physical quantities as complex functions (11), so that two dimensions are available. A *complex number* is written $z = x + jy$, where $j = \sqrt{-1}$, and x and y are called the *real* and *imaginary components*. One can think of z as a point in the *complex plane* representing all possible combinations of x and y, or it could be a vector in a cartesian coordinate system, as shown in Figure A.5.1*a*. The component x is plotted along the *real axis*, and y is plotted versus the *imaginary axis*. Two complex numbers $z_1 = x_1 + jy_1$ and $z_2 = x_2 + jy_2$ are equal only if $x_1 = x_2$ and $y_1 = y_2$.

Functions of a complex variable can also be defined, and they are always separable into real and imaginary parts. That is, the function $w(z)$ can always be written for $z = x + jy$ as

$$w(z) = u(x, y) + jv(x, y) \tag{A.5.1}$$

where $u(x, y)$ and $v(x, y)$ are wholly real. For example, one might have

$$w(z) = x^2 + y^2 - 2jxy \tag{A.5.2}$$

where the *real part* of $w(z)$, that is, $\mathrm{Re}[w(z)]$, is $u(x, y) = x^2 + y^2$ and the *imaginary part*, $\mathrm{Im}[w(z)]$, is $-2xy$. Two functions $w_1(z) = u_1(x, y) + jv_1(x, y)$ and $w_2(z) = u_2(x, y) + jv_2(x, y)$ are equal only if $u_1(x, y) = u_2(x, y)$ and $v_1(x, y) = v_2(x, y)$.

A valuable alternative notation for a complex number is to specify its position in polar coordinates, as shown in Figure A.5.1*b*. The length of the vector is

$$r = (x^2 + y^2)^{1/2} \tag{A.5.3}$$

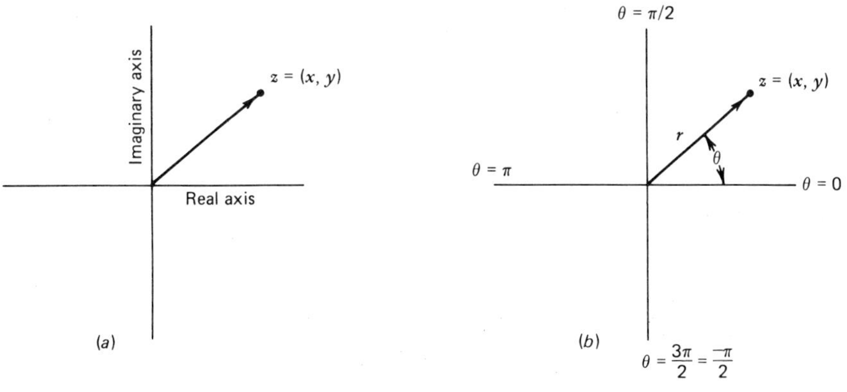

Figure A.5.!
Points in the complex plane. (*a*) Cartesian system. (*b*) Polar system.

and the phase angle θ is

$$\theta = \tan^{-1}\left(\frac{y}{x}\right) \tag{A.5.4}$$

An important function relates directly to this notation. We *define* the complex exponential as

$$\exp(z) = e^x(\cos y + j \sin y) \tag{A.5.5}$$

Thus,

$$e^z = e^x e^{jy} \tag{A.5.6}$$

where

$$\boxed{\exp(jy) = \cos y + j \sin y} \tag{A.5.7}$$

Note that e^{jy} always has a magnitude of unity, so that all values of the function lie on the circle of unit radius about the origin, as shown in Figure A.5.2, and y is the phase angle of the vector. Very often in science, sine and cosine terms are carried through derivations in terms of functions like (A.5.7).

The exponential function gives us a convenient way to express a complex number z in its polar form; that is,

$$z = x + jy = re^{j\theta} \tag{A.5.8}$$

Often the polar notation is the more useful form for applications, and functions may be written in terms of r and θ, instead of x and y.

Finally, we note that any function of a complex variable $w = u + jv$ can always be converted into a wholly real function by multiplying it by its *complex conjugate* $w^* = u - jv$:

$$ww^* = u^2 + v^2 \tag{A.5.9}$$

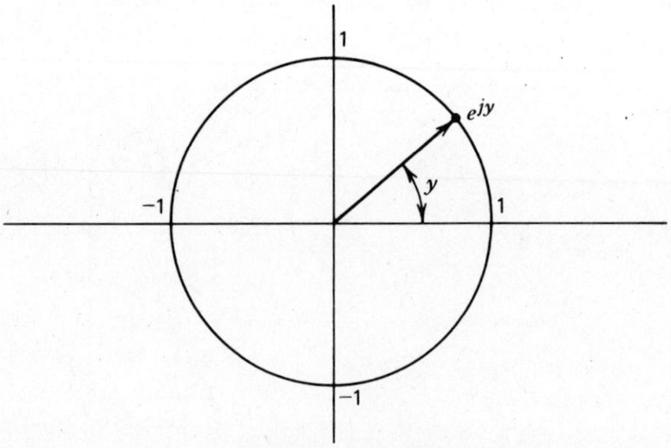

Figure A.5.2

Locus of values of $\exp(jy)$, that is, a circle at unit radius. Vector shows one value, with argument y as phase angle.

This feature is very useful in algebraic manipulation, for example, in removing imaginary components from denominators in fractions.

A.6 FOURIER SERIES AND FOURIER TRANSFORMATION

Any periodic waveform, such as the square wave in Figure A.6.1*a*, can be represented as a superposition of sinusoidal components (12–15) comprising a *fundamental frequency* $f_0 = 1/T_0$, where T_0 is the period of the waveform, plus the *harmonics* of f_0. That is,

$$y(t) = \frac{a_0}{2} + \sum_{n=1}^{\infty} [a_n \cos(2\pi n f_0 t) + b_n \sin(2\pi n f_0 t)] \tag{A.6.1}$$

or, alternatively,

$$y(t) = A_0 + \sum_{n=1}^{\infty} A_n \sin(2\pi n f_0 t + \phi_n) \tag{A.6.2}$$

where A_n is the *amplitude* of the component with frequency nf_0 and ϕ_n is its *phase angle*. The term A_0 is the *dc level*. This series is called a *Fourier series*, and the signal is a *Fourier synthesis* of the components. A few such components of the square wave are shown in Figure A.6.1*b*, and one can see how their sum begins to approximate the square wave itself.

The existence of the Fourier series makes it possible to represent a signal either in the *time domain*, as signal level versus time, or in the *frequency domain*, as the set of

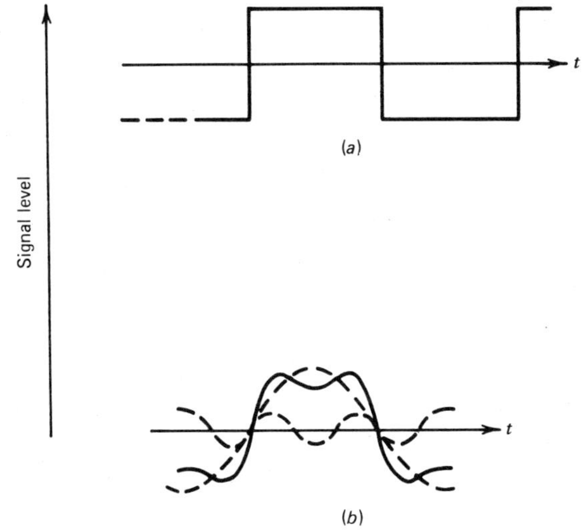

Figure A.6.1

(*a*) A square wave. (*b*) Two components (dashed) of the square wave, $\cos(2\pi f_0 t)$ and $-1/3\cos(6\pi f_0 t)$, and their sum (solid).

amplitudes and phase angles of the component sinusoids. Sometimes it is useful to dissect a time-domain signal into its components or to synthesize the time-domain signal from its components. Section 9.8 provides excellent illustrations of both cases. Our concern here is with the mechanism of interdomain conversions.

The *Fourier integral* (12–15) creates a frequency-domain function $H(f)$ from the time-domain relation $h(t)$:

$$H(f) = \int_{-\infty}^{\infty} h(t)\, e^{-j2\pi ft}\, dt \qquad \text{(A.6.3)}$$

This operation is called *Fourier transformation*, and $H(f)$ is the *Fourier transform* of $h(t)$. *Inverse transformation* by the integral:

$$h(t) = \int_{-\infty}^{\infty} H(f)\, e^{j2\pi ft}\, df \qquad \text{(A.6.4)}$$

allows generation of the time-domain function, given $H(f)$.

In most applications of interest to us, we must deal with waveforms that have been digitized (sampled) at a constant rate. For example, we might represent the current through a cell as a list of data points taken at evenly spaced times. Then a numeric algorithm, based essentially on the integral (A.6.2), would be used to generate the frequency-domain information, which would be a list of amplitudes and phase angles. The input to the algorithm is one period T_0 of the waveform represented as n points (usually 128, 256, 512, . . .). The output is a list of the amplitudes and phase angles for $n/2$ frequencies comprising the dc level, $1/T_0$, $1/2T_0$, . . ., $1/[(n/2) - 1]T_0$. The algorithm would probably be based on the *Fast Fourier transform* (FFT) technique (14, 15), and it might be implemented either in software in a computer or in a peripheral hardware device.

The FFT algorithms also have provision for inversion. One supplies amplitudes and phase angles for $n/2$ components at the frequencies given above, and the output is n evenly spaced time-domain points comprising one period.

The Fourier transformation has many properties that can be applied extremely usefully to signal processing. A simple illustration is noise reduction, or smoothing. Suppose one has a signal in which the information is at a few low frequencies, but there is noise at high frequencies. Transformation of signal yields the frequency spectrum, which one alters by setting the amplitudes of high-frequency components to zero. Inverse transformation yields smoothed time-domain data. Other operations allow integration and differentiation, correlations between two signals, or correlation of a signal with itself. Their details are beyond our scope, but they are covered thoroughly in several specialized sources (13–17).

A.7 REFERENCES

1. R. V. Churchill, "Operational Mathematics," 2nd ed., McGraw-Hill, New York, 1958.
2. G. Doetsch, "Laplace Transformation," Dover, New York, 1953.

3. H. Margenau and G. M. Murphy, "The Mathematics of Physics and Chemistry," 2nd ed., Van Nostrand, New York, 1956.
4. A. Erdelyi, W. Magnus, F. Oberhettinger, and F. Tricomi, "Tables of Integral Transforms," McGraw-Hill, New York, 1954.
5. "Handbook of Chemistry and Physics," 43rd ed., Chemical Rubber Publishing Company, Cleveland, Ohio, 1961.
6. G. E. Roberts and H. Kaufman, "Table of Laplace Transforms," Saunders, Philadelphia, 1966.
7. T. S. Peterson, "Calculus," Harper, New York, 1960.
8. W. Kaplan, "Advanced Calculus," 2nd ed., Addison-Wesley, Reading, Mass., 1973.
9. F. S. Acton, "Numerical Methods That Work," Harper & Row, New York, 1970, Chap. 1.
10. M. Abramowitz and I. Stegun, "Handbook of Mathematical Functions (AMS 55)," National Bureau of Standards, Washington, D.C., U.S. Government Printing Office, 1964.
11. R. V. Churchill, J. W. Brown, and R. F. Verhey, "Complex Variables and Applications," 3rd ed., McGraw-Hill, New York, 1976.
12. R. V. Churchill, "Fourier Series and Boundary Value Problems," McGraw-Hill, New York, 1941.
13. R. Bracewell, "The Fourier Transform and its Applications," McGraw-Hill, New York, 1965.
14. E. O. Brigham, "The Fast Fourier Transform," Prentice-Hall, Englewood Cliffs, N.J., 1974.
15. P. R. Griffiths, Ed., "Transform Techniques in Chemistry," Plenum, New York, 1978.
16. G. Horlick and G. M. Hieftje in "Contemporary Topics in Analytical Chemistry," Vol. 3, D. M. Hercules, G. M. Heiftje, and L. R. Snyder, Eds., Plenum, New York, 1978, Chap. 4.
17. J. W. Hayes, D. E. Glover, D. E. Smith, and M. W. Overton, *Anal. Chem.*, **45**, 277 (1973).

A.8 PROBLEMS

A.1 Show from the definition that

$$L\{\sin at\} = a/(a^2 + s^2)$$

A.2 Derive (A.1.17).

A.3 Find $L\{\sin at\}$ using (A.1.14).

A.4 Use convolution to find the inverse transform of $1/[s^{1/2}(s - 1)]$.

A.5 Use the Laplace transform method to solve for Y in the following cases:

(a) $Y'' + Y' = t^2$ with $Y(0) = 5$, $Y'(0) = -1$.

(b) $Y = 2\cos(t) - 2\int_0^t Y(\tau)\sin(t - \tau)\,d\tau$.

(c) $Y''' - Y'' - Y' + Y = \cos(t)$ with $Y(0) = Y'(0) = 0$, $Y''(0) = 5$.

The primes denote differentiation by t.

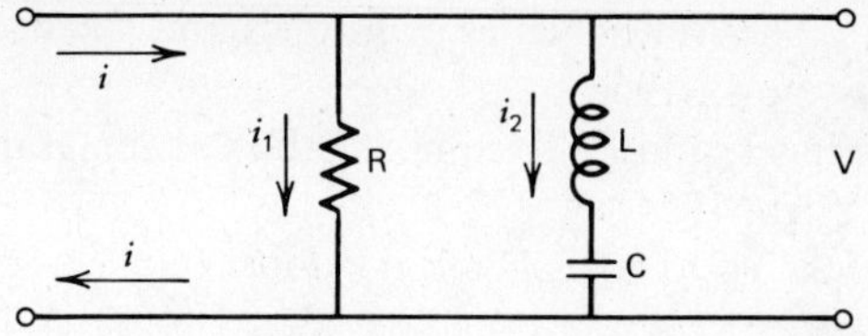

Figure A.8.1

A.6 A constant current i, is supplied to the network in Figure A.8.1 after $t = 0$. Before then $i = 0$ and $V = 0$. Find $V(t)$ for $t > 0$. Different combinations of R, L, and C will provide markedly different responses. Why?

A.7 Obtain a Taylor expansion of $\exp(ax)$ about $ax = 1$. Obtain the Maclaurin series. What approximations could be made for $\exp(ax)$ near $ax = 1$ and $ax = 0$?

A.8 Derive (A.3.3).

Appendix B
Digital Simulations of Electrochemical Problems

We have seen that the rate of an electrochemical process is affected by the rates at which reactants can be supplied to the electrode and products can be dispersed from it. Often the overall process is governed completely by these rates of mass transport and homogeneous chemical reaction. Nearly always, one can write the exact differential equations describing the transformations and movements of material, but many times they are solved in closed form with difficulty or not at all.

In recent years such problems have been attacked by variants of the *method of finite differences*. In one approach, a numerical model of the electrochemical system is set up within a digital computer, and the model is allowed to evolve by a set of algebraic laws derived from the differential equations defining material flow. In effect, one carries out a *digital simulation* of the experiment, and one can extract from it numeric representations of current functions, concentration profiles, potential transients, and so on.

We will outline the basic features of digital simulation here because the method has been very useful in a number of complex electrochemical problems involving complicated kinetic schemes, nonuniform current distributions at the working electrode, or spectroscopic-electrochemical interactions. Explicit simulation is a numerical approach to the solution of partial differential equations, but it is conceptually simpler than other numerical techniques. In addition, it often offers an aid to the intuitive grasp of the important processes in an electrochemical system. Several reviews have covered this topic, and they are recommended to the reader interested in more detail than we present below (1–5).

B.1 SETTING UP THE MODEL

B.1.1 The Discretized Model

By resorting to a simulation we are admitting our inability to cope with an electrochemical system that is described by continuous functions. Our mathematical skills cannot handle the calculus. So, we move backward one stage in sophistication and consider the electrolyte solution in terms of small, discrete volume elements. Throughout any element, the concentrations of all substances are regarded as uniform, but they vary from element to element. In general, we wish to study electrochemical experiments featuring linear diffusion to a planar electrode of area A. If edge diffusion is prevented, then even in the real system the concentration of any substance is constant over any plane parallel to the electrode. The concentration can vary only normal to its surface.

Thus, we construct a model featuring a sequence of volume elements extending away from the interface as shown in Figure B.1.1. The electrode surface is usually envisioned as being in the center of the first box, and each box j is taken to characterize the solution at a distance $x = (j - 1)\Delta x$ from the interface. If species A, B, . . . are present, their concentrations are $C_A(j)$, $C_B(j)$, ... What we have created, then, is a discrete model of the solution comprising arrays of concentrations, whose properties approximate those of the continuous system. Since we choose the size of Δx, it is said to be a *model variable*. The smaller we set Δx, the more elements will be needed in the arrays and the more refined our model will be.

It is also clear that if C_A at x_1 differs from that at a nearby point x_2 in any solution, diffusion will tend to equalize them. In addition, homogeneous reactions such as A + B → C might occur. Thus, the arrays $[C_A(j)]$, $[C_B(j)]$,. . . can represent the chemical system only for a limited period. There are laws of diffusion and reaction by which the concentrations interact with each other to transform themselves into different arrays that we could view as applying to the system at a later time.

Since we cannot digitally represent time in a continuous fashion, it must be broken into segments, too. Let one of them have a duration Δt. To model the system's evolution, we cast the reaction and mass transport laws into algebraic relations that will describe the changes these processes would bring about over one interval Δt. We first apply the relations to a set of starting concentration arrays, which characterize the initial condition of the system. This first application transforms the arrays to a different set that can be viewed as a picture of the system at time Δt. Applying the laws again on the new arrays yields a picture for $t = 2\Delta t$, and so on. The kth *iteration* of the laws of transformation therefore yields the model for $t = k\Delta t$. The time evolution of the continuous system is approximated by that of the model, and the discrepancy grows smaller as Δt, a second model variable, is reduced.

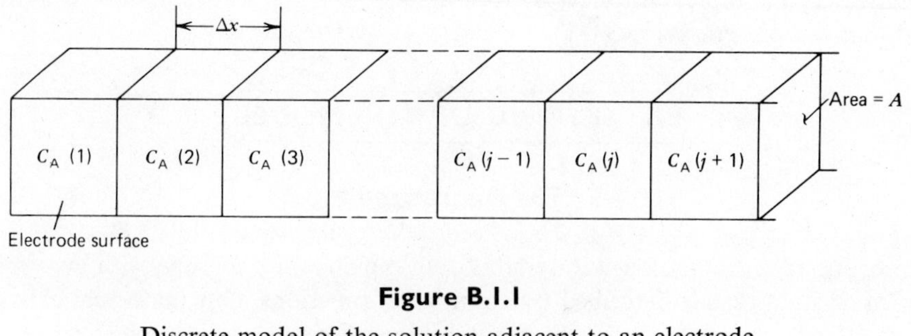

Figure B.1.1
Discrete model of the solution adjacent to an electrode.

B.1.2 Diffusion

The most common interactions between array elements are defined by Fick's laws of diffusion (see Section 4.3). The first law is

$$J(x, t) = -D\frac{\partial C(x, t)}{\partial x} \tag{B.1.1}$$

but the derivative is defined so that (B.1.1) can be recast:

$$J(x, t) = \lim_{\Delta x \to 0} -D\frac{[C(x + \Delta x, t) - C(x, t)]}{\Delta x} \tag{B.1.2}$$

The essence of the finite difference method is an assertion that Δx in a practical discrete model (a finite difference) can be made small enough that one can take

$$J(x, t) = -D\frac{[C(x + \Delta x, t) - C(x, t)]}{\Delta x} \tag{B.1.3}$$

or

$$J(x, t) = -\frac{D}{\Delta x}[C(x + \Delta x/2, t) - C(x - \Delta x/2, t)] \tag{B.1.4}$$

Now consider the second law:

$$-\frac{\partial C(x, t)}{\partial t} = \frac{\partial J(x, t)}{\partial x} \tag{B.1.5}$$

In finite difference form, it is

$$-\frac{C(x, t + \Delta t) - C(x, t)}{\Delta t} = \frac{J(x + \Delta x/2, t) - J(x - \Delta x/2, t)}{\Delta x} \tag{B.1.6}$$

Using (B.1.4) to substitute for the fluxes, one finds

$$C(x, t + \Delta t) = C(x, t) + \frac{D\Delta t}{\Delta x^2}[C(x + \Delta x, t) - 2C(x, t) + C(x - \Delta x, t)] \tag{B.1.7}$$

If one refers equation B.1.7 back to the model in Figure B.1.1, it is easy to see that this relation allows one to calculate the concentration in any box at $t + \Delta t$ from the concentrations in that box and its immediate neighbors at time t. Put in the jargon of the simulation, the meaning of (B.1.7) is that for any box j, the concentration resulting from iteration $k + 1$ is calculated from the concentrations produced in boxes $j - 1$, j, and $j + 1$ by iteration k. Thus,

$$C(j, k + 1) = C(j, k) + \frac{D\Delta t}{\Delta x^2}[C(j + 1, k) - 2C(j, k) + C(j - 1, k)] \tag{B.1.8}$$

Equation B.1.8 is the general law defining diffusion effects on any species in any box, except the first box. There the electrode boundary condition creates special circumstances that we will treat later.

These ideas are probably best consolidated by considering just how one might model a step experiment. The first stages are shown in Figure B.1.2. Initially the

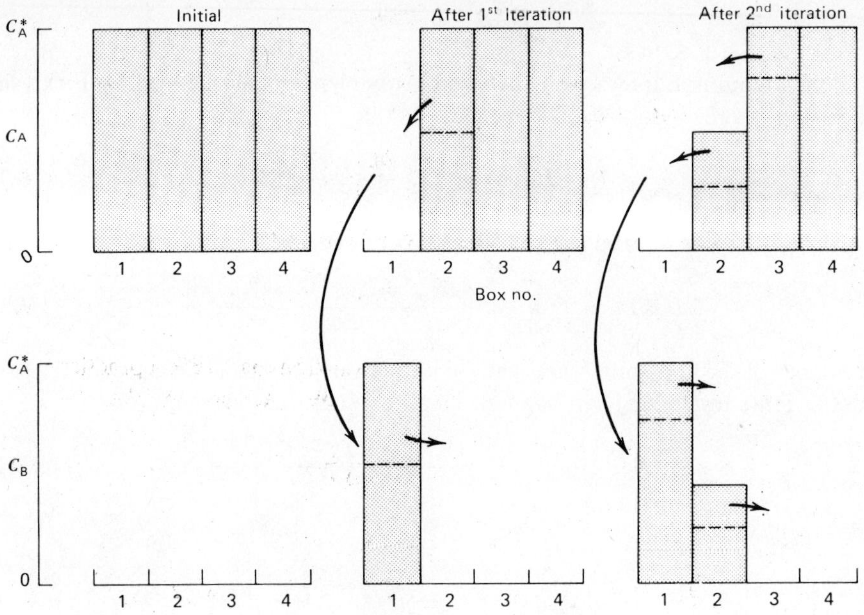

Figure B.1.2

Developing concentration profiles for a system undergoing the electrode reaction A + $ne \rightarrow$ B. Arrows show mass flow.

solution is uniform, and hence every box has a concentration $C_A(j, 0) = C_A^*$. If B is initially absent, $C_B(j, 0) = 0$. Suppose the step magnitude is large enough that the surface concentration of A is zero. Then in the first iteration, all the A in the first box is converted to B. Diffusion does not occur because the concentrations for $k = 0$ are uniform. Thus $C_A(1, 1) = 0$ and $C_B(1, 1) = C_A^*$, but for $j > 1$, $C_A(j, 1) = C_A^*$ and $C_B(j, 1) = 0$. In the second iteration, diffusion will alter the concentration in box 2, and fluxes of both A and B will cross the boundary between boxes 1 and 2. In order to maintain the interfacial condition that $C_A(1, k) = 0$ $(k > 0)$, the flux of A must be converted into B. Obviously this yields the current for the second iteration. Continuing the process will generate the concentration profiles and the current as functions of time.

B.1.3 Dimensionless Parameters

Note that if one wanted results for several starting concentrations C_A^*, an equal number of simulations would be needed. Consider, though, the effects of dividing (B.1.8) by C_A^*. If $f(j, k) = C(j, k)/C_A^*$, then

$$\boxed{f(j, k + 1) = f(j, k) + \mathbf{D}_M[f(j + 1, k) - 2f(j, k) + f(j - 1, k)]} \quad \text{(B.1.9)}$$

The constant $\mathbf{D}_M = D\Delta t/\Delta x^2$ is called the *model diffusion coefficient*. More will be said about it later. The f's, which are called *fractional concentrations*, are examples of *dimensionless parameters*.

Now suppose we simulate the step experiment again, but in place of concentrations we will substitute f_A's and f_B's. Equation B.1.9 describes the way in which these parameters are altered by diffusion, and the boundary condition is simply $f_A(1, k) = 0$ $(k > 0)$. Initially, $f_A = 1$ and $f_B = 0$ everywhere. Carrying out the simulation is straightforward, and one obtains the time evolutions of the fractional concentration profiles. The difference in using the dimensionless parameters is that *these profiles from a single simulation describe the characteristics of the experiment for every possible value of* C_A^*. To obtain the dimensioned profiles for a specific starting concentration, one need only multiply the f's by that value of C_A^*.

As an example of the usefulness of dimensionless parameters in the compact display of theoretical results, consider the homogeneous reaction $A \rightarrow B$ with rate constant k. The solution for the concentration of A at any time t is $C_A = C_A^* \exp(-kt)$ where C_A^* is the concentration at $t = 0$. Since the real variables are C_A and t, one might at first glance think of making a graphical display of the results by plotting C_A *vs.* t for different values of C_A^* and k. This involves a family of $m \times n$ curves, where m is the number of values of C_A^*, and n is the number of k values for each C_A^* (see Figure B.1.3). If one recognizes that the combination kt is a dimensionless parameter that always accounts wholly for the effects of k and t in this problem, then one could plot C_A *vs.* kt and represent the same information as before in m curves for different C_A^* values, as shown in Figure B.1.3*b*. If one further realizes that C_A/C_A^* is also a useful dimensionless parameter, then the equation can be written $f_A = \exp(-a)$, where $f_A = C_A/C_A^*$ and $a = kt$. Hence, a plot of f_A *vs.* a shows all of the desired information in a single curve (Figure B.1.3*c*), which is the essential shape function of the system. It is an example of a *working curve*.

Solving differential equations in terms of dimensionless parameters generally does yield solutions that characterize whole families of specific experimental situations. This is a magnificent asset, especially where numerical solutions are required, and therefore the use of dimensionless variables has become very common.

A confusing aspect of these parameters is their tendency to combine the effects of more than one observable, so that one has difficulty in mentally separating the effects of a single observable individually. For example, the abscissa of Figure B.1.3*c* deals with changes in k, or in t, or in both. It is easier to understand the working curve by thinking in terms of a given experiment, which always involves holding some variables constant.

One might study the reaction $A \rightarrow B$ by looking at the decay of A continuously with time. A given experiment would involve constant C_A^* and k values; hence the working curve can be regarded as a scaled decay function of concentration C_A *vs.* time t. We might even call f the *dimensionless concentration* and a the *dimensionless time*.

Alternatively, we might study $A \rightarrow B$ by measuring C_A at a single time t, which has a fixed known value. In that experimental context, the working curve in Figure B.1.3*c* would be seen as a plot of the concentration of A remaining (relative to the initial concentration) at the sampling time t *vs.* a scaled rate constant k. For that purpose, a could be called a *dimensionless rate constant*. Thus, the interpretation we apply to a working curve depends on the specific experiment at hand.

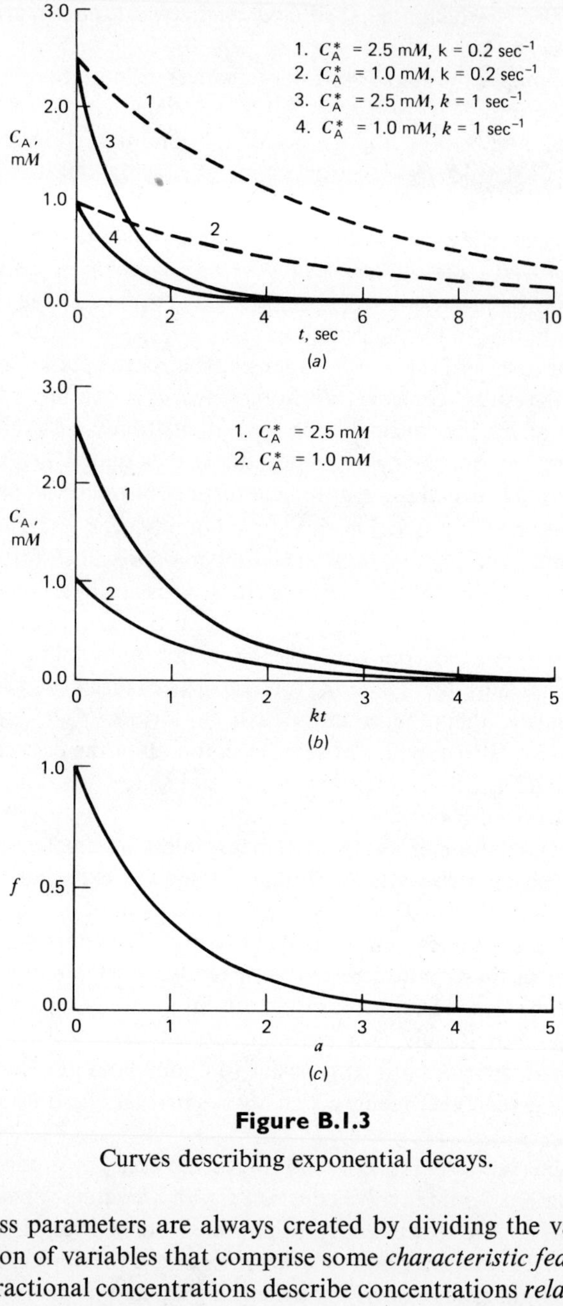

Figure B.1.3
Curves describing exponential decays.

Dimensionless parameters are always created by dividing the variables of interest by a combination of variables that comprise some *characteristic feature* of the system. For example, fractional concentrations describe concentrations *relative* to a characteristic concentration C^*_A. The parameter a can be understood in this way by recognizing that $1/k$ is the average lifetime of A (see Problem 3.6). Thus a is the ratio of the observation time to the lifetime of species A. Understood in this way, a is more than a numerical value; it is a guide to the design and interpretation of experiments. This capacity for expressing relationships between observables and characteristic features is a powerful aspect of dimensionless parameters, and with practice it can aid one's intuition in very useful ways.

B.1.4 Time

Time, t, measured into the simulation is $k\Delta t$, where k is the iteration number. Since Δt is a model variable, it is our choice, and in choosing its value we are making the equivalent statement that some known characteristic time t_k (which might be a step width, a scanning time, or some similar characteristic experimental duration) will be broken into $\mathbf{l}$ iterations in our model. Thus,

$$\Delta t = t_k/\mathbf{l} \tag{B.1.10}$$

Either $\mathbf{l}$ or Δt can be selected arbitrarily as the model variable, though it usually is more convenient to work in terms of $\mathbf{l}$. Thus, we might carry out simulations of a step experiment with duration t_k by breaking t_k into 100, 1000, or 10,000 iterations at will. All the simulations will yield equivalent results within their individual abilities to approximate events, but the larger the value of $\mathbf{l}$, the higher is the quality of the simulation. However, a larger $\mathbf{l}$ requires more computations too, so a compromise is chosen. Typically $\mathbf{l}$ is 100 to 1000 iterations/t_k.

Note also that time is easily evaluated in dimensionless terms by expressing it as the ratio t/t_k. Thus,

$$\boxed{\frac{t}{t_k} = \frac{k}{\mathbf{l}}} \tag{B.1.11}$$

B.1.5 Distance

The center of box j is at a distance $(j - 1)\Delta x$ from the electrode surface, and we have noted earlier that Δx is also our choice. However, there is a limit to its smallness. It turns out that $\mathbf{D}_M$, the model diffusion coefficient, cannot exceed 0.5. Otherwise the finite difference calculation will not be stable. A physical reason for this is that Δx and Δt are not independent. In our treatment of diffusion, we implicitly assumed that within a period Δt, material could diffuse only between neighboring boxes. If we try to set Δx too small for a given Δt, this assumption becomes inadequate, and the simulation diverges from reality.

Note also that, given Δt, $\mathbf{D}_M$ is an equivalent model variable to Δx. Instead of specifying Δx, it usually is more convenient to specify the dimensionless $\mathbf{D}_M$. Then,

$$\Delta x = \left(\frac{D\Delta t}{\mathbf{D}_M}\right)^{1/2} \tag{B.1.12}$$

The larger we choose $\mathbf{D}_M$, the smaller is Δx and the better is the model; hence $\mathbf{D}_M$ is usually set at a high constant value such as 0.45. Substituting for Δt, one obtains

$$\boxed{\Delta x = \left(\frac{Dt_k}{\mathbf{D}_M\mathbf{l}}\right)^{1/2}} \tag{B.1.13}$$

and it becomes clear that $\mathbf{l}$ is really the determinant of both temporal and spatial resolution for any simulation.

The distance of the center of box j from the electrode can now be written

$$x(j) = (j - 1)\left(\frac{Dt_k}{\mathbf{D}_M\mathbf{l}}\right)^{1/2} \tag{B.1.14}$$

The convenient dimensionless distance is obtained by placing the real system variables on the left and the model variables on the right:

$$\boxed{\chi(j) = \frac{x(j)}{(Dt_k)^{1/2}} = \frac{j - 1}{(\mathbf{D}_M\mathbf{l})^{1/2}}} \tag{B.1.15}$$

The expression $(j - 1)/(\mathbf{D}_M\mathbf{l})^{1/2}$ allows one to calculate $\chi(j)$ easily from the simulation parameters, and the expression $x(j)/(Dt_k)^{1/2}$ allows one to correlate the properties of box j with the properties of the experimental solution segment situated at a real distance x from a physical electrode. Note that $\chi(j)$ is the ratio of distance to the diffusion length $(Dt_k)^{1/2}$ for the characteristic time t_k.

Recalling the step simulation we considered above, we see that the most efficient way to use the calculation is to report the concentration profiles as functions of f_A and f_B *vs.* χ for various values of t/t_k. These curves would then fully characterize every possible electrochemical experiment satisfying the initial conditions and the boundary conditions. Given specific values of C_A^*, t_k, and D, it is a simple matter to convert the curves into functions of C_A and C_B *vs.* x for various values of t.

B.1.6 Current

In general, there is a flux of electroactive species across the boundary between boxes 1 and 2. For species A in iteration $k + 1$, it is

$$-J_A^{1,2}(k + 1) = DC_A^* \frac{[f_A(2, k) - f_A(1, k)]}{\Delta x} \tag{B.1.16}$$

In the absence of other processes this flux will effect a concentration change in box 1. However, there is in every experiment a boundary condition that dictates the circumstances at the electrode surface, and it must be maintained. Consider the example we have taken above, where $f_A(1, k) = 0$ at all $k = 0$. Since the surface concentration of A is zero, a flux of A will always move toward the electrode across the boundary between boxes 1 and 2. Maintenance of the surface condition demands that this flux be eliminated, but the only way to eliminate it is to convert it electrochemically to B. For iteration $k + 1$, then, the current is defined by $J_A^{1,2}(k + 1)$. We can write it:

$$i(k + 1) = \frac{nFADC_A^* f_A(2, k)}{\Delta x} \tag{B.1.17}$$

Substituting for Δx yields

$$i(k + 1) = \frac{nFAD^{1/2}C_A^* f_A(2, k)(\mathbf{D}_M\mathbf{l})^{1/2}}{t_k^{1/2}} \tag{B.1.18}$$

We can obtain a dimensionless current $Z(k)$ by following the standard recipe: Rearrange (B.1.18) so that experimental variables are on the left and model variables are on the right:

$$Z(k+1) = \frac{i(k+1)t_k^{1/2}}{nFAD^{1/2}C_A^*} = (\mathbf{D}_M\mathbf{l})^{1/2}f_A(2, k) \tag{B.1.19}$$

This definition of Z relates the actual current to the Cottrell current expected at time t_k (see Problem B.1).

The current for the first iteration is calculated differently because there is no flux. Instead the current flows because we first establish the surface condition by eliminating A from box 1 at that time. The number of moles electrolyzed in the Δt interval is $\Delta x A C_A^*$; hence the current is

$$i(1) = \frac{nFAC_A^*\Delta x}{\Delta t} = \frac{nFAC_A^*D^{1/2}\mathbf{l}^{1/2}}{t_k^{1/2}\mathbf{D}_M^{1/2}} \tag{B.1.20}$$

Thus,

$$Z(1) = (\mathbf{l}/\mathbf{D}_M)^{1/2} \tag{B.1.21}$$

To what time should we assign $Z(k)$? Since our current calculation really involves dividing the integral charge passed during an iteration by the duration of the iteration, it probably is more appropriate to assign the current to the midpoint, rather than to the end, of the iteration. Thus, we say that the dimensionless current $Z(k)$ flowed at $t/t_k = (k - 0.5)/\mathbf{l}$.

B.1.7 Thickness of the Diffusion Layer

In doing these calculations one needs to know how many boxes are needed in the simulation. A rule of thumb provides a safe answer: Any experiment that has proceeded for time t will alter the solution from its bulk character for a distance no larger than about $6(Dt)^{1/2}$. Thus,

$$j_{\max} - 1 \cong j_{\max} \cong \frac{6(Dt)^{1/2}}{\Delta x} \tag{B.1.22}$$

and

$$j_{\max} \cong 6(\mathbf{D}_M k)^{1/2} \tag{B.1.23}$$

Since $\mathbf{D}_M \leq 0.5$, one never needs to make calculations for any more than $4.2k^{1/2}$ boxes during iteration k.

B.1.8 Diffusion Coefficients

Note that the parameter $\mathbf{D}_M$ exists in diffusion expressions for each species. Each $\mathbf{D}_M$ contains the diffusion coefficient for the pertinent species, but since Δx and Δt are constants and since $D_A \neq D_B \neq D_C \ldots$, the $\mathbf{D}_M$'s cannot all be equal. We should write them as $\mathbf{D}_{M,A}$, $\mathbf{D}_{M,B}$, ... This obviously complicates the model, therefore, one frequently makes the assumption that all diffusion coefficients are equal. Then a single value of $\mathbf{D}_M$ suffices.

When this procedure is not satisfactory, one must take explicit account of the differences in D values by using different $\mathbf{D}_M$ values for each species. We have already seen that one of these parameters is equivalent to a model variable, and hence it can be chosen at will. The others then are determined by the fact that

$$\mathbf{D}_{M,i}/\mathbf{D}_{M,\mathrm{A}} = D_i/D_{\mathrm{A}} \tag{B.1.24}$$

This procedure ensures that the model will behave diffusively as the real system does.

B.2 AN EXAMPLE

Figure B.2.1 is a FORTRAN listing of an actual simulation. The treated problem is the Cottrell experiment which was solved analytically in Section 5.2.1. An electroreactant A is uniformly distributed initially, but at $t = 0$, a potential step is applied to force the surface concentration of A to zero by converting it faradaically to species B.

The simulation starts by setting up arrays to represent the fractional concentrations of A and B in each box. There are "old" and "new" arrays for each species that are related by rules discussed later. In addition, an array for the current-time curve is declared, and the model variables $\mathbf{l}$, $\mathbf{D}_{M,\mathrm{A}}$, and $\mathbf{D}_{M,\mathrm{B}}$ are fixed. Since $\mathbf{l}$ is only 100, this simulation has relatively low resolution. At most, 42 boxes will represent the diffusion layer. The concentration arrays are initialized to reflect the uniform starting concentration of A and the absence of B.

Upon beginning the first iteration, the new concentration arrays are calculated from the old according to the laws of diffusion. The boundary conditions require that species A be zero in the first box; hence **FANEW**(1) is then reset to zero and **FBNEW**(1) is incremented by an equal amount to reflect the faradaic conversion. The current $Z(k)$ is calculated from the amount of A converted. These operations conclude the chemical activity for the iteration.

If $k \neq 50$, the new arrays are reassigned to the old ones in preparation for the next iteration, then k is incremented and the chemical processes are applied again. When $k = 50$ the concentration profiles are typed out. The distance parameter $\chi(j)$ is calculated and printed, along with the fractional concentrations in each box.

When $k = \mathbf{l}$, the program exits the iteration loop and prints the current-time curve. The parameter T is the value of t/t_k associated with a given $Z(k)$, and **ZCOTT** is the dimensionless current calculated from the Cottrell equation (5.2.11). It is an exact solution to this problem, and is easily shown by rearrangement of (5.2.11) to be

$$Z_{\mathrm{Cott}} = \left[\pi^{1/2}\left(\frac{t}{t_k}\right)^{1/2}\right]^{-1} \tag{B.2.1}$$

The ratio $\mathbf{R} = Z/Z_{\mathrm{Cott}}$ is a comparison that allows an evaluation of the simulation.

Figure B.2.2 is a display of this ratio as a function of time through the simulation. The numbers by the points are the corresponding values of k. Ideally, $\mathbf{R}$ is exactly unity always. Figure B.2.2 shows that large errors occur in the first few iterations, as one must expect from the coarse nature of the model at that stage. However, by the tenth iteration, the error is only a few percent, and it falls steadily. For $t/t_k = 0.995$, there is only an error of 0.2%. Better results could have been obtained with a larger value for $\mathbf{l}$.

```
C     SET UP ARRAYS AND MODEL VARIABLES

      DIMENSION FAOLD(100),FANEW(100),FBOLD(100),FBNEW(100),Z(100)
      L=100
      DMA=0.45
      DMB=DMA


C     INITIAL CONDITIONS

      DO 10 J=1,100
      FAOLD(J)=1
      FANEW(J)=1
      FBOLD(J)=0
      FBNEW(J)=0
   10 CONTINUE
      K=0


C     START OF ITERATION LOOP

 1000 K=K+1


C     DIFFUSION BEYOND THE FIRST BOX

      JMAX=4.2*SQRT(FLOAT(K))
      DO 20 J=2,JMAX
      FANEW(J)=FAOLD(J)+DMA*(FAOLD(J-1)-2*FAOLD(J)+FAOLD(J+1))
      FBNEW(J)=FBOLD(J)+DMB*(FBOLD(J-1)-2*FBOLD(J)+FBOLD(J+1))
   20 CONTINUE


C     DIFFUSION INTO THE FIRST BOX

      FANEW(1)=FAOLD(1)+DMA*(FAOLD(2)-FAOLD(1))
      FBNEW(1)=FBOLD(1)+DMB*(FBOLD(2)-FBOLD(1))


C     FARADAIC CONVERSION AND CURRENT FLOW

      Z(K)=SQRT(L/DMA)*FANEW(1)
      FBNEW(1)=FBNEW(1)+FANEW(1)
      FANEW(1)=0


C     TYPE OUT CONCENTRATION ARRAYS FOR K=50

      IF (K.NE.50) GO TO 100
      TYPE
      TYPE
      DO 30 J=1,JMAX
      X=(J-1)/SQRT(DMA*L)
      TYPE X,FANEW(J),FBNEW(J)
   30 CONTINUE
      TYPE
      TYPE


C     SET UP OLD ARRAYS FOR NEXT ITERATION

  100 DO 40 J=1,JMAX
      FAOLD(J)=FANEW(J)
      FBOLD(J)=FBNEW(J)
   40 CONTINUE


C     RETURN FOR NEXT ITERATION IF K<L

      IF (K.LT.L) GO TO 1000


C     TYPE OUT CURRENT--TIME CURVE

      TYPE
      TYPE
      DO 50 K=1,L
      T=(K-0.5)/L
      ZCOTT=1/SQRT(3.141592*T)
      R=Z(K)/ZCOTT
      TYPE T,Z(K),ZCOTT,R
   50 CONTINUE
      END
```

Figure B.2.1

A digital simulation program for the Cottrell experiment.

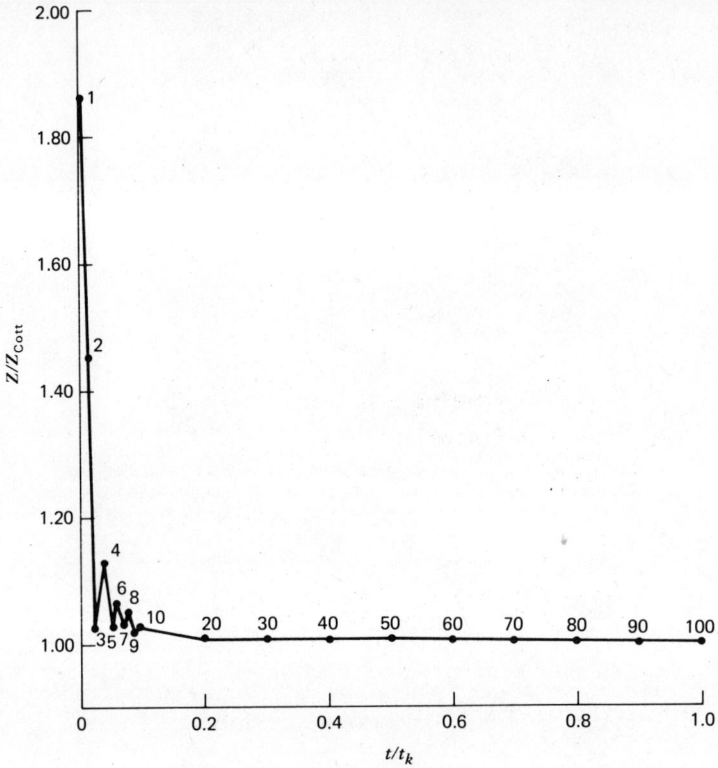

Figure B.2.2

Results of the simulation in Figure B.2.1 for $\mathbf{l} = 100$ and $\mathbf{D}_M = 0.45$. Simulated current parameter Z divided by analytical solution. Numbers by points are iteration numbers.

Figure B.2.3 offers the concentration profiles for $t/t_k = 0.5$. The points are from the simulation, and the curves are the analytical results from equation 5.2.13 and its complement. The agreement is clearly quite good.

Note that the sum of fractional concentrations for A and B is unity in each box. This rule always holds when $\mathbf{D}_{M,\mathrm{A}} = \mathbf{D}_{M,\mathrm{B}}$ (see equation 5.4.28); and it is useful for diagnosing programming errors leading to losses or gains of material. One of the problems with digital simulations is that errors are difficult to detect, so it is important to use every possible safeguard.

B.3 INCORPORATING HOMOGENEOUS KINETICS

If the only homogeneous dynamics of concern were the diffusion processes, simulation would find much less use than it does. Its utility is most appreciated when the electrochemical process is coupled to one or more homogeneous chemical reactions. Then the differential equations describing the system can easily become too difficult for an analytical solution.

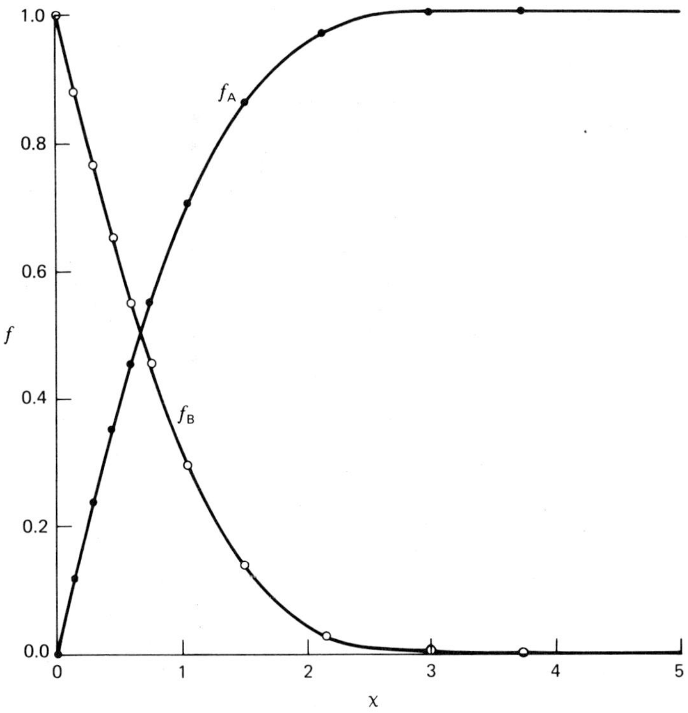

Figure B.2.3

Concentration profiles from the simulation program in Figure B.2.1 for $\mathbf{l} = 100$, $\mathbf{D}_M = 0.45$, $t/t_k = 0.5$. Points are simulated; curves are analytical.

B.3.1 Unimolecular Reactions

Consider the system in which an electrode reaction is followed by a unimolecular conversion:

$$\mathrm{A} + e \longrightarrow \mathrm{B} \text{ (at the electrode)} \tag{B.3.1}$$

$$\mathrm{B} \xrightarrow{k_1} \mathrm{C} \text{ (in solution)} \tag{B.3.2}$$

The differential equations describing B and C must account for both diffusion and reaction (see Chapter 11). For example,

$$\frac{\partial C_{\mathrm{B}}(x, t)}{\partial t} = D_{\mathrm{B}} \frac{\partial^2 C_{\mathrm{B}}(x, t)}{\partial x^2} - k_1 C_{\mathrm{B}}(x, t) \tag{B.3.3}$$

The first term on the right is just Fick's second law, and we have seen that its finite difference representation is given by (B.1.6). Thus, we can immediately write the finite difference analog to (B.3.3) as

$$C_{\mathrm{B}}(x, t + \Delta t) = C_{\mathrm{B}}(x, t) + \mathbf{D}_{M,\mathrm{B}}[C_{\mathrm{B}}(x + \Delta x, t) - 2C_{\mathrm{B}}(x, t) + C_{\mathrm{B}}(x - \Delta x, t)] - k_1 \Delta t \cdot C_{\mathrm{B}}(x, t) \tag{B.3.4}$$

Dividing by C_A^* and introducing the notation of the simulation, we obtain

$$f_B(j, k + 1) = f_B(j, k) + \mathbf{D}_{M,B}[f_B(j + 1, k) - 2f_B(j, k) + f_B(j - 1, k)]$$
$$- \frac{k_1 t_k}{\mathbf{l}} f_B(j, k) \tag{B.3.5}$$

Equation B.3.5 allows a one-step accounting of diffusion and kinetic effects for the B array during iteration $k + 1$.

In practice, the simultaneous calculation of diffusion and kinetic manifestations is not usually made. Instead, one ordinarily calculates the effects of diffusion first, without considering the reaction, and then one allows the reaction to modify the concentration arrays that were created just beforehand by the diffusion equations. We could diagram the operation as follows:

$$[f_B(j, k - 1)] \;\Big|\; \xrightarrow{\text{Diffusion}} [f'_B(j, k)] \xrightarrow{\text{Kinetics}} [f_B(j, k)] \;\Big|\; \rightarrow$$
$$\longleftarrow \text{Iteration } k \longrightarrow$$

Thus,

$$f'_B(j, k) = f_B(j, k - 1) + \mathbf{D}_{M,B}[f_B(j + 1, k - 1) - 2f_B(j, k - 1) + f_B(j - 1, k - 1)] \tag{B.3.6}$$

and

$$\boxed{f_B(j, k) = f'_B(j, k) - \frac{k_1 t_k}{\mathbf{l}} f'_B(j, k)} \tag{B.3.7}$$

This stepwise calculation is not rigorous, of course, but neither is (B.3.5). Still, either approach will yield good results if the model is sufficiently refined, that is, if $\mathbf{l}$ is large enough. There are two advantages to the ordered calculation:

1. Since kinetic effects appear separately from diffusion in the computer program, it is easier to make programming changes to account for different mechanistic details.
2. Equation B.3.5 can easily produce negative values of $f_B(j, k + 1)$ if kinetic effects are very important. Sometimes it is difficult to predict this behavior, and it is hard to apportion the available mass when it does happen. Mass allocation is more straightforward in the sequential approach because the total destruction of a reactant can only happen in a kinetic step.

Equation B.3.7 shows that the dimensionless kinctic parameter is $k_1 t_k$. Since it must be given a numeric value for any particular simulation, the results of that simulation are valid for all those experiments, but only those experiments, for which the product of t_k and k_1 is equal to the predetermined constant. Following the argument of Section B.1.3, we see that this dimensionless parameter is the ratio of the characteristic time t_k to the lifetime of B, which is $1/k_1$. In general, the effects of the unimolecular decay of B will hardly be felt in the experiment if $k_1 t_k$ is much less than unity, and they will be completely manifested for $k_1 t_k$ much greater than unity. The finite difference method is based on an approximation of true derivatives, and hence one can expect (B.3.7) to supply an accurate accounting of the kinetic effect only when $k_1 t_k/\mathbf{l}$ is not too large. Otherwise, the extent of decay per unit of time resolution is excessive. The upper limit of the most useful modelling range is therefore $k_1 t_k \sim \mathbf{l}/10$. The lower limit is reached when the kinetic perturbation no longer registers an experimentally significant impact.

B.3.2 Bimolecular Reactions

Now consider a bimolecular complication to an electrode process:

$$\text{A} + e \longrightarrow \text{B (at the electrode)} \tag{B.3.8}$$

$$\text{B} + \text{B} \xrightarrow{k_2} \text{C (in solution)} \tag{B.3.9}$$

For B, we have

$$\frac{\partial C_{\text{B}}(x, t)}{\partial t} = D_{\text{B}} \frac{\partial^2 C_{\text{B}}(x, t)}{\partial x^2} - k_2 C_{\text{B}}(x, t)^2 \tag{B.3.10}$$

Transforming this to the notation of the simulation exactly as above, we obtain the analog to (B.3.5):

$$f_{\text{B}}(j, k+1) = f_{\text{B}}(j, k) + \mathbf{D}_{M,\text{B}}[f_{\text{B}}(j+1, k) - 2f_{\text{B}}(j, k) + f(j-1, k)] - \frac{k_2 t_k C_{\text{A}}^*}{\mathbf{l}} f_{\text{B}}(j, k)^2 \tag{B.3.11}$$

Again, it is useful to handle diffusion and homogeneous reaction sequentially, so we split (B.3.11) into two parts. The diffusion effects are registered by (B.3.6) and the changes in concentration due to reaction are given by

$$\boxed{f_{\text{B}}(j, k) = f_{\text{B}}'(j, k) - \frac{k_2 t_k C_{\text{A}}^*}{\mathbf{l}} f_{\text{B}}'(j, k)^2} \tag{B.3.12}$$

The dimensionless parameter pertaining to the second-order process is clearly $k_2 t_k C_{\text{A}}^*$. Its value must be fixed for a given simulation, and several successive simulations must be carried out to show the effects of variations in $k_2 t_k C_{\text{A}}^*$. For reasons equivalent to those outlined above, the most useful modelling range for $k_2 t_k C_{\text{A}}^*$ is the interval below $\mathbf{l}/10$.

B.4 BOUNDARY CONDITIONS FOR VARIOUS TECHNIQUES

So far, we have considered only a step to a potential where the electroreactant is brought to the electrode at the mass-transfer-limited rate. We use a particularly simple boundary condition in that case: $C_{\text{A}}(0, t) = f_{\text{A}}(1, k) = 0$. In the example of Figure B.2.1, this boundary condition is enforced in the section headed "Faradaic Conversion and Current Flow." Other situations demand other conditions, and we outline some of them here.

B.4.1 Potential Steps in a Reversible System

Suppose the electrode reaction

$$A + ne \rightleftharpoons B \tag{B.4.1}$$

is reversible, so that the equation:

$$E = E^{0'} + \frac{RT}{nF} \ln \frac{C_A(0, t)}{C_B(0, t)} \tag{B.4.2}$$

always applies. In terms of fractional concentrations,

$$E = E^{0'} + \frac{RT}{nF} \ln \frac{f_A(1, k)}{f_B(1, k)} \tag{B.4.3}$$

which can be rearranged to give a dimensionless potential parameter:

$$\mathbf{E}_{\text{norm}} = \frac{(E - E^{0'})nF}{RT} = \ln \frac{f_A(1, k)}{f_B(1, k)} \tag{B.4.4}$$

or

$$\boxed{\frac{f_A(1, k)}{f_B(1, k)} = \exp(\mathbf{E}_{\text{norm}})} \tag{B.4.5}$$

To simulate, for example, an experiment in which the initial system was a uniform solution of A and a step was made to potential E, one would set up the initial conditions just as we have done before; then the ratio of $f_A(1, k)$ and $f_B(1, k)$ would be maintained at the value dictated through (B.4.5) by a value of $\mathbf{E}_{\text{norm}}$ corresponding to the step potential E. This value of $\mathbf{E}_{\text{norm}}$ is therefore a model variable, and a separate simulation would have to be carried out for each desired value of E. Note that the normalized potential E_{norm} is simply the value of the potential E expressed as the energy difference between E and $E^{0'}$ in units of kT.

Maintaining condition (B.4.5) usually will cause a diffusive flux across the boundary between boxes 1 and 2. This flux alters the ratio $f_A(1)/f_B(1)$ after the diffusive step in the simulation; therefore, the ratio must be reestablished by converting species A into B or vice versa. The amount converted gives rise to a current that is calculated as the dimensionless parameter $Z(k)$ by the methods discussed in Section B.1.6.

B.4.2 Potential Sweeps in a Reversible System

If we want to apply the program:

$$E = E_i + vt \tag{B.4.6}$$

to the system expressed in (B.4.1), then we have

$$\mathbf{E}_{\text{norm}} = \frac{(E_i - E^{0'})nF}{RT} + \frac{nFvt}{RT} \tag{B.4.7}$$

and (B.4.5) still holds. The first term is a normalized initial potential $\mathbf{E}_{i,\text{norm}}$ which would have to be specified as a model variable to the simulation. The second term describes the effects of the sweep, and its value would change as the simulation evolved; that is, the second term is a function of the iteration number k.

The specific function is obtained by evaluating it via (B.1.11):

$$\mathbf{E}_{\text{norm}} = \mathbf{E}_{i,\text{norm}} + \frac{nFvt_k}{RT} \cdot \frac{k}{\mathbf{l}} \tag{B.4.8}$$

However, we have yet to define the known time t_k corresponding to $\mathbf{l}$ iterations. Several choices could be made, but probably the most convenient one is to let t_k be the time required to scan RT/nF volts. Then $t_k = RT/nFv$, and

$$\boxed{\mathbf{E}_{\text{norm}} = \mathbf{E}_{i,\text{norm}} + \frac{k}{\mathbf{l}}} \tag{B.4.9}$$

At the end of each iteration, the value of $\mathbf{E}_{\text{norm}}$ is calculated from (B.4.9) and the corresponding ratio $f_{\text{A}}(1, k)/f_{\text{B}}(1, k)$ is computed from (B.4.5). Then A and B are interconverted to establish that ratio, and the current parameter $Z(k)$ is computed from the amount of material converted. To simulate a sweep from an initial potential E_i to a final potential E_f, the total number of iterations required is $|E_f - E_i|nF\mathbf{l}/RT$, since $\mathbf{l}$ iterations deals only with a potential change of RT/nF volts.

B.4.3 Accounting for Heterogeneous Kinetics

For an electrode reaction:

$$\text{A} + ne \underset{k_b}{\overset{k_f}{\rightleftharpoons}} \text{B} \tag{B.4.10}$$

where

$$k_f = k^0 \exp\left[\frac{-\alpha nF(E - E^{0\prime})}{RT}\right] \tag{B.4.11}$$

$$k_b = k^0 \exp\left[\frac{(1 - \alpha)nF(E - E^{0\prime})}{RT}\right] \tag{B.4.12}$$

the current is always given by

$$\frac{i}{nFA} = k_f C_{\text{A}}(0, t) - k_b C_{\text{B}}(0, t) \tag{B.4.13}$$

In terms of simulation variables, we obtain

$$\frac{it_k^{1/2}}{nFAD_{\text{A}}^{1/2}C_{\text{A}}^*} = Z(k) = \left(\frac{k_f t_k^{1/2}}{D_{\text{A}}^{1/2}}\right) f_{\text{A}}(1, k) - \left(\frac{k_b t_k^{1/2}}{D_{\text{A}}^{1/2}}\right) f_{\text{B}}(1, k) \tag{B.4.14}$$

where the clusters $(k_f t_k^{1/2}/D_{\text{A}}^{1/2})$ and $(k_b t_k^{1/2}/D_{\text{A}}^{1/2})$ are dimensionless rate constants. The dimensionless current can be calculated for any iteration from (B.4.14) if those parameters are specified.

To see how to make the specification, let us recast (B.4.11) and (B.4.12) into the appropriate forms:

$$\frac{k_f t_k^{1/2}}{D_A^{1/2}} = \left(\frac{k^0 t_k^{1/2}}{D_A^{1/2}}\right) \exp(-\alpha \mathbf{E}_{\text{norm}}) \tag{B.4.15}$$

$$\frac{k_b t_k^{1/2}}{D_A^{1/2}} = \left(\frac{k^0 t_k^{1/2}}{D_A^{1/2}}\right) \exp[(1 - \alpha)\mathbf{E}_{\text{norm}}] \tag{B.4.16}$$

Now we see that both rate constants can be calculated from $\mathbf{E}_{\text{norm}}$ when two dimensionless model variables are supplied to the simulation. One must specify the value of the *transfer coefficient* α, which is dimensionless in itself, and one must supply a value for $(k^0 t_k^{1/2}/D_A^{1/2})$, which is a *dimensionless intrinsic rate constant*. The normalized potential $\mathbf{E}_{\text{norm}}$ is treated exactly as in Sections B.4.1 and B.4.2, for steps and sweeps, respectively.

In a real simulation, the diffusion process would be allowed to take place within an iteration, and it would yield values for temporary $f_A(1)$ and $f_B(1)$. The current parameter $Z(k)$ would be calculated from (B.4.14), and an amount of species A equivalent to $(k_f t_k^{1/2}/D_A^{1/2}) f_A(1)$ would be converted to B, while a quantity of species B equivalent to $(k_b t_k^{1/2}/D_A^{1/2}) f_B(1)$ would be converted to A. These conversions would yield the final values $f_A(1, k)$ and $f_B(1, k)$ for the iteration.

B.4.4 Controlled Current

For the electrode reaction (B.4.1), the application of controlled current is equivalent to exercising control over the gradient in the concentration of A at the electrode surface, since

$$\frac{i}{nFA} = -J_A(0, t) = D_A\left(\frac{\partial C_A(x, t)}{\partial x}\right)_{x=0} \tag{B.4.17}$$

To convert this expression to the finite difference notation relevant to the simulation, we make the assumption that the concentration profile is linear from the center of box 1 (the electrode surface) to the center of box 2. Then,

$$\frac{i}{nFA} = D_A C_A^* \frac{[f_A(2, k) - f_A(1, k)]}{\Delta x} \tag{B.4.18}$$

Controlling the current in a real experiment is therefore equivalent to controlling the difference in fractional concentrations between boxes 1 and 2 in the model.

Now we rearrange (B.4.18) to obtain the usual current parameter:

$$Z = \frac{i t_k^{1/2}}{nFAD_A^{1/2} C_A^*} = \frac{D_A^{1/2} t_k^{1/2}}{\Delta x}[f_A(2, k) - f_A(1, k)] \tag{B.4.19}$$

Substituting from (B.1.13) gives

$$Z = (\mathbf{D}_{M,A}\mathbf{l})^{1/2}[f_A(2, k) - f_A(1, k)] \tag{B.4.20}$$

If the current has a constant magnitude, it is most convenient in this treatment to define the known time t_k as the transition time given by the Sand equation (7.2.14) for species A. Thus, **l** iterations correspond to

$$\tau^{1/2} = t_k^{1/2} = \frac{nFAD_A^{1/2}C_A^*\pi^{1/2}}{2i} \qquad \text{(B.4.21)}$$

and the current parameter is

$$Z = \frac{\pi^{1/2}}{2} = (\mathbf{D}_{M,A}\mathbf{l})^{1/2}[f_A(2, k) - f_A(1, k)] \qquad \text{(B.4.22)}$$

In carrying out an actual simulation, one must hold the difference in fractional concentration between the first two boxes at a constant value. With $t_k = \tau$, that required difference is given from (B.4.22) as

$$\boxed{f_A(2, k) - f_A(1, k) = \frac{\pi^{1/2}}{2(\mathbf{D}_{M,A}\mathbf{l})^{1/2}}} \qquad \text{(B.4.23)}$$

In each iteration, one allows diffusion to occur; then the value of $f_A(1)$ is adjusted downward so that (B.4.23) is maintained. Since this adjustment corresponds to a faradaic conversion, $f_B(1)$ must be adjusted upward by an equal amount. These steps give the final values $f_A(1, k)$ and $f_B(1, k)$ for iteration k.

If the system is the uncomplicated case of (B.4.1), the transition time, which is found when $f_A(1, k) = 0$, will be reached ideally in the **l**th iteration. Deviations from this result will occur when complications such as homogeneous kinetics are introduced into the electrode process.

The potential-time curve for a reversible system can be obtained by reporting the value of $\mathbf{E}_{\text{norm}}$ calculated with equation B.4.4 at each iteration. Equivalent data for a quasi-reversible system would require the specification of heterogeneous rate parameters in a fashion like that outlined in Section B.4.3.

B.5 SIMULATIONS IN CONVECTIVE SYSTEMS

Convective effects, such as those that occur with the rotating disk and rotating ring-disk electrodes (RDE and RRDE) can also be simulated (3, 4, 6). In this case, the flow of solution is taken into account by movement of the contents of the boxes from one location in solution to another. For example, near the surface of the RDE, the solution flow normal to the disk is given by (8.3.9); that is,

$$-\frac{dy}{dt} = -v_y = 0.51\ \omega^{3/2}\nu^{-1/2}y^2 \qquad \text{(B.5.1)}$$

The solution of this equation for movement of a solution element from y_1 to y_2 during the time increment Δt is

$$\frac{1}{y_2} - \frac{1}{y_1} = 0.51\ \omega^{3/2}\nu^{-1/2}\ \Delta t \qquad \text{(B.5.2)}$$

If the distances y_2 and y_1 are measured from the electrode surface, then as in (B.1.14), if we now let Δy play the role earlier held by Δx,

$$y_1 = (j - 1)\Delta y = (j - 1)\left(\frac{Dt_k}{\mathbf{D}_M \mathbf{l}}\right)^{1/2} \tag{B.5.3}$$

Then y_2, the position in the array where this solution element will reside at the end of the time increment Δt, written in terms of simulation parameters is $(j' - 1)\Delta y$, with j' representing the position in terms of box number. It can be obtained from the equation

$$\left[\frac{1}{(j' - 1)\Delta y}\right] - \left[\frac{1}{(j - 1)\Delta y}\right] = 0.51\ \omega^{3/2}\nu^{-1/2}\ \Delta t \tag{B.5.4}$$

or

$$\boxed{j - 1 = \frac{(j' - 1)}{1 - (j' - 1)\mathbf{V}_N}} \tag{B.5.5}$$

where $\mathbf{V}_N$ is a dimensionless constant:

$$\mathbf{V}_N = 0.51\ \omega^{3/2}\nu^{-1/2}\ \Delta x\ \Delta t = \mathbf{D}_M^{-1/2}\mathbf{l}^{-3/2} \tag{B.5.6}$$

Δx is given by (B.1.13), and

$$\Delta t = \frac{t_k}{\mathbf{l}} = \frac{\omega^{-1}\nu^{1/3}D^{-1/3}(0.51)^{-2/3}}{\mathbf{l}} \tag{B.5.7}$$

Here, t_k is defined in terms of convenient time parameters for the RDE. Thus the contents (i.e., concentrations, C_A, C_B, . . .) of each box at the end of a time period are replaced by those at locations calculated with (B.5.5); this procedure results in a new array representing the effect of convection normal to the electrode. In practice better accuracy at smaller values of $\mathbf{l}$ is obtained in the computation if (B.5.5) is modified slightly (6):

$$\boxed{j - 1 = \frac{(j' - 1)}{1 - [1.11(j' - 1)\mathbf{V}_N]}} \tag{B.5.8}$$

Diffusion normal to the electrode is also taken into account, by the procedures described in Section B.1.2.

For the RRDE, convection in the radial (r) direction is treated in a similar manner. In this case, volume elements or boxes are also constructed in the radial direction, measured from the center of the electrode and indexed by r_i, so that

$$r = r_i \Delta r \tag{B.5.9}$$

where Δr is defined in terms of the dimensions of the electrodes. Note that in this case each solution element is indexed by two simulation distance parameters, j and r_i. The application of the equation for solution velocity in the radial direction, (8.3.10), ultimately yields (6)

$$r_i = r'_i \exp[-1.03(j')\mathbf{V}_N] \tag{B.5.10}$$

where r_i is the radial position of a volume element that will be in position r', j' at the end of Δt. Radial diffusion is taken into account in the usual manner. Results of such two-dimensional simulations were described in Section 8.5.

Again, an especially valuable aspect to these digital simulations is the straightforward incorporation of kinetic effects, as described in Section B.3, into the treatment. This allows rather complicated problems, such as those involving dimerizations of the disk-generated species, to be solved for the RDE and RRDE (see Section 11.4) (7–9). Similar simulations have been carried out for the rotating double-ring system (10–12). The simulation approach discussed here involves the assumption that the solution flow velocity profiles follow the analytical solutions. Digital simulation of the solution flow itself, for example, to a rotating disk, is also possible (13). This technique might be especially valuable in electrochemical systems involving more complicated solution flow situations. Digital simulation of the hydrodynamics occurring at the expanding spherical surface of the DME has also been described (14).

B.6 MISCELLANEOUS DIGITAL SIMULATIONS

B.6.1 Electrical Migration and Diffuse Double-Layer Effects

The flux of a species A arising from an electrical field $\mathscr{E} = \partial\phi/\partial x$ is [see (4.1.11)]

$$J_A(x, t) = -\left(\frac{F}{RT}\right) z_A D_A C_A(x, t)\mathscr{E}(x) \tag{B.6.1}$$

Transfer of a species from box to box is treated in a manner similar to that used for diffusion, except that the change in concentration in this case depends on the field. The calculation of the field distribution is based on the usual electrostatic considerations (15), using, for example, the Poisson equation (12.3.5):

$$\frac{d^2\phi}{dx^2} = \frac{d\mathscr{E}(x)}{dx} = \frac{-\rho(x)}{\varepsilon\varepsilon_0} \tag{B.6.2}$$

Feldberg (16) treated the problem of the relaxation of the diffuse double layer following a coulostatic injection of charge by such a procedure. When a charge density σ^M is injected into the electrode, assumed to be initially at E_z, the field at any time is given by the Gauss law:

$$\mathscr{E}(x) = \left(\frac{-1}{\varepsilon\varepsilon_0}\right)[\sigma^M + F\int_0^x \sum_{i=1}^{n} z_i C_i(x)\, dx] \tag{B.6.3}$$

where i indexes the different species. At the instant of charge injection, before any ionic movement occurs, electroneutrality exists at all points in the solution, so that

$$\sum_{i=1}^{n} z_i C_i(x) = 0 \qquad \text{(for all } x\text{)} \tag{B.6.4}$$

and the field is constant, given by

$$\mathscr{E}(x) = \frac{-\sigma^M}{\varepsilon\varepsilon_0} \qquad \text{(for all } x\text{)} \tag{B.6.5}$$

This initial field induces a flux of ions that can be calculated by use of the finite difference form of (B.6.1), as well as the diffusional flux, calculated in the usual manner. At any time, the potential $\phi(x)$ can be calculated by integration of the field; thus ϕ_2 is given by

$$\phi_2 - \phi_\infty = \int_\infty^{x_2} \mathscr{E}(x)\,dx \tag{B.6.6}$$

Eventually the system relaxes to the distribution and ϕ_2 value obtained by the GCS approach (Section 12.3). A similar digital simulation approach was taken to calculate the charge and potential distributions in the space charge region that form inside the electrode at a semiconductor electrode/electrolyte interface (17).

B.6.2 Thin-Layer Cells and Resistive Effects

In the treatment of thin-layer cells (see Section 10.7), two additional effects must be considered. The distance between the electrode and the cell wall is small, and the total number of simulation boxes taken, $\mathbf{n}_L$, now represents the cell thickness. Moreover, the counter electrode is usually placed outside the thin-layer portion and an appreciable resistive drop occurs between each segment of the working electrode and the counter electrode. This effect is taken into account by dividing the working electrode into $\mathbf{n}_M$ segments, each at a different potential with respect to the reference electrode, and each with a different current density. An iterative procedure is used to determine the potential and current distribution at any given time in an electrochemical experiment (18). The resistive effects in thin-layer cells can be significant in determining the electrochemical response and are of importance in the interpretation of the behavior of thin-layer (e.g. electrochemical-ESR) cells and thin-film electrodes (19).

B.6.3 Advanced Simulation Methods

The simulation methods described here are quite straightforward and have been employed for a variety of electrochemical problems. Nevertheless, there are cases when the computation times required for accurate solutions become excessive, and more efficient numerical methods are appropriate. Generally, this is the case when very rapid coupled homogeneous reactions occur. As discussed in Section B.3, when k_1 is large, the reaction layer thickness $(D/k_1)^{1/2}$ may be small compared to the box thickness Δx. The requirement for accurate simulation can be given as $(D/k_1)^{1/2} \gtrsim \Delta x$ or, from (B.1.13), $(1/k_1) \gtrsim (t_k/\mathbf{l})$. Several approaches have been suggested. One can abandon the model of boxes with equal widths and use space elements of variable dimensions (20). Thus thin boxes are used near the electrode where the fast reactions are occurring, with the elements becoming wider farther out into the solution. A second approach involves using implicit methods for obtaining $f(j, k+1)$ [e.g., the Crank-Nicolson method (21)] rather than the explicit solution in (B.1.9) (20, 22, 23). More advanced numerical procedures, such as the orthogonal collocation technique, which has been applied to heat transfer problems and involves solution of simultaneous differential equations, have also been used (24, 25). These methods can result in appreciable savings of computer time, but require more mathematical sophistication and more difficult computer programming. Their use will probably only be justified when the computation times required for the accurate solution of an electrochemical problem by the simpler methods become unacceptable.

B.7 REFERENCES

1. S. W. Feldberg, *Electroanal. Chem.*, **3**, 199 (1969).
2. S. W. Feldberg in "Computers in Chemistry and Instrumentation," Vol. 2, "Electrochemistry" J. S. Mattson, H. B. Mark, Jr., and H. C. MacDonald, Jr., Eds., Marcel Dekker, New York, 1972, Chap. 7.
3. K. B. Prater, *ibid.*, Chap. 8.
4. J. T. Maloy, *ibid.*, Chap. 9.
5. J. T. Maloy in "Laboratory Techniques in Electroanalytical Chemistry," P. T. Kissinger, Ed., Marcel Dekker, New York, in press.
6. K. B. Prater and A. J. Bard, *J. Electrochem. Soc.*, **117**, 207 (1970).
7. K. B. Prater and A. J. Bard, *J. Electrochem. Soc.*, **117**, 335, 1517 (1970).
8. V. J. Puglisi and A. J. Bard, *J. Electrochem. Soc.*, **119**, 833 (1972).
9. L. S. R. Yeh and A. J. Bard, *J. Electrochem. Soc.*, **124**, 189 (1977).
10. J. Margarit and M. Levy, *J. Electroanal. Chem.*, **49**, 369 (1974).
11. J. Margarit, G. Dabosi, and M. Levy, *Bull. Soc. Chim. France*, **1975**, 1509.
12. J. Margarit and D. Schuhmann, *J. Electroanal. Chem.*, **80**, 273 (1977).
13. S. Clarenbach and E. W. Grabner, *Ber. Bunsenges, Phys. Chem.*, **80**, 115 (1976).
14. I. Ruzic and S. W. Feldberg, *J. Electroanal. Chem.*, **63**, 1 (1975).
15. K. J. Binns and P. J. Lawrenson, "Analysis and Computation of Electric and Magnetic Field Problems," Macmillan, New York, 1963.
16. S. W. Feldberg, *J. Phys. Chem.*, **74**, 87 (1970).
17. D. Laser and A. J. Bard, *J. Electrochem. Soc.*, **123**, 1828, 1837 (1976).
18. I. B. Goldberg and A. J. Bard, *J. Electroanal. Chem.*, **38**, 313 (1972).
19. I. B. Goldberg, A. J. Bard, and S. W. Feldberg, *J. Phys. Chem.*, **76**, 2250 (1972).
20. T. Joslin and D. Pletcher, *J. Electroanal. Chem.*, **49**, 171 (1974).
21. G. D. Smith, "Numerical Solutions of Partial Differential Equations," Oxford University Press, 1969.
22. N. Winograd, *J. Electroanal. Chem.*, **43**, 1 (1973).
23. T. B. Brumleve and R. P. Buck, *J. Electroanal. Chem.*, **90**, 1 (1978).
24. L. F. Whiting and P. W. Carr, *J. Electroanal. Chem.*, **81**, 1 (1977).
25. S. Pons and A. Bewick, Private communication, 1978.

B.8 PROBLEMS

B.1 Show that $Z(t)$ is proportional to the ratio of the current at time t to the Cottrell current at time t_k. What is the proportionality factor?

B.2 From concepts developed in Section 5.2, justify the statement that the diffusion layer in any experiment is contained in a thickness $6(Dt)^{1/2}$ measured from the electrode surface.

B.3 Set up a digital simulation of the Cottrell experiment and work through the first 10 iterations with a hand calculator. Use $\mathbf{l} = 50$ and $\mathbf{D}_M = 0.40$. Calculate $Z(k)$ for each iteration and compare it to $Z_{\text{Cott}}(k)$. Calculate the χ values corresponding to the first 12 boxes, and plot the concentration profiles f_A and f_B *vs.* χ for $t/t_k = 0.2$. Derive the functions describing f_A and f_B *vs.* χ and t/t_k from (5.2.13) and draw the analytical curves on your graphs of concentration profiles. Comment on the agreement between your model and the known solution.

B.4 Suppose one desired a simulation of chronocoulometry. Derive a dimensionless charge parameter analogous to $Z(k)$. In carrying out a simulation, to what time should the charge parameter calculated for iteration k be assigned?

B.5 Consider the following mechanism:

$$A + e \rightleftharpoons B \qquad \text{(at the electrode)}$$

$$B + C \xrightarrow{k_2} D \qquad \text{(in solution)}$$

Derive the diffusion-kinetic equations analogous to (B.3.11) and (B.3.12) and identify the dimensionless kinetic parameter involving k_2.

B.6 If computer time is available, carry out simulations of cyclic voltammetry for a quasi-reversible system. Let $\mathbf{l} = 50$ and $\mathbf{D}_M = 0.45$. Take $\alpha = 0.5$ and let the diffusion coefficients of the oxidized and reduced forms be equal. Cast your dimensionless intrinsic rate parameter in terms of the function ψ defined in (6.5.5), and carry out calculations for $\psi = 20$, 1, and 0.1. Compare the peak splittings in your simulated voltammograms with the values in Table 6.5.2.

B.7 To the simulation program devised for Problem B.6, add provision for first-order homogeneous decay of the reduction product B, that is,

$$A + ne \rightleftharpoons B \qquad \text{(quasi-reversible)}$$

$$B \xrightarrow{k_1} C \qquad \text{(in solution)}$$

Run a simulation for $\psi = 20$ and $k_1 t_k = 1$. Compare the results with those predicted by R. S. Nicholson and I. Shain, *Anal. Chem.*, **36**, 706 (1964).

Appendix C
Reference Tables

Table C.1

Selected Standard Electrode Potentials in Aqueous Solutions at 25° in V *vs.* NHE[a]

Reaction	Potential, V
$Ag^+ + e = Ag$	0.7996
$AgBr + e = Ag + Br^-$	0.0713
$AgCl + e = Ag + Cl^-$	0.2223
$AgI + e = Ag + I^-$	−0.1519
$Ag_2O + H_2O + 2e = 2Ag + 2OH^-$	0.342
$Al^{3+} + 3e = Al$(0.1 *M* NaOH)	−1.706
$Au^+ + e = Au$	1.68
$Au^{3+} + 2e = Au^+$	1.29
p-benzoquinone + $2H^+ + 2e$ = hydroquinone	0.6992
$Br_2(aq) + 2e = 2Br^-$	1.087
$Ca^{2+} + 2e = Ca$	−2.76
$Cd^{2+} + 2e = Cd$	−0.4026
$Cd^{2+} + 2e = Cd(Hg)$	−0.3521
$Ce^{4+} + e = Ce^{3+}$(1 *M* H_2SO_4)	1.44
$Cl_2(g) + 2e = 2Cl^-$	1.3583
$HClO + H^+ + e = \frac{1}{2}Cl_2 + H_2O$	1.63

(continued)

[a] More extensive tables can be found in the following sources: (1) W. M. Latimer, "The Oxidation States of the Elements and Their Potentials in Aqueous Solutions," Prentice-Hall, Englewood Cliffs, N.J., 1952 [Note that the signs used by Latimer are opposite to those used here. We follow the convention recommended by the International Union of Pure and Applied Chemistry (IUPAC)]. (2) A. J. Bard and H. Lund, Eds., "The Encyclopedia of the Electrochemistry of the Elements," Marcel Dekker, New York, 1973–. (3) G. Milazzo and S. Caroli, "Tables of Standard Electrode Potentials," Wiley-Interscience, New York, 1977.

An extensive compilation of standard potentials and thermodynamic data under the auspices of the Electrochemistry and Electroanalytical Chemistry Commissions of IUPAC is being prepared (A. J. Bard, J. Jordan, and R. Parsons, Eds.).

Table C.1—(continued)

Reaction	Potential, V
$Co^{2+} + 2e = Co$	−0.28
$Co^{3+} + e = Co^{2+} (3\ M\ HNO_3)$	1.842
$Cr^{2+} + 2e = Cr$	−0.557
$Cr^{3+} + e = Cr^{2+}$	−0.41
$Cr_2O_7^{2-} + 14H^+ + 6e = 2Cr^{3+} + 7H_2O$	1.33
$Cu^+ + e = Cu$	0.522
$Cu^{2+} + 2CN^- + e = Cu(CN)_2^-$	1.12
$Cu^{2+} + e = Cu^+$	0.158
$Cu^{2+} + 2e = Cu$	0.3402
$Cu^{2+} + 2e = Cu(Hg)$	0.345
$Eu^{3+} + e = Eu^{2+}$	−0.43
$\frac{1}{2}F_2 + H^+ + e = HF$	3.03
$Fe^{2+} + 2e = Fe$	−0.409
$Fe^{3+} + e = Fe^{2+} (1\ M\ HCl)$	0.770
$Fe(CN)_6^{3-} + e = Fe(CN)_6^{4-} (1\ M\ H_2SO_4)$	0.69
$2H^+ + 2e = H_2$	0.0000
$2H_2O + 2e = H_2 + 2OH^-$	−0.8277
$H_2O_2 + 2H^+ + 2e = 2H_2O$	1.776
$2Hg^{2+} + 2e = Hg_2^{2+}$	0.905
$Hg_2^{2+} + 2e = 2Hg$	0.7961
$Hg_2Cl_2 + 2e = 2Hg + 2Cl^-$	0.2682
$Hg_2Cl_2 + 2e = 2Hg + 2Cl^-$(sat'd. KCl)	0.2415
$HgO + H_2O + 2e = Hg + 2OH^-$	0.0984
$Hg_2SO_4 + 2e = 2Hg + SO_4^{2-}$	0.6158
$I_2 + 2e = 2I^-$	0.535
$I_3^- + 2e^- = 3I^-$	0.5338
$K^+ + e = K$	−2.924
$Li^+ + e = Li$	−3.045
$Mg^{2+} + 2e = Mg$	−2.375
$Mn^{2+} + 2e = Mn$	−1.029
$Mn^{3+} + e = Mn^{2+}$	1.51
$MnO_2 + 4H^+ + 2e = Mn^{2+} + 2H_2O$	1.208
$MnO_4^- + 8H^+ + 5e = Mn^{2+} + 4H_2O$	1.491
$Na^+ + e = Na$	−2.7109
$Ni^{2+} + 2e = Ni$	−0.23
$Ni(OH)_2 + 2e = Ni + 2OH^-$	−0.66
$O_2 + 2H^+ + 2e = H_2O_2$	0.682
$O_2 + 4H^+ + 4e = 2H_2O$	1.229
$O_2 + 2H_2O + 4e = 4OH^-$	0.401
$O_3 + 2H^+ + 2e = O_2 + H_2O$	2.07
$Pb^{2+} + 2e = Pb$	−0.1263
$Pb^{2+} + 2e = Pb(Hg)$	−0.1205
$PbO_2 + 4H^+ + 2e = Pb^{2+} + 2H_2O$	1.46
$PbO_2 + SO_4^{2-} + 4H^+ + 2e = PbSO_4 + 2H_2O$	1.685
$PbSO_4 + 2e = Pb + SO_4^{2-}$	−0.356
$Pd^{2+} + 2e = Pd$	0.83
$Pt^{2+} + 2e = Pt$	~1.2
$PtCl_4^{2-} + 2e = Pt + 4Cl^-$	0.73
$PtCl_6^{2-} + 2e = PtCl_4^{2-} + 2Cl^-$	0.74
$S + 2e = S^{2-}$	−0.508
$Sn^{2+} + 2e = Sn$	−0.1364

Table C.1—(*continued*)

Reaction	Potential, V
$Sn^{4+} + 2e = Sn^{2+}$	0.15
$Tl^{+} + e = Tl$	−0.3363
$Tl^{+} + e = Tl(Hg)$	−0.3338
$Tl^{3+} + 2e = Tl^{+}$	1.247
$U^{3+} + 3e = U$	−1.8
$U^{4+} + e = U^{3+}$	−0.61
$UO_2^{+} + 4H^{+} + e = U^{4+} + 2H_2O$	0.62
$UO_2^{2+} + e = UO_2^{+}$	0.062
$V^{2+} + 2e = V$	−1.2
$V^{3+} + e = V^{2+}$	−0.255
$VO^{2+} + 2H^{+} + e = V^{3+} + H_2O$	0.337
$VO_2^{+} + 2H^{+} + e = VO^{2+} + H_2O$	1.00
$Zn^{2+} + 2e = Zn$	−0.7628
$ZnO_2^{2-} + 2H_2O + 2e^{-} = Zn + 4OH^{-}$	−1.216

Table C.2

Estimated Electrode Potentials in Aprotic Solvents, in V *vs* aq SCE[a,b]

Substance	Reaction	Conditions[c]	Potential, V
Anthracene (An)	$An + e = An^{\cdot -}$	DMF, 0.1 *M* TBAI	−1.92
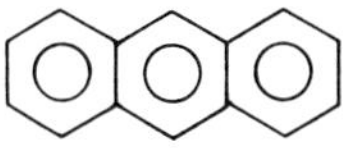	$An^{\cdot -} + e = An^{2-}$	DMF, 0.1 *M* TBAI	−2.5
	$An^{\cdot +} + e = An$	MeCN, 0.1 *M* TBAP	+1.3
Azobenzene (AB)	$AB + e = AB^{\cdot -}$	DMF, 0.1 *M* TBAP	−1.36
Ph—N=N—Ph	$AB^{\cdot -} + e = AB^{2-}$	DMF, 0.1 *M* TBAP	−2.0
	$AB + e = AB^{\cdot -}$	MeCN, 0.1 *M* TEAP	−1.40
	$AB + e = AB^{\cdot -}$	PC, 0.1 *M* TBAP	−1.40
Benzophenone (BP)	$BP + e = BP^{\cdot -}$	MeCN, 0.1 *M* TBAP	−1.88
O	$BP + e = BP^{\cdot -}$	THF, 0.1 *M* TBAP	−2.06
‖	$BP + e = BP^{\cdot -}$	NH_3, 0.1 *M* KI	−1.23[e]
PhCPh	$BP^{\cdot -} + e = BP^{2-}$	NH_3, 0.1 *M* KI	−1.76[e]
1,4 Benzoquinone (BQ)	$BQ + e = BQ^{\cdot -}$	MeCN, 0.1 *M* TEAP	−0.54
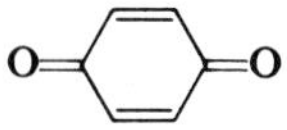	$BQ^{\cdot -} + e = BQ^{2-}$	MeCN, 0.1 *M* TEAP	−1.4
Ferrocene (Cp_2Fe)	$Cp_2Fe^{+} + e = Cp_2Fe$	MeCN, 0.2 *M* $LiClO_4$	+0.307
Nitrobenzene (NB)	$NB + e = NB^{\cdot -}$	MeCN, 0.1 *M* TEAP	−1.15
Ph—NO_2	$NB + e = NB^{\cdot -}$	DMF, 0.1 *M* $NaClO_4$	−1.01
	$NB + e = NB^{\cdot -}$	NH_3, 0.1 *M* KI	−0.42[e]
	$NB^{\cdot -} + e = NB^{2-}$	NH_3, 0.1 *M* KI	−1.24[e]
Oxygen	$O_2 + e = O_2^{\cdot -}$	DMF, 0.2 *M* TBAP	−0.87
	$O_2 + e = O_2^{\cdot -}$	MeCN, 0.2 *M* TBAP	−0.82
	$O_2 + e = O_2^{\cdot -}$	DMSO, 0.1 *M* TBAP	−0.73
$Ru(bpy)_3^{n+}$ (RuL_3^{n+})	$RuL_3^{3+} + e = RuL_3^{2+}$	MeCN, 0.1 *M* $TBABF_4$	+1.32
(bpy = 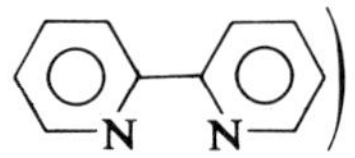)	$RuL_3^{2+} + e = RuL_3^{+}$	MeCN, 0.1 *M* $TBABF_4$	−1.30
	$RuL_3^{+} + e = RuL_3^{0}$	MeCN, 0.1 *M* $TBABF_4$	−1.49
	$RuL_3^{0} + e = RuL_3^{-}$	MeCN, 0.1 *M* $TBABF_4$	−1.73

(*continued*)

Table C.2—*(continued)*

Substance	Reaction	Conditions[c]	Potential, V
Tetracyano-quinodimethane (TCNQ)	$TCNQ + e = TCNQ^{\cdot-}$	MeCN, 0.1 *M* $LiClO_4$	+0.127
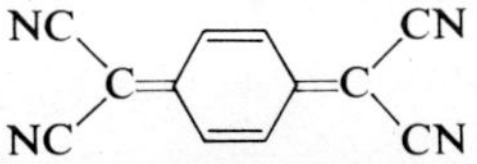	$TCNQ^{\cdot-} + e = TCNQ^{2-}$	MeCN, 0.1 *M* $LiClO_4$	−0.291
N,*N*,*N*′,*N*′-Tetramethyl-*p*-phenylenediamine (TMPD) 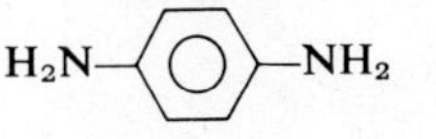	$TMPD^{\cdot+} + e = TMPD$	DMF, 0.1 *M* TBAP	+0.21
Tetrathiafulvalene (TTF)	$TTF^{\cdot+} + e = TTF$	MeCN, 0.1 *M* TEAP	+0.30
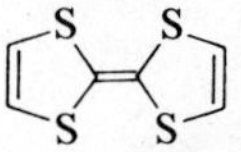	$TTF^{2+} + e = TTF^{\cdot+}$	MeCN, 0.1 *M* TEAP	+0.66
Thianthrene (TH)	$TH^{\cdot+} + e = TH$	MeCN, 0.1 *M* $TBABF_4$	+1.23
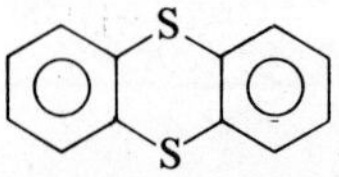	$TH^{2+} + e = TH^{\cdot+}$	MeCN, 0.1 *M* $TBABF_4$	+1.74
	$TH^{\cdot+} + e = TH$	SO_2, 0.1 *M* TBAP	+0.30[d]
	$TH^{2+} + e = TH^{\cdot+}$	SO_2, 0.1 *M* TBAP	+0.88[d]
Tri-*N*-*p*-tolylamine (TPTA) 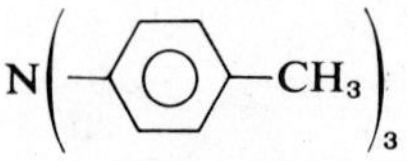	$TPTA^{\cdot+} + e = TPTA$	THF, 0.2 *M* TBAP	+0.98

[a] See footnote in Table C.1.

[b] Problems arise in reporting potentials in nonaqueous solvents. The frequent practice of using an aqueous SCE as a reference electrode introduces an unknown and sometimes irreproducible liquid junction potential. Sometimes reference electrodes made up in the solvent of interest (e.g., Ag/$AgClO_4$) or quasi-reference electrodes (QRE) are employed. Results here are reported versus an *aq* SCE unless noted otherwise. While there has not yet been an adopted convention for reporting potentials in nonaqueous solvents, a frequent practice is to reference these to the potential of a particular reversible couple in the same solvent. This couple (sometimes called the "reference redox system") is usually chosen on the basis of the extrathermodynamic assumption that the redox potential of this system is only slightly affected by the solvent system. Suggested reference redox systems include ferrocene/ferrocenium, Rb/Rb^+, $Fe(bpy)_3{}^{3+}/Fe(bpy)_3{}^{2+}$ (bpy = 2,2′-bipyridine), and aromatic hydrocarbon/radical cation. For further information concerning these problems, the following references can be consulted: (1) O. Popovych, *Crit. Rev. Anal. Chem.*, **1**, 73 (1970); (2) D. Bauer and M. Breant, *Electroanal. Chem.*, **8**, 282 (1975); (3) A. J. Parker, *Electrochim. Acta*, **21**, 671 (1976).

[c] See Standard Abbreviations, p. xvii.

[d] *vs.* $Ag/AgNO_3$(sat'd) in SO_2 at −40 °C.

[e] *vs.* Ag/Ag^+ (0.01 *M*) in NH_3 at −50 °C.

Index

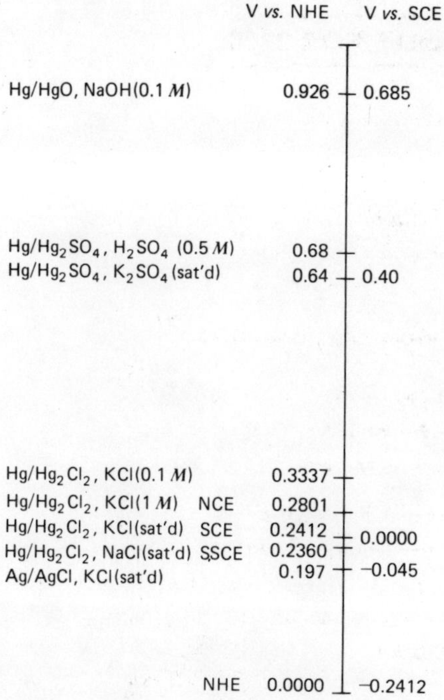

Figure E.1 Potentials of reference electrodes in aqueous solutions at 25 °C[a].

[a] See D. J. G. Ives and G. J. Janz, "Reference Electrodes," Academic Press, New York, 1961.

At other temperatures (t), °C:

SCE $$E = 0.2412 - 6.61 \times 10^{-4}(t - 25) - 1.75 \times 10^{-6}(t - 25)^2 - 9.0 \times 10^{-10}(t - 25)^3$$

NCE $$E = 0.2801 - 2.75 \times 10^{-4}(t - 25) - 2.50 \times 10^{-6}(t - 25)^2 - 4 \times 10^{-9}(t - 25)^3$$

PHYSICAL CONSTANTS

c	Speed of light in vacuo	2.99792×10^{8} m/sec
e	Elementary charge	1.60219×10^{-19} C
F	Faraday constant	9.64846×10^{4} C/equiv
h	Planck constant	6.62618×10^{-34} J-sec
k	Boltzmann constant	1.38066×10^{-23} J/K
N_A	Avogadro's number	6.02205×10^{23} mol^{-1}
R	Molar gas constant	8.31441 J mol^{-1}-K^{-1}
ε_0	Permittivity of free space	8.85419×10^{-12} C^2 N^{-1} m^{-2}